PARASITOLOGY
The Biology of Animal Parasites

Preface

The third edition of *Parasitology* constitutes a considerable revision of the previous editions. Chapters 1 and 2 of the second edition have been combined into a single, introductory chapter, with much rearrangement and shifting of various sections to later chapters. Our definition of parasitism has been changed. A new introductory approach has been made, which briefly describes a fish and its possible parasites as a model for parasite-host communities. A short general account of immunity is included in the introduction, and more specific statements on immunity have been added to the discussions of each major group of parasites in later chapters.

Our conviction that parasites and their hosts should be studied together if one is to form a true conception of parasitism has led us to combine the former chapters on effects on the host and effects on the parasite into one chapter — "Parasite-Host Responses." Throughout the text more physiologic material has been added. The taxonomy of the Protozoa, Nematoda and some smaller groups has been modified, and the chapter on "Miscellaneous Phyla" has been expanded. A number of new illustrations appear throughout the book, and some old ones were deleted or improved by labeling.

We have omitted references to the treatment of diseases because this book does not attempt to duplicate a text in clinical parasitology. Rather, we offer parasitism as an important, ubiquitous way of life.

Finally, this edition, like the first, "is written primarily for the advanced undergraduate student who has completed at least one year of introductory zoology or biology, and at least one course in elementary chemistry at the collegiate level."

Of the many persons who have aided in the preparation of this volume, the following have given particularly valuable advice and criticism, for which we are exceedingly grateful: Dr. John N. Belkin, University of California, Los Angeles; Dr. Thomas Bowman, Smithsonian Institution; Dr. Wilbur L. Bullock, University of New Hampshire; Dr. MayBelle Chitwood, Agricultural Research Service, Beltsville, Maryland; Dr. Sneed B. Collard, University of West Florida; Dr. Roger Cressey, Smithsonian Institution; Dr. Frederick Hochberg; Dr. B. M. Honigberg, University of Massachusetts; Dr. Kenneth C. Kates, Agricultural Research Service, Beltsville, Maryland; Dr. Marietta Voge, University of California, Los Angeles; and Dr. Martin Ulmer, Iowa State University.

SANTA BARBARA, CALIFORNIA *Elmer R. Noble*
SAN LUIS OBISPO, CALIFORNIA *Glenn A. Noble*

Contents

Section I
INTRODUCTION

Chapter 1

Introduction to Parasitology

During the millions of years that animals and plants have competed among themselves for food and space, parasites have invaded practically every kind of living body. These bodies, called **hosts,** also provide protection, and since hosts generally furnish different kinds of space in the form of external surfaces, organs, tissues and fluids, they usually acquire more than one kind of parasite. Today, most animals have on or within their bodies several species of parasites, sometimes totalling hundreds or even millions of individuals. There are, therefore, more kinds and numbers of animal parasites than free-living animals. The major groups of animal parasites are found among the Protista, the helminths (flatworms and roundworms) and the arthropods. The host and its parasites constitute a community of organisms living in close intimacy and exerting a profound effect upon each other.

To illustrate the kinds of parasites that an animal might support and some of the inter-relationships and problems that are involved in parasitism, we will select the fish as a typical vertebrate host. Almost any vertebrate would be satisfactory, but fish may be easily secured by students the world over for purposes of dissection. This book is concerned only with animal parasites, so we shall disregard the numerous bacterial, fungal and viral infections that plague freshwater and marine fishes. A careful dissection of a single fish usually reveals three or four species of animal parasites, seldom more than five or six, and sometimes only one.

An examination of any host for its parasites should start with the outside. The skin and scales of a fish are commonly the home of copepods and other Crustacea, encysted larval stages of digenetic trematodes (flukes), adult monogenetic trematodes, leeches, and several kinds of Protozoa. Copepods often have sharp claws that enable them to cling to the skin, or anchoring devices that are deeply embedded under scales. Fish lice (Branchiura) are temporary parasites of the skin. The encysted larvae of digenetic trematodes are called metacercariae, and the fish must be eaten by another host (e.g., a bird or another fish) before the larval flukes can develop into adult worms. Monogenetic trematodes may damage the host's skin by means of clawlike hooks on the posterior ends of their bodies. Leeches suck the fish's blood and may thereby transmit blood parasites from one fish to another. Protozoan parasites on the skin occasionally cause so much damage that the fish dies.

The next place to look is inside the mouth and on the gills. Here may be found the same kinds of worms as on the skin, plus additional kinds of parasites. Isopod crustaceans often cling to the gills or mouth lining; sometimes a single parasite is so large that it almost fills the mouth cavity. Hundreds of copepods may be partly embedded in the gills. Cysts of the protozoan order Myxosporida may appear as white spots or lumps. Thousands of metacercariae may be embedded in the gill filaments, as well as monogenetic trematodes.

If the fish is alive or freshly killed, blood smears should be made, and the fresh blood scrutinized for such parasites as the flagellated trypanosomes (related to those that cause

African sleeping sickness in man), and another protozoan form, the haemogregarines, which live within red cells. Fish often have more internal parasites than external ones, and any organ may be infected. Many parasites from the digestive tracts of fish have been described, especially nematodes and flukes. Thorny-headed worms, called acanthocephalans, are common in the intestine. Several kinds of Protozoa may be mixed with the fecal material. Coiled, larval nematodes are easily seen in the mesenteries and walls of the coelom as well as in the muscles. Larval tapeworms of several kinds inhabit a variety of organs. Myxosporida are very common in the gallbladder, urinary bladder, kidneys, muscles and other organs. Another group of protozoan parasite, called Microsporida, may infect the cells of most organs of the fish. Both Myxosporida and Microsporida may cause fatal diseases although, generally, death and disease seldom occur as the result of parasitic infections.

One of the first questions that a student usually asks is, "How is it possible for a fish (or any host) to live in apparent good health with so many parasites crowding its body?" The answer is a complicated one, involving a consideration of the results of gradual adaptations between hosts and parasites during their evolution together. After all, it is not to the advantage of the parasite to kill or even injure its host, because a healthy host means a healthy environment for the parasite.

Another question often asked by students is, "How do these parasites get into the host and what are their life cycles?" The answers are numerous and can be found throughout the pages of this book, where the various kinds of parasite and host relationships are described in detail. Many parasites have a relatively simple, direct life cycle whereby the infective stage (such as a cyst, spore or motile larva) released by one host is directly taken up (often eaten) by another host, in which the parasite grows and develops. Other species of parasites may have a complicated, indirect life cycle, requiring one or more intermediate hosts (such as a mosquito) to complete their development.

More specific questions that may be asked are: What are the important morphologic and physiologic features of parasites? How do parasites live within a host? How does a host respond to parasites? What are the nutritional requirements of infective stages? Upon what factors in a host do parasites depend? Do parasites provide anything of value to a host? How does the life cycle and behavior of a host affect its parasites? How do the parasites of one species affect those of another species in the same host? What factors trigger each developmental change during the life cycle of a parasite? What genetic and developmental factors have particular significance in parasitism? We have enough information to answer some of these questions in part, but all of them and many others need to receive much more attention, especially by experimental parasitologists.

SYMBIOSIS

The term "symbiosis" was proposed in 1879 by de Bary[6] to mean the "living together" of two species of organisms. This term came to be used in a more restricted sense to connote mutual benefit, as exemplified by the termite and its gut protozoans. Indeed, de Bary used a lichen as the clearest example of symbiosis. O. Hertwig defined symbiosis as "the common life, permanent in character, of organisms that are specifically distinct and have complementary needs." However, a cursory examination of symbiotes (= symbionts)—those organisms living together symbiotically—reveals a wide variation in permanency of the association, degree of intimacy, and degree of pathogenicity.

Textbooks on parasitology frequently distinguish the following three general kinds of symbiosis: (a) commensalism, (b) mutualism, and (c) parasitism. **Commensalism** occurs when one member of the associating pair, usually the smaller, receives all the benefit and the other member is neither benefited nor harmed. The basis for a commensalistic relationship between two organisms may be space, substrate, defense, shelter, transportation or food (Fig. 1–1). If the association is merely a passive transportation of the commensal by the host, it is called **phoresy.**

Mutualism occurs when each member of the association benefits the other. For example, in the association between termites and their flagellates and between ungulates and their ciliates, the parasites digest the food (cellulose) of the host in return for free board and lodging (see pp. 497 to 500 for details). Similar to this kind of symbiosis is the ubiquitous association between animals and such parasites as bacteria, yeasts and other fungi. These symbionts provide essential vitamins for their hosts. They are parasites, as defined below, and rightly belong in a textbook on parasitology.

Parasitism. An older definition of parasite, still seen in some textbooks, is the following: an organism that lives at the expense of its host and often inflicts some degree of injury or damage. One difficulty with this definition lies in its emphasis on harm or lack of benefit. How can we be certain that a symbiont does *not* affect its host in a way more subtle than causing obvious physical damage or changes in behavior? Numerous parasites apparently act as commensals most of the time, but become pathogenic when their numbers become unusually high. *Entamoeba histolytica*, a well known parasite of man, can cause dysentery, but most of the time it lives in the small intestine as a nonpathogen, and becomes pathogenic only when certain physiologic changes take place in the host, and probably in the parasite as well.

The original meaning of the word parasite was "situated beside," which referred to being near food; there was no reference to pathogenicity. Textbooks on parasitology are not necessarily restricted to a study of pathogenic parasites, and a "parasitologist" is frequently not concerned with pathogenic parasites at all. Should we therefore go back to the original term symbiosis to include *all* of our categories of associations, and entitle our text *Animal Symbiosis*, and call ourselves "symbiontologists"? While discussing an approach toward a course in the principles of parasitism, Huff *et al.*[11] concluded that "Although there is a clear trend to a biology of symbiosis, which would include parasitism, so that the basic course in the subject may eventually be one in symbiosis, present practice seems to call for a course in parasitism. For the present it seems useful to follow this practice, while continually bringing the wider range of relations into the picture." With these conclusions we agree.

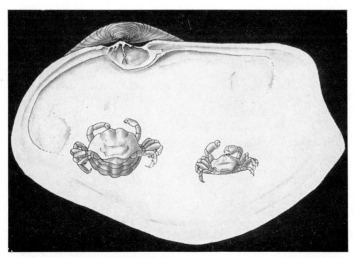

Fig. 1-1. Commensal male (small) and female pea crab, *Pinnixa faba*, in shell of gaper clam, *Schizothaerus nuttalii*.

A more accurate concept of parasitism, one that clearly separates it from other categories of symbiosis, is based on biochemical relationships between host and parasites. We have not been able to describe commensalistic, mutualistic, and parasitic relationships with greater precision because we do not understand enough of the economics of these various associations. Such an understanding requires precise knowledge of the biochemistry involved. However, making precise distinctions among the kinds of symbiotic associations is not really important. As our knowledge increases and the details of each partnership are analyzed and understood, the arbitrary boundaries of each category tend to disappear.

PARASITISM

If a species of parasite has lived with its host species for millions of years, each partner must have had to adapt itself to the other in many ways. Among the morphologic and functional changes that a free-living organism must undergo to become a parasite are metabolic changes that require the presence of host tissues or fluids. Parasites, therefore, are metabolically dependent upon their hosts.

This concept of parasitism has been developed by several parasitologists during the past 12 or 14 years. Cameron[2] stated that a parasite is "an organism which is dependent for some essential metabolic factor on another organism which is always larger than itself." Smyth[28] also described parasitism as an intimate association between two organisms in which the dependence of the parasite is metabolic. His definition, however, included an important addition, when he said that in parasitism "some metabolic by-products of the parasite are of value to the host." The following quotations are from Lincicome,[16] who reviewed the problem of defining parasitism and parasite:

Parasites are dependent on their environments for: 'a) developmental stimuli; b) nutritional materials; c) digestive enzymes; d) control of maturation.' [From Smyth.] It is not food alone therefore that is the basis of parasitism. The

organism and the whole of its process are involved. . . . Parasitism must be a two-way affair. . . . Parasitism fundamentally then has a chemical basis . . . This is the emphasis that has been lacking in all attempts to view parasitism philosophically to the present time. Parasitism is a great pattern of life on this planet. Fundamentally, it is an intimacy between two organisms that is basically a chemical, metabolic dependency in which there is a mutual exchange of chemical substances.

In the two earlier editions of this book we equated parasitism with symbiosis. We still accept the term symbiosis as a broad, general concept including commensalism, mutualism and parasitism, but the latter is a unique kind of symbiosis. *Parasitism may be defined, then, as an intimate association between two organisms in which the dependence of the parasite on its host is a metabolic one involving mutual exchange of chemical substances.*

As yet, we have little experimental evidence to support this concept of mutual chemical exchange. However, the evidence is accumulating (see, for example, Lincicome and Shepperson,[18] and Lincicome et al.[17]).

COMPARISON BETWEEN PREDATORY AND PARASITIC MODES OF LIFE

The tabulation on page 7 was prepared by Dr. Ralph Audy for classroom use (1966, unpublished) and is printed here with his permission. Each mode of life has its own kind of feeding relation "between one partner that is dependent on the other for its food supply, and the other partner (the victim) that provides its living substance or at least parts of itself to the other. Most of the statements should be qualified by 'customarily' or 'characteristically'."

IMMUNITY

Immune Reactions in Vertebrates

The body of a vertebrate animal normally reacts against the entry of foreign material. The material may be part of another animal or plant, a whole parasite (alive or dead), or any metabolic by-product of a living organism. The reactions of the host may

	DEPENDENT PARTNER	
	Predator	Parasite
Victim is known as	Prey	Host
Manner of feeding on individual victim	Destructive consumption	Cropping or sampling
Food provided by victim	Whole of body or fragment	Selected tissues, body fluids, excretions, etc.
Lethal to victim?	Yes, usually	Exceptional
Habitat	Co-exists with prey	Usually lives in or on host ("host as habitat") in its parasitic stage(s)
Size, compared with victim	Larger (or stronger)	Smaller or much smaller
Numbers, compared with victim	Less numerous	More or much more numerous
Encounter with prey or host	Momentary	Prolonged and/or repeated
Effect of grouping or crowding of victims	May protect prey	Encourages parasitism
Disease caused?	Rare and indirect through stress	Common, direct; may be great variety
Rate of multiplication compared with victim	Slower	Much faster

consist of the walling-off or encapsulation of the invading material, or the production of chemical substances called **antibodies,** which are definite protein entities. Each kind of foreign protein, called an **antigen,** stimulates the formation of one kind of antibody. The union of the antigen with the antibody initiates a series of complex events; this antigen-antibody reaction is usually specific. That is, each antibody normally combines only with the antigen (or one very similar to it) that engendered its production. As Stauber[29] has pointed out, the metabolic products of the parasite function to maintain the invader in its host; if the products have a physiologic activity, it is possible that antibodies to them will inhibit this activity. If inhibition occurs, the parasite will be unable to maintain itself in the host; thus a state of **immunity** will be established.

If the antibodies are unsuccessful in inhibiting the parasite, the host may be injured or killed. Usually, however, **a balance of activity** is reached. In parasitism,

this balance results from mutual cooperation between parasite and host, or mutual countervailing defense mechanisms of parasite and host. The host protects itself and tolerates the parasite. The parasite is allowed to live and, in turn, does little or no harm to the host, and often contributes something of value.

Compared with the immune responses caused by bacteria and viruses, the antigen-antibody manifestations involved with metazoan parasites are exceedingly complex. This complexity stems from the multiplicity of the antigen systems of each metazoan parasite. Because of the great variety of cells and tissues in the parasite body, many kinds of antigens are produced. During the development of a helminth parasite, especially one that goes through two or more stages of development and requires one or more intermediate hosts, biochemical and physiologic changes constantly occur that add to the complexity of the antigenic mosaic.[31] Heyneman[10] has stated, "This is the essential differ-

ence between microbial and helminth immunity—the worm's size and its antigenic
complexity. . . . An individual nematode larva,
passing through various growth stages as it
migrates through its host, presumably undergoing metabolic phases as well, sheds antigens
not only as successive larval cuticles, but more
importantly as a spewing out of metabolic
waste products and a variety of other secreted
and excreted antigenic substances."

Seven specific antigens have been demonstrated[15] by agar gel analysis from adult *Schistosoma mansoni* (a blood-dwelling trematode
parasite of man), as well as three from cercarial
larvae and five from the eggs of this worm.
Another analysis[3] of antigens from the same
species of parasite produced 21 in extracts
from adult worms, 11 shared by adults and
eggs, 14 shared with cercariae, and 12 in
excretions and secretion products.

Hosts and parasites may have antigens in
common. At least four were found to occur
in both *Schistosoma mansoni* and the laboratory
mouse.[5] Kagan[13] reported that only nine of
19 components in hydatid fluid (from a larval
tapeworm) were of parasite origin. The
concept of "eclipsed antigens" has been
proposed by Damian.[4] When an antigen of a
parasite resembles an antigen of its host, the
host does not recognize the parasite antigen
as being foreign. Thus the parasite antigen
is "eclipsed" and the host does not produce
antibodies against it. Such a relationship
would obviously be disadvantageous to the
host.

Parasite antigens are usually protein or
protein complexes on the surfaces of protozoa,
eggs or larvae, in excretions, secretions or
other fluid, in or on shed cuticles of worms, or
released by injured parasite cells or tissues.
In addition to the **ES antigens** (produced
from excretions and secretions), antigens are
elaborated by the entire, intact parasite
body, called **somatic antigens.** Somatic
antigens may be identified experimentally
in ground-up whole worms, in extracts made
from whole worms, or in specific tissues such
as cuticle or muscles. Protective immunity is
apparently best achieved by the presence of

whole, living organisms rather than dead
organisms or extracts. Examples of antigenic
components of parasite cells are lipoprotein,
glucoprotein, nucleoprotein, mucoprotein and
glucolipid. Antigens with broad specificity
are the polysaccharides, which exhibit blood-
group activity.[22] (See Kagan[14] for a review
of the characterization of parasite antigens.)

Antibodies are usually associated with the
host's serum globulin fractions. These numerous globulins with antibody activity are
called **immunoglobulins,** and as yet we
have little understanding of their biologic
functions. They are often considered as
belonging to two related groups: (1) those
that circulate in the blood **(humoral antibodies),** and (2) those that occur within cells
(cellular antibodies). Antibodies originate
in lymphoid cells (plasma cells) that occur in
the liver, spleen, lymph nodes and lymph
nodules. These cells are part of the circulatory system, and together with the phagocytes
of the blood, they comprise the **reticuloendothelial system.**

Details of antigen-antibody reactions may
be found in the several reviews and symposia
listed at the end of this chapter. In general,
host immune responses to metazoan parasites
are "manifested by changes in their [the
parasite] numbers, rate or extent of growth,
morphogenesis, extent of migration and
reproduction rate."[33] Antigen-antibody complexes initiate a rapid inflammatory response
in immune hosts. Proteolytic and other
enzymes are activated, and the increased
glycolytic processes result in greater acidity
at the site of inflammation. A variety of tissue
lesions may be produced. The complexity
of responses is a logical corollary to the
heterogeneity of antibodies.

When a host has made an immunologic
response to an antigen, the host is **sensitized**
to the antigen. Such a sensitization is
illustrated by the well known phenomenon
of allergic reaction (e.g., the rash, urticaria
and asthma that commonly occur in children
as the result of infection with the nematode,
Ascaris lumbricoides). A **delayed-type hypersensitivity** "is characterized by a state of

hypersensitivity to the antigen that does not depend upon demonstrable circulating antibodies and is not transferable by serum from one animal to another. It owes its name to the fact that the visible effects of antigen injected locally into a sensitized animal are relatively slow in onset and die away more gradually than the effects of immediate-type hypersensitivities. Immediate and delayed-type hypersensitivities commonly, but not always, coexist."[33]

Host red blood cells may become altered by parasitic infection to the extent that they are rendered antigenic toward the host. **Autoantibodies** against the host's own red cells augment their destruction.

The phenomenon of **cross reaction** occurs when antibodies formed from the action of antigens from one species of parasite react with the antigens of another species of parasite. For example, a cross reaction has been demonstrated between *Schistosoma mansoni* antigens and sera from patients with *Trichinella spiralis* infections. Another example was described by Hunter, *et al.*[12] when immunization in albino mice by larvae of the nematode, *Nippostrongylus brasiliensis*, was followed by a challenge infection with *Schistosoma mansoni* cercariae. Recovery of schistosomes was significantly lower as compared with controls. In **reciprocal cross immunity,** two different species of parasites are able to stimulate the production of antibodies that can react with the antigens of either species. In **nonreciprocal cross immunity,** only one of two different species can stimulate antibody formation that will affect both species.

If the parasite or parasitic substance used to reinfect (i.e., challenge) a host is of the same species as that which started the original infection (or is a vaccine made from the original species), the experiment is a **homologous** one; if the parasite or substance is of a different species, the experiment is a **heterologous** one, and is used to demonstrate cross reaction.

Concurrent infections (see p. 495) with two or more species of parasites in one host body are common. Therefore, the similarities and dissimilarities of antigens of the different parasites must be considered, as well as the immune responses of the host. Schad[26] has proposed a hypothesis stating that, "when co-occurring parasites are likely competitors, cross immunity may be a device evolved to limit the abundance of a competing species." In this theory, parasite species A produces an antigen that elicits an immunologic response against parasite species B, but not against A.

Premunition is a special kind of immunity to some parasites that do not provoke a lasting immunity by a single attack upon the host. Continued reinfection, however, maintains a state of relative immunity that protects the host from disease, with few if any symptoms. Premunition may be considered a compromise situation between parasite and host wherein both are able to remain alive.

Dineen[7] has proposed that the immune response creates an environment for the selection of genetic variants during the evolution of the parasite-host relationship. He described the factors that might determine the mean threshold level of parasitic infection as: (1) the degree of antigenic disparity between host and parasite, and (2) the rate of flow of antigenic information. If an antigen does not stimulate a response influencing the survival (or "fitness") of the parasite, it is immunologically impotent. There is no immunologic selective pressure to modify the parasite that produced the antigen, and such a parasite may remain highly antigenic. This situation might explain the presence of antibody with little or no resistance to infection. Dineen concluded that, "the role of the immunological response in the 'adapted' host/parasite relationship is to control the parasite burden rather than to cause complete elimination of the infection."

Although we have presented a few facts and theories relating to immune responses between vertebrates and their animal parasites, the underlying mechanisms of immunity remain obscure. Among helminth infections, a useful immunization technique for cattle that uses irradiated vaccine has been developed only against the nematode *Dictyocaulus viviparus*. A protective antibody has been characterized in terms of its biologic function only in the *Trypanosoma lewisi* parasite-host system. In 1967, Smithers[27] summed up our paucity of information by stating that "It follows that parasitologists are still faced with two basic problems in spite of the large amount of work done in this field: first, for the great majority of parasitic infections it is not yet possible to induce resistance by a safe and effective vaccine; second, although acquired immunity is common in parasitic infections, we know almost nothing of how it acts."

Immune Reactions in Invertebrates

Invertebrate hosts, like the vertebrates, are able to discriminate between normal body constituents and foreign matter, although antigen-antibody reactions have not been demonstrated in the invertebrates. While discussing protozoan parasites in arthropod hosts, Garnham[9] stated that the invertebrate host provides the most interesting features of infection, "tantalizing because most of the phenomena are inexplicable . . . the essence of the problem is the physical and chemical basis of susceptibility, or its converse, resistance. We have scarcely reached the stage of being able to define this problem."

Amebocytes (sometimes called haemocytes or lymphocytes) in the body cavity of metazoan invertebrates are responsible for intracellular digestion and also for removing, by phagocytosis, a wide variety of foreign particles of living or inert matter.[25] If the particles are large, such as larval trematodes in an unsuitable snail, the amebocytes respond by accumulating in layers around the parasite, forming an encapsulating nodule in which fibrils may be discerned. Ciliates sometimes parasitize the blood of a crab; if the ciliates are experimentally injected into the blood of an abnormal crab host, the parasites are agglutinated. Pan[23] reported that *Schistosoma mansoni* in the snail, *Australorbis glabratus*, often causes death of the host, and that death is probably due to "disruption of the tissues by migrating cercariae, and to the presence of

intense proliferative tissue reaction." A number of defense mechanisms in mosquitoes are known to result from protozoan and microfilarial infections. Microfilaria (nematodes) are commonly encapsulated or chitinized at all stages of the infection. The age of the host is an important factor—the older hosts become more heavily parasitized, but there are exceptions to this general phenomenon.[20]

Immune reactions attributed to acquired resistance in invertebrates have rarely been reported. Studies[21] on the responses of snails with the trematode, *Cotylurus flabelliformis*, demonstrated that snails infected with sporocysts were more resistant to penetration by homologous cercariae and to subsequent metacercarial development than were noninfected snails. Michelson[19] found that if the snail, *Australorbis glabratus*, is inoculated with *Schistosoma mansoni* eggs and an extract is made from the infected snails, the extract possesses a substance that immobilizes miracidia of *S. mansoni*. Such observations do not permit us to conclude that there are humoral as well as cellular responses in snails to foreign materials. Much more experimental work must be done, particularly with regard to the specificity of the reactions, and the use of tissue culture techniques.[30]

ZOONOSES

Zoonoses are diseases or infections that are naturally transferable between animals and man. In a broad sense, all animals are included in the definition, but most studies of zoonoses involve only diseases of vertebrates. The term **anthroponoses** means human diseases that are transmissible to animals.

The overall concept of zoonoses is complex. It involves man, another vertebrate, often an arthropod, the agent that causes the disease, and the environment—all forming a biologic whole. The interaction of these parts involves more than just a sum of the parts. A serious study of zoonoses should thus include the ecology of all organisms involved —parasite, animal, vector and man.

Many zoonoses, such as balantidiasis

(caused by an intestinal ciliate), fascioliasis hepatica (liver fluke disease) and tongue-worm infection, are found almost exclusively in animals and only rarely in man. Others, such as leishmaniasis (Oriental sore), flea infestation, African sleeping sickness and clonorchiasis (Chinese fluke infection), are common in both animals and man. Well over 100 zoonoses are known, and they may be grouped on the basis of the causative organisms: viruses, rickettsiae, bacteria, fungi, protozoa, nematodes, trematodes, cestodes and arthropods.

Hydatidosis is an example of a parasitic zoonosis with worldwide distribution. Hydatid disease is caused by a larval stage of the minute tapeworm, *Echinococcus granulosus* (see pages 244 to 248 for a description of the life cycle). Figure 1–2 illustrates the major factors involved in the study of this disease. Animal infection involves various farm and wild animals, but centers around dogs. The

fight against the spread of this infection is a public health problem and is based on the treatment of all dogs in infected areas, prevention of reinfection and elimination of stray dogs. For a thorough review of the zoonoses, see Beaver,[1] Van der Hoeden,[32] and Fiennes.[8]

AN ECOLOGIC APPROACH TO THE STUDY OF PARASITISM

The whole assemblage of parasites associated with a host population may be called the **parasite-mix.** Such an assemblage is a small biocoenose, and it includes all the viruses, bacteria, protozoa, molds, rickettsiae, worms and arthropods that live on or in another organism. The small biocoenose is a biologic entity that is constantly changing as it reacts with the environment. Parasitology is thus a study in ecology. Such an approach has been emphasized only in recent years.

When we label morphologic or physiologic

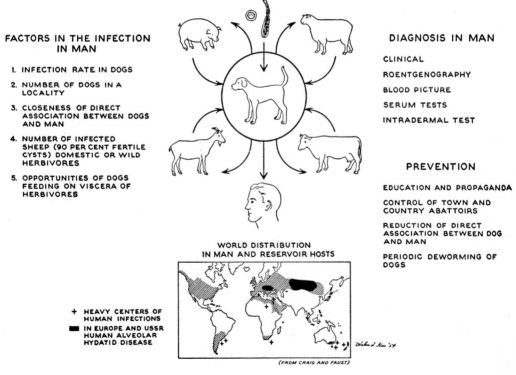

FACTORS IN THE INFECTION
IN MAN

1. INFECTION RATE IN DOGS

2. NUMBER OF DOGS IN A
   LOCALITY

3. CLOSENESS OF DIRECT
   ASSOCIATION BETWEEN DOGS
   AND MAN

4. NUMBER OF INFECTED
   SHEEP (90 PER CENT FERTILE
   CYSTS) DOMESTIC OR WILD
   HERBIVORES

5. OPPORTUNITIES OF DOGS
   FEEDING ON VISCERA OF
   HERBIVORES

DIAGNOSIS IN MAN

CLINICAL

ROENTGENOGRAPHY

BLOOD PICTURE

SERUM TESTS

INTRADERMAL TEST

PREVENTION

EDUCATION AND PROPAGANDA

CONTROL OF TOWN AND
COUNTRY ABATTOIRS

REDUCTION OF DIRECT
ASSOCIATION BETWEEN DOG
AND MAN

PERIODIC DEWORMING OF
DOGS

WORLD DISTRIBUTION
IN MAN AND RESERVOIR HOSTS

+ HEAVY CENTERS OF
HUMAN INFECTIONS
■ IN EUROPE AND USSR
HUMAN ALVEOLAR
HYDATID DISEASE

(FROM CRAIG AND FAUST)

Fig. 1-2. Epidemiology of hydatidosis, a zoonotic disease caused by the minute tapeworm, *Echinococcus granulosus*. (Meyer, courtesy of Univ. California Press.)

features as specific adaptations to parasitism we must bear in mind the universal need to adapt to the **environment.** Many characteristics that are described as hallmarks of the parasitic habit are also to be found among free-living species. The hallmark is sometimes present in only one or two species, or it may even disappear during a phase in the life cycle of an individual parasite. For example, cyst formation, so characteristic of parasitic protozoa, is common among free-living protozoa and metazoa. The complicated and significant alternation of sexual with asexual generations during life cycles of sporozoa, trematodes and other parasites is duplicated in foraminifera, hydroids and many other free-living species. The saprozoic form of nutrition can be illustrated abundantly among soil-dwelling organisms as well as among parasites.

In order to understand more completely the ecology of parasitism, we must thoroughly review environmental variables. We must avoid the promulgation of too many broad generalizations inadequately supported by specific data. Although generalizations must be synthesized and elaborated, they must emerge from detailed long-term studies, preferably with experimental work.

Ratcliffe *et al.*[24] have given an excellent example of how the conceptual approach known as **systems analysis** can help to overcome the difficulty of formulating a biologic model. Their study involved the nematode, *Haemonchus contortus*, and its sheep host. This approach is new to parasitology, and is a "method of model-building and analysis by which the selection of critical hypotheses can be made in a logical and systematic way."

GENERAL PRINCIPLES

In biology, a principle is a fundamental doctrine, theory or belief. Understanding the basic principles of ecology, evolution, genetics, morphogenesis, physiology and immunology is tantamount to understanding the basic principles of parasitology. These principles, however, must be adapted to the needs of parasitologists because parasitism is a great

deal more than a combination of parasites and hosts. Associations of these organisms create a SYSTEM that is unique. The components of the system can effectively be examined separately, but if principles of symbiosis are to be developed, the interrelations among all components of the system must be understood. The generalizations and hypotheses stated below could constitute the beginnings of a statement of "principles" of parasitism. Other principles may be found throughout the book, especially at the ends of the last three chapters.

Parasites have lost the capacity for free-living and have become dependent for their existence upon one or more other living species. They have, in general, lost sense organs, locomotor abilities, and certain metabolic functions, such as the elaboration of some digestive enzymes. These losses are compensated by various gains: a habitat that provides abundant food, shelter and some protection, a long individual life, specialized modes of reproduction and life cycles, and specialized organs of attachment.

The host has also lost some freedom. It must share its body with the parasite. The loss of food and the functions of resistance result in the diversion of energy. However, the host gains from the exchange of chemical substances with the parasite. In addition, the presence of one species of parasite often prevents the establishment of another, perhaps injurious, species.

Parasites and their hosts must struggle to keep these gains. They must cooperate so that the host remains in a healthy state and the parasite is not rejected. They must tolerate each other and resist each other, thereby becoming mutually adaptive and mutually beneficial. In this situation, the environment (the host) adjusts to the parasites. Since the host is the environment, the parasite must find a means of transport from environment to environment because a single host body provides limited space and it eventually dies. To satisfy this need, parasites depend upon the food and habits of the host. Appropriate triggering mechanisms initiate the change from infective stages to parasitic stages. Once the parasite has begun its existence in a new host body, other triggering mechanisms initiate each change of the parasite during its development.

REFERENCES

1. Beaver, P. C.: Zoonoses, with particular reference to parasites of veterinary importance. *In* Soulsby, E. (ed.): Biology of Parasites. pp. 215–227. New York, Academic Press, 1966.
2. Cameron, T. W. M.: Parasites and Parasitism. New York, Wiley, 1956.
3. Capron, A., Biguet, J., Rosé, F., and Vernes, A.: Les antigènes de *Schistosoma mansoni*. II. Étude immunoelectrophorétique comparée de divers stades larvaires et des adultes de deux sexes. Aspects immunologiques des relations hôte-parasite de la cercaire et de l'adulte de *S. mansoni*. Ann. Inst. Pasteur, *109*:798–810, 1965.
4. Damian, R. T.: Molecular mimicry: Antigen sharing by parasite and host and its consequences. Amer. Naturalist, *98*:129–149, 1964.
5. ———: Common antigens between adult *Schistosoma mansoni* and the laboratory mouse. J. Parasit., *18*:255–265, 1967.
6. de Bary, H. A.: Die Erscheinung der Symbiose. Strassburg, Karl J. Tübner, 1879.
7. Dineen, J. K.: Immunological aspects of parasitism. Nature (Lond.), *197*: 268–269, 1963.
8. Fiennes, R.: Zoonoses of Primates. The Epidemiology and Ecology of Simian Diseases in Relation to Man. Ithaca, N.Y., Cornell University Press, 1967.
9. Garnham, P. C. C.: Factors influencing the development of protozoa in their arthropodan hosts. *In* Taylor, A. E. R. (ed.): Host-Parasite Relationships in Invertebrate Hosts. Oxford, Blackwell Sci. Pub., 1964.
10. Heyneman, D.: Host-parasite resistance patterns—some implications from experimental studies with helminths. Ann. N.Y. Acad. Sci., *113*:114–129, 1963.
11. Huff, C. G., *et al.*: An approach toward a course in the principles of parasitism. J. Parasit., *44*:28–45, 1958.
12. Hunter, G. W., III., Velleca, W. M., and Crandall, R. B.: Studies on schistosomiasis. XXII. Cross resistance in *Schistosoma mansoni* and *Nippostrongylus brasiliensis* infections in mice. Exp. Parasit., *21*:15, 1967.
13. Kagan, I. G.: Seminar on immunity to parasitic helminths. VI. Hydatid disease. Exp. Parasit., *13*, 57–71, 1963.
14. ———: Characterization of parasite antigens. *In* Immunologic Aspects of Parasitic Infections. Sci. Pub. No. 150, pp. 25–36. Washington, D.C., Pan American Health Organization, 1967.
15. Kagan, I. G., and Norman, L.: Analysis of helminth antigens (*Echinococcus granulosus* and *Schistosoma mansoni*) by agar gel methods. Ann. N.Y. Acad. Sci., *113*:130–153, 1963.
16. Lincicome, D. R.: Chemical basis of parasitism. Ann. N.Y. Acad. Sci., *113*:360–380, 1963.
17. Lincicome, D. R., McLean, M. R., and Nelson, B. D.: Host serum malic dehydrogenase activity associated with *Trypanosoma lewisi* infection. Exper. Parasit., *20*:9–15, 1967.
18. Lincicome, D. R., and Shepperson, J. R.: Experimental evidence for molecular exchanges between a dependent trypanosome cell and its host. Exper. Parasit., *17*:148–167, 1965.
19. Michelson, E. H.: Development and specificity of miracidial immobilizing substances in extracts of the snail *Australorbis glabratus* exposed to various agents. Ann. N.Y. Acad. Sci., *113*:486–491, 1963.
20. Nelson, G. S.: Factors influencing the development and behaviour of filarial nematodes in their arthropodan hosts. *In* Taylor, A. E. R. (ed.): Host-Parasite Relationships in Invertebrate Hosts. pp. 75–119. Oxford, Blackwell Sci. Pub., 1964.
21. Nolf, L. O., and Cort, W. W.: On immunity reactions of snails to the penetration of the cercariae of the strigeid trematode, *Cotylurus flabelliformis* (Faust). J. Parasit., *20*:38–48, 1933.
22. Oliver-González, J.: Immunological properties of polysaccharides from animal parasites. Ann. Rev. Microbiol., *8*:353–361, 1954.
23. Pan, C-T.: Generalized and focal tissue responses in the snail, *Australorbis glabratus*, infected with *Schistosoma mansoni*. Ann. N.Y. Acad. Sci., *113*:475–485, 1963.
24. Ratcliffe, L. H., Taylor, H. M., Whitlock, J. H., and Lynn, W. R.: Systems analysis of a host-parasite interaction. Parasitol., *59*: 649–661, 1969.
25. Salt, G.: The defense reactions of insect to metazoan parasites. Parasit., *53*:527–642, 1963.
26. Schad, G. A.: Immunity, competition, and natural regulation of helminth populations. Amer. Naturalist, *100*:359–364, 1966.
27. Smithers, S. R.: The induction and nature of antibody response to parasites. *In* Immunologic Aspects of Parasitic Infections. Sci.

Pub. No. 150, pp. 43–49. Washington, D.C., Pan American Health Organization, 1967.

28. Smyth, J. D.: Introduction to Animal Parasitology. London, English Universities Press, 1962.

29. Stauber, L. A.: Some aspects of immunity to intracellular protozoan parasites. J. Parasitol., *49*:3–11, 1963.

30. Tripp, M. R.: Cellular responses of mollusks. Ann. N.Y. Acad. Sci., *113*:467–474, 1963.

31. Tromba, F. G.: Immunology of nematode diseases. J. Parasit., *48*:839–954, 1962.

32. Van Der Hoeden, J. (ed.): Zoonoses. Amsterdam, Elsevier, 1964.

33. WHO: Research in Immunology. Tech. Rep. Series No. 286. Geneva, WHO, 1964.

34. ———: Immunology and Parasitic Diseases. Report of a World Health Organization Expert Committee. Tech. Rep. Series No. 315. Geneva, WHO, 1965.

Section II
PHYLUM PROTOZOA

Chapter 2

Introduction; Subphylum Sarcomastigophora

INTRODUCTION

Protozoa abound in oceans, fresh water, soil and the bodies of other organisms. They are generally microscopic in size, and although they consist of a single cell with one or more nuclei, they are amazingly complex in structure, physiology and behavior (Fig. 2–1). Indeed, their complexity is so great that they are sometimes called "acellular," to distinguish them from the individual cells that make up a metazoan animal or plant.[2] Although some protozoan cells may group together in a colony, each cell maintains its independent function.

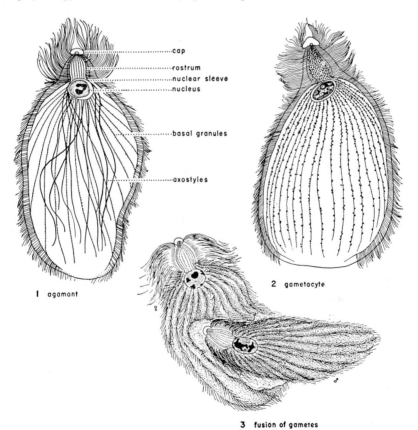

Fig. 2-1. *Eucomonympha imla* in the wood-feeding roach *Cryptocercus punctulatus.* Part of the process of gamete fusion is illustrated. Flagella are omitted from the upper surface of the rostrum in *1* (only rows of basal granules are shown). × 430. (Cleveland, courtesy of J. Morphology.)

Characteristics of Protozoa

No major morphologic features, outside of those common to cells of all plants and animals, can be listed for *all* the protozoa. Each subdivision of the phylum possesses its own distinctive morphology, as described in the classification on pages 20 to 25. The most conspicuous protozoan cytoplasmic inclusions, seen with an ordinary light microscope, are the various **vacuoles,** stored food in the form of **granules,** and the **chromatoid bodies** in some amebic cysts. **Mitochondria** appear as granules, rods or threads.

For detailed reviews of the physiology and biochemistry of protozoan parasites, see Danforth,[3a] Honigberg,[14] and von Brand.[38]

Nutrition

Protozoans exhibit three basic methods of obtaining nutrition. **Saprozoic** nutrition is the absorption of nutrients through the cell membrane. This method may be utilized with one of two other methods that follow. **Holozoic** nutrition is the ingestion of particulate food through the cytostome (cell mouth) or directly into the body wall through a temporary opening. Food vacuoles, in which digestion occurs, are formed in the cytoplasm. **Holophytic** nutrition is the synthesis of carbohydrates by means of chlorophyll (typical of green plants). This occurs in all unicellular species containing chloroplasts, such as *Euglena*. These species are usually classified as plants, and are generally placed in a separate phylum, but some may lose their chloroplasts and behave like animals. The boundary line between plants and animals is thin indeed, and there are logical grounds for using Ernst Haeckel's term "Protista" for all single-celled organisms.

Excretion

In the protozoa, excretion is accomplished primarily by diffusion through the cell membrane. Whether contractile vacuoles represent a primary excretory system can be questioned. These vacuoles, which are said to represent an osmoregulatory system, are absent from most marine and parasitic flagellates and amebas, but are present uniformly in marine and parasitic ciliates.

Undigested matter in ciliates is eliminated through a pore called a **cytopyge** or **cytoproct,** or through a temporary opening in the body wall.

Respiration

In parasitic species, respiration is usually termed aerobic (e.g., the malarial parasite, *Plasmodium*) or anaerobic (e.g., the amebic dysentery parasite, *Entamoeba histolytica*). In general, however, parasites carry on aerobic or anaerobic fermentations rather than complete oxidation, even when oxygen is plentiful. These fermentations are more varied than those occurring in vertebrate tissues, in which lactic acid is formed almost exclusively. Most parasitic protozoa, and also metazoan parasites, ferment carbohydrates to succinate. Lactate is the primary, if not the exclusive product of the malarial parasite, and is produced in lesser quantitites in some other species.

The host intestine is not completely devoid of oxygen. The oxygen tension in the intestinal gases of a pig has been estimated to average about 30 mm. Hg, or about one-fifth that of air. It should not be assumed that the oxygen tension of any one habitat is a uniform phenomenon—it varies considerably. Apparently, there is an oxygen gradient in the intestine, as evidenced by the fact that large numbers of protozoa may be found near the intestinal mucosa, with decreasing numbers toward the center of the lumen. A similar gradient exists in tissues near small blood vessels. However, these distributions of parasites might be due to other factors, such as differences in the pH of the different areas. Moreover, bacteria in the intestine may alter the conditions near a parasite by using the available oxygen themselves. For respiratory metabolism, see Danforth,[3a] and Ryley.[29a]

Reproduction and Life Cycles

Protozoan life cycles may consist of a series of simple binary fissions, multiple fissions (**schizogony**), or elaborate and precise integrations of sexual and asexual reproductions. Schizogony is a form of asexual multiplication. Since no gametes are involved, the process is sometimes called **agamogony,** in contrast to gamete production, which may be called **gamogony.** In schizogony, the nucleus undergoes repeated division; each nucleus then becomes surrounded by a separate bit of cytoplasm, and the original cell membrane ruptures, liberating as many daughter cells as there were nuclei. These daughter cells are **merozoites.** The parent cell undergoing nuclear division is called a **schizont.** If a multinucleate cell divides into portions that are still multinucleate, the process is called **plasmotomy** (as occurs in some Myxosporida). When a syncytium (many nuclei within one cell membrane) is produced (as in the Myxosporida), the process is called **nucleogony.**

Budding is another method of reproduction by some unicellular parasites. Essentially, the process is simply mitosis with unequal cellular division. This kind of exogenous budding commonly occurs among the Myxosporida (see p. 107). Endogenous budding within the cell membrane occurs in other protozoan parasites, including the Cnidospora.

Sexual reproduction occurs in parasitic protozoa in various forms. If two cells come together and exchange nuclear material, the process is **conjugation** (common in ciliates). When they separate, each cell may be called an **exconjugant.** If sex cells (**gametes**) are produced, they unite by **syngamy** to form a **zygote,** which is the first cell of a new individual. Gametes may be similar in appearance, in which case they are known as **isogametes,** or they may be dissimilar in appearance, resembling the eggs and sperm of higher organisms, in which case they are called **anisogametes.** For example, in the body of a mosquito, the gametocytes of malarial parasites form gametes of markedly different size and shape. The large variety (female) is a **macrogamete,** whereas the smaller one (male) is the **microgamete.** *Babesia bigemina,* the causative agent of cattle tick fever, has isogametes.

Coprophilic Protozoa

Coprophilic protozoa normally live in feces outside of the body, and this group does not include the usual intestinal protozoa. Often, however, cysts of coprophilic protozoa get into the gut of animals or man and are mistaken for parasites. Thirty or more species of such soil, sewage and fecal forms have been reported from various animals. They include flagellates, amebas and ciliates. The name **coprozoic** is also used to describe these organisms, but it should be restricted to those species that normally do not live in feces outside the body but that can be cultivated in feces.

If feces from any mammal remain at room temperature or even in a refrigerator for a few days, motile amebas, flagellates and cili-

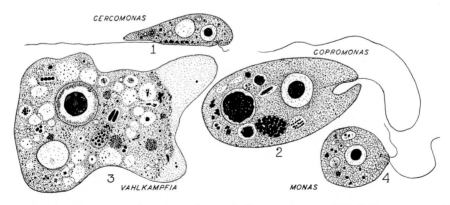

Fig. 2-2. Coprophilic protozoa from cattle feces. *1, Cercomonas* sp. × 5000. *2, Copromonas* sp. × 6000. *3, Vahlkampfia* sp. × 1600. *4, Monas communis* × 5000. (Noble, courtesy of J. Parasitol.)

ates of the coprophilic varieties may be found. Motile coprophilic amebas can be distinguished from parasitic species by the presence in the former of a contractile vacuole

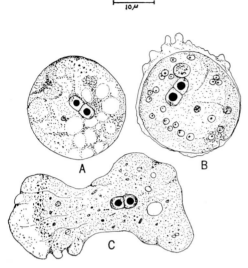

Fig. 2-3. *Sappinia diploidea*, a binucleate coprophilic ameba found in feces of large mammals. *A*. Probably a cyst of *S. diploidea*. *B*. A motile form entering the precystic stage. Note the parasites in the cytoplasm. *C*. Typical motile *S. diploidea* with a large, almost clear pseudopodium, a contractile vacuole, and ectoplasmic lines. There is a crescent-shaped mass of granules between each endosome and the nuclear membrane. (Noble, courtesy of J. Protozool.)

and, usually, a large endosome in the nucleus, although much variation exists. The ciliates are usually clearly freshwater species. The cyst stage of the Protozoa is the one that causes much of the confusion in identification.

A gradual increase in numbers of coprophilic protozoa normally occurs in stored feces at room temperature or in a refrigerator at 4° C. The order of appearance of the protozoa (Figs. 2–2 and 2–3) in mammal feces from pastures is usually as follows: flagellates (*Cercomonas, Copromonas, Monas*), amebas ((*Vahlkampfia, Sappinia*), and ciliates. The latter may not appear for several weeks, some of them are large holotrichs, and they tend to appear after the other protozoa have flared up in numbers and then died down.

Classification

The classification of the Protozoa is constantly being revised and improved. The system used here is that adopted by the Society of Protozoologists in 1964.[15] Only those taxa that contain parasites are listed, therefore the numbers before the taxa are not always consecutive. Representative families are mentioned on the following pages with descriptions of some of the more important genera and species.

Phylum PROTOZOA Goldfuss

Subphylum I. SARCOMASTIGOPHORA
  Flagella, pseudopodia, or both types of locomotory organelles; single type of nucleus except in developmental stages of certain Foraminiferida; typically no spore formation; sexuality, when present, essentially syngamy.
  Superclass *I.* MASTIGOPHORA Diesing
    One or more flagella typically present in trophozoites; solitary or colonial; asexual reproduction basically by symmetrogenic binary fission; sexual reproduction unknown in many groups; nutrition phototrophic, heterotrophic, or both.
    Class 1. PHYTOMASTIGOPHOREA
      Typically with chromatophores; if chromatophores lost secondarily relationship to pigmented forms clearly evident; commonly only one or two emergent flagella; amoeboid forms frequent in some groups; sexual reproduction known with certainty in few orders; mostly free-living.
      Order *6.* DINOFLAGELLIDA
        Two flagella, typically one transverse and one trailing; body usually grooved transversely and longitudinally, forming girdle and sulcus, each containing a flagellum; chromatophores usually yellow or dark brown, occasionally green or blue-green; many species thecate; food reserves starch and lipids.
    Class 2. ZOOMASTIGOPHOREA
      Chromatophores absent; one to many flagella; additional organelles may be present in mastigonts;

amoeboid forms, with or without flagella, in some groups; sexuality known in a few groups; species predominantly parasitic.

Order 3. RHIZOMASTIGIDA

Pseudopodia and one to four flagella.

Order 4. KINETOPLASTIDA

One to four flagella; kinetoplast argentophobic and Feulgen-positive, present as self-replicating organelle with mitochondrial affinities; most species parasitic.

Suborder (1) BODONINA

Typically two unequal flagella, one directed anteriorly, other posteriorly; no undulating membrane; kinetoplast absent secondarily in some species; free-living or parasitic.

Suborder (2) TRYPANOSOMATINA

One flagellum, either free or attached to body by means of undulating membrane; all species parasitic.

Order 5. RETORTAMONADIDA

Two to four flagella, one turned posteriorly and associated with ventrally located cytostomal area; cytostome bordered by fibril; parasitic.

Order 6. DIPLOMONADIDA

Body bilaterally symmetrical, with two karyomastigonts, each with four flagella and set of accessory organelles; most species parasitic.

Order 7. OXYMONADIDA

One or more karyomastigonts, each with four flagella, typically in two pairs, in motile stages; one or more flagella may be turned posteriorly, adhering for some distance to body surface; one to many axostyles; division spindle intranuclear; no Janicki-type parabasal apparatus; sexuality in some species; all parasitic.

Order 8. TRICHOMONADIDA

Typically four to six flagella, of which one is recurrent, per mastigont system; undulating membrane, if present, associated with recurrent flagellum; axostyle and non-dividing argentophilic Janicki-type parabasal apparatus (an organelle of the Golgi type) in each mastigont; division spindle extranuclear; sexuality unknown; true cysts unknown; all or nearly all parasitic.

Order 9. HYPERMASTIGIDA

Mastigont system with numerous flagella and multiple parabasal apparatus; basal bodies (kinetosomes) distributed in complete or partial circle, in plate or plates, or in longitudinal or spiral rows meeting anteriorly in centralized structure; nucleus single; division spindle extranuclear; sexuality in some species; all parasitic.

Suborder (1) LOPHOMONADINA

Extranuclear organelles arranged in one system; typically resorption of all old structures in division, with formation de novo of daughter organelles.

Suborder (2) TRICHONYMPHINA

Organization basically bilateral, with two or occasionally four mastigont systems; typically equal separation of systems in division, with total or partial retention of old structures when new systems are formed.

Superclass II. OPALINATA

Numerous cilia-like organelles in oblique rows over entire body surface; cytostome absent: two to many nuclei of one type; nuclear division acentric, binary fission generally interkinetal, thus usually symmetrogenic: known life cycles involve syngamy with anisogamous flagellated gametes; all parasitic.

Order 1. OPALINIDA

With characters of the superclass.

Superclass III. SARCODINA

Pseudopodia typically present; flagella, when present, restricted to developmental stages; cortical zone of cytoplasm relatively undifferentiated in comparison with other major taxa; body naked or with external or internal tests or skeletons of various types and chemical composition; asexual reproduction by fission; sexual reproduction, if present, with flagellate or, more rarely, amoeboid gametes; most species free-living.

Class 1. RHIZOPODEA

Locomotion associated with formation of characteristic lobopodia, filopodia, or reticulopodia; nutrition phagotrophic.

Subclass (1) LOBOSIA

Pseudopodia typically lobose, rarely filiform or anastomosing.

Order *1*. AMOEBIDA
Naked; typically uninucleate; majority free-living, many parasitic.
Subclass (5) LABYRINTHULIA
Groupings of spindle-shaped individuals which glide along filamentous tracks forming a slime net; occurrence of amoeboid stage in life cycle not clearly established; on marine plants and in soil.
Order *1*. LABYRINTHULIDA
With characters of the subclass.
Class 2. PIROPLASMEA [This class is now assigned to the Sporozoa. See p. 102.]
Small, piriform, round, rod-shaped or amoeboid; spores absent; no flagella or cilia; locomotion by body flexion or gliding; asexual reproduction by binary fission or schizogony; pigment not formed from host cell hemoglobin; heteroxenous; parasitic in vertebrate erythrocytes; known vectors ticks.
Order *1*. PIROPLASMIDA
With characters of the class.
Class 3. ACTINOPODEA
Spherical, typically floating forms, some attached secondarily; pseudopodia typically delicate and radiose, either axopodia or with filose or reticulate patterns; naked or with test; test membranous, chitinoid, or of silica or strontium sulfate; reproduction asexual and sexual; gametes usually flagellated.
Subclass (4) PROTEOMYXIDIA
Without test; filopodia and reticulopodia formed in some species; flagellated swarmers and cysts present in some species; most species parasitic.
Order *1*. PROTEOMYXIDA
With characters of the subclass.
Subphylum II. SPOROZOA
Spores typically present; spores simple, without polar filaments and with one to many sporozoites; single type of nucleus; cilia and flagella absent except for flagellated microgametes in some groups; sexuality, when present, syngamy; all species parasitic.
Class 1. TELOSPOREA
Spores present; reproduction sexual and asexual; locomotion by body flexion or gliding; pseudopodia ordinarily absent, if present used for feeding not locomotion; flagellated microgametes in some groups.
Subclass (1) GREGARINIA
Mature trophozoites extracellular, large; parasites of digestive tract and body cavity of invertebrates.
Order *1*. ARCHIGREGARINIDA
Life cycle apparently primitive, characteristically with three schizogonies; parasites of annelids, sipunculids, enteropneustids, and ascidians.
Order *2*. EUGREGARINIDA
Schizogony absent; parasites of annelids and arthropods.
Suborder (*1*) ACEPHALINA
Trophozoite composed of single compartment.
Suborder (*2*) CEPHALINA
Trophozoite septate, composed of more than one compartment.
Order *3*. NEOGREGARINIDA
Schizogony present, presumably reacquired secondarily; parasites of insects.
Subclass (2) COCCIDIA
Mature trophozoites small, typically intracellular.
Order *1*. PROTOCOCCIDA
Schizogony absent; two species known, in marine annelids.
Order *2*. EUCOCCIDA
Schizogony present; asexual and sexual phases in life cycle; in epithelial and blood cells of invertebrates and vertebrates.
Suborder (*1*) ADELEINA
Macrogamete and microgametocyte associated in syzygy during development; microgametocyte usually produces few microgametes; sporozoites enclosed in envelope; monoxenous or heteroxenous.
Suborder (*2*) EIMERIINA
Macrogamete and microgametocyte develop independently; syzygy absent; microgameto-

cyte typically produces many microgametes; zygote non-motile; oocyst does not increase in size during sporogony; sporozoites typically enclosed in sporocyst; monoxenous or heteroxenous.

Suborder (3) HAEMOSPORINA

Macrogamete and microgametocyte develop independently; syzygy absent; microgametocyte produces moderate number of microgametes; zygote motile in some forms; oocyst increases in size during sporogony; sporozoites naked; heteroxenous; schizogony in vertebrate and sporogony in invertebrate host; pigment ordinarily formed from host cell hemoglobin.

Class 2. TOXOPLASMEA

Spores absent; reproduction asexual by binary fission, endodyogeny or schizogony; locomotion by body flexion or gliding; pseudopodia and flagella absent in all stages; cysts or pseudocysts with many naked trophozoites; monoxenous.

Order 1. TOXOPLASMIDA

With characters of the class.

Class 3. HAPLOSPOREA

Spores present; reproduction asexual only; schizogony present; flagella absent; pseudopodia may be present.

Order 1. HAPLOSPORIDA

With characters of the class.

Subphylum III. CNIDOSPORA

Spores with one or more polar filaments and one or more sporoplasms; all species parasitic.

Class 1. MYXOSPORIDEA

Spore of multicellular origin; one or more sporoplasms; with two or three (rarely one) valves.

Order 1. MYXOSPORIDA

Spore with one or two sporoplasms and one to six (typically two) polar capsules; each capsule with coiled polar filament; filament probably with anchoring function; spore membrane generally with two, occasionally up to six, valves; coelozoic or histozoic in cold-blooded vertebrates.

Suborder (1) UNIPOLARINA

One to six polar capsules at or near anterior end of spore; capsules sometimes widely separated or located in central area, but polar filaments attached near anterior end.

Suborder (2) BIPOLARINA

Two widely separated polar capsules, one located and opening at or near each end of spore.

Order 2. ACTINOMYXIDA

Spore with three polar capsules, each enclosing polar filament; membrane with three valves; several to many sporoplasms; in invertebrates, especially annelids.

Order 3. HELICOSPORIDA

Spore with three sporoplasms surrounded by spirally coiled, thick filament; spore membrane with one valve; histozoic in insects.

Class 2. MICROSPORIDEA

Spore of unicellular origin; single sporoplasm; single valve; one long, tubular polar filament through which sporoplasm emerges; cytozoic in invertebrates, especially arthropods, and lower (rarely higher) vertebrates.

Order 1. MICROSPORIDA

With characters of the class.

Suborder (1) MONOCNIDINA

Spores independent.

Suborder (2) DICNIDINA

Two spores each with one polar filament, fused at base.

Subphylum IV. CILIOPHORA

Simple cilia or compound ciliary organelles in at least one stage of life cycle; subpellicular infraciliature universally present even when cilia absent; two types of nucleus, except in few homokaryotic forms; binary fission basically homothetogenic and generally perkinetal; sexuality involving conjugation, autogamy and cytogamy; nutrition heterotrophic; most species free-living.

Class 1. CILIATEA

With characters of the subphylum.

Subclass (1) HOLOTRICHIA

Somatic ciliature often simple and uniform; buccal ciliature, present in only two orders, basically tetrahymenal and generally inconspicuous.

Order 1. GYMNOSTOMATIDA

Essentially no oral ciliature; cytostome opening directly to outside; cytopharyngeal wall containing rods; body morphology and ciliation usually simple; mostly large.

Suborder (*1*) RHABDOPHORINA

Somatic ciliature usually simple; cytostome apical or lateral; cytopharynx with expansible armature of toxic trichocysts; commonly carnivorous.

Suborder (*2*) CYRTOPHORINA

Often dorso-ventrally flattened with ciliation restricted to ventral surface; cytostome ventral in anterior half of body; cytopharynx with complex armature of nematodesmata; commonly herbivorous.

Order *2.* TRICHOSTOMATIDA

Somatic ciliature typically uniform, but highly asymmetrical in some forms, vestibular but no buccal ciliature in oral area.

Order *3.* CHONOTRICHIDA

Somatic ciliature absent in mature individuals; vestibular ciliature in apical "funnel" derived from field of ventral cilia present on migratory larval forms; adults vase-shaped, attached to crustaceans by non-contractile, non-scopula-produced stalk; reproduction by budding.

Order *4.* APOSTOMATIDA

Somatic ciliature of mature forms spirally arranged; typically with unique rosette near inconspicuous cytostome; polymorphic life cycles with marine crustaceans usually involved as hosts.

Order *5.* ASTOMATIDA

Somatic ciliature typically uniform; cytostome absent; often large; some species with endoskeletons and holdfast organelles; catenoid "colonies" typical of some groups; mostly parasitic in oligochaetes.

Order *6.* HYMENOSTOMATIDA

Somatic ciliature typically uniform; buccal cavity ventral, with ciliature fundamentally composed of one undulating membrane on the right and adoral zone of three membranelles on the left; often small.

Suborder (*1*) TETRAHYMENINA

Oral ciliature, usually inconspicuous, composed of undulating membrane and three membranelles; vestibulum seldom present; body typically small.

Suborder (*2*) PENICULINA

Oral ciliature dominated by presence of "peniculi" deep in buccal cavity; outer vestibulum with uniform vestibular ciliature often present; usually large.

Suborder (*3*) PLEURONEMATINA

Somatic ciliature usually sparse, but prominent caudal cilium common; oral ciliature dominated by conspicuous external undulating membrane; adoral zone of membranelles hardly recognizable; vestibulum absent; cytostome typically subequatorial.

Order *7.* THIGMOTRICHIDA

Tuft of thigmotactic somatic ciliature typically present near anterior end of body; buccal ciliature, if present, located subequatorially on ventral surface or at posterior end; usually parasitic in or on bivalve molluscs.

Suborder (*1*) ARHYNCHODINA

Somatic ciliature usually uniform; oral ciliature and cytostome present.

Suborder (*2*) RHYNCHODINA

Somatic ciliature often reduced, absent in some forms; cytostome replaced by anterior sucker, occasionally on end of tentacle.

Subclass (*2*) PERITRICHIA

Somatic ciliature essentially absent in mature form; oral ciliature conspicuous, winding around apical pole counterclockwise to cytostome; body often attached to substrate by contractile, scopula-produced stalk or by prominent adhesive basal disc; colonial organization common; migratory larval form with aborally located ciliary girdle.

Order *1.* PERITRICHIDA

With characters of the subclass.

Suborder (*1*) SESSILINA

Predominantly sessile; with contractile or non-contractile stalk; some loricate; solitary or colonial.

Suborder (*2*) MOBILINA

Motile; without stalk; oral-aboral axis shortened; commonly with prominent adhesive basal disc at aboral end of body; often ectoparasitic on aquatic hosts.

Subclass (3) SUCTORIA
Mature stage without external ciliature of any kind; typically sessile forms, attached to substrate by non-contractile, scopula-produced stalk: ingestion through few to many suctorial tentacles; astomatous migratory larval stage, produced by budding, with some somatic cilia.
Order 7. SUCTORIDA
With characters of the subclass.
Subclass (4) SPIROTRICHIA
Somatic ciliature sparse in all but one order; cirri dominant feature of one order; buccal ciliature conspicuous, with adoral zone, typically composed of many membranelles, winding clockwise to cytostome; body often large.
Order 7. HETEROTRICHIDA
Somatic ciliature, when present, usually uniform; body frequently large; some species pigmented; a few species loricate with migratory larval forms.
Suborder (7) HETEROTRICHINA
With characters of the order *sensu stricto*.
Suborder (2) LICNOPHORINA
Somatic ciliature absent; conspicuous buccal membranelles on prominent oral disc; elaborate basal disc used for attachment; almost exclusively marine; ectoparasitic.
Order 2. OLIGOTRICHIDA
Somatic ciliature sparse or absent; buccal membranelles conspicuous, often extending around apical end of body; typically small; mostly marine.
Order 3. TINTINNIDA
All loricate, but motile; lorica exhibiting variety of shapes, sizes, and composition; oral membranelles conspicuous when extended from lorica; typically marine, pelagic.
Order 4. ENTODINIOMORPHIDA
Simple somatic ciliature absent; oral membranelles, functional in feeding, restricted to small area; other membranellar tufts or zones present in many species; pellicle firm, often drawn out posteriorly into spines; parasitic in herbivores.
Order 5. ODONTOSTOMATIDA
Somatic ciliature usually sparse; oral ciliature reduced to eight membranelles; body small, wedge-shaped, laterally compressed; pellicle sometimes with spines.
Order 6. HYPOTRICHIDA
Cirri arranged in various patterns on ventral body surface; adoral zone of membranelles prominent; body flattened dorso-ventrally.
Suborder (7) STICHOTRICHINA
Ventral somatic ciliature composed of variable number of rows of cirri; cirri relatively fine, generally numerous, and regularly distributed along their rows.
Suborder (2) SPORADOTRICHINA
Ventral somatic ciliature composed of six rows of fronto-ventral and transverse cirri, two rows of lateral cirri, and several caudal cirri; cirri heavy, typically few in number.
For general references to the protozoa, see Grell,[8] Hall,[10] Hyman,[19] Jahn and Jahn,[20] Kudo,[21] Levine,[22] MacKinnon and Hawes,[24] Manwell,[25] Pitelka,[29] Wenyon,[40] and Taylor and Baker.[34]

SUBPHYLUM SARCOMASTIGOPHORA
Superclass Mastigophora

The Mastigophora possess one or more long whip-like "tails" used for locomotion. In some species this structure is anterior in position, pulling the animal along rather than pushing it (e.g., *Copromonas ruminatium* in cattle feces).

ORDER DINOFLAGELLIDA

These free-swimming organisms, a few of which are parasitic, produce food reserves of starch and lipids and contain chromatophores. They are abundant in all oceans. The body possesses two grooves, one longitudinal (**girdle**) and one transverse (**sulcus**). A flagellum usually lies in each groove (Fig. 2–4).

Members of the Dinoflagellida can be found in numerous invertebrates. *Blastodinium* lives in the gut of the crustacean, *Cyclops;* a parasite of proportionate size in man would be as large as his liver. Parasitic dinoflagellates may also be found on other

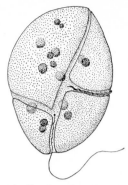

Fig. 2-4. A dinoflagellate showing the body grooves containing two flagella.

crustacea, in annelids and salps (ascidians), and on the gills of some freshwater and marine fish. These organisms can even be found within cells of siphonophores (colonial coelenterates) and in Protozoa (tintinnioid ciliates and Radiolaria). *Coccodinium*, for example, parasitizes other dinoflagellates and *Duboscquella tintinnicola* is a large (100 μ) parasite in tintinnioid ciliates. *Paradinium poucheti* lives in the body cavity of copepods, whereas copepod eggs may be inhabited by *Chytriodinium parasiticum* and by *Trypanodinium ovicola*.

ORDER RHIZOMASTIGIDA

The order Rhizomastigida shows affinities with the amebas by possessing ameboid bodies that often produce pseudopodia. In general, they possess one to four flagella, although members of one family, Multiciliidae, have many flagella. Most are freeliving. *Rhizomastix gracilis* is a species found in crane-fly larvae and in the salamander known as "axolotl." *Mastigina hylae* occurs in the intestines of many species of frogs; its nucleus is situated at the extreme anterior end of the body.

The genus *Histomonas*, commonly placed with this order, has strong affinities with the order Trichomonadida. A description of one species may be found on page 51.

ORDER KINETOPLASTIDA

Family Trypanosomatidae

This family consists of elongated leaf-like or sometimes rounded protozoa that possess a single flagellum at the anterior end. The flagellum serves as an organ of attachment as well as locomotion. However, it is absent in certain attached stages and in the amastigote stage. All parasites in this family are internal, and most are apparently harmless commensals, usually found in arthropods or vertebrates. The flagellated stages range in length from 4 μ to 130 μ.

The flagellum arises from a **kinetosome** at the floor of an invagination that usually occurs at the anterior end of the body but sometimes near the posterior end or in between (see first basal body in Fig. 2–5). This invagination is called the **flagellar pocket,** and it may be shallow or long enough to reach almost to the posterior end of the body. A contractile vacuole, and occasionally more than one, opens into the flagellar pocket. In at least one species (*Trypanosoma ariae*), a tube-like **cytopharynx** opens through a **cytostome** at the beginning of the pocket.

The kinetosome (sometimes called a **bleph-**

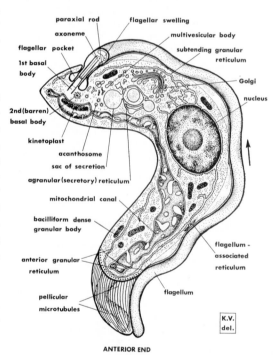

Fig. 2-5. The fine structure of *Trypanosoma congolense* in its bloodstream phase as revealed by phase contrast observations and electron microscopy. (Vickerman, courtesy of J. Protozool.)

aroplast, **basal body** or **centriole**) is a hollow cylinder composed of nine equally spaced fibrils; it averages about 1.5 μ in overall diameter. The fibrils are connected to those of the axoneme within the flagellum. There is some evidence that the kinetosomes are self-duplicating, but there is good evidence that they arise de novo each time the cell divides. They apparently contain DNA, although its origin is obscure (possibly from the nucleus), and its function in the kinetosome is unknown.

Just posterior to the kinetosome, which often cannot be seen, lies the **kinetoplast,** a spherical, rod- or disc-shaped structure peculiar to the family Trypanosomatidae. It is usually deeply stained in prepared specimens. The kinetoplast contains DNA that possesses different genetic properties from that of the nucleus. From the kinetoplast arises a large posterior mitochondrion (**posterior chondriome**), and often an even larger anterior mitochondrion (**anterior chondriome**). The sizes and shapes of these mitochondria vary with the species of trypanosome and the stage in its life cycle.

The kinetoplast is usually large, with the fibrillar DNA component well developed in cultivation forms of all genera of the Trypanosomatidae and in the blood forms of the *lewisi* group. The blood forms of the pathogenic trypanosomes belonging to the *brucei*

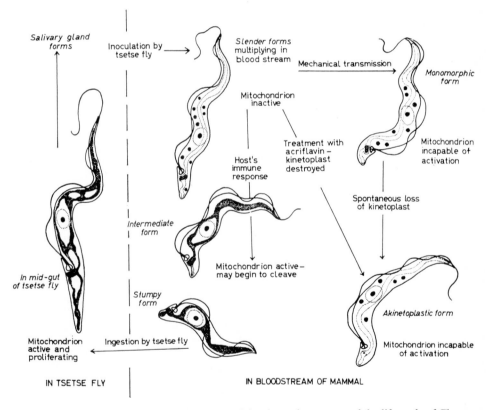

Fig. 2-6. Scheme showing activity of mitochondrion in various stages of the life cycle of *Trypanosoma rhodesiense* and its derivatives. The mitochondrion is inactive in the slender forms multiplying in the blood, but becomes active as the trypanosomes transform into stumpy forms, and remains active and proliferates in the first part of the cycle in the fly. What happens in the fly salivary glands is not yet known. If bloodstream forms are passaged by syringe in laboratory rodents, they lose the ability to undergo mitochondrial activation and can no longer transform or be transmitted by tsetse; such forms may spontaneously lose the kinetoplast. Slender bloodstream forms can also be rendered akinetoplastic artificially by treatment with certain drugs. (Vickerman, courtesy Sci. Prog.)

and *congolense* groups possess a smaller, more compact kinetoplast, with a less pronounced DNA element. The kinetoplast represents a specialized mitochondrion. Vickerman[35] has described kinetoplastic respiration in developing stages in insect vectors, where the kinetoplast is large and active, and akinetoplastic respiration in the narrow forms in mammalian blood, where the kinetoplast is reduced and inactive (Fig. 2–6). Presumably the kinetoplast, and more precisely the DNA contained therein, is responsible for the development of the mitochondrial cristae in the tubular extensions, and thus for the production of enzymes and cofactors of the tricar-

boxylic acid cycle and of the cytochrome-mediated electron-transport system.[33]

The six morphologic stages found in the life cycles of Trypanosomatidae are named mainly on the basis of the arrangement of the flagellum: its starting point (indicated by by the kinetoplast), its course through the body, and its point of emergence.[11]

1. **Amastigote** (leishmanial): rounded, no external flagellum.
2. **Promastigote** (leptomonad): antenuclear kinetoplast, flagellum emerging from the anterior end of the body.
3. **Opisthomastigote** (trypanosome or trypanomorphic): postnuclear kinetoplast,

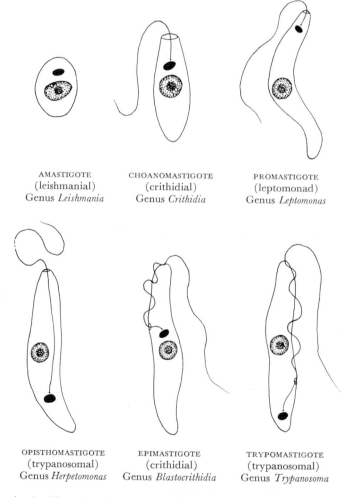

AMASTIGOTE
(leishmanial)
Genus *Leishmania*

CHOANOMASTIGOTE
(crithidial)
Genus *Crithidia*

PROMASTIGOTE
(leptomonad)
Genus *Leptomonas*

OPISTHOMASTIGOTE
(trypanosomal)
Genus *Herpetomonas*

EPIMASTIGOTE
(crithidial)
Genus *Blastocrithidia*

TRYPOMASTIGOTE
(trypanosomal)
Genus *Trypanosoma*

Fig. 2-7. Stages in the life cycles of trypanosomatid flagellates. The words in parentheses are those formerly used for these stages. (Modified from Hoare and Wallace, courtesy Nature.)

flagellum emerging from the anterior end (in the genus *Herpetomonas* only).

4. **Epimastigote** (crithidial): juxtanuclear kinetoplast, flagellum emerging from the side of the body and continuing to the anterior end as an undulating membrane.

5. **Trypomastigote** (true trypanosome): postnuclear kinetoplast, flagellum emerging from side, long undulating membrane.

6. **Choanomastigote** (crithidial): usually anterior kinetoplast, flagellum emerging from a wide funnel at the anterior end.

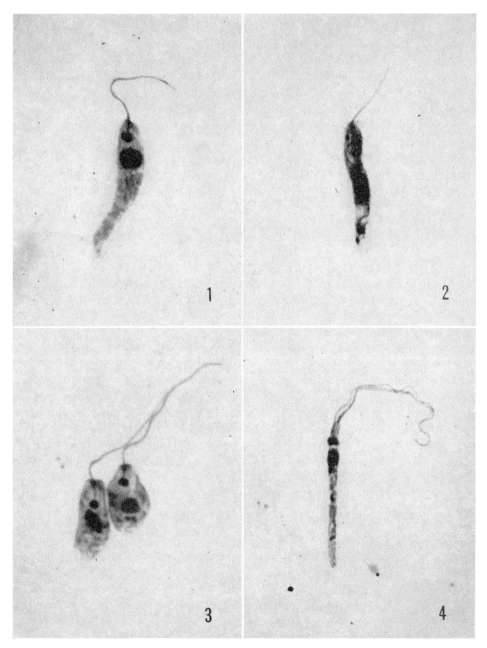

Fig. 2-8. Four genera occurring in the family Trypanosomatidae. (1) *Leptomonas;* (2) *Herpetomonas;* (3) *Crithidia;* (4) *Blastocrithidia.* (Wallace, courtesy of Exp. Parasitol.)

These stages are illustrated in Figure 2–7; each is typical of the mature stage of the genus listed with it. The life cycles of these genera may include several of the morphologic stages.

The following descriptions are of the main genera of Trypanosomatidae. Those of *Leptomonas*, *Herpetomonas*, *Crithidia* and *Blastocrithidia* (Fig. 2–8) are from Wallace[39]; the description of *Trypanosoma* is from Guttman and Wallace.[9] The genus *Phytomonas* is morphologically indistinguishable from *Leptomonas*, but because the various species live in plants, they are not described here. *Leptomonas* is so similar to *Herpetomonas* that some authorities doubt the existence of two separate genera.

Genus Leptomonas. Elongate trypanosomatids, these organisms are pointed at the posterior end, and pointed or narrowly rounded at the anterior end. A reservoir opens narrowly at the anterior end; the depth of the reservoir does not exceed one-fifth of the body length. The kinetoplast occurs below the base of the reservoir. The flagellum emerges from the bottom of the reservoir, passes out the mouth of the reservoir, and continues freely; there is no undulating membrane. The nucleus occurs in the middle third of the body length. Parasites of protozoa, nematodes, molluscs, and insects. Sample species: *L. bütschlii*.

Genus Herpetomonas. Opisthomastigote parasites of insects. The body is elongate and relatively slender, pointed or truncate at the posterior end, and pointed or rounded at the anterior end. The nucleus occurs in the middle or posterior third of the body length. The reservoir opens narrowly in the anterior end, and extends a variable distance into the body. The kinetoplast is situated anywhere from a point one-fifth to about nine-tenths of the body length. The flagellum originates near the kinetoplast, passes through the reservoir, and emerges through the mouth of the reservoir. There is no undulating membrane. Sample species: *H. muscarum*.

Genus Crithidia. Relatively small (4 to 10 μ long) choanomastigote parasites of insects, with a body usually shorter and wider than in other genera. The anterior end is truncate. The body is often constricted slightly in the anterior third, to produce a vase-like shape. The reservoir is wide, with its mouth (cytostome) occupying most of the truncate anterior end. The kinetoplast is large, lateral, and relatively far posterior, sometimes near or beside the nucleus. The flagellum moves in a circular motion. These organisms are often clustered together or attached like the pile of a carpet to the intestinal wall. Sample species: *C. fasciculata*.

Genus Blastocrithidia. Insect epimastigote flagellates (15 to 33 μ long) drawn out to a long point anteriorly. The reservoir opens along the side. The kinetoplast is situated just posterior to the base of the reservoir, and usually anterior to or beside the nucleus, but in occasional individuals, it occurs behind the nucleus. The flagellum arises near the kinetoplast, passes out through the reservoir and along an undulating membrane to the anterior end, whence it continues as a free flagellum. Flagellar cysts occur in some species. Sample species: *B. gerridis*.

Genus Trypanosoma. Elongate trypomastigote parasites (12 to 130 μ long) with a morphologically complex life cycle. In the form characteristic of the genus, the reservoir is considerably posterior to the nucleus. The kinetoplast is posterior to the base of the reservoir; a mass of electron-dense, anteroposterior fibers form a compact, sharply defined zone across the kinetoplast. The flagellum is attached to the body by an undulating membrane. The life cycle usually involves a vertebrate and an invertebrate host. Sample species: *T. cruzi*.

Genus Leishmania. Rounded or oval, parasites (1.5 to 3 μ) of invertebrates and vertebrates with a morphologically complex life cycle. Body forms include a promastigote with a free flagellum, and the characteristic amastigote, in which the flagellum is never fully emergent. The fine structure of the kinetoplast is like that of *Trypanosoma*. The life cycle usually involves two hosts. Sample species: *L. donovani*.

Trypanosomes

Trypanosomes are usually found in the body fluids of vertebrates, especially in the plasma, and in the digestive tracts of arthropod or leech vectors. Trypanosomes, however, may occur in any body organ and some have a "preference" for certain organs such as the heart. The numbers of these parasites in one host may be enormous. For example, 20 million to four billion can be recovered from the blood of an animal 100 hours after infection. However, the relation between numbers and infectivity of trypanosomes varies significantly during the normal development of these flagellates in the vector and in the mammal host. Host specificity also varies considerably. Some species, especially *Trypanosoma congolense*, occur in practically all domestic animals; others are found almost exclusively in one type of host, like *T. lewisi* in rats. Our knowledge of the structure of these organisms has been vastly increased by electron microscopy. One surprising discovery was the presence of a cytostome in the toad trypanosome, *T. mega*. This structure appears as a long, funnel-shaped gullet that extends well into the cytoplasm. Feeding apparently occurs by **pinocytosis** ("cell drinking"), which involves the passage of food down the cytostome and absorption through an enlargement or vacuole that forms at the end of the cytostome.

Identification of the many species and strains of trypanosomes is often exceedingly difficult, and usually requires a combination of methods to determine various characteristics. These usually involve a study of the location and behavior of the parasite in the insect vector; the morphologic types that occur in the vertebrate; method of transmission; and physiologic and biochemical characteristics. Trypanosomes are commonly placed in one of four groups: *lewisi, vivax, congolense,* and *brucei*. For a detailed classification of mammalian trypanosomes, see Hoare.[11]

Representative Species of Trypanosoma.
1. *Lewisi Group.* Posterior end pointed; kinetoplast large; not terminal; free flagellum always present; division occurs in amastigote, epimastigote or trypomastigote stages; transmission contaminative through feces (*T. rangeli* also inoculative).

TRYPANOSOMA LEWISI (Fig. 2–9). This common blood parasite of rats occurs throughout the world. It is a slender flagellate, pointed at both ends and averaging about 25 μ long. The life cycle of this nonpathogenic organism is similar to that of *T. cruzi*, but the intermediate host is the rat flea, *Ceratophyllus fasciatus*. *T. lewisi* has commanded considerable attention because it has been the subject of numerous physiologic studies. For example, rats can be immunized against *T. lewisi* infections by injections of metabolic products of the flagellate. A substance called **ablastin** is produced by the rat; this apparently inhibits reproduction of the trypanosomes and causes them to agglutinate, thereby making the task of the host white blood cells easier (see Chapter 21).

TRYPANOSOMA THEILERI (Fig. 2–9). This species is a large blood parasite of cattle. This flagellate may reach 70 μ in length in the ordinary blood stage, and up to 120 μ in length in animals suffering from a chronic infection of the disease. The lowest limit of unaided human vision is about 100 μ, so this relatively huge trypanosome can be seen without a microscope. The life cycle is similar to that of *T. cruzi*, and the insect vector is probably the horsefly (*Tabanus*).

TRYPANOSOMA (=SCHIZOTRYPANUM) CRUZI. South American trypanosomiasis (Chagas' disease) is found mainly in Central and South America and is caused by this species. A World Health Organization report estimates that at least seven million people have the disease. The incidence of infection may be as high as 50 per cent, especially in children. The adult flagellate lives in the blood and reticuloendothelial tissues of man and of dogs, cats, rats, monkeys, armadillos, opossums, and other mammals. Within the mammalian host, *T. cruzi* enters tissue cells and changes to amastigote forms that multiply rapidly. Amastigotes develop into promastigote and epimastigote stages, and finally to

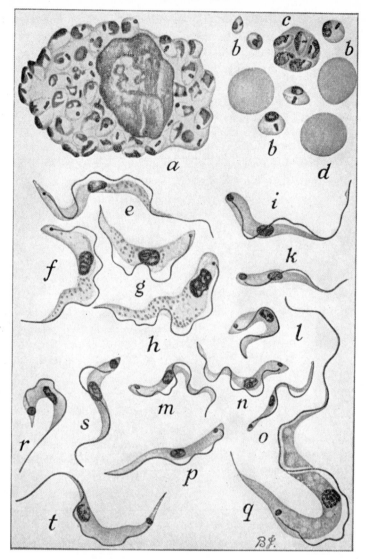

Fig. 2-9. *Leishmania* and *Trypanosoma* from various animals. *a,b,c, Leishmania* in tissue smears. *a.* Macrophage packed with parasites. *b.* Parasites scattered by rupture of host-cells. *c.* Detached portion of host-cell with parasites. *d.* Erythrocytes. *e–t.* Trypanosomes. *e–h. T. brucei. e.* Slender; *f,* intermediate; *g,* stumpy. *h.* Posterior-nuclear forms. *e,f. T. evansi* and/or *T. equiperdum. i. T. vivax. k. T. uniforme. l,p. T. congolense. l,m,n,o. T. simiae. q. T. theileri. r,s. T. cruzi. t. T. lewisi.* (Richardson and Kendall, courtesy Oliver and Boyd, 1957.)

trypomastigote forms that re-enter the blood and lymphatic vessels.

A common insect vector of *Trypanosoma cruzi* is *Panstrongylus megistus* in Brazil, or other members of the insect family Reduviidae. With a meal of blood the trypanosomes are taken to the posterior part of the insect gut (**posterior station**), where they develop into the amastigote and epimastigote forms and multiply. The epimastigote forms develop into small trypomastigotes. These "metacyclic" individuals occur in the rectum and represent the infective stage. The cycle in the insect requires about two weeks, but it varies with temperature and other factors. When man is bitten by the bug, the insect usually defecates while feeding and thus deposits infective parasites on the skin. Often, the irritated host rubs the bitten area, so there is considerable opportunity for the parasites to be rubbed into the wound made by the bug. Animals are sometimes infected by eating bugs or by eating other infected animals. Infected bugs may infect "clean"

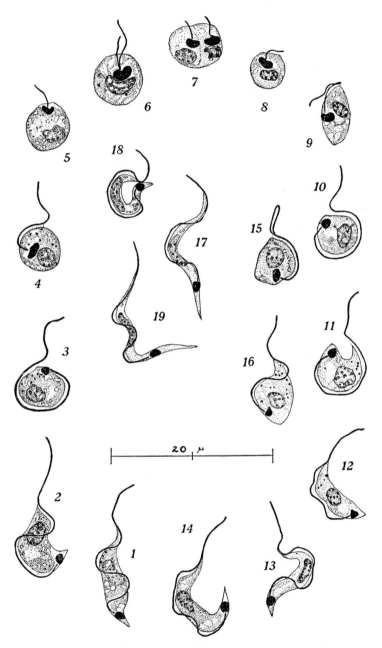

Fig. 2-10. Intramuscular development of Brazilian *Trypanosoma cruzi*. All figures were drawn with the aid of a camera lucida from slides stained with Jenner-Giemsa. The developmental cycle is illustrated by Figures *1* through *14*, exemplifying indirect development (Figures *9, 10, 11*, and *12*) of progressive parasites. Less frequently, direct development was observed (Figures *9, 15, 16* and *12*). (Wood, courtesy of Amer. J. Trop. Med. Hyg.)

bugs when they are kept together in the same chamber.[26] The insect is apparently infective for life.

Symptoms of Chagas' disease are so varied that a clear diagnostic picture is difficult to present. Often, however, the bug bites the area of the eye, especially in children. During the acute stage there occurs a unilateral swelling of the face (Romaña's sign), apparently due to the bite of the bug, not to the trypanosomes. The eye, usually the left one, becomes puffy and is often closed. Both eyes and even the whole face may become involved. As parasites invade body organs, enlargements of the spleen, lymph nodes and liver occur, with headaches, fever, anemia, and prostration. The "mega" condition of enlargement of the esophagus or colon appears to be related to this disease.[5] The heart, striated muscles, and nervous system may also be affected. Intramuscular forms (Fig. 2–10) are mostly amastigote stages. Distinct immunologic types exist that may explain clinical diversity.[28] The disease is diagnosed by a complement fixation test, fluorescent antibody test, and finding the flagellates in the blood. Treatment is of little value, and the mortality rate is high, especially in children.

In some parts of the southwest United States, wood rat nests are inhabited by triatomid bugs that are parasitized with *Trypanosoma cruzi*, and the disease in man has been reported a few times from this area. The flagellate has been found in raccoons and in other mammals in the United States, and it has been successfully established, experimentally, in sheep, goats, and pigs. Obviously, a potential source of considerable danger to man exists in North America. See von Brand[36,37] for a review of the physiology of *T. cruzi* and other Protozoa.

TRYPANOSOMA RANGELI (T. ARIARII). Man, monkeys (*Cebas fatuellus*), dogs, and possibly opossums, in Venezuela, Brazil, Colombia, Guatemala, Chile, and French Guiana, harbor this species. The insect vector is the triatomid bug, *Rhodnius prolixus*, which transmits the flagellate during the act of biting the host. To get into the salivary glands of the insect, the parasite must penetrate the midgut, pass through the hemocoel and migrate to the head region, where it leaves the vector through the hypopharynx. In the vertebrate host, the parasite multiplies only in the trypomastigote stage. This trypanosome is apparently not pathogenic to vertebrates, but damages its insect hosts.

TRYPANOSOMA MELOPHAGIUM. This common blood parasite of sheep in England is transmitted by the sheep ked (*Melophagus ovinus*), which is a wingless, bloodsucking fly. Probably up to 90 per cent of British sheep are infected with this flagellate. A similar parasite, *T. theodori*, occurs in goats.

2. *Vivax Group*. Monomorphic forms; free flagellum always present; rounded posterior end of body; development in the fly, *Glossina*, in proboscis (**anterior station**) only; *T. vivax* also transmitted mechanically by tabanid flies.

TRYPANOSOMA VIVAX. This active blood parasite occurs in practically all domestic animals in Africa, the West Indies and parts of Central and South America. Dogs and pigs are not easily infected. In Africa, the insect vector is the tsetse fly, but in other countries transmission is by mechanical means, indicating that the parasites have been introduced to these countries and have been able to maintain themselves in spite of the lack of a suitable intermediate host (see p. 555). These parasites average 22 μ in length, with a range of 20 to 26 μ. The posterior end of the body is characteristically wider than the anterior end. *T. vivax* does not have a stomach phase in the insect vector; it develops only in the proboscis.

TRYPANOSOMA UNIFORME. This species is shorter than *T. vivax* and occurs in cattle, sheep, goats and antelope. It is transmitted by tsetse flies.

3. *Congolense Group*. Monomorphic or polymorphic forms; free flagellum present or absent; kinetoplast typically marginal; medium in size; development in *Glossina* in midgut and proboscis. Monomorphic forms: free flagellum short or absent, undulating membrane inconspicuous; length usually 12 to 17 μ.

TRYPANOSOMA CONGOLENSE. This short flagellate lacks a free anterior flagellum. It averages about 13 μ in length, with a range of 9 to 18 μ. It is carried by the tsetse fly, from which infection occurs through the feces. A wide variance in degree of virulence suggests the existence of several strains of *T. congolense*. Usually the infection is serious in cattle, in which it causes anemia and other symptoms common to other forms of trypanosomiasis and, with some strains, is almost always fatal. These parasites may also be found in horses, sheep, goats and camels.

TRYPANOSOMA SIMIAE. A species similar to *T. congolense* but longer, *T. simiae* averages 14 to 24 μ. It is a common and virulent parasite of African monkeys, and is also found in sheep, goats, pigs and camels. It is the most important trypanosome of domestic swine, but apparently does not affect horses, cattle or dogs. It is transmitted by tsetse flies, and can be transferred mechanically by blood-sucking flies.

4. *Brucei Group*. Monomorphic or polymorphic forms; free flagellum present or absent; kinetoplast small, subterminal (lacking in *T. equinum*); undulating membrane conspicuous; development in *Glossina* in midgut and salivary glands (except in *evansi* subgroup).

TRYPANOSOMA GAMBIENSE. One of the causative agents of human sleeping sickness that is endemic in the west and central portions of Africa is *T. gambiense*. This disease should not be confused with the virus-caused sleeping sickness found in the United States and elsewhere. *T. gambiense* in the blood of man may be long and slender, measuring about 25 × 2 μ with a flagellum, or short and broad without a flagellum, or intermediate in shape. The parasite multiplies by longitudinal splitting, and it migrates through the body by way of the blood. Normal habitats are the blood plasma, cerebrospinal fluid, lymph nodes and spleen.

The vector host of this flagellate is usually the tsetse fly, *Glossina palpalis*, which bites man and feeds on his blood. In the intestine of the fly, the parasite reproduces and forms both epimastigotes and trypomastigotes. After two weeks or more in the gut of the fly, the flagellates migrate to the salivary glands (**anterior station**) where they become attached to the epithelium and develop into the infective stage.

The Gambian or chronic form of sleeping sickness primarily involves the nervous system and the lymphatic system. After an incubation period of one or two weeks, fever, chills, headache, and loss of appetite usually occur, especially in non-natives. As time goes on, enlargement of the spleen, liver and lymph nodes occurs, accompanied by weakness, skin eruptions, disturbed vision and a reduced pulse rate. As the nervous system is invaded by the parasites, the symptoms include weakness, apathy, headache, and definite signs of "sleeping sickness." A patient readily falls asleep at almost any time. Coma, emaciation and often death complete the course of the disease, which may last for several years. The mortality rate is high.

TRYPANOSOMA RHODESIENSE. Closely related to *T. gambiense* is *T. rhodesiense*. It is identical in appearance and has the same type of life cycle, but it also occurs in antelope and cattle. The insect vectors are the tsetse flies, *Glossina morsitans*, *G. pallidipes*, and others. *T. rhodesiense* causes the Rhodesian or acute and more rapid type of human sleeping sickness, which usually results in death within a year. Because of its relatively rapid course, Rhodesian trypanosomiasis rarely if ever causes the symptoms normally associated with sleeping sickness. The incidence of infection is less than that with *T. gambiense*, and the parasite is restricted to a much more limited area, being almost confined to the high tablelands of southeast Africa.

There is some justification in considering *Trypanosoma rhodesiense* as a virulent type of *T. gambiense*.[41] Certainly the distinctions between the two types of sleeping sickness are difficult to recognize. Much more knowledge about *Glossina* host relationships is needed, as well as more infectivity measurements and more research on the immunology of trypanosomiasis. For a review of the biologic aspects of trypanosomiasis research, see Lumsden.[23]

TRYPANOSOMA BRUCEI. This flagellate causes **nagana** in livestock. It is morphologically identical with *T. gambiense* and *T. rhodesiense*, and is also transmitted by the tsetse fly; however, it is unquestionably a different species.[3] *T. brucei* is widely distributed in Africa in dogs, sheep, goats, horses, mules, donkeys and camels. Cattle and pigs suffer mild infections, whereas laboratory animals may be severely infected. The parasites range from 25 to 35 μ in length by 2 to 3 μ in width. Long, short and intermediate forms may appear in the blood at the same time. As the parasites spread throughout the body, host reaction is shown by edema, anemia, fever, nervous symptoms, conjunctivitis, keratitis, blindness, paralysis and, especially in horses, death. *T. brucei* is present as a natural infection of many wild mammals in Africa. A related species, *T. suis*, is highly pathogenic to pigs.

TRYPANOSOMA EVANSI (Fig. 2–9). Another serious parasite of horses and camels is *T. evansi*, which causes the disease known as **surra.** A small percentage of this species develops without a kinetoplast. The organism can be found in dogs, in which the disease is often fatal, and in donkeys, cattle and elephants. It is widespread throughout Asia and occurs in parts of Africa and certain parts of Central and South America. There is no cyclic development in the insect host. The principal vector is the horse fly, *Tabanus*, but other flies also mechanically transmit the flagellate. Ticks (*Ornithodoros*) and even the vampire bat have been implicated. The pathology of the disease is similar to that of *T. brucei*.

TRYPANOSOMA EQUINUM. This trypanosome infects horses in South and Central America, causing a disease known as **mal de Caderas.** It may also infect bats, laboratory animals and a South American rodent called **capybara.** The flagellate resembles *T. evansi* but appears to lack a kinetoplast. This body is actually present, but it is not revealed by

staining, and it lacks the central structure, a condition called **dyskinetoplastic.**

TRYPANOSOMA EQUIPERDUM. The disease **dourine** in horses is caused by this flagellate (Fig. 2–9). It is transmitted during copulation, and thus is a venereal disease. The flagellate is morphologically identical to *T. evansi*, and its length is about 25 μ. It is a typical trypanosome, but it does not exhibit the usual changes in morphology during its life cycle. The parasite first affects the sexual organs of male and female horses, causing swelling and ulcers. Enormous numbers of parasites can be found in these ulcers. The nervous system may become involved, as evidenced by paralysis of the legs or parts of the face. The mortality rate is high, and there is no effective treatment. Dogs, mice, rats, and rabbits may become infected with *T. equiperdum*, but the parasites remain largely in the blood of these hosts. A complement fixation test has been developed to identify the disease in horses.

Trypanosomes of Other Vertebrates. Trypanosomes of birds have been reported primarily from domestic species, but undoubtedly many wild birds harbor as yet undescribed forms. *Trypanosoma avium* occurs in various birds, *T. gallinarum* is a parasite of chickens, and *T. hannai* may be found in pigeons. The names of other bird forms may be found in texts on protozoology.

Trypanosomes of fish are transmitted by leeches. Many fish trypanosomes, such as *Trypanosoma giganteum* in the ray and *T. percae* in perch, are unusually large (e.g., up to 130 μ long). *T. caulopsettae*, found in the blood of a sand flounder and in another fish ("witch") in New Zealand, is illustrated in Figure 2–11.

Trypanosoma mega. The toad, *Bufo regularis*, is host to *T. mega*. This parasite has been found to be a convenient and suitable flagellate for experimental studies. Steinert and Novikoff[33] reported the existence of a cytostome in this species.

Endotrypanum schaudinni. This interesting flagellate was found in the blood of a sloth in 1908 and was described as apparently an

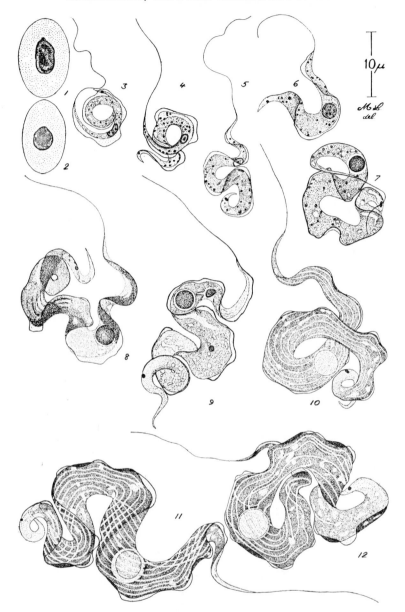

Fig. 2-11. *Trypanosoma caulopsettae* from a New Zealand sand flounder. *1* and *2* are red blood cells. (Laird, courtesy of Proc. Zool. Soc. London.)

endoerythrocytic form. Recent study[30] shows it to be truly within the red blood cell, where it usually occurs in the trypomastigote stage. Epimastigote stages may also occur. The vector is presumed to be the sandfly *Lutzomyia*. *Endotrypanum* seems to be clearly an independent genus and probably is best placed between *Trypanosoma* and *Leishmania*.

Leishmania

Among the flagellates that comprise the family Trypanosomatidae, the genus *Leishmania* is characterized by two different stages in its life cycle, each of which occurs in a distinct host. The amastigote stage, usually called a **Leishman-Donovan** body or **L-D** body, is found in the cytoplasm of reticulo-

endothelial cells, monocytes and other phagocytic cells of the vertebrate host. In this stage
the parasite measures 2 to 5 μ in diameter and
may be rounded or oval in shape. Its cytoplasm is often vacuolated. A few to 100 may
occupy one cell and their reproduction is by
binary fission. The other stage takes place in
the gut of phlebotomine sandflies, where the
parasite develops as a promastigote.

Details of the life cycle are not completely
known. Parasitized host cells rupture, liberating amastigotes that are engulfed by other
phagocytic cells. Biting sandflies pick up
both free amastigotes and free parasitized
cells. After active reproduction of the parasite in the midgut of the insect, the infection
is established in the anterior part (anterior
station) of the sandfly gut. When the infected sandfly bites again, promastigotes
eventually reach the blood or other tissues of
the vertebrate host, change into amastigotes
and are engulfed by phagocytic cells. It is
possible that the promastigotes are engulfed
first and then assume the amastigote form.

Authorities disagree on the number of valid
species and strains. Most of these flagellates
are morphologically indistinguishable. Differences, however, are frequently found in the
reaction of the host, the epidemiology of the
disease, reservoir hosts, species of sandflies involved in transmission, and serology. Different names have been given to the diseases
produced, but infection with a *Leishmania*
species is technically called **leishmaniasis.**
Leishmaniasis is an example of a zoonosis that
reaches man through an insect. Mammalian
reservoir hosts,[7] especially various rodents,
dogs, and certain carnivores, maintain the
infection in nature.

Of the various species of *Leishmania* that
have been described as infecting man, we
shall discuss the four most widely accepted.

Leishmania tropica. A serious cutaneous disease called **Oriental sore** that occurs in
tropic and subtropic regions of the world is
caused by *L. tropica.* Its distribution is, in
general, similar to that of kala-azar, but the
specific areas of occurrence frequently do not
coincide. It may be found in numerous

parts of China, Asia Minor, southeast Russia,
Greece, southern Europe, northern and
western Africa and in the New World. The
transmitting agents include phlebotomine
flies. Skin sores vary from a small pimple to
large ulcerated areas. The parasites rarely
invade adjacent mucocutaneous areas. Development from the bite of a sandfly to the
appearance of skin sores requires a few weeks
to a few months. The infection passes from
rodent to rodent, the natural hosts.

Leishmania mexicana. This species occurs in
some countries of central and south America,
principally Mexico and British Guiana and,
to a lesser extent, the Amazon region. It
causes a cutaneous lesion and does not spread
to mucous areas. The lesion that often occurs
on the ear is known as **chiclero ulcer.** In
experimental animals, the infection may lead
to visceralization, especially to involvement
of the spleen and liver. The vector seems to
be *Lutzomyia flaviscutellata.* Several species of
forest rodents are found infected in nature.

Leishmania braziliensis. This complex of
species or varieties is present in most parts of
the tropic and subtropic regions of the New
World, ranging at least from Panama to
Argentina. The pathology, epidemiology,
reservoir hosts and species of sandflies involved in transmission vary considerably from
one place to another, so the organism is often
called *Leishmania braziliensis* sensu lato.

The most severe form of this infection,
called **espundia,** is endemic in the jungles of
Brazil, Bolivia, Peru and some other South
American countries. The clinical form frequently involves the mucous membranes of
mouth and nose, and the lesions are resistant
to treatment. In certain arid areas of
western Peru, however, a benign form called
uta is present. In these areas involvement of
the respiratory membranes has not been observed, and the skin lesions usually heal spontaneously. Between espundia and uta occur
all the transitional forms of cutaneous leishmaniasis in the New World. In addition, an
anergic form of cutaneous leishmaniasis that
produces extensive skin infection resembling

leprosy has also been found. The etiologic agent is named *Leishmania braziliensis pifanoi.*

Throughout areas where *Leishmania braziliensis* is endemic, several mammals, both domestic and wild, have been found naturally infected. The list includes dogs, horses, wild rodents and certain forest carnivores. The importance of such animals as reservoir hosts in relation to the disease in man is still to be determined. Among the several species of phlebotomine sandflies that could be incriminated as vectors are *Lutzomyia flaviscutellata, L. intermedius* and *L. trapidoi.* Other species have been found naturally infected with promastigote flagellates, but further investigations are necessary in order to determine their relation, if any, to leishmaniasis.

Leishmania donovani. Visceral leishmaniasis is produced by this species. Infection occurs chiefly in the spleen and liver, secondarily in bone marrow, intestinal villi and other areas. The disease is called **kala-azar** and **dum dum fever.** It occurs in east India, Assam, the Mediterranean area, southern Russia, north Africa,[18] central Asia, northeast Brazil, Colombia, Argentina, Paraguay, El Salvador, Guatemala and Mexico. It is a rural disease with reservoirs of infection in dogs, foxes, rodents and other mammals. Promastigotes may be found in flies (e.g., *Phlebotomus argentipes, P. orientalis*) three days after feeding. They occupy the anterior portion of the gut by the fourth or fifth day. On the seventh to ninth day after the fly has fed a second time, promastigotes are in the proboscis and become infective. The fly lives 14 to 15 days.

The first response to a bite is a small, nonulcerated skin sore. Infection metastasizes to the viscera, and the spleen and liver enlarge. Mortality is particularly high when the disease occurs in new geographic areas. Headache, fever and weakness often accompany infection. This disease occurs as three types or varieties, separated geographically. The first two types, Indian and Sudanese, are found mainly in human adults, rodents, and possibly lizards, but not in dogs. The third type, Mediterranean, is much more widespread, occurring in southern Europe, middle Asia, China, Central and South America. In this type, dogs are important hosts, and jackals and other mammals have been found to be infected. Transmission of the disease may occur from vertebrate to vertebrate by direct contact, thus omitting the vector.

About two years after the acute stage of kala-azar occurs in the viscera, post-kala-azar leishmaniasis may appear. Manifestations range from depigmented areas of the skin to pronounced nodular lesions (Fig. 2–12).

Dogs also show pathogenic effects of infection. Loss of hair is a characteristic manifestation of severe visceral leishmaniasis in these animals. Ulcers on the lips and eyelids, weakness and enlargement of the liver are common symptoms. Positive diagnosis relies on discovery of the parasite in infected tissues and in peripheral blood. Stray dogs should be eliminated and insecticides used to control the sandflies.

Probably, many other species of *Leishmania* occur in animals but have not yet been reported. Those that have been described include *L. enrietti* in laboratory guinea pigs, *L. canium* in dogs, *L. chamaeleonis* in lizards, and *L. denticis* in the silverfish, *Dentex argyrozona.*

Physiology

The development of the chondriome (see p. 27) that accompanies the transition of the trypomastigote blood stages to the insect or cultivation phases may be responsible not only for the biochemical transformation of the parasites, but also for the changes in body form.[35] No adequate evidence is available of toxin production by the trypanosomes.

Carbohydrates. All members of the family Trypanosomatidae are basically aerobic, exhibit a high rate of carbohydrate metabolism, and depend upon exogenous glucose as an energy source.

The different oxidative pathways employed in carbohydrate metabolism by different genera and species and by the invertebrate and vertebrate phases in the life cycles of individual genera appear to be related directly to the host environment.

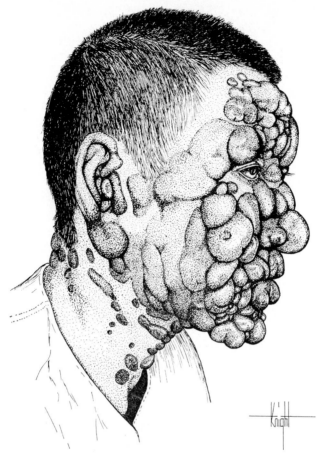

Fig. 2-12. A severe dermal, post-visceral case of "kala-azar" infection caused by
Leishmania donovani.

The blood forms of the pathogenic *brucei* group and the *evansi* group of trypanosomes lack cytochrome pigments, cyanide-sensitive respiration and a functional Krebs cycle. These components are normally associated with mitochondria. Such blood forms are, therefore, incapable of oxidizing carbohydrates by way of the citric acid cycle. On the other hand, all the culture forms of the trypanosomes hitherto studied, including *Leptomonas* and *Crithidia* (restricted to invertebrates), have an operative citric acid cycle and active mitochondrial kinetoplasts. The same is true of the blood forms of the *lewisi* group, and probably also of the pathogenic *Trypanosoma vivax* and *T. congolense.*

The blood forms of the *brucei* and *evansi* groups may be highly dependent upon dehydrogenase systems.[36] In pathogenic forms, a respiratory enzyme may exist that is capable of being oxidized as fast as or perhaps faster than the cyanide-sensitive species.[27]

Proteins. Although trypanosomes derive energy primarily from glucose, there is evidence (e.g., from cultural studies of *Leishmania*) that these parasites are capable of utilizing proteins for energy. Some amino acids, such as glutamic acid, appear to be derived from exogenous sources, but numerous others are formed by transamination.

Lipids. Trypanosomes apparently do not store lipid reserves, and few if any fat droplets have been found in these cells. When grown in defined media, they have not demonstrated a lipid requirement under normal conditions of laboratory cultivation.[29] In a study of lipid composition and mitochondrion associated lipid metabolism, Dixon[4] found that blood forms of *Trypanosoma lewisi* (which possess active mitochondria) have a rate of fatty acid oxidation 20 times greater than blood stages of *T. rhodesiense* (which do not possess active mitochondria). The levels of fatty acid oxidation and synthesis in *T. lewisi*, however, are comparatively low.

Nucleic Acid Metabolism. Trypanosomes depend primarily, or perhaps exclusively, upon their hosts for preformed purine rings, of which adenine appears to be the most important, and for pyrimidines as precursors of nucleic acids and of many coenzymes. This conclusion is based primarily upon a study of such cultivation forms as *Trypanosoma cruzi* and *T. mega*, as well as the blood forms of *T. lewisi* and *T. equiperdum*.

Immunology

Immunity to Oriental sore has been shown to follow supposedly complete recovery from a primary infection with *Leishmania tropica*. Vaccine treatment has been successful, and artificial immunity to this disease was practiced long before the etiologic agent was known. Natives of the Near East usually inoculate material from a sore into an unexposed part of the body to prevent the occurrence of a sore with its resulting scar on the face. Throughout the course of all clinical types of leishmaniases, there is a striking increase in immunologically competent cells. This increase is generalized in visceral leishmaniasis and local in the cutaneous and mucotaneous varieties.[1]

When dead and washed promastigotes of any species of *Leishmania* are inoculated into the skin of a person with Oriental sore, a delayed hypersensitivity reaction (**Montenegro reaction** or **leishmanin reaction**) occurs. The response also occurs with antigens prepared from *Trypanosoma cruzi*, and can be evoked long after spontaneous cure. The reaction itself is no indication of immunity, but is a mechanism for mobilizing immunolog-ically competent cells to a site containing *Leishmania* antigens where the cells are effective after immunity is established.

Permanent immunity occurs after the skin lesion heals, but if the sore is surgically removed before healing, immunity does not follow. Classic antibodies have not yet been positively demonstrated in Oriental sore, although antibodies against the promastigote form have been established. Cross immunity may occur between *Leishmania tropica* and *L. mexicana*, but not between these species and *L. donovani*.

Cellular reactions in visceral leishmaniasis are essentially the same as those in cutaneous leishmaniasis. All the mammalian leishmanias have common antigens in addition to specific ones. With visceral leishmaniasis, circulating antibodies can definitely be demonstrated by complement fixation using antigens from animal and human tissues. A patient with Chagas' disease, however, may respond positively to this test. It is of interest to note that hamsters infected with *Leishmania donovani* or with *L. tropica* are far more resistant to the malarial parasite, *Plasmodium berghei*, than are normal hamsters.

Although a complement fixation test for the presence of *Trypanosoma equiperdum* has been used successfully in Canada to eliminate dourine in horses, serologic diagnosis of trypanosome infections of wild mammals has had limited success. Immunologically induced changes in serum proteins can be detected by protein recognition techniques, but ignorance of the pattern of the antigenic variation makes reliable interpretation of results difficult.

Twenty or more antigenic variants may occur within a given strain of trypanosome. These variants are apparently correlated with periodicity in the infectivity of the blood, i.e., with successive parasitemic waves. An infected animal may survive the first such wave but succumb to the second. Laboratory animals may be immunized against a given antigenic trypanosome by the injection of antigenic serum or inoculation of living organisms, but such immunized animals are pro-

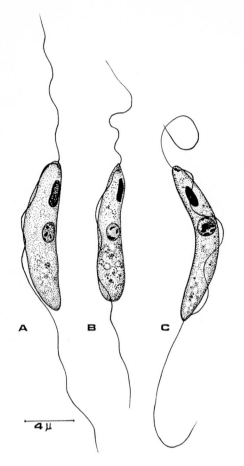

A   B   C

4 μ

Fig. 2-13. *Cryptobia stilbia* from the stomach of a deepsea fish. The two kinetosomes in *C* probably indicate the start of mitosis. (Noble, courtesy of J. Parasitol.)

Family Cryptobiidae

Many members of this family of flagellates appear superficially like trypanosomes but they all possess two anterior flagella instead of one. One of these flagella is usually trailing while the other extends anteriorly. The trailing one is the outer margin of an undulating membrane in most species. A large and usually elongate kinetoplast is characteristic of the family. The majority of described species inhabit the blood of freshwater fish, and utilize leeches as intermediate hosts. Other cryptobias live in the stomachs of marine fish, and in the gonads of pulmonate snails, chaetognaths, siphonophores and a few other animals. *C. borreli* in the blood of fish, and *C. helicis* in the seminal vesicles of the snail, *Helix*, are typical examples. The length of the body varies in different species from about 20 to 120 μ.

The kinetosome and kinetoplast are located anterior to the nucleus (Fig. 2–13). There is some justification for separating these flagellates into two genera: *Trypanoplasma*, which exist in the blood of fish, and *Cryptobia*, which exist in the stomach of fish and in various organs of invertebrates. Transmission of the species that do not inhabit the blood of their hosts is probably accomplished by ingestion of the flagellates, or, in snails, by way of spermatophores and fertilized eggs.

ORDER RETORTAMONADIDA

Family Retortamonadidae

Chilomastix mesnili. This flagellate is the best known member of the family. It is probably harmless and lives in the cecum and colon of man and other primates and the pig. The trophozoite, or motile form, is usually pear-shaped and averages about 12 μ in length. The spiral groove in the body is frequently not seen in the usual hematoxylin-stained preparations, and the three anterior flagella are also difficult to see. Probably the most diagnostic feature is the location of the large nucleus very near the anterior end of the body. The cyst is about 9 μ long, oval in shape, with a broad extension of the cyst wall at one end to form a spade-like or cone-shaped

tected only against the homologous antigenic type.

Trypanocidal properties from an immune serum have been removed by adsorption with *Trypanosoma lewisi*, leaving the reproduction-inhibiting property. The *in vivo* manifestations of these two properties are strikingly different, and one may be formed independently of the other. The reproduction-inhibition is caused by the antibody **ablastin.** Immune serum from rabbits immunized with *T. lewisi* contains only the trypanocidal property, and not the reproduction-inhibiting one.

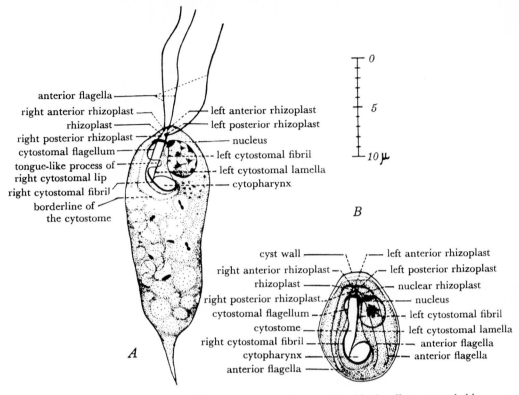

anterior flagella

right anterior rhizoplast

rhizoplast

right posterior rhizoplast

cytostomal flagellum

tongue-like process of
right cytostomal lip

right cytostomal fibril

borderline of
the cytostome

left anterior rhizoplast

left posterior rhizoplast

nucleus

left cytostomal fibril

left cytostomal lamella

cytopharynx

0

5

10 μ

B

cyst wall

right anterior rhizoplast

rhizoplast

right posterior rhizoplast

cytostomal flagellum

cytostome

right cytostomal fibril

cytopharynx

anterior flagella

left anterior rhizoplast

left posterior rhizoplast

nuclear rhizoplast

nucleus

left cytostomal fibril

left cytostomal lamella

anterior flagella

anterior flagella

A

Fig. 2-14. *Chilomastix intestinalis* from the guinea pig. *A.* Trophic flagellate, ventral side, showing body structures. *B.* Cyst, ventral side. (Nie, courtesy of J. Morph.)

projection. The large nucleus, large cytostome, and various filaments can usually be seen in stained cysts. Other species of *Chilomastix* occur in the rectum of amphibians, fish, and many mammals (Fig. 2–14).

Retortamonas intestinalis. This parasite is another harmless small flagellate of man (Fig. 2–15). It measures about $6 \times 4\,\mu$. Like *Chilomastix*, it possesses a cytostomal groove but has only two anterior flagella, one extending anteriorly and the other passing backward through the cytostome and then trailing from the body. Other members of the genus have been found in insects [e.g., *Embadomonas* (*Retortamonas*) *gryllotalpae* in the mole cricket].

ORDER DIPLOMONADIDA

Family Hexamitidae

Giardia lamblia (Figs. 2–16 and 17). This universal and common inhabitant of the

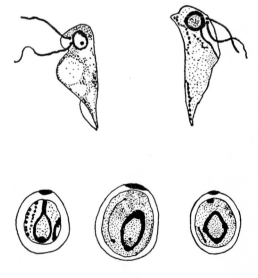

Fig. 2-15. *Retortamonas intestinalis* from man showing two motile forms and three variations of the cyst. (Hogue, courtesy of J. Hyg.)

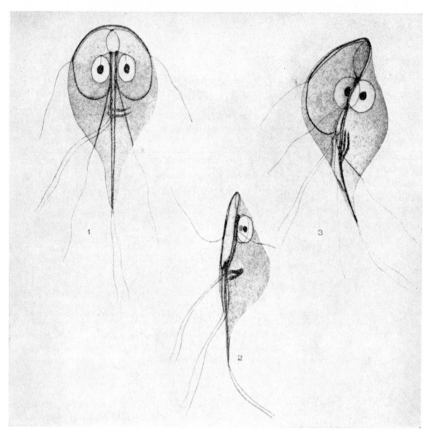

Fig. 2-16. *Giardia lamblia*, motile form in various aspects. (Kofoid and Swezy, courtesy of University of California Publications in Zoology.)

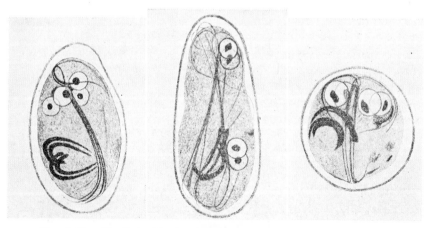

Fig. 2-17. *Giardia lamblia* showing cysts and various stages of division (Kofoid and Swezy, courtesy of University of California Publications in Zoology.)

upper small intestine of man, monkeys and pigs is probably the best known member of the family. Present in at least ten per cent of the people of the United States, it is usually harmless, except in rare instances. Motile *Giardia* average about $7 \times 14\,\mu$ and they possess two nuclei and four pairs of flagella. The body is unusual for a protozoan cell because it possesses a suction disc or sucker on its ventral surface. The central pair of flagella apparently work as a pump, taking fluid out of the cavity of the sucker. The mature cyst is usually $8 \times 11\,\mu$ in size and possesses four nuclei. A comma-shaped "parabasal" body and deeply stainable strands, often occurring in pairs, give the cyst its distinctive appearance.

Transmission of *Giardia* from host to host occurs by the ingestion of cysts in contaminated food or drinking water, or from an infected person who may have cysts on his hands, body or clothes. When the parasite is pathogenic it causes diarrhea, abdominal pain and loss of weight. Apparently it interferes with normal fat absorption in the small intestine. Pathogenicity occurs more often in children than in adults. Diagnosis is confirmed by finding cysts or motile forms in feces.

Similar *Giardia* have been reported from many other animals. Rabbits and dogs may be heavily invaded, and rats can be infected in the laboratory. *G. muris* in mice is distinct from *G. duodenalis* and other species in rats. Pathogenicity in these hosts is doubtful, but, as in man, infection is sometimes associated with diarrhea.

Hexamita meleagridis. Another well-known member of the family Hexamitidae, this organism causes infectious catarrhal enteritis in turkeys. The disease is also known as **hexamitiasis** and is characterized by diarrhea, nervousness, loss of weight and death, especially in younger birds. The infectious organism, *H. meleagridis*, averages about $9 \times 4\,\mu$, is pear-shaped and possesses eight flagella, two of which are attached to the posterior end of the animal, while six arise from the anterior end. Other members of the

genus infect other birds. *Hexamita salmonis* (Fig. 2–18) is a species that develops in epithelial cells of the cecum of trout.

ORDER TRICHOMONADIDA

A diagnostic feature of this order is the parabasal body that is composed of a granule or one or more elongated bodies, often attached by fibrils to the kinetosomes of flagella. The parabasal body is probably homologous with the Golgi apparatus of metazoan cells. The **axostyle** (Fig. 2–20) is a filament or hyaline rod that passes lengthwise through the body, often projecting from the posterior end. There may be a few or many flagella (typically four to six). A trailing flagellum is characteristic, and there is often an undulating membrane. Sexual stages are unknown. In addition to the two families described below, there are the Devescovinidae and Calonymphidae,

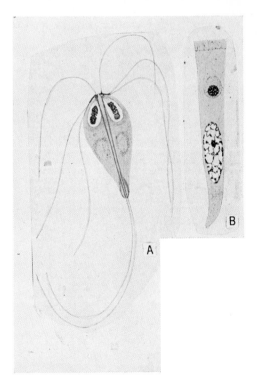

Fig. 2-18. *Hexamita (Octomitus) salmonis,* a parasitic flagellate of trout.

A. Typical flagellate, $\times$ 2360. *B.* Epithelial cell from cecum of trout containing early stage of parasite, $\times$ 1230. (Davis, Bull. Bureau of Fisheries.)

both restricted to termites (see pp. 497 and 556). For a detailed discussion of the evolutionary and systematic relationships of this order, see Honigberg.[13]

Family Monocercomonadiae

In this family are placed many flagellates that live in the digestive tracts of insects and vertebrates. There may be several flagella, but usually no undulating membrane. A representative genus is *Monocercomonas* (Fig. 2–19) found in insects and vertebrates.

Family Trichomonadidae

This family is well known for its important parasites of man and of animals, both vertebrate and invertebrate. The flagellates are characterized by the presence of several anterior flagella, a **pelta** that lies at the anterior margin of the body and occurs in all trichomonads, an undulating membrane, a deeply staining **costa** that extends along the base of

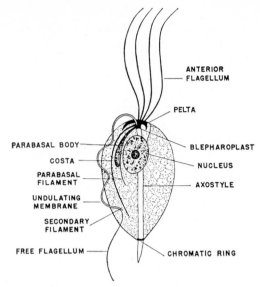

Fig. 2-20. Structures of *Trichomonas*. (Levine, courtesy of Burgess Publishing Company.)

the undulating membrane, and an axostyle.[13] There is only one nucleus.

The name *Trichomonas* was first given to a genus that was believed to possess three anterior flagella. Unfortunately, confusion has arisen because species have been found with more than three flagella. The original group actually possesses four flagella, but the name *Trichomonas* remains. The following genera are now recognized: *Tritrichomonas*, with three anterior flagella (Fig. 2–21*M*); *Trichomonas*, with four anterior flagella (Fig. 2–20); and *Pentatrichomonas*, with five anterior flagella (Fig. 2–21).

Representative Types of Trichomonads

Pentatrichomonas hominis. This species lives in the cecum and adjacent intestinal area (Fig. 2–21) and is one of the three trichomonads that parasitize man; it is also found in monkeys, cats, dogs and rats. It is oval in shape, measuring about $10 \times 4\,\mu$, with five anterior flagella (four of which are grouped together at their bases), and a free posterior flagellum extending from the end of a long undulating membrane. At its anterior end, it has a costa and a parabasal apparatus composed of a filament and one or more granules.

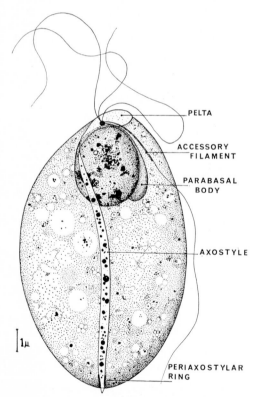

Fig. 2-19. A composite line diagram of *Monocercomonas molae* from the sunfish, *Mola mola*. (Noble and Noble, courtesy of J. Protozool.)

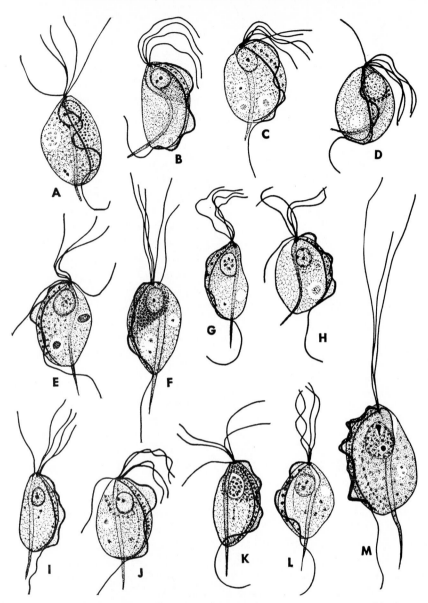

Fig. 2-21. *Pentatrichomonas hominis* from the intestine of man, *A* to *D*, compared with the same species from the monkey, *E* to *G*. *H* is from the cat, *I*, from the dog, *J* and *K* from the rat. *L* is *Trichomonas gallinarum* from the pheasant. *M* is *Tritrichomonas fecalis*. (Wenrich, courtesy of J. Morphology.)

No cysts have been found. The geographic distribution is worldwide, but the incidence of infection is heaviest in the warmer countries. Although it is frequently associated with pathogenic conditions, there is no evidence that the organisms actually are responsible for the symptoms.

Trichomonas tenax. *T. tenax* is somewhat smaller than *Pentatrichomonas hominis*, averaging about 7 μ in length. It is similar to *P. hominis* in general appearance, but the undulating membrane extends about two-thirds the length of the body, and there is no trailing flagellum. The parabasal body is elongate and prominent. This flagellate lives in the human mouth and is a harmless

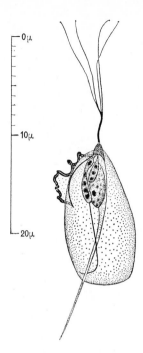

0μ

10μ

20μ

Fig. 2-22. *Trichomonas vaginalis.* The parabasal body is V-shaped and the pelta appears as a dark-staining triangular membrane at the anterior end of the body. Note the double nature of the outer margin of the undulating membrane. (Honigberg and King, courtesy of J. Parasitol.)

commensal. Its habitat provides the basis for the name *T. buccalis*, by which it is sometimes called.

Trichomonas vaginalis (Fig. 2–22). The largest of the three trichomonads of man is *T. vaginalis*. It averages about 13 μ in length, but the range is from 10 to 20 μ. The undulating membrane usually reaches to the middle of the body, but it may be shorter. There is no trailing flagellum. The parabasal body, with a parabasal filament, is large. The axostyle characteristically curves slightly around the nucleus.[17] The most significant difference between this species and the other two is its habitat. It lives in the vagina, urethra, or in the prostate gland. The parasites are usually transferred from one host to another through sexual intercourse.

Trichomonas vaginalis is worldwide in distribution, and the incidence of infection in women varies from 20 to 40 per cent, whereas in men it varies from 4 to 15 per cent. In women with abnormal vaginal secretion (usually a white discharge), the incidence of infection may be as high as 70 per cent.[12] The flagellates can be found in abundance in this leukorrheic discharge. As with the other species, no cysts are produced. The parasite may even be found in newborn babies and children.

Pathogenic strains exist.[6] *T. vaginalis* and *T. gallinae* (Fig. 2–23) have been found within the cytoplasm of host cells such as macrophages and epithelial cells.[17] These two flagellates were found to be 80 to 100 per cent viable after long storage at −196° C. (*T. gallinae* after nearly six years).[16]

Trichomonas gallinarum (Fig. 2–21). Avian trichomoniasis in chickens, turkeys, and other domestic birds is caused by this parasite, which is pear-shaped, like other members of the genus, and averages about 7 × 10 μ in size. The flagellate lives in the lower intestines of the birds and is the cause of diarrhea, loss of appetite and weight, ruffled feathers and lesions in the intestinal wall. Cysts are not produced, and the method of transmission from bird to bird is unknown. *Trichomonas gallinae* (Fig. 2–23) is a similar parasite that infects the upper intestinal tract of poultry, especially of pigeons.

Tritrichomonas foetus (Fig. 2–24). Three anterior flagella characterize this species, which reaches a length of 24 μ. There is a long undulating membrane which is bordered on its outer margin by a flagellum that becomes free posteriorly. The axostyle is unusually wide, and it projects a short distance from the posterior end of the animal; the parabasal body is large. The parasites live in the preputial cavities of bulls and are thus readily transferred to the vaginas of cows during coitus. In the vagina, the trichomonads multiply for about three weeks. After the cow is in heat again, the parasites become reduced in numbers or disappear completely. This cycle continues for three to four months, after which time the infection is usually lost. These flagellates may invade other reproductive organs and are apparently able to carry on their metabolic activities both aerobically (not very effectively) and anaerobically.

If trichomonads get into the uterus, as they easily do, they may cause temporary infertility or abortion. They may even get into the unborn young, appearing in the fetal membranes, the amniotic and allantoic fluids and often in the stomach of the fetus. The parasites seem to have little effect on bulls. Some bulls apparently have a natural resistance to infection, and cows are able to develop some degree of immunity.

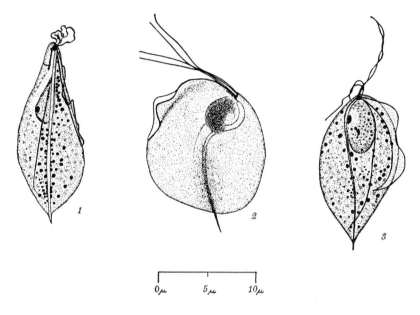

Fig. 2-23. *Trichomonas gallinae* from intestines of birds. *1.* Domestic pigeon. Note concentration of granules in axial area. *2.* Domestic turkey. Dried smear, stained with Giemsa. Note double marginal filament. (Courtesy of Dr. Hawn.) *3.* Red-tailed hawk. (Stabler, courtesy of J. Morphology.)

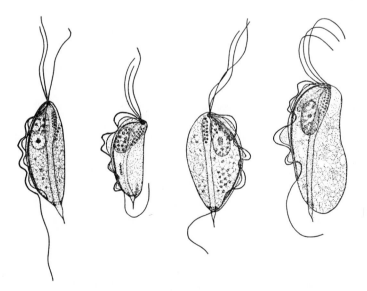

Fig. 2-24. *Tritrichomonas foetus*, a parasite that can cause serious disease in cattle. Side views. × 400. (Wenrich and Emmerson, courtesy of J. Morphology.)

Physiology

The physiology of trichomonads has received a great deal of study during the past 15 years. These investigations were reviewed by Shorb,[31] from whom the following brief summary was adapted. Much of the work was devoted to enzyme systems and carbohydrates used for energy and growth.

All species of trichomonads seem to utilize effectively glucose, fructose, galactose and maltose. The pentoses and sugar alcohols do not support growth. Lactose, sucrose, trehalose, glycogen, starch and dextrin support growth poorly. Other carbohydrates also support only fair to poor growth, and species of trichomonads vary in the number of carbohydrates they use. Different sugars have different effects on growth and length of life. Trehalose in *Tritrichomonas foetus*, for example, could produce a population peak only 55 per cent of that produced by glucose, but it lasted almost twice as long as did the glucose control.

Anaerobic studies have yielded varying results due to variations in techniques and in components of the experiments. Energy metabolism and motility of *Tritrichomonas foetus* was maintained aerobically or anaerobically at 37° C. by fermentation of intracellular glycogen reserves and any extra substrates available. The substrates were degraded to CO_2, H_2, succinic acid and acetic acid.

Aerobic respiration has been studied by several investigators. As usual, results varied. For example, oxygen uptake by *Tritrichomonas foetus* was stimulated within 60 minutes by pure oxygen in the absence of glucose. In the presence of this sugar, the uptake was slightly inhibited after 120 minutes. In contrast, the high oxygen tension was toxic to *Trichomonas vaginalis*. It increased endogenous respiration but almost completely inhibited oxygen uptake with glucose. Energy metabolism seems to be centered around the intracellular glycogen reserves supplemented by any extracellular substrates. In general, carbohydrate utilization for respiration is similar to that for growth and anaerobic metabolism. Oxygen uptake and anaerobic fermentation in most trichomonads are inhibited by iodoacetate, arsenite, fluoride and possibly azide. The presence of catalase is suggested by the lack of inhibitory action of arsenate, malonate and hydrogen peroxide. Cyanide in general did not inhibit oxygen uptake except for *Tritrichomonas foetus* and *Trichomonas vaginalis*. Some species, e.g., *Trichomonas suis*, were not inhibited by most of the compounds studied. The work on enzyme inhibitions and the sugars used indicated that the trichomonads probably break down carbohydrates through the usual Embden-Meyerhof scheme of glycolysis, but the steps to terminal respiration are not through the usual Krebs cycle. *T. vaginalis* apparently lacks all of the citric acid cycle enzymes except malic dehydrogenase and pyruvate dehydrogenase.

Proteolytic enzymes and those involved with lipid metabolism are undoubtedly present but very few studies have been made. The field is wide open for research. Students are referred to Shorb[31] for a review of trichomonad requirements for vitamins, minerals, components of nucleic acids, amino acids, fatty acids and sterol.

Immunology

Serologic tests for antibodies against trichomonads have become more reliable with the development of more precise testing methods and a better understanding of antigen-antibody relationships. A complement-fixation test was positive in 80 per cent of adult female and 40 per cent adult male patients infected with *Trichomonas vaginalis*. Agglutination and agglomeration reactions vary, depending on the strains of *T. vaginalis* present. This is also true of the complement-fixation test, so a comparison of the two tests is valid only if the same strain of parasite is used. Immunity conferred on the host as the result of infection by *T. vaginalis* is apparently temporary. The intradermal or skin test is also used and is reliable for the diagnosis of genitourinary trichomoniasis.

ORDER HYPERMASTIGIDA

These protozoa possess many flagella as the name indicates, and they may also possess many parabasal bodies and axostyles. There is a single nucleus and no cytostome. All the species are found in the alimentary tracts of termites, cockroaches and wood roaches. Only two out of at least eight families will be mentioned here.

Family Hoplonymphidae

The members of this family are found in termites and wood roaches. *Hoplonympha natator*, in the intestine of the termite, *Kalotermes simplicicornis*, is a narrow, spindle-shaped organism with two tufts of flagella on its anterior end. The nucleus is near the anterior end and spiral grooves occur on the pellicle. *Barbulanympha ufalula* (Fig. 2–25) is an irregularly rounded flagellate with two anterior groups of flagella, below which extends a ring of elongated parabasal bodies surrounding the nucleus. The axostyle consists of a group of filaments that extend into the cytoplasm from the nuclear area. *Rhynchonympha* and *Urinympha talca* are other examples of this family in the roach.

Family Trichonymphidae

Members of this family possess a projection at the anterior end of the body. This "rostrum," which may be cone-shaped, is covered with long flagella except for its tip,

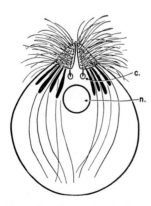

Fig. 2-25. *Barbulanympha ufalula*, resting organism prior to division. (Cleveland, courtesy of J. Protozool.)

while other flagella cover much of the rest of the body in longitudinal rows. The base, or more, of the body may be bare.

Trichonympha corbula is typical of the genus. It possesses circular bands of flagella, each band of a different length. The nucleus is surrounded by the narrow rods of the parabasal body, and the cytoplasm usually contains bits of wood that the animal has ingested. Several other species of this genus live in the intestines of termites and in wood roaches. The sexual cycle of *Trichonympha* has been thoroughly investigated and is partially illustrated in Figure 2–26. *Pseudotrichonympha grassii* in the termite, *Coptotermes formosanus;* and *Mixotricha, Deltotrichonympha,* and *Eucomonympha imla* (Fig. 2–1), from the wood roach are other examples of the family.

FLAGELLATES OF UNCERTAIN AFFINITIES

Until further morphologic and life history studies are made, we will consider the following genera of flagellates as **incertae sedis.**

Costia (Fig. 2–27). This genus has two short anterior flagella and two trailing flagella (all arising within a slightly spiral longitudinal groove), and a cytostome that attaches to the skin of fish. *C. necatrix* causes serious damage and often a high rate of mortality in salmon, trout, and pond fishes.

Enteromonas hominis (Fig. 2–28). This tiny flagellate, about $4 \times 7 \mu$, has been reported from the intestine of man in the warmer countries of the world, but it is not pathogenic. There are three anterior flagella and one that arises anteriorly but adheres to the body as it passes backwards, then emerges as a free flagellum posteriorly. A cytostome is said to be present. The cyst possesses four nuclei, two at each pole.

Histomonas meleagridis (Fig. 2–29). Although often placed in the order Rhizomastigida, a group of protozoa with permanent flagella as well as pseudopodia, this parasite has strong trichomonad affinities and probably belongs in the order Trichomonadida.

This flagellate ranges in diameter from 4 to 30μ, and it may possess up to four flagella or none. The parasite usually lives

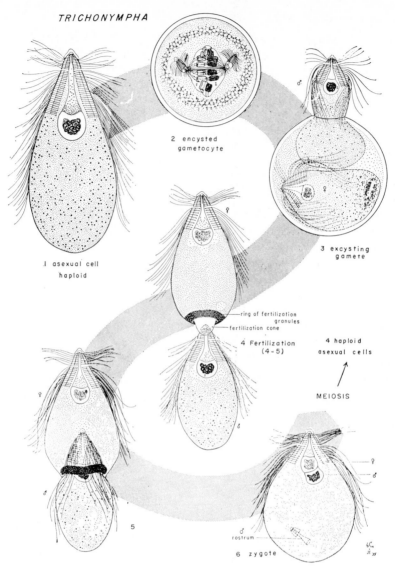

Fig. 2-26. *Trichonympha* and some steps in its sexual cycle.
(Cleveland, courtesy of J. Protozool.)

in the liver of poultry, causing infectious enterohepatitis, but it may also infect the ceca, kidneys or spleen. The bird's head characteristically turns a dark, almost black color—hence the common name of the disease, "blackhead." However, darkening does not always occur, or it may result from other diseases. No protective cyst is formed, so when the parasite is eliminated from the host's body with feces, it soon dies. The

protozoa, however, may become enclosed within the egg shell of the poultry cecal worm, *Heterakis gallinae*, while still in the host intestine. When the embryonated eggs of the worm are eaten by poultry, the enclosed *Histomonas* are liberated.

Blackhead is one of the most serious diseases of turkeys, and it costs the American poultry farmer approximately three and a half million dollars a year. The disease is not as serious

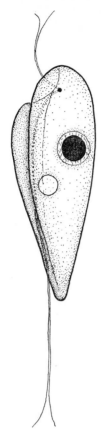

Fig. 2-27. *Costia necatrix*, showing ventral body groove, basal granule, flagella, nucleus, and contractile vacuole. $\times$ 7,500. (Tavolga and Nigrelli, courtesy of Trans. Amer. Micro. Soc.)

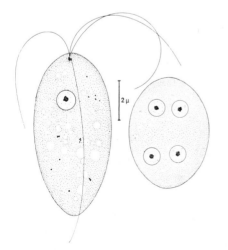

Fig. 2-28. *Enteromonas hominis*, motile and cyst.

for chickens, which usually recover after a mild reaction.

Resistant chickens and turkeys may harbor both the protozoon, *Histomonas meleagridis*, and the worm, *Heterakis gallinae*, at the same time. Such birds are reservoirs of infection for all new chicks. Chickens and turkeys obviously should not be raised together. Even raising young turkeys with older turkeys is apt to expose the young ones to serious protozoan and worm diseases.

Symptoms of histomoniasis are similar to those of many other poultry diseases. The birds become listless, their wings droop and their eyes are partly or completely closed much of the time. With turkeys especially, the head may become darkened. Young birds may die within two or three days, and older birds may last a week or two before death, or they may recover after a few days. Cleanliness and isolation are the keys to prevention.

Superclass Opalinata

ORDER OPALINIDA

Members of this group oare called **opalinids.** There are about 150 species, found mainly in the intestines of amphibia but occasionally in fish, snakes and in at least one lizard (*Varanus*). These saprozoic parasites are flattened and are covered with rows of cilia. Usually numerous nuclei are scattered throughout the cytoplasm. The geographic distribution of these protozoa is worldwide, but each has its own special distribution. The largest genus, *Opalina*, for example, can be found in almost any continent except Australia and South America.

The life cycles of opalinids are not completely known. In some species the body apparently divides into small organisms that encyst and pass from the host with feces. These cysts are eaten by new hosts (tadpoles) within which they encyst and develop to maturity. Other opalinids apparently undergo gametogenesis. The gametes unite and the resulting zygote is liberated from the body as a cyst that is eaten by another host.

The genus *Protoopalina* is considered to be

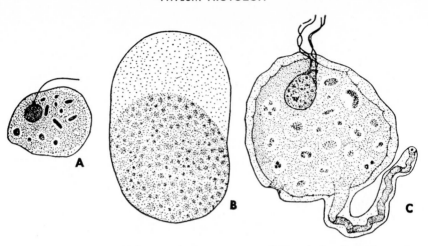

Fig. 2-29. *Histomonas* from chickens and pheasants.   × 1300.   *A.* Small specimen from chicken cecum.   *B.* Sketch of living specimen.   *C.* Specimen from pheasants showing the peculiar "tube" that is characteristic of these flagellates.   (Wenrich, courtesy of J. Morphology.)

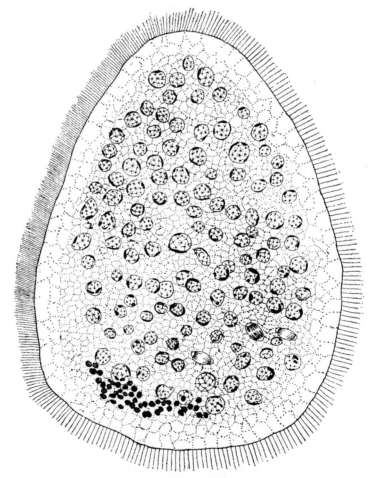

Fig. 2-30. *Opalina ranarum.* × 460.   Shows many nuclei, some of which are in the process of division. Most of the endospherules have been omitted.   (Metcalf, courtesy of U.S. Nat. Mus. Bull.)

primitive. Its members are cylindrical or spindle-shaped and range in size from about 100 to over $300\,\mu$ long by about 20 to $70\,\mu$ wide. *P. saturnalis* occurs in the marine fish, *Box boops*, while *P. intestinalis* and *P. mitotica* occur in the intestines of amphibia.

Opalina. This genus contains flattened, multinucleate forms that are found in amphibia. The parasites range in size from 90 to over $500\,\mu$ by 30 to $180\,\mu$, with a thick-ness of 20 to $40\,\mu$. Among the several known species are *O. ranarum* (Fig. 2–30), *O. hylaxena*, *O. oregonensis* and *O. spiralis*.

Zelleriella. This genus contains individuals that have only two nuclei, and whose bodies are considerably more flattened than are those of other genera. These parasites live in the intestines of amphibia. *Z. hirsuta* is about $130 \times 60 \times 22\,\mu$, and inhabits the gut of the toad, *Bufo cognatus*. *Z. opisthocarya*

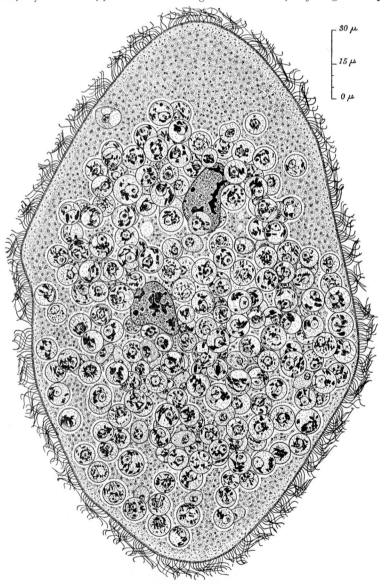

Fig. 2-31. *Zelleriella opisthocarya*, an opalinid containing over 200 cysts of *Entamoeba*. (Stabler and Chen, courtesy of Biol. Bull.)

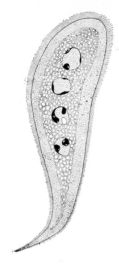

Fig. 2-32. *Cepedea lanceolata.* $\times$ 850. (Metcalf, after Bezzengerger, U.S. Nat. Mus. Bull.)

is itself parasitized by an ameba, *Entamoeba* sp.[32] (Fig. 2–31).

Cepedea (Fig. 2–32). This genus is also found in amphibians. The protozoa are cylindrical or pyriform in shape and their range in size is about the same as that indicated for the genus *Opalina*. *C. cantabrigensis* is a representative species from the toad, *Bufo lentiginosus.*

REFERENCES

1. Adler, S.: Leishmania. *In* Dawes, B.: Advances in Parasitology. Vol. 2, pp. 35–96. New York, Academic Press, 1964.
2. Boyden, A.: Are there any "acellular animals"? Science, *125*:155–156, 1957.
3. Cunningham, M. P., and Vickerman, K.: Antigenic analysis in the *Trypanosoma brucei* group, using the agglutination reaction. Trans. Roy. Soc. Trop. Med. Hyg., *56*, 48–59, 1962.
3a. Danforth, W. F.: Respiratory metabolism. *In* Chen, T. T. (ed.): Research in Protozoology. Vol. 1, p. 201. Long Island City, Pergamon Press, 1967.
4. Dixon, H.: Lipid metabolism in trypanosomes. Trans. Roy. Soc. Trop. Med. Hyg., *61*, 135, 1967.
5. Ferreira-Santos, R.: Aperistalsis of the esophagus and colon (megaesophagus and megacolon) etiologically related to Chagas' disease. Symposium by South American Authors. I. Amer. J. Digest. Dis., *6*, 700–726, 1961.
6. Frost, J. K.: *Trichomonas vaginalis* and cervical epithelial changes. *In* The cervix. Ann. N.Y. Acad. Sci., *97*:792–799, 1961.
7. Garnham, P. C. C.: The leishmanias, with special reference to the role of animal reservoirs. Am. Zoologist., *5*:141–151, 1965.
8. Grell, K. G.: Protozoologie. Heidelberg, Springer-Verlag, 1956.
9. Guttman, H. N., and Wallace, F. G.: Nutrition and physiology of the Trypanosomatidae. *In* Hutner, S. H. (ed.): Biochemistry and Physiology of Protozoa. Vol. III, pp. 459–494. New York, Academic Press, 1964.
10. Hall, R. P.: Protozoology. New York, Prentice-Hall, 1953.
11. Hoare, C. A.: The classification of mammalian trypanosomes. Ergebnis. Microbiol. Immunitätsf. Exper. Therapie., *39*:43–57, 1966.
12. Honigberg, B. M.: Comparative pathogenicity of *Trichomonas vaginalis* and *Trichomonas gallinae* to mice. I. Gross pathology, quantitative evaluation of virulence and some factors affecting pathogenicity. J. Parasitol., *47*:545–571, 1961.
13. ————: Evolutionary and systematic relationships in the flagellate order Trichomonadida. J. Protozool., *10*:20–62, 1963.
14. Honigberg, B. M.: Chemistry of parasitism among some Protozoa. *In* Florkin, M., and Sheer, B. J. (eds.): Chemical Zoology. pp. 695–814. New York, Academic Press, 1967.
15. Honigberg, B. M. (chairman) and Committee: A revised classification of the phylum Protozoa. J. Protozool., *11*:7–20, 1964.
16. Honigberg, B. M., Farris, V. K., and Livingston, M. C.: Preservation of *Trichomonas vaginalis* and *Trichomonas gallinae* in liquid nitrogen. Progress in Protozool., No. *236*:199–200, International Congress Series No. 91, 1965.
17. Honigberg, B., and King, V.: Structure of *Trichomonas vaginalis* Donné. J. Parasit. *50*:345–364, 1964.
18. Hoogstral, H., and Heyneman, D.: Leishmaniasis in the Sudan Republic. Amer. J. Trop. Med. Hyg., *18*:1091–1210, 1969.
19. Hyman, L. H.: The Invertebrates. I. Protozoa through Ctenophora. New York, McGraw-Hill, 1940.
20. Jahn, T. L. and Jahn, F. F.: How to Know the Protozoa. Dubuque, Iowa, William C. Brown, 1949.
21. Kudo, R. R.: Protozoology, 4th ed. Springfield, Ill., Charles C Thomas, 1954.
22. Levine, N. D.: Protozoan Parasites of Domestic Animals and of Man. Burgess Pub. Co. 412 pp., 1961.

23. Lumsden, W. H. R.: Biological aspects of trypanosomiasis research. *In* Dawes, B.: Advances in Parasitology. Vol. 3, pp. 1–57. New York, Academic Press, 1965.

24. Mackinnon, D. L., and Hawes, R. S. J.: An Introduction to the Study of Protozoa. Oxford, Clarendon Press, 1961.

25. Manwell, R. D.: Introduction to Protozoology. New York, St. Martin's Press, 1961.

26. Marinkelle, C. J.: Direct transmission of *Trypanosoma cruzi* between individuals of *Rhodnius prolixus* Stal. Rev. Biol. Trop.; *13*:55–58, 1965.

27. Moulder, J. W.: The oxygen requirements of parasites. J. Parasit. *36*:193–200, 1950.

28. Nussenzweig, V., Deane, L. M., and Kloetzel, J.: Acquired immunity in mice infected with strains of immunological types A and B of *Trypanosoma cruzi*. Exptl. Parasit. *14*:233–239, 1963.

29. Pitelka, D. R.: Electron-microscopic structure of protozoa. New York, Macmillan, 1964.

29a.Ryley, J. F.: Carbohydrates and respiration. *In* Kidder, G. W. (ed.): Chemical Zoology. Vol. 1. New York, Academic Press, 1967.

30. Shaw, J. J., and Bird, R. G.: The intracellular habitat of *Endotrypanum schaudinni*. Trans. Roy. Soc. Trop. Med. Hyg., *58*:2, 1965.

31. Shorb, M. S.: The physiology of trichomonads. *In* Hutner, S. H. (ed.): Biochemistry and Physiology of Protozoa. Vol. III, pp. 383–457. New York, Academic Press, 1964.

32. Stabler, R. M., and Chen, T.: Observations on an *Endamoeba* parasitizing opalinid ciliates. Biol. Bull., *70*:56–71, 1936.

33. Steinert, M., and Novikoff, A. B.: The existence of a cytostome and the occurrence of pinocytosis in the trypanosome, *Trypanosoma mega*. J. Biophysic. Biochem. Cytol., *8*:563–569, 1960.

34. Taylor, A. E. R., and Baker, J. R.: The Cultivation of Parasites *in vitro*. Oxford, Blackwell Sci. Pub., 1968.

35. Vickerman, K.: Polymorphism and mitochondrial activity in sleeping sickness trypanosomes. Nature, *208*:762–766, 1965.

36. von Brand, T.: The physiology of blood flagellates. *In* Parasitic Infections in Man. pp. 90–113. New York, Columbia University Press, 1951.

37. ———: Metabolism of Trypanosoma cruzi. A review of recent developments. Medicina Tropical, 261–275, 1968.

38. ———: Biochemistry of Parasites. New York, Academic Press, 1966.

39. Wallace, F. G.: The trypanosomatid parasites of insects and arachnids. Exp. Parasitol., *18*:124–193, 1966.

40. Wenyon, C. M.: Protozoology. London, Bailliere, Tindall & Cox, 1926.

41. Willett, K. C.: Some observations on the recent epidemiology of sleeping sickness in Nyanza Region, Kenya, and its relation to the general epidemiology of gambian and rhodesian sleeping sickness in Africa. Trans. Roy. Soc. Trop. Med. Hyg., *59*:374–394, 1965.

Chapter 3

Superclass Sarcodina

ORDER AMOEBIDA

The Sarcodina are usually microscopic, and float or creep in fresh or salt water, although occasionally they may be sessile. The cytoplasm is usually divided into an outer, homogeneous, more-or-less clear ectoplasm and an inner, more granular endoplasm that makes up the bulk of the cytoplasm. The endoplasm contains fluid-filled vacuoles that may or may not contract, food vacuoles, mitochondria, other organelles and the nucleus. Within the cytoplasm of young cysts of some species are **chromatoid bodies.** These structures are deeply stained with certain dyes and are convenient objects for taxonomic purposes. They develop in the motile stage of amebas and mature in the cyst, but disappear as the cyst ages. They are composed of masses of ribosomes and thus function in protein metabolism.[27]

Pseudopodia are temporary projections of the body surface and cytoplasm that function for locomotion and feeding. They may be finger or tongue-like (lobopodia), filamentous (rhizopodia), or stiff, narrow structures containing an axial rod (axopodia). Axopodia are semipermanent. Only lobopodia are commonly found among the parasitic sarcodina; they are thrust out and withdrawn in an unpredictable fashion, and the amebas may move rapidly or sluggishly.

Food capture occurs when a pseudopodium flows around a suitable object and engulfs it, i.e., takes it into a temporary vacuole. There are several variations of this method. Nutrients may also be absorbed directly through the cell membrane. Food is stored as glycogen in a vacuole in some species.

The nucleus contains an **endosome** (nucleolus) that may be large or small, and may consist of a group of granules. It may be centrally located or eccentric in position. Chromatin material can be seen as small masses or minute granules on the inner nuclear membrane, and thus it appears as a distinct ring in optical section. Chromatin may also be scattered in the nucleus, or it may form a ring or "halo" around the endosome.

Reproduction usually occurs by binary fission, multiple fission, plasmotomy or budding, but sexual processes are involved in some species. Cyst formation is common among many of the parasitic forms. The motile organism (vegetative or trophic stage, or **trophozoite**) becomes rounded and secretes a resistant cyst wall around itself. Within this cyst, the nucleus may divide several times.

Parasites of this order are usually much smaller than the well-known free-living *Amoeba proteus*, many averaging 10 μ in diameter, and many of them are usually seen only in the cyst stage. The motile stage moves by means of relatively broad pseudopodia, and most of the species have only one nucleus, except in the cyst forms. Many species live in the digestive tracts of vertebrates and of invertebrates.

Family Dimastigamoebidae

This family of soil and freshwater inhabitants clearly shows affinities with flagellates. These protozoans possess an ameboid phase as well as a flagellated phase (usually biflagellated). Common genera are *Dimastig-*

amoeba, *Naegleria* and *Trimastigamoeba*. Sometimes found in old feces of insects and of vertebrates, they are coprophilic (see p. 19). *Naegleria* and *Hartmannella* are both potential pathogens of mammals, including man. *Naegleria* has been isolated from human spinal fluid. A few of these amebas in water dropped into the nose of mice or rats cause paralysis and often death.[2]

Family Amoebidae

These amebas are usually free-living or coprophilic and do not possess a flagellate phase. *Vahlkampfia patuxent* is a uninucleate ameba that lives in the gut of oysters. It has a large, broad pseudopodium and it may reach a diameter of 140 μ when cultured in vitro; its usual diameter in the host is about 20 μ. The nucleus contains a large endosome. Reproduction is by fission and cyst formation. Other species are commonly found in the intestines of mammals as coprophilic protozoans (p. 20). *Acanthamoeba*, which resorbs its endosome during division, and *Hartmannella* are small forms similar to *Vahlkampfia*, and they may also be found as coprophilic species. Species of *Hartmannella* have been found as intracellular parasites of grasshoppers and freshwater snails. *Acanthamoeba* is seriously pathogenic to mice and monkeys if inoculated into these animals. It has been isolated as a contaminant from trypsinized monkey kidney cells.[13] (See p. 556 for description of the pathology of *Acanthamoeba*.) All of these amebas, including the well-known genus *Amoeba*, possess large endosomes in the nucleus, and all may be found in the soil. *Sappinia diploidea* (Fig. 2–3), sometimes found in great numbers in old feces, has two nuclei.

Family Entamoebidae

The genus *Entamoeba* is usually found in the intestines of invertebrates and vertebrates. The vesicular nucleus possesses an endosome which is usually small and located at or near the center of the nucleus. Granules may occur on the inner wall of the nuclear membrane or around the endosome. The nuclei of the cysts are similar but may number from one to eight. Chromatoid bodies are often present in young cysts.

The genus *Endamoeba* was originally created for *E. blattae* in cockroaches. There is no endosome but numerous granules are located near the thick nuclear membrane. Cysts may contain many nuclei. Species of this genus are restricted to invertebrates.

Two relatively recent methods of studying amebas (as well as other protozoa) involve axenic cultivation and cryogenic preservation. Axenic cultivation is the maintenance of these organisms in the absence of all other cells. Axenic cultivation is essential for an understanding of nutritional requirements, growth behavior, rates of multiplication and metabolism of the parasites. Cryogenic preservation is freeze-preservation at ultralow temperatures. Using liquid nitrogen as the refrigerant, amebas may be stored at −196° C. for several years with little or no chemical or physical change. Long-term preservation eliminates most of the costly and time-consuming requirements of cultivation maintenance. For details on these two technical processes see Diamond.[4,5]

REPRESENTATIVE AMEBAS

Amebas in Invertebrates

Entamoeba sp.,[25] which are morphologically similar to *E. ranarum* in frogs, live in *Zelleriella* opalinids, which parasitize toads and frogs. The motile or trophozoite stage of the amebas averages 8 μ in diameter. They live in pockets in the opalinids, sometimes in such large numbers that there is little room for anything else (Fig. 2–32). Cysts averaging 9.4 μ occur in the opalinid host. Some of the amebas are parasitized by dot-like organisms which are probably *Sphaerita* (see p. 63). Thus the frog gut contains opalinids that harbor amebas that are host to *Sphaerita*.

Endamoeba blattae. This ameba lives in the intestines of cockroaches and termites. The parasite varies from 10 to 150 μ in diameter, and it forms multinucleate cysts. Pseudopodia are usually broad, and striations often

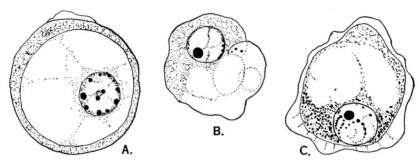

Fig. 3-1. *Entamoeba phallusae*, an ameba from the intestine of an ascidian at Plymouth, England. (Mackinnon and Ray, courtesy of J. Marine Biol. Assn. of the United Kingdom.)

appear in the body. The organism is probably harmless to the insects.

Entamoeba phallusae. This parasite (Fig. 3–1) has been found in the intestine of the ascidian, *Phallusia mamillata*, at Plymouth, England.[15] It measures 15 to 30 μ in the motile stage; cysts average 21 $\times$ 19 μ. The nucleus is prominent and eccentric. Small, unidentified amebas have also been found in the ascidian, *Clavellina lepadiformis*.

Entamoeba aulastomi lives in the gut of a horseleech, *Haemopsis sanguisuagae*. The cysts contain four nuclei. *Endamoeba philippinensi* and *Endamoeba javanica* both may be found in the intestine of the wood-feeding roach, *Panesthia javanica*. The bee, *Apis mellifica*, is the host of *Endamoeba apis*, and crane fly larvae have been found to harbor *Endamoeba minchini*.

Hydramoeba hydroxena. This pathogenic parasite lives on the coelenterate, *Hydra*. The ameba ranges in size from 60 to 380 μ in diameter, and it may kill a hydra in a few days by eating its epithelial cells.

Amebas in Fish

Not more than eight amebas have been reported from fish, and some of these may be the same species. *Amoeba mucicola* may be found in the branchial mucus and on the skin of the marine fish, *Symphodus tinca*, whereas the mucus lining the stomach of the rainbow trout, *Salmo shasta*, contains *Schizamoeba salmonis*. *Entamoeba ctenopharyngodoni* occurs in the rectum of the freshwater fish, *Ctenopharyngodon idellus*, in China. The sunfish,

Mola mola, harbors *Entamoeba molae* in its hindgut. This ameba is somewhat similar to *E. histolytica*, but the nucleus contains a subperipheral crescent of granules. A careful search would undoubtedly reveal other species of amebas in both freshwater and marine fish.

Amebas in Amphibia

Entamoeba ranarum. This parasite (possibly pathogenic) lives in the large intestine of frogs. The motile stages usually vary in size from 10 to 50 μ in diameter. There is a small endosome in the nucleus, and the cysts usually possess four nuclei, although as many as sixteen have been reported. Similar amebas have been found in salamanders.

Amebas in Reptiles

Entamoeba invadens (Fig. 3–2). The best-known *Entamoeba* parasite from these hosts is *E. invadens*. It lives in the intestine, is

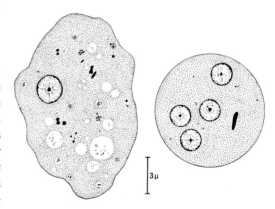

Fig. 3-2. *Entamoeba invadens*, motile and cyst forms.

similar in many respects to *E. histolytica* of man, and may be highly pathogenic. This ameba has become a favorite for *in vitro* studies of metabolism. A prerequisite for mass encystation of *E. invadens* within its host is the ingestion by the protozoon of minute particles of carbohydrate. The ameba is harmless in herbivorous turtles, as compared with its effects in experimental snakes. This difference is probably related to the abundance of particulate carbohydrate in the turtle's intestine, and the absence of it in the gut of the snake.[17] *E. invadens* has been found to be pathogenic to chicks. See Balamuth[1] for a consideration of the effects of some environmental factors upon growth and excystation.

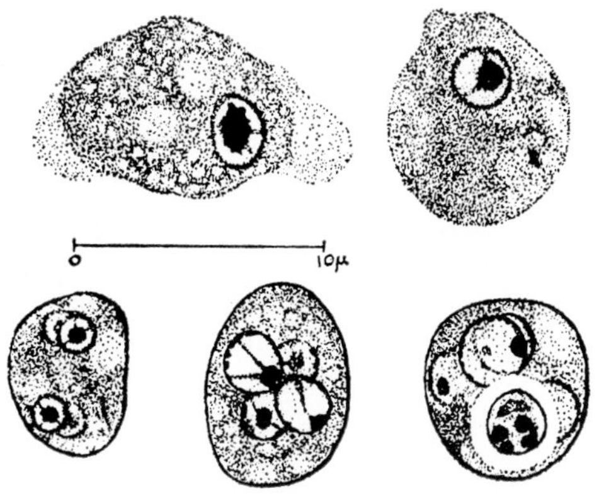

Endolimax clevelandi. This ameba (Fig. 3-3) lives in the intestines of the turtle, *Pseudemys floridana mobilensis.* The small trophic stage averages about 7 μ in diameter. A finely granular ectoplasm encloses a coarser endoplasm. Vacuoles may or may not be present and some of them may contain bacteria or other food. The nucleus is vesicular in appearance with a large endosome. Cysts are often elliptical (an *Endolimax* trait), and they range from 4.5 to 10 μ in diameter. The mature cyst contains four nuclei, which are similar to those in the motile stage.

Amebas in Birds

The freedom from amebas enjoyed by birds as compared with their burden of other parasites is difficult to explain. Probably a more thorough search of birds for intestinal amebas would be productive. The few amebic parasites reported from birds include *Entamoeba lagopodis* (*E. gallinarum?*) and *Endolimax gregariniformis* (*E. janisae*) in the cecum of the grouse, *Lagopus scoticus; Entamoeba anatis* in the duck; and *E. gallinarum* in various fowl.

Amebas in Mammals

Entamoeba bovis. In the intestines of cattle, *E. bovis* (Fig. 3-4) has a motile form that averages 5.3 $\times$ 7.5 μ, with a smoothly

Fig. 3-3. *Endolimax clevelandi* from the turtle, showing motile stages (top row) and cysts (bottom row). (Gutierrez-Ballesteros and Wenrich, courtesy of J. Parasitol.)

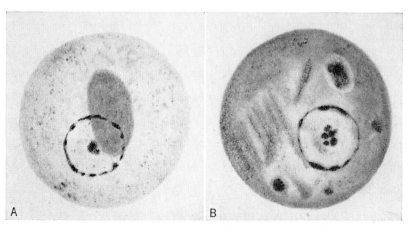

Fig. 3-4. *Entamoeba bovis* cysts from the intestine of cattle. (Noble, E., courtesy of Univ. Calif. Pub. Zool.)

granular cytoplasm filled with vacuoles. The large nucleus has conspicuous peripheral chromatin and a large central endosome made up of compact granules. Cysts average 8.8. μ, possess a single nucleus and, in young forms, irregular chromatoid bodies that stain deeply with hematoxylin. Glycogen vacuoles are common in the cysts. *E. histolytica* has also been reported from the intestine of cattle.

Entamoeba ovis. Sheep are host to *E. ovis* in the intestines. Examination of host feces rarely discloses motile forms. Cysts average 7.2 μ in diameter, and they contain chromatoid bodies of irregular shapes. The nucleus is often eccentric in position and is occasionally surrounded by a clear area. Peripheral chromatin is moderate in amount, and the endosome is composed of a relatively large compact mass of granules. Many variations of these characteristics exist. *E. caprae* has been reported from the stomach of sheep.

Entamoeba debliecki. Goats and pigs are normal hosts to this intestinal ameba. The motile stage averages 13 $\times$ 16 μ, and it possesses a finely granular ectoplasm and a coarser endoplasm. The nucleus is large, usually pale, and the peripheral chromatin appears to be a homogeneous ring. The central endosome is large and often indistinct. Cysts range from 4 to 17 μ in diameter, averaging 8 μ. Highly variable chromatoid bodies occur in large cysts. The endosome varies from a very large mass to a small one surrounded by a "halo" of granules. Three other species have been recorded from the goat intestine. They are *E. wenyoni*, *E. dilimani* and *E. caprae*. The latter has also been found in the stomach.

Considerable confusion exists with regard to the correct names for entamoebas of domestic animals. They all possess essentially the same morphologic features. Hoare has stated that "there are definite indications that clinical amoebiasis caused by *Entamoeba histolytica*—both acquired from human sources and spontaneous—does occur in certain mammals, and it is conceivable that they might serve as reservoirs of human infection. Furthermore, the possibility of other species

of amoebae being pathogenic cannot be excluded." See Noble and Noble[20] and Hoare[10] for discussions of the problem and a review of the literature.

Entamoeba intestinalis. This ameba has been found in the cecum and colon of horses and is said to occur in other animals. Trophic forms are common whereas the cysts are rare or absent. The nuclei are of the *E. coli* type, and the eccentric endosome is surrounded by a ring of granules. *E. gedoelsti* has been suggested as a better name for this ameba. *E. equi* (*intestinalis?*) has also been found in the feces of horses.

Dogs harbor amebas, but the identification of species is uncertain. *Entamoeba venaticum* (*E. histolytica?*, *E. caudata?*) has been described from the large intestine, and *E. gingivalis* (*E. canibuccalis?*) has been reported from the mouth. *E. histolytica* was identified in 8.4 per cent of the dogs in Memphis, Tennessee. *E. coli* and *Endolimax nana* have also been reported from the intestine of the dog.

Cats apparently do not have their own amebas, but they can easily be infected with *Entamoeba histolytica* from man. For this reason, kittens are often used in experiments with this important human parasite.

Rabbits may harbor *Entamoeba cuniculi* (*E. muris*), which is similar to *E. coli* of man. *Endolimax* sp. also live in the cecum.

Guinea pigs are host to *Entamoeba cobayae* (*E. caviae* and *E. muris?*) which resembles *E. coli* of man. This species occurs in the cecum, whereas the intestine harbors *Endolimax caviae*.

Rats and mice, like the two hosts above, possess an intestinal ameba that resembles *Entamoeba coli*. It is named *E. muris* (Fig. 3–5). Rats naturally infected with *E. histolytica* have been reported from England, Iraq, Indonesia and the United States. *Endolimax ratti* occurs in the rat colon. It is morphologically indistinguishable from human *E. nana*.

Macaque monkeys may harbor several amebas that are morphologically identical with those of man. Those species reported from the large intestine are: *Entamoeba*

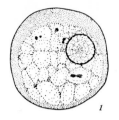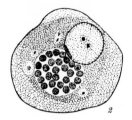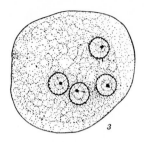

Fig. 3-5. *Entamoeba muris*, found in rats and mice. The figures show nuclear structure. *1.* Ameba with highly vacuolated endoplasm and endosome unstained (hemalum). *2.* With included parasite, *Sphaerita*. *3.* With four nuclei. (Wenrich, courtesy of J. Morphology.)

histolytica (*E. nuttalli, E. duboscqi, E. cynomolgi?*); *E. coli* (*E. legeri*); *E. chattoni; Endolimax nana* (*E. cynomolgi*); *Iodamoeba bütschlii* (*I. kueneni*).

Amebas of Man

The genera and species of amebas in man are *Entamoeba histolytica, E. hartmanni, E. coli, E. gingivalis, Endolimax nana, Iodamoeba bütschlii* and *Dientamoeba fragilis.* As with the amebas of animals, the trophozoites of these species are usually uninucleate, and the cysts are commonly multinucleate. Most amebas of man are not pathogenic, but *E. histolytica* causes one of the more serious parasitic diseases (amebiasis) of tropic and temperate countries.

Entamoeba moshkovskii. *E. moshkovskii* can easily be mistaken for *E. histolytica* when sewage-contaminated water is examined. The four-nucleate cysts of each species appear practically identical. Although their relationship is probably only phylogenetic,[3] there is a possibility that *E. moshkovskii* is a reduced-temperature strain of *E. histolytica.* This possibility is supported by the existence of some strains (e.g., Laredo type) of *E. histolytica* that can be cultivated easily at both 20° C. and 37° C.[8]

Entamoeba histolytica. This ameba is common in man, apes and monkeys, and may also be found in pigs, dogs, cats and rats. For a long time it was believed that all the varieties of *Entamoeba histolytica* belonged to a single species. It now seems clear that there are two distinct species: *E. histolytica* and *E. hartmanni.* Differentiation is based on structural features, size ranges, cultivation differences, and antigenic constitution. *E. hartmanni* is small and nonpathogenic. *E. histolytica* occurs as two races or strains: a smaller, common, nonpathogenic form, and a much larger, virulent form. For details of these species and varieties, see Hoare,[11] Sarkisyan,[24] Goldman and Gleason,[9] and Neal.[19]

About 600 million people are infected with the large strain of *Entamoeba histolytica*, but only a small fraction of this number have any clinical symptoms.[12] Certain aspects of the host-parasite relationship in amebic infections remain obscure in spite of the enormous amount of study that has been carried on for well over 80 years. The role of the host in the *initiation* of pathogenicity is particularly elusive. The intestinal mucosa, host diet, bacterial flora, and concurrent infections with other parasites are factors that influence amebiasis. See p. 66 for a discussion of pathogenicity.

Mature motile *Entamoeba histolytica* (Fig. 3–6) averages about 25 μ in diameter (range 10 to 60 μ). The cytoplasm consists of a clear ectoplasm and a finely granular endoplasm. The endoplasm may include food vacuoles filled with red blood cells in various stages of digestion. Clear vacuoles may also be present. Movement is irregular and is associated with pseudopodia that are usually broad but may sometimes be finger-like. The type and rapidity of movement varies, depending on the consistency of the surrounding medium, age of the parasite, temperature,

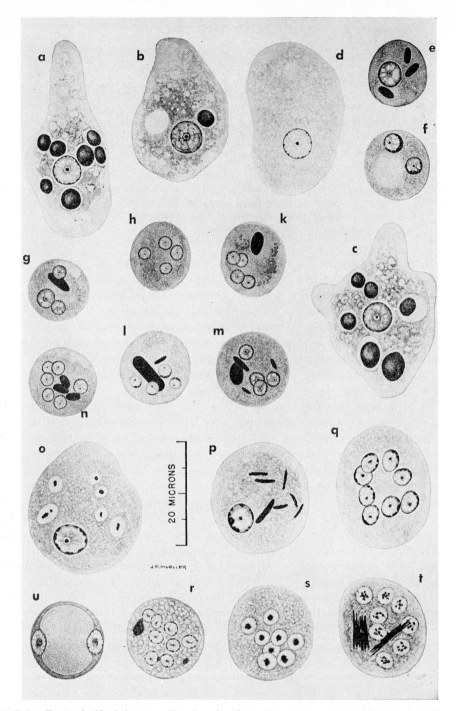

Fig. 3-6. *Entamoeba histolytica.* *a–c.* Trophozoites from a single case, showing ingested red blood cells. *d.* Trophozoite from another case showing delicate cytoplasm and typical nucleus. *e–n.* Various cysts, with and without chromatoid bodies, showing 1, 2, 3, 4 and 6 nuclei of varying character. Chromatoid bodies are commonly present in young cysts, absent in older cysts. In exceptional cases, eight nuclei are present.

Entamoeba coli. *o.* Trophozoite with ingested bacteria, showing blunt pseudopod, typical nucleus. *p.* Cyst with chromatoids, single nucleus. *q–t.* Cysts with eight nuclei, showing varying conditions of cytoplasm, chromatoids, and nuclear structure. *u.* A common form of coli cyst containing a large central glycogen vacuole and two large nuclei. (Mueller, *The Story of Amebiasis,* courtesy of Winthrop Laboratories.)

variety or race of ameba, stage of treatment of the host, and many other factors. Food for these parasites usually consists of bacteria or other organic material found in the intestine. Red cells are found only in pathogenic forms. Food vacuoles appear to be formed by invagination of the plasma membrane, and usually a different vacuole is formed for each kind of food particle.[7]

The vesicular nucleus measures about 3 to 5 μ in diameter. Within it lies the endosome. This structure may appear as a single granule or a closely packed cluster of minute granules. Often a ring or halo appears to surround the endosome, and sometimes spoke-like lines radiate from the endosome to the nuclear membrane. Other granules, usually of uniform size, lie against the inner wall of the nuclear membrane. Considerable variation occurs in these morphologic features.

Pathogenic forms of this ameba live in lesions of the host's large intestine. They divide by mitosis. The lumen-dwelling forms engulf bacteria, and after a period of feeding and reproduction, vacuoles disappear and the amebas become rounded. Soon a cyst wall begins to form; these uninucleated stages are **precysts.** One or more deeply-staining bar-shaped chromatoid bodies characteristically develop in the cytoplasm of the cyst. Cysts (Fig. 3–6) are about 12 μ in diameter (range 5 to 20 μ) and usually have four nuclei. **Metacystic stages** involve the division of the quadrinucleate organism into four small uninucleate stages, which subsequently grow and divide. Cysts pass out of the body with feces and become the infective stage. Forty-five million cysts may be discharged in the feces of one infected person in one day.

These cysts are killed by drying. The cyst wall[16] of *Entamoeba histolytica* consists of one layer, 0.5 μ in diameter, whereas that of *E. coli* (see below), consists of two layers totaling 1 μ in thickness. Of the two amebas, *E. coli* is far more resistant to drying.

When eaten by a new host, a cyst is carried to the small or large intestine where it escapes the confinement of the cyst wall. The process of escape is called **excystation.** The actual process of excystation starts with increased activity of the ameba within its cyst. Clear ectoplasmic pseudopodia are formed at various points within the cyst and they travel rapidly around the periphery. Frequently they press against the wall at certain spots, as though the imprisoned organism were searching for an exit. Soon a pseudopodium is seen to be applied to one point on the internal surface of the cyst wall. Shortly afterwards the tip of the pseudopodium squeezes through a minute and previously invisible pore at this spot and it appears like a tiny hernia on the outside. Just how the perforation is made is not clearly understood, but probably a cytolytic substance is involved.

When mature, the amebas usually remain in the lumen of the intestine. Not only is the pH near the mucosa more stable than that of the rest of the lumen contents, but gases and organic materials in the paramucosal lumen (that portion of the lumen immediately adjacent to the mucosa) differ from those in the center of the lumen. These physiologic strata are obviously of importance in the physiology and distribution of intestinal parasites. They help to explain the different ecologic niches of such related species as *Entamoeba histolytica* and *E. coli*—the former appears to favor a closer contact with the mucosa.

Entamoeba histolytica can be cultivated, collected in suspension, and a sterile extract prepared that shows a high proteolytic activity against a casein or gelatin substrate. Although this capacity was found by Neal[18] in two out of five invasive strains, he concluded that "the present evidence does not convincingly demonstrate that high proteolytic activity is required for tissue invasion by amoebae, but may accompany another factor." Possibly the parasites merely enter lesions associated with colitis, abrasions, temperature fluctuations or abnormal diet, or damaged areas produced by other agents, such as bacteria.

The relationship between bacteria and amebas is important and little understood.

Normally, the presence of living bacteria or of other living cells is essential not only for in vitro cultivation of *Entamoeba histolytica* in the laboratory, but for the life of these protozoa in the intestine of their host. The pathogenicity of this ameba is influenced by both pathogenic and nonpathogenic bacteria.[22] (For details on pathogenicity, see Neal[18] and Stamm.[26]) Amebas from ameba-trypanosome cultures were unable to establish themselves in germ-free guinea pigs without bacteria, but the addition of either of two species of bacteria (*Aerobacter aerogenes* or *Escherichia coli*) not only permitted the establishment of the amebas, but promoted invasion of host tissues. Differences in experimental invasiveness and in specific nutritional requirements of *E. histolytica* may be due to bacterial components of the culture, rather than variations in the protozoa. Experiments now in progress with axenic cultures may throw light on this problem.

Since motile amebas are delicate, they die quickly outside the body. Transmission depends on the cysts, which must be kept moist or they, too, will die. If kept moist, they can live for a few weeks to a few months, depending on the temperature. Cysts normally enter a new host in drinking water or food. They may be carried mechanically by such insects as flies and cockroaches, or by people with unclean hands. Diagnosis is confirmed by finding cysts or trophozoites in feces.

Entamoeba histolytica may not be confined to the intestine. Once a lesion is made in the intestinal wall, the amebas have an opportunity to travel through the body by way of the blood or lymph and to invade any other tissue. In fact, they have been found in almost all soft tissues, but the most common extra-intestinal locus of infection is the liver, particularly the right lobe. Here the infection may be mild, or a large abscess may develop. Abscesses may also occur in the skin[14] (Fig. 3–7), lungs, or brain. Tenderness in the hepatic region is probably caused by infection of the hepatic loop of the intestine rather than a liver abscess.[23]

The pathology of amebiasis varies a great

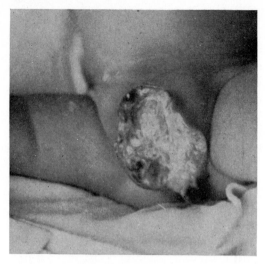

Fig. 3-7. Cutaneous amebiasis. (Courtesy of Dr. Francisco Biagi F., Ciudad University, Mexico.)

deal. Symptoms include abdominal pain, nausea, vomiting, mild fever, diarrhea, blood and mucus in feces, tenderness over the sigmoidal region of the colon, and hepatitis. Serious infections involve ulceration of the colon, appendicitis, abscess of the liver and, rarely, abscess of the brain. The virulency of *Entamoeba histolytica* can be attenuated by a series of subcultures, and increased by repeated passage of the amebas through laboratory animals. Repeated daily stool examinations should be made before proclaiming a negative diagnosis.

Prevention of the disease involves proper disposal of sewage, avoidance of fecal contamination of food, water and eating utensils, and by education. The immunology of *Entamoeba histolytica* has been studied extensively, but "so far, biochemical and physiological studies have not resulted in a test which will distinguish species of *Entamoeba*."[23]

Entamoeba coli. Another ameba found in the human large intestine is the nonpathogenic *E. coli* (Fig. 3–6). This organism has a worldwide distribution, and the incidence of infection varies greatly. The overall incidence is probably 28 per cent, making it the most common species of intestinal ameba in man. Ten to 50 per cent of the population has been reported to be infected in various areas. The type and thoroughness of examination has greatly influenced the accuracy of the reports.

The life cycle of this parasite differs from that of *Entamoeba histolytica* in that *E. coli* does not enter tissues. The trophozoite of *E. coli* ranges from 20 to 30 μ in diameter. It is more sluggish than *E. histolytica*, and the cytoplasm usually appears much more dense and crowded with food vacuoles. Although there is little clear ectoplasm, a thin clear area is generally seen around the nucleus of the trophozoite in stained specimens. The nucleus possesses heavier peripheral chroma-

tin than does *E. histolytica* and the endosome is eccentrically placed. Young cysts are apt to have a large glycogen vacuole, and they may possess few nuclei; mature cysts possess eight nuclei, each being similar to the nucleus of the motile stage. The chromatoid bodies of the cysts are usually slim with pointed or irregular ends, thus appearing different from the cigar-shaped or bar-shaped bodies in *E. histolytica*. Unlike those of *E. histolytica*, the cysts of *E. coli* are not readily killed by drying, which probably accounts for the high incidence of infection. Possibly, *E. coli* cysts may be air-borne, but this method of infection has

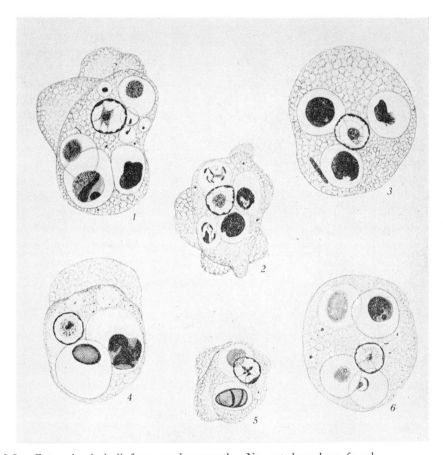

Fig. 3-8. *Entamoeba gingivalis* from monkey mouth. No cysts have been found.
1. The ameba is thrusting out three pseudopodia. *2.* Many small pseudopodia present, compressed nucleus. The fragmented remains of salivary corpuscles are visible within the vacuoles. *3.* The endoplasm contains one large rod-shaped bacterium. The peripheral chromatin of the nucleus is heavily beaded and the endosome eccentric. *4.* One large endosomal granule, one broad pseudopod, and large vacuoles. *5.* Six granules (chromosomes?) within the nucleus. They are attached by fibers to the peripheral chromatin. Fibers mainly attached to one large peripheral blob. *6.* One small, irregular, central endosomal granule. (Kofoid and Johnstone, courtesy of Univ. Calif. Pub. Zool.)

not been proved. Monkeys and apes also share this parasite with us; pigs apparently have it, but rats seem to be free from it.

Entamoeba gingivalis. An ameba frequently available in a classroom is the mouth form, *Entamoeba gingivalis* (Fig. 3–8). As with *Dientamoeba fragilis* (p. 70), only the trophic stage is known. The ameba lives in the gingival areas around the teeth of man, other primates, dogs and cats. It can be gathered by gentle probing around the bases of the teeth with a toothpick. The incidence of infection is high, probably around 50 per cent, but the organisms are found more frequently associated with diseased conditions than with healthy mouths. There is no evidence, however, that the amebas cause disease. Related species occur in horses and pigs. Since there are no cysts and no intermediate hosts are involved, *E. gingivalis* represents the simplest kind of life cycle.

The size of *Entamoeba gingivalis* ranges from 5 to 35 μ in diameter, with an average of about 15 μ. In general these amebas appear similar to *E. histolytica*, but usually they contain many more large food vacuoles that often enclose remnants of ingested host white blood cells. Chromatin of the nuclei of these host cells takes a deep stain which often gives the cytoplasm of the ameba the appearance

of containing several large black nuclei. Sometimes a definite ectoplasm may appear to be separated from the endoplasm by a deeply staining ring.

The nucleus is lined with beaded peripheral chromatin, and it contains an endosome that may consist of a single granule or, more commonly, of several closely grouped granules. Often, spoke-like fibrils connect the endosome with the nuclear membrane.

Endolimax nana. This intestinal ameba of man, apes, monkeys and pigs averages about 10 μ in diameter in the motile stage (Fig. 3–9). It lives in the cecum and in the large intestine of 10 to 20 per cent of the human population of most countries of the world. Although it is not pathogenic, its presence indicates that the host has been contaminated with fecal material from somebody else. Therefore the presence of *Endolimax nana*, or of any other intestinal parasite whose method of transmission is directly from intestine to mouth, is an indicator that a pathogenic parasite may be present.

Trophozoites of *Endolimax nana* usually appear about one-half the size of those of *Entamoeba histolytica*, although there may be larger individuals. The cystoplasm often looks pale and vacuolated, and the pseudopodia are usually short and broad, with ends

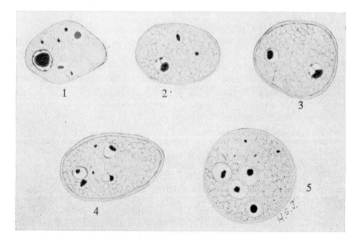

Fig. 3-9. *Endolimax nana*, a small intestinal ameba of man and monkeys. $\times$ 2500.

1. Trophozoite with a few endoplasmic inclusions. *2,3,4,5.* One, two and four nucleated cysts. Note the characteristic chromatoid bodies in *2. 5.* May be mistaken for a small cyst of *Entamoeba histolytica.* (Anderson *et al., Amebiasis,* courtesy of Charles C Thomas.)

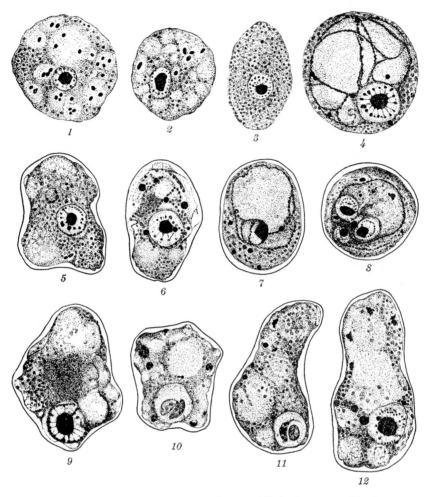

Fig. 3-10. *Iodamoeba bütschlii* from the intestine of man. All the figures on this plate were drawn by Mr. R. L. Brown from slides of a single infection.

1 to *3*. Trophozoites. *1*. Typical individual; nucleus with large deeply-stained endosome surrounded by a row of granules attached to it by radial fibrils. *2*. Small individual with larger nucleus; endosome elongated, granular ring not quite complete. *3*. Probably precystic individual; food bodies mostly absent; endosome of nucleus relatively small; granules more numerous on one side of endosome. *4*. Large early cyst. Large nucleus with granules still in a ring around endosome; apparently several glycogen vacuoles not yet coalesced; dark chromatoid strands at periphery of glycogen vacuoles. *5*. Cyst of irregular contour; granules of periendosomal ring uneven in size.

6. Normal-sized cyst; note volutin-like granules in cytoplasm; one large intranuclear granule attached to endosome. *7*. Cyst with small nucleus; endosome a lateral plaque against the nuclear membrane, an unusual condition considered to be abnormal. *8*. Round cyst with three and possibly four nuclei; fourth nucleus not clearly distinguishable. *9*. Large cyst with periendosomal granules of nucleus arranged as a peripheral ring as in species of *Entamoeba*. *10* and *11*. Cysts with lightly staining endosomes in nuclei; granules massed into crescents. *12*. Large cyst with endosome of nucleus staining more darkly; granules not confined to a lateral crescent. (Wenrich, courtesy of Proc. Amer. Phil. Soc.)

showing hyaline ectoplasm. Food vacuoles are often present. The single nucleus, usually not readily seen in unstained specimens, contains a large endosome that is often eccentric in position and may even lie against the nuclear membrane. Sometimes minute threads appear to connect the endosome with this membrane. Since the nuclear membrane generally lacks the chromatin clusters or granules common to *E. coli* and to *E. histolytica*, this membrane appears to be an exceedingly fine line and often is barely discernible.

Cysts of *Endolimax nana* average 9 μ in diameter and are often so pale that they are difficult to see, even in stained material. Four nuclei are usually present and they are similar to those of the motile stage, but only one or two nuclei may be present. In stained material the cysts frequently appear to be no more than indistinct oval bodies containing four tiny, dark dots (the endosomes).

Iodamoeba bütschlii (=I. williamsi). This intestinal ameba (Fig. 3–10) gets its name from its large glycogen vacuole, which readily stains brown with iodine. The organism normally has a diameter of 9 to 14 μ in the motile stage in which the ectoplasm is not clearly differentiated from the endoplasm.[21] The nucleus has little peripheral chromatin and the large endosome is surrounded by a mass of refractile granules. In stained specimens the endosome is dark but the surrounding granules are usually lighter.

The cyst (about 9 μ long) is unusual because of its rectangular, oval or variable shape. It, too, normally possesses a large glycogen vacuole. The single nucleus contains a deeply stainable endosome that usually lies against the nuclear membrane, and the

granules may be clustered in the shape of a crescent near the endosome. There are many variations in the appearance of these cysts and occasionally two nuclei are present. The ameba is probably nonpathogenic, but reports of pathogenicity exist. It has also been reported from the large intestine of monkeys, apes and pigs, and is probably the same as *Iodamoeba suis*.

Dientamoeba fragilis. This ameba (Fig. 3–11), as its name indicates, is characterized by the presence of two nuclei. Cysts have never been found. It lives in the intestine of man and is probably present in at least 20 per cent of the population of most countries; an incidence of over 51 per cent, however, may occur in local areas. The ameba has been found in monkeys and was once reported from sheep. Examination techniques that are especially suited for the identification of cysts would, of course, fail to reveal *Dientamoeba fragilis*. Since the ameba is often difficult to see, most surveys report a very low incidence of infection.

The size of the ameba varies considerably, ranging from 3 to 22 μ with an average of about 8 μ. The cytoplasm, usually pale in appearance even in stained specimens, frequently contains small food vacuoles. Although most specimens possess two nuclei, mononucleate forms are not uncommon, and there may even be several nuclei. Each nucleus contains a cluster of closely grouped granules, frequently four, which form the endosome. Frequently a fibril (*paradesmose*) may be seen connecting the two nuclei. There is practically no peripheral chromatin on the nuclear membrane. This ameba resembles a flagellate in some ways.

Sometimes, a tube-like extension may be

 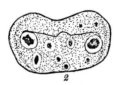 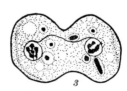

Fig. 3-11. *Dientamoeba fragilis.* *1.* Typical binucleate individual. *2* and *3.* Stages in division. *4.* Uninucleate stage. (Wenrich, courtesy of J. Morphology.)

seen projecting from the body of *Dientamoeba fragilis*. This organelle seems to be a food-gathering apparatus, and is suggestive of a similar structure found on the flagellate, *Histomonas meleagridis* (p. 51). The mode of transmission is uncertain, but a suggestion has been made that the ameba may be carried by an intestinal worm such as *Trichuris* or *Enterobius*.[6]

IMMUNITY

Most animals and most people apparently possess a natural immunity against amebic infection, as is evidenced by the fact that only a small per cent of infected hosts have amebiasis. This protection against disease is probably due to adjustment between parasite and host and does not necessarily depend on antigen-antibody responses. Certainly antibodies are produced, especially against amebas that are in parenteral contact with tissues. If amebas remain in the intestinal lumen, the host may not produce the antibodies. The presence of antibodies, however, does not prove the presence of amebas, because antibodies may remain long after the amebas have disappeared.

Serologic diagnosis of amebiasis has been successful using various techniques, especially the gel-diffusion precipitin test. This test is helpful in that it reveals nonspecific systems that might be present. Hemagglutination and precipitation tests yield a high percentage of positives. Tests must be interpreted with regard to the background of antibodies ("noise"), which varies in different geographic areas.

The fluorescent antibody test is becoming increasingly popular with workers in serologic techniques. It involves coupling a fluorescent dye to an antibody, thereby making the antibody-antigen reaction visible. The relationships of species of *Entamoeba* have been clarified by this technique.

CLASS ACTINOPODEA

Order Proteomyxida

This group is poorly defined as to its proper taxonomic position and much study is needed to clarify the life cycles of most of the known species. Families of the order include the Labyrinthulidae, Pseudosporidae and Vampyrellidae. Most species attack algae but *Labyrinthula macrocystis* parasitizes eel grass (*Zostera marina*). The labyrinthulids are uninucleate but they join to form a network of individuals. The pseudosporids possess ameboid and flagellate stages. In the vampyrellids the mature stage is a plasmodium.

REFERENCES

1. Balamuth, W.: Effects of some environmental factors upon growth and encystation of *Entamoeba invadens*. J. Parasitol., 48:101–109, 1962.
2. Culbertson, C. G., Ensminger, P. W., and Overton, W. M.: Pathogenic *Naegleria* sp.—Study of a strain isolated from human cerebrospinal fluid. J. Protozool., 15:353–363, 1968.
3. de Carneri, Ivo: Studi su *Entamoeba moshkovskii*. III. Isolamento di 17 ceppi a Milano. Caratteristiche morfoalogiche e adattamento termico all'-ambiente di 4 ceppi milanese e di 1 ceppo Brasiliano. RiVista di Parassitologia, 27:73–88, 1966.
4. Diamond, L. S.: Recent advances in the cultivation and cryogenic preservation of *Entamoeba histolytica*. Amer. J. Gastroent., 41:336–370, 1964.
5. Diamond, L. S., Meryman, H. T., and Kafig, E.: Preservation of parasitic protozoa in liquid nitrogen. *In* Martin, S. M. (ed.): Culture Collections: Perspectives and Problems. pp. 189–192. Toronto, University of Toronto Press, 1963.
6. Dobell, C.: Researches on the intestinal protozoa of monkeys and man. X. The life history of *Dientamoeba fragilis*—observations, experiments and speculations. Parasitology, 32:417–459, 1940.
7. Fletcher, K., Maegraith, B. G., and Jarmilinta, R.: Electron microscope studies of trophozoites of *Entamoeba histolytica*. Ann. Trop. Med. Parasitol., 56:496–499, 1962.
8. Goldman, M., and Cannon, L. T.: Antigenic analysis of *Entamoeba histolytica* by means of fluorescent antibody. Amer. J. Trop. Med. Hyg., 16:245–254, 1967.
9. Goldman, M., and Gleason, N. N.: Antigenic analysis of *Entamoeba histolytica* by means of fluorescent antibody. IV. Relationship of two strains of *E. histolytica* and one of *E. hartmanni* demonstrated by cross-absorption techniques. J. Parasitol., 48:778–783, 1962.

10. Hoare, C. A.: The enigma of host-parasite relations in amebiasis. The Rice Institute Pam., *45*:23–35, 1958.

11. ———: Amoebic infections in animals. Vet. Rev. Annot., *5*:91–102, 1959.

12. ———: Considerations sur l'étiologie de l'amibiase d'apres le rapport hôte-parasite. Bull. Soc. Path. Exot., *54*:429–441, 1961.

13. Jahnes, W. G., Fullmer, H. M., and Li, C. P.: Free living amoebae as contaminants in monkey kidney tissue culture. Proc. Soc. Exper. Biol. Med., *96*:484–488, 1957.

14. León, L. A.: Amibiasis Cutánea. Rev. Med. Mex., *42* (899):375–384, 1962.

15. Mackinnon, D. L., and Ray, H. A.: An amoeba from the intestine of an ascidian at Plymouth. J. Marine Biol. Ass. of the U.K., *17*(2):583–586, 1931.

16. McConnachie, E. W.: The morphology, formation and development of cysts of *Entamoeba*. Parasitology, *59*:41–53, 1969.

17. Meerovitch, E.: On the relation of the biology of *Entamoeba invadens* to its pathogenicity in snakes. J. Parasitol., *43*(Abstracts):41, 1957.

18. Neal, R. A.: Enzyme proteolysis by *Entamoeba histolytica*; biochemical characteristics and relationship with invasiveness. Parasit., *50*:531–550, 1960.

19. ———: Experimental studies on *Entamoeba* with reference to speciation. *In* Dawes, B. (ed.): Advances in Parasitology. Vol. 4, pp. 1–51. New York, Academic Press, 1966.

20. Noble, G. A., and Noble, E. R.: Entamoebae in farm mammals. J. Parasitol., *38*:571–595, 1952.

21. Pan, Chia-Tung: 1959. Nuclear division in the trophic stages of *Iodamoeba bütschlii* (Prowazek, 1912) Dobell, 1919. Parasitol., *49*: 543–551, 1959.

22. Phillips, B. R., *et al.*: Studies on the ameba-bacteria relationship of amoebiasis. Comparative results of the intracecal inoculation of germfree, monocontaminated and conventional guinea pigs with *Entamoeba histolytica*. Amer. J. Trop. Med. Hyg., *4*:675–692, 1955.

23. Powell, S. J., Wilmot, A. J., and Elsdon-Dew, R.: Hepatic amoebiasis. Trans. Roy. Soc. Trop. Med. Hyg., *53*:190–195, 1959.

24. Sarkisyan, M. A.: 1957. Observations on *Entamoeba hartmanni*, Prowazek, 1912. Med. Parasitol. i Parazitarn Bolezni, *26*:618–623, 1957. (Referat. Zhur. Biol., 91033, 1958.)

25. Stabler, R. M., and Chen, T.: Observations on an endamoeba parasitizing opalinid ciliates. Biol. Bull., *70*(1):62–71, 1936.

26. Stamm, W. P.: *Entamoeba histolytica* in man. *In* Taylor, A. E. R. (ed.): The Pathology of Parasitic Diseases. pp. 1–14. Oxford, Blackwell Sci. Pub., 1966.

27. Svihla, G., and Barker, D. C.: Ultraviolet microscopy of *Entamoeba invadens*. J. Parasit., *49*(Sect. 2):37–38, 1963.

Chapter 4

Subphyla Sporozoa and Cnidospora

SUBPHYLUM SPOROZOA

A wide and diversified assemblage of protozoan parasites characterized by the presence of spores has generally been placed, as a matter of convenience, in a single class or subphylum, Sporozoa, in spite of the absence of any clear-cut features that bind the assemblage into an obvious taxonomic unit. The complicated spores of the order Myxosporida, with their polar filaments, are totally unlike spores of the genus *Plasmodium*, which produce infective sporozoites. The electron microscope has shown that some forms without spores (e.g., *Toxoplasma*), which were formerly placed as appendages of uncertain affinities, are structurally similar to some of the sporozoan group. Thus, many of these groups have been reclassified according to the scheme found on page 22.

In the subphylum Sporozoa, organelles of locomotion are not evident to the casual observer, but these protozoa are able to move about readily. We have observed, in some species, active pseudopodial projections of the body. Electron microscopy reveals minute papillae covering the body of several species; others are covered with "hairs" that look like cilia (e.g., *Rhynchocystis pilosa* in earthworms). Knobs, filaments, sucker-like depressions, bristles, internal fibrils and myonemes all may play a part in locomotion as well as other functions. Both sexual and asexual reproduction occur in many species, as in the malarial organism, *Plasmodium*.

With few exceptions, spores are produced containing one to many motile sporozoites.

There are no polar capsules, polar filaments, cilia or flagella (except for flagellated microgametes in some species). Vertebrates and invertebrates are hosts.

Class Telosporea

SUBCLASS GREGARINIA

The gregarines comprise a very large group of parasites limited to invertebrates, in which they occupy both tissue cells and body cavities. Gregarines within cells are only a few microns in diameter, but species in body cavities may be much larger, the maximum

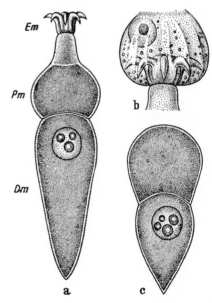

Fig. 4-1. *Corycella armata* a gregarine showing the epimerite, *Em;* protomerite, *Pm;* and deutomerite, *Dm.* *a.* Entire parasite. *b.* Epimerite anchored in host cell. *c.* Gamont detached from epimerite. (Grell, *Protozoologie*, courtesy of Springer-Verlag.)

being about 10 mm. in length. In some spe-cies the body is divided into two main parts: an anterior **protomerite,** and a larger, poste-rior part, the **deutomerite,** which contains the nucleus. These types are **cephaline** gregarines. In some species, the protomerite possesses an anterior anchoring device, the **epimerite** (Fig. 4–1), which is left in host tissue when the parasite breaks away. The **acephaline** gregarines lack the septum divid-ing the body, but sometimes possess an anchoring device (known as the **mucron**), which remains attached to the parasite when the parasite becomes free from the host tissue. Hosts become infected by swallowing spores.

Order Archigregarinida

These parasites are found in the intestinal tracts of marine invertebrates, particularly polychaete annelids. *Selenidium* is the best-known genus.

In the primitive life cycles of these organ-isms, the **trophozoites** undergo schizogony. The nucleus divides a few or many times without division of the cell itself. When the cell finally divides, there are as many daugh-ter cells as there were nuclei. This type of trophozoite is called a **schizont.** The daughter cells that appear as the result of schizogony are **merozoites,** which eventually grow into **gametocytes.** Spore formation is commonly completed in the sea.

Order Neogregarinida

These gregarines are parasitic in insects. They have a schizogonic phase in their life cycles and each spore forms eight sporozoites. *Ophryocystis* is a typical genus.

Order Eugregarinida

Sporozoites enter the host and develop into mature trophozoites without schizogony. The trophozoites become gamonts that unite in pairs (a form of union called **syzygy**), a cyst develops, and the unit is known as a **gameto-cyst.** Sexual reproduction takes place and the zygote undergoes sporogony. In the sub-order Cephalina, the gregarines often do not enter host tissue cells, but the acephaline

sporozoites are usually intracellular parasites. They often grow to a large size and protrude from the host cells. If they are in intestinal cells, they finally leave the gut lining and move freely in the intestinal cavity.

Suborder Acephalina

This suborder consists of nonseptate para-sites living mostly in the coelom of the host.

Monocystis lumbrici. This parasite (Fig. 4–2) lives in the seminal vesicles of the earth-worm, *Lumbricus terrestris,* and in related worms. Trophozoites, or sporadins, are elongated, measuring about 200 by 65 μ. Infection of the worm occurs by ingestion of mature spores. In the gut of the worm sporozoites are liberated and they penetrate the gut wall, migrate through the body and enter the seminal vesicles. Here the young trophozoites enter tissue cells and increase in size. Older trophozoites become free from host tissue and migrate to the sperm funnel, where they become attached again to host cells. Two of them join and secrete a wall around themselves, becoming a cyst. Within this cyst gametes are formed and produce zygotes (sometimes called sporoblasts). The sporoblasts become spores and the cycle is completed.

Acephaline gregarines have been found in many oligochaete annelids, sea cucumbers, nemerteans, insects, and many other inverte-brates. *Lankesteria,* for example, lives in the gut of insects, planaria and ascidians. A typical life cycle is that of *Lankesteria culicis,* a parasite of the mosquito, *Aëdes aegypti.* The cycle starts with the zygote, formed by gametes, occurring in the malpighian tubes of the mosquito larva (pupa). Many such zygotes occur in a large cyst. Each zygote develops into a spore within which eight sporozoites are formed. When the pupa becomes an adult mosquito the sporozoites are liberated into the water. Larval mos-quitoes eat the sporozoites, which enter cells lining the intestine. If a spore is eaten, it ruptures in the intestine, liberating the eight sporozoites. These tiny, motile, spindle-shaped organisms enter gut cells and

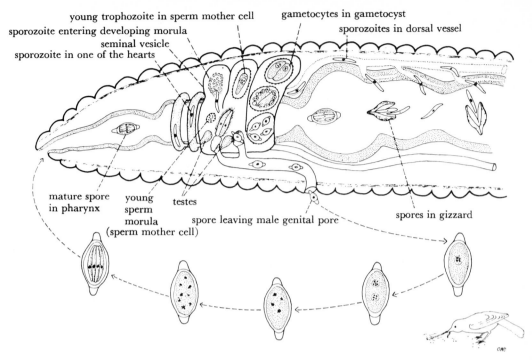

young trophozoite in sperm mother cell   gametocytes in gametocyst
sporozoite entering developing morula   sporozoites in dorsal vessel
seminal vesicle
sporozoite in one of the hearts

mature spore   young   testes
in pharynx   sperm
                          morula   spore leaving male genital pore   spores in gizzard
                  (sperm mother cell)

Fig. 4-2.   Earthworm (*Lumbricus terrestris*), host of *Monocystis*, showing endogenous phase of life cycle of parasite.   (Olsen, Animal Parasites, 1967, courtesy of Burgess Publishing Co.)

develop into trophozoites, which leave host cells and become gametocytes. When the larval mosquito pupates, the gametocytes make their way into the malpighian tubes. Here the parasites unite in pairs, develop a cyst wall and divide into many gametes that join to produce new zygotes.

Suborder Cephalina

Twelve families belong to this suborder, which is found primarily in arthropods and worms. The body of the trophozoite consists of more than one segment or compartment. A typical life cycle is that of *Stylocephalus longicollis* (Fig. 4–3), a parasite of the beetle *Blaps mortisaga*.

Gregarina garnhami (Fig. 4–4). These parasites occupy the intestinal ceca and midgut of the migratory locust *Schistocerca gregaria*. Sometimes the protozoa destroy considerable areas of the cecal epithelium. The parasites may be present by the hundreds and occur in such masses that there is a barrier between the food material in the lumen of the gut and the gut wall. The gregarines occur in both nymph and adult locusts.

Gregarina blattarum. A common species found in cockroaches (*Blatta orientalis* and others) is *G. blattarum*. The trophozoites are relatively enormous, reaching 1100 μ; larger ones are easily seen with the naked eye. Cyst contents reach the outside through eight to ten sporoducts. The life cycle is basically similar to that described for *Stylocephalus longicollis*.

SUBCLASS COCCIDIA

These intracellular parasites occur in both vertebrates and invertebrates. They enter epithelial cells and may cause considerable damage, often ending in the death of the host. Their distribution is worldwide. Only a few representative families are described here.

The life of a typical coccidian parasite begins with the zygote, which is a fertilized **macrogamete.** The parasite at this stage lies embedded in the intestinal wall or other epithelial surface, from which it soon breaks

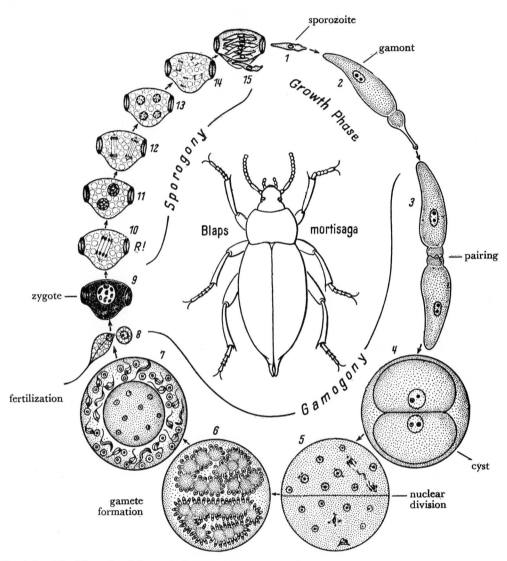

Fig. 4-3.   The life cycle of the cephaline gregarine, *Stylocephalus longicollis,* a parasite of the intestine of the beetle, *Blaps mortisaga.*   (Grell, courtesy of Springer-Verlag.)

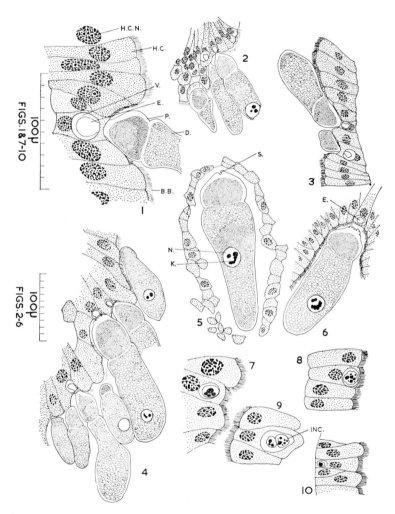

Fig. 4-4. *Gregarina garnhami*, a eugregarine parasite of the intestinal ceca and midgut of the migratory locust, *Schistocerca gregaria*, showing stages in the intestinal ceca. *1* to *4*. Cephalonts (adults) attached to epithelial cells. *5* to *6*. Sporonts (destined to develop into spores) free in the ceca. *7* to *10*. Hyaline inclusions in host cells. (Canning, courtesy of J. Protozool.).

away, and thus lies free in the lumen of the gut. It develops a protective cyst wall and emerges from the host as a unicellular **oocyst.** The cyst wall has two main layers. The exterior layer is probably protein, whereas the inner layer is lipid in nature and is associated with a protein lamella. The viability of an oocyst depends primarily on temperature, but it is also influenced by moisture, oxygen, bacteria and fungal action.

After a few hours or days, depending on the species, the oocyst undergoes a type of division known as **sporogony.** The nucleus divides into two, then four and finally eight nuclei. Each nucleus, with its accompanying cytoplasm, becomes a sporozoite. In one genus (*Eimeria*) two sporozoites lie within an enclosing membrane, the **sporocyst.** The sporocyst with its sporozoites is called a **spore.** In another genus (*Isospora*), there are four sporozoites in each of two spores. At this stage the oocyst is infective to another host, and, if eaten, the eight sporozoites are liberated in the intestine, and each enters a cell of the mucosa. Each sporozoite becomes rounded and forms a **schizont.** Schizogony occurs, during which each schizont multiplies into merozoites that form in a manner similar to that depicted in Figure 4–15 (p. 93) for the malarial parasite, *Plasmodium.*

Merozoites break out of host cells and enter new epithelial tissue. They can develop in the nucleus of host cells as well as in the cytoplasm. Eventually, macro- and microgametocytes are formed. The macrogametocytes enlarge into macrogametes that remain in host epithelial cells. Microgametocytes develop into motile microgametes that are liberated from host cells and swim to the cells containing macrogametes. Fertilization then takes place, thus completing the cycle. Characteristically, coccidia show a high degree of host specificity. See Davies,[14] Hammond,[26] Levine,[42] Levine and Ivens,[43] and Pellérdy[58] for reviews of the coccidia.

Order Eucoccida

The life cycle involves both sexual and asexual phases. Schizogony is present. These parasites are found in blood cells and epithelial cells of vertebrates and invertebrates.

Suborder Adeleina

Family Adeleidae. Members of this family parasitize intestinal epithelium and associated glands of invertebrates. Occasionally they occur in vertebrates, as in the kidneys of mammals.

Adelina deronis. This species lives in peritoneal cells of the annelid worm, *Dero limnosa.* The oocyst contains about 12 sporocysts, each of which contains two sporozoites. Oocysts are taken into the digestive tract of the worm, where its sporozoites are liberated. The sporozoites make their way to peritoneal cells, where they become trophozoites. The free-living trophozoites may become attached end-to-end, thus forming a chain of individuals.

Other genera of the family include *Adelea, Karyolysus* (Fig. 4–5), *Orcheobius, Chagasella,* and *Klossia* (Fig. 4–6). Each species varies in the numbers of sporocysts and sporozoites that are produced from the oocysts, and in the hosts and tissues that are parasitized.

Family Haemogregarinidae. This group of coccidia possesses life cycles involving two hosts: a vertebrate host in which the parasite lives in the circulatory system, and an invertebrate host involving the digestive system. Oocysts are small, and there are no sporocysts. Until more complete life histories are known, the identity and taxonomic position of many described species will remain in doubt. Some of them may belong to the genus *Hepatozoon*[30] (p. 82).

Haemogregarina stepanowi (Fig. 4–7). This haemogregarine begins its life cycle as a zygote (ookinete) in the gut of the leech, *Placobdella catenigera.* The zygote divides three times and forms eight sporozoites. When an infected leach takes blood from the turtle, *Emys orbicularis,* the sporozoites pass to the vertebrate host and enter red blood cells. Here schizogony occurs and produces merozoites that are liberated from the host cells and infect new red blood cells. Eventually some of the merozoites that enter these cells

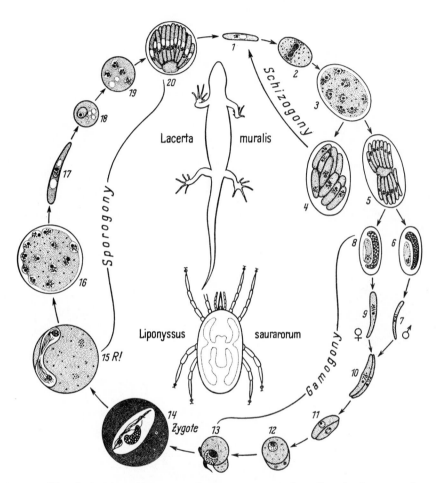

Fig. 4-5. The life of *Karyolysus lacertarum*. Sporogony occurs in a lizard whereas schizogony and gamogony occur in a mite. (Grell, courtesy of Springer-Verlag)

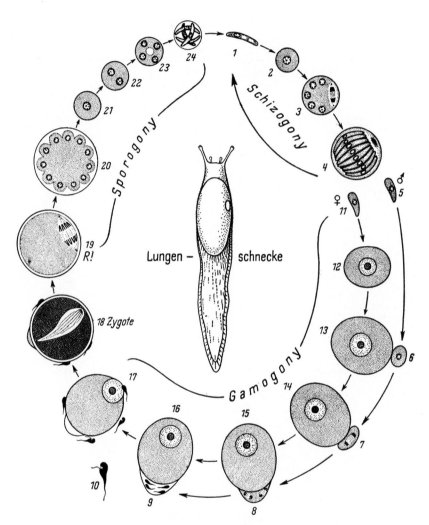

Fig. 4-6. The life cycle of the gregarine *Klossia*, which parasitizes a slug. (Grell, courtesy of Springer-Verlag.)

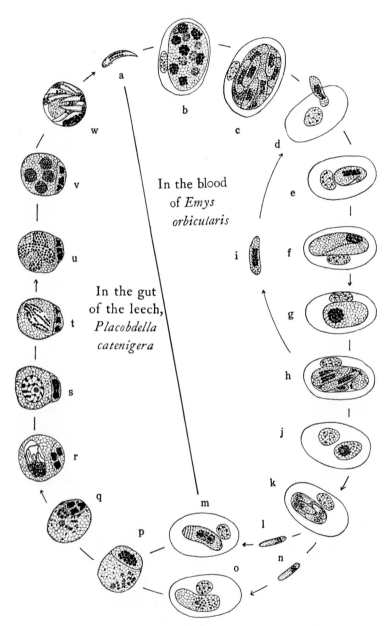

In the blood
of *Emys
orbicularis*

In the gut
of the leech,
*Placobdella
catenigera*

Fig. 4-7. Development of *Haemogregarina stepanowi* (after Reichenow) × 1200. *a*, sporozoite; *b* to *i*, schizogony; *j* to *k*, gametocyte-formation; *l*, *m*, microgametocytes; *n*, *o*, macrogametocytes; *p*, *q*, association of gametes; *r*, fertilization; *s* to *w*, division of the zygote nucleus to form eight sporozoites. (Kudo, courtesy of Charles C Thomas.)

develop into gametocytes. If these cells are ingested by a leech, the gametocytes are liberated, mature to gametes, and produce zygotes.

Haemogregarines of fish may enter leukocytes as well as erythrocytes, and they can be separated into two broad categories: haemogregarines (without asexual multiplication in red cells) and schizohaemogregarines (with schizogony occurring in red cells).[41] In spite of the large numbers of described species, however, the method of transmission from one fish to another remains a mystery. Haemogregarines of reptiles are transmitted by the ingestion of infected arthropod vectors.

Family Hepatozoidae. Many species of *Hepatozoon* occur in the red blood cells or leukocytes of fish, reptiles, birds and mammals. In the vector, the oocyst is large and contains numerous sporocysts in which there are many sporozoites. The parasite is transmitted by flies, lice, ticks and mites, in which the sexual stages occur.

Suborder Eimeriina

In this suborder, there is no syzygy.

Family Lankesterellidae. The genus *Lankesterella* occurs in amphibia, reptiles and birds; the genus *Schellackia*, in reptiles. These two genera may be synonymous. Schizogony, gametogony and sporogony all take place in the tissue cells (e.g., macrophage cells of spleen, liver, lungs, kidneys) of the vertebrate host. Sporozoites enter red blood cells, lymphocytes and monocytes, which may be ingested by the transport host (leech or mite). The parasites do not develop in the transport host. *L. adiei* is common in English sparrows throughout the world. *L. paddae* (Fig. 4–8) occurs in passerine birds. For more details of this genus see Stehbens.[77]

Family Aggregatidae. In this family, schizogony takes place in the cells of one host and most of the gamogony and sporogony in the cells of another host. Marine annelids, crustaceae and mollusks serve as hosts. Oocysts typically contain many sporozoites.

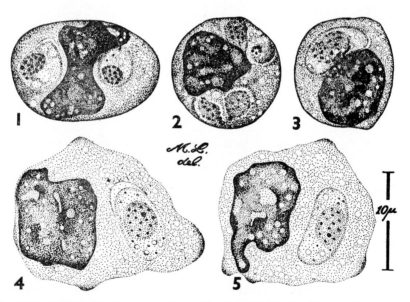

Fig. 4-8. *Lankesterella paddae* from several species of birds (family Zosteropidae, or Silver Eyes) from the South Pacific.

1 and *2*. Division stages, lymphocytes of *Gallirallus australis scotti*, New Zealand. *3*. Nondividing form, lymphocyte of *Zosterops lateralis lateralis*, New Zealand. *4* and *5*. Nondividing forms, monocytes of *Zosterops lateralis griseonota*, New Caledonia, and *Z. flavifrons flavifrons*, Futuna, respectively. (Laird, courtesy of J. Parasitol.)

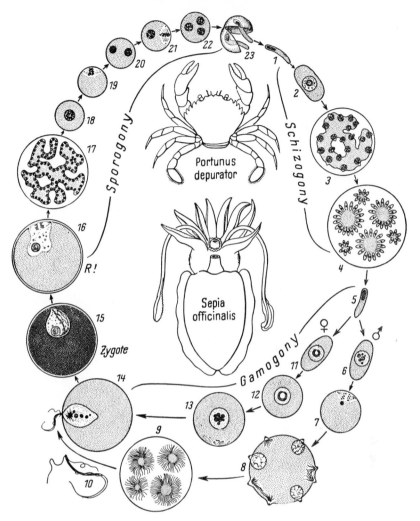

Fig. 4-9. *Aggregata eberthi* showing its life cycle, which includes a crab and the cuttlefish. Schizogony occurs in the intestine of the crab whereas sporogony and gamogony occur in the submucosal cells of the intestine of the cuttlefish. (Grell, courtesy of Springer-Verlag.)

Aggregata eberthi (Fig. 4–9). Schizogony occurs in the crab, *Portunus depurator*, and both sporogony and gamogony in the cuttlefish, *Sepia officinalis*. Sporozoites develop into merozoites in the intestinal connective tissue cells of the crustacean host. When a cuttlefish eats an infected crab, the merozoites develop into gametocytes in the gut wall of the mollusk. Gametes are released, unite, and form zygotes. From the zygotes develop the sporoblasts, spores, and finally sporozoites. Crabs eat the spores, and the sporozoites enter intestinal cells.

Family Eimeriidae. This family contains a large number of genera and species, many of which have considerable economic importance. Infections with the eimerid sporozoa are referred to as a *coccidiosis*. Gametes, each possessing three flagella, are produced independently and are of two sizes (macro- and microgametes). The oocyst (Fig. 4–11) usually contains one to many sporocysts, although in a few species sporocysts may be lacking. The number of sporozoites in each sporocyst varies with the species. The oocyst, which is the stage transferred from one host to an-

another, is usually immature when found in fresh feces. A diagram of a generalized life cycle is given in Figure 4–10. Note that schizogony and gametogony take place within host cells, and that sporogony occurs outside of the host body. (See Fig. 4–23.)

Eimeria. The genus *Eimeria* is characterized by the presence of four spores in each oocyst and two sporozoites in each spore. The many species occur primarily in the intestinal cells of vertebrates, but may also be found in epithelial cells of the liver, bile duct or other organs. Some species infect invertebrates. *Eimeria schubergi*, for example,

inhabits the intestinal epithelium of the centipede, *Lithobius forficatus*. Cross infection experiments indicate that *Eimeria* species are highly host specific.

Coccidiosis is a self-limiting disease, because after a given number of schizont generations all merozoites are transformed into gametocytes. Under such circumstances the life cycle of the parasite is completed. Thus in coccidiosis the severity of the disease depends largely upon the number of organisms that initiate the infection.[26] The effect of coccidiosis varies widely with the species of host, species of coccidia, age and resistance of host,

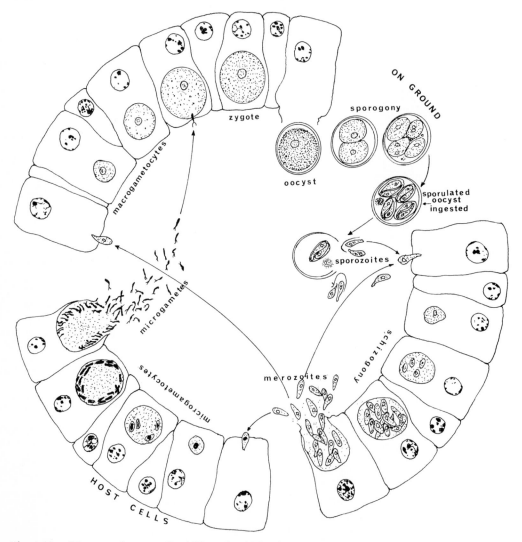

Fig. 4-10.   Diagram of a generalized life cycle of *Eimeria*. The same pattern occurs in the genus *Isospora*.

degree of infection, and many other factors. The following description of the effect of coccidiosis in calves, however, gives a general picture of the symptoms that might be seen in most animals with heavy infections.

The symptoms of coccidiosis may include rough coat, weakness, listlessness, nervousness, poor appetite, diarrhea, and loss of weight or poor gains in weight. The general weakness may cause the calf to defecate without rising.

When standing, the calf may attempt to defecate and not be able to pass feces; the intense straining results in an arched back, raised tail, and a "pumping" of the sides. Diarrhea may be watery or only slightly liquid, being quite unlike the "white scours" of calves less than three weeks old. Diarrhea caused by coccidiosis may contain many strands of gelatinous mucus and splotches or streaks of blood. In infections with *Eimeria zurnii* and *E. bovis*, it may be extremely bloody and even include shreds of intestinal tissue or occasionally short lengths of the tubular lining of the damaged intestine.[15]

Coccidiosis is rare in horses, but cattle are infected by at least 15 species of *Eimeria*. Common parasites are *E. zurnii* and *E. bovis*. Treatment has not been completely successful. Cleanliness is the best preventive, since the disease is carried from animal to animal by oocysts in manure.

At least ten species of *Eimeria* occur in sheep and goats. *E. arloingi*, *E. parva*, *E. faurei* and *E. ninae-kohl-yakimova* are common species. They infect the tissues of the digestive tract and cause diarrhea, destruction of epithelial tissues, weakness, loss of weight and death. As with calves, clean, dry, uncrowded living conditions are strongly recommended.

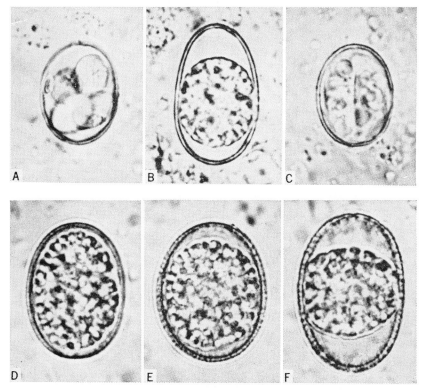

Fig. 4-11. Oocysts of species of *Eimeria* occurring in swine. *A* to *C. E. debliecki*. *A*. Sporulated cyst of smaller form with polar granule at one side of cyst. *B*. Cyst of larger form with rounded cytoplasm. Clear cyst wall. *C*. Mature cyst.

D to *F. Eimeria scabra. D*. Immature form with cytoplasm completely filling cyst. *E*. Cyst with cytoplasm beginning to contract. *F*. Cyst with cytoplasm almost entirely contracted, nearly "ball" stage. × 1500. (Henry, courtesy of Univ. Calif. Pub. Zool.)

Hogs may become infected with five species of *Eimeria*. *E. debliecki* and *E. scabra* (Fig. 4–11) are the most pathogenic. Dogs and cats are hosts to two genera of coccidia, being infected with *E. canis* and *E. felina*. Some of the same coccidia occur in foxes. Rabbits may be infected with at least ten species of *Eimeria*. A well-known species is *E. stiedae*. Most wild mammals and birds have their own species of coccidia, and they are far too numerous even to mention here. Undoubtedly, many species are yet to be found and described.

Coccidia occur abundantly in domestic and wild birds. The parasites cause a serious disease in chickens, geese and turkeys, but apparently not in ducks. In chickens and turkeys, the parasites mainly occur in the small intestine, but they may also be found in other parts of the gut. Twelve or more species infect domestic poultry, and among the most common in chickens in the United States are *E. necatrix*, *E. acervulina*, *E. maxima* and *E. tenella*. Turkeys are infected with *E. meleagridis*, *E. meleagrimitis* and other species. Young birds are more susceptible than older ones, and immunity is built up against species harbored. In a bird with heavy infection the intestine may be enlarged and thickened; droppings may be bloody, greenish, brownish or watery. Birds may be obviously ill, and many of them die. One interesting result of infection is a rise in the blood sugar level of the bird.

There is some evidence[25] that *Eimeria tenella* within macrophages causes enhanced ribonucleoprotein production by the host cells, and that this new protein is incorporated into the newly formed merozoites.

Clean, uncrowded, dry living areas for poultry, as for mammals, are of great importance in prevention and control.

Isospora. The genus *Isospora* is characterized by the presence of two spores in each oocyst and four sporozoites in each spore. Dogs and cats are parasitized by *Isospora bigemina*, *I. revolta* and *I. felis*. *Isospora suis* occurs in hogs. Frogs may have *I. lieberkuhni*. Birds also are parasitized by this genus of coccidia: *Isospora lacazii* is found in the small intestine of sparrows, blackbirds and other passerine birds, and *I. buteonis* occurs in the gut of hawks.

Isospora hominis and *I. belli* (the latter is more common) are found rarely in man, probably invading the ileum and cecum. Details of the life cycles have not been worked out, but undoubtedly they follow the same general pattern as outlined for all coccidia. The parasites have been found in man in the Mediterranean area, the Balkan countries, India, the Far East, Hawaii, Mexico, Cuba and South America.

Oocysts of *Isospora hominis* (Fig. 4–12) average about 28 by 14 μ in size. In freshly passed stools the oocysts usually contain zygotes. If the specimen is kept in a covered

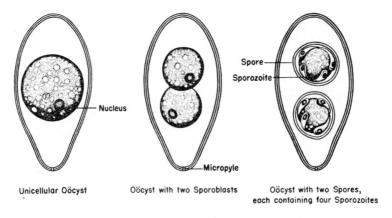

Unicellular Oöcyst   Oöcyst with two Sporoblasts   Oöcyst with two Spores, each containing four Sporozoites

Fig. 4-12. *Isospora hominis.* Oocysts in various stages of development. (U.S. Medical School Manual)

dish at room temperature for one or two days, the oocysts mature and contain the typical two spores, each enclosing four sporozoites.

Symptoms of this infection include abdominal pain, diarrhea, flatulence, digestive disturbances, abdominal cramps, nausea, loss of appetite, lassitude, loss of weight and fever. After the initial infection, the incubation period lasts about a week, then symptoms may or may not appear for about a month. During this time oocysts are discharged. When oocysts are no longer found, the disease has run its course and reinfection apparently does not occur. "The more recent work showing that a generalized systematic response and circulating antibodies do in fact exist as a result of coccidial infections in chickens, rabbits, turkeys, and cattle has not as yet provided a solution to this problem, because there is no proof that these antibodies are actually protective."[26]

See evidence on p. 105 that *Toxoplasma* is an isosporan.

Suborder Haemosporina

Relationships between the haemosporina and coccidia are indicated by the general similarity in schizogony and sporogony. The name "haemosporina" indicates that the parasites live in the blood. Sexual reproduction, which occurs in the blood-sucking invertebrate host, results in sporozoites that must be introduced into the vertebrate host by the "bite" of the invertebrate. Schizogony occurs in red blood cells and, in some species, in other cells of the vertebrate body. Microgametes and macrogametes develop independently and the zygote is motile. Pigment (hematin) is usually formed from host hemoglobin.

Family Plasmodiidae. This family contains protozoa that live in vertebrate tissue and blood cells and are transmitted by an insect vector. Schizogony occurs in vertebrate hosts, whereas sporogony occurs in insects. Zygotes are motile and the sporozoites are naked. Opinion differs as to which genera should be placed within the family,

and there is much need for further research on structure, physiology and life cycles.

Hepatocystis. The members of this genus are commonly found in arboreal mammals, especially in lower monkeys of the Old World tropics. The parasite is transmitted by biting midges (family Ceratopogonidae). A common species is *H. kochi*, a parasite of African monkeys. It lives in parenchyma cells of the liver, where merozoites are formed. Characteristically, only gametocytes appear in monkey blood. In the midge, *Culicoides adersi*, oocysts develop between the eyes and brain or in thoracic muscles.

Plasmodium. This is the only genus in which schizogony occurs in red blood cells, and the sexual phase develops almost always in mosquitoes.

MALARIA. Of all the diseases of mankind, malaria is one of the most widespread, best known and most devastating. It has always been a disease associated with forests and with migrant populations such as armies and nomads. It has undoubtedly played a role in the history of civilization because large areas of the earth have been subjected to the ruinous effects of the illness, and to debilitation and death caused by the parasites. "The full picture of malaria is rarely seen in stable, partially immune populations but comes into painful evidence when nonimmunes are introduced."[81] An analysis of the status of malaria today shows that the disease is under control in most areas of the world, or at least the prospects for control are encouraging. It is no longer a serious public health problem in the colder regions of the globe. Recent estimates indicate that during the past ten or 12 years the annual malaria deaths have been reduced from 2.5 million to less than one million, and that over 500 million days of illness have been prevented each year.

In most tropic regions of the world, however, malaria "keeps its place in the front rank of great plagues of the human species and can be called the King Herod of the diseases of infants and children in underdeveloped areas."[9] Why should this be so, when we know the causes of the disease, and the details of the parasite life cycle? We know a great deal about the mosquito that transmits malaria. We possess drugs that can cure the disease, so why do we permit so many people to die from it? The problem is not primarily biologic nor medical—it is economic and sociologic. Masses of people in tropic and semitropic countries do not have the economic resources with which to combat the mosquito, nor to identify and treat the millions of patients. Nor, in many areas, do patients and their families possess enough social awareness of the problem to assist adequately the public health authorities.

In spite of the tremendous advances in malaria research that have been made over the past decade, many basic problems must still be solved, such as the problem of host specificity among vertebrates and mosquitoes, the biochemical response to infection, immune responses, the infective stage and the nature of its activation, the origin of relapses, the intrinsic factors that help to determine whether a mosquito is a suitable vector, and the problem of drug resistance.

For a consideration of the ecology of malaria, see May.[52] For a review of research in malaria, see Garnham,[22] Sadun and Osborne,[66] and Sadun and Moon.[65] For a detailed account of terminology of malaria and of malaria eradication, see the Report of a Drafting Committee, WHO.[88] For a review of cultivation methods, see Taylor and Baker.[78] For immunology, see WHO.[89] For some aspects of human malaria, see McGregor.[53] For microscopic diagnosis, see Walker.[85]

THE LIFE CYCLE OF PLASMODIUM. The genus *Plasmodium* is the cause of malaria, and three species are common in man: *P. vivax*, *P. malariae* and *P. falciparum*. These plasmodia are worldwide in distribution, occurring largely in the warmer countries but also in such cold areas as North Korea, Manchuria, and southern Russia. A fourth species, *P. ovale*,[63] appears to be confined to tropic Africa and islands of the western Pacific. Table 4–1, p. 90, gives the comparative characteristics of these four species of *Plasmodium*, and Plates I, II, III and IV

LEGEND FOR PLATE I.

Stages of *Plasmodium vivax* in human erythrocytes from thin-film preparation.
1. Normal sized red cell with marginal ring form trophozoite.
2. Young signet ring form trophozoite in a macrocyte.
3. Slightly older ring form trophozoite in red cell showing basophilic stippling.
4. Polychromatophilic red cell containing young tertian parasite with pseudopodia.
5. Ring form trophozoite showing pigment in cytoplasm, in an enlarged cell containing Schüffner's stippling.*
6, 7. Very tenuous medium trophozoite forms.
8. Three ameboid trophozoites with fused cytoplasm.
9, 11, 12, 13. Older ameboid trophozoites in process of development.
10. Two ameboid trophozoites in one cell.
14. Mature trophozoite.
15. Mature trophozoite with chromatin apparently in process of division.
16, 17, 18, 19. Schizonts showing progressive steps in division ("presegmenting schizonts").
20. Mature schizont.
21, 22. Developing gametocytes.
23. Mature microgametocyte.
24. Mature macrogametocyte.

* Schüffner's stippling does not appear in all cells containing the growing and older forms of *P. vivax* as would be indicated by these pictures, but it can be found with any stage from the fairly young ring form onward.

(From the Manual for the Microscopical Diagnosis of Malaria in Man, National Institutes of Health Bulletin No. 180. By Aimee Wilcox.)

PLATE I

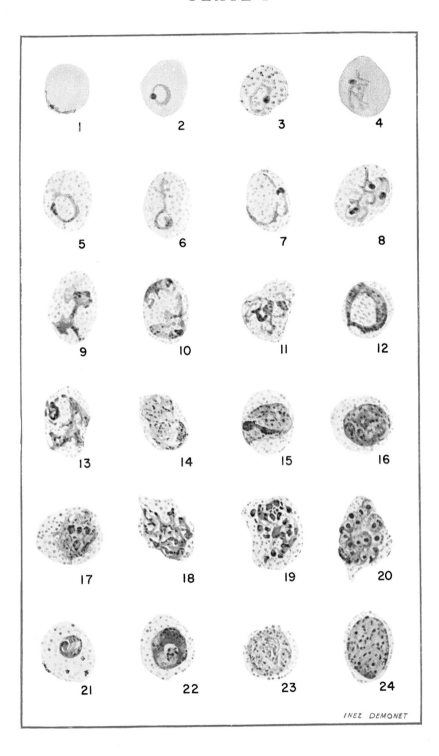

INEZ DEMONET

Stages of *Plasmodium malariae* in human erythrocytes from thin-film preparation.
1. Young ring form trophozoite of quartan malaria.
2, 3, 4. Young trophozoite forms of the parasite showing gradual increase of chromatin and cytoplasm.
5. Developing ring form trophozoite showing pigment granule.
6. Early band form trophozoite—elongated chromatin, some pigment apparent.
7, 8, 9, 10, 11, 12. Some forms which the developing trophozoite of quartan may take.
13, 14. Mature trophozoites—one a band form.
15, 16, 17, 18, 19. Phases in the development of the schizont ("presegmenting schizonts").
20. Mature schizont.
21. Immature microgametocyte.
22. Immature macrogametocyte.
23. Mature microgametocyte.
24. Mature macrogametocyte.

(From the Manual for the Microscopical Diagnosis of Malaria in Man. National Institutes of Health Bulletin No. 180. By Aimee Wilcox.)

PLATE II

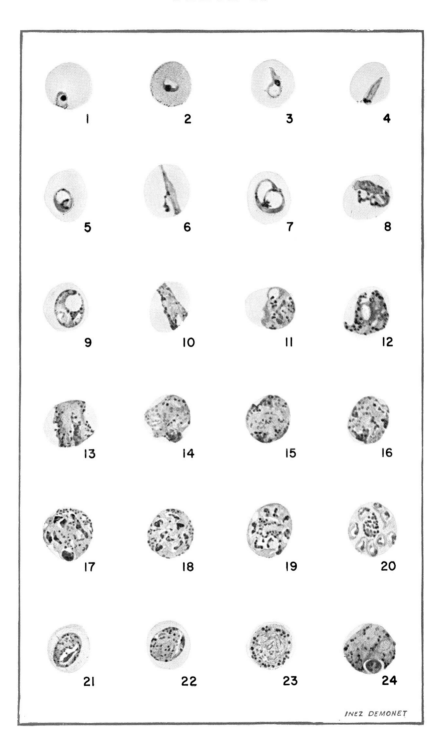

INEZ DEMONET

Stages of *Plasmodium falciparum* in human erythrocytes from thin-film preparation.

1. Very young ring form trophozoite.
2. Double infection of single cell with young trophozoites, one a "marginal form," the other, "signet ring" form.

3, 4. Young trophozoites showing double chromatin dots.

5, 6, 7. Developing trophozoite forms.

8. Three medium trophozoites in one cell.

9. Trophozoite showing pigment, in a cell containing Maurer's spots.

10, 11. Two trophozoites in each of two cells, showing variation of forms which parasites may assume.

12. Almost mature trophozoite showing haze of pigment throughout cytoplasm; Maurer's spots in the cell.

13. Estivo-autumnal "slender forms."

14. Mature trophozoite, showing clumped pigment.

15. Parasite in the process of initial chromatin division.

16, 17, 18, 19. Various phases of the development of the schizont ("presegmenting schizonts").

20. Mature schizont.

21, 22, 23, 24. Successive forms in the development of the gametocyte—usually not found in the peripheral circulation.

25. Immature macrogametocyte.

26. Mature macrogametocyte.

27. Immature microgametocyte.

28. Mature microgametocyte.

(From the Manual for the Microscopical Diagnosis of Malaria in Man. National Institutes of Health Bulletin No. 180. By Aimee Wilcox.)

PLATE III

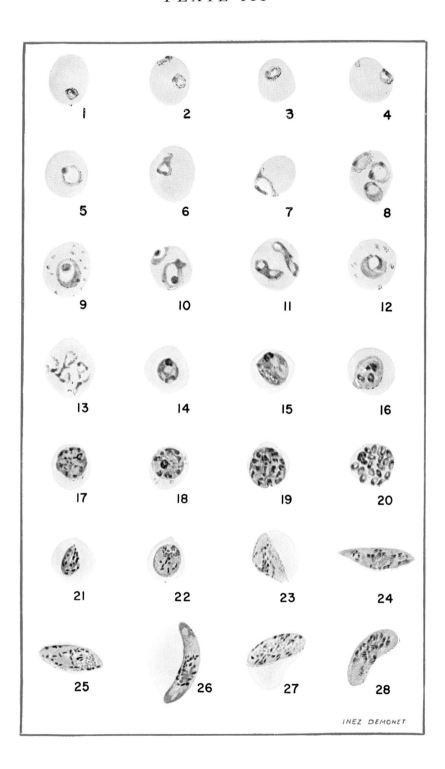

INEZ DEMONET

PLATE IV

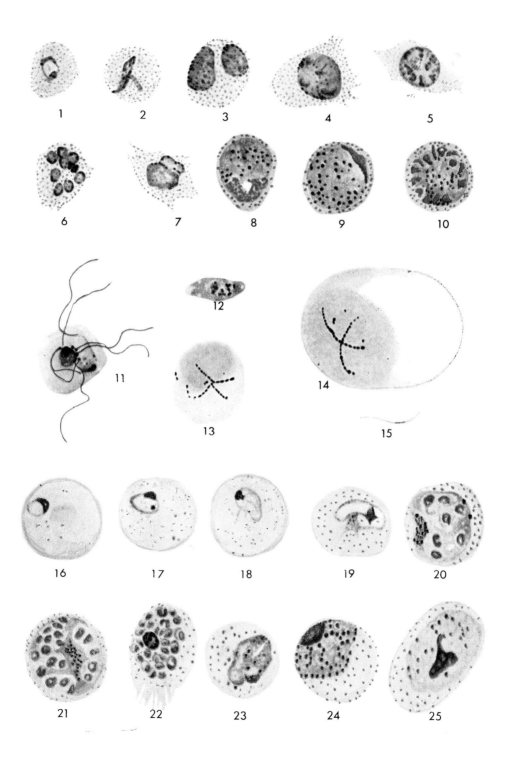

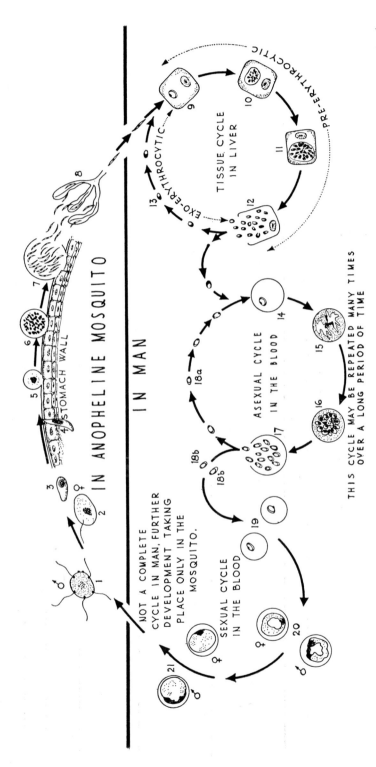

Fig. 4-13. The life cycles of *Plasmodium* in a mosquito and in man. (Blacklock and Southwell, courtesy of H. L. Lewis & Company, London.)

IN ANOPHELINE MOSQUITO

IN MAN

NOT A COMPLETE
CYCLE IN MAN, FURTHER
DEVELOPMENT TAKING
PLACE ONLY IN THE
MOSQUITO.

STOMACH WALL

PRE-ERYTHROCYTIC

EXO-ERYTHROCYTIC

TISSUE CYCLE
IN LIVER

ASEXUAL CYCLE
IN THE BLOOD

SEXUAL CYCLE
IN THE BLOOD

THIS CYCLE MAY BE REPEATED MANY TIMES
OVER A LONG PERIOD OF TIME

Table 4-1. Comparative Characters of Plasmodia of Man*

(STAINED THIN SMEARS)

Stage or Period	Plasmodium vivax	Plasmodium ovale	Plasmodium malariae	Plasmodium falciparum
Early trophozoite or ring	Relatively large; usually one prominent chromatin dot, sometimes two; often two rings, sometimes more, in one cell	Compact; one chromatin dot; double infection uncommon	Compact; one chromatin dot; double cell infections rare	Small, delicate; sometimes two chromatin dots; multiple red cell infection common; appliqué forms frequent
Large trophozoite	Large; markedly ameboid; prominent vacuole; pigment in fine rodlets; abundant chromatin	Small; compact; not ameboid; vacuole inconspicuous; pigment coarse	Smaller than vivax: compact; often band-shaped; not ameboid; vacuole inconspicuous; pigment coarse	Medium size; usually compact rarely ameboid; vacuole inconspicuous; rare in peripheral blood after half grown; pigment granular
Young schizont or pre-segmenter	Large; somewhat ameboid; dividing chromatin masses numerous; pigment in fine rodlets	Medium size; compact; chromatin masses few; pigment coarse	Small; compact; chromatin masses few; pigment coarse	Small; compact; chromatin masses numerous; single pigment mass; rare in peripheral blood
Mature schizont or segmenter	Schizonts and merozoites large; pigment coalescent	Merozoites larger than in P. malariae; irregular rosette	Schizonts smaller but merozoites larger	Smaller merozoites; single pigment mass
Number of merozoites	12 to 24, usually 12 to 18	6 to 12, usually 8	6 to 12, usually 8	8 to 26, usually 8 to 18
Microgametocytes (usually smaller and less numerous than macrogametocytes)	Spherical; compact; no vacuole; undivided chromatin; diffuse coarse pigment; cytoplasm stains light blue	Similar to P. vivax but somewhat smaller; never abundant	Similar to P. vivax but smaller and less numerous	Crescents usually sausage-shaped; chromatin diffuse; pigment scattered large grains; nucleus rather large; cytoplasm stains paler blue

Macrogametocytes	Spherical; compact; larger than microgametocyte; smaller nucleus; pigment same; cytoplasm stains darker blue	Similar to P. vivax but somewhat smaller; never abundant	Similar to P. vivax but smaller and less numerous	Crescents often longer and more slender; chromatin central; pigment more compact; nucleus compact; cytoplasm stains darker blue
Pigment except in mature schizonts	Short, rather delicate rodlets irregularly scattered; not much tendency to coalesce	Similar to but somewhat coarser than P. vivax; sometimes clumped or in lateral bands	Seen in very young rings; granules rather than rods; tendency toward peripheral scatter	Pigment granular; early tendency to coalesce; typical single solid mass in mature trophozoite; coarse scattered "rice grains" in crescents
Alterations in the infected red cell	Enlarged and decolorized; Schüffner's dots usually seen	Enlarged; decolorized; prominent Schüffner's dots appear early; infected cells may be oval-shaped with fimbriated ends	Cell may seem smaller; fine stippling (Ziemann's dots) occasionally seen	Normal size but may have "brassy" appearance; Maurer's dots (or "clefts") common; Garnham's bodies occasionally seen
Length of asexual phase	48 hours	49 to 50 hours	72 hours	36–48 hours, usually 48
Prepatent period, minimal	8 days	9 days	14 days, average 28 to 37	5 days, average 8 to 12
Usual incubation period	8 to 31 days, average 14	11 to 16 days	28 to 37 days, average 30	7 to 27 days, average 12
Interval between parasite patency and gametocyte appearance	3 to 5 days	5 to 6 days; appearance irregular and numbers few	10 to 14 days; appearance irregular and numbers few	8 to 11 days
Developmental period in mosquito	30 days or more at 17.5° C.; 16 to 17 days at 20° C.; 10 days at 25° C.; 10 days at 28° to 30° C. (82.4–86° F.)	16 days at 25° C., 14 days at 27° C.	30 to 35 days at 20° C.; 25 to 28 days at 22–24° C. (80° F.)	22 to 23 days at 20° C.; 10 to 12 days at 27° C. (80° F.)

* From Faust, E. C., Russell, P. F. and Jung, R. C.: Clinical Parasitology, ed. 7. pp. 206–207. Philadelphia, Lea & Febiger, 1970.

show the stages of each species in human erythrocytes as they appear in stained blood smears.   The mosquito vector is discussed with other arthropods on page 406.

The life cycle of *Plasmodium* (Fig. 4–13) starts with a zygote in the stomach of a female mosquito (many species of *Anopheles* and, for lizard and bird malaria, culicine mosquitoes). The zygote is active and moves through the stomach or midgut wall.   The parasite at this stage is called an *ookinete*.   Under the lining of the gut, the ookinete becomes rounded, forms a cyst, and is called an *oocyst*.

Sporozoites invade the entire mosquito— many of them get into the salivary glands and so are in a favorable position to enter the next host when the mosquito bites man.

Warren and Bennett[86] have emphasized that:

the mosquito is not just a passive medium in which sporogony occurs but is a complicated biological mechanism which actively confronts the parasite at three major stages of its sporogonic development. . . . Many species of mosquitoes completely fail to support the parasite . . . others feebly . . . The success of the parasite in its invertebrate host is also limited by the feeding and breeding habits of the mosquito. Not only must the parasite find an *Anopheles* in which it can develop, but the mosquito must have a high rate of returning to re-feed on the same vertebrate. . . . The situation is further complicated by recent evidence that some culicine mosquitoes are partially susceptible to primate plasmodia. . . . The relationship between the parasite and its dipteran host is so specialized as to severely limit the parasite's distribution.

When the sporozoites are introduced into the blood of man by the bite of a mosquito, a series of cycles begins that involves different cells and tissues.   The sporozoites mark the end of the sexual cycle.   They promptly enter various tissue cells, such as liver parenchyma and fixed macrophages.   This is known as the *exoerythrocytic* cycle, because the organisms have not yet entered red blood cells.   The sporozoites may now be called *cryptozoites*. The cryptozoites undergo schizogony, rupture, and liberate *metacryptozoites*, which may enter other tissue cells and repeat the schizogony cycle.   This repetition apparently does not

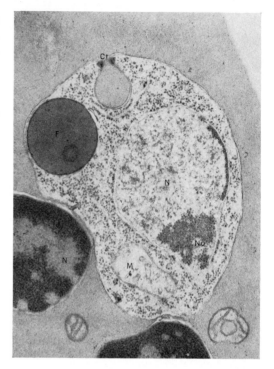

Fig. 4-14. An erythrocytic trophozoite of *Plasmodium gallinaceum* showing a cytostome (Ct) that is ingesting host cell cytoplasm. *N.* Nucleus. *Nu.* Nucleolus. *F.* Food vacuole. *M.* Mitochondrion. (Aikawa, Huff, and Sprinz, Military Medicine, courtesy of Association of Military Surgeons, U.S.)

occur in the cycle of *Plasmodium falciparum*, and, consequently, relapses seldom occur in *falciparum* malaria.   If one considers the exoerythrocytic cycle as the original or primitive condition, then the short duration of this cycle in *P. falciparum* would indicate that this species is the most highly advanced of those plasmodia that infect man.

Metacryptozoites gain access to the blood, enter the erythrocytes and start the *erythrocytic phase* of the life cycle.   Within the red blood cells, young plasmodia have a red nucleus and a ring-shaped blue cytoplasm in typically stained smears.   This appearance gives rise to the name *signet ring* for the parasite at this stage.   The ring configuration alters as the plasmodia begin to grow within the blood cells.   The parasites at this stage may be active, and are trophozoites.

A cytostome has been described for both erythrocytic and exoerythrocytic stages of several species of *Plasmodium* of mammals and birds (Fig. 4–14). Ingestion of erythrocyte cytoplasm occurs by a process called *pinocytosis*, wherein droplets of host cytoplasm are engulfed by the parasites through invaginations of the plasma membrane, with subsequent formation of food vacuoles. This process has been described as occurring through the wall of the cytostome[2] and at other regions on the surface of the parasite.[64]

Schizogony occurs again. Because the schizont is characterized by dividing or segmenting nuclei, it is sometimes called a *segmenter* when mature. The cytoplasm of the red blood cell may contain various pigment granules (e.g., Schüffner's dots). The plasmodium divides into merozoites (Fig. 4–15), which are comparable to the metacryptozoites formed earlier in the life cycle. Merozoites (Fig. 4–16) break out of the red cells,

and each may enter another erythrocyte and repeat the process of schizogony. Ladda[39] showed that during the rapid penetration of red cells by merozoites of rodent malarial parasites, "the paired organelles and toxonemes associated with the conoid region are completely depleted suggesting that these structures contain substances which are used up in penetration." During penetration the host red cell membrane is not disrupted and "the parasite is enclosed in a vacuole within the host cell and is separated from direct contact with the host cell cytoplasm by the vacuolar membrane . . . and the material contained in the growth vacuole." After penetration the merozoite loses its complex structural organization, then begins a period of growth and enlargement, and the formation of polysomes and proliferation of endoplasmic reticulum that "indicates an intensive synthetic activity." (See Fig. 4–17, p. 95.)

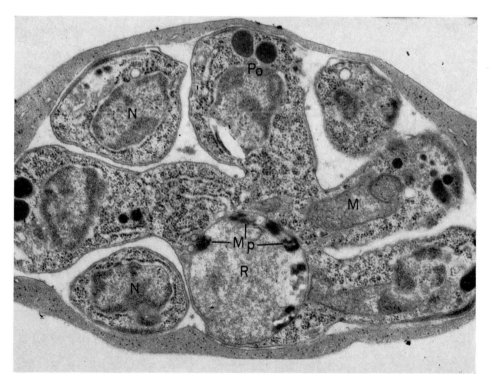

Fig. 4-15. A schizont of *Plasmodium cathemerium* with budding new merozoites. *Mp.* Malaria pigment. *N.* Nucleus. *M.* Mitochondrion. *Po.* Paired organelles. *R.* Residual body. (Aikawa, Huff, and Sprinz, Military Medicine, courtesy of Association of Military Surgeons, U.S.)

Significant increases in the osmotic fragility of infected red cell populations and a volume increase of red cells accompanying erythrocytic growth of the parasite have been reported by several workers. In a summary of his study of the quantitative cytochemistry of malaria-infected erythrocytes, Bahr[7] stated that: "It appears then that the expansion of the erythrocyte membrane uses existing material, which will, undoubtedly, change permeability, mechanical and osmotic fragility, surface charge and sedimentation properties, all of which may favor the accelerating growth of the parasite and the subsequent release of merozoites."

Some of the merozoites in the blood cells develop into sexual forms that grow into male microgametocytes or female macrogametocytes. When a mosquito bites man at this stage of the life cycle, the gametocytes are taken into the insect's stomach and there they mature into the microgametes and macrogametes. The microgametes are flagellate-like outgrowths from the microgametocytes. They become detached and behave like the sperm cells of higher animals. Fertilization now takes place and the zygote thus formed completes the life cycle.

A summary of the reproductive cycles just outlined may be stated as follows: Once the

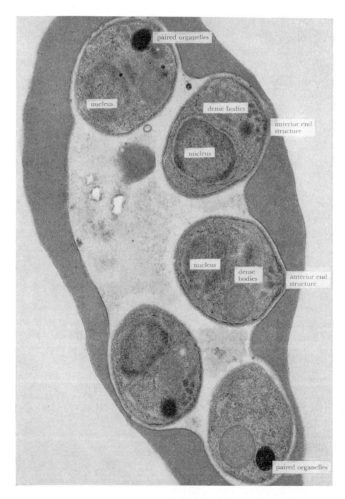

Fig. 4-16. Five merozoites of *P. floridense* in red blood cell. Note poorly developed endoplasmic reticulum, abundant ribosomes and a pellicular complex. $\times$ 28,000. (Aikawa and Jordan, courtesy of J. Parasitol.)

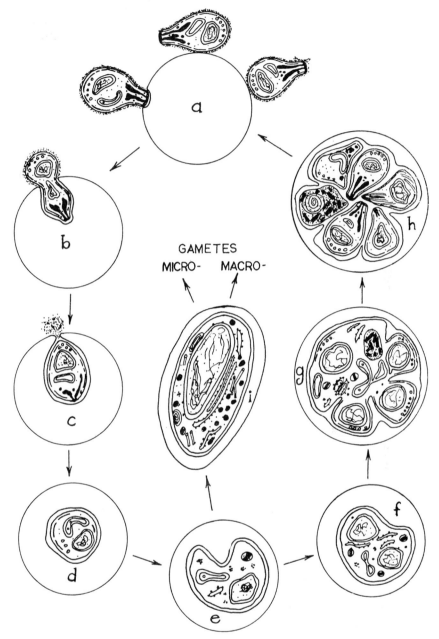

Fig. 4-17. Erythrocytic cycle of rodent malarial parasites. *a.* Free merozoites contact many red cells, but penetration only occurs when merozoite becomes properly oriented toward the host cell, *b* and *c.* The parasite creates a depression in the host cell membrane which expands into the red cell. forming a cavity as the parasite advances. The site of initial merozoite contact on the red cell membrane remains relatively constricted, forming a tight ring that deforms the penetrating parasite. *c* and *d.* Edges of the tight ring fuse as the parasite completely enters the red cell. The parasite becomes enclosed in a vacuole in the host cell. *d* and *e.* Merozoite undergoes complete dedifferentiation and begins the maturation process. *f, g,* and *h.* Trophozoite grows and finally undergoes segmentation into highly differentiated merozoites, which are subsequently released to infect other red cells, thus repeating the cycle. *i.* A small number of trophozoites show no evidence of nuclear division and develop large numbers of dense cytoplasmic bodies (toxonemes) and microtubules. These parasites give rise to gametocytes. (Ladda, courtesy Association of Military Surgeons, U.S.)

infection is well started in man, schizogony normally occurs regularly in the liver cells (exoerythrocytic cycle) and in the blood cells (erythrocytic cycle). At intervals of five to 12 days, depending on the species, multiplication takes place in the liver, and each generation then produces a large number of parasites that invade the blood. A smaller number of parasites usually continues the cycle in the liver. Cyclic multiplication in the blood is manifested in cyclic onset of chills and fever. The length of a single cycle in the blood varies with the species of *Plasmodium*.

PHYSIOLOGY OF PLASMODIUM. Our rapidly increasing knowledge of the physiology and biochemistry of protozoan parasites is clearly indicated by Honigberg's review[28] and the papers in Sadun and Moon.[65] Much of the following information and speculation is based on these reviews.

PROTEINS. The nitrogenous requirements of the parasites are largely derived from the host cells. The erythrocytic stages of several species of *Plasmodium* have been found to take in the cytoplasm of the host red cell by means of pinocytosis and the hemoglobin provides the chief source of nitrogen. The insoluble residual product of the reaction whereby the parasite utilizes hemoglobin is *hemozoin*. Malarial parasites apparently depend upon the host erythrocytes for ATP,[55] an excellent source of purines.

In the red cells parasitized by *Plasmodium* there is a large free amino acid pool that consists largely of glutamate, aspartate, alanine and glycine. The free amino acid pool in the parasites that develop within the erythrocytes apparently depends upon the CO_2-fixing capacities of the parasites, as well as on degradation of hemoglobin or uptake of amino acids from the plasma.[82] The growing plasmodium is responsible for almost all of the alteration of metabolism of the parasitized cell.

Little is known about the kinds of amino acids required by the malarial parasites, but we do know that methionine is essential for growth. The folic acid group of compounds may play an important role in the transmethylation reactions involved in the conversion of cystein to methionine, as well as in the nucleic acid synthesis. Experiments on the in vitro growth of *Plasmodium* indicate that the following compounds, in addition to methionine, are required: glutathione, panthothenate and para-aminobenzoic acid (PABA).

On the basis of cultivation experiments, it has been shown that only aspartic acid, glutamic acid and alanine are synthesized from glucose by *P. knowlesi*. All the other required amino acids, particularly isoleucine and methionine, must, therefore, be derived from digestion of the host hemoglobin.

LIPIDS. Relatively little is known about lipid metabolism of malarial parasites. Growth of *Plasmodium* in erythrocytes is characterized by a pronounced increase in lipid content of the host cell. In vitro studies have shown that *P. berghei* within red blood cells actively synthesizes phospholipids, chiefly cephalins and lecithins, from glucose carbon and free oleic acid, and it apparently depends upon the host cells for the major part of the fatty acids required.

CARBOHYDRATES. Suspensions of red cells parasitized by *Plasmodium* utilize more glucose than the noninfected erythrocytes and show a higher metabolic rate. The collaborative nature of this phenomenon is indicated by the impaired metabolic competence of *Plasmodium* freed from host cells as compared with the efficient utilization of glucose by the erythrocyte-parasite system. There is evidence,[55] based on a study of metabolism of *Plasmodium berghei*, that the parasite utilizes the reserve ATP of the host cells, and thus depends on host cells for energy supply.

We know little about the enzymes involved in the tricarboxylic acid cycle. Malic dehydrogenase activity is much greater in free parasites than in uninfected cells. Sherman[71] has shown that the malarial parasites *Plasmodium lophurae* and *P. berghei* synthesize their own dehydrogenase systems. The parasite-host "complex contains activities which are the sum of the content of the growing parasites and the residium of the host cell." The dehydrogenases studied "seem to provide examples of the subtle nature of the differences in the host and the parasite at the molecular level and suggest that perhaps the parasite grows at the expense of the host by certain small but significant catalytic advantages."

The dependence upon the host cell for the synthesis of a single coenzyme essential in protein, carbohydrate or lipid metabolism undoubtedly is a major factor in restricting the parasite to intracellular existence.

HOST RESPONSES. Malaria, especially the *falciparum* type, is characterized by anoxemia (deficiency in the oxygen content of the blood) and anoxia in adjacent tissues. There may be increased permeability of capillary walls, constriction of blood vessels and stagnation of blood in capillaries, particularly in the brain.[51] The anemia accompanying malarial infections results in part from direct destruction of the red cells by the intraerythrocytic parasites, and in part by the phagocytosis of nonparasitized erythrocytes.

In both human and animal malarials the adrenals are probably affected, and hypoglycemia occurs in the hosts, at least during the terminal stages of the infections.[84] Serious disturbances of liver function apparently occur in all infections during the course of both naturally acquired and artificially induced *falciparum* malaria. These disturbances occur even when there is little or no detectable invasion of peripheral red blood cells, and they include congestion, degeneration, necrosis and atrophy of parenchymal cells. In *falciparum* malaria, infected red cells become viscid and adhere together in various internal organs, causing symptoms that vary depending on which organ is infected.

There is some evidence[51] that plasmodia may produce a toxin that profoundly affects the basic metabolic processes of host tissue cells by inhibiting respiration and oxidative phosphorylation. These events begin during the erythrocytic phase of the cycle.

The most characteristic response of human hosts to malarial infection is the paroxysm, which begins with chills and a gradually mounting fever. The patient's teeth chatter, he gets "goose pimples" and he tries to cover himself with all the blankets he can find. After 20 to 60 minutes he ceases shivering and quickly begins to feel intensely hot. His fever may reach 106° F., accompanied by headache and nausea, and his temperature may remain high for one to four hours. Perspiration then starts, may become profuse, and lasts a few hours as his temperature subsides. The entire paroxysm lasts from six to ten hours and occurs every third day (with *Plasmodium vivax*) for one to two weeks.[27]

An abnormal condition in man, called *sickle-cell anemia*, may provide protection against malaria. Allison[3] has stated that: "The sickle-cell gene, which is common in some southern European and Asian and many African populations, is responsible for the production of an abnormal haemoglobin. The condition is so severe that very few sickle-cell homozygotes survive to adulthood . . . the heterozygous state confers protection against falciparum malaria." This phenomenon and others involving the protective value of inherited blood abnormality traits are subjects of considerable controversy, and definite conclusions concerning them are hazardous to make.

IMMUNOLOGY. Human populations exposed to frequent infections with *Plasmodium* acquire a strong immunity which, to *P. falciparum* at least, is basically serologic in nature. Field studies have resulted in the successful passive transfer of immunity to man, and have demonstrated the existence, in the blood of immune persons, of a humoral factor or factors in the serum globulin complex that are capable of reducing parasitemia. Naturally acquired immunity in man is apparently effective only against the asexual stages of the parasite that develop in the red blood cells.[89]

If, as seems highly probable, immunity in man is not effective against developing or mature gametocytes or against sporozoites, inoculation of viable sporozoites will lead to successful penetration of and development in hepatic cells, even in immune hosts.[53] This situation is particularly significant with *Plasmodium falciparum*, which has such a high multiplication rate in the liver and a high merozoite content of its erythrocytic schizonts. The density of viable sporozoites is thus of major importance, but we need more research on this point, as well as on variations among the mosquito hosts.

Plasmodium produces soluble antigens, recoverable in sera of animals with acute infections, and antibodies to these antigens appear during recovery from the infections. There is some evidence for autoimmunity in malaria,[90] but "except for the fairly clear demonstration of an antibody-antigen basis of immunity demonstrated for *P. lophurae* and *P. knowlesi*, we have very little information concerning what the mechanisms of immunity to malaria might be."[13] In all probability, frequent antigenic variation occurs in all four human species of *Plasmodium*, as well as in species in other mammals and birds. Acquired immunity against malaria appears to be strain specific, and, as such, constitutes a barrier against the development of a successful program of artificial immunization against the disease.

Acquired immunity to *Plasmodium berghei* infections in rats controls the course of primary infection and confers resistance to later homologous challenge. Cellular immunity may be involved, but humoral factors, presumably antibodies, are more likely to be implicated. Antibodies probably function by diminishing the parasite's ability to penetrate the red cell. The greatest antiparasitic or protective effect is obtained in sera from rats that have received repeated immunizing infections.[76] X-irradiated sporozoites of *P. berghei* have been used to produce complete protection of mice against an otherwise 100 per cent lethal sporozoite inoculum.

All efforts to develop an active, dependable malarial immunization have met with failure to date because protective antigens have not yet been sufficiently isolated, purified or characterized. Most workers have been using intact or irradiated *Plasmodium*, and they

have been concerned with circulating antibodies reacting with plasmodial antigens. Much work needs to be done on the production of specifically sensitized lymphocytes that are involved with cell-mediated immunity, including delayed hypersensitivity. Attempts at vaccination using preparations of whole parasites have produced only partial protection, probably because immune sera contain small proportions of specific malarial antibody.

The whole problem of malarial pathology and host defenses is being considered constantly by many workers, and new results are being published every month. For further details, see Moulder,[54] Geiman,[24] Garnham,[22] Sadun and Osborne,[66] and Sadun and Moon.[65]

PLASMODIA IN ANIMALS OTHER THAN MAN. REPTILIAN MALARIA. *Plasmodium* sp. are commonly found in lizards of North and South America, Africa, East Indies, Pacific islands and Australia, but much rarer in Europe and Asia. *P. minasense* is probably the most widely distributed species. Mosquitoes apparently are the vectors. The vertebrate tissue forms of the parasites and the erythrocytic schizogony stages often vary in character.[29] These stages suggest that lizard malaria is the earliest form of the disease to evolve. The first report of a malaria parasite undergoing sporogonic development in an insect other than a mosquito appeared in 1970.[6a] *Plasmodium mexicanum* is an inhabitant of western fence lizards, *Sceloporus occidentalis*, in California. Sporogonic stages of this parasite were found in sandflies, *Lutzomya vexatrix occidentis* and *L. stewarti*. These nocturnal flies feed on the blood of reptiles and terrestrial amphibians and they inhabit burrow systems of ground squirrels during the day. Few reports of *Plasmodium* in snakes have been made. For a comprehensive review of reptilian malaria, see Garnham.[22]

AVIAN MALARIA. The wide geographic range of bird malaria is easily explained by the migratory habits of the hosts. Mosquitoes of the genera *Aedes* and *Culex* are the most common vectors; several other genera are involved, but seldom *Anopheles*. The subgenus *Haemamoeba* includes the common *Plasmodium relictum*, infecting mostly passerine birds (especially sparrows); *P. cathemerium*, also in passerine birds (type host, the domestic sparrow); and *P. gallinaceum*, in the jungle fowl and chickens. In other subgenera are a number of species including *P. lophurae*, which infect chickens and other gallinaceous birds; and *P. elongatum*, in finches, sparrows and other passerine birds.

In contrast to *Plasmodium* in mammals, the avian species always have a nucleolus and typical mitochondria.[1] Garnham[22] has listed the chief differences between avian and primate exoerythrocytic stages as follows:

The exoerythrocytic stages of avian parasites are found in mesodermal tissues; those of the primate occur in the liver parenchyma. The primary cycle occupies two generations in the avian and one in the primate parasites. The avian EE parasites are infective on subinoculation, the primate are non-infective; the former can arise from the inoculation of blood stages, the latter arise only from sporozoites or their direct descendants in hepatic tissue. Avian exoerythrocytic schizonts produce a maximum of a thousand merozoites, primates produce a minimum of a thousand, and normally many more.

Terzakis *et al.*[79] have studied the development of the avian *Plasmodium gallinaceum* in the mosquito, *Aëdes aegypti*, with the electron microscope. They found that the oocyst cytoplasm rapidly segregates itself into sporoblasts, each with a number of nuclei, by a process of vacuole formation. Sporoblasts develop into sporozoites by a budding process (Fig. 4–18). Sporozoites, each with a single nucleus and a minute pore (*micropyle*) at one end, become distributed throughout the body of the mosquito after the enlarged oocyst bursts, and are transferred to the vertebrate host through the bite of the insect. The production of sporozoites in the mosquito is known as *sporogony*, but the oocysts become polysporoblastic without the formation of a cyst wall or sporocysts.

The pathology in avian malaria caused by *Plasmodium gallinaceum* is greatly influenced by the site of secondary exoerythrocytic schizogony. The growth of tissue stages in capillary endothelium of various organs, especially in the brain, results in blockage of the vessels, and the host dies of "cerebral malaria." If the disease is caused by *P. elongatum*, the parasites invade the bone marrow, and the birds may die of aplastic anemia.

SIMIAN MALARIA. *Plasmodium cynomolgi* of lower monkeys was first described about 65 years ago, but now it is recognized as a species complex comprising at least eight species or subspecies among monkeys in Malaya, Taiwan, Cambodia, south India and Ceylon. *P. bastianellii* and *P. cynomolgi*, both from Malaya, are well-known species. *P. knowlesi* of monkeys in Malaya, *P. inui* in the Philippines, and *P. gonderi* in Africa are also being extensively studied at the present time. In the New World monkeys, *P. simium* resembles *P. ovale*, and *P. brasilianum* resembles *P. malariae*. *P. reichenowi* is a natural parasite of chimpanzees. In addition to *P. hylobati* from gibbons of Java, three other species have been isolated from gibbons in Malaya.

Plasmodium cynomolgi has been successfully transmitted from monkeys to man and back again. Experimental infections of man can be achieved through the bites of mosquitoes infected with *P. brasilianum*, *P. bastianellii* and *P. inui*. *P. ovale*, *P. vivax*, and *P. falciparum* of man can develop at least partially in chimpanzees, while *P. malariae* can develop in gorillas and chimpanzees. Gibbons are excellent experimental hosts for *P. falciparum*. Recent reports include a case of natural infection of man by *P. knowlesi*.[10] These examples suffice to emphasize the zoonotic nature of malaria, and the possibility of monkeys and apes serving as reservoir hosts for human infection. An informative report on malaria in man and monkeys was made by Coatney.[12] See also Garnham.[22] For a statement on mosquito vectors of simian malaria, see p. 410.

Simian malarial parasites, like those of man, have a complex of antigens. "Immunity is conferred on one another by all homologous strains of simian malaria parasites. The various *P. cynomolgi* subspecies show no

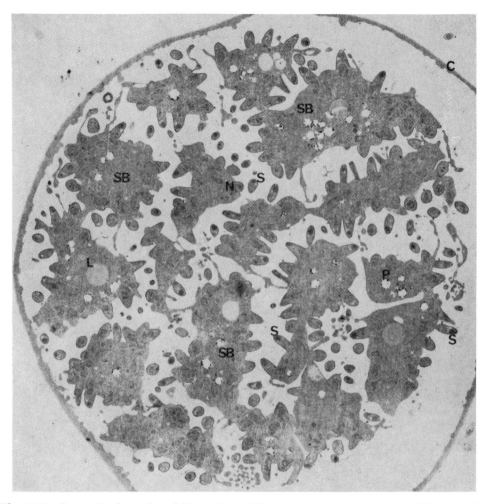

Fig. 4-18. Sporozoite formation of *Plasmodium gallinaceum.* Individual sporozoites (*S*) emerge from the sporoblast cytoplasm (*SB*) by budding. *N.* Nuclei. *P.* Pigment inclusions. *L.* Lipid droplets. × 4500. (Terzakis, Sprinz, and Ward, courtesy of Military Med.)

cross immunity either to each other or to different species. . . . The African parasite, *P. gonderi*, was able to infect monkeys immune to all oriental species tested, including *P. knowlesi.* . . . The three species, *P. knowlesi*, *P. coatneyi* and *P. fragle*, showed considerable cross immunity amongst themselves and against other oriental species, suggesting that they possess a broad spectrum of common antigens."[83] A related genus, *Hepatocystis*, is largely confined to arboreal mammals of the tropics.

A new experimental simian host for malarial parasites of man is the Colombian owl monkey or night monkey, *Aotus trivigatus*, which is susceptible to *Plasmodium falciparum*, *P. vivax* and *P. malariae*, as well as to a strain of *P. knowlesi* being maintained in the rhesus monkey, *Macaca mulatta*. Experimentally produced malaria in *Aotus* is being successfully used for biochemical, cultivation and immunologic studies.

RODENT MALARIA. *Plasmodium berghei* was discovered in the African tree rat in 1948 in Katanga, and four years later, *P. vinckei* was reported from the same area in rodents. Both species are transmitted by the same mosquito, *Anopheles dureni*. *P. chabaudi* has also been found to be a natural parasite of rodents. *P. berghei* exhibits several ecologic races, and their study suggests an evolutionary adaptation to different climatic zones. The erythrocytic forms of *P. berghei* have a strong predilection for reticulocytes or younger red cells.

The growth and division synchrony in *Plasmodium berghei* is abolished by experimental removal of the host's pineal body.[5] The pineal also mediates the vascular capture and release of late growth forms of this parasite. The mechanism by which the pineal is

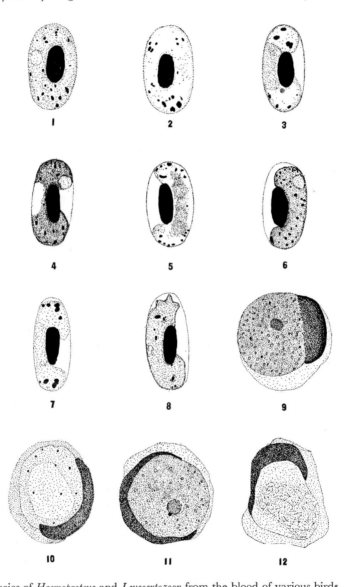

Fig. 4-19. Species of *Haemoproteus* and *Leucocytozoon* from the blood of various birds. *1* to *10* are × 1942; *11* and *12* are × 1750. *1* and *2. Haemoproteus archilochus* from the ruby-throated humming bird, showing a macro- and microgametocyte entirely enclosing the host cell nucleus. *3* and *4. Haemoproteus quiscalus* from the bronzed grackle. A micro- and macrogametocyte. *5* and *6. Haemoproteus* sp. from the Baltimore oriole. A micro- and macrogametocyte. *7* and *8. Haemoproteus* sp. from the blue jay. A micro- and macrogametocyte. *9* and *10. Leucocytozoon coccyzus* from the yellow-billed cuckoo. A macro- and microgametocyte. *11* and *12. Leucocytozoon sakharoffi* from the eastern crow. A macro- and microgametocyte. (Coatney and West, courtesy of Amer. Midland Nat.)

activated and by which it carries out its part in regulating growth and division synchrony is unknown. It may be the junction point for a neural and chemical pathway. Evidence for this suggestion comes from the fact that the pineal hormone, ubiquinone, and the related compounds vitamin K and vitamin E, play a role in augmenting growth and division synchrony of *P. berghei* infections in mice.[7]

Ladda *et al.*[40] have listed the similarities between trophozoites of rodent and human *Plasmodium:* (1) both lack a nucleolus; (2) pigment and smooth membrane vesicles are similar; (3) both exhibit double limiting membranes; (4) both lack mitochondrial forms; and (5) both exhibit micropyle-type structures that are apparently specialized sites of absorption.

Family Haemoproteidae. This family is similar in many respects to some of the Plasmodiidae. Only gametocytes appear in vertebrate blood. The intermediate host is a fly and vertebrate hosts are reptiles and birds.

Haemoproteus. This genus occurs in domestic ducks and turkeys and in wild birds and reptiles. Gametocytes (Fig. 4–19) occur in host erythrocytes, and schizogony takes place in endothelial cells of blood vessels. Unlike *Plasmodium*, members of the genus *Haemoproteus* do not produce pigment in infected host cells. Vectors are midges (Ceratopodonidae) and louse flies (Hippoboscidae).

Haemoproteus columbae is a familiar parasite of the red blood cells of the pigeon, *Columba livia*, and other birds. The sexual phase occurs in the hippoboscid flies of the genera *Lynchia, Pseudolynchia* and *Microlynchia*. Fertilization takes place in the stomach of the fly, and the zygote enters the stomach wall. Asexual reproduction produces a cyst full of sporozoites. The cyst ruptures and the liberated sporozoites enter the bird host with the bite of the fly.

The sporozoites now start the schizogonic cycle by entering endothelial cells of blood vessels of various organs and dividing. Merozoites are produced and are released by the rupture of the infected host cells. These merozoites may enter other endothelial cells or red blood cells. In the latter they become somewhat sausage-shaped, and develop into gametocytes that mature to gametes in the stomach of the fly.

Leucocytozoon (Fig. 4–19). This genus, whose vectors are black flies (Simuliidae) and midges (e.g., Culicoides), has relatively few species, most of which, if not all, occur in birds. The life cycle lacks erythrocytic schizogony, but schizogonic stages occur in the parenchyma cells of the liver or in various other organs, and the schizonts may reach 150 μ in length. As the name indicates, the parasites (gametocytes) may be found in white blood cells, but erythrocytes are also involved. Pigment is not formed in the infected red cells. The zygote in the vector develops into a motile ookinete that eventually forms sporozoites.

The two best known species, *L. simondi* and *L. smithi*, of ducklings and turkeys respectively, are pathogenic. The course of the disease in ducklings may be so rapid that the young birds appear in good health one day and are dead the next day. The mortality rate varies a great deal and if a duck recovers it remains a carrier of the disease. There is no effective treatment.

The nature of the disease in turkeys parallels that in ducks with minor modifications. Infected birds are weak and nervous and may die in a few hours after symptoms appear. Treatment is of little value, if any. Keeping blackflies away from the birds is a sure preventive but difficult for the average farmer. Screened pens for young birds are recommended. See papers by Fallis *et al.*,[17,18] and Liu.[44]

Class Piroplasmea

Members of this class are minute, pyriform, rod-shaped or ameboid parasites of erythrocytes and, occasionally, of leukocytes or histiocytes of vertebrates. Locomotion is by gliding or body flexion. Reproduction is by binary fission or schizogony; sexual stages have not been observed. Ticks are commonly used as intermediate hosts. The presence of a conoid,[73] similar to that in

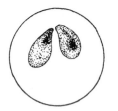

Fig. 4-20. *Babesia bigemina* in cattle red blood cells.

Toxoplasma, Eimeria, and other genera, is the reason for placing this class with the Sporozoa rather than with the amebas.

Order Piroplasmida

Family Babesiidae. *Babesia.* Considerable damage to livestock and to other animals is caused by members of this genus; the disease is known as *babesiasis* or *piroplasmosis.* The common name for the disease in cattle is *cattle tick fever.* In the red blood cells of cattle, the parasite typically divides, apparently only by budding, to form two pear-shaped bodies (trophozoites); hence the species name, *bigemina* (Fig. 4-20). Within the red cells the trophic forms engulf large portions of the erythrocyte cytoplasm, as does *Plasmodium.* The pigment hemozoin is not formed (see page 96). The arthropod vector is the tick, *Boophilus margaropus,* which feeds on cattle blood. Babesiasis is primarily a disease of older animals. Cattle may suffer fever, loss of appetite, constipation or diarrhea, bloody urine and anemia. Experimental infections of cattle and dogs have revealed exoerythrocytic parasites in the lumen of capillaries of almost all internal organs, including the brain.[37] The mortality rate is about 90 per cent, and death may occur in one week. Treatment is of little value, but control is successful when the tick can be destroyed. Other species in cattle are *B. bovis* and *B. argentina.*

With *Theileria parva*[36] of cattle, schizogony takes place chiefly in the cells of the lymphatic system. After schizogony, minute forms, 1 to 2 μ in diameter, appear in the red blood cells. Since the red cells are not normally destroyed, there is no anemia or blood in the urine, but

other symptoms are similar to those of babesiasis. Mortality in African vectors may be as high as 95 per cent. A sexual cycle in tick vectors has been described, but it needs confirmation. *Gonderia* is a related genus that infects lymphocytes and red cells of cattle and other ungulates, especially in Africa and Asia.

The relationships of the species described under this order are uncertain, and several of them may be synonyms. *Babesia, Piroplasma, Babesiella* and *Nuttallia* have been described from mammals; *Aegyptianella* from birds, and *Babesiosoma (-Haemohormidium)* from amphibians and fish. Final determination concerning the relationships among these genera must await more detailed studies of immune reactions, electron microscopic photographs, and life history investigations. The genus *Dactylosoma,* found in the blood of reptiles and amphibians where schizogony and gametogony take place in red blood cells, probably is a piroplasmid. For a detailed account of the development of *Babesia* and *Theileria* in ticks, see Riek,[61] and for a consideration of the cycle of both genera in the vertebrate hosts, see Ristic.[62] Arthur[6] has discussed the ecology of ticks with reference to the transmission of piroplasms. See Ball[8] for an account of some confusing relationships among these protozoa and other bodies to be found in the blood of reptiles and amphibians.

Babesia also attacks sheep, goats, dogs, pigs, horses, birds, and probably many other animals.[73] Equine babesiasis is manifested by progressive anemia.[72] Dog piroplasmosis is caused by *Babesia canis.* As with the cattle form, these protozoa enter red blood cells and cause an increase in temperature, pulse rate,

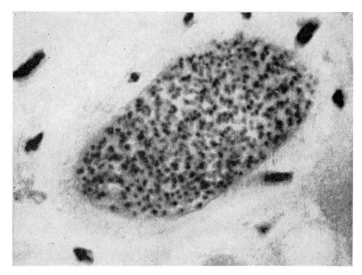

Fig. 4-21. *Sarcocystes lindemanni* in left ventricular wall of a Panamanian child. The cysts contained about 170 cells. Except for enlargement and slight hyaline degeneration of the muscle fibers, there were no myocardial lesions. × 1500. (Gilmore, Kean and Posey, Am. J. Trop. Med.)

and respiration rate. Dogs become weak and anemic, and they often die, but generally they acquire an immunity after an infection.[67] The disease is transmitted by the brown dog tick, *Rhipicephalus sanguineus*. At least one human case of piroplasmosis has been reported.[20] A man whose spleen had been removed some years earlier acquired the infection and died with symptoms of blackwater fever.

Class Toxoplasmea

Spores are absent, but cysts or pseudocysts are formed with naked trophozoites. Reproduction is by binary fission or internal budding. Schizogony has been described, but its presence is questionable. Locomotion is by gliding or body flexion. There are no intermediate hosts.

Order Toxoplasmida

Family Sarcocystidae. *Sarcocystis* (Fig. 4–21). This genus is frequently found in the muscles of animals slaughtered for human consumption and in wild herbivores. Cattle may be 100 per cent infected. The organisms are also found in birds and reptiles and, rarely, in dogs, cats and man. Among the many *Sarcocystis* that have been described are

S. meischeriana in pigs, *S. muris* in mice, *S. harvathi* in chickens and *S. lindemanni* in man. Only 15 human cases have been reported. In striated muscle the organisms occur in elongate compartmented cysts called *Miescher's sacs*, which may be so small that they cannot be seen without a lens, or so large that they appear as short white streaks.[49] Cysts range in size from 25 μ to 5 cm. Motile crescent-shaped trophozoites, 6 to 15 μ in length, are crowded within the compartments. The parasites may also be found in heart muscle.

Transmission possibly occurs by the ingestion of trophozoites from feces or from infected muscles, but the evidence is not clear. Pathogenicity, if it occurs, is mild. Normally,

Fig. 4-22. *Toxoplasma gondii*. The line at the lower right represents one micron in length.

meat found to be infected is condemned for human consumption.

Family Toxoplasmatidae. *Toxoplasma gondii* (Fig. 4–22). Several species have been described, but *gondii* is the only one generally accepted. *Toxoplasma gondii* is one of the most common parasites of man and other vertebrate animals. It is crescentic or oval in shape, about $3 \times 6 \mu$ in size, although individuals 12μ long have been reported. Often, the nucleus is situated in the blunter end of the body. Fibrils from an anterior truncated cone (*conoid*) extend over the anterior two-thirds of the body.[34] Five to 18 cylindrical or club-shaped *toxonemes* run longitudinally within the body. (These details cannot be seen in slides usually available to students.)

A revealing electron microscopic study of

the trophozoites of *Toxoplasma gondii* compared with merozoites of *Eimeria perforans* was made by Scholtyseck and Piekarski.[68] They showed that the fine structures of these two organisms are remarkably similar (Fig. 4–23) and indicate a close relationship. Comparable studies[4] indicate a close systematic relationship among *Eimeria, Lankestrella, Eucoccidium, Plasmodium, Sarcocystis, Besnoitia* and *Toxoplasma*.

Evidence indicates that *Toxoplasma gondii* is an intestinal coccidian of cats.[18a,31a,70a] Oocysts of the *Isospora* type appeared in cat feces after infection with *Toxoplasma*, and were used to infect other cats. Cats were also infected by ingesting *Toxoplasma* from infected mice. Sporozoites from excysted oocysts were infective for monkey kidney cell cultures. Ultrastructural examinations and other stud-

EIMERIA PERFORANS

TOXOPLASMA GONDII

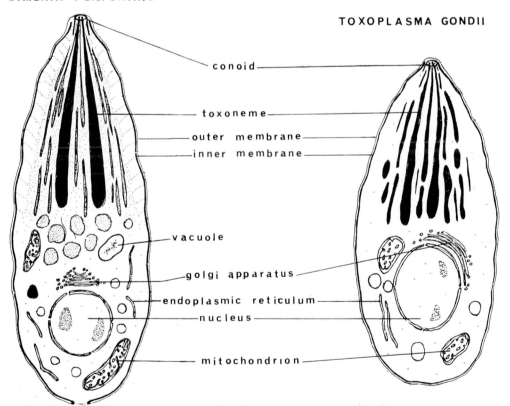

- conoid

- toxoneme
- outer membrane
- inner membrane

- vacuole

- golgi apparatus
- endoplasmic reticulum
- nucleus

- mitochondrion

Fig. 4-23. A merozoite of *Eimeria perforans* compared to a trophozoite of *Toxoplasma gondii*, based on electron microscope studies. The similarity is striking and indicates that the genus *Toxoplasma* belongs to the Sporozoa. (Redrawn from Scholtyseck and Piekarski, Ztschr. Parasitenk.)

ies showed that *T. gondii* has typical coccidian gametocytes and schizonts. Transmission occurs by ingesting infective oocysts from host feces and possibly by eating muscle or other tissue infected with mature *Toxoplasma*.

Toxoplasma utilizes oxygen and glucose and produces CO_2. The main source of energy appears to be glucose. An active hexokinase has been demonstrated, and lactic, acetic and another unidentified acid have been detected as metabolic products of glucose. *T. gondii* can oxidize reduced DPN (diphosphopyridine nucleotide), and it possesses a cytochrome system.[19] For a consideration of *Toxoplasma* antigens, see Chordi *et al.*[11] Excellent reviews of *Toxoplasma gondii* and its distribution and pathogenicity are available (e.g., see Levine[42]). Portions of the following discussion have been adapted from Jacobs.[32] See also Nozik and O'Connor.[57]

TOXOPLASMOSIS. Toxoplasmosis is most common in warm, moist areas but is worldwide in distribution; *Toxoplasma* is probably the most common pathogenic parasite of man, infecting 30 to 50 per cent of the human race. Serologic evidence indicates that 90 per cent of the population of some areas are affected. Responses to *Toxoplasma gondii* vary widely, but, fortunately, disease occurs much less frequently than infection. It is necessary to learn the reasons for this variation in response. In order to explain the clinical forms of toxoplasmosis and to establish the etiologic relationship between it and other disease forms, it is necessary to understand the natural history of the infection.

In laboratory animals, *T. gondii* proliferate locally at the site of entry, then quickly disseminate throughout the body by way of the circulation. These organisms are obligately intracellular, and they are probably carried in the circulation within infected leukocytes. The occasional organisms that might be found free in the plasma or other extracellular fluids probably had been recently released from cells. The *Toxoplasma* organisms can invade many different types of cells and tissues, such as cardiac and skeletal muscle, alveoli, cells of the kidney tubules, intestinal mucosa, liver parenchyma, endothelial and reticular cells, and neurons.

The organisms reproduce within the various tissues by means of *endodyogeny*, in which two daughter cells are produced within the organism, or by schizogony. Eventually the mother cell ruptures, liberating the trophozoites, which then invade neighboring cells to create multiple foci of necrosis. Circulating antibodies are built up as the infection develops, and the proliferation of trophozoites diminishes and almost ceases, except in neural tissues. However, parasite cysts grow and develop, beginning as early as eight days after the initiation of the infection. These persist in muscle and other tissues. Eventually, cysts are the predominant form of the parasite in neural and extraneural tissues, although trophozoites may continue to produce an infection of the retinal neurons years after the acute infection has passed.

The cysts are usually dormant, residing in tissues without provoking cellular reactions or antibody production. However, *Toxoplasma* may occasionally show extraneural activity late in the course of the disease, as evidenced by parasitemia or congenital transmission in chronically infected mice. Whether this activity can be attributed to residual trophozoites or to organisms released from cysts is difficult to determine.

The most common symptom of acute infection is lymphadenopathy, involving peripheral lymph nodes in the cervical, supraclavicular, and inguinal regions, and sometimes in the mesentery. Myositis, anemia, hepatosplenomegaly, fever and headache may be associated in varying degrees with adenopathy. Mild pulmonary infiltration may sometimes occur. In the fatal form of the disease, which is much rarer, pneumonitis, myocarditis, and encephalitis occur.

Toxoplasma produce interferon, which is inhibitory to viruses in mammalian cells. In spite of this, toxoplasma organisms and viruses were found to be able to cohabit the same cell. In a laboratory experiment, human cancer-derived cells could be infected with *Toxoplasma* even though they were already infected with any one of the viruses causing the following diseases: measles, pseudorabies, adenovirus, Newcastle disease or bovine parainfluenza.

Human infections often occur in veterinarians, slaughter-house workers, rabbit dressers and rabbit trappers. It is probably transmitted by the oral route through infected meat or animal products. Dubey[16] fed mice infected with *Toxoplasma* to cats infected with the nematode worm *Toxocara cati*. From these cats he recovered worm eggs that contained *Toxoplasma*. The infected eggs were fed to other cats, which then acquired toxoplasmosis. (See Hutchinson.[31]) A recent outbreak of the disease in man was traced to the eating of contaminated hamburgers.

Cysts in carnivores may derive from infected prey, but cysts in herbivores must come from other sources not yet determined. In experimental mice, the kidney, lungs and intestines may serve as portals of exit for cysts.[33] *Toxoplasma* is capable of passing into the external environment in the feces of experimentally infected cats; these forms remain viable for at least a year. Congenital transmission takes place in mammals, and in human beings the fetus apparently is infected only during the acute stage of the disease.

Besnoitia. This genus has been found in a few mammals (cattle, horses, deer, rodents) and lizards. It resembles *Toxoplasma* morphologically and in its type of reproduction, but immunologically it is different, although the two genera appear to have antigens in common.[50] *Besnoitia* is motile at room temperature while *Toxoplasma* is not. The two genera differ in type of cyst formation, serologic reactions and kind of pathology produced. The morphologic and antigenic similarities justify placing them in the same family.

Pols[59] has described in detail the morphology and pathogenicity of *Besnoitia besnoiti* in cattle, in which the parasites may be common in subcutaneous and connective tissue.

More recently, Garnham[21] has described the life cycle and morphology of *B. sauriana* from a lizard in British Honduras. He described the cyst as macroscopic with a very thick cyst wall, characterized by the presence of numerous hypertrophied host-cell nuclei. "The parasites ('spores') multiply by binary fission, enclosed in a fine membrane, within this cell, which eventually they entirely fill. The 'spores' easily contaminate the blood and may be mistaken for *Schellackia* or *Toxoplasma* in blood smears of infected lizards."

Class Haplosporea

Order Haplosporida

The affinities of the Haplosporida have long been disputed. *Ichthyosporidium*, common in cells, tissues and body cavities of invertebrates and lower vertebrates, has a schizogony phase during its development. It is without flagella or cilia, but sometimes has pseudopodia. *Haplosporidium*, *Nephridiophaga* (in roaches), *Minchinia*, *Sporozoon* and *Coleospora* are among the genera placed in this order. Sprague[74] has pointed out that the haplosporidians are much more like microsporidans than any other spore-forming group, and he suggested that these two orders be placed in the class Microsporea. *Haplosporidium* is found in aquatic annelids and tunicates, and its spores are truncated with a lid at one end. *Minchinia nelsoni* is pathogenic in oysters.[75]

SUBPHYLUM CNIDOSPORA

Spores contain polar filaments.

Class Myxosporidea

Order Myxosporida

These parasites are usually found in the gallbladder, urinary bladder or in other hollow organs of both marine and freshwater fish; occasionally they infect amphibians and reptiles. They are also found in the liver, spleen, kidneys, gills, skin or other organs, in which they are frequently enveloped by host tissue. In body cavities the parasites may be ameboid.

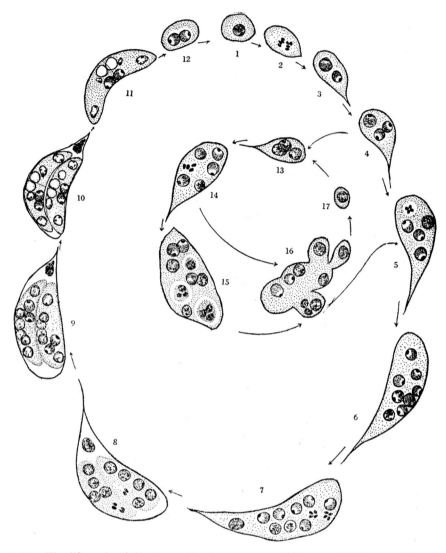

Fig. 4-24. The life cycle of *Ceratomyxa blennius*, a myxosporidan parasite from a tidepool blenny. *1*, Zygote; *2* to *11*, sporogony; *8*, two sporoblasts showing reduction division; *9*, young spores; *11*, mature spore; *12*, sporoplasm with gametes; *13* to *17*, nucleogony with cytoplasmic growth and budding; *15*, internal budding; *16*, external budding; *17*, uninucleate bud. (Noble, courtesy of J. Morphology.)

Reproduction occurs by the production of spores that are characterized by the presence of two valves and one or more polar capsules. Each spore develops into a multinucleate organism. The life cycle of a myxosporidan parasite is illustrated in Figure 4–24. Infection does not usually harm the host, except occasionally, when the parasites may be so destructive as to produce up to 100 per cent mortality.

The two nuclei of the sporoplasm unite when the spore opens its valves upon gaining entrance to the host, by ingestion or other means. The freed sporoplasm thus becomes a zygote, and the entire life cycle of the large polysporous species normally occurs within the original zygote membrane. Budding or plasmotomy may take place. Multiplication in the polysporous species is accomplished largely by nucleogony (the formation of a syncytium). The diploid phase of the cycle lasts until the formation of a sporoblast occurs, which starts the sporogony part of the cycle. The sporoblast (or pansporoblast, if more than two spores are formed) is derived from a specialized cell that begins as an ordinary vegetative cell or nucleus. Sporogony may be monosporous, disporous, or polysporous. Sporoblasts of six to eight nuclei, or pansporoblasts of 14 or more nuclei, may be found. Each sporoblast normally contains six generative and two somatic, residual nuclei. Of the six generative nuclei, two form the thin-walled shell valves, two give rise to the polar capsules, each containing a polar filament, and two, by a reduction division, become the haploid sporoplasm nuclei or gametes. Spores of some species (as well as some microsporidia) are surrounded by characteristic mucous envelopes that can be revealed by the addition of India ink to material under investigation.[47]

The order Myxosporida consists of a highly specialized group of protozoan parasites that have become modified so as to possess some basic metazoan characteristics. Common genera of Myxosporidia are *Leptotheca*, *Ceratomyxa*, *Myxidium*, *Henneguya* and *Myxobolus* (Fig. 4–25). *Myxosoma* (= *Myxobolus*) *squamalis* is a parasite in rainbow trout, silver salmon, and chum salmon. In the trophozoite stage (growth stage), the parasite attacks the scales of fish and makes its way inside the scales and skin, where it sporulates. Scales and surrounding tissue may be damaged. For further details on life cycles, see Noble.[56] For a detailed account of the ultrastructure of Myxosporida, see Lom and de Puytorac.[45] For a general review (in Russian) of the Myxosporida, see Shulman.[69]

Order Helicosporida

Helicosporidium parasiticum is the single species belonging to this genus and order. It

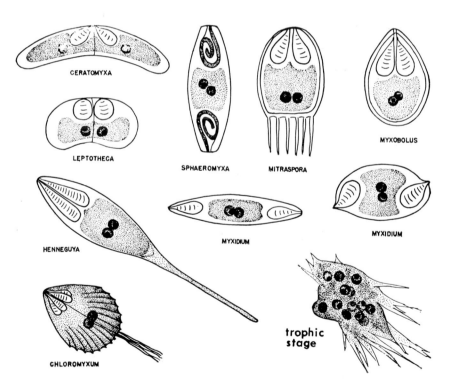

Fig. 4-25. Sample spores of the order Myxosporida. Only about half of the described genera are represented. At the lower right corner is a trophic stage with slim, pointed pseudopodia that are typical of several genera.

lives in the body cavity, the fat body and the nerve tissue of *Dasyhelea obscura*, a dipteran insect whose larvae are found in the sap in wounds of elm and horse chestnut trees. Each spore contains three uninucleate sporoplasms and a long, spirally-coiled filament. Other fly larvae associated with the wounds in these trees may be parasitized.

Order Actinomyxida

The exact taxonomic position of this group of organisms is disputed. Each spore is surrounded by a membrane possessing three valves. There are also three polar capsules, each enclosing a polar filament. In the mature stage eight spores develop within the sporocyst. There are two families: Tetractinomyxidae and Triactinomyxidae. All species occur in sipunculids and in tubificid annelids. *Tetractinomyxon intermedium*, possessing spores 7 to 8 μ in diameter, is found in the coelom in the sipunculid, *Petalostoma minutum*. *Triactinomyxon legeri* lives in the worm, *Tubifex tubifex*. Its spore contains 24 uninucleate sporoplasms. These organisms unite in pairs that are liberated as binucleate amebulae, each of which divides several times and forms a cyst (*pansporocyst*) containing cells of two sizes. These cells are the gametes. Each small cell fuses with a larger one (anisogamy) and forms a zygote, which develops into the spores that are transmitted from host to host. For further information and a description of sporogony in *Siedleckiella*, see Janiszewska.[35]

Class Microsporidea

Order Microsporida

Microsporidans (Figs. 4–26 and 27) are intracellular parasites, mostly of arthropods (commonly in gut epithelium and fat bodies) and fish (common in skin and muscles); they are also found in all invertebrate phyla including Protozoa. Spores are 2.0 to 25.0 μ long, have walls that contain chitin, and possess a single polar filament. There is no intermediate host.

Within the host digestive tract the sporo-plasm of the spore is injected into host tissue through the hollow polar filament, which everts like a finger of a glove. The energy for ejection is supplied by a special organelle called a "polaroplast," which imbibes water and produces pressure.[48] The parasite then enters the host gut epithelium and, by way of the blood or body cavity, is carried to the skin, muscles or other tissue. It enters host cells and undergoes asexual division and sporogony. If enough host cells are entered, the host dies because of the degeneration of these cells.

The following examples illustrate the wide variety of hosts that may become infected by microsporidians. *Nosema* and *Glugea* (Fig. 4–28) commonly occur in mammals. The brains and hearts of rabbits, dogs and humans may be invaded. The mammalian forms were formerly known as *Encephalitozoon*.[87] *Nosema bombycis* is a destructive parasite of the silkworm, while *N. apis* causes serious disease in honey bees. *N. stegomyiae* occupies various organs, especially midgut tissue, of the mosquito, *Anopheles gambiense*. Some bryozoa have parasitic microsporidans in their germ cells and body cavity. *Cyclops fuscus* is inhabited by *N. cyclopi*, whereas the cytoplasm of the cephaline gregarine, *Frenzelina conformis*, may be parasitized by *N. frenzelinae*. The urinary bladder of the fish, *Opsanus tau*, is parasitized by the myxosporidan, *Sphaerospora polymorpha* which, in turn, is parasitized by the microsporidan, *N. notabilis*, which lives in the trophozoite of the protozoan host. The myxosporidans may be killed by their parasites.

Another hyperparasite is the microsporidan, *Perezia lankesteriae*, which lives in the cytoplasm of the gregarine, *Lankesteria ascidiae*. This microsporidan also may be found in the intestine of the tunicate, *Ciona intestinalis*. Other genera and species of Microsporida inhabit mosquitoes, mice, termites, mayflies, crustaceans, annelids, flies and hemipterans. For detailed descriptions of morphology, see Kudo and Daniels,[38] and Lom and Várva.[46] For a list of microsporidans that infect insects, see Thomson.[80] For a synopsis of spe-

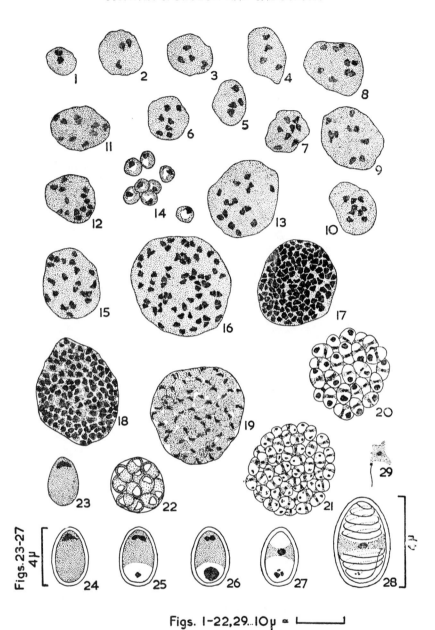

Figs. 1-22,29..10μ ≈ |_____|

Fig. 4-26. *Plistophora culicis*, a microsporidan that parasitizes the Malpighian tubules of larvae of *Culex pipiens* and of the adult mosquito, *Anopheles gambiae*.

1. Binucleate schizont. *2 to 13.* Development of the schizonts. *14.* Merozoites produced by segmentation of 8-nucleate schizont. *15 to 17.* Development of the sporont. *18.* Cytoplasmic cleavage within sporont; nuclei large. *19.* Fully developed sporoblasts within sporont; nuclei compacted.

20. Group of macrospores; sporoplasm binucleate, metachromatic granule large. *21.* Group of microspores at later stage of development. *22.* Fresh preparation of small macrospore cyst. *23 to 27.* Development from sporoblast to mature spore. *28.* Diagrammatic representation of the structure of the mature spore. *29.* Sporoplasm after emergence from spore. (Canning, courtesy of Rivista di Malariologia.)

cies from freshwater and euryhaline fish, see Putz et al.[60]

MORPHOLOGIC RELATIONSHIPS AMONG THE SPOROZOA AND CNIDOSPORA

The electron microscope has disclosed a remarkable similarity of fine structure among many species of Sporozoa. Garnham et al.[23] showed that *Lankesterella* appears very much like *Toxoplasma*, *Laverania*, *Eimeria*, and *Sarcocystis*. They concluded that ". . . in fact the possession of a number of common features seems to draw the few organisms we have studied closer together rather than to separate them." The features in common among

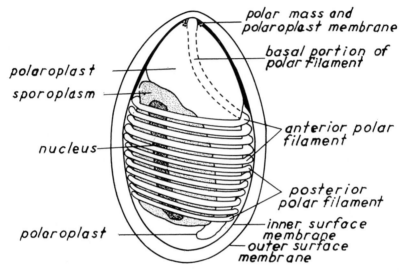

Fig. 4-27. Schematic drawing showing the structure of the spore of a microsporidan, *Thelohania californica*, as revealed by electron microscopy. (Kudo and Daniels, courtesy J. Protozool.)

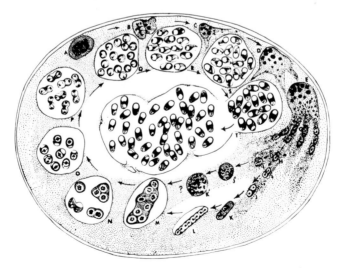

Fig. 4-28. Intracellular development of *Glugea*. *A.* Host cell nucleus that divides repeatedly by amitosis; the endoplasmic reticulum of the nucleus is continuous with the plasma membrane of plasmodial stages of the microsporidan parasites. *H* to *L* and *I* to *K.* Parasite plasmodia that expand and break up to form spores. *T.* Mature spores. (Sprague and Vernick, courtesy of J. Protozool.)

all of these species are usually the conoid, peripheral fibers, paired organelles, typical mitochondria and Golgi bodies (Fig. 4–23). A conoid has been found in the ookinete of *Plasmodium* as well as in merozoites of these parasites. For a short discussion of some of these similarities among *Besnoitia, Toxoplasma, Sarcocystis, Eimeria, Lankesterella, Haemamoeba, Laverania* and M-organism, see Sheffield.[70]

REFERENCES

1. Aikawa, M.: The fine structure of the erythrocytic stages of three avian malarial parasites, *Plasmodium fallax, P. lophurae,* and *P. cathemerium.* Amer. J. Trop. Med. Hyg., *15*(4):449–471, 1966.

2. Aikawa, M., Huff, C. G., and Sprinz, H.: Comparative feeding mechanisms of avian and primate malarial parasites. *In* Sadun, E. H., and Osborne, H. (eds.): Research in Malaria. pp. 969–983. Suppl. to Milit. Med., *131*:847–1272, 1966.

3. Allison, A. C.: Inherited factors in blood conferring resistance to Protozoa. *In* Garnham, P. C. C., Pierce, A. E., and Riott, I. (eds.): Immunity to Protozoa. pp. 109–122. Philadelphia, F. A. Davis, 1963.

4. Andreassen, Jørn, and Behnke, O.: Fine structure of merozoites of a rat coccidian *Eimeria miyairii,* with a comparison of the fine structure of other sporozoa. J. Parasit., *54*(1):150–163, 1968.

5. Arnold, J. D., Berger, A. E., and Martin, D. C.: Role of the endocrine system in controlling growth and division synchrony of *Plasmodium berghei* in mice. J. Parasit., *55*:956–962, 1969.

6. Arthur, D. R.: The ecology of ticks with reference to the transmission of Protozoa. *In* Soulsby, E. J. L. (ed.): Biology of Parasites. pp. 61–84. New York, Academic Press, 1966.

6a. Ayala, S. C., and Lee, D.: Saurian malaria: development of sporozoites in two species of phlebotomine sandflies. Science, *167*, 891–892, 1970.

7. Bahr, G. F.: Quantitative cytochemistry of malaria infected erythrocytes (*Plasmodium berghei, Plasmodium chabaudi* and *Plasmodium vinckei*). Milit. Med., *134*(10):1013–1025, 1969.

8. Ball, G. H.: Some protozoa and other bodies found in the blood of reptiles and amphibians. Progress in Protozool., Internat. Congr. Ser. No. 91, Excerpta Medica Foundation, Abstract, pp. 127–128, 1965.

9. Bruce-Chwatt, L. J.: Comments on biology of human malaria. *In* Sadun, E. H., and Osborne, H. (eds.): Research in Malaria. pp. 963–966. Supp. to Milit. Med., *131*(9): 847–1272, 1966.

10. Chin, W., Contacos, P. G., Coatney, G. R., and Kimball, H. R.: A naturally acquired quotidian-type malaria in man transferable to monkeys. Science, *149*:865, 1965.

11. Chordi, A., Walls, K. W., and Kagan, I. G.: Analysis of *Toxoplasma gondii* antigens by agar diffusion methods. J. Immunol., *93*:1034–1044, 1964.

12. Coatney, G. R.: Simian malarias in man: facts, implications, and predictions. Amer. J. Trop. Med. Hyg., *17*:147–155, 1968.

13. Cox, H. W.: Comments on autoimmunity in malaria. Amer. J. Trop. Med. Hyg., Supp., *13*:225–227, 1964.

14. Davies, S. F. M., Joyner, L. P., and Kendall, S. B.: Coccidiosis. London, Oliver & Boyd, 1963.

15. Davis, L. R., and Bowman, G. W.: Bovine Coccidiosis. *In* Animal Diseases. The Yearbook of Agriculture. Washington, D.C., U.S.D.A., pp. 314–317. 1956.

16. Dubey, J. P.: Studies on *Taxocara cati* larvae infected with *Toxoplasma gondii.* J. Protozool., Supp., Abstract No. 167, *14*, 1967.

17. Fallis, A. M., and Bennett, G. F.: On the epizootiology of infections caused by *Leucocytozoon simondi* in Algonquin Park, Canada. Can. J. Zool., *44*:101–112, 1966.

18. Fallis, A. M., Davis, D. M., and Vickers, M. A.: Life History of *Leucocytozoon simondi* Mathis and Leger in natural and experimental infections and blood changes produced in the avian host. Can. J. Zool., *29*:305–328, 1951.

18a. Frenkel, J. K., Dubey, J. P., and Miller, N. L.: *Toxoplasma gondii* in cats: fecal stages identified as coccidian oocysts. Science, *167*, 893–896, 1970.

19. Fulton, J. D., and Spooner, D. F.: Metabolic studies on *Toxoplasma gondii.* Exp. Parasit., *9*:293–301, 1960.

20. Garnham, P. C. C.: Parasitological problems in tropical medicine. Radioisotopes in Trop. Med., IAEA, 305–321, 1962.

21. ———: *Besnoitia* (Protozoa: Toxoplasmea) in lizards. Parasitology, *56*:320–334, 1966.

22. ———: Malaria Parasites and Other Haemosporidia. Oxford, Blackwell Sci. Pub., 1966.

23. Garnham, P. C. C., Baker, J. R., and Bird, R. G.: The fine structure of *Lankesterella garnhami.* J. Protozool., *9*:107–114, 1962.

24. Geiman, Q. M.: Comparative physiology: mutualism, symbiosis, and parasitism. Ann. Rev. Physiol., *26*:75–108, 1964.

25. Gresham, G. A., and Cruickshank, J. G.: Protein synthesis in macrophages containing *Eimeria tenella*. Nature, Supplement 15, *184*:1153, 1959.

26. Hammond, D.: Coccidiosis of Cattle. Logan, Utah, Utah State University Publication, 1964.

27. Hawking, F., Worms, M., and Gammage, K.: 24- and 48-hour cycles of malaria parasites in the blood; their purpose, production and control. Trans. Roy. Soc. Med. Hyg., *62*: 731–760, 1968.

28. Honigberg, B. M.: Chemistry of parasitism among some Protozoa. *In* Florkin, M., and Sheer, B. J. (eds.): Chemical Zoology. pp. 695–814. New York, Academic Press, 1967.

29. Huff, C. G.: Exoerythrocytic stages of avian and reptilian malarial parasites. Exp. Parasit., *24*:383–421, 1969.

30. Hull, R. W. and Camin, J. H.: Haemogregarines in snakes: The incidence and identity of the erythrocytic stages. J. Parasitol., *46*:515–523, 1960.

31. Hutchinson, W. M.: The transmission of *Toxoplasma gondii*. Trans. Roy. Soc. Trop. Med. Hyg., *61*:80–89, 1967.

31a. Hutchinson, W. M., Dunachie, J. F., Chr. Siim, J., and Work, K.: Coccidian-like nature of *Toxoplasma gondii*. Brit. Med. J., *1*, 142–144, 1970.

32. Jacobs, L.: Toxoplasmosis. Israel J. Med. Sci. *1*:511–513, 1965.

33. Jacobs, L., and Melton, L.: Demonstration of *Toxoplasma* in the kidneys of mice with chronic infections. J. Parasit., Supp. 2, Abstract No. 4, *51*:17, 1965.

34. Jadin, J. M., and Creemers, J.: Ultrastructure et Biologie des Toxoplasmes. III. Observations de Toxoplasmes Intraerythrocytaires chez un Mammifere. Acta Tropica, *25*(3): 267–270, 1968.

35. Janiszewska, J.: *Siedleckiella antonii* sp. n. Uwagi nad sporogeneza u. rodzaju *Siedleckiella* i u innych Actinomyxidia. (*Siedleckiella antonii* sp. n. Remarks on the sporogenesis in the genus *Siedleckiella* and in other Actinomyxidia). Zool. Polon., *6*:88–100, 1955.

36. Jarrett, W. F. H., Crighton, G. W., and Pirie, H. M.: *Theileria parva*: kinetics of respiration. Exp. Parasit., *24*:9–25, 1969.

37. Krylov, M.: Exoerythrocytal development stages of *Piroplasma bigeminum*. Acta Protozool., *2*:97–102, 1964.

38. Kudo, R. R., and Daniels, E. W.: An electron microscopic study of the spore of a microsporidian, *Thelohania californica*. J. Protozool., *10*:112–120, 1963.

39. Ladda, R. L.: New insights into the fine structure of rodent malarial parasites. Milit. Med., *134*(10):825–865, 1969.

40. Ladda, R., Arnold, J., and Martin, D.: Electron microscopy of *Plasmodium falciparum*. I. The structure of trophozoites in erythrocytes of human volunteers. Trans. Roy. Soc. Trop. Med. Hyg., *60*:369–375, 1966.

41. Laird, M.: New haemogregarines from New Zealand marine fishes. Trans. Roy Soc. N. Z., *79*:589–600, 1952.

42. Levine, N. D.: Protozoan Parasites of Domestic Animals and of Man. Minneapolis, Burgess, 1961.

43. Levine, N. D. and Ivens, V.: The Coccidian Parasites (*Protozoa, Sporozoa*) of Rodents. Urbana, University of Illinois Press, 1965.

44. Liu, S. K.: (The pathology of leucocytozoon disease in chicks.) In Chinese, with Eng. summ. Mem. Coll. Nat'l Taiwan Univ., *5*:74–80, 1958.

45. Lom, J., and de Puytorac, P.: Studies on the myxosporidian ultrastructure and polar capsule development. Protistologica, *1*:53–65, 1965.

46. Lom, J. and Vávra, J.: Contribution to the knowledge of Microsporidian spore. I. Electron microscopy. II. The sporoplasm extrusion. Abstracts of Papers Presented at the International Conference of Protozoologists, Praha, pp. 259–260, 1961.

47. ———: Mucous envelopes of spores of the subphylum Cnidospora (Doflein, 1901). Věstnik Československ. Spol. Zool. (Acta Soc. Zool. Bohemoslov.), *27*:4–6, 1963.

48. ———: The mode of sporoplasm extrusion in microsporidian spores. Acta Protozool., *1*:81–89, 1963.

49. Ludvik, J.: The electron microscopy of *Sarcocystis miescheriana* Kuhn 1865. J. Protozool., *7*:128–135, 1960.

50. Lunde, M. N., and Jacobs, L.: Antigenic relationship of *Toxoplasma gondii* and *Besnoitia jellisoni*. J. Parasit., *51*:273-276, 1965.

51. Maegraith, B. G.: Pathogenic processes in malaria. *In* Taylor, A. E. R. (ed.): The Pathology of Parasitic Diseases. pp. 15–32. Oxford, Blackwell Sci. Pub., 1966.

52. May, J. M.: The ecology of malaria. *In* May, J. M. (ed.): Studies in Disease Ecology. New York, Hafner, 1961.

53. McGregor, I. A.: Considerations of some aspects of human malaria. Trans. Roy. Soc. Trop. Med. Hyg., *59*:145–152, 1965.

54. Moulder, J. W.: The Biochemistry of Intracellular Parasitism. Chicago, University of Chicago Press, 1962.

55. Nagarajan, K.: Metabolism of *Plasmodium berghei*. II. ^{32}Pi incorporation into high-

energy phosphate. Exp. Parasit., 22:27–32, 1968.

56. Noble, E. R.: Life cycles in the Myxosporidia. Quart. Rev. Biol., 19:213–235, 1944.

57. Nozik, R. A., and O'Connor, G. R.: The so-called toxin of toxoplasma. Amer. J. Trop. Med. Hyg., 78:511–515, 1965.

58. Pellérdy, L. P.: Coccidia and Coccidiosis. Budapest, Akadémiai Kiadó, 1965.

59. Pols, J. W.: Studies on bovine besnoitiosis with special reference to the aetiology. Onderstepoort J. Vet. Res. 28:265–356, 1960.

60. Putz, R. E., Hoffman, G. L., and Dunbar, C. E.: Two new species of Plistophora (Microsporidea) from North American fish with a synopsis of Microsporidea of freshwater and euryhaline fishes. J. Protozool., 12:228–236, 1965.

61. Riek, R. F.: The development of Babesia spp. and Theileria spp. in ticks with special reference to those occurring in cattle. In Soulsby, E. J. L. (ed.): Biology of Parasites. pp. 15–32. New York, Academic Press, 1966.

62. Ristic, M.: The vertebrate developmental cycle of Babesia and Theileria. In Soulsby, E. J. L. (ed.): Biology of Parasites. pp. 127–141. New York, Academic Press, 1966.

63. Rousset, J.-J., Couzineau, P., and Baufine-Ducrocq, H.: Plasmodium ovale Stephens 1922. Ann. Parasit. Hum. Comp. 44:273–328, 1969.

64. Rudzinska, M. A., Trager, W., and Bray, R. S.: Pinocytic uptake and the digestion of hemoglobin in malaria parasites. J. Protozool., 12:563–576, 1965.

65. Sadun, E. H., and Moon, A. P. (eds.): Experimental Malaria. Milit. Med., 134(10): 729–1306, 1969.

66. Sadun, E. H., and Osborne, H. (eds.): Research in Malaria. Supp. to Milit. Med., 131(9):847–1272, 1966.

67. Schindler, R.: Serological and immunological investigations in babesiosis. Progress in Protozool., Second Internat. Conf. on Protozool., Abstract No. 19, p. 34, 1965.

68. Scholtyseck, E., and Piekarski, G.: Elektronenmikroskopische Untersuchungen an Merozoiten von Eimerien (Eimeria perforans und E. stiedae) und Toxoplasma gondii. Ztschr. Parasitenk., 26:91–115, 1965.

69. Schulman, S. S.: (Myxosporidia in the Fauna of the USSR.) In Russian. Moscau-Leningrad, Academia Nauk USSR Zool. Institute, 1966.

70. Sheffield, H. G.: Electron microscope study of the proliferative form of Besnoitia jellisoni. J. Parasit., 52:583–594, 1966.

70a. Sheffield, H. G., and Melton, M. L.: Toxoplasma gondii: the oocyst, sporozoite, and infection of cultured cells. Science, 167, 892–893, 1970.

71. Sherman, I.: Malic dehydrogenase heterogeneity in malaria (Plasmodium lophurae and P. berghei). J. Protozool. 13:344–349, 1966.

72. Sibinovic, S., Sibinovic, K. H., Ristic, M. and Cox, H. W.: Physical and serologic properties of an antigen prepared from erythrocytes of horses with acute babesiosis. J. Protozool., 13:551–553, 1966.

73. Simpson, C. F., Kirkham, W. W., and Kling, J. M.: Comparative morphologic features of Babesia caballi and Babesia equi. Amer. J. Vet. Res., 28:1693–1697, 1967.

74. Sprague, V.: Suggested changes in "A revised classification of the Phylum Protozoa," with particular reference to the position of the haplosporidans. Systematic Zool., 15:345–349, 1966.

75. Sprague, V., Dunnington, E. A., Jr., and Drobeck, E.: Decrease in incidence of Minchinia nelsoni in oysters accompanying reduction of salinity in the laboratory. Nat. Shellfish Assoc., 59:23–26, 1969.

76. Stechschulte, D. J., Briggs, N. T., and Wellde, B. T.: Characterization of protective antibodies produced in Plasmodium berghei infected rats. Milit. Med., 134(10):1140–1146, 1969.

77. Stehbens, W. E.: Observations on Lankesterella hylae. J. Protozool. 13:59–62, 1966.

78. Taylor, A. E. R., and Baker, J. R.: The Cultivation of Parasites in Vitro. Oxford, Blackwell Sci. Pub., 1968.

79. Terzakis, J. A., Sprinz, H., and Ward, R. A.: Sporoblast and sporozoite formation in Plasmodium gallinaceum infection of Aedes aegypti. In Sadun, E. H., and Osborne, H. (eds.): Research in Malaria. pp. 984–992. Supp. to Milit. Med., 131(9):847–1272, 1966.

80. Thomson, H. M.: A list and brief description of the Microsporidia infecting insects. J. Insect Path., 2:346–385, 1960.

81. Tigertt, W. D.: Present and potential malaria problem. In Sadun, E. H., and Osborne, H. (eds.): Research in Malaria. pp. 853–856. Supp. to Milit. Med., 131(9): 847–1272, 1966.

82. Ting, I. P., and Sherman, I. W.: Carbon dioxide fixation in malaria. I. Kinetic studies in Plasmodium lophurae. Comp. Biochem. Physiol., 19:855–869, 1966.

83. Voller, A., Garnham, P. C. C., and Targett, G. A. T.: Cross immunity in monkey malaria. J. Trop. Med. Hyg., 69:121–123, 1966.

84. von Brand, T.: Biochemistry of Parasites. New York, Academic Press, 1966.

85. Walker, A. J.: Manual for the microscopic diagnosis of malaria. Sci. Pub. No. 87. Washington, D.C., Pan American Health Organization, 1963.

86. Warren, M., and Bennett, G. F.: Vectoral relationships between the haemosporidia of lower primates. Progress in Protozool., Abstract, pp. 175–176, 1965. New York, London, International Congress Series 91, Excerpta Medica Foundation.

87. Weiser, J.: On the taxonomic position of the genus *Encephalitozoon* Levaditi, Nicolau and Schoen, 1923 (Protozoa: Microsporidia). Parasitol. *54*:749–751, 1964.

88. WHO: Terminology of Malaria and of Malaria Eradication. Report of a Drafting Committee. Geneva, WHO, 1963.

89. WHO: Immunology of Malaria. Techn. Rep. Series No. 396. Geneva, WHO, 1968.

90. Zuckerman, A.: Recent studies on factors involved in malarial anemia. *In* Sadun, E. H., and Osborne, H. (eds.): Research in Malaria. pp. 1201–1216. Supp. to Milit. Med., *131*(9):847–1272, 1966.

Chapter 5

Subphylum Ciliophora

CLASS CILIATEA

Ciliates move by means of hair-like projections of the cytoplasm, called **cilia,** each of which is much like a little flagellum in appearance and structure. The cilia are arranged in various ways in the different groups. The Holotrichida possess simple cilia that may cover the entire surface or be limited to certain areas, whereas other groups possess cilia in limited areas. Cilia may become specialized in function and be grouped into bundles (**cirri**), membranes or tight clusters resembling spines. Zones of cilia around the mouth or cytostome are **oral** in position; zones near the opposite end are **aboral** in position. Some organelles formed by modified cilia are known as **membranelles.**

The order Heterotrichida consists of ciliates usually covered with cilia, but the cilia tend to be reduced on the dorsal surface; an undulating membrane usually exists along the edge of the cytostome. Oligotrichs possess relatively few cilia, but they do have membranelles. The order Peritrichida contains ciliates usually attached to some surface, the attachment being either direct or by means of a stalk. Some stalked forms are colonial. A sucker-like organelle is common with the parasitic or commensal *Trichodina*, whereas the peculiar *Caliperia brevipes* develops arms reaching around the gill filaments of its host (a skate), thus enabling the ciliate to maintain its position by a clasping mechanism (Fig. 5–6).

Beneath the outer surface of the body in all ciliates is a system of granules and fibrils that together are known as the **infraciliature.** Two types of nuclei exist: the large **macro-** nucleus (usually single), and one to many smaller **micronuclei.** Reproduction occurs by both asexual and sexual means. No fusion of independent gametes occurs, but two entire protozoa come together temporarily (**conjugation**) and exchange micronuclei. Most groups of ciliates contain at least a few parasitic species, and some groups are entirely parasitic. Ciliates may be found in or on mollusks, annelids, crustacea, echinoderms, and vertebrates. Only representative families are discussed here.

Subclass Holotrichia

Cilia often cover the entire body and are of equal length; buccal ciliature, if present, is inconspicuous. Encystment is common, and reproduction is by conjugation or transverse fission. This group includes few parasitic forms.[1,3]

ORDER GYMNOSTOMATIDA
Family Bütschliidae

These ciliates have special interest for parasitologists because they live in the alimentary tract of herbivorous mammals. Their size varies considerably, but averages about $35 \times 55\,\mu$. They are oval, barrel-shaped or pear-shaped, covered with cilia, possess a "mouth" or **cytostome** at the anterior end and an "anus" or **cytopyge** at the posterior end. Contractile vacuoles are common. If a drop of fluid from the cecum or colon of horses or from the rumen of cattle, camels or other herbivores is placed on a slide and examined immediately through a microscope, one might see swarming masses of these ciliates. For a discussion of the role these organisms play in the digestive tract of their hosts, see p. 498; see also Hungate.[2] *Bütschlia* is found in cattle, and among the many genera to be found in horses are *Blepharoconus*, *Bundleia*, and *Holophyroides*.

Family Entorhipidiidae

The members of this family are flattened and have a lobe-like anterior end and a tapered or pointed posterior end. They live in the gut of sea urchins, which seem to be especially favored by ciliates. The sea urchins, *Strongylocentrotus*, *Echinus*, *Toxopneustes* and other genera, harbor in their intestines species of *Entodiscus* (Fig. 5–1), *Biggaria*, and *Entorhipidium*.[5]

Among the many ciliates belonging to other families or orders that live in sea urchins are *Colpidium* (Tetrahymenidae), *Anophrys* and *Uronema* (Cohnilembidae), *Plagiopyla* (Plagiopylidae), *Colpoda* (Colpodidae), the hypotrich, *Euplotes*, and the peritrich, *Trichodina*.

Family Isotrichidae

This family is characterized by the possession of a cytostome at or near the apical end of the body and by uniform and complete surface ciliation. The ciliates are characteristically found in the stomachs of ungulate ruminants. There are three genera: *Isotricha*, *Dasytricha* and *Protoisotricha*. *Isotricha prostoma* and *I. intestinalis*, both about $120 \times 65\,\mu$ in size, and *Dasytricha ruminantium* are examples of species. One species of *Isotricha* may be found in the gut of a cockroach.

ORDER TRICHOSTOMATIDA

The cytostome of ciliates in this order is usually situated at the base of an oral groove or pit, the wall of which bears dense cilia.

Family Balantidiidae

The best known member of the genus *Balantidium* is *B. coli*. Other species are *B. suis*, in the pig; *B. caviae*, in the cecum of the guinea pig; *B. duodeni*, in the gut of the frog; and *B. praenucleatum*, a large species reaching $127\,\mu$ in length, found in the colon of the cockroach, *Blatta orientalis*.

Balantidium coli is the only ciliate parasite of man, although coprophilic species may be found occasionally. It is practically worldwide in distribution, lives in the large intestine, and is definitely pathogenic, causing the disease **balantidiasis.** *B. coli* may also be found in monkeys.

The motile form of *Balantidium coli* (Fig. 5–2) is roughly oval in shape and averages about $75 \times 50\,\mu$ in size. The length ranges from 50 to $100\,\mu$, and the breadth, 40 to $70\,\mu$. Rows of cilia cover the body and a peristomal region of slightly longer cilia guards a cleft leading to the cytostome. The macronucleus is a sausage-shaped structure, and the small, dot-like micronucleus lies so close to the macronucleus that it often cannot be seen. The cytoplasm may contain many food vacuoles and two contractile vacuoles, one large and one small. Asexual reproduction occurs by transverse fission. Sexual reproduction is by conjugation. A host diet rich in starch apparently favors the growth of this parasite.

The cyst of *Balantidium coli* is approximately round in outline and measures about $55\,\mu$ in diameter. Inside the cyst wall lies the ciliated

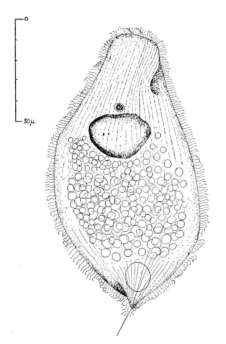

Fig. 5-1. *Entodiscus borealis*, a ciliate from a sea urchin. (Powers, courtesy of Biol. Bull.)

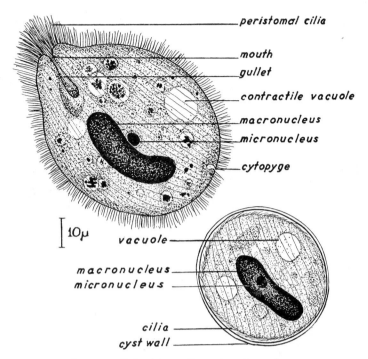

Fig. 5-2. *Balantidium coli*, trophozoite and cyst.

parasite in which can be seen the large macronucleus and usually a vacuole but often little else. The living cysts are pale yellow or greenish in color.

When cysts are eaten, they excyst in the host intestine and the released ciliates begin to feed on cell fragments, starch grains, fecal material and other organic matter. Often they invade the mucosa and submucosa of the large intestine or cecum, causing ulceration. Symptoms of infection include diarrhea, abdominal pain, dysentery, nausea, vomiting, weakness and loss of weight. Diagnosis is confirmed by finding the cysts in stool specimens.

ORDER CHONOTRICHIDA

These peculiar ciliates usually possess the general shape of a vase, and are often stalked. They range in length from about 30 to over 100 μ, and may be found attached to various aquatic animals, especially crustacea. Most of the described species are attached to marine animals, but one species, *Spirachona*

gemmipara, occurs on the gill plates of fresh-water gammarids (amphipods). Young amphipods are heavily infected. Marine genera include *Stylochona*, *Kentrochona* and *Chilodochona*. A marine species, *Oenophorachona ectenolaemus*, is illustrated in Fig. 5–3.

ORDER APOSTOMATIDA

Family Foettingeriidae

Foettingeria actinarum. This European ciliate feeds on material within the gastro-vascular cavity of a sea anemone. It leaves the host and encysts on some object in the sea while undergoing cell division. Products of this division are relased and come to rest on crustacea (copepods, ostracods, and amphipods) as secondary hosts. Here they encyst. Sea anemones receive the ciliate when they feed on infected crustacea.

Chromidina. This genus lives on the renal epithelium and pancreas of squids. It may reach a length of 3 mm. Its macro-nucleus has been described as appearing like

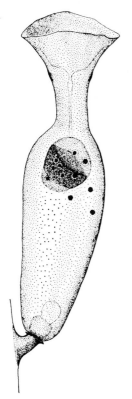

Fig. 5-3. Mature *Oenophorachona ectenolaemus*, a marine chonotrichous ciliate. (Matsudo and Mohr, courtesy J. Parasitol.)

known during the last few years because of studies on protozoan physiology. Some parasitic species show some degree of pathogenicity. The genus *Colpidium* has a representative living in sea urchins. Various species of the genus *Tetrahymena* have been experimentally established in larval and adult insects, guppies, tadpoles, and embryo chicks. In insects, the protozoans live as facultative parasites in the hemolymph. *T. corlissi* was established in guppies and tadpoles through artificially produced wounds. In experimental chicks, the circulatory system, body cavity and muscle tissue may be invaded by two strains of these parasites, whereas the yolk sac or allantoic sac may be invaded by six strains.

Tetrahymena limacis. This parasite lives in the renal organ of the European gray garden slug, *Deroceras reticulatum*. The ciliate can be grown axenically. *Tetrahymena* is also a normal endoparasite of mosquitoes, millipeds, slugs and other animals.

Glaucoma. Members of this genus are usually free-living, but they may occur as facultative parasites of arthropods and in the central nervous system of fish and amphibians. The ciliate's tolerance of CO_2 favors the parasitic habit.

the aimless markings made by a person trying out a new pen.

ORDER ASTOMATIDA

As the name implies, this group is characterized by the absence of a mouth. The organisms are somewhat larger than some of the other groups, usually being between 250 and 350 μ in maximal length, but ranging from 100 to 1,200 μ. Transverse fission and budding are common methods of reproduction and result in the formation of a chain of individuals. They are found in fresh and salt water invertebrates, especially in oligochaete worms.

ORDER HYMENOSTOMATIDA

Family Tetrahymenidae

Most of the members of this family are free-living and the group has become well

Family Ophryoglenidae

Ichthyophthirius multifiliis. This common parasite[4] (Fig. 5-4) lives in the skin of many species of freshwater fish. It is 0.1 to 1.0 mm. long, and may cause serious eruptions that often lead to the death of the fish. The parasites leave the lesions, drop to the bottom of the stream, lake or aquarium, secrete a gelatinous capsule and reproduce by binary fission. Hundreds of new ciliates are formed in each capsule; when released, they are ready to attack new fish.

ORDER THIGMOTRICHIDA

Family Ancistrocomidae

The most interesting characteristic of this group is the presence of a tentacle that enables these ciliates to attach themselves to the body of the host and ingest food. The

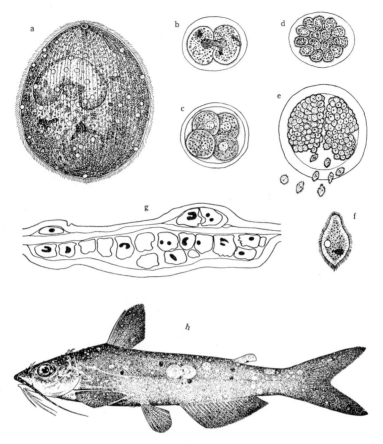

Fig. 5-4. *Ichthyophthirius multifiliis*, a parasite of the skin of fish. *a*. Free-swimming individual. $\times$ 75. (Bütschli) *b* to *e*. Development within cyst. *f*. A young individual. $\times$ 400. (Touquet) *g*. Section through a fin of infected carp showing numerous parasites. $\times$ 10. (Kudo) *h*. A catfish, *Ameurirus albidus* heavily infected with the ciliate. (Stiles) (Kudo, courtesy of Charles C Thomas)

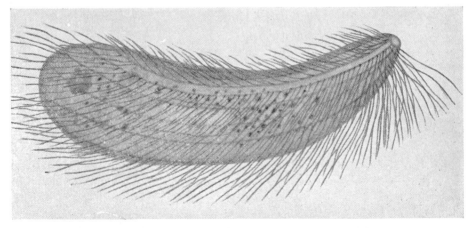

Fig. 5-5. *Ancistrocoma (Parachaenia) myae*, a ciliate from the clam, *Mya arenaria*. (Kofoid and Bush, courtesy of Bull. du Musie Royal d'Histori Naturelle de Belgique.)

attachment is made more secure by the action of thigmotactic cilia, hence the name of the order. The family Ancistrocomidae has many genera, all species of which may be found attached to the gills and palps of mollusks. The ciliates are, in general, oval to elongate with a heavy covering of cilia of fairly uniform size. The anterior tips of the organisms are usually slightly projected. They are comparatively small, ranging from about 15 to 60 μ in the greatest diameter.

Ancistrocoma myae. This transparent, pale green, holotrichous ciliate lives in the excurrent siphon and pericardial cavity of the clam, *Mya arenaria*. It is 40 to 100 μ long (Fig. 5–5).

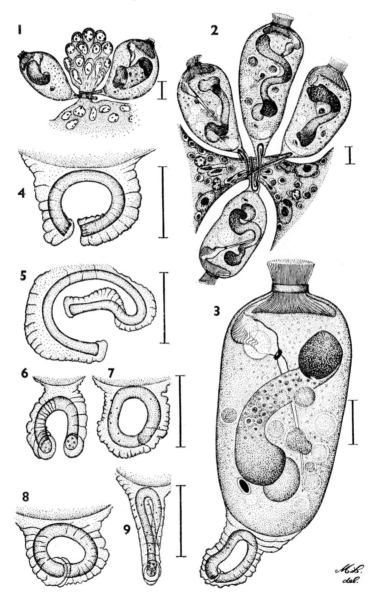

Fig. 5-6. *Caliperia brevipes* from the gills of the skate, *Raja erinacea*. The scale lines to the right of the figures represent 10 μ taken at the same magnification. *1* and *2*. Habit. × 580. *3*. Whole animal, detached from host. × 1450. *4* to *9*. The attachment organelle. × 2340. (Laird, courtesy of Can. J. Zool.)

Subclass Peritrichia

ORDER PERITRICHIDA

The oral surface of these ciliates is flattened and forms a disc that has a counterclockwise spiral of one or more rows of cilia. Many of the free-living species are attached to a substrate by a stalk.

Family Scyphidiidae

Ellobiophrya donacis. This ciliate is so modified that it has developed limb-like posterior projections that join around the trabecula of a molluscan gill and hold the ciliate in place. *Scyphida acanthoclini* similarly attaches itself to the gills of *Acanthoclinus quadridactylus;* and *Caliperia brevipes* (Fig. 5–6) is attached to the gills of skates (Rajidae).

Fig. 5-7. *Trichodina parabranchicola,* a ciliate from the gills of various intertidal zone fish. × 880. The scale line represents 10 μ at the same magnification. (Laird, courtesy Trans. Royal Soc. of New Zealand.)

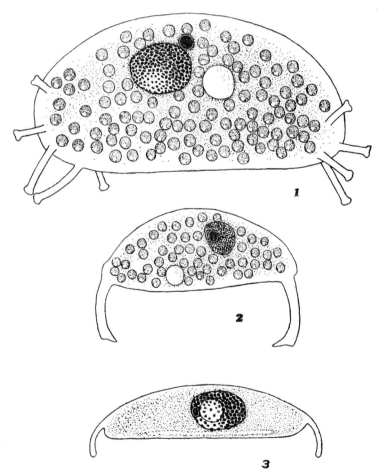

Fig. 5-8. Three species of *Allantosoma,* parasitic suctorea from the large intestine of the horse. × 1707. *1. A. intestinalis. 2. A. dicorniger. 3. A. brevicorniger.* (Hsiung, courtesy of Iowa State Coll. J. Sci.)

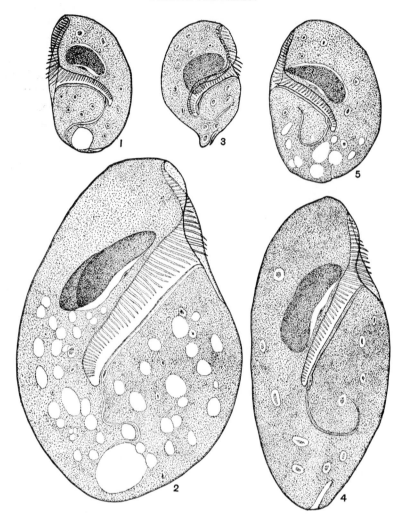

Fig. 5-9. *Nyctotherus cordiformis*, a ciliate found in frog and toad tadpoles.

1. Typical condition of *Nyctotherus cordiformis*, as seen in adult Anuran hosts. Length, 140μ; width, 96μ. *2. Nyctotherus* from tadpole of *Rana clamitans*. Length, 340μ; width, 230μ. Note bilobed nucleus and anterior position of mouth and oblique pharynx. *3. Nyctotherus* from tadpole of *R. catesbeiana*. Length, 132μ; width, 90μ. Note curved pharynx and constricted posterior end.

4. Nyctotherus from tadpole of *R. catesbeiana*. Length, 300μ. Note narrowness, with macronucleus and pharynx almost longitudinal in position. *5. Nyctotherus* from tadpole of *R. clamitans*. Length, 170μ; width, 120μ. Body broadly oval in shape; cytostome forward. (Higgins, courtesy of Trans. Amer. Micros. Soc.)

Family Urceolariidae

Trichodina. Species of this genus (Fig. 5–7) are found in or on fish, amphibians, mollusks, bryozoans, turbellarians, sponges, hydroids, various worms, crustaceans and echinoderms. The aboral end is a flattened disc equipped with rings of cuticular teeth-like skeletal elements. Locomotor organelles consist of posterior membranelles; cirri and undulating velum are present. Most species are ectoparasitic. One of the most well-known species is *T. pediculus*, which "skates" on the surface of the coelenterate, *Hydra*. Economically important species live on various fish, one species being lethal to goldfish. *T. urinicola* inhabits the urinary bladder of amphibians. Another entozoic species, *Urceolaria urechi*, may be found in the intestine of the echiuroid marine worm, *Urechis caupo*, on the United States Pacific coast. Members of this family often feed on bacteria.

Subclass Suctoria

ORDER SUCTORIDA

The name of this group indicates one of its major characteristics: suctorial tentacles. Cilia are absent in the adult stages and most of the species are attached to the substrate by a stalk. *Allantosoma intestinalis* (Fig. 5–8), one of the few parasitic species, occurs in the large intestine and cecum of the horse. In this host the parasite becomes attached to other ciliates. *A. brevicorniger* may also be found attached to ciliates in the colon or cecum of the horses, while *A. dicorniger* lives unattached in the colon of these hosts. The last two species are about 30 μ long.

Another suctorian, *Syphaerophyra sol*, may occasionally be found as parasitic in the ciliates *Stylonychia mytilus*, *Epistylis plicatilis* and *Paramecium aurelia*. In the host cytoplasm the suctorian is rounded, but when it leaves the host it rapidly produces radiating tentacles. *P. aurelia* is seriously affected by the parasite. The cytoplasm becomes vacuolated and the nucleus tends to disintegrate.[6]

Subclass Spirotrichia

The cilia around the cytostome (adoral zone of membranelles) of this group are arranged in a spiral fashion, passing from the right side of the peristome into the cytopharynx.

ORDER HETEROTRICHIDA

Family Plagiotomidae

The peristome of these organisms contains an undulating membrane, and the entire body is ciliated. Parasitic forms may be found in many invertebrates and vertebrates. *Nyctotherus* is one of the best-known genera (Fig. 5–9). These ciliates are, in general, kidney-shaped; the indentation of the body contains a zone of cilia and leads to a cyto-

stome that opens to a ciliated "esophagus." One species, *N. faba*, has been reported from the intestine of man. It is probably not pathogenic, nor are the species that live in animals noticeably harmful. *N. velox* can be found in the milliped, *Spirobolus marginatus; N. ovalis* inhabits the cockroach, *Blatta orientalis; N. parvus* and *N. cordiformis* (Fig. 5–9) live in amphibians. The latter ciliate can often be found in large numbers in the colons of frogs, toads, and tadpoles. This ciliate conjugates in the intestine of tadpoles at the time of metamorphosis of the host. *Plagiotoma lumbrici*, a related species, may be found in the coelom of earthworms, and *Metopus circumlabens* (Fig. 5–10) is an inhabitant of Bermuda sea urchins.

ORDER ENTODINIOMORPHIDA

Throughout this survey of parasitic ciliates several groups have been mentioned as being found in the digestive canals of herbivores. The Entodiniomorphida are especially well-

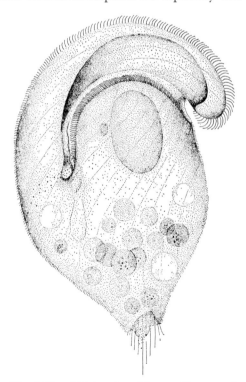

Fig. 5-10. *Metopus circumlabens*, a ciliate parasite in Bermuda sea urchins. (Biggar, courtesy of J. Parasitol.)

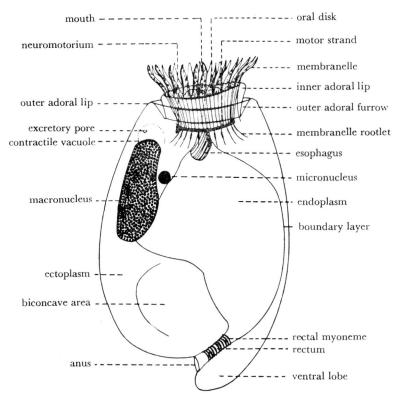

mouth
oral disk
neuromotorium
motor strand
membranelle
inner adoral lip
outer adoral lip
outer adoral furrow
excretory pore
membranelle rootlet
contractile vacuole
esophagus
micronucleus
endoplasm
macronucleus
boundary layer
ectoplasm
biconcave area
rectal myoneme
rectum
anus
ventral lobe

Fig. 5-11. *Entodidium biconcavum*, semidiagrammatic lateral view. (Kofoid and MacLennan, courtesy of University of California Press.)

known as rumen ciliates in cattle, sheep and related hosts. The family Ophryoscolecidae has many species existing in the rumen of cattle and sheep. Representative genera are: *Isotricha* (family Isotrichidae), *Entodinium* (Fig. 5–11), *Diplodinium* and *Ophryoscolex*. Cycloposthiidae are found more often in horses. The elephant, rhinoceros and even gorilla and chimpanzee may play host to members of the latter family. See page 498 for a discussion of these and other ciliates in ruminants. See Hungate[2] for a complete review of the rumen and its microbes.

REFERENCES

1. Berger, J.: Speciation among cryptochilid hymenstome ciliates inhabiting the sea urchin gut. J. Parasit., *13*(Supp. 15):9, 1966.

2. Hungate, R. E.: The Rumen and Its Microbes. New York, Academic Press, 1966.
3. Kozloff, E. N.: Morphological studies on holotrichous ciliates of the family Hysterocinetidae. II. *Craticuloscuta escobari* Gen. Nov., Sp. Nov. and *Epicharocotyle kyburzi* Gen. Nov., Sp. Nov. J. Protozool., *12*:335–339, 1965.
4. Mosevich, T. N.: Electron microscopic study of the structure of the contractile vacuole in the ciliate *Ichthyophthirius multifiliis* (Fouquet). Acta Protozool., *3*:61–68, 1965.
5. Powers, P. B. A.: Studies on the ciliates of sea urchins. A general survey of the infestations occurring in Tortugas echinoids. Papers of the Tortugas Laboratory, Carnegie Institute of Washington, *29*:293–326, 1935.
6. Reyes, R. P., and Lopez-Ochoterena, E.: *Sphaerophrya sol* (Ciliata: Suctoria) parasitic in some Mexican ciliates. J. Parasit., *49*:697, 1963.

Section III
PHYLUM PLATYHELMINTHES

Chapter 6

Introduction; Classes Turbellaria, Trematoda (Subclasses Monogenea, Aspidobothria)

INTRODUCTION

Members of this phylum are usually flattened dorsoventrally without segmentation, and are bilaterally symmetrical. They possess an incomplete digestive tract (except in Cestoidea), lack a body cavity, and are without special skeletal, circulatory, or respiratory structures; they have a flame-cell type of excretory system. The nervous system consists of a pair of anterior ganglia with one to three pairs of longitudinal nerve cords connected by transverse commissures. Both sexes are contained in one individual (with few exceptions), and the organisms undergo direct or indirect development. For a discussion of digestion see Jennings.[4]

Flatworms are divided into three chief classes: **Turbellaria,** which are almost all free-living; **Trematoda,** all parasitic and known as flukes; and **Cestoidea,** all parasitic and known as tapeworms (Chap. 9).

CLASS TURBELLARIA

These flatworms are mostly free-living. They usually have a ciliated epidermis, an undivided body, and a simple life cycle. They are grouped into five orders with many suborders, but only the order Rhabdocoela contains the few turbellarians that are parasitic.

To have one flatworm living as a parasite in another flatworm is rather unusual, but the rhabdocoel *Oekiocolax plagiostomorum* lives in the mesenchyme of a free-living turbellarian, *Plagiostomum.* It causes degeneration of the ovaries of its host. *Plaravortex gamellipara*

is parasitic in the clam, *Modiolus.* Sea cucumbers, sea lilies, sea urchins, sea stars, sipunculids and marine crustacea have all been known to become infected with various rhabdocoels.

Ectocommensal rhabdocoels generally live on the gills or on the body surface of freshwater crustaceans, or more rarely on gastropod mollusks or turtles. Their diet resembles that of free-living species—they are carnivorous and feed on protozoa, rotifers, nematodes, crustacea, etc. *Syndesmis* (Fig. 6–1) represents the relatively few internal parasites found in this order. It lives in the intestine of sea urchins and other echinoderms, and is worldwide in its distribution. Its length is commonly 3 to 4 mm.

Suborder Temnocephalidia

This suborder is composed of small worms with tentacles and sucker-like structures (Fig. 6–2); they live on crustaceans and a few other animals. *Fecampia* is a genus that is pathogenic to its host, a European marine crustacean. The ciliated, immature parasite enters the hemocoel of its host and undergoes profound modification, losing most of its viscera but retaining its intestine. In the crustacean it reaches sexual maturity, then leaves its host, produces eggs and dies.

CLASS TREMATODA

These flatworms are all parasitic, some living on the surface of their hosts and some inside. The body is covered by a tegument that is usually smooth but may be **spiny.**

129

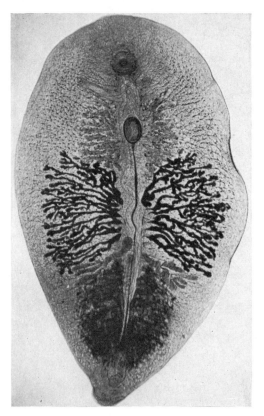

Fig. 6-1. *Syndesmis* sp., a turbellarian parasite from the intestine of the sea urchin *Strongylocentrotus purpuratus*. (Preparation and photograph by Anthony T. Barnes.)

There are one to several suckers on almost all species. The life cycle of the first subclass, Monogenea, involves one host, whereas that of the second subclass, Digenea, involves two or more hosts. Almost all digenetic trematodes are internal parasites, but a few are ectoparasitic (e.g., adult *Transversotrema* are found under the scales of scorpid marine fish in Australia).

Fish bear the heaviest burden of fluke parasites. Birds come next, having about three times as many kinds of flukes as occur in amphibians, and reptiles, and mammals. A single host may harbor only one species, but hundreds of individuals.

The study of trematodes of fish is a field wide open for investigators, and many of these flukes have yet to be described. Adult flukes of numerous species occur in the intes-

tines or other organs of fish, and larval forms are commonly embedded in skin, gills, mesentery, muscles, liver and other organs. In addition, monogenetic species may be attached to the skin, gills, or mouth parts. The monogenetic species apparently do little harm. The Digenea may or may not be injurious, depending on the numbers of individual worms and the organ infected. Penetration and encystment of some larvae elicit the production of melanophores that surround the parasites. This host reaction is probably a defense mechanism.

Subclass Monogenea

Although these flukes are usually found as ectoparasites on the lower vertebrates, especially fish, some inhabit the gill chambers, mouth cavity, urinary bladder, cloaca, ureters or body cavity. Contrary to the browsers on the skin, gill dwellers are sedentary because their food, the host blood, is abundantly available. The life cycles of monogenetic trematodes do not involve more than one host. Adult parasites are attached to the host by a modification of the posterior end known as a **haptor** or, more accurately, **opisthaptor** (the haptor at the anterior end is the **prohaptor**). The opisthaptor normally possesses suckers or hooks or both, and effectively holds the parasite to the skin or gills of the host. In general, an egg hatches into a larva that swims to its host, attaches to the skin, gills or elsewhere and gradually changes by metamorphosis into the adult.

The morphology of monogenetic trematodes is fundamentally similar to that of digenetic trematodes; however, there are differences that arise from the peculiar mode of life of the monogenetic trematodes (Fig. 6-3). These differences are sufficiently great to consider seriously placing these worms with the cestodes. (See page 561 for a discussion of this question.)

Adult Monogenea range in length from a few millimeters to 2 or 3 cm.; the body outline varies from spindle-shaped to circular. The most striking feature is the opisthaptor. This organ usually has one or more suckers or

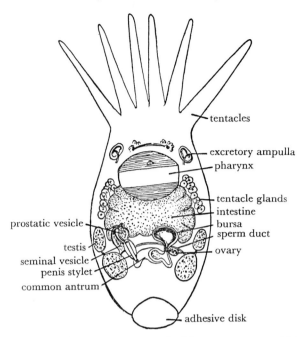

Fig. 6-2.   *Temnocephala* (after Haswell, 1893). (Hyman, courtesy of McGraw-Hill.)

cups used to attach the worm to the host. The Greek name for cup is **cotyl,** and many of the nomenclatural terms for members of this group incorporate this word. For example, the genus *Cyclocotyla* includes monogenetic trematodes that possess a ring of suckers on the opisthaptor. The simpler forms of Monogenea may not have a fully-formed sucker, but only an expanded posterior end of the body. As the number of suckers increases, the posterior part of the body becomes larger and forms a disc. This disc may be divided radially by septa, and the whole organ, including septa and suckers, is activated by muscles. Added to this complex are hooks, also activated by muscles, which can be extended or withdrawn somewhat like cat claws. These organs form an adhesive structure or holdfast that helps the parasite maintain its precarious perch on the host.

The anterior end of the parasite also usually possesses one or more suckers, ordinarily not as well developed as the oral sucker of digenetic trematodes. Sometimes, various other head organs, such as lappets, glandular areas, and extensions, exist instead of suckers.

The function of these organs is primarily adhesive, probably to hold the mouth to the region of feeding.

The mouth is normally located anteriorly, but it may be near the middle of the body. The shape of the opening ranges from slit-like to circular. It may be in, near or removed from the anterior sucker or suckers. On the basis of histochemical examination, Uspenskaya[8] has shown that most monogenes feed principally on host blood.

Male and female genital pores may open together or separately, but they usually occur closely associated on the ventral surface toward the anterior end of the animal. The tegument of the body is generally smooth and is pierced by a few other openings, such as the excretory ducts and the ducts from secretory glands.

Inside the parasite (Fig. 6-4) there is no body cavity, the various organs lying in a sort of packing tissue, the **parenchyma.** Like the digenes, the Monogenea carry on excretion with the aid of flame cells, tubules and vesicles. The digestive tract consists of the mouth, pharynx, esophagus, and gut.

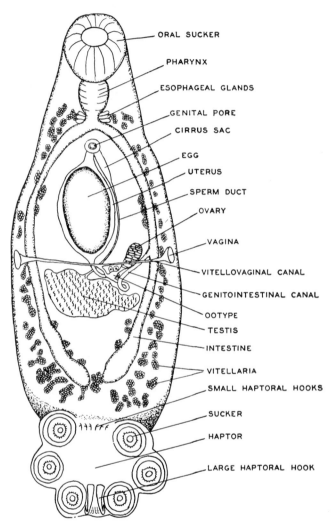

ORAL SUCKER

PHARYNX

ESOPHAGEAL GLANDS

GENITAL PORE

CIRRUS SAC

EGG

UTERUS

SPERM DUCT

OVARY

VAGINA

VITELLOVAGINAL CANAL

GENITOINTESTINAL CANAL

OOTYPE

TESTIS

INTESTINE

VITELLARIA

SMALL HAPTORAL HOOKS

SUCKER

HAPTOR

LARGE HAPTORAL HOOK

Fig. 6-3. *Polystomoidella oblongum,* a monogenetic trematode from the urinary bladder of turtles. (Cable, courtesy of Burgess Publishing Co.)

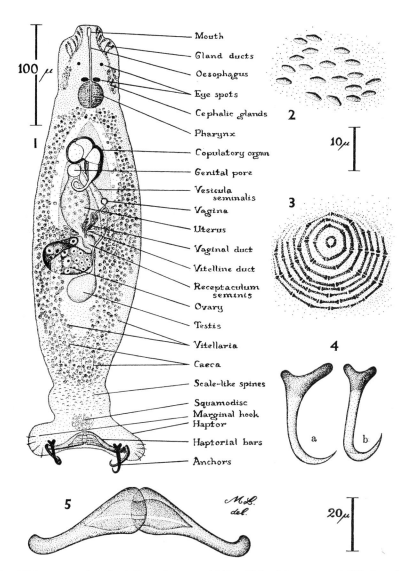

Mouth
Gland ducts
Oesophagus
Eye spots
Cephalic glands
Pharynx
Copulatory organ
Genital pore
Vesicula seminalis
Vagina
Uterus
Vaginal duct
Vitelline duct
Receptaculum seminis
Ovary
Testis
Vitellaria
Caeca
Scale-like spines
Squamodisc
Marginal hook
Haptor
Haptorial bars
Anchors

100 μ

1

2

10μ

3

4

a   b

5

M.L.
del.

20μ

Fig. 6-4. *Diplectanum melanesiensis* from the serranid fish, *Epinephelus merra*. *1.* Whole animal, ventral view. *2.* Scale-like spines of posterior part of body. *3.* Squamodisc. *4.* (*a*) Dorsal and (*b*) ventral anchors. *5.* Haptorial bars. (Laird, courtesy of Can. J. Zool.)

The gut may be simple or branched, sometimes with innumerable small blind pouches or **ceca.** There is no anus. The worms are hermaphroditic and the reproductive systems are basically the same as those of digenetic trematodes (Chap. 7). There may be a single testis or many testes. Sperm pass through the vasa efferentia to the vas deferens and are assisted from the body by a copulatory organ. Female structures are more complicated. The ovary is often branched or folded, and several organs providing nourishment and shell material to the egg and ensuring its fertilization are present. Figure 6–5 presents a diagrammatic representation of the genitalia and egg of a monogenetic trematode. Egg formation is essentially similar to that described for digenetic trematodes. The term **oncomiracidia** refers to larvae of all monogenetic trematodes.[6] Larvae find their host by chemical attraction.[5]

For general reference to monogenes, see Bychowsky,[2] Baer and Euzet,[1] Sproston,[7] and Yamaguti.[10] For a bibliography of monogene literature, see Hargis *et al.*[3]

Order Monopisthocotylea

An oral sucker is lacking or weakly developed in this group. The anterior end often possesses two or more lobes formed by clusters of adhesive glands. The opisthaptor consists of a prominent disc armed with two or three pairs of large hooks (anchors) and up to 16 marginal hooklets.

Gyrodactylus. This viviparous member of the group (Fig. 6–6) will serve as a representative species. It lives on the surface of freshwater fish and frogs, infesting host organs of locomotion and respiration especially. The larvae develop within the uterus of the parasite and may contain clusters of embryonic cells. The opisthaptor of the adult carries no suckers, but has a row of 16 small hooks along its edge and a large pair of hooks in the center.

Marine invertebrates may also be parasitized by monogenes belonging to this group. Squids are host to the genus *Isancistrum*, and the copepod *Caligus* may have clumps of *Udonella* attached to it. The latter parasite possesses two anterior suckers and a simple opisthaptor. The copepod is itself a parasite on marine fish.

Order Polyopisthocotylea

In this group, the mouth is surrounded by the prohaptor, which consists of one or two suckers or two pits. The opisthaptor may or may not be armed, but always has suckers or sucker-like bodies containing clamps. These parasites are almost exclusively blood feeders.

MONOGENETIC TREMATODE

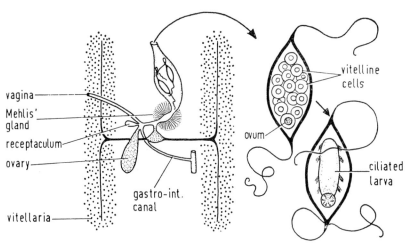

Fig. 6-5. Diagrammatic representation of genitalia and egg of a monogenetic trematode. (Smyth and Clegg, courtesy of Exper. Parasitol.)

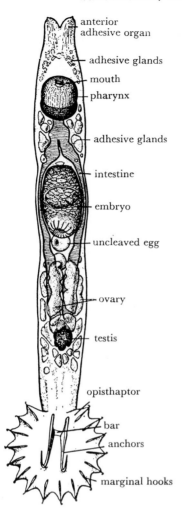

anterior
adhesive organ

adhesive glands

mouth

pharynx

adhesive glands

intestine

embryo

uncleaved egg

ovary

testis

opisthaptor

bar

anchors

marginal hooks

Fig. 6-6. *Gyrodactylus.*
(Hyman, courtesy of McGraw-Hill.)

Polystoma integerrimum. Endoparasitism is not common among the Monogenea, but there are some well-known examples. *P. integerrimum* lives in the urinary bladder of frogs and toads. Larval flukes usually attach themselves to the gills of tadpoles and remain in arrested growth until host metamorphosis takes place, when they migrate to the intestine and into the urinary bladder of the adult frog. The parasite's eggs are produced in the spring of the fourth year.[9] In some instances, the larval fluke attaches itself to a very young tadpole and matures as a neotenic (sexually mature) worm, and begins to

lay eggs by the end of the fifth week. These worms are smaller than the endoparasitic ones, and they die when their tadpole hosts metamorphose. This phenomenon of two possible generations, a neotenic one on young tadpoles and an endoparasitic one in older tadpoles and adult frogs, is reminiscent of the two possible life histories of the roundworm, *Strongyloides stercoralis* (p. 282).

Diplozoon ghanense. This species (Fig. 6–7) lives on the gills of the fish *Alestes macrolepidotus*. Its greatest distinction is that, although it is hermaphroditic, a permanent union of the worms occurs. During the larval stage of this fluke a small fleshy knob appears on the dorsal surface. Eventually this knob becomes fitted into a ventral sucker of another larval worm. The two worms become securely fused together and cannot be separated. The gonads then begin to develop, and finally the vagina of one individual opens in the region of the uterus and vas deferens of the other. This arrangement is reciprocal and cross fertilization is made easy.

The family Microcotylidae is represented in Figures 6–8 and 6–9 by three species from Gulf of Mexico fish. Some peculiarities of this group are well illustrated. Sucker-like clamps on the ends of four pairs of lateral peduncles occur in *Choricotyle louisianensis*, a parasite of fish.

Subclass Aspidobothria

This subclass holds an uncertain position between the monogenetic and the digenetic trematodes, but it is closer to the Digenea. The group is divided into two families, Aspidogastridae and Stichocotylidae. The aspidogastrids are characterized by an enormous circular or oval sucker occupying the greater part of the ventral surface (Fig. 6–10). They are marine or freshwater parasites of the mantle and pericardial and renal cavities of clams and snails, of the gut of fish and turtles, and of the bile passages of fish. These parasites develop directly from free-swimming larvae, but details of their development are obscure. The stichocotylids, comprising the single genus *Stichocotyle*, are elon-

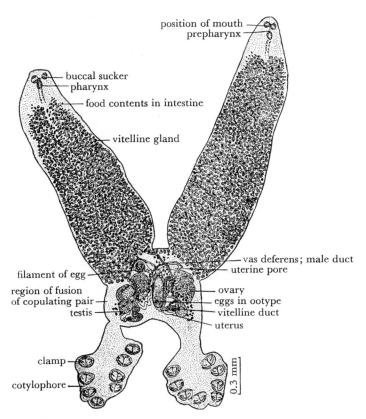

Fig. 6-7. Two *Diplozoon ghanense* in permanent copulation.
(Thomas, courtesy J. West. African Sci. Assoc.)

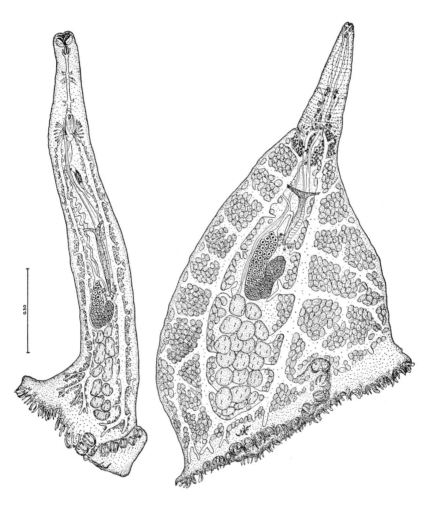

Fig. 6-8. *Axinoides raphidoma* on the left, *A. truncatus* on the right. Both are monogenetic trematodes from Gulf of Mexico fish. (Hargis, courtesy of Proc. Helmin. Soc. Wash.)

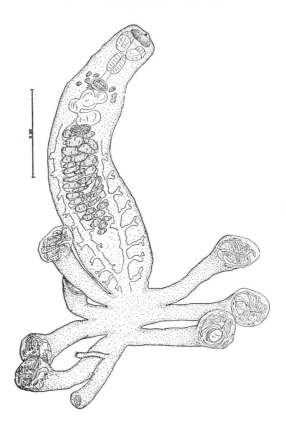

Fig. 6-9. *Choricotyle louisianensis*, from the gills of Southern Whiting. (Hargis, courtesy of Trans. Amer. Micros. Soc.)

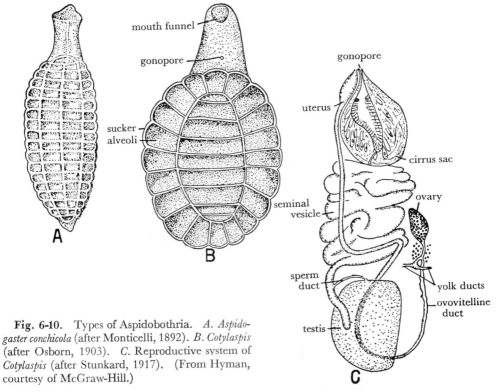

Fig. 6-10. Types of Aspidobothria. *A. Aspidogaster conchicola* (after Monticelli, 1892). *B. Cotylaspis* (after Osborn, 1903). *C.* Reproductive system of *Cotylaspis* (after Stunkard, 1917). (From Hyman, courtesy of McGraw-Hill.)

gate slender worms, about 10 cm. in length, parasitic in the bile passages or spiral valve of skates. The subclass is sometimes called Aspidocotylea or Aspidogastrea. For a discussion of the evolutionary relationships of this group, see pages 561 and 562. For a systematic account, see Yamaguti.[10]

REFERENCES

1. Baer, J. G., and Euzet, L.: Classe des Monogènes. Monogenoidea Bychowsky. *In* Grassé, P.: Traité de Zoologie. Tome IV, pp. 243–325. Paris, Masson et Cie Editeurs Libraires de l'Acad. Med., 1961.

2. Bychowsky, B. E.: Monogenetic Trematodes. Their Systematics and Phylogeny. (Edited by W. J. Hargis, Jr., trans. by P. C. Oustinoff.) Washington, D.C., American Institute of Biological Sciences, 1957.

3. Hargis, W. J., Lawler, A. R., Morales-Alamo, R., and Zwerner, D. E.: Bibliography of the Monogenetic Trematode Literature of the World. Virginia Institute Marine Science, Special Scientific Report 55, Gloucester Point, Virginia.

4. Jennings, J. B.: Digestion in flatworms. *In* Florkin, M., and Scheer, B. (eds.): Chemical Zoology. pp. 308–326. New York, Academic Press, 1968.

5. Kearn, G. C.: Experiments on host-finding and host-specificity in the monogean skin parasite *Entobdella soleae*. Parasitology, *57*: 585–605, 1967.

6. Llewellyn, J.: The larvae af some monogenetic trematode parasites of Plymouth fishes. J. Mar. Biol. Assoc. U. K., *36:* 243–259, 1957.

7. Sproston, N. G.: A Synopsis of the Monogenetic Trematodes. London, Zoological Society, 1946.

8. Uspenskaya, A. V.: Nutrition of monogenetic trematodes (In Russian). Dokladi Akademii Nauk SSSR, *142*:1212–1215, 1962.

9. Williams, J. B.: The dimorphism of *Polystoma integerrimum* (Frölich) Rudolphi and its bearing on relationships within the polystomatidae: Part III. J. Helminthol., *35*:181–202, 1961.

10. Yamaguti, S.: Systema Helminthum. Vol. IV. Monogenea and Aspidocotylea. New York, Interscience, 1963.

Chapter 7

Class Trematoda, Subclass Digenea (Introduction)

GENERAL CONSIDERATIONS

Morphology

This subclass is characterized by cuplike muscular suckers, usually without hooks or other accessory holdfast organs; by genital pores that normally open on the ventral surface between the suckers; and by a single, posterior excretory pore.

The classic shape of a digenetic trematode, or fluke, is that of a thick, oval leaf. However, many variations exist, from those resembling a short piece of pencil, tapered at both ends, to narrow ribbons elongated to 40 feet (about 12 m.) or more (see p. 202). In contrast, some species (*Euryhelmis*) are wider than they are long. Although the presence of two suckers is typical, only one occurs on some flukes and a few species lack both. Normally one sucker surrounds the mouth and is called the **oral** or **anterior sucker.** The other sucker is called the **acetabulum, posterior sucker,** or **ventral sucker. Monostomes** are flukes with one sucker; **distomes** possess two suckers. The general pattern of fluke anatomy is illustrated in Figure 7–1, which can serve to identify the organs used in diagnosis.

Tegument

This is a metabolically active, cellular unit with an outer layer, the **epicuticle,** which is anuclear, syncytial, and connected by narrow cytoplasmic tubes to the nucleated regions of the tegument and through the basal membrane and muscle layers of the body to cytoplasmic masses lying in the parenchyma. Microvilli, as well as pinocytotic vesicles, have been identified on the outer surface of larvae and adults. The tegument of the liver fluke, *Fasciola hepatica*, contains both mitochondria and Golgi bodies.[43] All of these structural features suggest secretory and excretory functions.[13] The absorption of glucose takes place through the tegument, thus assisting in the general nutritional activities of the worm. The tegument is also associated with respiratory and sensory functions. It is usually resistant to the action of pepsin and trypsin, probably because of the presence of acid mucopolysaccharides and polyphenols. This resistance is probably a main factor in protecting the worms from being digested by the host.

A histochemical analysis of the tegument of *Fasciola hepatica* and *Echinostoma revoltum* demonstrated that it "is formed by a lipoprotein complex. The characteristic granules in the proper cuticular layer consist of an acid mucopolysaccharide. . . . The scales and spines of these trematodes are formed by a scleroproteid with a characteristically high content of only [sic] cysteine and also with a multitude of lipoid substances."[40] For an account of the fine structure of the tegument of developmental and adult stages of trematodes, see Bils and Martin.[3]

Respiration

Adults in the intestine or other organs where little or no oxygen exists are anaerobic. However, flukes can utilize oxygen when it is available, so these parasites are called **facultative anaerobes.** The respiration rate of some digenetic trematodes was found by Vernberg[44] to be directly proportional to the

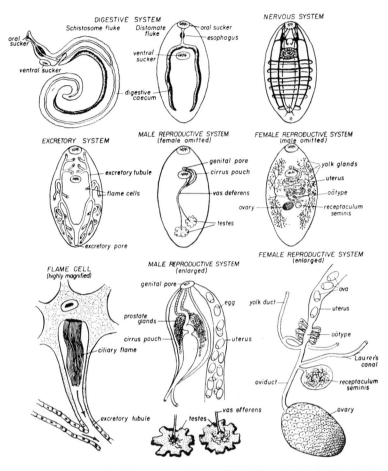

Fig. 7-1. Morphology of adult human flukes. (U.S. Navy Medical School Manual.)

oxygen tension down to a partial pressure of three per cent or lower. Below this level, the parasite was able to regulate the rate of respiration.

Nervous System

Paired ganglia in the anterior end of the body serve as a brain, and from it lead various main nerve trunks which, in turn, lead to branches innervating all parts of the body.

Digestive System

The mouth leads to the pharynx, esophagus and gut. As in the Monogenea, the gut may be simple or highly branched with many ceca. Working with rediae of *Cryptocotyle lingua*, Krupa *et al.*[25] found many highly flex-

ible ribbon-like folds extending from the epithelial cells of the intestine into the lumen. Nutritive phagocytosis was suggested. There usually is no anus, although in a few species (Echinostomatidae) an opening exists between the ceca and the excretory vesicle. Some flukes of fish have one or two anal pores that open to the outside.

The mode of feeding is suctorial, associated with the attachment process of the oral sucker and the muscular pharynx.[14,17] Enzymatic secretions of the parasites that act on host tissues might play a role. Gut-dwelling species may feed on epithelial tissues and associated mucoid secretions of the host. Apparently there are differences in the digestive processes related to differences in the cells and microvilli lining the gut of the parasite. The gastrodermis is capable of both secretion and absorption. Digestion is primarily an extracellular process.

Erasmus and Öhman[14] found that in the fluke *Cyathocotyle bushiensis*, parasitic in the cecum of ducks, the ventrally situated adhesive organ contains gland cells rich in protein and RNA. These cells produce enzymes consisting of alkaline phosphatase, esterases of several kinds, and leucine aminopeptidase. The enzymes are effective in freeing the columnar epithelium of the host, and the cells thus freed become reduced to a granular material that is taken into the oral sucker of the fluke and digested intracellularly within its cecum. The adhesive organ of the parasite thus serves for attachment and for extracorporeal digestion.

Excretory System

Typical of all flatworms, this system consists of flame cells (flame bulbs) connected by tubules uniting to form larger ducts that either open independently to the outside or join to form a urinary bladder that opens at or near the posterior end of the animal (Fig. 7–2). The flame cells and ducts function not only for excretion, but also for water regulation and possibly to keep body fluids in motion. The position of flame cells is indicated by a formula. For example, $2[(3+3+3) \quad (3+3+3)]$ indicates that on each side of the body there are three groups of three flame cells both in the anterior portion of the body and in the posterior portion (Fig. 7–3). These distribution patterns of flame cells are of value in species diagnosis (see Komiya[24]).

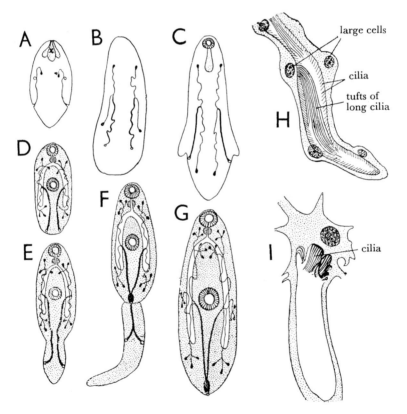

Fig. 7-2. The excretory system of Digenea. *A.* Miracidium. *B.* Sporocyst. *C.* Redia. *D,E,F.* Stages in development of the cercaria. *G.* Metacercaria. *H.* Tufts of long cilia and large cells forming the ciliated wall of the canal (not seen in the adult). *I.* Young stage flame cell from *Dicrocoelium dendriticum.* (Dawes, courtesy of University Press, Cambridge.)

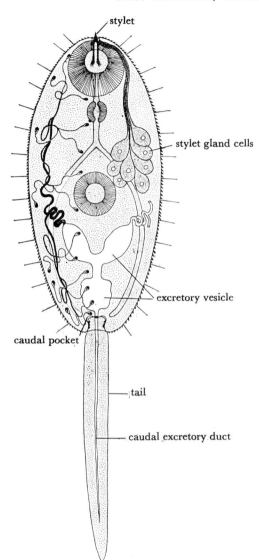

stylet

stylet gland cells

excretory vesicle

caudal pocket

tail

caudal excretory duct

Fig. 7-3. Cercaria of *Plagiorchis* (= *Multiglandularis*) *megalorchis* from the turkey. Dorsal view, showing stylet glands on right and excretory system on left.

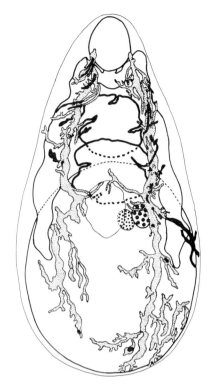

Fig. 7-4. *Paramphistomum bathycotyle*. Morphology of the lymphatic system. Dorsal view. (Lowe, courtesy Zeit. Parasitenk.)

The excretory systems of certain cercariae and metacercariae contain concretions composed chiefly of calcium carbonate and a trace of phosphate. These concretions may be active in the fixation of carbon dioxide and in the buffering of acids, but their function is not known with certainty.[31]

"Packing tissue" of the body is a mass of parenchymal cells that function in secreting the interstitial material and may also be responsible for storing glycogen and, in the absence of a circulatory system, for transporting metabolic products. A "lymphatic" system[30] occurs in the digenetic family Paramphistomatidae and ten others, and in the monogenetic family Sphyranuridae. A tubular system distinct from the excretory system, the lymphatic system consists of one pair of longitudinal canals branching repeatedly to supply the various tissues and organs (Fig. 7-4). For development of the excretory system, see p. 156.

Reproductive System

The testes and ovary may be rounded or branched. Usually there are two testes, but some digenetic flukes may have more than 100 (e.g., in some avian schistosomes). Most species in the family Monorchiidae possess a single testis. An enlarged detail of the distal

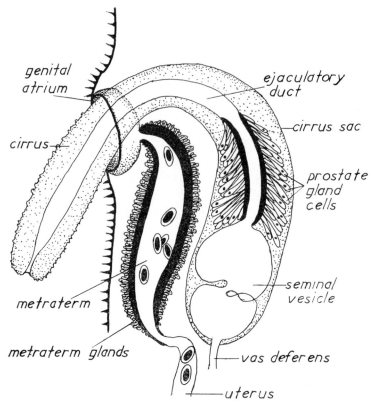

Fig. 7-5. Genital opening in a fluke showing the point where eggs emerge from the body and an extended male organ, the cirrus, which transmits sperm to another worm.

end of the male reproductive system (Fig. 7-5) shows the apparatus used to transfer sperm to the female part of another worm or perhaps to the female part of the same worm. The **cirrus** is analogous to the penis of higher animals. In the female system, **Laurer's canal** may be a vestigial duct corresponding to the vaginal canal in tapeworms. When it functions as a sperm storage organ, a seminal receptable or **receptaculum** is also usually present. In a few flukes, this canal has been shown to be a copulatory structure.

Vitellaria (vitelline glands) consist of yolk and eggshell producing cells. The large amount of this gland is probably associated with considerable protein demand for production of eggshells. The vitellaria may be dispersed (Fig. 7-6) or clumped; they are connected by minute ducts to larger vitelline ducts, which eventually form main channels that come from each side of the organism. These channels meet near the midline, forming the **vitelline reservoir.** The common vitelline duct coming from the reservoir joins the oviduct, and beyond this region it enlarges to become the **ootype.** The ootype is surrounded by a mass of minute gland cells known together as **Mehlis' gland** (Fig. 7-7). For a long time this gland was considered the main contributor of the materials that make up the eggshell, and therefore was often called the "shell gland." Now, however, there is ample evidence that Mehlis' gland cannot be directly concerned with secreting the bulk of the eggshell material.[41] It has been suggested that secretions of Mehlis' gland may lubricate the uterus and thus aid the passage of eggs along this tube, and also that the secretions might activate spermatozoa. A more recent study[18] showed that this gland

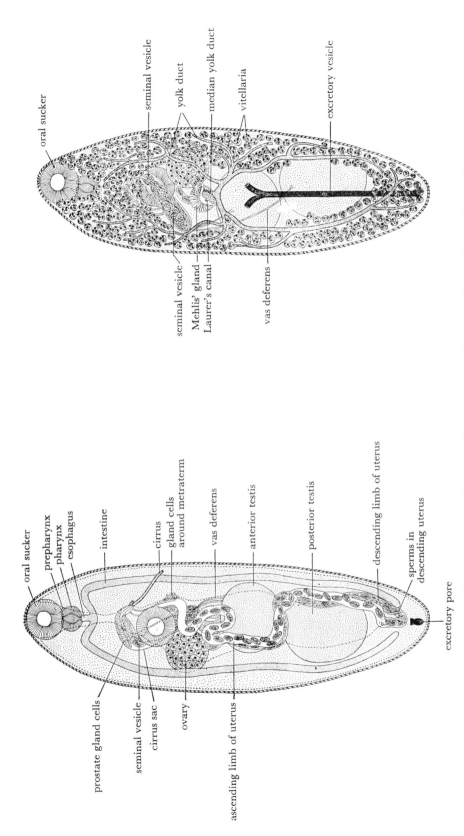

Fig. 7-6. *Plagiorchis (Multiglandularis) megalorchis*, showing the dispersed distribution of the vitella-ria and vitelline ducts. (Rees, courtesy of Parasitology)

in tapeworms secretes a phospholipid that may effect the release of eggshell precursors from vitelline cells. Undoubtedly, the function of the gland in cestodes is similar to its function in trematodes.

The diagrammatic representations of the genitalia of digenetic trematodes (Fig. 7–8) and of eggshell formation (Fig. 7–7) are taken from Smyth and Clegg,[41] as is the following description of the method of shell formation. Mature eggs leave the ovary and pass down the short oviduct, being fertilized by sperm that have been stored in the receptaculum or in Laurer's canal. The vitelline reservoir is filled with unusual cells that have been produced by the vitelline glands. These vitelline cells possess a nucleus, a mass of yolk material, and granules that are destined to produce the eggshell. These granules have been called **vitelline globules** but are better named **shell globules.** As the egg moves through the ootype a group of vitelline cells surrounds it, the number of vitelline cells being constant for each species of fluke. Shell globules are extruded from the vitelline cells and coalesce, thus forming

a thin membrane. This membrane is the outer portion of the eggshell. More shell globules are released and the eggshell is built up from within. Completion of the shell occurs in the lower part of the oviduct. The eggshell material is protein and it soon becomes brownish in color and hardens.

A different method of eggshell formation occurs in *Syncoelium*. In this fluke, the egg may possess a thin membrane formed by Mehlis' gland and/or the vitelline gland, but the bulk of the eggshell is furnished by the uterus, which has large gland cells containing precursors to the eggshell. The vitelline cells of *Parastrigea* also contribute copious amounts of glycogen and RNA.[9]

LIFE CYCLES

In vertebrate hosts, adult flukes live in the digestive tract, in ducts associated with the alimentary canal, in the blood, lungs, gallbladder, urinary bladder, or in almost any other organ of the body. The first cleavage of the zygote in an adult worm produces a germinal cell and a somatic cell. The latter divides further to form the soma of the next

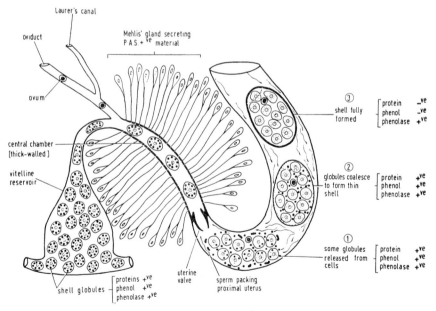

Fig. 7-7. Diagrammatic representation of eggshell formation in a trematode, based on *Fasciola hepatica*. The process is essentially the same in tapeworms. (Smyth and Clegg, courtesy of Exper. Parasitol.)

DIGENETIC TREMATODES

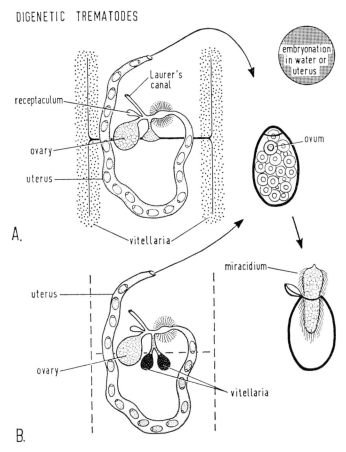

Fig. 7-8. Diagrammatic representation of female genitalia of digenetic trematodes. *Type A,* with extensive vitellaria; *Type B,* with condensed vitellaria. (Smyth and Clegg, courtesy of Exper. Parasitol.)

generation, and the germinal cell forms reproductive organs. This germinal lineage is preserved throughout the life cycle of trematodes.[37] Eggs usually leave the host through the intestine. If they get into water, they are either eaten by a snail, in which they hatch, or they hatch in the water and become free-swimming, ciliated **miracidia** that penetrate the snail. Miracidia are equipped with penetration glands, an excretory system, germ cells, and in some cases, an eye spot (Fig. 7–10).

Miracidia freed from their capsules (eggshell) dart about randomly. They show great flexibility and continual peristaltic changes of form. Their general behavior suggests that they are "seeking something with

feverish haste."[2] They find a snail, clam, copepod or, rarely, an annelid host by chemotactic attraction or by some other means. There is good evidence that at least some species perceive and are attracted by the mucus of a snail.[16,23,46] Many miracidia, however, reach their intermediate hosts purely by chance. Miracidia often demonstrate a high degree of selectivity for their hosts. Age of the mollusk is probably a controlling factor in the selection and entry by miracidia.

In a study of the compatibility and incompatibility concept as related to trematodes and mollusks, Cheng[6] concluded that "ambient environmental conditions as well as innate taxes, when such occur, are generally stronger determinants of miracidial migration and behavior than are chemotactic forces. Thus it is only when the other factors serve to bring miracidia into the immediate vicinity of the mollusc that chemotaxis becomes an effective attractant. There is also evidence which suggests that materials extruded from molluscs may inhibit rather than serve to enhance post-parasite contact."

Entry of miracidia into a snail host is either active or passive. If the egg is small, as is true with members of the fluke families Heterophyidae, Opisthorchiidae, Brachylaimidae, and Plagiorchiidae, it is usually eaten by the snail, whether aquatic or terrestrial, or sometimes by a bivalve mollusk. The miracidium emerges from the egg in the snail's intestine. Since snails readily eat

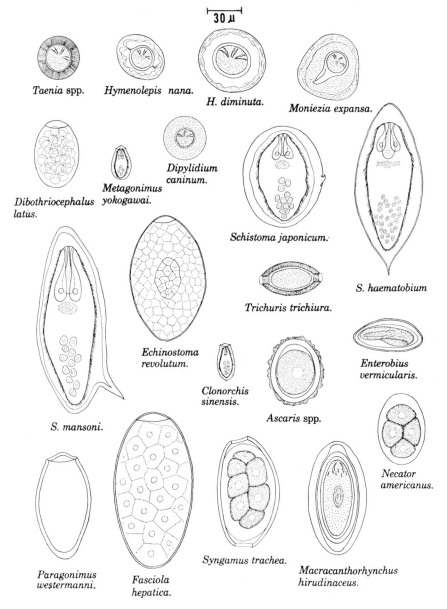

Fig. 7-9. Representative parasitic worm eggs of man and of animals. (Schell, courtesy of John Wiley and Sons)

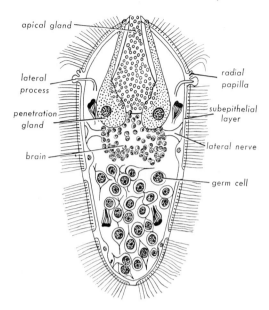

apical gland

lateral process

penetration gland

brain

radial papilla

subepithelial layer

lateral nerve

germ cell

Fig. 7-10. Miracidium of *Schistosomatium douthitti* to show internal anatomy. (After Price) (Smyth, The Physiology of Trematodes. University Reviews in Biology. Courtesy of Oliver and Boyd Ltd.)

feces, there is considerable opportunity for the passage of fluke eggs into their intestines. If the egg is of medium size or large (e.g., Schistosomatidae, Fasciolidae, Troglotrematidae), it hatches in water or moist soil and the miracidium must find the right species of snail to enter. It penetrates through the body surface in the region of the mantle, "foot," head or tentacles. According to some investigators, miracidia probe with their anterior end and attempt to bore into almost any object, including various species of unsuitable snails and other animals such as planarians. They may keep trying to enter an unsuitable host or an unsuitable part of the snail host, such as the shell, until they die of exhaustion. They seem to find the right host by accident, and then succeed in penetrating it simply because their boring mechanism is adapted to succeed only with that host. Most snails, however, probably secrete a substance that attracts the parasite.

A somewhat different description of the method of entry has been given by Dawes

for *Fasciola gigantica*.[11] He stated that a preliminary attachment of the parasite to the host's body occurs by suctorial action. Attachment takes place at an unciliated region of the miracidium. The parasite produces a cytolytic substance that breaks down host tissue and makes a perforation in the skin. The cytolytic substance is produced by penetration glands, formerly called a "gut." The miracidium usually sheds its cilia and thrusts itself into the snail's body.

Within the snail, one or more developmental stages occur: **sporocyst, redia** and **cercaria** (Fig. 7–11). The first two stages can be considered germinal sacs (Fig. 7–12), and the first division of the germinal cell in the germinal sac produces a somatic cell and a germinal cell. The sporocyst usually produces a second generation of sporocysts before rediae are developed. Likewise, each redia may produce a second redial generation before cercariae are formed. Sometimes the redial generation is suppressed and sporocysts give rise to cercariae (Figs. 7–13, 7–14). Cercariae leave the snail host and usually develop into encysted **metacercariae** in a second intermediate host. Metacercariae develop into adult worms. In a few groups another stage, an unencysted **mesocercaria,** is formed in a third intermediate host. Rarely, metacercariae are formed on aquatic vegetation.

Thus, we can divide the digenetic life cycle into three generations: the miracidium mother-sporocyst generation, the daughter-sporocyst/redia generation, and the cercaria/metacercaria/adult generation. James and Bowers[22] have reviewed numerous facts and theories concerning digene life cycles and they suggest that the cycles include a cyclic alternation of homologous generations. Some details of these developmental stages follow.

A sporocyst may be short or branched, or resemble a long, irregularly coiled tube (Fig. 7–14). A few sporocysts (e.g., *Heronimus*) may pulsate. (See *Leucochloridium*, p. 175.) In some species, at least temporary contact with a snail is apparently all that is necessary for a miracidium to metamorphose into a sporo-

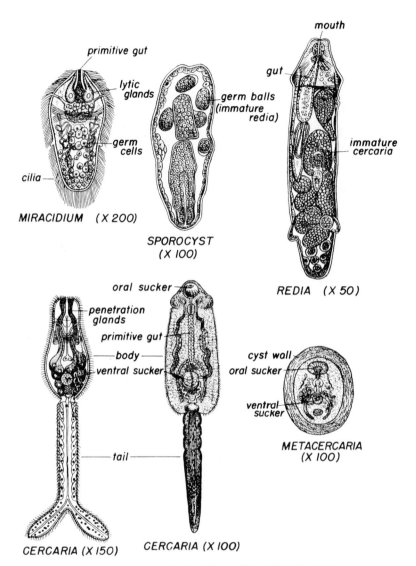

Fig. 7-11.   Larval types of human flukes.   (United States Naval Medical School Laboratory Manual.)

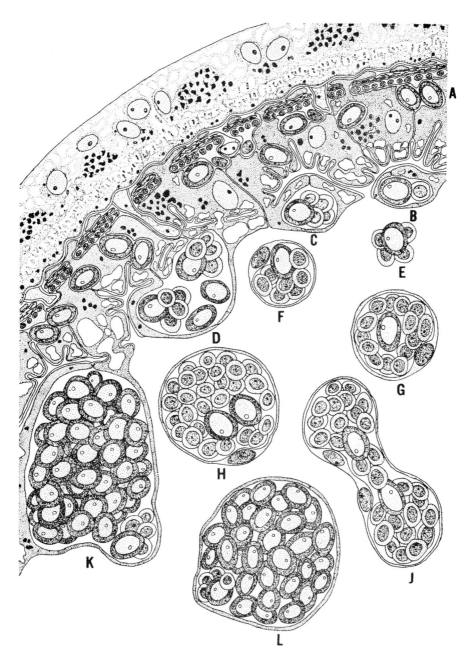

Fig. 7-12. Diagram of body wall of daughter sporocyst of *Cercaria bucephalopsis haimeana*, reconstructed from electron-micrographs and transverse sections seen under the light microscope. *A*. Germinal cell in subtegument. *B*. Early germinal ball in germinal cyst, consisting of one germinal and one somatic cell. *C*. Germinal ball in germinal cyst, consisting of one germinal and four somatic cells. *D*. Germinal cyst containing four germinal cells, two of which have divided to produce a five-celled germinal ball. *E*. Germinal ball in body cavity, consisting of 1 germinal and 4 somatic cells. *F*, Germinal ball in body cavity consisting of one germinal, seven somatic and one investing cell. *G*. Germinal ball in body cavity consisting of one germinal, 14 somatic and one investing cell. *H*. Germinal ball in body cavity consisting of two germinal, 28 somatic and one investing cell. *J*. Late stage of division of advanced germinal ball into two. *K*. Germinal mass enclosed in germinal cyst. *L*. Germinal mass free in body cavity. (James and Bowers, courtesy of Parasitology)

151

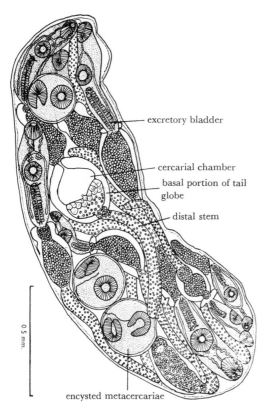

excretory bladder

cercarial chamber

basal portion of tail
globe

distal stem

0 5 mm.

encysted metacercariae

Fig. 7-13. Sac-like daughter sporocyst of *Phyllodistomum simile;* the adult fluke is found in the urinary bladder of brown trout, *Salmo trutta.* Within the sporocyst can be seen cercariae and metacercariae. (Thomas, courtesy of Proc. Zool. Soc., London)

another species in the same snail. Larval trematodes apparently utilize their host serum proteins, either directly or indirectly from the hepatopancreatic cells.[7]

During a study of metamorphosis of miracidia into cercariae, Muftic[33] extracted with methanol a crystalline substance from the haemolymph of the snail host (*Biomphalaria glabrata*) of schistosomes. This substance, possibly composed of steroids derived from the ovotestis of the snail, allowed miracidia in vitro to develop through the sporocyst stages into cercariae that were infective to mice. Steroids extracted from snail tissues have chemical and biologic properties similar to ecdysones of insects. Muftic postulated that parasites such as flagellates and trematodes may use steroids (designated "morphogenetic vitamins") produced by their hosts as morphogenetic factors.

James[21] has suggested that in most digenes the daughter-sporocyst/redia generations are pedogenetic embryos. Rediae show the beginnings of adult characteristics, each having developed a sucker and an embryonic gut. Most of the internal tissue is germinative, and within the redia the cercariae are formed.

Mature cercariae possess two suckers, a branched gut and a tail (Fig. 7–15). They also contain several kinds of gland cells, including penetration cells and cystogenous types. The cystogenic cells participate in the formation of the metacercarial cyst wall.[38,48] Like the miracidia, cercariae may have eye spots or photoreceptors consisting of sensory and pigment-containing cells.[35] The tail is considered homologous with the posterior region of the other generations of the life cycle, being a secondary aquatic adaptation.

Cable[4] has summarized some concepts of digene development in snails as follows. The ability of certain sporocysts to produce miracidia, and the fact that some cercariae and metacercariae may revert to the function of germinal sacs, demonstrate that the trematode life history is a series of polymorphic but fundamentally equivalent generations. Reproduction in germinal sacs is principally a type of parthenogenesis in which

cyst.[5] That snails produce a substance initiating the body transformation of the parasite is evident by the fact that once the process has started, it continues even if the developing sporocyst is removed from the snail.

A study of the ultrastructure of the rediae of *Cryptocotyle lingua* showed ribbon-like folds extending from the epithelial cells of its intestine into the lumen. Entrapped food globules in the folds suggested nutritive phagocytosis. The cytoplasmic folds, projections or flaps of the tegument also suggested a nutritive function by phagocytosis (or pinocytosis).[25] Rediae of other trematodes are known to ingest particulate matter as well as to absorb nutrients.[32] Lie *et al.*[28] have described active ingestion of rediae of one species by rediae of

meiosis is not involved (apomictic partheno-genesis). Polyembryony may be operative in the miracidium/mother-sporocyst and daughter-sporocyst/redia generations of some species, if an embryo that arises parthenogenetically dissociates to produce more than one offspring. The type and sequence of generations in the molluscan host are not fixed, but may depend on temperature, the extent of differentiation that occurred before the parasite entered the mollusk, or the intimacy of the parasite-host relationship. (See pp. 155 and 472 for some effects of larval trematodes on the snail host.)

Cercariae may migrate through the body before leaving the snail. Various reports show that migration occurs by way of the blood, genital ducts or digestive tract. The larvae may rupture the wall of the gut and enter the visceral hemocoel, the main venous vessels, the heart and the mantle sinuses, then emerge through the inner surface of the mantle.[36] (In cyclocoelids and a few other groups, cercariae encyst in the rediae that produce them and never have a migratory phase.) Once outside the snail, the cercariae then may penetrate their next host directly by burrowing into the skin. They

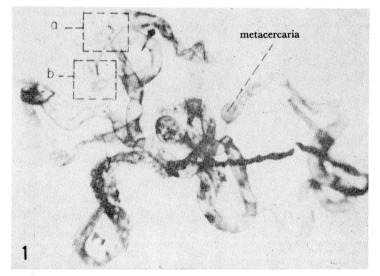

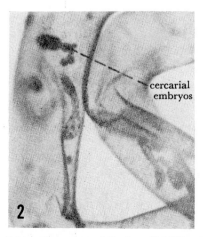

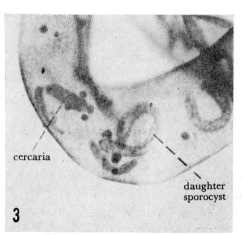

Fig. 7-14. Sporocyst of *Cotylurus flabelliformis* in the snail *Lymnaea stagnalis appressa.* 1. Entire mother sporocyst. 2. Detail from area "a." 3. Detail of area "b." (Hussey, Cort and Ameel, courtesy of J. Parasitol.)

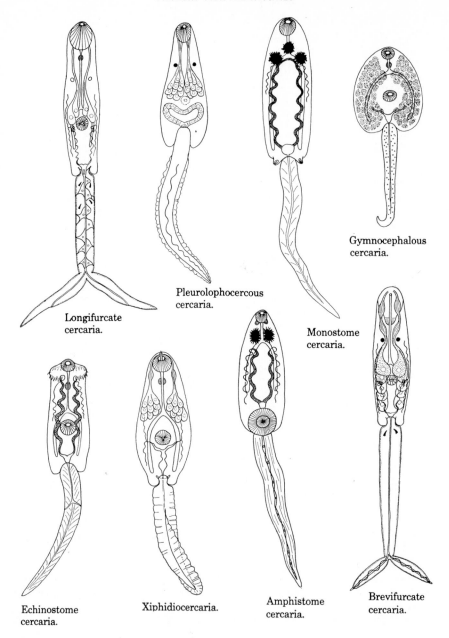

Longifurcate
cercaria.

Pleurolophocercous
cercaria.

Monostome
cercaria.

Gymnocephalous
cercaria.

Echinostome
cercaria.

Xiphidiocercaria.

Amphistome
cercaria.

Brevifurcate
cercaria.

Fig. 7-15. Representative types of cercariae. (Schell, courtesy of John Wiley and Sons)

do so more readily in those animals whose acellular elements of dermal connective tissue are least resistant to enzymatic action. Such susceptible hosts include young individuals, old individuals with youthful connective tissue, and individuals exposed to stresses.[27]

Cercariae often do not enter the definitive host directly but invade a transport host (e.g., fish, insect, crustacean, or almost any invertebrate), lose their tails, become rounded and may secrete a cyst-like wall about themselves. They are now called metacercariae. Some groups of cercariae possess an excretory epithelium that may immobilize wastes and thus permit the encysted metacercariae to grow and develop, even to functional maturity.[4] (See p. 180 for a description of the metacercarial cyst wall of the liver fluke, *Fasciola hepatica*.[12])

The final host eats vegetation with metacercariae attached or devours the prior intermediate host and so becomes infected. Excystation of the young fluke may be mediated by digestive enzymes of the host. Reducing and surface active agents and a carbon dioxide atmosphere may stimulate excystment.[31] The released metacercariae make their way to the site of preference, mature, and produce eggs, thus completing the cycle.

Snails may serve as both primary and secondary intermediate hosts. For example, the echinostome, *Echinoparyphium dunni*, utilizes freshwater snails for its sporocyst, redial and cercarial stages; the emerged cercariae enter the same species of snail or another species to become metacercariae.[29] Snails containing sporocysts can develop immunity to penetration by such cercariae (e.g., the strigeid fluke of ducks, *Cotylurus*).[34] Sexual maturity and egg production in metacercarial stages have been described by several authors.[8] Stunkard[42] has suggested that this kind of progenesis "appears to be a relict, the survival of an earlier developmental method."

Certain members of the family Bucephalidae use a freshwater mussel (*Lampsilis siliquoidea*) as the intermediate host. *Cercaria tiogae* inhabits the freshwater unionid clam, *Alasmidonta varicosa;* and a related species, *C. catatonke*, has been seen emerging from the clam, *Strophitus undulatus quadriplicatus. Cercaria milfordensis* is a small, stout-bodied distome that lives in the gonad and digestive gland of the common mussel, *Mytilus edulis* from Long Island, New York. The sporocysts are orange in color and give the mantle lobes of the mussel an orange shade. *Phyllodistomun bufonis*, a fluke living in the urinary bladder of toads, *Bufo boreas*, produces fully embryonated eggs whose miracidia penetrate clams (*Pisidium adamsi*). Mother and daughter sporocysts occur in the gill lamellae.

Cercariae encyst in nymphs of dragonflies. Another species of this genus lives in topminnows (*Fundulus sciadicus*) and also uses clams as intermediate hosts. The cercariae encyst within the daughter sporocysts. Minnows acquire the infection by eating infected clams.

The marine annelid worm, *Eupomatus dianthus*, serves as the first intermediate host of the fluke, *Cercaria loossi*, which is probably a fish blood fluke. In the Antarctic, available hosts would seem to be few and far between; the life of the larval trematode, *Cercaria hartmanae*, encourages this assumption because these larvae develop in other species of rediae that occur in the coelom of the marine annelid, *Lanicides vayssierei*, at Ross Island. *Allocreadium alloneotenicum* is found as a precociously mature adult fluke in the hemocoel of the caddis fly larvae belonging to the genus *Limnephilus*. Fluke eggs are liberated from disintegrating dead caddis fly larvae. Miracidia emerge and penetrate the bodies of freshwater clams (*Pisidium abditum*). The life cycle within the clam involves one sporocyst stage, two rediae and the cercariae. The latter leave the clam, penetrate caddis fly larvae, and mature in the hemocoel.[45] Immature flukes have been found in larval, pupal and adult mosquitoes.

"Cercariae" of members of the subfamily Gymnophallinae are ordinarily found in the snail, *Littorina saxatilis*. The parasite is either a cercaria that reproduces like a sporocyst or redia, or it is a sporocyst or redia that looks like a cercaria. Because of this peculiar combination of types, it is called a "parthenita." The rounded organism appears to be a cercaria, but is filled with small tailless cercariae.[20]

Larval stages of trematodes, especially the early stages, are definitely host specific, in contrast to many adult flukes, which may parasitize several species of vertebrate hosts. *Fasciola hepatica*, for example, may inhabit the livers of man, sheep, cattle, pigs, deer, rabbits, and other animals.

The effect of parasitism on the intermediate hosts is often not known. Trematodes, how-

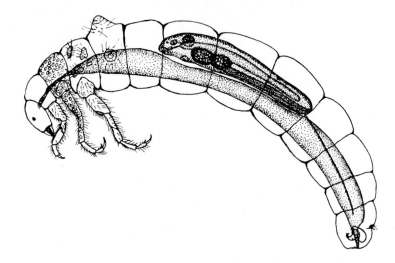

Fig. 7-16. The caddis fly larva *Limnephilus* shown in lateral view with the adult fluke *Allocreadium alloneotenicum* in the hemocoel. Fluke eggs are shown only in the middle thoracic segment. (Wootton, courtesy of Biol. Bull.)

ever, usually inhibit the growth of the gonads or even cause castration. Thus, parasitized snails are unable to reproduce. Etges and Gresso[15] found that when the snail, *Biomphalaria glabrata*, is infected with larvae of the blood fluke, *Schistosoma mansoni*, inhibition of the snail's egg laying occurs during the fourth week of infection, before the onset of cercarial emergence. Limited egg production is resumed well before the loss of infection, about 90 to 100 days after miracidial penetration. The rate of egg production by infected snails is about ten per cent of normal, but the eggs develop and hatch nearly normally. Normal and infected snails show a marked difference in thermal resistance, the latter succumbing readily to increased temperature. One study[7] demonstrated that damage to the hepatopancreas often occurs. Rediae may ingest host cells, cause sloughing of tissues, fat accumulation and depletion of glycogen. Heavy infections often destroy the liver of the snail.

CLASSIFICATION OF THE DIGENETIC TREMATODES

The development of the excretory system has a significant bearing on the correct classi-

fication of trematodes. At first, all species follow a fundamental or primitive plan. In the lateral folds of the young cercaria there develop simple, separated flame cells. They are connected to primary collecting tubules that traverse the embryo and open posteriorly through the body wall. As the cercaria matures, the primary collecting tubules meet and fuse along the midline, forming an excretory vesicle in the posterior region. This vesicle opens through a **bladder pore,** which possesses a sphincter. There may also be an excretory atrium. In digenetic trematodes the excretory bladder may be a primitive, thin-walled and relatively inefficient type, or an advanced, thick-walled efficient variety, but intermediate types occur. *Schistosoma* and other forked-tailed cercariae possess the simple, thin-walled type of bladder. In other families of trematodes, the excretory bladder begins as the thin-walled type and later develops thicker, more muscular walls. In some cercariae (Fig. 7–17), the primary excretory tubes enter the tail, while in others they do not.

Some of the families of digenetic trematodes are listed in the next chapter with a discussion of important or representative genera and species. For general references to the

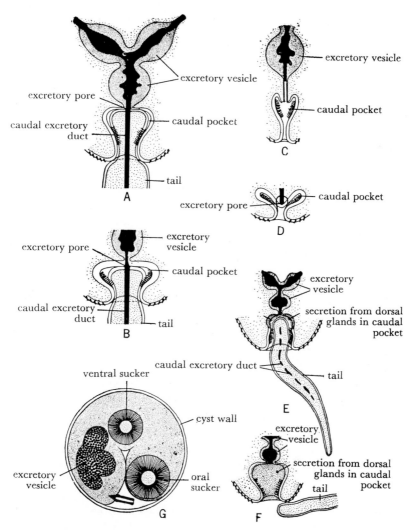

Fig. 7-17. The excretory vesicle of *Plagiorchis* (*Multiglandularis*) *megalorchis* and its relation to the tail in the cercaria. *A*. Excretory vesicle, tail root and caudal pockets of cercaria, expanded. *B*. Excretory vesicle (in part), tail root and caudal pockets of cercaria, contracted. *C*. Caudal pockets of cercaria after tail is shed, expanded. *D*. Caudal pockets of cercaria after tail is shed, contracted. *E*. Beginning of shedding of tail. *F*. Completion of shedding of tail. *G*. Fully formed cyst. (Rees, courtesy of Parasitology)

digenetic trematodes, see Baer and Joyeux,[1] Dawes,[10] Hyman,[19] Skrjabin,[39] and Yamaguti.[47]

La Rue's[26] system is based chiefly on larval life histories and morphology, especially the development of the excretory bladder. It is not a perfect system, but as La Rue himself said, it "may lead eventually to the development of a system which will portray with some degree of accuracy the genetic relationships within the Digenea." We have adopted this system.

Class TREMATODA

Subclass DIGENEA

Superorder ANEPITHELIOCYSTIDIA

Cercariae with thin excretory bladder, not epithelial; with forked or single tail.

Order STRIGEATOIIDA

Cercariae fork-tailed or modified from that condition; usually distomate; miracidia with one or two pairs of flame cells.

FAMILIES include: Strigeidae, Diplostomatidae, Cyathocotylidae, Clinostomatidae, Spirorchiidae, Aporocotylidae, Schistosomatidae, Azygiidae, Bivesiculidae, Transversotrematidae, Cyclocoelidae, Brachylaimidae, Fellodistomatidae, Bucephalidae.

Order ECHINOSTOMIDA

Cercariae develop in collared redia with stumpy appendages; life cycle usually involves three hosts.

FAMILIES include: Echinostomatidae, Cathaemasiidae, Fasciolidae, Psilostomatidae, Philophthalmidae, Haplosplanchnidae.

Suborder PARAMPHISTOMATA

Cercariae without penetration glands; bodies heavily pigmented; no collar in rediae; two-host life cycle; encyst on substrate. (Some authorities believe this should be a separate order.)

FAMILIES include: Paramphistomatidae, Gastrothylacidae, Cladorchidae, Diplodiscidae, Brumptidae, Gastrodiscidae, Heronimidae, Notocotylidae, Pronocephalidae.

Superorder EPITHELIOCYSTIDIA

Cercariae with thick-walled epithelial bladder; cercarial tail single, reduced in size or lacking.

Order PLAGIORCHIIDA

Cercariae without caudal excretory vessels: stylet present or lacking; encystment mostly in invertebrates, rarely in vertebrates.

FAMILIES include: Dicrocoeliidae, Eucotylidae, Haplometridae, Lecithodendriidae, Lissorchiidae, Macroderoididae, Microphallidae, Ochetosomatidae, Plagiorchiidae, Acanthocolpidae, Allocreadiidae, Lepocreadiidae, Megaperidae, Monorchiidae, Opecoelidae, Gorgoderidae, Troglotrematidae, Zoogonidae, Renicolidae.

Order OPISTHORCHIIDA

Cercariae with excretory vessels in tail; no stylet.

FAMILIES include: Opisthorchiidae, Heterophyidae, Acanthostomatidae, Cryptogonimidae

Suborder HEMIURATA

Nonciliated miracidia; second intermediate host a copepod. (Some authorities believe this should be a separate order.)

FAMILIES include: Hemiuridae, Halipegidae, Dinuridae, Didymozoidae.

REFERENCES

1. Baer, J. G., and Joyeux, C.: Classé des Trématodes (Trematoda Rudolphi). *In* Grassé, P.: Traité de Zoologie. Tome IV. Paris, Masson et Cie Editeurs, Libraires de l'Acad. Méd., 1961.

2. Barlow, C. H.: Life cycle of *Fasciolopsis buski*. Amer. J. Hyg., Monogr. Ser. 4, 1925.

3. Bils, R. F., and Martin, W. E.: Fine structure and development of the trematode integument. Trans. Amer. Micr. Soc. *85*:78–88, 1966.

4. Cable, R. M.: Thereby hangs a tail. J. Parasit., *51*:3–12, 1965.

5. Campbell, W. C., and Todd, A. C.: In vitro metamorphosis of the miracidium of *Fascioloides magna* (Bassi, 1875) Ward, 1917. Trans. Amer. Micr. Soc., 74:225–228, 1955.

6. Cheng, T. C.: The compatibility and incompatibility concept as related to trematodes and molluscs. Pacific Sci., *22*:141–160, 1968.

7. Cheng, T. C., and Snyder, R. W., Jr.: Studies on host-parasite relationships between larval trematodes and their hosts. I. A review. II. The utilization of the host's glycogen by the intramolluscan larvae of *Glypthelmins pennsylvaniensis* Cheng, and associated phenomena. Trans. Amer. Micr. Soc., *81*:209–228, 1962.

8. Chernogorenko-Bidulina, M. I., and Bliznyuk, I. D.: The Life-Cycle of the Trematode *Sphaerostoma bramae* Müller, 1776. Doklady Biol. Sci. Sect. (Doklady Akademii Nauk SSSR). Translation, *134*:780–782, 1960.

9. Coil, W. H.: Observations on the histochemistry of *Parastrigea mexicanus* (Strigeidae: Digenea) with emphasis on egg shell formation. Trans. Amer. Micr. Soc., *88*:127–135, 1969.

10. Dawes, B.: The Trematoda (With Special Reference to British and Other European Forms). New York, Cambridge Univ. Press, 1956.

11. ———: Penetration of *Fasciola gigantica* Cobbold, 1856 into snail hosts. Nature, *185*:51–53, 1960.

12. Dixon, K. E., and Mercer, E. H.: The fine structure of the cyst wall of the metacercaria of *Fasciola hepatica*. Quart. J. Micr. Sci., *105*:385–389, 1964.

13. Erasmus, D. A.: The host-parasite interface of *Cyathocotyle bushiensis* Khan, 1962 (Trematoda: Strigeoidea). II. Electron microscope studies of the tegument. J. Parasit., 53:703–714, 1967.

14. Erasmus, D. A., and Öhman, C.: The structure and function of the adhesive organ in strigeid trematodes. Ann. N.Y. Acad. Sci., *113*:7–35, 1963.

15. Etges, F. J., and Gresso, W.: Effect of *Schistosoma mansoni* upon fecundity in *Australorbis glabratus*. J. Parasit., *51*:757–760, 1965.

16. Etges, F. J., and Decker, C. L.: Chemosensitivity of the miracidium of *Schistosoma mansoni* to *Australorbis glabratus* and other snails. J. Parasit., *49*:114–116, 1963.

17. Halton, D. W.: Observations on the nutrition of digenetic trematodes. Parasitology, *57*:639–660, 1967.

18. Hanumantha-Rao, K.: The problem of Mehlis's gland in helminths with special reference to *Penetrocephalus ganapatii* (Cestoda: Pseudophyllidea). Parasitology, *50*:349–350, 1960.

19. Hyman, L. H.: The Invertebrates: Platyhelminthes and Rhynchocoela. Vol. II. New York, McGraw-Hill, 1951.

20. James, B. L.: A new cercaria of the subfamily Gymnophallinae (Trematoda: Digenea) developing in a unique 'parthenita' in *Littorina saxatilis* (Olivi). Nature, *184*:181–182, 1960.

21. ———: The life cycle of *Parvatrema homoeotecnum* sp. nov. (Trematoda: Digenea) and a review of the family Gymnophallidae Morozov, 1955. Parasitology, *54*:1–41, 1964.

22. James, B. L., and Bowers, E. A.: Histochemical observations on the occurrence of carbohydrates, lipids and enzymes in the daughter sporocyst of *Cercaria bucephalopsis haimaena* Lucoze-Duthiers, 1854 (Digenea: Bucephalidae). Parasitology, *57*:79–86, 1967.

23. Kawashima, K., Tada, I., and Miyazaki, I.: Host preference of miracidia of *Paragonimus ohirai* Miyzaki, 1939 among three species of snails of the genus Assiminea. Kyushu J. Med. Sci., *12*:99–106, 1961.

24. Komiya, Y.: The Excretory System of Digenetic Trematodes. Committee of Jubilee Publ. in Commemoration of 10th Anniversary of Dr. Y. Komiya as Chief of Dept. Parasitology. Tokyo, National Institute of Health, 1961.

25. Krupa, P. L., Bal, A. K., and Cousineau, G. H.: Ultrastructure of the redia of *Cryptocotyle lingua*. J. Parasit., *53*:725–734, 1967.

26. La Rue, G. R.: The classification of digenetic trematodes: a review and a new system. Exper. Parasit., *6*:306–344, 1957.

27. Lewert, R. M., and Mandlowitz, S.: Innate immunity to *Schistosoma mansoni* relative to the state of connective tissue. Ann. N.Y. Acad. Sci., *113*:54–62, 1963.

28. Lie, K. J., Basch, P. F., and Heyneman, D.: Direct and indirect antagonism between *Paryphostomum segregatum* and *Echinostoma paraensei* in the snail *Biomphalaria glabrata*. Zeit. f. Parasitenk., *31*:101–107, 1968.

29. Lie, K. J. and Umathevy, T.: Studies on Echinostomatidae (Trematoda) in Malaya. X. The life history of *Echinoparyphium dunni* sp. n. J. Parasit., *51*:793–799, 1965.

30. Lowe, C. Y.: Comparative studies of the lymphatic system of four species of amphistomes. Zeit. f. Parasitenk., *27*:169–204, 1966.

31. Martin, W. E., and Bils, R. F.: Trematode excretory concretions: formation and fine structure. J. Parasit., *50*:337–344, 1964.

32. McDaniel, J. S., and Dixon, K. E.: Utilization of exogenous glucose by the rediae of *Parorchis acanthus* (Digenea: Philophthalmidae) and *Cryptocotyle lingua* (Digenea: Heterophyidae). Biol. Bull., *133*:591–599, 1967.

33. Muftic, M.: Metamorphosis of miracidia into cercariae of *Schistosoma mansoni* in vitro. Parasitology, *59*:365–371, 1969.

34. Nolf, L. O., and Cort, W. W.: On immunity reactions of snails to the penetration of the cercariae of the strigeid trematode, *Cotylurus flabelliformis* (Faust). J. Parasit., *20*:38–48, 1933.

35. Pond, G. G., and Cable, R. M.: Fine structure of photoreceptors in three types of ocellate cercariae. J. Parasit., *52*:483–493, 1966.

36. Probert, A. J., and Erasmus, D. A.: The migration of *Cercaria X* Baylis (Strigeida) within the molluscan intermediate host *Lymnaea stagnalis*. Parasitology, *55*:77–92, 1965.

37. Rees, F. G.: Studies on the germ-cell cycle of the digenetic trematode *Parorchis acanthus* Nicoll. Part II. Structure of the miracidium and germinal development in the larval stages. Parasitology, *32*:372–391, 1940.

38. Rees, G.: The histochemistry of the cystogenous gland cells and cyst wall of *Parorchis acanthus* Nicoll, and some details of the morphology and fine structure of the cercaria. Parasitology, *57*:87–110, 1967.

39. Skrjabin, K. I.: Trematodes of Animals and Man. Elements of Trematodology. (A continuing series of volumes in Russian.) Moscow, Akademiya Nauk SSSR, 1947–1962.

40. Šlais, J., and Ždárská, Z.: Contribution to the histochemistry of the cuticle and cuticular structures of trematodes. Folia Parasit. (Praha), *14*:311–319, 1967.

41. Smyth, J. D., and Clegg, J. A.: Egg-shell formation in trematodes and cestodes. Exp. Parasit., *8*:286–323, 1959.

42. Stunkard, H. W.: Progenetic maturity and phylogeny of digenetic trematodes. J. Parasit. *45*:(Sect. 2) 15, 1959.

43. Threadgold, L. T.: The tegument and associated structures of *Fasciola hepatica*. Quart. J. Micr. Sci., *104*:505–512, 1963.

44. Vernberg, W. B.: Respiration of digenetic trematodes. Ann. N.Y. Acad. Sci., *113*:261–271, 1963.

45. Wootton, D. M.: Studies on the life-history of *Allocreadium alloneotenicum sp. nov.*, (Allocreadiidae-Trematoda). Biol. Bull., *113*:302–315, 1957.

46. Wright, C. A.: Host-location by trematode miracidia. Ann. Trop. Med. Parasit., *53*: 288–292, 1959.

47. Yamaguti, S.: Systema Helminthum. The Digenetic Trematodes of Vertebrates. Vol. I, parts 1 and 2. New York, Interscience Publishers, 1958.

48. Ždárská, Z.: The histology and histochemistry of the cystogenic cells of the cercaria *Echinoparyphium aconiatum* Dietz, 1909. Folia Parasit. (Praha), *15*:213–232, 1968.

Chapter 8

Class Trematoda, Subclass Digenea
(Representative Families)

SUPERORDER ANEPITHELIOCYSTIDIA

Cercariae possess a thin, nonepithelial excretory bladder and a forked or single tail.

Order Strigeatoida

Family Strigeidae

These intestinal flukes are found in reptiles, birds, and mammals. Usually the anterior part of the body is flattened and concave, and the other part is more cylinder-shaped (Fig. 8–1). The concave portion has an adhesive organ or "holdfast"; the posterior portion contains most of the reproductive system. There are two small suckers, and the overall length of the body ranges from a few millimeters to about 20 mm.

The excretory system of these flukes differs from that of others in that the excretory bladder enlarges and forms narrow tubules or a series of pockets or lacunae. These modifications are called the **reserve bladder system;** they may extend over the dorsal and lateral regions of the body. The system functions in excretion and may also serve as a hydrostatic skeleton and a vehicle for the circulation of dissolved nutrients. Lipid droplets occur in the lacunae.[11]

The life cycle usually includes the miracidium, sporocyst, daughter-sporocyst, cercaria and metacercaria, The metacercaria may be the **tetracotyle** type, which possesses two regular suckers and two pseudosuckers; the **diplostomulum** type, which has pseudosuckers, a leaf-like concave forebody and a small conical hindbody; or the **neascus** type, poles, frogs or other vertebrates, becoming

which is like the diplostomulum but lacks pseudosuckers. The metacercaria may be encysted, or free and active as in the Diplostomatidae. For a description of larval forms in fish, see Hoffman.[16]

Strigea elegans. Adults of these flukes live in snowy owls. Miracidia enter snails and develop into sporocysts, rediae and cercariae. The latter emerge and become mesocercariae in frogs and toads. A third intermediate host, the garter snake, harbors the tetracotyle stage, which possesses a definite cyst wall. The frogs and toads may be only paratenic hosts.

Cotylurus flabelliformis. Flukes may have their own fluke parasites. Cercariae of this strigeid, a common fluke of ducks, may enter planorbid and physid snails, where they develop into metacercariae only in those individuals that harbor sporocysts and rediae of other species of trematodes. They do not enter snails containing C. flabelliformis sporocysts. The strigeid cercariae enter other sporocysts and rediae where they are apparently protected from immune reactions of the abnormal snail hosts, and where, as hyperparasites, they are able effectively to utilize the nourishment that the sporocysts and rediae have secured from the snail host. The normal environment for metacercariae of C. flabelliformis is the hermaphroditic gland of lymnaeid snails.

Alaria arisaemoides. This is another fluke that has more than one intermediate host. From the snail, cercariae enter tad-

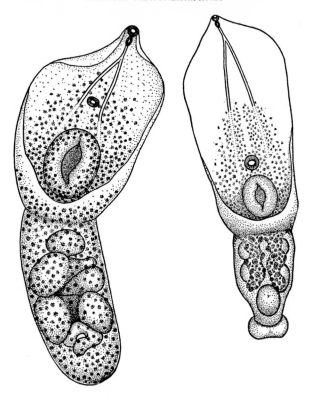

Fig. 8-1. Left, *Neodiplostomum paraspathula*, ventral view of entire animal. × 40. Right, *Neodiplostomum orchilongum*, ventral view of entire animal, from the intestine of a hawk. (Noble, courtesy of J. Parasitol.)

mesocercariae. The amphibians are eaten by raccoons, mice, rats and other mammals in which the metacercariae stage develops. The final host eats the mice, rats or raccoons. The parasites live as adults in the intestines of carnivores, such as dogs, cats, weasels, or minks. This would appear to be a four-host cycle, but since the amphibia are paratenic hosts, it is more truly a three-host cycle. (See Fig. 8–2.)

Family Spirorchiidae

This family is composed of monoecious flukes that are distomes or monostomes inhabiting primarily the heart, larger arteries and sometimes other blood vessels of turtles. Some parasites may occupy other tissues, such as the submucosa of the esophagus. The genus *Spirorchis* is characterized by a central row of testes; the genus *Haplorhynchus*

contains two testes; and the genus *Vasotrema* has a single testis that is spirally coiled. Eggs make their way from the host's blood vessels to the intestine, pass out of the turtle's body and hatch. The miracidia enter lymnaeid snails, develop into a first generation sporocyst, then another sporocyst, then into the furcocercous (forked-tail) cercariae that leave the snail and penetrate the soft tissues of the next turtle host. Other genera are *Haplotrema*, *Learedium*, *Neospirorchis* and *Amphiorchis*.

Family Aporocotylidae

This family is composed of elongated, worm-like parasites that live in blood vessels of fish. Suckers are weak or absent, and the body is covered with minute scales. The flukes are tiny, measuring less than 1 to 5 mm. long, and the sexes are separate.

There are four species, one of which, *Aporocotyle simplex*, is illustrated in Figure 8–3.

Family Schistosomatidae (= Bilharziidae)

Schistosomes are by far the best known of all blood flukes because of the serious disease (schistosomiasis or bilharziasis) that they cause in man. Animals are also hosts to these worms. *Schistosoma bovis* (Fig. 8–16) lives in the portal system and mesenteric veins of cattle and sheep. It occurs rarely in horses, antelopes and baboons, and may be found in Africa, southern Asia and southern Europe. The life cycle is essentially the same as that described for species in man. *S. intercalatum* may be a race of *S. bovis* that has become adapted to man. Snails belonging to the genus *Bulinus* serve as intermediate hosts for *S. bovis*. Another species parasitic in cattle, sheep, goats, carabao, antelope, and occasionally in dogs and horses is *S. spindale*, which may be the same species as *S. nasale*. *S. spindale* also occurs in Africa and southern Asia and uses snails of the genera *Planorbis*, *Indoplanorbis* and *Lymnea*. *S. incognitum* occurs in pigs in India.

Heterobilharzia americana. This blood fluke infests dogs in the United States. Its life cycle is similar to that of *Schistosoma mansoni*, described on page 172. As with other blood flukes, the most serious pathogenic response is the host reaction to the masses of eggs that are produced, which may disseminate to various parts of the body. Host tissue surrounding the eggs becomes infiltrated with epithelioid cells and leukocytes, producing a granular mass called a **granuloma**. The infection is common in several southern states and is associated with swamplands.

Other genera are *Ornithobilharzia*, *Bilharziella*, *Trichobilharzia* and *Pseudobilharzia*. The relationships of these flukes are not well understood and deserve much more study. Sexes are separate, with the female often longer than the male and more slender. Females range from 3 to 15 mm. (usually less than 10), but one species (*Gigantobilharzia acotyles*), which lives in the abdominal veins of the black-headed gull, may reach 165 mm.

Schistosomatium douthitti is a small blood fluke (1.9 to 6.3 mm. long) that lives in the hepatic portal system of muskrats and various mice. It is found in North America and ranges as far north as Alaska.

Blood Flukes of Man. It has been estimated that one person in 12 of the world's population is infected with schistosomes.[4] Almost all of these parasites belong to one of three major species.[30] Their general external anatomy is illustrated in Figure 8–4, and a comparison of the two sexes is shown in Figure 8–5.

Schistosoma japonicum. This fluke occurs in Japan, Korea, the Philippines, Sulawesi (Celebes), and China. The overall incidence of infection is from 10 to 25 per cent. In Taiwan the fluke is unusual in that it does not infect man, but only various other mammals. In every other respect the parasite appears to be identical with *Schistosoma japonicum* found elsewhere in the Far East. A reported case of human schistosomiasis from southern Taiwan has not been verified. The woman involved probably ate some meat heavily infected with *Schistosoma* eggs, which then passed through her intestine and were found during a routine examination. The most common host of the Taiwan species is the rat, but flukes may also be found in shrews and bandicoots. Various laboratory reservoir hosts include pigs, dogs, cats, carabao, cattle, goats, horses and rodents.

Male *Schistosoma japonicum* average 15 × 0.5 mm. in size. An oral sucker lies at the anterior end and an acetabulum is situated at the end of a short proejction from the body. The body has a groove along much of its length, thus appearing somewhat like a narrow boat. The female lies within this depression or fold, which is called the **gynecophoral canal**. Armstrong[2] has suggested that the male produces a pheromone that controls the maintenance and maturation of the female. The tegument is smooth, except for minute spines in the gynecophoral canal and on the suckers. There is no muscular pharynx, and the intestinal ceca unite posteriorly to form a single cecum. Usually,

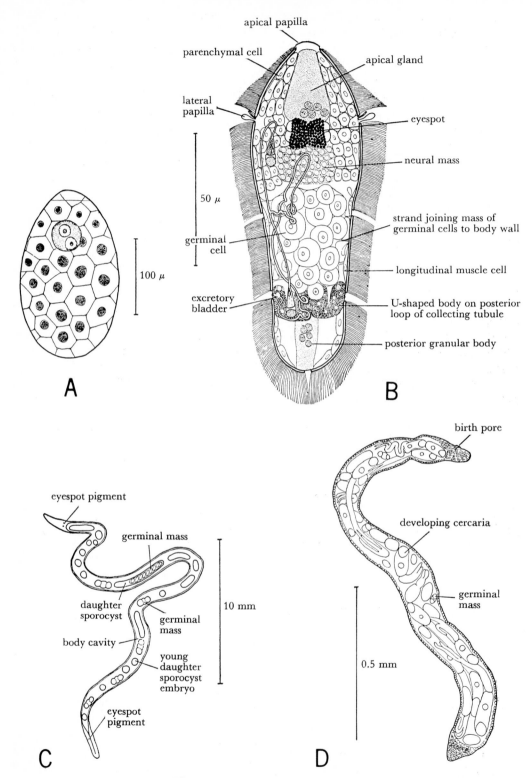

Fig. 8-2. See legend on facing page.

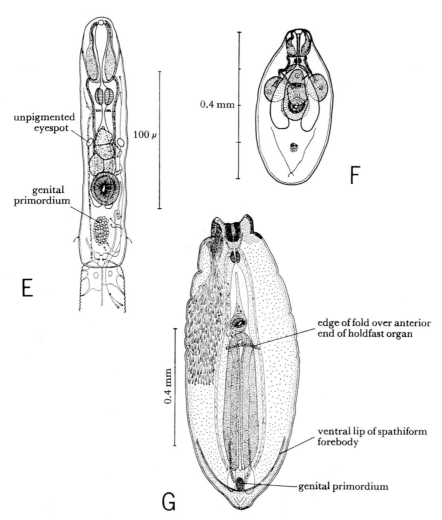

Fig. 8-2. Developmental stages of *Alaria arisaemoides* from the gut of dogs and foxes. (*A*) Freshly passed egg (camera lucida). (*B*) Miracidium in dorsal view (composite freehand; anterior loop of collecting tubule shown through neural mass; longitudinal and circular fibers, and dorsal row of nuclei not shown). (*C*) 18-day mother sporocyst containing developing daughter sporocysts (freehand). (*D*) Fully developed daughter sporocyst containing germinal masses and developing cercariae (slightly flattened; camera lucida). (*E*) Ventral view of body of cercaria (composite drawing based on camera lucida outline and details from living material). (*F*) Dorsal view of mesocercaria, showing gut and penetration glands (camera lucida). (*G*) Fully developed diplostomulum. Forebody glands and their ducts shown on right side. Note remnants of penetration gland duct in oral sucker and fold over anterior end of holdfast (camera lucida; ventral view). (Pearson, courtesy of Canad. J. Zool.)

seven testes are located near the anterior end of the body, and ducts from the testes join to form a seminal vesicle that leads to a short duct; this in turn opens at the genital pore

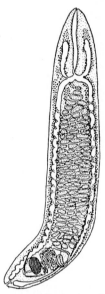

Fig. 8-3. *Aporocotyle simplex,* from the gills of flounder and other fish. About 4 mm. long. (Dawes, The Trematoda, courtesy of The University Press, Cambridge.)

situated just posterior to the pedunculate acetabulum. There is no muscular cirrus.

Females average 20 × 0.3 mm., thus being both longer and narrower than the males. The gynecophoral canal usually is not long enough to enclose the female, so loops of its body generally can be seen extending from the canal. The ovary is situated in the posterior half of the body and from it extends a long uterus that, in mature worms, is filled with 50 to 300 eggs. A genital pore opens just behind the ventral sucker. Numerous vitelline glands lie on either side of the long medium vitelline duct in the posterior quarter of the body.

These blood flukes live chiefly in the superior mesenteric veins, where the females extend their bodies from the male or leave the male to lay eggs (Fig. 8–7) in small venules of the mesenteries or of the intestinal wall. The oval to rounded eggs require a few days to develop into mature miracidia within the egg shell. Eggs measure $90 \times 60\,\mu$ and possess a minute lateral hook that is usually difficult to see. Masses of eggs cause pressure on the thin venule walls that are weakened

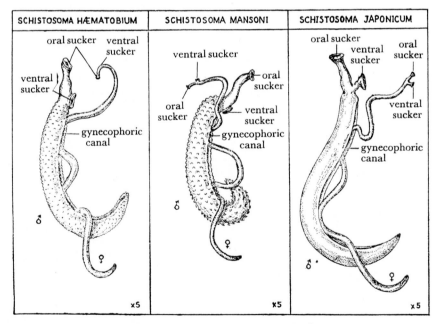

Fig. 8-4. Important schistosomes of man. (Belding, Textbook of Clinical Parasitology, courtesy of D. Appleton-Century Co.)

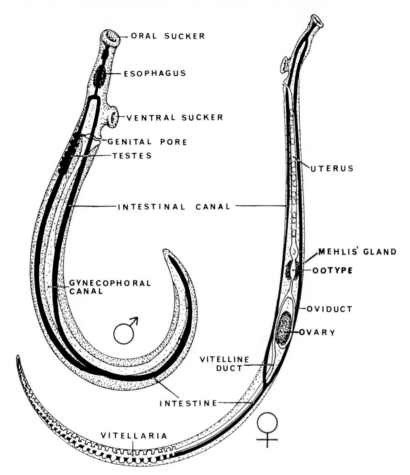

ORAL SUCKER

ESOPHAGUS

VENTRAL SUCKER

GENITAL PORE

TESTES

UTERUS

INTESTINAL CANAL

MEHLIS' GLAND

OOTYPE

GYNECOPHORAL CANAL

OVIDUCT

OVARY

VITELLINE DUCT

INTESTINE

VITELLARIA

Fig. 8-5. *Schistosoma*, illustrating the main differences between the adult male and female.

by secretions from the histolytic glands of the miracidia within the eggs. The walls rupture and the eggs get into the intestinal lumen and thus to the outside. Up to 3,500 eggs a day may be laid by a single worm, and in a heavy infection there may be thousands of worms in the blood vessels. See Figure 8–6 for the life cycle.

Hatching takes place in water. Although pH, salinity, temperature and other environmental factors are important, factors within the egg probably play a major role in the hatching process. The released miracidia can live for a few hours, during which time they actively swim about as though "in search" of their molluscan host. These hosts are various species of the minute snail *Oncomelania* (Fig.

8–8). Miracidia penetrate the snail's foot or body with the aid of histolytic gland secretions, the process requiring only a few minutes. Within the snail the cilia are shed and the miracidia become sporocysts. These sporocysts enlarge and the germ cells within them develop into a second generation of sporocysts, which mature in the digestive gland of the snail. Germ cells within the second generation sporocysts give rise to cercariae, thus skipping the usual redial generation. About a month is required from miracidial penetration of the snail to emergence of cercariae. These cercariae possess forked tails, hence the name "schistosoma" or "split body." They usually emerge during the early part of the night and are large enough (335 μ

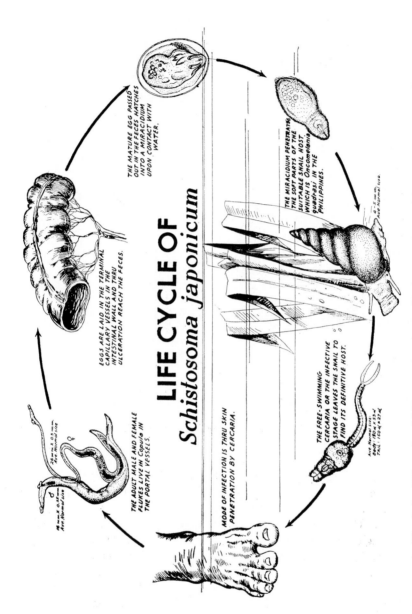

LIFE CYCLE OF *Schistosoma japonicum*

THE MATURE EGG PASSED OUT IN THE FECES HATCHES INTO A MIRACIDIUM UPON CONTACT WITH WATER.

THE MIRACIDIUM PENETRATES THE SOFT PARTS OF THE SUITABLE SNAIL HOST, WHICH IS *Oncomelania quadrasi* IN THE PHILIPPINES.

EGGS ARE LAID IN THE TERMINAL CAPILLARY VESSELS IN THE INTESTINAL WALL AND THRU ULCERATION REACH THE FECES.

THE FREE-SWIMMING CERCARIA OR THE INFECTIVE STAGE LEAVES THE SNAIL TO FIND ITS DEFINITIVE HOST.

THE ADULT MALE AND FEMALE FLUKES LIVE IN COPULA IN THE PORTAL VESSELS.

MODE OF INFECTION IS THRU SKIN PENETRATION BY CERCARIA.

Fig. 8-6. Life cycle of *Schistosoma japonicum*. (Pesigan, courtesy of Santo Tomas J. Med.)

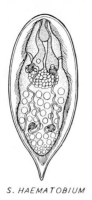

S. HAEMATOBIUM

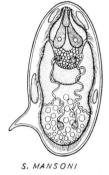

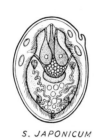

S. JAPONICUM

S. MANSONI

Fig. 8-7.   Eggs of *Schistosoma*.

overall length) to be seen with the unaided eye. They are active swimmers but may hang motionless from the surface of the water.

Penetration of man or other mammals usually occurs within 48 hours, and the life span of the cercariae is limited to a very few days. Penetration is probably similar to the following description of this process for *Schistosoma mansoni*.[44] When the cercariae come in contact with the skin of an appropriate host, they "loop" for variable periods of time by attaching themselves alternately with the oral and ventral suckers. When unattached, the oral end of the body constantly probes into every irregularity encountered. Points of entry include wrinkled areas, the bases of follicular eminences, points of scale attachment, distal hair-skin angles, orifices, follicular canals and entry sites used by previous cercariae. Apparently calcium and magnesium ions are essential for successful penetration,[24] which is strongly influenced by temperature.

Eventually the cercariae become closely attached by their oral suckers and they assume a vertical or oblique position in relation to the surface. At this time the body is elongated and slender, and the muscular oral sucker thrusts into the entry site. Penetrating cercariae apparently release an enzyme from glands at their anterior ends. Several pairs of large glands are usually located on each side of the midline, and they empty through long narrow ducts at the anterior end of the larva in the neighborhood of the oral sucker. Although the exact nature of the secretion of the penetration glands is not known, some cercariae apparently produce hyaluronidase, an enzyme that hydrolyzes hyaluronic acid, one of the principal substrates of connective tissue. Ramming motion and partial eversion of the sucker bring the openings of the ducts from the penetration glands into contact with the stratum corneum. Through these ducts the gland content is secreted into the host tissue. Alternate contraction and elongation of the body and energetic movements of the tail accompany thrusts of the oral sucker. Elongation of the body at the time the oral sucker initially probes into the entry site permits the cercariae to push into very small crevices. Thus attached and buried orally, the body is contracted and the diameter increased, with the result that the breach in the host tissue is enlarged. This sequence of slenderizing the body while the oral tip is being thrust deeper, and then contracting the body and pouring out glandular secretions seems to be the regular means of enlarging the entry pores for gradual penetration. Constant tail activity provides added forward thrust, but tail thrust is not a necessity since tailless cercariae have been seen penetrating the host. Exploratory time for the cercariae is about 1.8 to 2.3 minutes and entry time about three to seven minutes.

Migration of *Schistosoma japonicum* through the body commences with entry into minute lymphatic vessels, thence to the heart and circulatory system. These postpenetration cercariae are known as **schistosomules.** Young flukes leave the blood vessels in the liver,

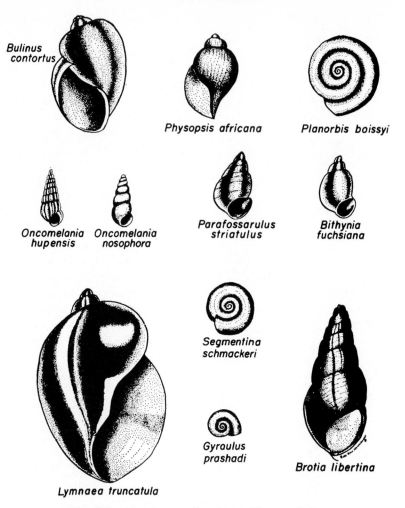

Fig. 8-8. First intermediate hosts of human flukes.
(United States Naval Medical School Laboratory Guide.)

where they develop to maturity. After several weeks they pass from the liver back into the circulatory system and remain in the superior mesenteric veins. Copulation takes place and the females enter the gynecophoral canals of the males. Usually one to two months elapse from the time of penetration by cercariae to the appearance of schistome eggs in human feces. The life span of adult worms is usually between five and 30 years, and several thousand individuals may exist in one host.

Migrating worms usually cause little or no damage or symptoms, but occasionally there are serious reactions, such as a pneumonia re-

sulting from the invasion of the lungs. The liver-inhabiting phase may be symptomless or marked by enlargement and tenderness of this organ and toxic reactions. Other symptoms include cough, fever, diarrhea, eosinophilia, enlargement of spleen and liver, anemia, ascites and an extreme wasting away, so that the arms, legs and body in general become alarmingly thin while the abdomen is huge with fluid (Fig. 8–9). Ulceration and necrosis of tissues in the intestine are characteristic.

Symptoms and pathology are related to the number of worms, the tissues infected and the sensitivity of the host.[27] Although toxins,

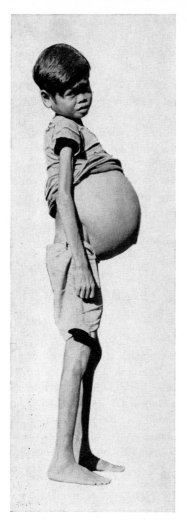

Fig. 8-9. Advanced case of schistosomiasis in a Filipino boy. (Courtesy of Dr. Robert E. Kuntz)

nourished persons who often expose themselves to the cercariae. Man and animals can develop some immunity after a previous infection.[33] Even the snail, *Biomphalaria glabrata*, shows immune reactions against *Schistosoma mansoni*. Promising research is being carried out by Hsü and Hsü[17] and others using the nonhuman Taiwan strain of *S. japonicum* as an antigen for building up immunity. These workers injected cercariae of this strain into monkeys, and they found that the animals were resistant to subsequent challenge infections with the human type strain from Japan.

Diagnosis should be based on finding eggs in the feces or in biopsied tissue. Complement fixation reaction, dermal tests and slide flocculation tests are of value.

Prevention and control involve educating the people who use infected waters, developing sanitary waste disposal measures, providing clean drinking water, and destroying the snail vectors.[31]

Schistosoma haematobium. About 40 million people are infected with this species of blood fluke in north, central and west Africa, Asia Minor, Cyprus and southern Portugal. The worms live in the veins of the urinary bladder and associated areas, rarely in the portal system, and occasionally in mesenteric veins. In chimpanzees, they have been reported from the urinary bladder, rectum, lungs, liver, appendix, and mesenteric veins.[9]

These flukes are intermediate in size between the other two blood flukes of man, with males reaching 15 mm. in length (average about 12 mm.), and ranging from 0.8 to 1 mm. in width. The body of the male is covered with small tubercular prominences, probably sensory in function. The slender, filiform females are about 2 cm. long, and when they have been feeding on blood, are dark in color. Their morphology is similar to that of *S. japonicum* and they have minute tubercles on the terminal portions of the body that usually project from the male gynecophoral canal. The female is able to extend her body in a quick snake-like manner from the canal into small veins where her eggs are

dead worms and malnutrition may be involved in hepatosplenic schistosomiasis in experimental mice and rats, there is clear evidence[43] that the schistosome ovum is the primary factor in pathogenesis. This evidence was based on electron microscope study of the liver parenchyma and schistosome pigment. Female worms do more damage than male worms because of the host reaction to the masses of worm eggs.

Prognosis is good in light cases when the patient does not become reinfected. It is grave in cases of long duration and in under-deposited, and then withdraw quickly into the canal.

Eggs average 170×40 to 70μ and are characterized by a pointed projection on the terminal end. Eggs penetrate the wall of the bladder or gut to the lumen and thence to the outside. The method of passing through tissue is somewhat obscure. Probably the miracidia within the egg secrete histolytic fluids that diffuse out of the egg and weaken the bladder wall sufficiently to permit the egg to be pushed through by various pressures on it. Possibly the spine helps.

The life cycle is similar to that of the other two flukes. The main intermediate hosts are snails belonging to the families Physidae and Bulinidae (Fig. 8–8). No rediae are produced, and cercariae develop directly within sporocysts. In man the infection is called **urinary schistosomiasis.** As the urinary bladder gradually becomes infiltrated with eggs, ulceration and bleeding may occur. Eventually the eggs become calcified. Symptoms usually do not appear for several months or more after infection. Positive diagnosis is based on finding eggs in urine.

Schistosoma mansoni. This species is the smallest of the three. Males average 1 cm. in length and females 1.6 cm. The disease is called **schistosomiasis mansoni** or **intestinal bilharziasis.** It occurs in Africa, Arabia, Madagascar, Brazil, Venezuela, Puerto Rico, Surinam and the Dominican Republic. Wild rodents may be reservoirs of infection.

Eggs measure 114 to $175 \mu \times 45$ to 68μ and they possess a spine that projects from the shell on one side near the end of the egg. Miracidia enter several types of snails but primarily those of the family Planorbidae. Responses of the miracidia to chemical attractants were studied by using agar and starch gel pyramids impregnated with chemicals.[29] Short-chain fatty acids, some amino acids and a sialic acid attracted the miracidia and stimulated them to attach and attempt to enter the agar. The solvent action of various chemicals (e.g., acetone) removed the attracting substances from the snail *Biomphalaria* (= *Australorbis*) *glabrata.* Subsequent addition of butyric or glutamic acids to these snail tissues restored the capacity to attract.

Penetration by miracidia of *Schistosoma mansoni* was described by Wajdi,[47] as follows: (1) A miracidium is pressed against the snail by action of its cilia. (2) Adhesive glands in the miracidium produce a mucoid substance causing the miracidium to adhere to the snail. The substance also functions for lubrication. (3) An anterior papilla elongates and works its way through the snail's epithelium like a drill. (4) The penetration gland secretion apparently digests the tissue of the snail, thereby aiding the papilla in enlarging the area of damage through which the miracidium enters. After infection, the snail produces a miracidia-immobilizing substance.[32] It is interesting to note that gerbils become infected with this fluke by eating infected snails.[28]

Sporocysts of *Schistosoma mansoni* survive longest in snail tissues of the head-foot, tentacles, pseudobranch and mantle collar. Cercariae survive longest in sinuses and veins as well as in the tissues listed above. Infected snails may show stunted growth and physiologic castration. Emergence of cercariae can be most frequently observed from the surface of the mantle collar and pseudobranch, and occasionally from the head-foot.[36]

Cercariae rapidly penetrate the skin of humans and develop into schistosomules as they make their way to the liver, where they mature. Adults feed on blood during their sojourn in the liver. After a week or more, they migrate to the inferior mesenteric veins and reach the capillaries of the sigmoidorectal area. Eggs penetrate the gut wall and pass from the body with feces. In mice and hamsters, mature male *Schistoma mansoni* in the mesenteric portal system are necessary for the sexual maturation of female worms. If male worms have not reached sexual maturity, the females do not mature. In man, symptoms include abdominal pains, tiredness and diarrhea.

Bueding[6] presented a review of studies of the glycolytic enzymes of *Schistosoma mansoni* to determine whether these enzymes of the parasite were the same as comparable ones in the host. The enzymes studied were phosphoglucose isomerase, lactic dehydrogenase, and hexokinases. Differences definitely existed in the nature of the enzymes catalyzing the same reactions in *S. mansoni* and in the host, and these differences may occur at different levels. Phosphoglucose isomerase of the parasites could be distinguished from that of host muscle (rabbit) only by the use of immunologic procedures. Lactic dehydrogenase of the worms and of host muscle differed from each other in their behavior toward specific antibodies, in their affinities for one of their substrates, and in the effect of the hydrogen ion concentration on their activities. Hexokinases of the parasite and of the host differed in kinetics as well as in substrate specificities. The selectiveness of the action of immune sera on enzymes of the worm has demonstrated the possibility of interfering with the functional integrity of the parasite without altering the normal functioning of similar enzymes of the host. The significance of this possibility for a more rational approach to the chemotherapy of schistosomiasis is obvious.

The genetics of parasites is an almost untouched field of study. Fundamental to any such study is a knowledge of the numbers and kinds of chromosomes that occur in these animals. The diploid number of chromosomes in *Schistosomatium douthitti* is 14, and 16 in *Schistosoma mansoni*, *S. haematobium* and *S. japonicum*. These genera of flukes have an XY pair of chromosomes in the female worm and an XX pair in the male.

Many species of flukes belonging to the family Schistosomatidae are parasitic in birds and in other animals (Fig. 8–16). Many snails harbor the developmental stages, and various species of cercariae may be found in large numbers in freshwater streams, ponds or lakes or in the intertidal region of the seashore. These cercariae are not as host-specific as are the earlier stages, and often try to penetrate the skin of vertebrates not normal to their life cycle. Studies on host recognition in schistosomes indicate that these cercariae apparently do not react to light, but to secretions from sebaceous glands, which cause them to attach themselves to almost any nearby skin. In certain areas, bird cercariae may thus attempt to enter, or actually do enter human skin. They do not mature in the human body, but the penetration of the skin causes a dermatitis and often intense itching—a reaction known as "swimmer's itch." Occasionally, in particularly susceptible persons, the reaction is severe enough to cause prostration. This dermatitis should not be confused with another irritation, also called "swimmer's itch," caused by certain blue-green algae.

At least one type of cercarial dermatitis results from the activity of cercariae from flukes that live in blood vessels of the noses of ducks. Eggs enter the nostrils of the duck and can be found in large numbers in mucus. From there they get into water and go through the usual stages in a snail. Cercariae are released into the water by the thousands, ready for the unwary swimmer. Marine schistosome dermatitis occasionally occurs among bathers along the seashore. The cercariae of *Austrobilharzia variglandis* from the snail, *Nassarius obsoletus*, has been shown to be the culprit in San Francisco Bay, California.[15]

Accidental infection with cercariae, metacercariae or even adult flukes of various species is not uncommon. Occasionally metacercariae become attached to the throat of man, causing considerable irritation. The parasites are taken into the mouth with food or drinking water and are usually species not normal to man.

In addition to the three main schistosomes found in man, at least six other species have been seen in human blood, or their eggs have been found in human feces or urine. These species, rare in man, are probably more common in monkeys, cattle, or other animals.

Immunity. The study of trematode immunity has been devoted primarily to the Schistosomatidae, with lesser emphasis given to Troglotrematidae, Opisthorchiidae and Fasciolidae. "To establish a classical antibody basis for schistosome immunity, three criteria must be fulfilled: (1) Antibodies must be demonstrably present when immunity is proven; (2) passive transfer of protective antibodies should be possible; and (3) the protective property should be absorbable or fractionated from the serum and in consequence the serum should lose its ability to transfer resistance. To date there has been no conclusive evidence for all three of the above postulates for schistosome immunity."[19] See Bruijning[5] for a general review of immunology of schistosomiasis.

Antibodies are readily demonstrated in host serum but this does not necessarily mean that the host is immune. For resistance to become established, the antibodies must act "on the parasite in such a way as to interfere with its general growth or metabolic processes or to upset its relation to the host tissue by enclosing it in a precipitate or by blocking its alimentary, excretory or reproductive orifices."[40]

Although the evidence for classic immunity in man has not been conclusive, it does indicate that immunization in schistosomiasis is possible.[37] In animals and man, antibody response occurs with the presence of any stage in the life cycle. Live schistosome eggs in antisera accumulate a precipitate of small globules on the egg shell after a few hours of incubation. The host becomes positive about 40 days after infection. Miracidia in host antisera become immobilized within five minutes due to a clumping of the cilia. Cercariae in antisera develop a plastic-like covering, the **pericercarial envelope,** or they become sticky and agglutinate.

Initial innate resistance to invasion of schistosome cercariae and to the subsequent success of the infection results from ". . . the state of organization of the acellular polysaccharide-protein substances forming the intercellular cement, subepithelial basement membranes, and ground substance of the host. These acellular connective tissue substances vary in their density and resistance to enzymatic alteration, in water content, and in water soluble components, the age of the host, its nutritional status, and with various hormonal influences and stress conditions."[25]

Antigens may be "somatic," obtained from the whole organism, or "metabolic," consisting of secretions and excretions. Fife et al.[12] have designated the latter as **exoantigens** and have isolated a serologically active one from *Schistosoma mansoni* cercariae. This antigen consists primarily of a polysaccharide, with galactose and galactosamine being the principal sugars; there is considerable linkage of the molecule through component amino groups. In contrast, the somatic antigen is protein or possibly glycoprotein, with little or no amino linkage. The miracidial cephalic glands and the egg shell have been identified as major sources of antigen.[46]

In laboratory animals, antigens from adult worms and cercariae are effective in the development of protection but eggs apparently play no role. Reports on host protection by artificial means conflict. Several workers have reported the induction of resistance against infection by injecting serum from experimental animals previously infected or artificially immunized with the flukes. According to Oliver-Gonzalez, however, "as to methods for producing protection or acquired immunity by artificial means, we are at a standstill. Very little or no protection can be given to experimental animals by injections of suspension of whole cercariae, adults, and eggs or by injections of their extracts."[35]

Diagnosis by serologic tests has been moderately effective. Antigens from adult worms and cercariae often give good results. Fluorescent antibody tests are especially rewarding. A skin test has been developed by which schistosome eggs are allowed to hatch in a small volume of water and then incubated for a short period. The water contains strongly antigenic metabolic products; if injected intradermally into an infected host, these produce a wheal. Vaccination experiments using homogenates, extracts or metabolic products have met with varying success. Experiments using life cycle stages of *Schistosoma japonicum* have been more successful than those with other species.

In experimental animals, infection with nonhuman strains of schistosomes (e.g.,

S. japonicum from Taiwan, and *S. bovis* from cattle) produces protection against human strains. A nonreciprocal cross-resistance between the roundworm *Nippostrongylus brasiliensis* and *Schistosoma mansoni* was shown by Hunter *et al.*[18] Albino mice infected with the roundworm were challenged with *S. mansoni* cercariae seven days later. There was a significant reduction in the number of adult flukes recovered in the treated mice compared to the controls. In monkeys, resistance to *Schistosoma* is produced by adult worms, but it is particularly effective against schistosomules. Both stages produce antigens, but the schistosomules are more vulnerable.

Rhesus monkeys tolerate an initial infection of *Schistosoma mansoni*, but become strongly resistant to a second infection. The first worm incorporates antigens into its body surface. Smithers *et al.*[39] suggest that the worm masquerades as its host, thereby preventing its rejection as foreign tissue. "This could explain the phenomenon of concomitant immunity; how the fully established adult worms escape the consequences of the immune response that they themselves engender. It may also explain their long persistence in the blood of man in what ought to be an immunologically hostile environment."

Few studies have been made on the immune responses of snails infected with larval flukes. These snails do possess some defense mechanism. For example, their "lymphocytes" may be primitive blood cells, which phagocytize foreign particles and eliminate them through the alimentary tract. Obviously, the cells could also engulf parasites of sufficiently minute size (e.g., microsporidia). Immunity to trematode infection is suggested by a study on the response of snails to the larvae of *Cotylurus flabelliformis*.[34] Snails infected with sporocysts of this fluke were more resistant to penetration by homologous cercariae and to subsequent metacercarial development than were noninfected snails. Aqueous tissue extracts prepared from *Schistosoma mansoni*-infected *Biomphalaria glabrata* were found to immobilze 76 to 100 per cent of homologous miracidia.[32]

Acquired immunity to schistosomes has been demonstrated in a large number of experimental animals and it is generally assumed that a similar resistance occurs in man. This assumption is somewhat hazardous because immunity is not always present and may be only partially effective. Different mechanisms may operate in different hosts and during different stages in the parasite life cycle.

See Warren and Newill[46a] for a bibliography of the world's literature on schistosomiasis from 1852 to 1962.

Family Brachylaimidae

These small or medium-sized distomes occur in the intestines of many different groups of vertebrates, especially in birds and mammals. The suckers are large and the genital pore opens in the posterior part of the body. The testes are roughly tandem in position with the ovary between them. The genera include *Brachylaimus*, *Itygonimus* and *Leucochloridium*. The latter genus is remarkable for its very unusual sporocysts. The unencysted metacercariae of some members of the family may damage their hosts. In the slug (*Limax*), for example, these parasites destroy the kidney tissue, and the results may be fatal.

Leucochloridium macrostomum. This European species is found in the rectum of the crow, sparrow, shrike, jay, nightingale and other birds. It is oval in outline and measures about 1.8 × 0.8 mm. and is about 0.45 mm. thick. Eggs pass from the intestine of birds and are eaten by aquatic or terrestrial snails (e.g., *Succinea*), in which the miracidia are liberated and penetrate host tissues. Sporocysts are formed which give rise to daughter sporocysts. These second generation sporocysts make their way through the snail host, become branched, and enter the head and tentacles. Here they become enlarged and colored, making the head and appendages of the snail appear to consist of elongated sacs ornamented with green, brown or orange bands. The sacs pulsate with the rhythm of a human heart beat and are noticeable to birds, which peck at them. Inside the sacs are encysted cercariae that are infective to the birds; since the snail tissue ruptures easily, the sacs are readily released and eaten. Even when released from the confinement of the snail, the sporocyst sacs pulsate. One wonders why it ever became necessary for a larval fluke to wave a colored flag at its next host and ask to be eaten!

Family Bucephalidae (= Gasterostomidae)

This family is characterized by the presence of the mouth near the middle of the body. The tiny adult flukes, less than 1 mm. long,

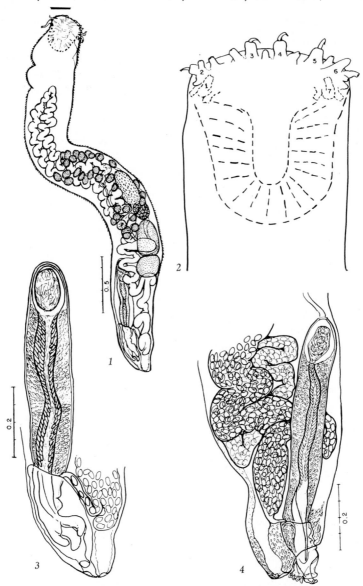

Fig. 8-10. Two species of gasterostomes from marine fish on the Pacific coast of Mexico. All figures were drawn with the aid of a camera lucida. The projected scale indicates millimeters in each figure. (*1*) Dorsal view of *Bucephalus heterotentaculatus*. (*2*) Composite diagram. Dorsal view of anterior sucker of *B. heterotentaculatus*. (*3*) Dorsal view of cirrus sac and genital atrium of *B. heterotentaculatus*. (*4*) Lateral view of cirrus sac and genital atrium of *Bucephaloides cybii* (Park, 1939) Hopkins, 1954. (Bravo-Hollis and Sogandares-Bernal, courtesy of J. Parasitol.)

live in the gut of carnivorous fish. The miracidia enter oysters or other marine bivalves or freshwater clams, thus differing markedly from the usual life cycle that uses a snail as the first intermediate host. Cercariae develop within the sporocyst without rediae and then leave the mollusk, swim about in the water, and become entangled in the fins of small fish. The entanglement is aided by the presence of two long tails, which give the name "oxhead" cercariae to this stage.

These cercariae encyst under the scales of the small fish and are eaten by larger carnivorous fish. Typical genera of these flukes are *Bucephalus* (Fig. 8–10), *Rhipidocotyle* and *Prosorhynchus*.

Order Echinostomida

Family Echinostomatidae

This group of flukes is characterized by the presence of a collar of spines at the anterior end. The body is elongated and covered with

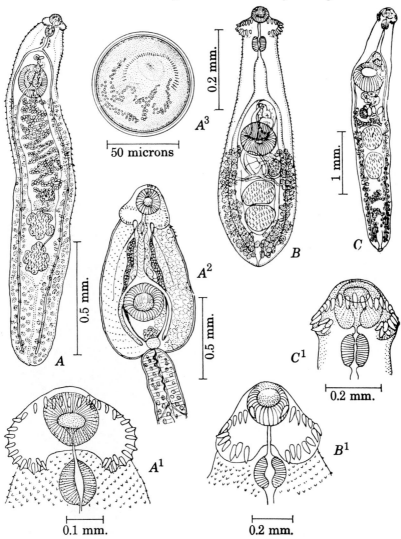

Fig. 8-11. Echinostomes. *A, Echinostoma lindoense; A¹*, head of same; *A²*, cercaria of same. (After Sandground and Bonne, Amer. J. Trop. Med., *20*, 1940.) *A³*, encysted metacercaria of *E. ilocanum*. (After Tubangui and Pasco, Philip. J. Sci., *51*, 1933.) *B, Echinochasmus japonicus; B¹*, head of same. (After Yamaguti.) *C, Euparyphium melis; C¹*, head of same. (After Beaver, J. Parasitol.) (Chandler and Read, An Introduction to Parasitology, courtesy of John Wiley and Sons)

spines, and the two suckers are usually close together; the acetabulum is larger than the oral sucker.

The ducts associated with the reproductive system of echinostomes differ from those of other groups. Excess sperm and vitelline granules pass through Laurer's canal and are discharged at the dorsal surface of the worm. Sperm are stored in a uterine seminal vesicle, instead of the usual type of seminal vesicle.[26] An interesting organ, the **ovicapt** or **oocapt,** occurs in echinostomes as well as in many other types of flukes. It is a muscular structure surrounding the base of the oviduct and its contractions cause eggs to be sucked from the ovary into the oviduct. The Germans have a descriptive word for this structure, the "Schluckapparat" or gulping apparatus. Laurer's canal, the ovicapt, and the part of the oviduct distal to the ovicapt are ciliated.

Echinostoma ilocanum. This fluke (Fig. 8–11) occurs in the intestines of men, dogs, cats, and rats in the Philippines. It averages about 5 × 1.5 mm. in size and possesses a ring of spines around the anterior sucker. Spine-like scales occur on the integument. Miracidia penetrate the snail (*Gyraulus*), within which they develop into sporocysts, rediae, another generation of rediae, and then cercariae. The encysted metacercariae occur in almost any snail or freshwater clam, so, if infected snails are used as food, the parasite gains entry into man.

Most of the other species of *Echinostoma* occur in birds. Fifteen to 20 such species have been described from the intestines of such birds as the cormorant, grebe, owl, thrush, domestic duck, goose, pheasant, partridge, stork, crane, and hawk. Other genera of the family include *Echinoparyphium, Euparyphium, Echinochasmus, Patagifer, Himasthla, Petasiger,* and *Parorchis.*

Family Fasciolidae

These broad, flat flukes are often seen in biology laboratories because of their large size and availability—some of them reach 100 mm. in length. They may be covered with small scales; the two suckers are close together, and the anterior end is often drawn out into a cone-like projection. The genital pore is located just anterior to the ventral sucker.

Fasciola hepatica (Fig. 8–12). This sheep liver fluke measures 20 to 30 mm. × 10 to 13 mm. and, as its common name indicates, is found in the liver, gallbladder, or associated ducts. The parasite may also be found in cattle, horses, goats, rabbits, pigs, dogs, squirrels, other animals and men, and its distribution is worldwide. The overall incidence of infection in man is less than one per cent. One of the distinguishing features of the adult fluke is the generous branching of most of the conspicuous organs of the body. In a poorly prepared specimen it is difficult to identify the testes, ovaries, digestive canals,

Fig. 8-12. *Fasciola hepatica* (photomicrography by Zane Price). (Markell and Voge, Diagnostic Medical Parasitology, courtesy of W. B. Saunders Co.)

and uterus because of the complicated branching or coiling of all of them.

The life cycle (Fig. 8–13) begins with fertilized eggs, which pass from the uterus of the worm into the gallbladder, bile or hepatic ducts of the host.[8] These eggs are about 140 μ long, are operculated and are carried to the intestine, leaving the body with feces. If they get into water, the opercula fly open and the ciliated miracidia are liberated. The miracidia in the eggs prepare for escape by producing an enzyme that digests the substance

that holds the operculum to the shell. Factors within the host as well as external environmental conditions seem to activate this enzyme of the egg. The substance binding the operculum of the egg to the shell probably consists primarily of a protein, thus a proteolytic enzyme would seem to be involved. This enzyme apparently functions from inside the shell, because it has no influence on hatching when it is applied to the outside. Other concepts to explain the mechanism of hatching have been proposed. One of the

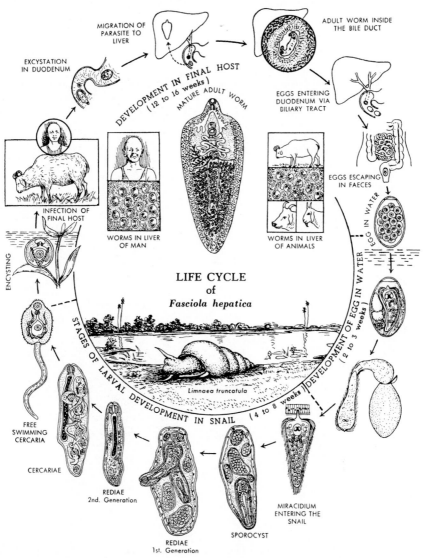

Fig. 8-13. Life cycle of *Fasciola hepatica*, a liver fluke of sheep, goats and cattle, and, occasionally, of man. (Chatterjee, Human Parasites and Parasitic Diseases, courtesy of Dr. K. D. Chatterjee.)

latest[48] postulated an initial response of the enclosed miracidium to light. Light activates the miracidium, which then alters the permeability of a viscous cushion that lies beneath the operculum. The cushion expands, internal pressure is built up until the operculum ruptures and flies back, liberating both cushion and miracidium.

The miracidium swims around for a few hours and then penetrates the body of a snail (e.g., *Lymnea*). Within the snail, each miracidium develops into a sporocyst that produces the first generation of rediae, which in turn produces the second generation of rediae. During the hot summer there is usually only one rediae generation. Rediae give rise to cercariae that leave the snail, often at night, and swim to aquatic vegetation and encyst as metacercariae. One or two months are re-

quired for development from the miracidial to the metacercarial stage. The cyst wall of a metacercaria of the liver fluke, *Fasciola hepatica*, is composed of four major layers[10]: (1) an outer tanned-protein layer with irregular meshwork containing cigar-shaped bodies; (2) a finely fibrous layer, composed mostly of mucopolysaccharide; (3) an inner layer, also mostly mucopolysaccharide; (4) a dense compact layer composed of numerous protein sheets stabilized by disulphide linkages and formed from tightly wound scrolls.

The vertebrate host acquires infection by ingesting the metacercariae with water plants or drinking water. In the intestine of man, the parasite excysts and migrates through the gut wall and the body cavity to the liver, where it takes up residence in bile passages. Two months are required to reach maturity

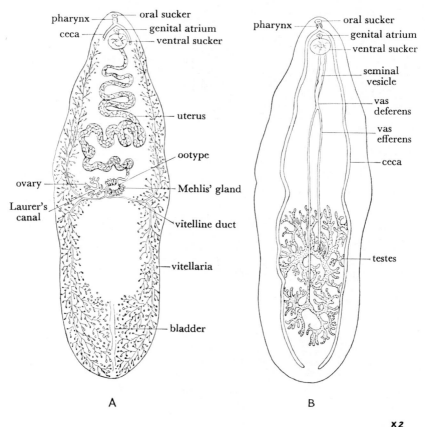

A   B

X2

Fig. 8-14. *Fasciolopsis buski.* (*A*) Female reproductive organs, ventral view. (*B*) Male reproductive organs, ventral view. (Adapted from Odhner, 1902.) (Belding, Textbook of Clinical Parasitology, courtesy of Appleton-Century-Crofts.)

in the liver or gallbladder. Eggs pass down the bile ducts into the intestine and out to soil or water. The entire life cycle may require as much as five months for completion. For an excellent review of the relationships between various species of *Fasciola* and their molluscan hosts, see Kendall.[20]

The degree of pathogenicity of liver flukes to man depends on many factors, particularly the number of worms present and the organs infected. Mechanical and toxic damage are characteristic. The parasite occasionally gets into the lungs, brain, or other tissues. Pain in the region of the liver, abdominal pain, diarrhea, fever, and anemia are some of the usual symptoms. A study of experimentally infected mice has shown that young flukes feed on hepatic cells and that "blood is inevitably included in the diet."[8] If infected raw liver is eaten, immature flukes may become attached to the buccal or pharyngeal membranes, causing pain, irritation, hoarseness or coughing. This type of *Fasciola* infections is known as **halzoun**. Diagnosis of fascioliasis depends primarily on finding the eggs in stool specimens.

Fasciola gigantica. This fluke has a life cycle similar to that of *F. hepatica* and is common in cattle and other ruminants in Asia, Africa and a few other areas. It is exceedingly rare in man. *F. gigantica* is not as broad as *F. hepatica* but is longer, measuring 25 to 75 mm. × 3 to 12 mm.

Fasciolopsis buski. This big relative of *Fasciola hepatica* may reach 75 mm. in length, but it is usually smaller (Fig. 8–14). Its life cycle is similar to that of *F. hepatica*, but the adult lives in the small intestine of man and pigs. Most cases of infection have been found in China, but there is a low incidence in India. Man acquires the fluke by eating water chestnuts (*Eliocharis*), the nuts of the red caltrop (*Trapa*) or other water plants on which the metacercariae attach themselves.

Fascioloides magna. This large liver fluke lives in cattle, sheep, moose, deer and horses. It is an oval fluke that may reach 26 × 100 mm. in size. The life cycle is similar to that of *Fasciola hepatica*.

Little work has been done on toxins or other injurious substances that these worms might produce. The lethal effects of extracts of some worm tissues on bacteria, however, can be demonstrated by grinding worms with sand, adding saline, filtering, and adding the filtrate to bacteria cultures. Filtrates thus obtained from *Fasciola gigantica* and *Ascaris vitulorum* show marked bacteriostatic action.[14]

SUBORDER PARAMPHISTOMATA

Family Paramphistomatidae (=Amphistomidae)

The anterior end of these flukes possesses a mouth but no sucker. The single sucker, or acetabulum, is located at the posterior end. The general shape of the body, which is often covered with papillae, is unlike the typical leaf-shape of other flukes—it frequently is rounded and sometimes looks more like a gourd or a pear with a hole in the top end. Common genera are *Homalogaster*, *Gastrodiscus*, *Watsonius* and *Gigantocotyle*. The life history of *Gigantocotyle explanatum*, a liver fluke of domestic ruminants in India, is illustrated in Figure 8–15.

Watsonius watsoni. This species is normally found in monkeys but has once been reported from man. It is about 9 mm. long and was found in a man's intestine in large numbers where it was apparently pathogenic.

Gastrodiscoides (=Gastrodiscus) hominis (Fig. 8–16). *G. hominis* is common in the intestine of man in some areas of India, Assam and a few other places. The worm is about 6 mm. long and possesses a large disc-like portion of the body which has a sucker.

Paramphistomum microbothrioides. This short, thick fluke possesses a sucker around the genital pore. It lives in the rumen of cattle and other ruminants, and may be found in South America and the United States, although the identity of the species in these two countries has been questioned. In appropriate snail hosts the usual sequence of sporocyst, redia, and cercaria appear. The cercariae leave the snail and encyst as metacercariae on vegetation. No evidence exists that the adult flukes are pathogenic, but the immature forms in the small intestine cause

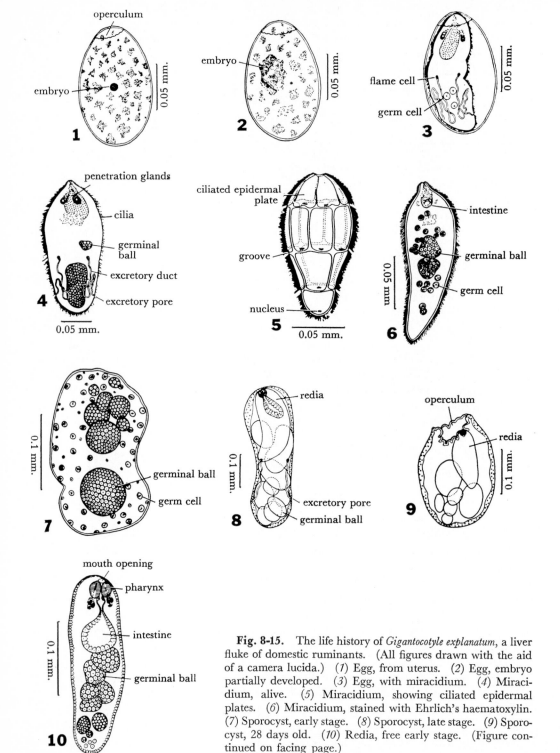

Fig. 8-15. The life history of *Gigantocotyle explanatum*, a liver fluke of domestic ruminants. (All figures drawn with the aid of a camera lucida.) (*1*) Egg, from uterus. (*2*) Egg, embryo partially developed. (*3*) Egg, with miracidium. (*4*) Miracidium, alive. (*5*) Miracidium, showing ciliated epidermal plates. (*6*) Miracidium, stained with Ehrlich's haematoxylin. (*7*) Sporocyst, early stage. (*8*) Sporocyst, late stage. (*9*) Sporocyst, 28 days old. (*10*) Redia, free early stage. (Figure continued on facing page.)

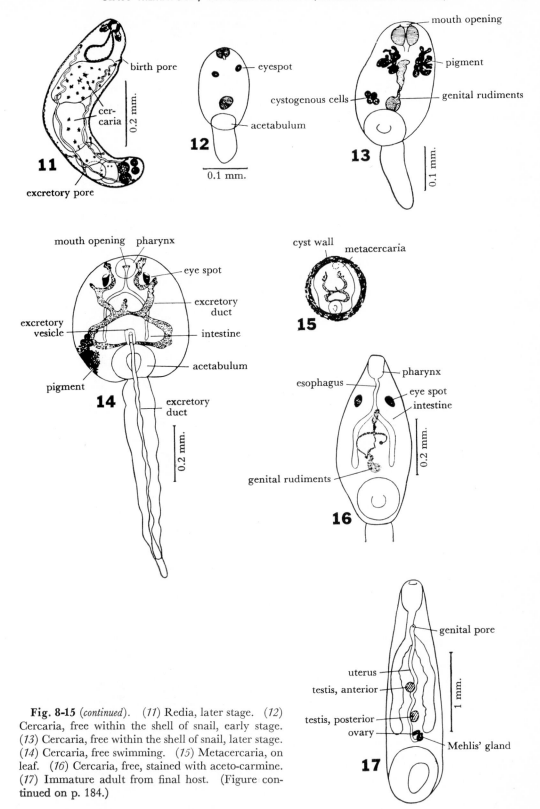

Fig. 8-15 (*continued*). (*11*) Redia, later stage. (*12*) Cercaria, free within the shell of snail, early stage. (*13*) Cercaria, free within the shell of snail, later stage. (*14*) Cercaria, free swimming. (*15*) Metacercaria, on leaf. (*16*) Cercaria, free, stained with aceto-carmine. (*17*) Immature adult from final host. (Figure continued on p. 184.)

inflammation, edema, hemorrhage, and destruction of the intestinal villi.

SUPERORDER EPITHELIOCYSTIDIA

The cercaria possesses a thick-walled epithelial bladder and a single tail that is reduced in size or lacking.

Order Plagiorchiida

Family Dicrocoeliidae

These flattened and somewhat translucent flukes occur in the gut, gallbladder, bile ducts, liver, or pancreatic ducts of amphibians, reptiles, birds, and mammals. The ovary is situated behind the testes and the uterus fills most of the posterior part of the body. Common genera are *Eurytrema*, *Dicrocoelium* and *Brachycoelium*.

Eurytrema pancreaticum. This is a parasite (Fig. 8–17) of the pancreatic ducts of hogs, water buffaloes and cattle in the Orient and of some camels and monkeys in the Old World. It has also been reported from the duodenum of sheep and goats in South America, and it was once found in man. The fluke measures about 12 × 7 mm. Cattle in Kuala Lumpur acquire the fluke by eating grasshoppers containing metacercariae.[3]

Dicrocoelium dendriticum (= D. lanceolatum). This elongated, narrow fluke measures about 8 × 2 mm. and is known as the **lancet fluke** (Fig. 8–18). It occurs in the bile ducts of cattle, sheep, goats, horses, camels, rabbits, pigs, deer, elk, dogs, and occasionally man; distribution is worldwide.

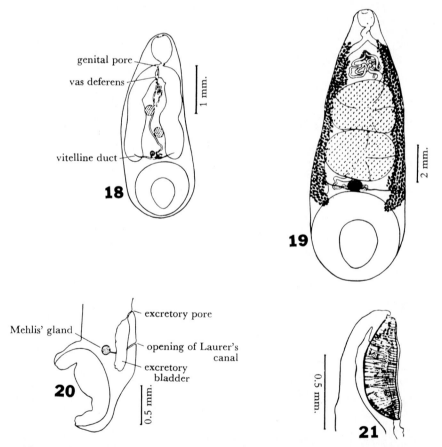

Fig. 8-15 *(continued).* *(18)* Immature adult from final host. *(19)* Gravid adult from final host, ventral view. *(20)* Adult, sagittal section of posterior half. *(21)* Adult, sagittal section of pharynx. (Singh, courtesy of J. Parasitol.)

The fluke has a well-developed ventral sucker that is larger than the anterior sucker, an unbranched ceca, slightly lobed testes that are almost tandem, no spines on the cuticle, and vitelline glands that occupy the middle third of the body. The eggs of *D. dendriticum* must be eaten by an appropriate snail before they will hatch. The snails are land forms (*Cionella lubrica*), so a cycle of larval growth without swimming stages had to develop. There are two sporocyst stages, and the cercariae aggregate in masses called slime balls in the respiratory chamber of the snail.

These slime balls are deposited on the soil or grass as the snail crawls. Although sheep or other final hosts may possibly acquire the infection by eating the slime balls or even the snails, ants (*Formica fusca*) serve as secondary hosts by eating these masses of cercariae, and the sheep acquire their infections by eating the ants.[23]

Family Fellodistomatidae

Sexually mature flukes belonging to the genus *Gymnophallus* inhabit the gallbladder, intestine, ceca and bursa-Fabricii of shore

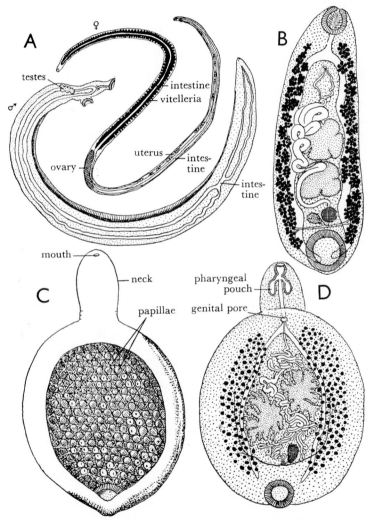

Fig. 8-16. Schistosomatidae (*A*) and Paramphistomatidae (*B–D*) of mammals. (*A*) *Schistosoma bovis*, male and female. (*B*) *Paramphistomum cotylophorum.* (*C,D*) *Gastrodiscus*, showing external characters in ventral view (*C*) and internal organs (*D*). (Dawes, The Trematoda, courtesy of the University Press, Cambridge.)

7

birds. Asexual generations of these parasites occur in bivalve mollusks, and they produce cercariae (from sporocysts) that belong to the Dichotoma group of furcocercous larvae. Some of the intermediate hosts from which cercarial stages and sporocysts of this genus have been reported are: *Mytilus edulus*, *Macoma balthica*, *Scrobicula tenuis*, *Cardium edule*, and *Hiatella arctica*. Upon emerging from the first intermediate host, the cercariae attach themselves to the mantle or body-wall of bivalve or gastropod mollusks, where they develop as unencysted metacercariae that almost reach definitive size and are easily mistaken for adults. These metacercariae produce lesions on the mantles of their hosts, sometimes accompanied by deposition of nacreous material. Such pearly formations in the mantle of the *Mytilus edulus*, *Donax truncatus* and other bivalves are not uncommon, and have often been reported as caused by a distome trematode.

Fellodistomum felis. This fluke inhabits the gallbladder of catfish. It has an unusual life cycle in that the sporocyst and cercariae develop in clams whereas the metacercariae develop in starfish. The final host acquires the infection by eating starfish.

Family Microphallidae

This group somewhat resembles the Heterophyidae. They are small, some species being less than 1 mm. long, although many are closer to 2 mm. in length. The body tends to be thick, pear-shaped and spiny. The two branches of the intestine are unusually short and there is a penis papilla. The usual host is a shore bird such as a sea gull, plover, godwit, sandpiper or tern in which the parasites are found in the intestine.

Representative genera are *Microphallus*, *Maritrema*, *Levinseniella*, *Spelotrema*, and *Spelophallus*.

Levinseniella minuta. This tiny fluke

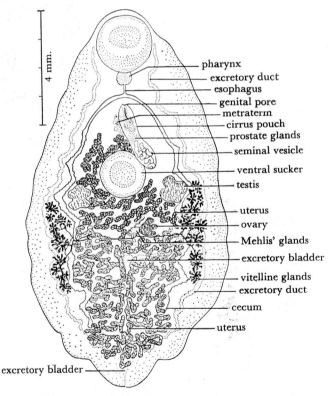

Fig. 8-17. *Eurytrema pancreaticum.* (Looss, *In* Chandler and Read, An Introduction to Parasitology, courtesy of John Wiley and Sons.)

averages only about 0.16 mm. long and 0.1 mm. wide. Its general appearance and life history is given in Figure 8–19.[45] The normal vertebrate hosts are scaups or other diving ducks, but white mice easily become infected in the laboratory. Snails eat the fluke eggs, which hatch into miracidia and then develop into sporocysts. Cercariae emerge from the sporocysts and encyst, forming the metacercaria stage in the snail. Ducks become infected by eating infected snails.

Family Plagiorchiidae

These flukes are found mainly in the intestines of frogs, snakes, and lizards, but they may occur in amphibians, birds and bats (Fig. 8–20). The family is large, containing hundreds of species. The flukes are small or only moderately large and are covered with spines; the two suckers are not far apart and there is a Y-shaped urinary bladder. The parasites also live in the respiratory tract, urogenital tract and sometimes the gallbladder. After leaving the snail, young worms (cercariae) penetrate aquatic insect larvae (Fig. 8–21), crayfish, or tadpoles. The vertebrate host acquires the infection by eating one of these transport hosts. Common genera are *Plagiorchis*, *Glypthelmins*, *Opisthoglyphe*, *Renifer*, *Haematoloechus* (= *Pneumoneces*, the common lung fluke of frogs), *Prosthogoni-*

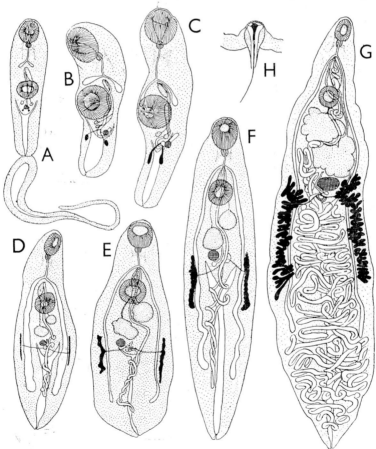

Fig. 8-18. Stages of *Dicrocoelium dendriticum* developing in the definitive host. (*A*) Cercaria vitrina, removed from cyst. (*B–F*) Juvenile flukes collected after infection: *B*, 8 days old; *C*, 9 days old; *D*, 12 days old; *E*, 14 days old, and *F*, 16 days old. (*G*) Full-grown fluke. (*H*) A touch papilla on the suckers of the adult. (After Neuhaus, 1939) The magnifications of *ABC/DEF/G* stand in the relation 5:1/2:4/1. (Dawes, The Trematoda, courtesy of the University Press, Cambridge.)

mus and *Sympetrum*. Most of the flukes are only a few millimeters long, but some may reach 20 mm. in length or more.

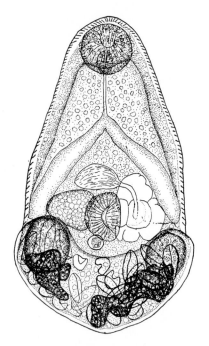

Fig. 8-19. *Levinseniella minuta.* Adult specimen, 0.17 mm. long, from a white mouse. (Stunkard, courtesy of J. Parasitol.)

Family Allocreadiidae

Common intestinal flukes of marine and freshwater game fish such as trout and bass belong to this family. *Crepidostomum, Bunodera, Allocreadium,* and *Pharyngora* are the more common genera. Usually they are small, often not longer than 1 or 2 mm. The cuticle is not spinous. In some genera (*Crepidostomum* and *Allocreadium*), the rediae are found in small bivalves and the metacercariae occur in other bivalves, crayfish, amphipods or mayfly nymphs. Other genera utilize snails as the first intermediate host and various other animals as the second intermediate host. *Helicometrina elongata* (Fig. 8–22), about 4 mm. long, lives in the small intestine of the tidepool fish *Gobiesox meandricus*, in Bodega Bay, California.

Family Troglotrematidae

Most of the genera belonging to this family are less than 10 mm. long and some may be only 1 or 2 mm. in length. In general, the group is composed of rather thick, spinous or scaly flukes that have an oval outline and live in various organs and sinuses of birds and mammals.

The genus *Paragonimus* has special interest

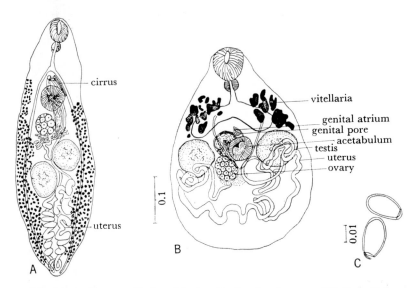

Fig. 8-20. (*A*) *Plagiorchis vespertilionis.* (*B*) *Acanthatrium jonesi;* eggs. (*C*) Both flukes are found in Korean bats. All figures drawn with the aid of a camera lucida. The projected scale indicates millimeters in each figure. (Sogandares-Bernal, courtesy of J. Parasitol.)

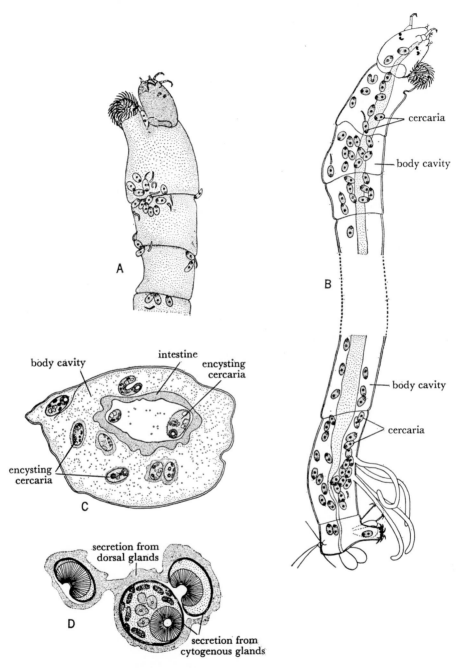

Fig. 8-21. Chironomid (midge) larvae infected with fluke larvae. (*A*) Anterior extremity of chirono-mid larva showing accumulation of cercariae on the surface in the junctions between the segments. (*B*) Anterior and posterior extremities of chironomid larva showing numerous cercariae in body cavity, not yet encysted. (*C*) Transverse section through chironomid larva showing encysting cercariae in body cavity and alimentary canal. (*D*) Three cercariae in process of encysting, from body cavity of chironomid larva. (Rees, courtesy of Parasitology.)

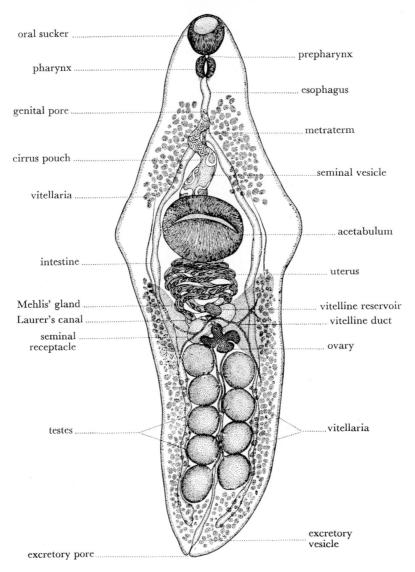

Fig. 8-22. *Helicometrina elongata* from a tidepool fish. × 42. Drawn from a stained whole mount with the aid of a camera lucida. Lateral bulges in the acetabular region are partially due to pressure during fixation. The vasa efferentia and the interconnections of the smaller female ducts were added to the drawing from studies on sectioned material. (Noble and Park, courtesy of Trans. Amer. Micr. Soc.)

because it contains the common lung fluke of man, *Paragonimus westermani*. The following account of the genus is from various sources, but primarily from the thorough review by Yokogawa.[50]

Species generally considered valid at the present time are: *Paragonimus westermani* (*P. ringeri*, *P. edwardsei*), found in man, the dog, cat, pig, and many other mammals; *P. kell-*

icotti, in the cat, dog, pig, mink, goat, fox and experimental animals; *P. ohirai*, in the pig, dog, weasel, badger, wild boar and experimental animals; *P. iloktsuensis*, in the rat and dog; and possibly *P. compactus*. These lung flukes all have the same general characteristics and life cycle.

Paragonimus westermani. This is one of the most important trematodes of man in

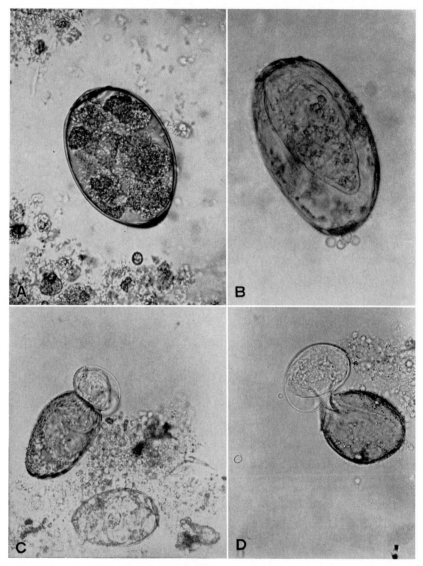

Fig. 8-23. Photographs showing hatching of the egg of the lung fluke *Paragonimus westermani.* (Courtesy of Dr. K. J. Lie.)

some areas of the world, especially in the Far East. It has been found in Korea, Japan, China mainland, Taiwan, Philippines, Indonesia, Africa, India, some Pacific islands, Peru and Ecuador. The incidence of infection ranges from 7 to 44 per cent.

Anatomic details of an adult *Paragonimus westermani* are shown in Fig. 8–24. There is no true cirrus or cirrus pouch. The fluke usually measures about $10 \times 5 \times 4$ mm. in size. It is plump, reddish in color and

covered with minute cuticular spines. It usually lives in the lungs, often paired with another fluke, and is enclosed within a host-produced cyst. Many other organs, however, may harbor the worms.

Eggs (Fig. 8–23), which average 87×50 μ, are brownish in color and sometimes are present in such numbers that they give the sputum a brown or rust colored tinge. These operculated eggs are coughed up from the lungs and expectorated, or swallowed and

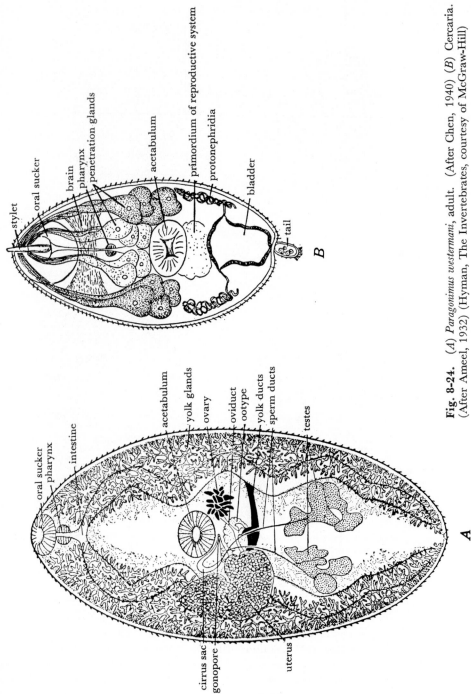

Fig. 8-24. (*A*) *Paragonimus westermani*, adult. (After Chen, 1940) (*B*) Cercaria. (After Ameel, 1932) (Hyman, The Invertebrates, courtesy of McGraw-Hill)

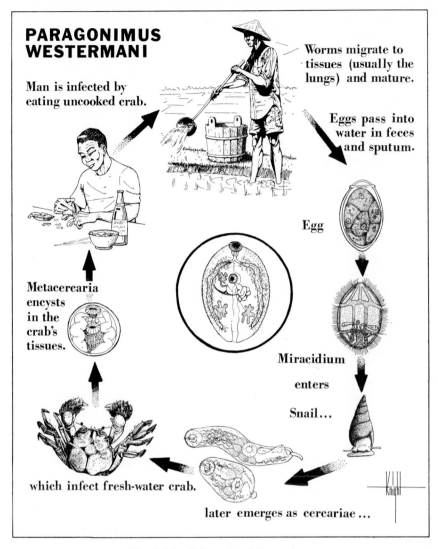

PARAGONIMUS WESTERMANI

Man is infected by eating uncooked crab.

Worms migrate to tissues (usually the lungs) and mature.

Eggs pass into water in feces and sputum.

Egg

Metacercaria encysts in the crab's tissues.

Miracidium

enters

Snail...

which infect fresh-water crab.

later emerges as cercariae...

Fig. 8-25. Life cycle of *Paragonimus*.

leave the body with feces. After two to several weeks of development in moist soil or water, they hatch and the miracidia swim about until a suitable snail appears. See Figure 8–25 for the life cycle.

Hatching is initiated by a sudden increase in activity of the larva inside the egg. The animal may at first lunge at the operculum but it soon flattens its anterior end against this structure. A series of violent contractions then occurs and the posterior portion of the body is extended until the operculum suddenly flies back as on a hinge. Water passes

through the exposed vitelline membrane causing it to swell and extend through the opercular opening. A bubble is produced which may become larger than the egg itself, forming a transparent sac into which the miracidium finally forces itself. The larva swims rapidly about inside this sac until the membrane is ruptured and the ciliated worm swims away.[1]

Mature miracidia average 0.08×0.04 mm. in size and are covered with four rows of ciliated ectodermal plates. *Oncomelania* and several other genera of snails are common first intermediate hosts. These snails are usually found in swift mountain streams, often away from towns or villages. Infection of the snails probably depends primarily on animal reservoir hosts rather than on man. The incidence in snails is surprisingly low. Out of hundreds of snails examined in endemic areas in the Philippines, we have found only two per cent infected. Age immunity apparently protects some snails.[41] Miracidia shed their ciliated plates during the act of penetrating the snail. The mother sporocyst generation develops free in the lymphatic system, usually along the esophagus, stomach or intestine. First generation rediae emerge from the sporocyst about four weeks after infection. Second generation rediae appear in the digestive gland 63 to 75 days after infection.

Microcercous cercariae emerge from the snail. They are ellipsoidal and possess a short tail with a spiny tip. The body is covered with small spines and a prominent stylet lies in the dorsal side of the oral sucker. There are two types of penetration glands plus two irregular rows of mucoid glands on each side of the body. Cercariae of *Paragonimus westermani* in the laboratory do not readily escape from snails.

Cercariae enter a crayfish or a crab. Crabs belonging to the genera *Potomon*, *Eriocheir*, *Sesarma* and *Parathelphusa* are the usual second intermediate hosts. Crabs probably ingest the cercariae. Possibly cercariae penetrate the crabs immediately after molting when the joint membranes are soft, although Yokogawa[49] believes this penetration does not occur since he found metacercariae only in blood vessels. However, metacercariae (0.34 to 0.48 mm.) in the crustaceans may be found in the gills, hearts, muscles of the legs and body, liver or reproductive organs. Infected crayfish usually belong to the genera *Cambaroides* and *Cambarus*. The time between infection of the crustacean and the appearance of mature metacercariae in its tissues may be several weeks to several months.

When a new host eats uncooked or poorly cooked crayfish or crabs, the metacercariae excyst in the intestine within an hour of ingestion. The minute worms are pinkish in color and possess a large, elongate, sac-like bladder. They penetrate the intestinal wall and usually go through the diaphragm to the pleural cavity. They may wander around the abdominal cavity for 20 days or more and enter various organs in which they may reach sexual maturity. Even after penetrating the diaphragm, they may never get to the lungs but enter the liver or other tissue. A heavy infection of the liver frequently is observed in experimental animals. In animals, usually two worms exist in a lung cyst, but this phenomenon rarely occurs in man, one worm being the rule. It has been observed, in experimental animals, that worms possibly locate one another by chemical attractants.[42] Eggs appear in the sputum $2\frac{1}{2}$ to three months after a man eats infected crustacea.

Excysted metacercariae have been observed to enter wounds and mucous membranes of experimental animals. Some vertebrate hosts may acquire infection by eating metacercariae freed from their crab hosts. In an aquarium in which infected crabs are kept, a few free metacercariae may be present. These cysts might have fallen from tissues when crabs fought. Crabs are often prepared for cooking on chopping blocks, and the juice left on the block may contain metacercariae that adhere to foods subsequently prepared on the same block or to the fingers of the cook. The juice of fresh crabs is used in various preparations in Korea and the the Philippines. Crabs are often "pickled" or salted or dipped in wine or vinegar before eating. These practices rarely kill all metacercariae. The encysted parasites can live several days in a dead crab, one or two days in diluted wine, or three weeks in an ice chest at $10°$ C. in ten per cent formalin.

Undoubtedly some immunity to paragonimiasis develops. The incidence of infection is much lower in children under ten years old than in older children or adults. A host sex factor seems to be operative, since in Korea 80 per cent of the cases in children are boys, although this may simply reflect a difference in the eating habits or other behavior of boys.

Symptoms of infection include a cough, profuse expectoration, occasional appearance of blood in sputum, pain in the chest, brown sputum and muscular weakness. Pathology involves bleeding spots where worms have penetrated tissues, leukocyte infiltration, tearing of muscles, scar tissue, host allergic reaction, and fibrous cyst formation around the worms. These cysts may contain blood-tinged material, purulent chocolate-colored fluid, worm eggs, living worms, dead worms or no worms. It is interesting to note that hemoglobin taken from the worms has been reported to be different from that in the red blood cells of the host. Apparently *Paragonimus* employs the heme moiety of the host directly in the formation of its own pigments.[22] Migration of the worms to other tissues or organs may be troublesome. Cerebral involvement can be extremely serious. Eggs distributed widely throughout the host cause inflammation, and secondary bacterial infection may occur.

Diagnosis consists primarily of finding eggs in sputum or feces, but the clinical picture also has value. Intradermal and complement fixation tests are useful.

Control should involve treatment of patients, elimination of reservoir hosts, disinfection of sputum and feces of patients, destruction of snails, crabs and crayfish, prevention of human infection by cysts that become free from the arthropod host, and education against eating raw or poorly cooked or preserved crabs or crayfish. See Yokogawa[50] for a comprehensive review of *Paragonimus* and paragonimiasis.

Nanophyetus (=Troglotrema) salmincola. This parasite is known by the curious name **salmon poisoning fluke.**[32a] Microbial pathogens are commonly transmitted to vertebrate hosts by ectoparasites, but some years ago parasitologists were surprised to discover that this trematode has taken the place of an arthropod vector of a rickettsial agent that causes a serious disease in dogs. The tiny fluke is roughly oval in shape, about a millimeter in length, and lives in the intestine. Eggs are about $40 \times 70\,\mu$ in size. Various

snails (e.g., *Goniobasis silicula*) serve as the first intermediate host within which cercariae similar to those of *Paragonimus westermani* develop. These cercariae penetrate salmon, other salmonid and nonsalmonid fish, and salamanders (*Dicamptodon ensatus*), in which they enter the kidneys.[13] The parasites become infected with a rickettsia, *Neorickettsia helminthoeca*, which causes a serious disease in the natural final host that eats the salmon. The dog, fox, raccoon, and other fish-eating mammals harbor the fluke and may thus become infected with the rickettsia. Fever, vomiting, and dysentery are some of the more severe symptoms. The mortality rate is fairly high, but recovery confers immunity.

Order Opisthorchiida

Family Opisthorchiidae

The habitat of these lanceolate distomes is the bile passages of vertebrates. Suckers are small, the ovaries are located in front of the testes, and a genital pore occurs just in front of the ventral sucker. The life cycle involves a snail and fish or amphibian intermediate hosts, and the usual stages of miracidia, sporocyst, redia, cercaria, and metacercaria.

Opisthorchis (= Clonorchis) sinensis. The Chinese liver fluke is the best known member of the family. Its prevalence in China is evident from the common name, but it is also found in Korea, Japan and other southeast Asian countries. The incidence of infection in man reaches 15 to 70 per cent in local areas and 100 per cent in reservoir hosts such as cats, dogs, tigers, foxes, badgers and mink. This fluke (Fig. 8–26) averages about 18×4 mm. in size. The eggs ($27 \times 16\,\mu$) are shaped somewhat like an electric light bulb with a distinct "door" or operculum on top. They are readily killed by desiccation, but can live six months at $0°$ C. if kept moist.

The life cycle (Fig. 8–27) includes a snail (e.g., *Parafossarulus* or *Bulimus*) that eats the eggs. Miracidia hatch in the snail rectum or intestine and penetrate the walls of these organs to become sporocysts. After leaving the snail, cercariae swim to freshwater fish where the larval parasites penetrate the skin

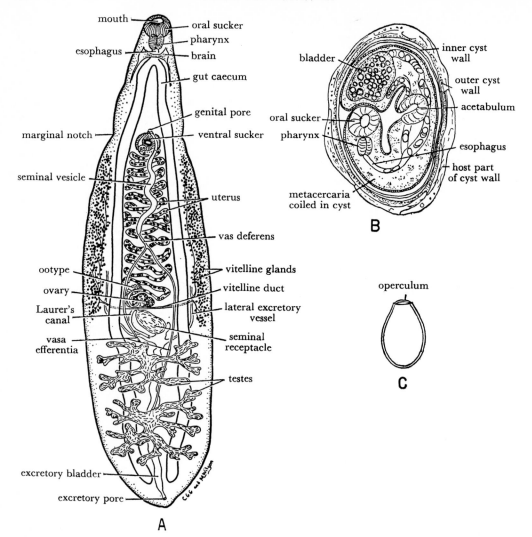

Fig. 8-26. *Opisthorchis sinensis.* (*A*) Dorsal view of adult. (Brown, Selected Invertebrate Types, courtesy of John Wiley & Sons.) (*B*) Metacercaria. (After Komiya and Tajimi, 1940.) (*C*) Capsule. (After Looss, 1907.) (*B, C* from Hyman, The Invertebrates, vol. II, courtesy of McGraw-Hill.)

and encyst as metacercariae. At least 80 species (ten families) of fish, most belonging to the family Cyprinidae in China, Korea and Japan, have been incriminated as second intermediate hosts, but only about 12 of them are chiefly responsible for human infection.[51]

Man gets the infection by eating uncooked, infected fish. The normal route of this fluke to the liver appears to be by way of the bile duct. The incubation period in man is about three weeks. Although a few dozen worms constitute an average infection, one case having 21,000 worms was reported! See Komiya[21] for a review of *Clonorchis* and clonorchiasis.

Symptoms, if any, include eosinophilia, nausea, epigastric pain, edema, diarrhea, vertigo, fluid in body tissues, and wasting of the body. Prevention rests mainly on cooking all freshwater fish from endemic areas. Diagnosis is confirmed by finding eggs in fecal samples.

Opisthorchis felineus (=O. tenuicollis). This species is found in much of Europe and the Orient. It averages about 10 × 2.5 mm. in size, and lives in the liver of man, cats, dogs, foxes and other mammals. The life cycle is similar to that of *O. sinensis*, eggs being eaten by the snail, *Bulimus*. Liberated cercariae possess penetration glands and large tails, and they are active swimmers. They become attached to various species of fish, drop their tails, penetrate the skin under scales, and become metacercariae in the skin or muscles. Man and other vertebrate hosts become infected by ingesting uncooked or partially cooked, infected fish. The mature metacercariae in fish sometimes reach a diameter of 300 μ, so they can be seen as tiny spots with the naked eye. In the vertebrate host they excyst and make their way to the bile ducts of the liver where they mature.

A few dozen of these worms in the human liver seem to cause no appreciable harm. When the number gets into the hundreds, there may be pain, congestion, liver enlargement, bile stones, and cirrhosis.

Family Heterophyidae

These small intestinal flukes live in birds and mammals including dogs, cats, and man. The parasites are usually only 1 or 2 mm. long, rarely reaching 3 mm. Some are distomes and some monostomes; the cuticle is scaly, at least on the anterior part of the body. The general shape of the body is ovoid or pyriform, and the ovary is located in front of the testes or testis. The metacercarial stage occurs in fish or amphibians.

Heterophyes heterophyes. This minute fluke (Fig. 8-28) averages 1.3 × 0.35 mm.; it may be pear-shaped, oval or somewhat elongated. The outer surface is covered with spine-like scales. The midventral surface seems to possess two adjacent suckers, but careful examination shows that one of them is the genital opening. This organ is retractile, and it may be called a genital sucker or a **gonocotyl**. The oval eggs average about 28 × 15 μ, thus they are smaller than many protozoan cysts. They resemble the eggs of

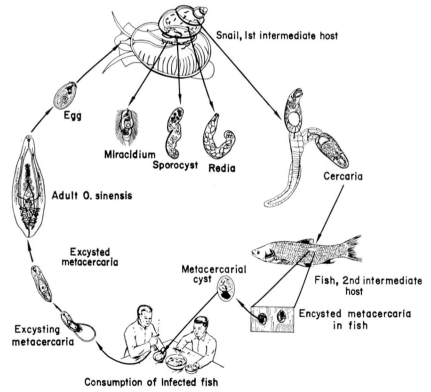

Fig. 8-27. Life cycle of *Opisthorchis sinensis*. (Yoshimura, courtesy of J. Parasitol.)

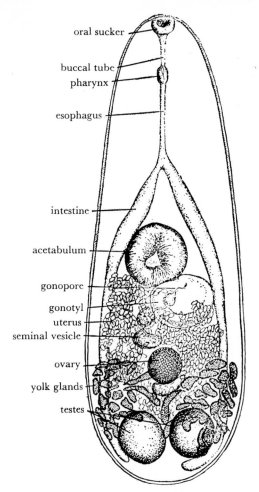

oral sucker

buccal tube
pharynx

esophagus

intestine

acetabulum

gonopore

gonotyl
uterus
seminal vesicle

ovary

yolk glands

testes

Fig. 8-28. *Heterophyes heterophyes*, intestinal fluke of man. (After Witenberg, 1929) (Hyman, The Invertebrates, vol. II, courtesy of McGraw-Hill)

related species and must be eaten by the right species of snail to survive. Metacercariae occur in various fish that may be ingested by man. The adult parasite lives in the intestine, normally causing little harm, but it may produce pain and diarrhea. Eggs of this species and related forms may also find their way into the blood and cause mild to serious trouble in organs into which they filter. The fluke is especially prevalent throughout the Orient. There are a number of related species that occasionally infect man (*Cryptocotyle lingua, Haplorchis yokogawai, Heterophyes katsuradai, H. brevicaeca,* and *Diorchitrema* sp.). *Cryptocotyle lingua* lives in dogs be-

tween the intestinal villi and in the lumen, and causes considerable damage due, in the main, to mechanical factors. A heavy infestation causes the production of a great deal of mucus, sloughing of tissue, pressure atrophy and necrosis.

Metagonimus yokogawai. A more important related fluke is *M. yokogawai* (Fig. 8–29). It, too, lives in the small intestine of man, dogs, and cats, and it has even been found in pelicans. It occurs in the Orient and a few other places. This fluke is only slightly larger than *Heterophyes heterophyes*, and has a similar life cycle, with the metacercariae encysting in freshwater trout and in other salmonid and cyprinoid fishes. A diagnostic character of adults is the eccentric position of the ventral sucker, which is located about half way between the anterior and posterior ends of the body but to the right of the midline. After ingestion of the metacercariae by man, the young worms grow to maturity in two to three weeks. Infection may produce colicky pains and diarrhea.

Family Hemiuridae

Hemiurus. These flukes are elongated, cylindrical worms with a nonspinous tegument. They are sometimes called **appendiculate** flukes, and they vary in length from a few to 15 mm. A characteristic feature of the body of some is its division into an anterior **soma** and a posterior **ecsoma.** These two parts may be telescoped together, the ecsoma being withdrawn into the soma. The dividing line between the thick-walled anterior part and the thin-walled posterior part is often clearly evident as a constriction (Fig. 8–30). Often the vitellaria consist of a few large bodies rather than many scattered particles. The flukes usually inhabit the gut, stomach, gallbladder, esophagus, or pharynx of marine fish, and they are worldwide in their distribution. Other genera include *Aphanurus, Sterrhurus* and *Lecithochirium.*

Nonappendiculate genera include *Hysterolecitha* and *Derogenes.* The latter (Fig. 8–31) is the most widely distributed digenetic trematode of marine fish, being found in cod,

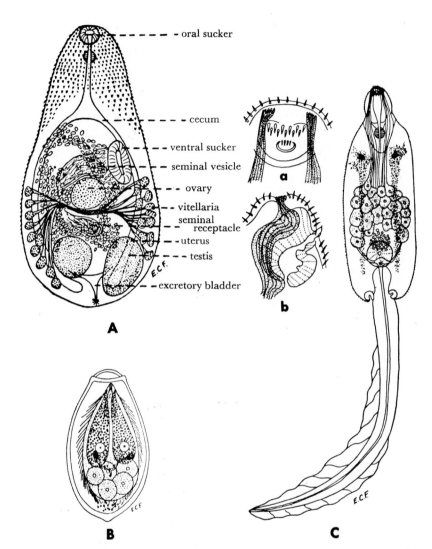

Fig. 8-29. (*A*) Adult specimen of *Metagonimus yokogawai*, ventral view. × 36.   (After Faust) (*B*) Egg of *M. yokogawai*, with mature miracidium. × 1300.   (After Faust) (*C*) Cercaria of *M. yokogawai*. ×200.   (*a, b*) Ventral and lateral aspects of anterior end, showing relationship of lytic gland ducts and integumentary spines to oral opening; greatly enlarged.   (Adapted in part from Faust, 1929, in part from Yokogawa, 1931) (Faust and Russell, Clinical Parasitology, Lea & Febiger)

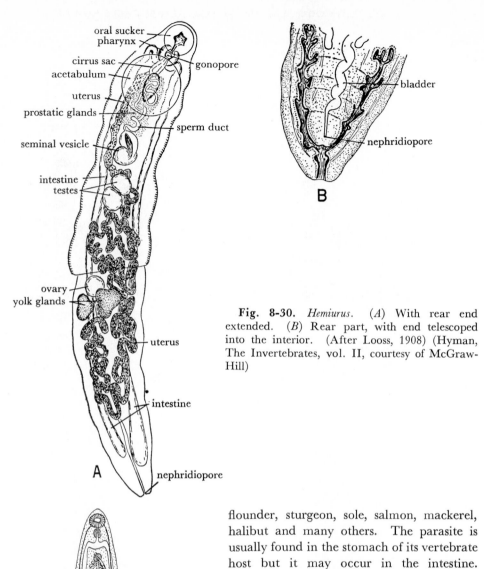

Fig. 8-30. *Hemiurus.* (*A*) With rear end extended. (*B*) Rear part, with end telescoped into the interior. (After Looss, 1908) (Hyman, The Invertebrates, vol. II, courtesy of McGraw-Hill)

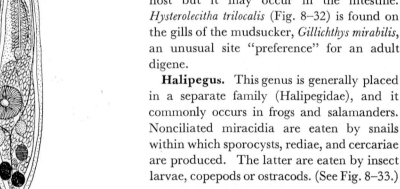

Fig. 8-31. *Derogenes varicus.* × 30. (Dawes, The Trematoda, courtesy of the University Press, Cambridge.)

flounder, sturgeon, sole, salmon, mackerel, halibut and many others. The parasite is usually found in the stomach of its vertebrate host but it may occur in the intestine. *Hysterolecitha trilocalis* (Fig. 8–32) is found on the gills of the mudsucker, *Gillichthys mirabilis,* an unusual site "preference" for an adult digene.

Halipegus. This genus is generally placed in a separate family (Halipegidae), and it commonly occurs in frogs and salamanders. Nonciliated miracidia are eaten by snails within which sporocysts, rediae, and cercariae are produced. The latter are eaten by insect larvae, copepods or ostracods. (See Fig. 8–33.)

Family Didymozoidae

These peculiar flukes usually live in pairs in cysts or cavities of their marine fish hosts. They may occur in the body cavity, kidney, body surface, esophagus, gut, musculature, subcutaneous tissue, pharynx or other places. The worms may have a broad posterior portion of the body (*Didymocystis*) or they may be thread-like (*Nematobothrium*), or have more bizarre shapes. Most of them are hermaphroditic, but in some (*Wedlia*, *Köllikeria*) the sexes are separate. The ventral sucker is absent in most members of this family, the pharynx is reduced or absent and the life histories are not known (Fig. 8–34). In at least one cyst-dwelling species (*Kollikeria filicollis*), the sexes are not entirely separate but the functional "males" and "females" are dimorphic.

A most unusual fluke, probably belonging to this family, may be found under the skin and between muscles of the sunfish, *Mola mola*. We have tried to remove the entire worm from the fish but find it an almost im-

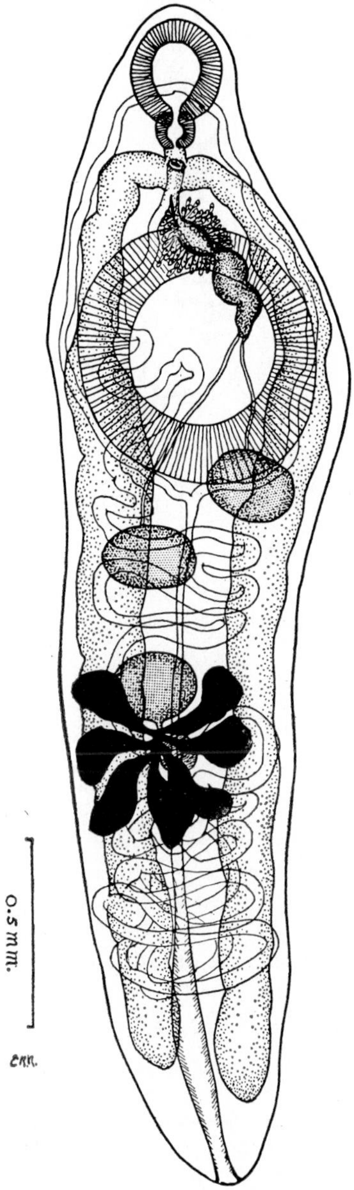

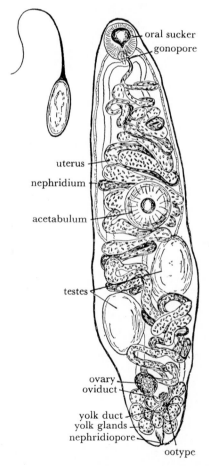

Fig. 8-32. *Hysterolecitha trilocalis.* Composite drawing of mature adult. The uterine coils have been reduced to numbers below average conditions and eggs omitted in order not to obscure other organs. Ventral view. (King and Noble, courtesy of J. Parasitol.)

Fig. 8-33. *Halipegus*, nonappendiculate hemiurid from the oral cavity of frogs. Capsule is shown at upper left. (After Krull, 1935) (Hyman, The Invertebrates, vol. II, courtesy of McGraw-Hill)

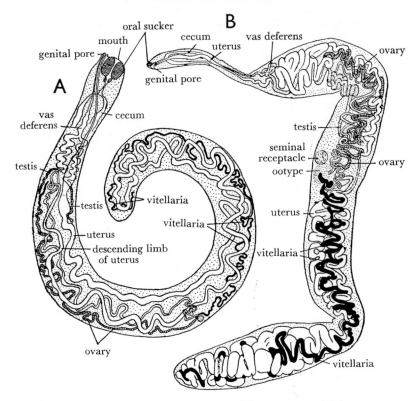

Fig. 8-34. Two species of *Didymozoon*. (*A*) *D. scombri*. (*B*) *D. faciale*. (Dawes, The Trematoda, courtesy of University Press, Cambridge.)

possible task. The parasite may be over 40 feet (13 meters) long and follows a tortuous course extending from one side of the host to the other, winding in and out through fin supports and dipping down between muscles and becoming tangled with a few to many other individuals.

REFERENCES

1. Ameel, D. J.: *Paragonimus*, its life history and distribution in North America and its taxonomy (Trematoda-Troglotrematidae). Amer. J. Hyg., *19*:279–317, 1934.

2. Armstrong, J. C.: Mating behavior and development of Schistosomes in the mouse. J. Parasitol., *51*:605–616, 1965.

3. Basch, P. F.: Completion of the life cycle of *Eurytrema pancreaticum* (Trematoda: Dicrocoeliidae). J. Parasitol., *51*:350–355, 1965.

4. Blanc, F., and Nosny, Y.: Nosographie des schistosomoses. Ann. Soc. Belge Méd. Trop., *47*:17–34, 1967.

5. Bruijning, C. F. A.: Immunology of schistosomiasis. Ann. Soc. Belge Méd. Trop., *47*: 117–126, 1967.

6. Bueding, E.: Carbohydrate metabolism of *Schistosoma mansoni*. J. Gen. Physiol., *33*:475–495, 1950.

7. Dawes, B.: The Trematoda (With Special Reference to British and Other European Forms.) Cambridge, at the Univ. Press, 1956.

8. ———: On the growth and maturation of *Fasciola hepatica* L. in the mouse. J. Helminthol., *36*:11–38, 1962.

9. De Paoli, A.: *Schistosoma haematobium* in the chimpanzee—a natural infection. Amer. J. Trop. Med. Hyg., *14*:561–565, 1965.

10. Dixon, K. E., and Mercer, E. H.: The fine structure of the cyst wall of the metacercaria of *Fasciola hepatica*. Quart. J. Micr. Sci., *105*:385–389, 1964.

11. Erasmus, D. A.: Ultrastructural observations on the reserve bladder system of *Cyathocotyle bushiensis* Kahn, 1962 (Trematoda: Strigeoidea) with special reference to lipid excretion. J. Parasitol., *53*:525–536, 1967.

12. Fife, E. H., Jr., Sleeman, H. K., and Bruce, J. I., Jr.: Isolation and characterization of a serologically active exoantigen of *Schistosoma mansoni* cercariae. Exper. Parasitol., *20*:138–146, 1967.

13. Gebhardt, G. A., Millemann, R. E., Knapp, S. E., and Nyberg, P. A.: "Salmon poisoning" disease. II. Second intermediate host susceptibility studies. J. Parasitol., *52*:54–59, 1966.

14. Gharib, H. M.: A preliminary note on the bacteriostatic properties of some helminths of animals. J. Helminthol., *35*:225–232, 1961.

15. Grodhaus, G., and Keh, B.: The marine, dermatitis-producing cercaria of *Austrobilharzia variglandis* in California. (Trematoda: Schistosomatidae). J. Parasitol., *44*:633–638, 1958.

16. Hoffman, G. L.: Synopsis of Strigeoidea (Trematoda) of fishes and their life cycles. U.S. Fish and Wildlife Serv. Fish Bull., *60* (175):439–469, 1960.

17. Hsü, S. Y. L., and Hsü, H. F.: New approach to immunization against *Schistosoma japonicum*. Science, *133*:766, 1961.

18. Hunter, G. W., III, Velleca, W. M., and Crandall, R. B.: Studies on schistosomiasis XXII. Cross resistance in *Schistosoma mansoni* and *Nippostrongylus brasiliensis* infections in albino mice. Exper. Parasitol., *27*:9–15, 1967.

19. Kagan, I. G.: Mechanisms of immunity in trematode infections. *In* Soulsby, E. J. L. (ed.): Biology of Parasites. p. 277–299. New York, Academic Press, 1966.

20. Kendall, S. B.: Relationships between the species of *Fasciola* and their molluscan hosts. *In* Dawes, B. (ed.): Advances in Parasitology. Vol. 3, pp. 59–98. New York, Academic Press, 1965.

21. Komiya, Y.: *Clonorchis* and clonorchiasis. *In* Dawes, B. (ed.): Advances in Parasitology. Vol. 4, pp. 53–106. New York, Academic Press, 1966.

22. Kruidenier, F. J., and Windsor, D. A.: Pigment of *Paragonimus kellicotti* Ward, 1908. J. Parasitol., *50*(Sect. 2, Supp.): 53, 1964.

23. Krull, W. H.: Experiments involving potential definitive hosts of *Dicrocoelium dendriticum* (Rudolph, 1891) Looss, 1899: Dicrocoeliidae. Cornell Vet., *46*:511–525, 1956.

24. Lewert, R. M., Hopkins, D. R., and Mandlowitz, S.: The role of calcium and magnesium ions in invasiveness of schistosome cercariae. Amer. J. Trop. Med. Hyg., *15*:314–323, 1966.

25. Lewert, R. M., and Mandlowitz, S.: Innate immunity to *Schistosoma mansoni* relative to the state of connective tissue. Ann. N.Y. Acad. Sci., *113*:54–62, 1963.

26. Lie, K. J.: Studies on Echinostomatidae (Trematoda) in Malaya. IX. The Mehlis' gland complex in echinostomes. J. Parasitol., *51*:789–792, 1965.

27. Lincicome, D. R. (ed.): Frontiers in research in parasitism: I. Cellular and humoral reactions in experimental schistosomiasis. Exper. Parasitol., *12*:211–240, 1962.

28. Luttermoser, G. W.: Infection of rodents with *Schistosoma mansoni* in ingestion of infected snails. J. Parasitol., *49*:150, 1963.

29. MacInnis, A. J.: Responses of *Schistosoma mansoni* miracidia to chemical attractants. J. Parasitol., *51*:731–746, 1965.

30. Malek, E. A.: The ecology of schistosomiasis. *In* May, J. M. (ed.): Studies in Disease Ecology. New York, Hafner Pub. Co., 1961.

31. McMullen, D. B.: Schistosomiasis control in theory and practice. Amer. J. Trop. Med. Hyg., *12*:288–295, 1963.

32. Michelson, E. H.: Miracidia-immobilizing substances in extracts prepared from snails infected with *Schistosoma mansoni*. Amer. J. Trop. Med. Hyg., *13*:36–42, 1964.

32a. Millemann, R. E., and Knapp, S. E.: Biology of *Nanophyetus salmincola* and "salmon poisoning" disease. *In* Dawes, B. (ed.): Advances in Parasitology. Vol. 8, pp. 1–41. New York, Academic Press, 1970.

33. Newsome, J.: Problems of fluke immunity: with special reference to schistosomiasis. Trans. Roy. Soc. Trop. Med. Hyg., *50*:258–274, 1956.

34. Nolf, L. O., and Cort, W. W.: On immunity reactions of snails to the penetration of the cercariae of the strigeid trematode, *Cotylurus flabelliformis* (Faust). J. Parasitol., *20*:38–48, 1933.

35. Oliver-Gonzalez, J.: Our knowledge of immunity to schistosomiasis. Amer. J. Trop. Med. Hyg., *16*:565–567, 1967.

36. Pan, Chia-Tung: Studies on the host-parasite relationship between *Schistosoma mansoni* and the snail *Australorbis glabratus*. Amer. J. Trop. Med. Hyg., *14*:931–976, 1965.

37. Sadun, E. H.: Immunization in schistosomiasis by previous exposure to homologous and heterologous cercariae by inoculation of preparations from schistosomes and by exposure to irradiated cercariae. Ann. N.Y. Acad. Sci., *113*:418–439, 1963.

38. Senft, A. W.: Recent developments in the understanding of amino acid and protein metabolism by *Schistosoma mansoni in vitro*. Ann. Trop. Med. Parasitol., *59*:164–168, 1965.

39. Smithers, S. R., Terry, R. J., and Hockley, E. J.: Do adult schistosomes masquerade as their hosts? Trans. Roy. Soc. Trop. Med. Hyg., *62*:466–467, 1968.

40. Smyth, J. D.: The Physiology of Trematodes. San Francisco, W. H. Freeman, 1966.

41. Sogandares-Bernal, F.: Studies on American paragonimiasis. I. Age immunity of the snail host. J. Parasitol., *57*:958–960, 1965.

42. ———: Studies on American paragonimiasis. IV. Observations on pairing of adult worms in laboratory infections of domestic cats. J. Parasitol., *52*:701–703, 1966.

43. Stenger, R. J., Warren, K. S., and Johnson, E. A.: An electron microscopic study of the liver parenchyma and schistosome pigment in murine hepatosplenic schistosomiasis mansoni. Amer. J. Trop. Med. Hyg. *16*:473–482, 1967.

44. Stirewalt, M. A., and Hackey, J. R.: Penetration of host skin by cercariae of *Schistosoma mansoni*. I. Observed entry into skin of mouse, hamster, rat, monkey, and man. J. Parasitol., *42*:565–580, 1956.

45. Stunkard, H. W.: The morphology and life history of *Levinseniella minuta* (Trematoda: Microphallidae). J. Parasitol., *44*:225–230, 1958.

46. von Lichtenberg, F. C.: The bilharzial pseudotubercle: a model of the immuno-pathology of granuloma formation. *In* Immu-nologic Aspects of Parasite Infections. Pan American Health Organization, Washington, D. C., WHO, 1967.

46a. Warren, K. S., and Newill, V. A.: Schisto-somiasis: A Bibliography of the World's Literature from 1852–1962. Cleveland, The Press of Case Western Reserve Univ., 1967.

47. Wajdi, N.: Penetration of the miracidia of *S. mansoni* into the snail host. J. Helminth. *40*:235–244, 1966.

48. Wilson, R. A.: The hatching mechanism of the egg of *Fasciola hepatica* L. Parasitol., *58*:79–89, 1968.

49. Yokogawa, M.: Studies on the biological aspects of the larval stages of *Paragonimus westermani*, especially the invasion of the second intermediate hosts. (II). Jap. J. Med. Sci. Biol., *5*:501–515, 1952.

50. ———: *Paragonimus* and paragonimiasis. *In* Dawes, B. (ed.): Advances in Parasitology. Vol. 3, pp. 99–158. New York, Academic Press, 1965.

51. Yoshimura, H.: The life cycle of *Clonorchis sinensis*: a comment on the presentation in the seventh edition of Craig and Faust's *Clinical Parasitology*. J. Parasitol., *57*:961–966, 1965.

Chapter 9

Class Cestoidea, Subclass Cestodaria

INTRODUCTION

General Anatomy

The class Cestoidea consists of parasites, commonly called **tapeworms,** that live as adults in the intestines of vertebrates. Their bodies are usually white or yellowish, ribbon-like, and divided into short segments called **proglottids** (Figs. 9–1, 9–4, 9–5). There are no digestive, circulatory, respiratory or skeletal organs; it would seem that everything possible has been eliminated in order to produce a thoroughly efficient reproductive machine. Larval stages live in a wide variety of invertebrate and vertebrate animals.

The tapeworm head is a holdfast called a **scolex,** which is armed with hooks, suction organs ("suckers"), or both (Figs. 9–2, 9–3). There are three kinds of suckers:

1. **Bothria** (from the Greek word meaning "hole" or "trench") are usually slit-like grooves with weak suction powers, as in *Dibothriocephalus latus*, the fish tapeworm (Fig. 9–2);

2. **Phyllidia** (meaning "like a leaf") are ear-like or trumpet-like and have thin, flexible margins. The order Tetraphyllidea (p. 221), as the name indicates, possesses four phyllidia;

3. **Acetabula** are suction cups that occur, four in a circle, around the head of the order

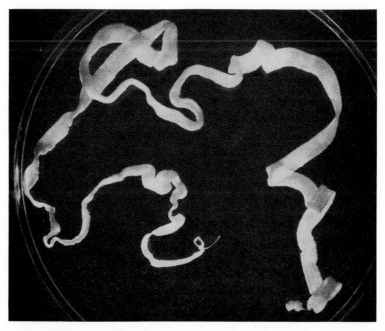

Fig. 9-1. Entire tapeworm, *Hymenolepis diminuta.* (Photograph by Zane Price.) (Markel and Voge, Medical Parasitology. Courtesy of W. B. Saunders.)

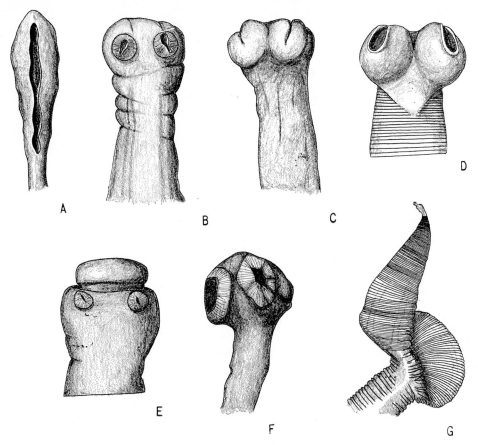

Fig. 9-2. Cestode scoleces. (*A*) *Dibothriocephalus*. Note bothria or grooves. (*B*) Mesocestoides. Note absence of hooks and rostellum. (*C*) *Moniezia*. (*D*) *Thysanosoma*. Note absence of hooks and rostellum. (*E*) *Raillietina*. Note presence of rostellum. (*F*) *Hymenolepis*. Note retracted rostellum. (*G*) *Fimbriaria*. Note special holdfast organ. (Whitlock, Illustrated Laboratory Outline of Veterinary Entomology and Helminthology. Courtesy of Burgess Publishing Co.)

Cyclophyllidea (p. 229), which includes the common tapeworms of man.

Below the scolex is the neck (**zone of proliferation**) and the body (**strobila**). Whereas each larval stage of a cestode is unquestionably an individual organism, the adults have sometimes been considered a "linear colony of highly specialized zooids."[23] However, "the preponderance of evidence appears to support the opinion that the adult cestode is an individual rather than a colony."[17]

The proglottids, which number from 3 to 3,000, become progressively more mature toward the posterior end of the tapeworm. The terminal segments, especially in the more primitive families of cestodes, may sometime become detached at an early stage in their development to live and mature independently in the intestine of their host. More commonly, however, the terminal gravid or ripe proglottids become little more than sacs filled with eggs.

Tegument. The tegument of cestodes is largely proteinaceous, containing certain polysaccharides, glycoprotein, mitochondria, vacuoles, and membranes. The parasite-host interface has been defined as the "region of chemical juxtaposition of regulatory mechanisms of both host and parasite."[7] The peripheral layer, the surface of which projects into numerous minute brush-like or digitiform processes (revealed by the electron microscope), is an anucleate syncytium morphologically specialized for absorption, and it overlies a muscular layer containing secretory cells.[4,9]

High levels of protein synthetic activity occur in the subtegumental cells, and these proteins are presumably secreted into the tegumental matrix.[5] The tegument of these

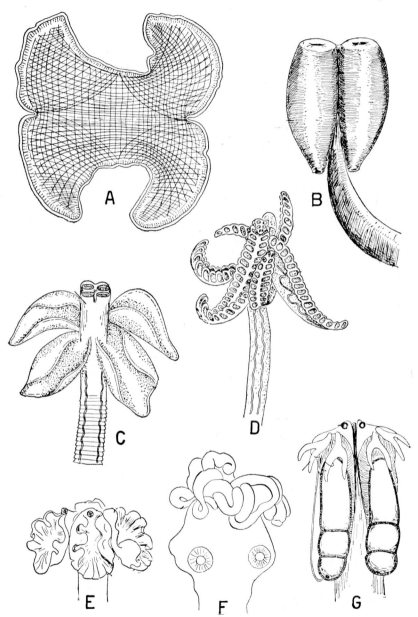

Fig. 9-3. Cestode scoleces. (*A*) Section through the scolex of a pseudophyllid to show the bothria and musculature (after Benham, 1909). (*B*) *Bothridium pythonis*, from the python, showing bothridia fused to tubes (after Southwell, 1930). (*C*) *Mysophyllobothrium* with four bothridia and a myzorhynchus with four suckers (after Shipley and Hornell, 1906). (*D*) *Echeneibothrium* with areolae (after Linton, 1887). (*E*) *Phyllobothrium*, with ruffled bothridia, each with an accessory sucker, from a skate. (*F*) *Polypocephalus* (after Linton, 1891). (*G*) *Acanthobothrium*, with areolae, hooks and accessory suckers (after Southwell, 1925). (Hyman, The Invertebrates. Vol. II. Permission of McGraw-Hill Book Co.)

parasites is basically the same as that described for trematodes on page 140, but the functions are consistent with the nutritional requirements of a parasite that does not have a digestive tract.

Reproductive System. Each mature proglottid usually contains one set of female organs and one set of male organs. A few species, such as the dog tapeworm (p 231), have duplicate sets of male and female organs in each proglottid. Only a few tapeworms are dioecious.[21] Occasionally the male organs disappear before the female organs become functional. These organs are fundamentally similar to those of most flukes, but a vaginal canal opens to the outside. Scattered testes

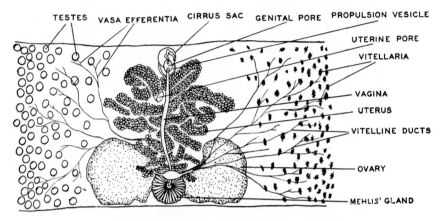

Fig. 9-4.   Proglottid of the fish tapeworm *Dibothriocephalus latus*.   (Cable, An Illustrated Laboratory Manual of Parasitology.   Courtesy of Burgess Publishing Co.)

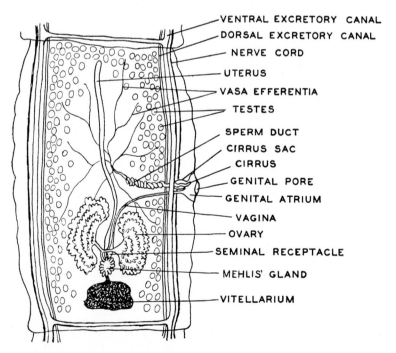

Fig. 9-5.   A mature proglottid of *Taenia pisiformis*, found in the dog and in other mammals.   (Cable, An Illustrated Laboratory Manual of Parasitology.   Courtesy of Burgess Publishing Co.)

occur throughout each proglottid, and the vitellaria, containing shell globules and yolk material, are commonly grouped in one clump, but they may be dispersed along the lateral margins of the worm. Some proglottids in a mature strobila may be sexually undifferentiated. These and other reproductive organs are illustrated in Figures 9–4 and 9–5.

Self fertilization involving the sexual structures of a single proglottid, or reciprocal fertilization between proglottids of the same worm may occur. Reciprocal fertilization between proglottids of different worms may also take place if more than one tapeworm is present in the same region of the intestine. A single tapeworm may lay thousands of eggs a day, and one to two million eggs during its lifetime.

The organization of the reproductive organs may be used to divide cestodes into two groups.[14] Group I includes those worms with vitellaria scattered throughout the peripheral region of the proglottid (Pseudophyllidea and Tetrarhynchidea), or those with vitellaria occurring in broad or narrow lateral masses (Tetraphyllidea, Proteocephalidea and Lecanicephalidea). In some species (Pseudophyllidea), the eggs are thick-walled, oval and operculated, as in flukes, whereas in others (Tetraphyllidea), the eggs are thin-walled, round and nonoperculated.

Group II includes those cestodes with compact vitelline glands usually located in the midline of the proglottid, as typified by the Cyclophyllidea. Two small median glands occur in the Mesocestoididae, but some Anoplocephalidae do not have vitelline glands. Embryos of this group are usually protected by the **embryophore** (Fig. 9–6). Figures 9–7 and 9–8 illustrate these two groups.

The reproductive organs and egg shell formation of the family Pseudophyllidae are closely related to those of digenetic trematodes, as described on page 144. Globules released from the vitelline cells coalesce around the mass of these cells, which surround the ovum. These globules form the egg shell (Fig. 7–7). Cyclophyllidean ces-

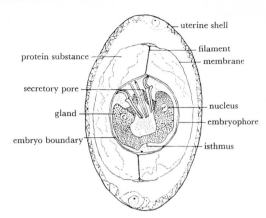

Fig. 9-6. Ventral view of oncosphere (or embryo) of *Raillietina cesticillus*, enclosed in embryonic membranes. (Reid, courtesy of Trans. Amer. Micr. Soc.)

todes have small vitellaria, and the ovum is surrounded by a few cells, occasionally only one. Ova of *Taenia* and other genera are surrounded by a thick embryophore (Figs. 9–8 and 9–9).

Nervous System. This system consists of cerebral ganglia in the anterior end of the worm, from which extend two main lateral trunks that span the length of the body. There are several other longitudinal nerves, and connecting all of these are many branching nerves that extend to all parts of the body. A heavier concentration of nerves and neurons occurs in such organs as suckers and excretory vesicles (Fig. 9–10).

Excretory System. The excretory or osmoregulatory system is similar to that of flukes, and consists of flame cells (flame bulbs) that connect with transverse and longitudinal collecting tubules (Fig. 9–5 and 9–11). These longitudinal vessels are located along each side of the body. Fluid flows anteriorly in the smaller, dorsal vessel and posteriorly in the larger, ventral vessel. These vessels generally open directly to the surface but a caudal excretory vesicle and a caudal excretory pore might be present.

The exact function of the excretory vessels is in doubt, but it seems clear that in addition to the usual role of ridding the body of certain metabolic wastes, these tubules help to main-

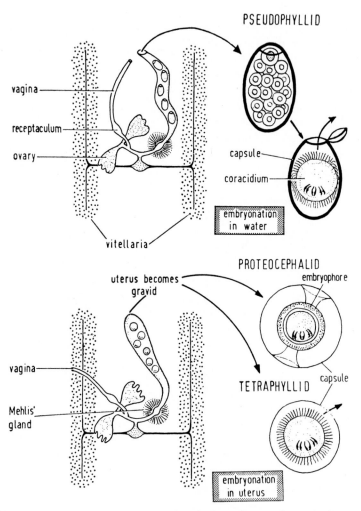

Fig. 9-7.   Diagrammatic representation of genitalia and eggs of cestodes in Group I:
with extensive vitellaria.   (Smyth and Clegg, courtesy of Exp. Parasitol.)

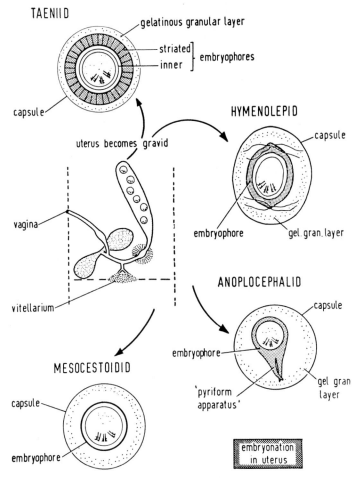

Fig. 9-8. Diagrammatic representation of genitalia and eggs of cyclophyllidean cestodes in Group II. (Smyth and Clegg, courtesy of Exp. Parasitol)

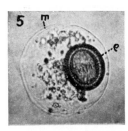

Fig. 9-9. *Taeniarhynchus saginatus.* Photomicrograph of egg. (*m*) External membrane of the egg. (*e*) Embryophore containing oncosphere or six-hooked embryo. (Gradwohl and Kouri, Clinical Laboratory Methods and Diagnosis, courtesy of C. V. Mosby Co.)

tain hydrostatic pressure. The study of excretory products in vivo is extremely difficult, so one has to rely either on analogies with free-living organisms or on studies with parasites in cultivation.

Muscle System. Movements of the body and locomotion are accomplished by various sets of muscles. A thin circular muscle layer and one or more longitudinal muscle layers occur close to the tegument. Within the parenchyma are well-developed transverse, longitudinal, dorsoventral and, occasionally, diagonal muscles. The scolex is especially well endowed with muscles associated with the suckers and hooks (Fig. 9-3). The transmission of waves of contraction is apparently

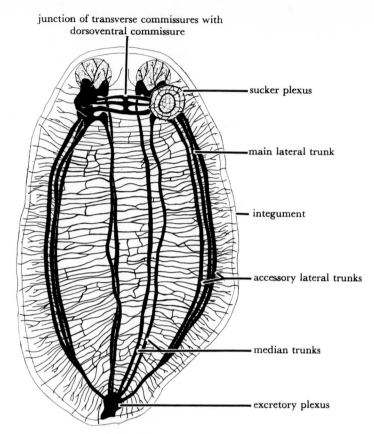

Fig. 9-10. Nervous system of *Mesocestoides* sp. based on histochemical and histologic preparations. (Hart, courtesy of J. Parasit.)

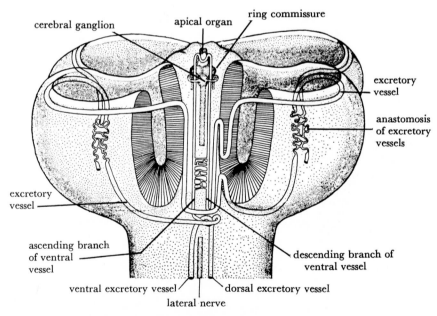

Fig. 9-11. Scolex of *Tetrabothrius affinis*. Reconstruction of the left half of the excretory system with part of the bothridia removed. (Rees, courtesy of Parasitology.)

inherent in the muscle fibers; detached, ripe segments often move actively.

Physiology

The physiology of tapeworms has received much attention, but we are only beginning to understand the complex biochemical nature of these organisms. There is a linear gradient in metabolic rate along the strobila, so that observations made on proglottids at one point in the body may differ from those made on other proglottids a short distance away. Extreme care must be exercised in analyzing the results of experiments, and definite notice must be taken of the region of the tapeworm under analysis.

Numerous species of cestodes (e.g., *Taenia taeniaeformis*, *Cysticercus fasciolaris*) contain large amounts of inorganic substances localized in **calcareous corpuscles.** The composition of this material varies considerably, and it is especially common in larval stages. Such substances apparently serve to buffer body acids, and thus protect the parasite from the harmful effects of the host's stomach.

Tapeworms depend on the breakdown of glycogen for their energy source. In general, the kinds of carbohydrates that can be utilized by tapeworms in vitro are limited; most of the tapeworms studied are capable of fermenting only certain monosaccharides.[6] Anophocephalids, however, use certain disaccharides.

Because of differences between the physiology of the host and that of its tapeworm, the quality of carbohydrate ingested by the host has a direct bearing on the growth and reproduction rates of the worm. Some of the effects of carbohydrates on the worm may be due to preliminary effects on the physiochemical characteristics of the habitat. Variations in the quantity of carbohydrate ingested by the host modify the competitive effects of one tapeworm species on another, because different species are not affected proportionally. Seasonal changes in the behavior of tapeworms and in their carbohydrate content may be correlated with the feeding behavior of the host. The imposition of the dynamics of gut physiology on the specific carbohydrate requirement of cestodes may be of importance in the distribution of the parasites and in the selection of species of parasites that appear and prosper in a particular host.[6]

The rapid advances that have been made in our understanding of the biochemistry and physiology of tapeworms have been reviewed by Read and Simmons[8] and Smyth.[13] Some physiologic modifications may be correlated with the obligate parasitism of tapeworms. These modifications include:

Nutritional dependence on carbohydrate for growth and reproduction coupled with the limitation effected by the capacity to use only simple sugars; limited ability to perform transamination reactions; a general incapacity to digest complex organic molecules from the surrounding medium; very active membrane transport systems associated with the outer surface and showing stereospecificity for various classes of small organic molecules such as monosaccharides and amino acids; an incapacity of many species to osmoregulate but with obvious adaptations to the osmotic pressures and ionic strengths of vertebrate body fluids; and specific reactions of oncospheres and later larval stages to physiochemical factors involved in establishment within a host. The limited information available suggests that, except for carbohydrate, the nutritional requirements of tapeworms may not be much more complicated than those of vertebrates. The physiological specializations thus far delineated in tapeworms strongly support the hypothesis that the broad determinant of obligate parasitism in this group of organisms is dependence on the chemical regulatory mechanisms possessed by their vertebrate hosts. Tapeworms seem to lack many functions usually associated with maintenance of a steady state.[8]

The nutrition of cestodes has been studied rather extensively in recent years, and the results of these studies were reviewed by Smyth in 1969. The following was adapted from his paper:[13]

Evidence gathered from in vitro and metabolic studies suggests that a complex nutritional relationship exists between hosts and cestodes. Two facts of special interest have come to light through these studies: (1) Cestodes have the ability to "fix" CO_2, as do plants and some microorganisms. This demonstrates that cestodes can use waste metabolic products efficiently from the host's intestine. (2) Some species of cestodes seem to take in nutrients through direct contact with the mucosa. In vitro, *Echinococcus granulosus* develops strobilarly only when a solid or semisolid proteinaceous substrate is provided, with which it makes intimate contact. A larval protoscolex will grow in a cystic direction when such a substrate is lacking.

If it can be shown that other cestodes with a penetrative type of rostellum are able to absorb materials directly from the host mucosal surface, "our general ideas of how a cestode obtains its nutrition will require revision."[13] It is possible that enzymes excreted by the tegument of the cestode digest proteins that are in contact or close to the worm's body, and the resulting amino acids are absorbed by the tegument. "It is now known that large molecules—such as dipeptides or even whole protein molecules—may be ingested. For this to occur, the cestode tegument must be capable of pinocytosis and we have no unequivocal experimental evidence (as yet!) that this is the case although ultra-structure studies are suggestive."[13]

Several factors related to the physiology of the host's intestine probably influence the rate of growth, the associated form of organogeny, and the beginning of maturation in the cestode; the following factors are probably the most important:

physico-chemical:	temperature, pH, pCO_2, pO_2, E_h
biochemical:	composition of bile, concentration and ratio of amino acids
morphological:	micromorphology of mucosa
nutritional:	diet of host, especially regarding carbohydrate level
immunological:	host species or strain
hormonal:	sex of host
behavioural:	whether host is stressed or normal.[13]

An example of studies on cestode growth is shown in Figure 9–12. Note the initial lag phase of about 24 hours, after which there is a

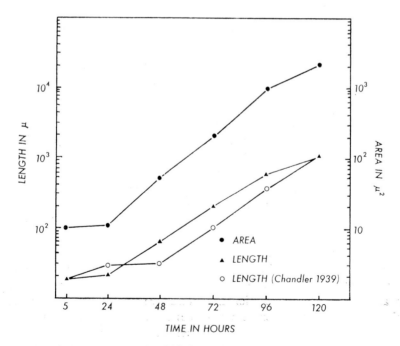

Fig. 9-12. Growth of *Hymenolepis diminuta* during the first five days in the final rat host, semi-log scale. (After Goodchild and Harrison, 1961.) (Smyth, The Physiology of Cestodes, courtesy of W. H. Freeman and Co.)

sharp rise leading to a length of well over 10 mm. in five days.

Immunity

Most investigations of acquired immunity to tapeworms involve those cestodes that have a mammalian tissue phase during larval development. Larvae of *Hymenolepis nana*, *Hydatigera taeniaeformis*, several species of *Taenia* and *Echinococcus* are commonly studied because they make intimate contact with host tissues. (These forms are described on the following pages.) A high degree of acquired resistance is common for these species, and it may approach absolute immunity.[24] The immune responses to early stages of the tapeworm life cycle tend to inhibit invasion of the parasite and to curtail its establishment within the host.

The rat tapeworms *Hymenolepsis nana* and *H. diminuta* have been used extensively for immunologic investigations. When a suitable host eats an egg, the embryo (oncosphere) is released and penetrates rapidly into host tissues, developing into a cysticercoid larva. Factors that cause larval release from eggs include bile salts, trypsin and probably pepsin. The host apparently reacts to this infection by changing the intestinal epithelium in some manner or by producing some substance in the intestine that reacts with larvae from a second infection. This reaction is usually successful in preventing a second infection— most oncospheres from a second infection are not able to penetrate the intestinal epithelium. If they are successful they soon die, indicating a second wall of defense on the part of the host. Possibly some parasites are able to maintain themselves within their host by changing their antigenic structure from time to time and thus escape the host antibodies that would otherwise destroy them. Immunity to *Hymenolepis nana* in mice occurs after egg infection but not directly after infection with cysticercoids. This difference is understandable, because egg infection leads to the tissue phase of the parasite while cysticercoid (indirect) infection leads to the adult phase in the gut.

Antibodies may be produced against this tapeworm.[22] If rabbits are immunized by injections of homogenates of adult *H. nana*, then bled and the serum collected, this serum is strongly sensitized against the worms. An adult *H. nana* placed in such serum immediately becomes unusually active and withdraws its rostellum. After a few hours a precipitation occurs on the worm, beginning with the anterior end and progressing posteriorly along the strobila. Undoubtedly, some hosts possess a certain amount of natural immunity against some worm parasites.

In dual infections of mice and rats with *Hymenolepis nana* and *H. diminuta*, the *H. nana* are much more successful in maintaining themselves. The immunogenic tissue phase of the *H. nana* elicits an inhibitory reaction to infectivity and growth of the nonimmunogenic *H. diminuta*. The host reaction that develops against *H. nana* is just as strong against *H. diminuta*. Large numbers of common antigens seem to exist among helminths. Antibodies get into the lumen of the host gut from sloughed and disintegrated mucosa cells and also by passing through cell membranes.[22]

It is possible that complementary genetic systems have developed in some animal parasites and their hosts that involve the specific reaction of one gene in the parasite with one gene in the host, giving a common phenotype of resistance or susceptibility. Such a genetic relation has developed between the flax rust and its host.

Adult cestodes, like lumen-dwelling trematodes, are generally poorly immunogenic. "Clearly, unless the scolex contact with the intestinal mucosa is very intimate, sufficient antigenic stimulus to the host will not be provided."[13] Passive immunization, using antisera as a vaccine, has been achieved in laboratory animals, but vaccination against important human tapeworm infections has not been very successful. Before humans or domestic animals can be successfully vaccinated against diseases caused by cestodes (e.g., hydatidiasis), detailed analyses using advanced immunologic techniques must be formed to determine the origin and composi-

PSEUDOPHYLLIDEA

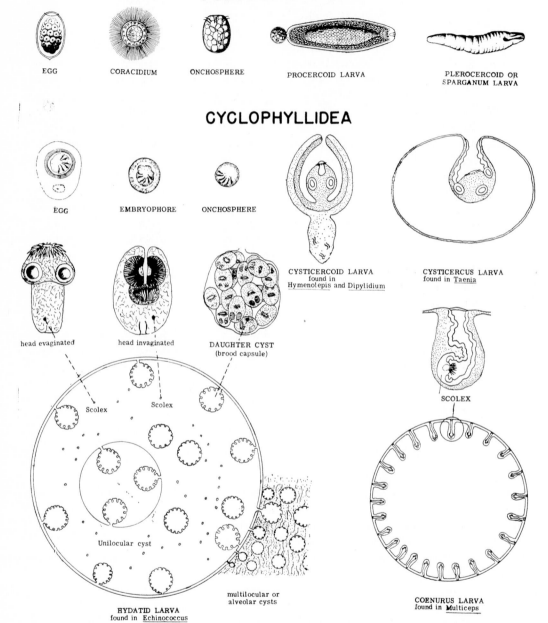

EGG   CORACIDIUM   ONCHOSPHERE   PROCERCOID LARVA   PLEROCERCOID OR SPARGANUM LARVA

CYCLOPHYLLIDEA

EGG   EMBRYOPHORE   ONCHOSPHERE

CYSTICERCOID LARVA
found in
Hymenolepis and Dipylidium

CYSTICERCUS LARVA
found in Taenia

head evaginated   head invaginated   DAUGHTER CYST
(brood capsule)

SCOLEX

Scolex   Scolex

Unilocular cyst

multilocular or
alveolar cysts

HYDATID LARVA
found in Echinococcus

COENURUS LARVA
found in Multiceps

Fig. 9-13. Tapeworms of man, immature stages. (Courtesy Navy Medical School,
National Navy Medical Center.)

tion of the metabolites and the antigens produced by all stages of cestodes. Cytologic studies of the histopathology of hosts would supplement these analyses. Moreover, in vitro cultivation of the parasites will play an important role in isolating antigens and determining the effects of antibodies.[13]

Life Cycles

The main types of cestode larval progression are represented in the Orders Pseudophyllidea and Cyclophyllidea (Fig. 9–13). In the Pseudophyllidea (e.g., the broad fish tapeworm, *Dibothriocephalus latus*, p. 224), the early embryo develops three pairs of hooks within the egg and thus is known as a **hexacanth embryo** or **oncosphere.** If this embryo hatches and becomes free-swimming, it is a ciliated **coracidium** (Fig. 10–4). Copepods (e.g., *Diaptomus*) eat the coracidia of *D. latus;* the cilia are shed and the young worms migrate into the body cavity (hemocoel), becoming **procercoid** larvae. The next host is a fish that eats the copepod. Salmon, trout, perch, pike and other fish thus become infected. The procercoid is freed in the intestine of the fish and makes its way into the muscles. Here it elongates into a **plerocercoid** larva (Figs. 10–3, 10–5). Because of an early mistake in identification, the plerocercoid larva was thought to be another type of worm, and was called a **sparganum.** The name persists, and infection of a fish or other animal with plerocercoids is called **sparganosis.** A plerocercoid may be minute or, in a few species, 5 to 10 cm. in length. Fish-eating mammals, including man, are the final hosts. Plerocercoids develop into mature tapeworms in the intestine, and eggs appear a few weeks after infection. (See Fig. 10–3, p. 225.)

It has been suggested that plerocercoid larvae of an elasmobranch tetraphyllidean cestode occur in most genuine oriental pearls.[1] Apparently the oyster feeds on small crustaceans that harbor the first intermediate stage of the tapeworm.

The order Cyclophyllidea, which includes most of the important cestodes of man, dis-plays three main types of immature tapeworms (Fig. 9–13): the **oncosphere,** the **cysticercoid,** and the **cysticercus.** The oncosphere is a six-hooked embryo like that described for the Pseudophyllidea. The three pairs of hooks may become vigorously active when the embryo hatches, enabling it to penetrate host tissue. The oncosphere moves with its hooked end in advance. In some species, the portion containing larval hooklets is constricted off as a **cercomere** and atrophies. The head of the tapeworm, or **scolex,** develops at the opposite end. Thus a definite metamorphosis occurs in the life cycle. In cysticercoid larvae, the scolex is retracted into a small bladder; the larvae may or may not have a solid tail (often the entire larva appears solid). This stage is frequently found in an arthropoid intermediate host. A cysticercus (bladderworm) larva has a rounded, fluid-filled cyst or bladder into which the scolex is invaginated. A **coenurus** is a larger bladderworm that has several or many invaginated scoleces attached to its inner, or germinative, layer. A **hydatid cyst** is a bladderworm in which daughter bladderworms develop, each similar to a coenurus. The term **metacestode** is used to indicate all of the stages between the egg and the adult that may be found in the intermediate host. See Voge[19] for a review of the biology of postembryonic stages of cestodes. See Rybicka[11] for a discussion of embryogenesis, and Smyth[12] for a review of the biology of cestode life histories.

In 1965, Stunkard[18] reaffirmed the following statement that he made in 1953[16]:

There is no precise basis for determining specificity in cestodes, and the same species under different physiological conditions, or in different hosts, may manifest striking morphological differences. Furthermore, these parasitic species, as a result of their complex life-cycles, hermaphroditism and self-fertilization, may readily develop divergent strains adapted to particular host-species. Under these circumstances, and in the absence of data from controlled experiments to determine the effects of different environmental conditions on development and adult structure, specific determination remains merely a matter of opinion.

Classification

The classification followed here is primarily that of Hyman, but modified slightly on the basis of the reviews by Stunkard.[16][17] Three minor orders (Aporidea, Disculicepitidea, Nippotaenidea) have been omitted. A few families are discussed but no attempt is made to list all of them, and only a few species are described. For details on classification, see the works of Hyman,[2] Spasskii,[15] Stunkard,[17] Voge,[20] Wardle and McLeod,[23] and Yamaguti.[25]

Phylum Platyhelminthes
Class Cestoidea
  Subclass Cestodaria
    Order Amphilinidea
    Order Gyrocotylidea
  Subclass Cestoda
    Order Tetraphyllidea
    Order Diphyllidea
    Order Lecanicephalidea
    Order Proteocephalidea
    Order Tetrarhynchidea (= Trypanorhyncha)
    Order Pseudophyllidea
      Families include: Haplobothriidae, Diphyllobothriidae
    Order Spathebothridea
    Order Cyclophyllidea
      Families include: Davaineiidae, Dilepididae, Hymenolepididae, Anoplocephalidae, Linstowiidae, Mesocestoididae.
    Order Caryophyllidea
      Families include: Caryophyllidae

SUBCLASS CESTODARIA

In this group[3] of worms there is no scolex, and the body is not segmented. These cestodes are therefore sometimes called **monozoic,** in contrast to the **polyzoic** forms that have proglottids. The two orders of Cestodaria given below are also characterized by larval stages that have ten hooks, in contrast to the six hooks possessed by practically all the orders of Cestoda. The general appearance of the Cestodaria resembles that of a fluke rather than a tapeworm. The worms are found in the intestines and body cavities of fish. The cestodarians have generally been considered the most primitive of the cestoides, but there is considerable evidence that they are progenetic plerocercoid larvae. See page 563 for a discussion of their phylogeny.

Order Amphilinidea

Amphilina foliacea. *A. foliacea* lives in the body cavity of sturgeons (*Acipenser*). It is an oval, flat worm without a scolex or digestive tract and with a protrusile proboscis. It is hermaphroditic, with the ovary scattered throughout the animal. There is a long, loosely coiled uterus, a vagina, and male and female openings, both at the posterior end of the body. Adults range in length from a few to about 40 mm.

The life history of *Amphilina foliacea* starts with the egg that develops into a ciliated larva while still in the uterus of the mother worm. The adult worm presumably penetrates the abdominal wall of the host before expelling eggs. The larva is called a **lycophore** or **decacanth,** and it does not emerge from the egg until it is eaten by the second host, an amphipod. In the crustacean host the larva makes its way into the body spaces and develops into a procercoid, then into a plerocercoid stage. The latter is infective to the sturgeon, which eats the amphipod. Sexually mature worms contain larval hooklets.

Order Gyrocotylidea

These elongated flatworms are nonsegmented, and they live in the spiral intestine of chimaeroid fish (Holocephali).

Gyrocotyle. *G. urna* and *G. fimbriata* are representative species. The genus *Gyrocotyle* consists of worms with bodies composed of one segment flattened dorsoventrally. At the anterior end is a muscular sucker, the **acetabulum.** The posterior end is roughly funnel-shaped with an anterior dorsal pore and a wide posterior attachment organ. The borders of this organ are thin and folded in a complex manner, forming the rosette. The lateral borders of the body, which has spines on its surface, are thin and undulant or ruffled. The vaginal pore is dorsal in position, the male genital pore and the uterine pore are ventral. All three pores occur in the anterior fourth of the body.

Little is known of the life cycles of these parasites. A ten-hooked ciliated larva emerges from the egg and may enter the host tissues directly without using an intermediate host.[10]

Gyrocotyle fimbriata lives in the intestine of the ratfish, *Hydrolagus colliei*, which can be caught along the western coast of the United States. The parasites average 32 mm. in length, but range from 13 to 63 mm. The general anatomy is that described for the genus, and other details may be seen in Figure 9–14.

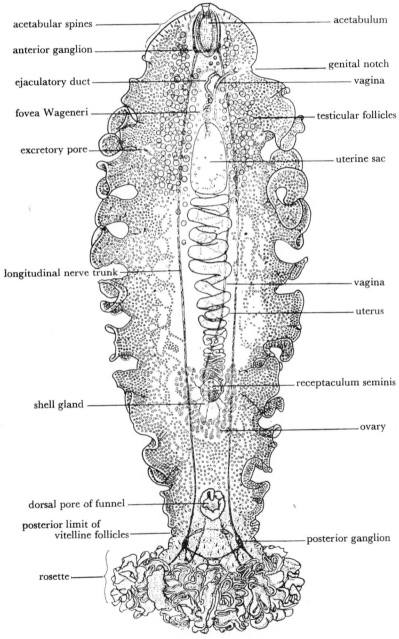

acetabular spines — acetabulum
anterior ganglion — genital notch
ejaculatory duct — vagina
fovea Wageneri — testicular follicles
excretory pore — uterine sac
longitudinal nerve trunk — vagina
— uterus
— receptaculum seminis
shell gland — ovary
dorsal pore of funnel — posterior ganglion
posterior limit of vitelline follicles
rosette

Fig. 9-14. *Gyrocotyle fimbriata,* dorsal aspect. × 57. (Lynch, courtesy of J. Parasitol.)

REFERENCES

1. Baer, J. G.: Ecology of Animal Parasites. Urbana, University of Illinois Press, 1951.
2. Hyman, L. H.: The Invertebrates: Platyhelminthes and Rhynchocoela. The Acoelomate Bilateria. Vol. II. New York, McGraw-Hill, 1951.
3. Joyeux, Ch., and Baer, J. G.: Classe des Cestodaires. Cestodaria Monticelli. In Grassé, P.: Traité de Zoologie. Vol. IV, pp. 327–346. Paris, Masson et Cie Éditeurs, Libraires de l'Acad. Méd. Paris, 1961.
4. Lumsden, R. D.: Cytological studies on the absorptive surfaces of cestodes: 1. The fine structure of the strobilar integument. Zeit. f. Parasitenk., 27:355–382, 1966.
5. ———: Cytological studies on the absorptive surfaces of cestodes: 2. The synthesis and intracellular transport of protein in the strobilar integument of Hymenolepis diminuta. Zeit. f. Parasitenk., 28:1–13, 1966.
6. Read, C. P.: The role of carbohydrates in the biology of cestodes. VIII. Some conclusions and hypotheses. Exp. Parasit., 8:365–382, 1959.
7. Read, C. P., Rothman, A., and Simmons, J. E.: Studies on membrane transport, with special reference to parasite-host integration. Ann. N.Y. Acad. Sci., 113:154–205, 1963.
8. Read, C. P., and Simmons, J. E., Jr.: Biochemistry and physiology of tapeworms. Physiol. Rev., 43:263–305, 1963.
9. Rothman, A. H.: Electron microscopic studies of tapeworms: The surface structure of Hymanolepis diminuta (Rudolphi, 1819) Blanchard, 1891. Trans. Amer. Micro. Soc., 82:22–30, 1963.
10. Ruszkowski, J. S.: Etudes sur le Cycle Evolutif et sur la Structure des Cestodes de mer. II. Sur les Larves de Gyrocotyle urna (Gr. et Wagen.). Bull. Intern. Acad. Polon. Sci., Sér B., pp. 629–641, 1932.
11. Rybicka, K.: Embryogenesis in cestodes. In Dawes, B.: Advances in Parasitology. Vol. 4, pp. 107–186, New York, Academic Press, 1966.
12. Smyth, J. D.: The Biology of Cestode Life-Cycles. Technical Communication No. 34, Commonwealth Bureau of Helminthology, St. Albans, Herts, England, 1963.
13. ———: The Physiology of Cestodes. San Francisco, W. H. Freeman, 1969.
14. Smyth, J. D., and Clegg, J. A.: Egg-shell formation in trematodes and cestodes. Exp. Parasit., 8:286–323, 1959.
15. Spasskii, A. A.: Essentials of cestodology. In Skrjabin, K. I. (ed.): Anoplocephalate Tapeworms of Domestic and Wild Animals. Vol. I. Moscow, Acad. Sci. U.S.S.R., 1951. Published in England by Israel Prog. Sci. Trans. Birron & Cole, 1961.
16. Stunkard, H. W.: Life histories and systematics of parasitic worms. Systematic Zool., 2:7–18, 1953.
17. ———: The organization, ontogeny, and orientation of the cestoda. Quart. Rev. Biol., 37:23–34, 1962.
18. ———: Variation and criteria for generic and specific determination of diphyllobothriid cestodes. J. Helminth., 39:281–296, 1965.
19. Voge, M.: The post-embryonic developmental stages of cestodes. In Dawes, B. (ed.): Advances in Parasitology. Vol. 5, pp. 247–297. New York, Academic Press, 1967.
20. ———: Systematics of cestodes—present and future. In Schmidt, G. D. (ed.): Problems in Systematics of Parasites. pp. 49–72. Baltimore, University Park Press, 1969.
21. Voge, M., and Rausch, R.: Observations on Shipleya inermis Fuhrmann, 1908 (Cestoda: Acoleidae). J. Parasitol., 42(5):547–551, 1956.
22. Von Brand, T., Mercado, T. I., Nylen, M. U., and Scott, D. R.: Observations on function, composition and structure of cestode calcareous corpuscles. Exper. Parasit., 9:205–214, 1960.
23. Wardle, R. A., and McLeod, J. A.: The Zoology of Tapeworms. Minneapolis, University of Minnesota Press, 1952.
24. Weinmann, C. J.: Immunity mechanisms in cestode infections. In Soulsby, E. J. L. (ed.): Biology of Parasites. New York, Academic Press, 1966.
25. Yamaguti, S.: Systema Helminthum. Vol. II. The Cestodes of Vertebrates. New York, Interscience, 1959.

Chapter 10

Class Cestoidea, Subclass Cestoda

These flatworms typically possess a scolex bearing hooks and suckers; the strobila consists of three to many proglottids, and the embryo usually has six hooklets.

ORDER TETRAPHYLLIDEA

These tapeworms are commonly found in the intestines of elasmobranch fish (e.g., sharks, rays); they are characterized by four phyllidia on the scolex (Fig. 9–3, p. 207). These "suction" organs of attachment are usually broad and leaf- or trumpet-like, and they may be relatively simple or complex. The worms are moderate in size, usually not exceeding 10 cm. in length, possessing at most only a few hundred proglottids. The ovary is bilobed, and each lobe is constricted horizontally; the vagina lies dorsal to the uterine sac, and the vitellaria occur as two marginal bands. The order contains two families: Phyllobothriidae and Oncobothriidae. Life cycles of the various genera have not been completely delineated, but they are basically similar to those of the pseudophyllids (p. 224). Urea, which circulates through the digestive tract of elasmobranchs, seems to play a special role in maintaining osmotic relations of these cestodes.[11]

ORDER LECANICEPHALIDEA

Like the tetraphyllids, these tapeworms live in the intestines of elasmobranch fish. The two groups of worms are similar in many respects, but the scolex of the lecanicephalids consists of two main parts in tandem. The anterior portion may be bulb-like (many possess tentacles or suckers), and the posterior part may bulge like a cushion and bear four suckers. The order Lecanicephalidea is composed of small tapeworms possessing relatively few proglottids and more-or-less cylindrical bodies. Complete life cycles have not yet been described. The oyster, *Crassostrea virginica*, is said to be the natural intermediate host of metacestode stages of members of the genus *Tylocephalum*.[4] Three families comprise this order: Lecanicephalidae, Cephalobothriidae, and Discocephalidae.

ORDER PROTEOCEPHALIDEA

The order includes only one family, Proteocephalidae, but it is a large group with many well-known species. The worms inhabit the intestines of amphibians, reptiles, and fish. The genus *Proteocephalus* occurs in many freshwater teleosts throughout the world. Figure 10–1 illustrates larval stages of *P. parallacticus*. *Lintoniella adhaerens* has been found in the hammerhead shark. The worms, in general, are only a few centimeters long; mature proglottids are longer than they are broad. The scolex is varied but usually has four simple suckers flush with the surface of the body and near the anterior tip. The scolex may or may not extend beyond the suckers. This extension sometimes possesses hooks. Vitellaria occur in two marginal bands.

The life cycle of proteocephalids involves an oncosphere that develops into a plerocercoid larva in the body cavity of a copepod (*Cyclops*). The copepod is eaten by fish or amphibians, in which the adult worm develops. Plerocercoids of *Proteocephalus ambloplitis* have been reported to migrate from the parenteral cavity of bass through the gut wall and into the gut lumen. Apparently they do this in response to an increase in temperature and with the aid of histolytic secretions.

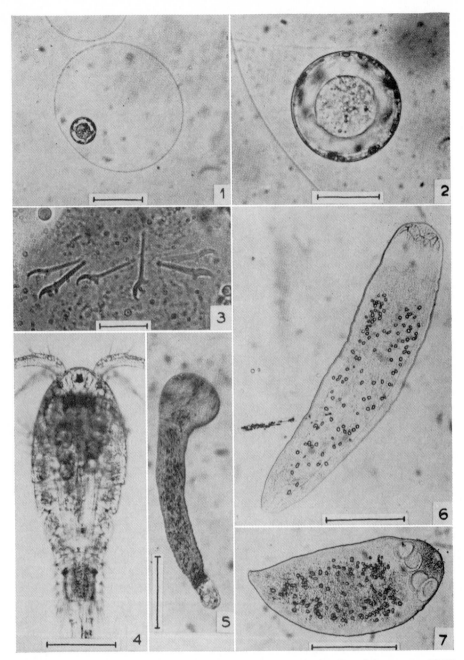

Fig. 10-1. Photomicrographs of the cestode *Proteocephalus parallacticus* from lake trout. (*1*) The egg. (*2*) Greater magnification of the egg. (*3*) Embryonic hooks. (*4*) *Cyclops bicuspidatus* with subspherical metacestodes in situ. (*5*) Cercomer stage. (*6*) Unfixed plerocercoid showing general morphology. (*7*) Same plerocercoid as in *6* during fixation with ten per cent formol-saline. (Value of scale for each figure is: *1, 4, 5* = 0.10 mm.; *2* = 0.03 mm.; *3* = 0.01 mm.; *6, 7* = 0.02 mm.) (Freeman, reproduced by permission of the National Research Council of Canada from the Canadian Journal of Zoology, *42*:393, 1964.)

Vertebrates may also serve as intermediate hosts when they are eaten by larger fish, amphibians or reptiles. In these cases the plerocercoid larvae usually inhabit the liver or other organ of the first vertebrate host and, when eaten, remain in the intestine of the second vertebrate host.

group is characterized by a scolex that possesses large hooks at its anterior end and two large, boat-like bothridia, each formed by a fusion of two of these sucker-like attachment organs (Fig. 9–3, p. 207). The long "neck" of the worm is spiny, and the entire worm is small, having fewer than 20 proglottids.

ORDER DIPHYLLIDEA

The single genus *Echinobothrium* contains few species. Adults occur in the intestines of elasmobranch fish and the larval stages inhabit marine mollusks and crustaceans. The

ORDER TETRARHYNCHIDEA (=TRYPANORHYNCHA)

The scolex of these tapeworms has four long tubes, within each of which lies a slender tentacle armed with rows of hooks (Fig.

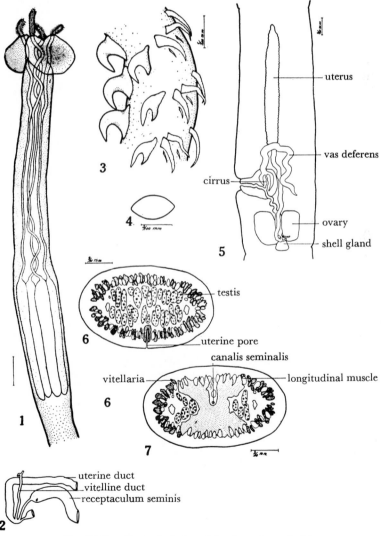

Fig. 10-2. *Tentacularia* found in elasmobranch fish.
(Hart, courtesy of Trans. Amer. Micr. Soc.)

10–2). These tentacles can readily be extended from the scolex and withdrawn into the tubes. In addition, the scolex possesses two to four phyllidia that usually are not well developed. The entire holdfast end of the worm is long and cylindrical. Yolk glands are usually distributed in a sleeve-like layer in the cortex of each proglottid; the testes extend into the region behind the ovary. The vagina and its opening are ventral to the cirrus pouch; the sperm duct does not cross the vagina before entering the cirrus pouch. Embryonation is completed after eggs have been expelled from the proglottid.

These tapeworms are usually less than 100 mm. in length and may be only a few millimeters long. They normally inhabit elasmobranchs, especially the spiral valve. As far as we know, the life cycle involves two intermediate hosts, the first a copepod and the second a teleost fish.

ORDER PSEUDOPHYLLIDEA

An important characteristic of this group of tapeworms is the presence of two **bothria** on the scolex instead of suction cups. In this order, the bothria usually are not specialized (Fig. 9–3, p. 207). The bothrium may be a short, longitudinal, slit-like groove in some species, and a wider depression in others. In either case the bothria possess weak holdfast properties. In some groups of worms the margins of the bothria join, forming a tube. The length of worms of this order varies from a few millimeters to 25 m. or more. Some forms are **monozoic,** that is, possessing a body without segmentation; but most of them are **polyzoic,** a term referring to the familiar divisions of the body (strobila) into proglottids.

Usually, only one set of reproductive organs occurs in each proglottid, but a few species of worms have two sets. The genital opening is often on the midventral surface and even middorsal, but it may also occur laterally. The ovary is bilobed, yolk glands are numerous and scattered, and the uterus opens to the outside on the ventral surface.

Family Haplobothridae

Haplobothrium globuliforme. This species (Fig. 25–6, p. 565) has four retractile, spined "tentacles" that suggest a relationship with the tetrarhynchs, but the anatomy of its gravid segments places it with the pseudophyllids. The life cycle of this cestode includes a coracidium that is eaten by *Cyclops*, within which it develops into a procercoid larva. A bony fish (e.g., the bullhead, *Ameiurus nebulosus*) eats the crustacean, and the procercoid is liberated and develops into a plerocercoid in the liver of the fish. A ganoid fish, *Amia calva*, eats the bullhead and thus becomes infected.

Family Diphyllobothriidae

Dibothriocephalus latus. This tapeworm is an important parasite of man, but most of the several genera of this family are parasites of marine mammals. *D. latus*, the fish tapeworm or broad tapeworm (formerly named (*Diphyllobothrium latum*), has a scolex that is almond-shaped, measuring 2 to 3 × 0.7 to 1 mm., with deep dorsal and ventral grooves. The anterior 20 per cent of the body is composed of small immature segments, while the rest consists of mature and gravid segments (Fig. 10–3). The entire worm may be 3 to 10 m. or more in length. Most of the segments are wider than long, but most of the posterior gravid segments are longer than wide. The testes are numerous, small, rounded bodies situated in the lateral folds or dorsal side of the proglottid. The vas deferens is much convoluted and proceeds anteriorly from the midplane at the beginning of the posterior third of the proglottid. It enlarges into a seminal vesicle and ends in a cirrus that is median in position, approximately between the first and second third of the body.

The ovary is symmetrically bilobed and is located in the posterior third of the proglottid with Mehlis' gland (see p. 208) between the lobes. Vitelline glands occupy the same lateral areas as do the testes, but the former are ventral to the latter. The uterus is in the form of a rosette, occupying the middle field of the proglottid. The vagina is a narrow, coiled tube, its coils interspersed with those of the uterus.

Dibothriocephalus latus eggs (Fig. 10–4) are broadly ovoid, yellowish to golden brown, operculated, 55 to 76 × 41 to 56 μ, and non-embryonated when voided with feces of the host. The life cycle is described on p. 217. Adult worms may be found in the small intestine of man, pigs, dogs, cats or many other

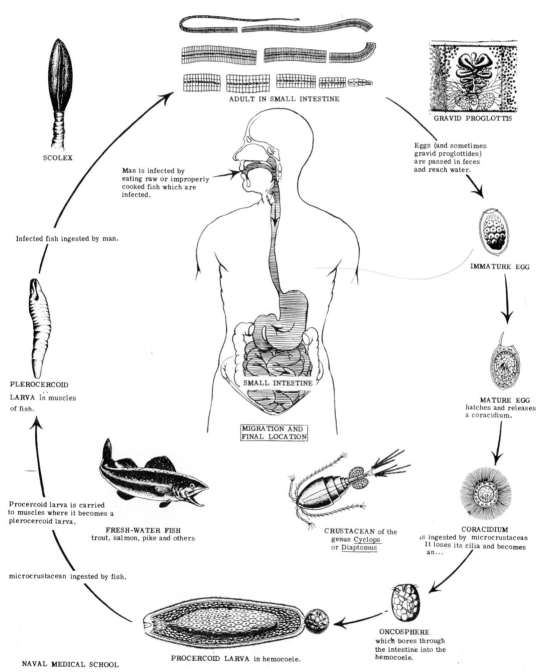

SCOLEX

ADULT IN SMALL INTESTINE

GRAVID PROGLOTTIS

Man is infected by eating raw or improperly cooked fish which are infected.

Eggs (and sometimes gravid proglottides) are passed in feces and reach water.

Infected fish ingested by man.

IMMATURE EGG

PLEROCERCOID LARVA in muscles of fish.

SMALL INTESTINE

MIGRATION AND FINAL LOCATION

MATURE EGG hatches and releases a coracidium.

Procercoid larva is carried to muscles where it becomes a plerocercoid larva.

FRESH-WATER FISH trout, salmon, pike and others

CRUSTACEAN of the genus Cyclops or Diaptomus

CORACIDIUM is ingested by microcrustacean It loses its cilia and becomes an...

microcrustacean ingested by fish.

NAVAL MEDICAL SCHOOL

PROCERCOID LARVA in hemocoele.

ONCOSPHERE which bores through the intestine into the hemocoele.

Fig. 10-3. Life cycle of *Dibothriocephalus latus*. (Courtesy of Naval Medical School, National Naval Medical Center)

mammals. Some of these parasites may have both birds and mammals as definitive hosts. The intermediate hosts are freshwater copepods and fish.

Symptoms of infection are often absent. Sometimes digestive discomfort, anemia, abdominal pains, nervous disorder or enteritis occur in man. If a mature tapeworm is situated in the proximal part of the small intestine, the host and the worm compete for vitamin B_{12}. If enough of this vitamin is removed by the worm, the host may suffer from

pernicious anemia.[18] If the scolex is not found after treatment, several months should elapse before a negative diagnosis is made. Prevention consists of thoroughly cooking fish before eating, or of freezing them at $-10°$ C. for 24 hours. Reservoir hosts should be eliminated if possible, and untreated sewage should be prevented from flowing into freshwater lakes or rivers.

Sparganosis (p. 217) may occur in man from infection by the plerocercoid larvae of tapeworms belonging to the genus *Spirometra*.

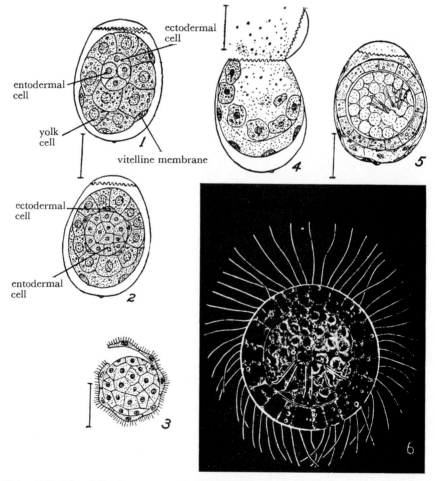

Fig. 10-4. *Dibothriocephalus latus* eggs and coracidium. (*1, 2*) Segmented eggs, showing origin of ectoderm and endoderm. Tissues shrunk away from the eggshell owing to fixation. (Modified after Schauinsland.) (*3*) Immature embryo, showing development of cilia. (After Schauinsland.) (*4*) Egg just after liberation of the coracidium. (*5*) Egg a few hours before hatching. (*6*) Vogel's figure of the coracidium by dark field illumination.

The lines near the figures represent 0.02 mm. (Vergeer, courtesy of Papers Mich. Acad. Sci., University of Michigan Press)

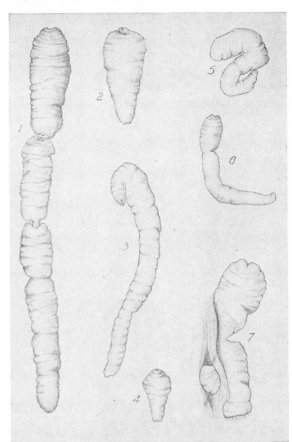

Fig. 10-5. Plerocercoids of *Dibothriocephalus latus*. (*1*) Old preserved plerocercoid of *D. latus* showing deep constrictions. × 5. (*2, 3, 4, 5, 6*) Young plerocercoids of *D. latus* preserved while in flesh to show normal positions. × 11¼. (*7*) The same, with part of the flesh of the fish still in position. × 11¼. (Vergeer, courtesy of J. Infectious Diseases)

The normal hosts for these larvae are frogs, snakes, or amphibious mammals. Southeast Pacific islanders occasionally used crushed fresh frogs as a poultice, and have been known to acquire plerocercoids from this practice. Figure 10–5 illustrates plerocercoids from fish.

Another member of the genus *Dibothriocephalus*, *D. erinacei*, is more strictly a parasite of dogs and cats, and it has a particularly interesting life cycle. Coracidia emerge from the eggs and are eaten by copepods, which in turn are eaten by young frogs. If procercoid larvae have developed within the copepods, frogs become infected and plerocercoid stages occur in the adult amphibians. The number of plerocercoids in a single frog may range from one to 25, located mainly in the hind legs and abdomen. Oral transfer to cats and dogs results in the establishment of adult tapeworms in the intestine.

ORDER CARYOPHYLLIDEA

Family Caryophyllaeidae

Adults of these little worms live in freshwater teleost fish or in tubificid oligochaetes. Species that live in annelid worms may be larval forms that have become sexually mature (paedogenetic). *Archigetes* (Fig. 10–6) is the best-known genus. It is apparently capable of maturing in fish as well as in tubificids, and some species do not require a fish host.[5]

Pliovitellaria wisconsinensis. This worm averages 5 mm. × 1 mm. in size. It lives in the intestine of the shiner, *Notemigonus crysoleucas auratus*, and in the minnow, *Hyborhynchus notatus*, in Wisconsin. It has a poorly defined scolex and an elongated body that is oval in cross section.

ORDER SPATHEBOTHRIDEA

These small worms were formerly included with the order Pseudophyllidea. They never have true bothria or suckers. There is no external segmentation, but some internal proglottisation, and the uterus opens between the male and female apertures. Medullary testes occur in two oval lateral bands; the ovary is rosettiform or bilobed. The operculated eggs have thick shells. Adults have been described as neotenic procercoids, and they occur in the more ancient groups of fish.

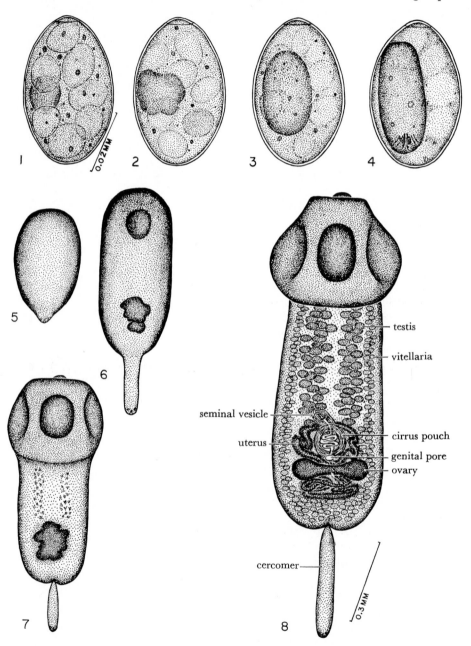

Fig. 10-6. *Archigetes iowensis.* (*1–4*) Oncosphere development. (*4*) Egg after 13 days in water, oncosphere fully formed. (*5* to *8*) Procercoids. (*8*) At 60 days (experimentally reared), procercoid nearly complete. (Calentine, courtesy of J. Parasitol.)

ORDER CYCLOPHYLLIDEA

Most of the important tapeworms of man and domestic animals belong to this order. The order is also well represented among the adult tapeworms of birds, but not so well among those of the amphibians and reptiles. Like the pseudophyllids, members of this order have a wide range in length. Some of them are only a few millimeters long, while others may reach a length of 30 m. An important characteristic of the group is the presence of four well-developed suckers on the scolex. The anterior tip of the scolex usually projects as a *rostellum*, which may or may not bear hooks and may be retractable (Fig. 10–7). Proglottids are usually flattened, and the genital apertures are located marginally on one or both sides. The ovary is typically bilobed or fan-shaped and the testes consist of scattered granules. The yolk gland is nor-mally compact, lying posterior to the ovary. The gravid uterus may be branched or sac-like and contains the embryos enclosed by embry-onic membranes. In some species, the uterus forms egg capsules containing one or several eggs. Almost all of the species are hermaph-roditic, but the genus *Dioecocestus* (and a few others) is dioecious. The male and female strobila of the latter can be recognized by their difference in shape.

Arthropods, annelids, mollusks and verte-brates serve as intermediate hosts; and am-phibians (rarely), reptiles, birds, and mam-mals harbor the adult tapeworms.

Family Davaineidae

The scolex of these small to moderately large tapeworms has hooks, suckers, and a cushion-shaped rostellum. An important species is *Raillietina* (=*Skrjabinia*) *tetragona*, a

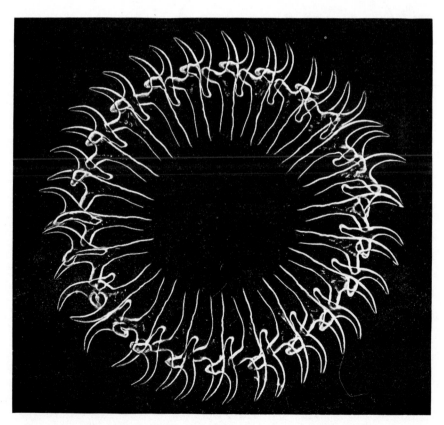

Fig. 10-7. Hooks on the scolex of the tapeworm *Taenia pisiformis*. (Courtesy of Amer. Inst. Biol. Sci. Drawing by D. W. C. Marquardt.)

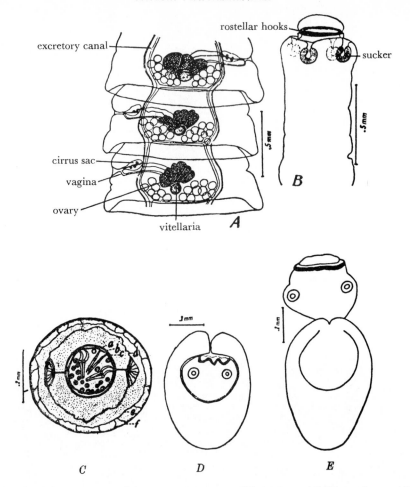

Fig. 10-8. *Raillietina cesticillus.* (*A*) Mature segments. (*B*) Scolex. (*C*) Oncosphere. (*D*) Cysticercoid, invaginated. (*E*) Cysticercoid, evaginated. (*a*) Surface layer of oncosphere. (*b, c, d, e, f*) The five membranes of the oncosphere. (Lapage, Veterinary Parasitology, courtesy of Oliver and Boyd Ltd.)

common tapeworm of domestic fowl. This worm may reach 25 cm. in length. Larval stages occur in ants or in maggots of the house fly.

Raillietina bonini. The definitive host for this tapeworm is the pigeon. Eggs from the intestine of a bird are eaten by a snail or slug in which the cysticercoid develops. Pigeons become infected by eating the mollusks (Fig. 10–9).

Raillietina cesticillus. Probably the best-known member of the family is *R. cesticillus* because it is a common tapeworm of poultry. Chickens, pheasants, guinea fowl and wild birds are often infected. The adult worms

may reach 130 mm. in length and are about 2 mm. wide. The four suckers are small and there are 400 to 500 minute hooks that circle the scolex. Eggs are passed from the host with feces and are eaten by the intermediate host, which may be one of several species of beetles or even the house fly. In the insect, cysticercoids are formed, and birds become infected by eating insects. Figure 10–8 illustrates the stages in the life cycle. Figure 9–6 (p. 209) shows the oncosphere within its embryonic membrane. This tapeworm has been grown in bacteria-free (gnotobiotic) chicks without showing abnormal growth.

Raillietina loeweni. This tapeworm is a

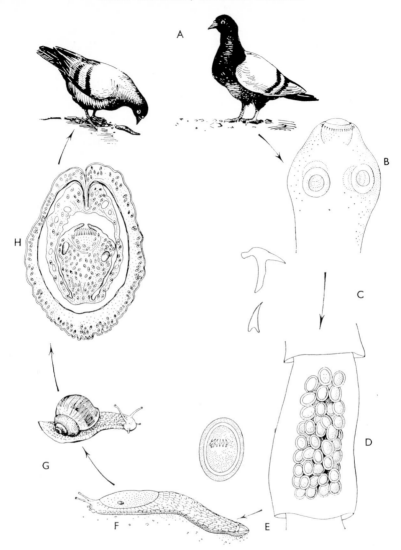

Fig. 10-9. *Raillietina bonini*, a tapeworm of pigeons. The illustration shows a scolex, gravid proglottid, the two possible intermediate hosts and an enlarged view of the cysticercoid stage that is found in snails or slugs. (Joyeux and Baer, *In* Grassé, Traité de Zoologie, courtesy of Masson et Cie, Editeurs)

common parasite of the hare, *Lepus californicus melanotis*. The intermediate host is the harvest ant belonging to the genus *Pheidole*, which the hare inadvertently ingests while eating vegetation.[1]

Davainea meleagridis. This small member of the family is found in the turkey, *Meleagris gallopavo*. Mature specimens are only 5 mm. long by 950 μ wide. The scolex is about 165 μ wide. When one remembers that the limit of vision with the unaided eye

is about 100 μ, he can appreciate the difficulty encountered in hunting for such tiny tapeworms (Fig. 10–10).

Family Dilepididae

Dipylidium caninum. *D. caninum* is worldwide in distribution and is common in the small intestines of dogs, cats and other carnivores but rare in man. The few infections that do occur in man are usually in children. The tapeworm averages about a

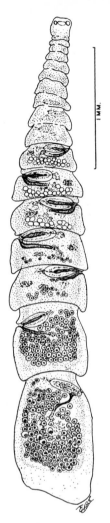

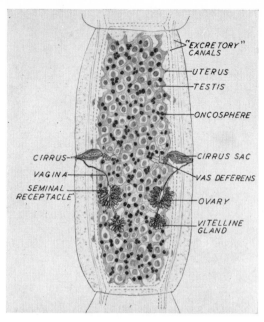

Fig. 10-11. *Dipylidium caninum*, mature proglottid. The uterus forms a network around the testes. (After various authors.)

Fig. 10-10. *Davainea meleagridis*, a tapeworm from the turkey. Whole specimen, whole mount. (Jones, after Fuhrmann, Proc. Helminth. Soc. Wash.)

foot in length (10 to 70 cm.) and can be recognized by the elongated, somewhat almond-shaped mature proglottids (Fig. 10–11). The rhomboidal scolex possesses an anterior projection, the **rostellum,** which is armed with several transverse rows of hooks and can be retracted into a rostellar sac. Below the rostellum are four prominent suckers. A mature proglottid (Fig. 10–11) contains two sets of reproductive organs with an opening on each side of the body. The uterus develops as a network of canals or cavities.

Hooked eggs (Figs. 10–12, 10–13), 24 to 40 μ in diameter, occur in oval packets containing five to 20 eggs each. Ripe proglottids containing these packets, or balls of eggs, break loose from the strobila and look like active little worms about the size and shape of a pumpkin seed. When they reach the outside, they rupture and the eggs may be ingested by larvae of the fleas *Ctenocephalides canis* and *C. felis,* or the human flea, *Pulex irritans,* or adult biting lice, *Trichodectes canis.* Within these, the eggs hatch and larvae migrate to the body cavity. By the time the insect has reached maturity, the tapeworm has developed into a cysticercoid stage. Dogs and cats get the fleas or bits of them into their mouths and, if swallowed, the parasite is carried to the intestine where the cysticercoid stage matures into an adult tapeworm. Children apparently get infected fleas or parts of them under their fingernails and become infected by putting their fingers into their mouths. The life cycle is illustrated in Fig. 10–14.

Symptoms in children are absent to mild.

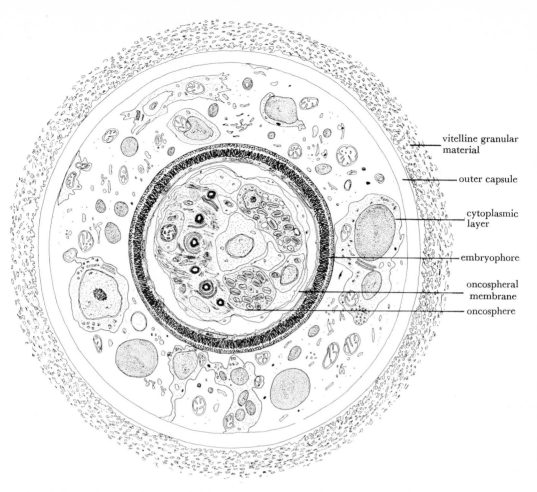

Fig. 10-12. Schematic drawing of the mature egg of *Dipylidium caninum*, based on electron microscope photographs. (Pence, courtesy of J. Parasitol.)

vitelline granular material

outer capsule

cytoplasmic layer

embryophore

oncospheral membrane

oncosphere

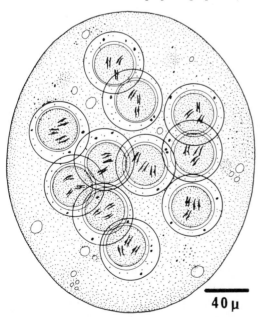

40 μ

Fig. 10-13. *Dipylidium caninum* packet of eggs.

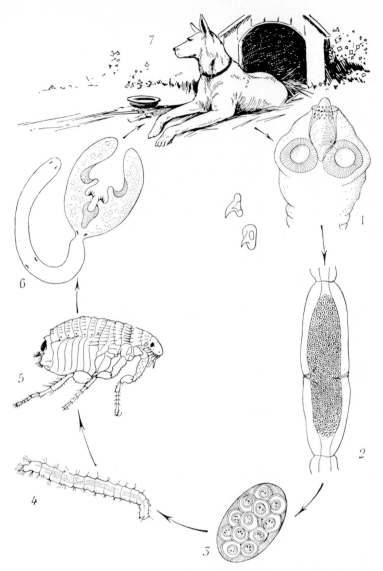

Fig. 10-14. Life cycle of *Dipylidium caninum*. (*1*) Scolex with partially retracted rostellum. (*2*) A gravid proglottid. (*3*) An egg packet containing 12 eggs. (*4*) A larval flea that eats the eggs. (*5*) An adult flea. (*6*) The infective cysticercoid stage found in the flea. (*7*) A dog, the definitive host. (Joyeux and Baer, *In* Grassé, Traité de Zoologie, courtesy of Masson et Cie, Editeurs)

Diagnosis involves finding egg packets or entire proglottids in feces. Dogs and cats and their sleeping quarters should be kept as clean as possible, their bodies "wormed" often, and treated frequently with insecticides.

Family Hymenolepidae

Hymenolepis nana. This common cosmopolitan species is found in man as well as rats and mice. Since the parasite averages only about 32 mm. (usually 25 to 40 mm.) in length, it is called the dwarf tapeworm (Fig. 10–15). The rostellum of *H. nana* is retractable, like that of *Dipylidium caninum*, but it possesses a single ring of hooks. Proglottids are wider than long, and they contain one set of reproductive organs (Fig. 10–16).

The life cycle[17] is unusual for tapeworms

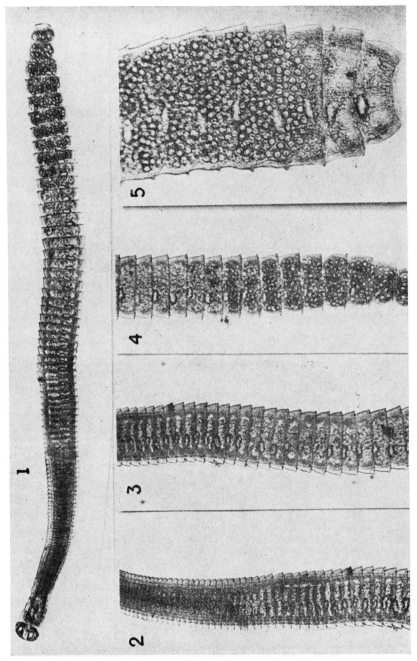

Fig. 10-15. *Hymenolepis nana.* (Originals of Kouri.) (*1*) Complete specimen. × 26. (*2, 3, 4*) Anterior, center, and posterior thirds of the parasite, respectively. × 40. (*5*) Posterior fourth of the parasite. × 100. The gravid segments are filled with eggs. Most of the caudal segments are completely emptied. The third from last has partially lost its ova. (Gradwohl and Kouri, Clinical Laboratory Methods and Diagnosis, courtesy of C. V. Mosby Co.)

(Fig. 10–17). Eggs (Fig. 6–5, p. 134) are 30 to 50 μ in diameter. They reach the outside and may be ingested by grain beetles or fleas in which the oncospheres develop into tailed cysticercoids (Fig. 10–18). Often these eggs are swallowed directly by men or mice, and when the oncospheres are liberated they develop into tailless cysticercoids in the intestinal villi. Thus the intermediate host in the life cycle can be eliminated. Cysticercoids mature, drop into the lumen of the small intestine and develop into adult tapeworms in one to two weeks. The life span is short and worms are evidently eliminated.

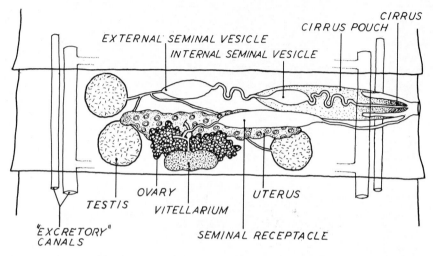

Fig. 10-16. *Hymenolepis nana*, mature proglottid.

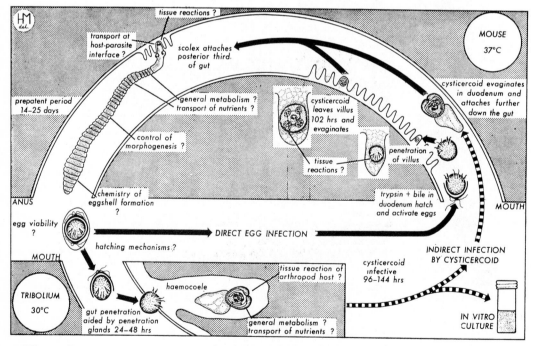

Fig. 10-17. Life cycle of the rodent cestode, *Hymenolepis nana*, and the physiologic problems associated with it. (Smyth, The Physiology of Cestodes, University Reviews in Biology, courtesy of Oliver and Boyd.)

The incidence of infection in man ranges from less than one per cent to 28 per cent. The numbers of worms in one host may be high—7,000 were taken from one human patient. Symptoms of infection, if any, are usually mild, but toxic reactions such as nervous disorders, sleeplessness, diarrhea and intestinal pain may occur. Diagnosis is best made by finding *Hymenolepis nana* eggs in stool specimens. Prevention measures include personal cleanliness, destruction of rats and mice and a well-balanced diet to promote resistance to infection.

Hymenolepis diminuta. The common

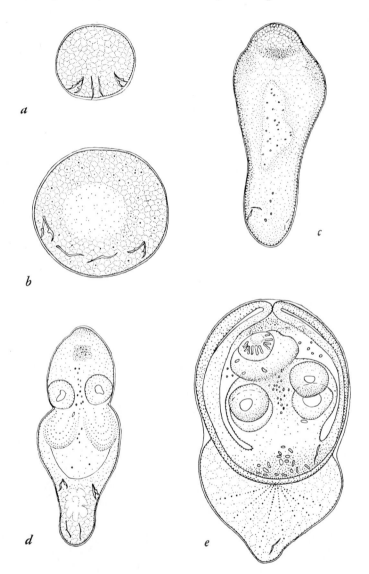

Fig. 10-18. Five stages in the growth of *Hymenolepis nana*. (*a*) Stage 1. Solid sphere, showing paired hooks and external membrane. (*b*) Stage 2. Cavity and dispersal of oncosphere hooks. (*c*) Stage 3. Two body divisions, elongation of cavity and anterior zones of sucker and rostellum primordia.

(*d*) Late stage 3. Process of withdrawal, showing separation of "neck" tissue, which will become layer immediately enveloping scolex. (*e*) Stage 4. Withdrawn scolex, enlarged rostellum with partly developed hooks, and clearly demarcated suckers. All drawings freehand, relative size indicated by oncosphere hooks (10–12 μ). (Voge and Heyneman, courtesy of Univ. Calif. Publ. Zool.)

species in rats is *H. diminuta*, which only occasionally get into man. The incidence of infection in man is usually less than one per cent, but in favorable localities (e.g., a few areas of India) it may run as high as six per cent. The maximal number of worms recorded from one man is 19. The worm averages 45 cm. in length, thus being considerably larger than *H. nana*. Its size, however, is partly a function of the age of the rat host, but the basis for this relationship is not well understood (see p. 468).

With the aid of the electron microscope, Rothman[13] has described minute projections covering the surface of *Hymenolepis diminuta*. He called these projections **microtriches** (singular **microthrix**) and suggested that they serve to increase the absorptive area of the body surface, to help hold the worm next to the gut lining, and to agitate gut fluids.

Eggs of *Hymenolepis diminuta*, unlike those of *H. nana*, have a sculptured shell and lack polar filaments. Adults of *H. diminuta*, in contrast to *H. nana*, have an indefinite life span and do not show aging. *H. diminuta* adults were maintained in rat hosts for 14 years by 13 successive surgical transplantations.[10] The worm obviously has a life span potential considerably longer than that of its host.

The life cycle requires an intermediate host in which the cysticercoids develop. Many kinds of insects serve as this host, e.g., grain beetles, earwigs, fleas, dung beetles and cockroaches. The intermediate layer of the cysticercoid body wall is a dynamic area containing a number of enzymes. The physiology and pattern of development of this larva is essentially similar to that of the indirect cycle of *H. nana*.

Hymenolepis diminuta tend to occupy the anterior part of the host small intestine during the early morning hours, then move to the posterior end of the small intestine, arriving in largest numbers by late afternoon. This circadian migratory behavior appears to be correlated with the periods of host feeding[12] (rats normally feed at night). Investigations of population distributions of other helminths

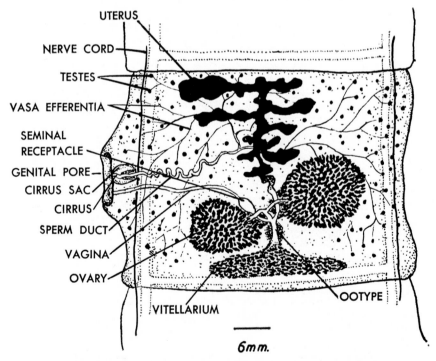

Fig. 10-19. *Taenia solium*, the pork tapeworm. Mature proglottid.
(Noble and Noble, Animal Parasitology Laboratory Manual)

in the vertebrate gut should include the possibility of similar behavior.

Hymenolepis has received a large share of the research on the physiology and biochemistry of tapeworms. The sizes and reproductive rates of this parasite are related to carbohydrate ingestion of the host (see p. 462). There is convincing evidence for mediated absorption of amino acids, sugars, purines and pyrimidines from the host gut (see review by Read[9]).

Kilejian *et al.*[6] found evidence that the intestinal contents of rats infected with *Hymenolepis diminuta* contained at least twice as much lipid as did the controls. Most of the fatty acids are undoubtedly supplied by bile, but others are probably supplied by the pancreas and the intestine itself. In addition to these endogenous sources of lipid, tapeworms probably absorb lipids brought to the intestine by the host's food. The synthesis of fat by the tapeworm may "buffer" the rapidly formed organic acids by chemical synthesis to neutral fats.[19] There seems to be no real evidence that the fat is used as a source of energy.

Symptoms of infection are mild. Control of the tapeworm consists of keeping rats away from stored fruits and grains, keeping insects away from such material and by being careful not to eat insect-contaminated food.

Family Taeniidae

Taenia solium. The pork tapeworm, *Taenia solium* (Fig. 10–19), is, as the common name indicates, one that man may acquire from eating uncooked pork. The incidence of infection in man varies with the locality from less than one per cent to about eight per cent, with a worldwide figure of two to three per cent. The figures may be inaccurate, however, because this species is easily confused with the beef tapeworm, *Taeniarhynchus saginatus* (formerly called *Taenia saginata*) (see p. 242).

The adult pork tapeworm ranges in length from 2 to 7 m. and has a scolex with a rounded rostellum. This structure is armed with large hooks that alternate with small hooks, giving the appearance of a double ring. There are four prominent round suckers. Microscopic, spine-like projections cover the body surface. The gravid proglottids (Fig. 10–20) are longer than wide and contain a single set of reproductive organs. The parasites are fairly common in Europe but are rare in parts of the Orient, United States and England. They are very common in Mexico where they cause much brain cysticercosis. Man is the only definitive host. Camels, dogs,

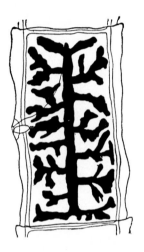

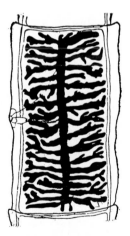

Fig. 10-20. *Taenia solium* on the top and *Taeniarhynchus saginatus* on the bottom. Gravid proglottids showing typical uterine patterns. Considerable variation in numbers of uterine branches occurs in both worms. (Roudabush, An Aid to the Diagnosis of Helminths Parasitic in Humans, courtesy of Wards Natural Science Establishment)

monkeys, and man may serve as intermediate hosts in addition to pigs.

The life cycle of *Taenia solium* (Fig. 10–21) starts with a thick-walled round egg averaging about 38 μ in diameter. It contains the characteristic embryo with its three pairs of hooks. Usually the eggs remain in proglottids that become isolated from the rest of the strobila

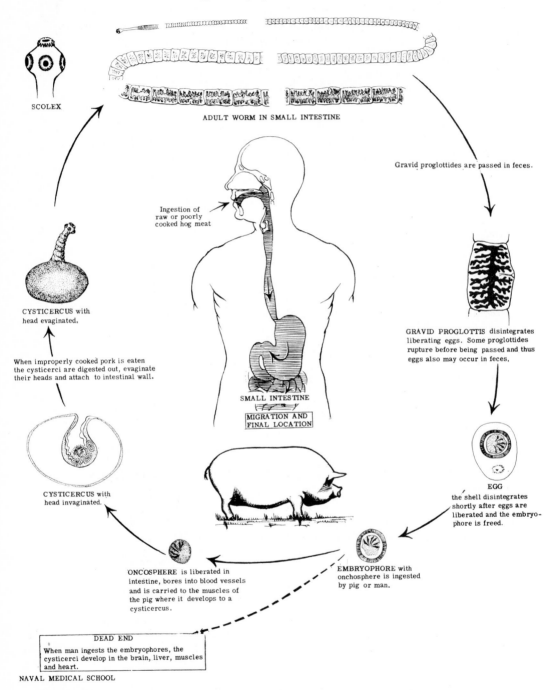

SCOLEX

ADULT WORM IN SMALL INTESTINE

Gravid proglottides are passed in feces.

Ingestion of raw or poorly cooked hog meat

CYSTICERCUS with head evaginated.

When improperly cooked pork is eaten the cysticerci are digested out, evaginate their heads and attach to intestinal wall.

SMALL INTESTINE

MIGRATION AND FINAL LOCATION

GRAVID PROGLOTTIS disintegrates liberating eggs. Some proglottides rupture before being passed and thus eggs also may occur in feces.

CYSTICERCUS with head invaginated.

EGG
the shell disintegrates shortly after eggs are liberated and the embryophore is freed.

ONCOSPHERE is liberated in intestine, bores into blood vessels and is carried to the muscles of the pig where it develops to a cysticercus.

EMBRYOPHORE with onchosphere is ingested by pig or man.

DEAD END
When man ingests the embryophores, the cysticerci develop in the brain, liver, muscles and heart.

NAVAL MEDICAL SCHOOL

Fig. 10-21. Life cycle of *Taenia solium*. (Courtesy of Naval Medical School, National Naval Medical Center)

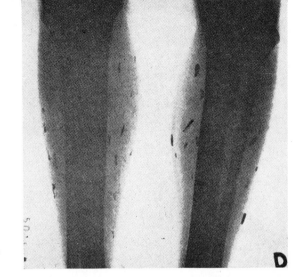

Fig. 10-22. X-ray of the lower limbs of a case of generalized cysticercosis, attended at the outpatient department of the Institute of Tropical Medicine (Director, Prof. P. Kouri) of the University of Havana. (Original photograph of Kouri, Basnuevo, and Sotolongo.) Note the calcified *Cysticercus* in the soft tissues. (Gradwohl and Kouri, Clinical Laboratory Methods and Diagnosis, courtesy of C. V. Mosby Co.)

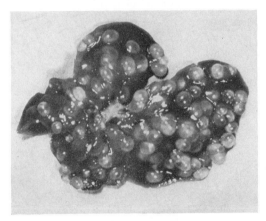

Fig. 10-23. *Cysticercus fasciolaris* (*Cysticercus taeniaeformis*) in an experimental rat liver. (Courtesy of Dr. Robert E. Kuntz)

and pass from the body of the host. Pigs, man, dogs, or other animals ingest these eggs in contaminated food. Oncospheres are liberated in the small intestine and make their way through the gut wall to blood vessels and are carried to all parts of the body. In various organs, especially muscles, the larval worms leave the blood and develop into the cysticercus or bladderworm stage (Figs. 10–22, 10–23). A study of cysticercosis in swine and bovines revealed the location of the parasites to be within the lymphatic capillaries of muscles.[14] The muscles of

hogs may sometimes become so filled with these parasites (sometimes called *Cysticercus cellulosae*) that the meat is called **measly pork.** The fluid within these cysticerci is composed largely of host blood plasma. Man usually acquires the adult tapeworm by eating uncooked or poorly cooked infected pork. The bladderworm becomes evaginated in the intestine, hook and suckers enable the young worm to become attached to the gut wall, and a new worm develops to maturity.

Symptoms of adult tapeworm infection may be absent or there may be mild general reactions. Rarely there is diarrhea, loss of weight, nervous symptoms and even perforation of the intestinal tract. Symptoms of cysticercus infection in man depend on extent of infection and location of the bladderworms. Cysticercosis of the brain would obviously produce symptoms different from cysticercosis of the forearm muscles. Diagnosis of intestinal infection is based on finding proglottids or eggs in stool specimens and is confirmed by finding the scolex. If the scolex is not recovered, four to six months are necessary to be sure the entire worm is no longer present. Treatment for cysticercosis, other than surgery, is of little value. Prevention consists in thoroughly cooking all pork before eating it. Proper sewage disposal is important.

As is apparent from the above discussion, bladderworms can be more serious to the host than can the adult worms. Some of these parasites were first discovered as larval stages and thought to be adults of new species or genera, and so were given new names. Later it was found that they were stages in the life cycles of other parasites, so two sets of names arose for the organisms. For example, *Cysticercus tenuicollis* is the thin-necked bladderworm of domestic ruminants, but a study of its complete life cycle shows it to be the larval stage of *Taenia hydatigena* of dogs. *Cysticercus ovis* causing sheep "measles" is the larva of *Taenia ovis* of dogs.

The larval stage of the tapeworm, *Taenia solium*, is only one of several types of bladderworms of pigs. *Cysticercus cellulosae* has also been reported from other domestic mammals. This parasite usually measures about 5×10 mm. when fully mature and infective to man. Sometimes the cysticerci are so numerous as to occupy more than one-half of the total volume of a piece of flesh (Fig. 10–23). The worms are characteristically located in the connective tissue of striated muscles but they may be found in any organ or tissue of the body. Various authorities disagree as to the site of "preference." Belding[2] lists possible infected tissue in the following order: tongue,

neck, shoulder, intercostal muscles, abdominal muscles, psoas, femoral, posterior vertebral. The U.S.D.A. Yearbook 1942[16] lists heart, head, diaphragm, abdomen, tongue. There are no definite symptoms of bladderworm infestation in animals, and there is no known practical treatment for removal of the parasites from swine.

Taeniarhynchus saginatus. This cosmopolitan "beef tapeworm" is also called the "unarmed tapeworm" because the scolex does not possess hooks (Fig. 10–24). It is more prevalent in man than is *Taenia solium*. It is longer than the pork species, usually measuring from 5 to 10 m., and it lives in the small intestines. One extreme specimen reached 25 m. in length, about three times as long as the entire human intestine. The

Fig. 10-24. *Taeniarhynchus saginatus*, scolex. The beef tapeworm. The line on the right represents 0.22 mm.

Table 10–1.   Differential Diagnosis

Character	Taenia solium	Taeniarhynchus saginatus
Rostellum	Present, armed	Absent
Testes	375—575, confluent posterior to vitellarium	880—1200, not confluent posterior to vitellarium
Cirrus pouch	Extends to excretory vessels	Does not extend to excretory vessels
Ovary	Three lobes	Two lobes
Vaginal sphincter	Absent	Present
Ova	Spherical	Oval
Gravid segments	Do not leave host spontaneously	Leave host spontaneously
Uterine branches	Less numerous	Numerous

The most reliable and easily assessed criteria for the differentiation of the two species in man are:
   1. The presence or absence of an armed rostellum
   2. The number of ovarian lobes
   3. The presence or absence of a vaginal sphincter

(Adapted from Verster, courtesy of Zeit. Parasit.)

entire strobila possesses from 1,000 to 2,000 proglottids. Table 10–1 summarizes major differences between the two species.

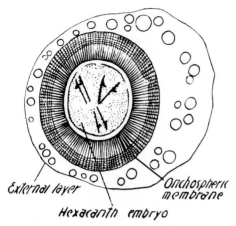

External layer

Onchospheric membrane

Hexacanth embryo

Fig. 10-25. Schematic drawing of egg of *Taeniarhynchus saginatus*. (Gradwohl and Kouri, Clinical Laboratory Methods and Diagnosis, courtesy of C. V. Mosby Co.)

The life cycle is essentially similar to that of *Taenia solium* and starts with an almost identical egg (Fig. 10–25). Ova are expelled only from detached proglottids that migrate to the perianal area. The intermediate hosts are cattle, buffalo, or other ungulates, in which the bladderworm stage is called *Cysticercus bovis*. Heavy infections cause the "measly" condition to occur primarily in jaw muscles and in the heart. Man is probably the only definitive host, and he acquires the infection by swallowing the cysticerci in uncooked, infected beef. Symptoms and treatment are the same as for the pork species.

Taenia pisiformis. This species (about 500 mm. long) possesses a life cycle much like that of *T. solium*. The intermediate hosts are rabbits, rats, squirrels or some other rodent that might be eaten by dogs. Oncospheres usually are found in the liver of these animals while the infective stage, the cysticercus, in-

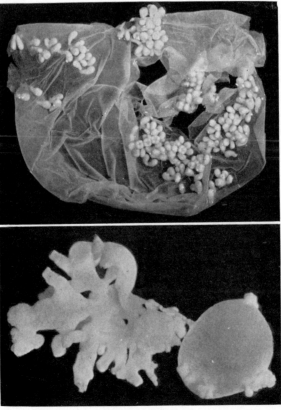

Fig. 10-26. Two types of coenurus of *Multiceps*. (Faust and Russell, Clinical Parasitology, Lea & Febiger)

habits the peritoneal cavity. The rodents are eaten by dogs, cats, wolves, foxes and other carnivores that thus become the definitive hosts.

Taenia taeniaeformis (T. crassicollis). This tapeworm occurs in domestic and wild cats. The adult worm apparently does little damage but the larval form, called *Cysticercus fasciolaris,* may cause considerable harm to the rat or mouse host in which it is normally found. The larva gets into the rodent liver and there becomes encapsulated. A serious cancer-like growth apparently may arise from this encapsulation.

Multiceps multiceps. Normally found in dogs and wolves, this worm is similar to *Taenia solium* in appearance. When the eggs are eaten by ruminant animals such as sheep, cattle and horses or related wild mammals, the larvae develop into the bladder stage, each of which contains many scoleces (Fig. 10–26). This stage is a **coenurus,** which resembles a brood capsule of *Echinococcus granulosus* (see p. 246). Although it may develop in almost any tissue, it often occurs in the brain, causing giddiness. Thus the common name "gid worm" or "gid tapeworm" (*Coenurus cerebralis*). It is most prevalent in sheep but has rarely occurred in man, probably due to accidental ingestion of eggs from dog feces. Prognosis in man is grave and there is no effective treatment. Dogs acquire the adult tapeworm by eating infected parts of sheep.

Echinococcus granulosus. One of the most serious larval tapeworm infections in man is caused by *E. granulosus.* It is called the **hydatid worm** (Fig. 10–27) because it forms hydatid cysts in various organs. The normal host for the adult parasite is the dog, in which hundreds of the worms may occur in the small intestine. Wolves and jackals also harbor the adult worm, and in some areas foxes are probably infected. The parasite is especially common in sheep-raising countries such as Australia, parts of South America and the Middle East (see Rausch and Nelson, 1963[8]). The entire worm consists of four segments, including the scolex,

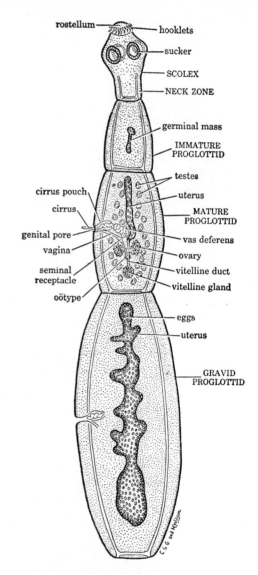

Fig. 10-27. *Echinococcus granulosus,* surface view of whole worm. (Brown, Selected Invertebrate Types, courtesy of John Wiley & Sons Inc.)

and it is only 3 to 5 mm. long. The scolex has a rectractable rostellum armed with a double circle of hooks, and there are four moderately sized suckers. Only a single ripe proglottid occurs at any one time and it, although tiny, somewhat resembles that of *Taenia solium.* The tegument is covered by minute spines (Fig. 10–28) and possesses sensory endings (Fig. 10–29). Apparently

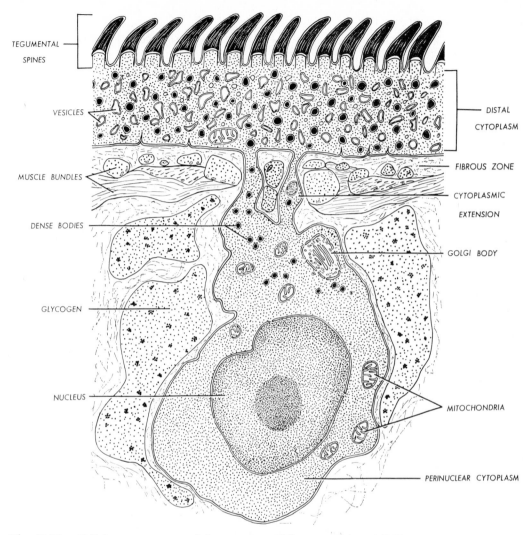

TEGUMENTAL SPINES

VESICLES

MUSCLE BUNDLES

DENSE BODIES

GLYCOGEN

NUCLEUS

DISTAL CYTOPLASM

FIBROUS ZONE

CYTOPLASMIC EXTENSION

GOLGI BODY

MITOCHONDRIA

PERINUCLEAR CYTOPLASM

Fig. 10-28. Cellular arrangement of the tegument of the protoscolex of *Echinococcus granulosus*. This arrangement is the same in the adult form. (Morseth, courtesy of J. Parasitol.)

the adult worms do little damage to their hosts.

The fine structure of the nervous system of *Echinococcus granulosus* was described by Morseth[7] as follows: lateral nerve trunks are composed of unmyelinated fibers without a cellular sheath. Mitochondria and vesicles of several sizes occur in the fibers. Clear vesicles accumulate on one side of synaptic junctions, whereas dense vesicles, possibly neurosecretory, occur at a presumed neuromuscular junction and in nerve processes.

The important part of the life cycle (Fig. 10–30), so far as its pathogenicity is concerned, is the larval stage, which may occur in man, cattle, sheep, camels, horses, moose, deer, pigs, rabbits, etc. These hosts ingest *Echinococcus granulosus* eggs, which are almost identical with those of *Taenia* sp. The oncospheres are liberated in the intestine and, as in other species, enter mesenteric veins and make their way to various organs, especially the lungs and liver. The spleen, kidneys, heart, brain, or even bone may be infected. In these organs the oncosphere develops into a spherical cyst or hydatid that may grow to

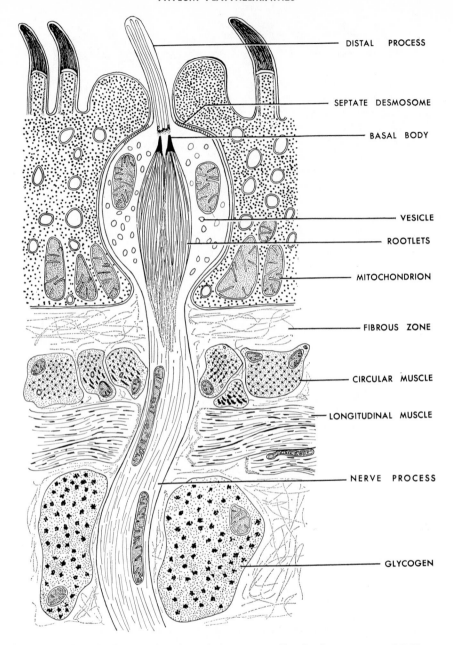

DISTAL   PROCESS

SEPTATE DESMOSOME

BASAL BODY

VESICLE

ROOTLETS

MITOCHONDRION

FIBROUS ZONE

CIRCULAR MUSCLE

LONGITUDINAL MUSCLE

NERVE   PROCESS

GLYCOGEN

Fig. 10-29.   Longitudinal section through a sensory ending in the tegument of *Echinococcus granulosus.*   (Morseth, courtesy of J. Parasitol.)

a diameter of 15 cm. (6 inches!). The size and shape of the cyst may be limited by the organ or space in which it develops. The inner lining of the cyst is germinative and gives rise to scoleces, brood capsules and daughter cysts (Fig. 10–31). Brood capsules and scoleces may become free from their attachments and form a loose mass on the floor of the cyst; this mass is known as **hydatid sand.** The cyst is filled with **hydatid fluid.** There may be as many as two million scoleces in a large cyst that might

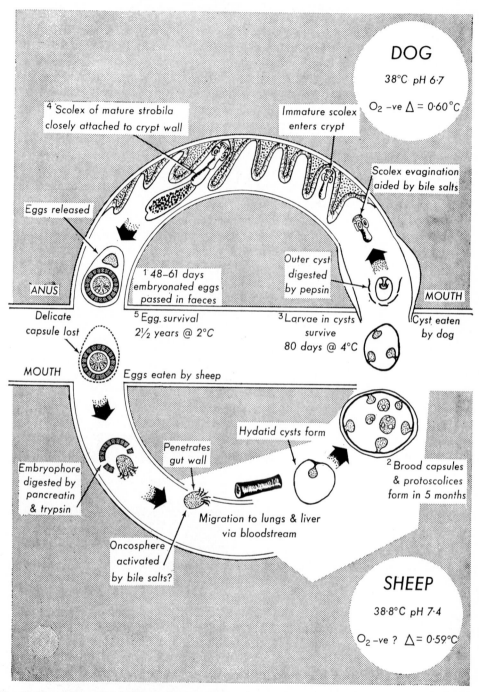

Fig. 10-30.   Life cycle of *Echinococcus granulosus* and some of the physiologic factors relating to it. (Smyth, *In* Dawes, Advances in Parasitology, courtesy of Academic Press)

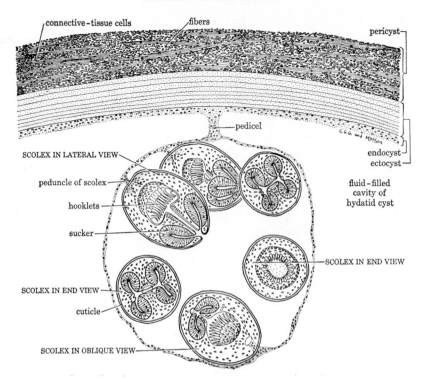

Fig. 10-31. *Echinococcus granulosus*, portion of the hydatid. (Brown, Selected Invertebrate Types, courtesy of John Wiley & Sons Inc.)

contain 2 L. of fluid. Old, enormous cysts have been reported to contain over 15 L. of fluid. These fluid-filled cysts are **unilocular** and they may persist for years. They may develop within bone, thus forming osseous hydatid cysts. Figure 10–32 shows a cyst-like mass, somewhat larger than a fist, taken from the abdomen of a woman in New York. The small (0.5 to 2.5 cm. in diameter) echinococcal cysts it contained are illustrated in Figure 10–33.

In heavily endemic areas, 50 per cent of the dogs are infected with adult worms, and up to 90 per cent of the sheep and cattle and 100 per cent of the camels may be infected with hydatid cysts. The incidence of hydatid disease (echinococcosis or hydatidosis) in man may run as high as 20 per cent (in a few areas in South America), but it is usually much lower.[3]

Echinococcus multilocularis. This worm produces an even more serious condition be-

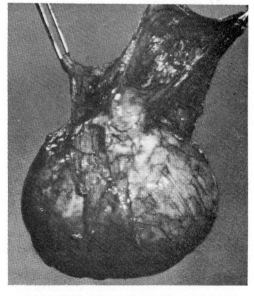

Fig. 10-32. *Echinococcus* cyst, somewhat larger than a fist. (Bohrod, courtesy of Medical Radiology and Photography)

Fig. 10-33. Small cysts taken from the mass illustrated in Figure 10–32. (Bohrod, courtesy of Medical Radiology and Photography.)

cause the cyst becomes irregular, filled with connective tissue and gelatinous masses, and grows like a malignant tumor with metastasis. This type is a **multilocular** or **alveolar** cyst. The tapeworm, found chiefly in Europe and northern Asia, looks much like a small *E. granulosus*. In length it ranges from 1.4 to 3.4 mm. The testes lie posterior to the cirrus sac, and the uterus does not possess lateral pouches. Although dogs, cats, or man may harbor the adult worm, the principal host is the fox. Intermediate hosts are usually mice in which the liver may become heavily infected. The many pockets that may be formed are the basis for the name **multilocular.**

Damage by any cyst is, of course, related to the size of the cyst and to its location. Simple cysts seem to do little harm to animals, although pressure on surrounding organs may be of consequence. Rupturing cysts may cause allergic responses, and enlarging cysts may destroy bone or impair the normal functioning of other organs. Large or migrating cysts can be serious. Although the adult worms do not live long, the cysts may remain alive for many years. Host tissue reactions,

toxemia, eosinophilia, pressure effects, obstruction of blood vessels and other factors indicate the presence of the parasites.

The only treatment for infection by the cysts of *Echinococcus* is surgery, and even this procedure is often unsuccessful or impractical. Prevention consists mainly of keeping dogs free from infection and avoiding accidental ingestion of tapeworm eggs. Obviously the best way to prevent dogs from becoming parasitized by those worms with intermediate hosts is to prevent dogs from eating the intermediate hosts. To prevent man from becoming infected with the cysts, personal hygiene is important. Keep dogs clean; don't pet infected dogs; never allow a dog to "kiss" your face.

Family Anoplocephalidae

Members of this family possess scoleces without hooks. Female reproductive organs may be single or double. The parasites live in the intestines of birds, herbivorous mammals and primates. The cysticercoid stage occurs in oribatid mites and insects. A few genera (e.g., *Avitellina*) lack yolk glands.

Adult worms that live in the intestines of domestic and wild ruminants include the genera *Anoplocephala*, *Moniezia* and *Thysanosoma*. Rabbits, hares and other rodents may harbor *Andrya* or *Cittotaenia*. Other genera may be found in birds, and *Bertiella* is a parasite of monkeys, apes and, occasionally, man. In 1937 Stunkard[15] worked out the life cycle of *Moniezia expansa* (Fig. 10–34), a common tapeworm of sheep. The worm eggs in host feces contaminate pastures. Minute, free-living oribatid mites creep over the soil, usually in the evening, and probably ingest the worm eggs accidentally with their natural food. Within two to five months the young tapeworms have developed into the cysticercoid stage in the mites. Sheep eat the mites, which often cling to forage grass. Cysticercoids are released in the sheep small intestines, where they mature in about 30 days. Adult *M. expansa* reach a length of 600 cm. (20 feet). The mature proglottids are wider than long.

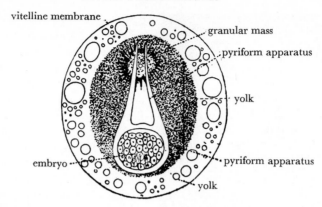

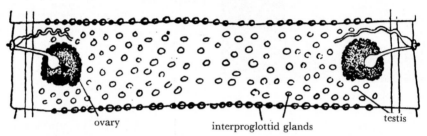

Fig. 10-34. *Moniezia expansa.* Proglottid (after Mönnig) and egg (after Railliet). (Lapage, Veterinary Parasitology, courtesy of Oliver and Boyd Ltd.)

Thysanosoma actinioides. This tapeworm is found in domestic and wild ruminants, and develops in psocid lice (order *Psocoptera*). These small insects may be found on pasture vegetation.

Family Linstowiidae

These worms are small to medium in size. The scolex is unarmed. They primarily inhabit insectivorous animals, and the larval stages occur in beetles. Because of the unarmed scolex and other features, the group is sometimes placed as a subfamily of the Anoplocephalidae. An important genus is *Inermicapsifer. I. madagascariensis* is unusual in that a few cases of human infections have been reported, although this species normally parasitizes rats. The adult is about 40 cm. in length. The life cycle is unknown. *Oochoristica* occurs in reptiles and mammals. *O. ratti,* in rats and mice, uses various insects as intermediate hosts.

Family Mesocestoididae

Members of this family are found as adults in birds and mammals. They are small to medium in size. The holdfast has four prominent suckers but no rostellum. The genital aperture is median and on the ventral surface of the body; eggs occur in a posterior parauterine organ.

Mesocestoides. This genus has a peculiar structure called the **parauterine** organ that is formed by parauterine cells arising in medullary tissue between anterior margins of the ovarian lobes. The cells form a reticular syncytium and the organ is apparently involved with the developmental processes of embryos.

REFERENCES

1. Bartel, M. H.: The life cycle of *Raillietina* (R.) *loeweni* Bartel and Hansen, 1964 (Cestoda) from the black-tailed jackrabbit, *Lepus californicus melanotis* Mearns. J. Parasitol., *51*: 800–806, 1965.
2. Belding, D. L.: Textbook of Clinical Parasitology. Ed. 2. New York, Appleton-Century-Crofts, 1952.
3. Cameron, T. W. M., and Webster, G. A.: The ecology of hydatidasis. *In* May, J. M. (ed.): Studies in Disease Ecology. New York, Hafner, 1961.

4. Cheng, T. C.: The coracidium of the cestode *Tylocephalum* and the migration and fate of this parasite in the American oyster, *Crassostrea virginica*. Trans. Amer. Micr. Soc., *85*(2):246–255, 1966.

5. Kennedy, C. R.: Taxonomic studies on *Archigetes* Leuckart, 1878 (Cestoda: Caryophyllaeidae). Parasitology, *55*:439–451, 1965.

6. Kilejian, A., Ginger, G., and Fairbairn, D.: Lipid metabolism in helminth parasites. IV. Origins of the intestinal lipids available for absorption by *Hymenolepis diminuta* (Cestoda). J. Parasitol., *54*:63–68, 1968.

7. Morseth, D. J.: Observations on the fine structure of the nervous system of *Echinococcus granulosus*. J. Parasitol., *53*:492–500, 1967.

8. Rausch, R. L., and Nelson, G. S.: A review of the genus *Echinococcus* Rudolphi, 1801. Ann. Trop. Med. Parasit., *57*:127–135, 1963.

9. Read, C. P.: Nutrition of intestinal helminths. *In* Soulsby, E. J. L. (ed.): Biology of Parasites. pp. 101–126, New York, Academic Press, 1966.

10. ———: Longevity of the tapeworm, *Hymenolepis diminuta*. J. Parasitol., *53*:1055–1056, 1967.

11. Read, C. P., Douglas, L. T., and Simmons, J. E., Jr.: Urea and osmotic properties of tapeworms from elasmobranchs. Exper. Parasit., *8*:58–75, 1959.

12. Read, C. P., and Kilejian, A. Z.: Circadian migratory behavior of a cestode symbiote in the rat host. J. Parasitol., *55*:574–578, 1969.

13. Rothman, A. H.: Electron microscope studies of tapeworms: The surface structure of *Hymenolepis diminuta* (Rudolphi, 1819) Blanchard, 1891. Trans. Amer. Micr. Soc., *82*: 22–30, 1963.

14. Šlais, J.: The location of the parasites in muscle cysticercosis. Folia Parasit., *14*:217–224, 1967.

15. Stunkard, H. W.: The development of *Moniezia expansa* in the intermediate host. Parasitol., *30*:491–501, 1937.

16. U.S.D.A. Yearbook (1942): Keeping Livestock Healthy. 1942 Yearbook of Agriculture, Washington, D.C.

17. Voge, M., and Heyneman, D.: Development of *Hymenolepis nana* and *Hymenolepis diminuta* (Cestoda: Hymenolepididae) in the intermediate host *Tribolium confusum*. Univ. Calif. Pub. Zool., *59*, 549–580, 1957.

18. Von Bonsdorff, B.: *Diphyllobothrium latum* as a cause of pernicious anemia. Exper. Parasitol., *5*:207–230, 1956.

19. Warren, M., and Daugherty, J.: Host effects on the lipid fraction of *Hymenolepis diminuta*. J. Parasitol., *43*:521–526, 1957.

Section IV
PHYLUM ACANTHOCEPHALA

Chapter 11

Phylum Acanthocephala

GENERAL CONSIDERATIONS

The phylum Acanthocephala is composed of thorny-headed worms, so called because of the many thorn-like hooks that occur on the proboscis (**acanth** means a thorn and **cephala** refers to the head). These cosmopolitan worms are all endoparasitic as adults in the digestive tracts of terrestrial and aquatic vertebrates, ranging from fish to man, but occurring especially in fish. They are whitish or slightly yellow parasites having wrinkled or smooth bodies that range in length from 1 mm. to over 1 m. Larval stages are found in invertebrates. Hundreds of young stages may be found in a single intermediate host, and the vertebrate host may harbor thousands of the adult worms. Damage to the intestinal epithelium and proliferation of host connective tissue are the most pronounced histopathologic effects.[2] For a review of the biology of the Acanthocephala, see Nicholas.[12]

The proboscis and associated structures are called the **presoma**. The rest of the body is the **trunk**. Small spines occur on the trunk in some genera (e.g., *Corynosoma strumosum* in seals). The proboscis, neck and trunk may become modified as accessory anchoring devices, such as a bulb-like inflation of the anterior end and a general covering of cuticular spines. In adults, the body cavity is a fluid-filled space between the viscera and the body wall, and is a **pseudocoel** rather than a true coelom. There is no digestive tract or true circulatory system in either adult or larval worms. The principal carbohydrate reserve is glycogen. Sexes are separate, females are almost always larger than males, and the posterior aperture, or **gonopore**, is the only body opening. In general, tissues have lost cellular identity and there are relatively few nuclei in the entire body.

The body wall consists of a cuticula, a thick, syncytial, fibrous, three-layered subcuticula (or hypodermis), a thin dermis, and a double layer of syncytial muscles (an inner layer and an outer circular layer). The inner layer of the epidermis contains a network of spaces or **lacunae** that contain nutritive fluid and function as a food-distributing mechanism. The number of nuclei in the body wall is approximately constant for each species, at least in the early stages and often throughout life.

The proboscis is an anterior structure, globular or cylindrical in shape, bearing rows of recurved hooks or spines that serve to attach the worm to the gut of its host. In some species the proboscis of the adult and that of the infective larva are identical in appearance; in others, the adult proboscis may become markedly modified. In some worms the proboscis is permanently anchored in host tissue, as in *Filivollis* in birds, *Polymorphus* in ducks and *Pomphorhynchus bulbocolli* in fish. The proboscis can usually be withdrawn into a muscular **proboscis receptacle** or **proboscis sac**. This sac extends into the body cavity from the neck region. The hooks on the proboscis vary according to size, shape, number and arrangement and thus have considerable taxonomic value. See Figure 11–1 for general anatomy.

Lemnisci are paired organs, usually elongated and pendulus, that extend into the body cavity from the neck region. They apparently serve as fluid reservoirs when the

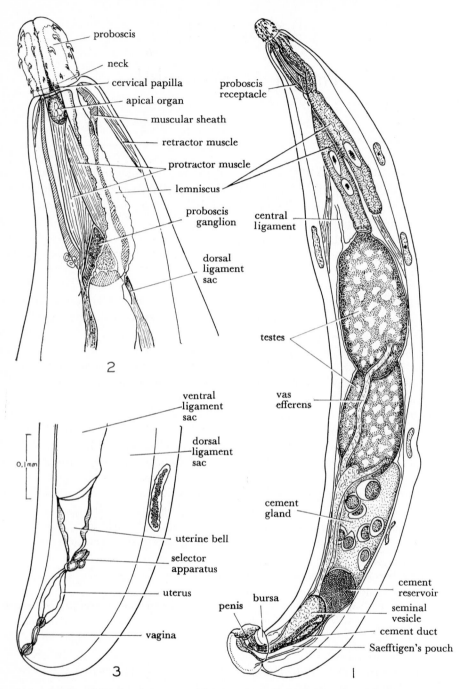

Fig. 11-1. Adults of *Paulisentis fractus*. (*1*) Mature male, lateral view. (*2*) Same, proboscis and anterior trunk region enlarged. (*3*) Posterior portion of female, lateral view. (Cable and Dill, courtesy of J. Parasitol.)

proboscis is invaginated. There is much histochemical evidence that the lemnisci may also serve a metabolic function—especially for fat metabolism. They are often surrounded by muscles at their basal portion.

Ligament sacs extend from the proboscis sheath or from the adjacent body wall and form tubes that surround the reproductive organs. They commonly do not persist in adults, and only one may be present. A **ligament strand** is attached to the gonads and extends the length of the ligament sacs.

Protonephridia serve as excretory organs. They consist of flame bulbs and collecting tubules that occur in some members of the class Archiacanthocephala. The flame bulbs are grouped into two masses attached to the reproductive organs and empty by way of a canal or bladder into the sperm duct or uterus.

The nervous system consists mainly of a ganglion in the proboscis sheath and of nerves that connect the ganglion to other organs and tissues of the body. In addition, a pair of genital ganglia, with nerves, occurs in the male. **Sense organs** are found in the proboscis and in the penis and male bursa.

Male reproductive organs consist of a pair of testes, one behind the other, and a common sperm duct formed by the union of a duct from each testis. The common duct leads through the penis to the outside. There is usually a cluster of large **cement glands** that empty into the common sperm duct, sometimes by way of a cement gland reservoir. The penis projects into an eversible bursa. This cup-like terminal structure with a thick, domed cap is used to hold the female during copulation.

Female reproductive organs consist of an ovary fragmented into **ovarian balls** that lie in the ligament sac or are free in the pseudocoel. A **uterine bell**, at the end of the ligament sac is a funnel or cup-like structure that receives ova from the ovarian balls. At the base of the uterine bell is a group of cells, the **selective apparatus**, and bell pockets. The reproductive tube continues as an

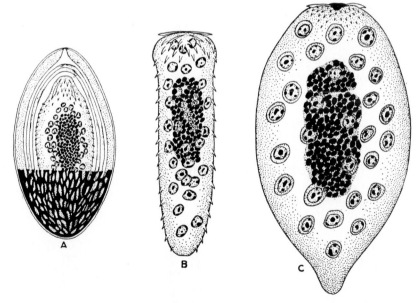

Fig. 11-2. Shelled embryo and acanthor stages of *Macracanthorhynchus hirudinaceus* (Pallas). Magnification approximately 500 diameters. (*A*) "Egg" or shelled embryo. Surface appearance shown only on lower pole; on remainder of drawing membranes are shown in optical section. (*B*) Acanthor, stage I, from lumen of midgut of beetle larva. Note the larval rostellar hooks and small body spines. (*C*) Acanthor, stage II, after penetrating the wall of the midgut of the beetle larva, about five to 20 days after artificial infection. (Redrawn from Kates, 1943.) (Van Cleave, courtesy of J. Parasitol.)

elongated uterus. A vagina connects the uterus to the terminal opening, or gonopore.

During copulation the male discharges sperm into the vagina. Then secretions from the cement glands block the exit of the vagina, thus preventing escape of sperm. In some species of worms, the cement gland secretions form a cap over the entire terminal end of the female. Fertilized eggs go through early embryologic development in the ligament sac or pseudocoel. When eggs emerge from the gonopore they contain a hooked larva called the **acanthor.** A host must eat the eggs before the embryos can hatch. See Figure 11-3 for representative eggs.

Hatching of the eggs of *Moniliformis dubius* (p. 262), an acanthocephalan of rats with larval stages in the cockroach, was studied in the laboratory by Edmonds.[4] Eggs hatched in certain electrolytes provided that the molarity of the solution was greater than 0.2 and the pH greater than 7.5. The addition of CO_2 lowered the required pH level and increased the percentage of eggs that hatched. The embryo released small amounts of chitinase, which possibly assisted in the decomposition of a chitinous membrane that surrounds the embryo. Activation of egg hatching was probably ionic rather than osmotic. The optimal hatching temperature seemed to be between 15 to 37° C. Eggs hatched in stimulating fluids at 20° C. after storage in distilled water at 5° C. for at least four weeks.

Little is known about details of the life cycles of most species, but undoubtedly they all follow the same general plan. Each female usually has only one reproductive cycle. At the time of copulation, all eggs are subject to fertilization, and the female may become greatly distended with stored embryonated eggs that she can discharge selectively over a long period of time. The eggs are normally spindle-shaped, often resemble diatoms, and are commonly eaten by aquatic insects and crustaceans.

The intermediate hosts for acanthocephalids of fish are usually benthic amphipods (*Gammarus, Pontoporeia*), isopods (*Asellus*), or ostracods (*Cypris*). Snails and aquatic insects sometimes act as either transport hosts, or as secondary intermediate hosts. Some species of acanthocephalids may require a second and even a third intermediate host, but usually only a single invertebrate host is involved. No superimposed multiplicative cycle occurs similar to that which occurs in the flukes, and there is no asexual or parthenogenetic development. One adult arises from each zygote. There is no free-living stage.

The following is a life cycle involving one invertebrate host: An egg eaten by an arthropod hatches into an **acanthor,** develops into an **acanthella,** and becomes a **juvenile** (which might progress to a **cystacanth**), and is eaten by the final (vertebrate) host, in which it becomes an adult. Transport hosts can be either vertebrates or invertebrates. Within the invertebrate host, the acanthor is liberated from the egg, bores through the gut

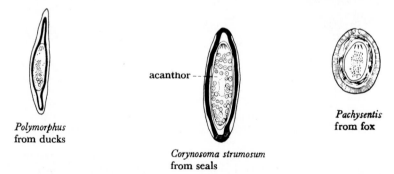

acanthor

Polymorphus
from ducks

Corynosoma strumosum
from seals

Pachysentis
from fox

Fig. 11-3. Representative eggs of Acanthocephala. (Olsen, Animal Parasites, Burgess Pub. Co., 1967.)

wall and develops into the acanthella. After several weeks or more, the acanthella becomes a juvenile, which is generally called a cystacanth (the second infective stage), differing from an adult primarily by being sexually immature. The term "juvenile" should be used only when the larva resemble the adult. The vertebrate host becomes infected by eating the arthropod intermediate host or a transport host. Sexual development occurs in the final host.

Because host specificity for the intermediate host is more limited than that for vertebrate hosts, the distribution of acanthocephalids is largely determined by invertebrate hosts. For example, some birds acquire *Corynosoma* by eating fish. This genus occurs in aquatic birds and marine mammals, and the fish is presumably a transport host. DeGiusti[3] has made a significant observation on the life cycle of *Leptorhynchoides thecatus*, a parasite of the rock bass, *Ambloplites rupestris*. An amphipod is the intermediate host, but if the fish swallows an amphipod containing a juvenile worm that has not quite completed its full development, the parasite does not mature in the intestine of the final host. Instead, it penetrates the gut wall and becomes encysted in the mesenteries.

The biochemistry of the Acanthocephala needs much more study, and only a few generalizations can be made at the present time. Nicholas[12] reviewed the information to date on this subject. Glycogen constitutes the principal carbohydrate reserve, and lipids accumulate in considerable amounts. Studies on *Moniliformis dubius* in rats have provided some information on intermediary metabolism. Glucose, fructose, mannose and mal-tose stimulate glycogen synthesis in worms taken from fasted rats. The Embden-Meyerhof scheme of glycolysis appears to operate in this species. *M. dubius* excretes organic acids under aerobic and anaerobic conditions in vitro, and the respiration appears to be adapted to semi-anaerobic or anaerobic environments.

REPRESENTATIVE SPECIES

Macracanthorhynchus hirudinaceus (Figs. 11–4, 11–5). This worm is exceedingly rare in man but is a common species in hogs, where it occurs in the small intestine attached to the gut wall. It is usually large, the females ranging from 20 to 65 cm. in length whereas the males range from 5 to 10 cm. The largest worms may be about the width of an ordinary pencil. The animal is flattened in normal life but it soon becomes cylindrical when collected and preserved. Irregular transverse folds give the worm a wrinkled appearance. The small protrusible proboscis is armed with hooks in six spiral rows. The body tapers from just behind the proboscis to the posterior end, where there is a bell-like bursa. Testes are located in the anterior half of the male worm.

Female worms produce eggs containing mature acanthors. Eggs (67 to 110 × 40 to 657 μ) leave the pig's intestine with feces and they become widely scattered. They are extremely resistant to adverse environmental conditions and remain viable for years. Birds may carry the eggs on their feet or may even eat them with contaminated food and pass them unharmed through their own bodies. The usual invertebrate host is a larval bettle that eats the eggs.

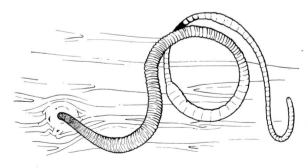

Fig. 11-4. *Macracanthorhynchus hirudinaceus* adult attached to the gut wall of a pig.

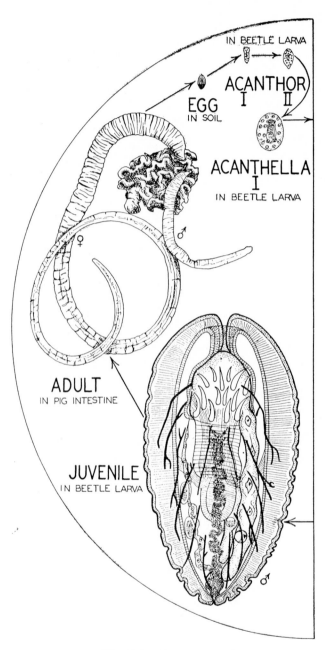

IN BEETLE LARVA

EGG
IN SOIL

ACANTHOR
I   II

ACANTHELLA
I
IN BEETLE LARVA

ADULT
IN PIG INTESTINE

JUVENILE
IN BEETLE LARVA

Fig. 11-5. *Legend on opposite page.*

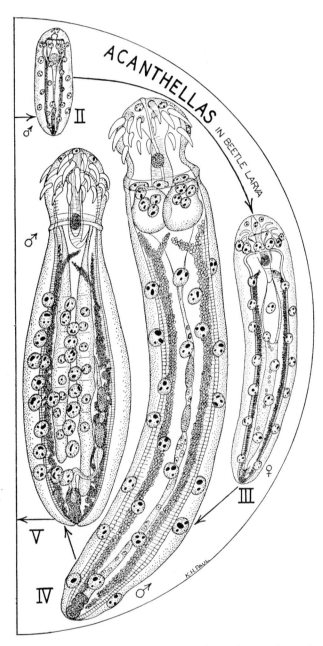

Fig. 11-5. The life cycle of *Macracanthorhynchus hirudinaceus* (Pallas), the thorny-headed worm of swine. Drawings of adult worms about natural size; all other stages at fairly uniform magnification of approximately 38 diameters. The elongated, branching, dark bodies in the juvenile are the nuclei which will become the giant nuclei of the subcuticula in the adult. (After Kates, 1943) (Van Cleave, courtesy of J. Parasitol.)

Cockchafer beetles (*Melolontha vulgaris*), rose chafer beetles (*Cetonia aurata*), water beetles (*Tropisternus collaris*), and various species of dung beetles are the common hosts. Eggs hatch in the beetle gut and the acanthor uses its hooks to bore through the intestinal wall to the hemocoel, where it develops into the acanthella that gradually develops most of the adult characteristics. By the end of six weeks to three months the juvenile (cyst-acanth) stage is reached. The beetle may be eaten by a second intermediate host but usually it is devoured by the final host. This juvenile is thus the infective stage for pigs or man. The life span of the adult in the pig is usually less than a year. Although 400 acanthors have been found in one naturally infected beetle, there are usually fewer than 30 worms in a pig. Occasionally dogs and other mammals that might eat beetles are parasitized by this worm.

Ulceration, necrotic areas, anemia and even penetration of the intestinal wall with subsequent peritonitis may occur in serious infections. Control consists mainly of sanitary measures in raising pigs, such as rearing them on concrete floors and promptly removing feces.

Moniliformis moniliformis. This species is a cosmopolitan acanthocephalan that lives in the small intestine of rats, mice, dogs, and cats. The worms have a beaded appearance because of annular thickenings. Males range from 4 to 13 cm. in length while females are 10 to 30 cm. long. The proboscis possesses 12 to 15 rows of hooks. Eggs, 85 to 120 × 40 to 50μ, are eaten by beetles (*Blaps gigas, Tenebrio molitor, Calandra orizae*) or by cockroaches (*Blatta orientalis, Blatella germanica*). The life history closely resembles that of *Macracanthorhynchus hirudinaceus*. Juveniles appear in four to six weeks and as many as 100

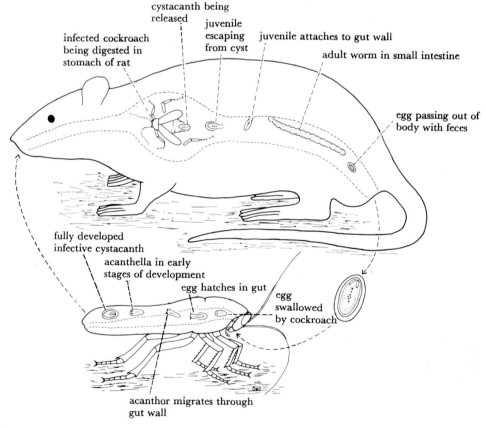

Fig. 11-6. Life cycle of *Moniliformis dubius*. (Olsen, Animal Parasites, Burgess Pub. Co., 1967.)

may inhabit one cockroach. The infection rate in rats varies, but may run as high as 20 per cent in *Rattus alexandrinus*. Frogs, toads, and lizards have been suggested as potential transport (paratenic) hosts. Growth of the worm ceases when the rat is placed on a diet devoid of carbohydrate.

Moniliformis dubius is also common in rats (Fig. 11–6) and has been reported from man. There is some question about the identity of this species with *M. moniliformis*. This problem has been "settled" several times by the taxonomic "lumpers" and "splitters." The developing acanthella in the cockroach, *Periplaneta americana*, is surrounded by a capsule consisting of closely packed vesicles that arises from the surface membrane of the insect's haemocytes.[11]

The following few examples are additional representatives of the hundreds of species that have been described from all groups of vertebrates.

Polymorphus minutus (family Polymorphidae) is a small bright orange acanthocephalan that lives in the small intestine of chickens, ducks and possibly geese and other water birds. The male is about 3 mm. long and the female grows up to 10 mm. It can cause serious disease in domestic and wild fowl. The intermediate host is the amphipod, *Gammarus pulex*. A closely related species, *Filicollis sphaerocephalus*, is also found in aquatic birds. The invertebrate host is a crustacean. *Filicollis anatis* is another small species that inhabits the intestines of ducks, geese and other water birds. The male is 6 to 8 mm. and the female, 10 to 25 mm. long. *Asellus aquaticus*, a freshwater isopod, is the intermediate host. *Neoechinorhynchus* is common in turtles and fish.[9]

CLASSIFICATION

Affinities of this phylum to other worms are not clear. Developmental stages are more similar to those of flatworms than to the nematodes, but the group is probably closest to the cestodes, although distinctly separate. There are about 500 known species, and probably many more have not yet been described.

The three main groups below are sometimes listed as classes instead of orders, but modern authorities point out that the similarities are too great to justify separation into classes.[2a] For further information on the Acanthocephala, see Baer,[1] Bullock,[2a] Golvan,[5,6,7,8] Hyman,[10] Nicholas,[12] Van Cleave,[13] and Yamaguti.[15] For the evolution of the group, see p. 565.

Order *1.* ARCHIACANTHOCEPHALA
"Parasites of terrestrial birds and mammals in which the proboscis hooks are arranged in concentric circles, trunk spination is lacking, the main lacunar channels are median in position, protonephridia are generally present, two persistent ligament sacs are found in the females and eight cement gland cells in the male."[14]
Family GIGANTORHYNCHIDAE
Elongate worms in birds and Central and South American anteaters; no protonephridia. Genus: *Mediorhynchus*.
Family OLIGACANTHORHYNCHIDAE
Short or elongate worms often curved, coiled or even spirally rolled; short spheroid or ovoid proboscis; protonephridia usually present; found in birds and mammals. Genera: *Macracanthorhynchus, Oncicola, Hamanniella*.
Family MONILIFORMIDAE
Elongate worms; body somewhat twisted; found in rodents and other terrestrial mammals; no protonephridia. Genus: *Moniliformis*.
Order *2.* PALAEOACANTHOCEPHALA
"Parasites of fish, aquatic birds and marine mammals in which the proboscis hooks are arranged in alternating radial rows, trunk spines are frequently present, the main lacunar channels are lateral, protonephridia are absent, the single ligament sac ruptures in females and the male usually has six cement gland cells."[14]
Family RHADINORHYNCIDAE
Usually with spiny trunk; elongate worms in fish. Genera: *Rhadinorhynchus, Telosentis, Illiosentis, Leptorhynchoides, Serrasentis, Gorgorhynchus*.
Family ECHINORHYNCHIDAE
No trunk spines; found in fish and amphibians. Genera: *Echinorhynchus, Acanthocephalus*.
Family POMPHORHYNCHIDAE
Long, slender, cylindrical neck, usually with prominent, bulbous swelling at anterior end; found in fish. Genus: *Pomphorhynchus*.
Family POLYMORPHIDAE
Relatively short, straight, often plump worms in aquatic birds and marine mammals; proboscis of various shapes; spines often on

anterior part of trunk; neck and foretrunk frequently sharply differentiated from hindtrunk. Genera: *Polymorphus, Filicollis, Corynosoma, Bolbosoma* (in whales).

Order 3. EOANCANTHOCEPHALA

With giant epidermal nuclei. Rather small parasites of many fish, including ganoids, amphibians and reptiles; presence in elasmobranchs and mammals is probably accidental. Intermediate hosts include ostracods, copepods, amphipods, annelids, insects and cyclostomes (Petromyzon). Proboscis hooks arranged in alternating radial rows; trunk spines may or may not be present; the main lacunar channels are dorsal and ventral; protonephridia are absent; two ligament sacs are found in the female but the ligament breaks down in sexually mature worms; a syncytial cement gland occurs in the male.

Family QUADRIGYRIDAE

With trunk spines; found in fish, primarily in Asia. Genera: *Quadrigyrus, Pallisentis, Acanthosentis*.

Family NEOECHINORHYNCHIDAE

Without trunk spines; found primarily in freshwater and marine fish, but several species occur in North American turtles. Genera: *Neoechinorhynchus, Octospinifer, Octospiniferoides* (or *Paulisentis*), *Tanaorhamphus*.

REFERENCES

1. Baer, J. G.: Embranchement des Acanthocephales. *In* Grassé, P. (ed.) Traité de Zoologie. Tome IV, pp. 731–782. Paris, Masson et Cie Editeurs, 1961.

2. Bullock, W.: Intestinal histology of some salmonid fishes with particular reference to the histopathology of acanthocephalan infections. J. Morph., *112*:23–44, 1963.

2a.———: Morphological features as tools and as pitfalls in acanthocephalan systematics. *In* Schmidt, G. D. (ed.): Problems in Systematics of Parasites. pp. 9–43. Baltimore, University Park Press, 1969.

3. DeGiusti, D. L.: The life cycle of *Leptorhynchoides thecatus* (Linton), an acanthocephalan of fish. J. Parasitol., *35*:437–460, 1949.

4. Edmonds, S. J.: Some experiments on the nutrition of *Moniliformis dubius* Meyer (Acanthocephala). Parasitology, *55*(2):337–344, 1964.

5. Golvan, Y. J.: Le Phylum des Acanthocephala. Première Note. Sa Place dans l'Echelle Zoologique. Ann. Parasit. Hum. Comp., *33*: 538–602, 1958.

6. ———: Le Phylum des Acanthocephala. Deuxième Note. La Classe des Eoacanthocephala (Van Cleave 1936). Ann. Parasit. Hum. Comp., *34*:5–52, 1959.

7. ———: Le Phylum des Acanthocephala. Troisième Note. La Classe des Palaeacanthocephala (Meyer 1931). Ann. Parasit. Hum. Comp., *35*:138–165, 1960.

8. ———: Le Phylum des Acanthocephala. Quatrième Note. La Classe des Archiacanthocephala (A. Meyer 1931). Ann. Parasitol. Hum. Comp., *37*:1–72, 1962.

9. Hopp, W. B.: Studies on the morphology and life cycle of *Neoechinorhynchus emyais* (Leidy). An acanthocephalan parasite of the map turtle, *Graptemus geographica* (Le Sueur). J. Parasitol., *40*:284–299, 1954.

10. Hyman, L. H.: The Invertebrates: Acanthocephala, Aschelminthes and Entoprocta. Vol. III. New York, McGraw-Hill, 1951.

11. Mercer, E. H., and Nicholas, W. L.: The ultrastructure of the capsule of the larval stages of *Moniliformis dubius* (Acanthocephala) in the cockroach *Periplaneta americana*. Parasitology, *57*(1):169–174, 1967.

12. Nicholas, W. L.: The biology of the Acanthocephala. *In* Dawes, B. (ed.): Advances in Parasitology. Vol. 5, pp. 205–246. New York, Academic Press, 1967.

13. Van Cleave, H. J.: Acanthocephala of North American Mammals. Illinois Biol. Monogr., *23*, Nos. 1–2, Urbana, University of Illinois Press, 1953.

14. Watson, J. M.: Medical Helminthology. London, Baillière Tindall & Cox, 1960.

15. Yamaguti, S.: Systema Helminthum. Vol. V. Acanthocephala. New York, Interscience, 1964.

Section V
PHYLUM NEMATODA

Chapter 12

Introduction

INTRODUCTION

The phylum Nematoda contains an almost unbelievable number of worms that are free-living in water and soil, and an impressive number of species that are parasitic in plants and animals. Biology students constantly find thread-like worms coiled in the muscles, connective tissue or other organs of laboratory-dissected animals. Fish especially are notorious for their burdens of worms. Hundreds of individual worms and several species often inhabit one host. Insects and other invertebrates have their own roundworm parasites, and also serve as intermediate hosts. Even fossil nematodes are known. The worm *Heydonius antiguus* was found projecting from the anus of the beetle *Hesthesis immortua* in Rhine lignite. Various other species have been found in amber.

Many nematodes have yet to be described, and the parasite-host relationships are just beginning to be understood. Morphologic or behavioral changes of the host sometimes occur after heavy infestation. Some parasitic nematodes possess their own parasites. Thus *Aphelenchoides parietinus*, a soil nematode parasitic in plants, may be infected with a microsporidian. Nematology is certainly an open field for anyone interested in investigating parasites.

Morphology

Nematodes are usually round in cross section; hence they are called "roundworms" in contrast to most "flatworms." They should not be confused with the segmented worms or annelids, represented by the earthworms. Compared with trematodes and ces-todes, the life histories of nematodes are less varied and their anatomy less adapted to parasitism. This relatively unspecialized development is undoubtedly related to the fact that, as a group, nematodes are not all parasitic, as are trematodes and cestodes.

Figures 12–1, 12–2, and 12–3 give a generalized picture of some of the organ systems of nematodes. The neck or tail region of these worms may bear cuticular fin or wing-like flanges. In the neck these structures are known as **cervical alae** (singular, **ala**), whereas those on the tail are **caudal alae.** The latter may be supported by fleshy papillae and are used as copulatory organs.

Labial papillae (around the mouth) are sensory, perhaps including a tactile function; others, such as a pair of cervical (**deirid**) papillae and various genital and caudal papillae, are definitely tactile.

Tegument

The published accounts of the numbers of layers, sublayers and functions of the nematode tegument are diverse. The two main layers are the cuticle and an inner hypodermis. The cuticle covers the external surface and lines the buccal cavity, esophagus, vagina, excretory pore, cloaca and rectum. This tough, protective membrane is usually smooth but may be covered with spines (Fig. 12–4) or have longitudinal and transverse ridges; it is resistant to the action of host digestive enzymes. It is noncellular and is secreted by the hypodermis.[4]

The cuticle consists basically of three layers that are usually subdivided and strengthened by fibrils often compacted to form fibers:

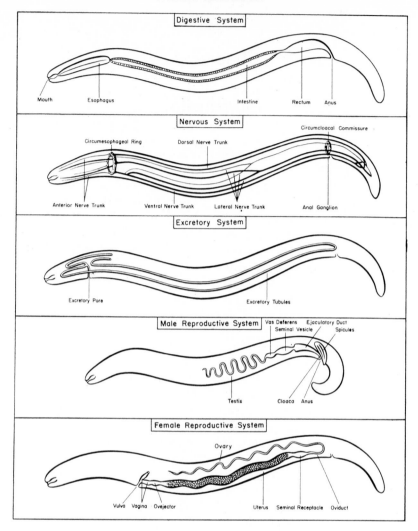

Fig. 12-1 Major organ systems of roundworms. (United States Navy Medical School Laboratory
Guide)

an outer **cortical layer,** a middle **matrix
layer,** and a **basal layer**. Associated with
the plastic matrix layer is a system of rods
that probably have a skeletal function.[17]
Inglis[12] considered the rods to be canals
("punctation canals"); he believed that all
the modifications of the cuticle are based on
this system of canals, and that these modifications result from the need for a cuticle that is
flexible longitudinally and strong radially.
Radial strength may also be achieved by the
presence of transverse alternately rigid and
flexible annules or by spiral fibers (as in

Ascaris, see p. 299), or by some other kind of
strongly patterned modification. Cuticular
canals have been demonstrated in a few
species (e.g., *Ascaris*), but they have not been
seen in most nematodes. For an account of
the ultrastructure of the cuticle of some larval
nematodes, see Eckert and Schwarz,[10] and
Jamuar.[13] For an account of histochemical
events that accompany formation of the adult
cuticle, see Kan and Davey.[14]

Twenty amino acids have been found in
the cuticle of several species of nematodes,
as well as small amounts of carbohydrates

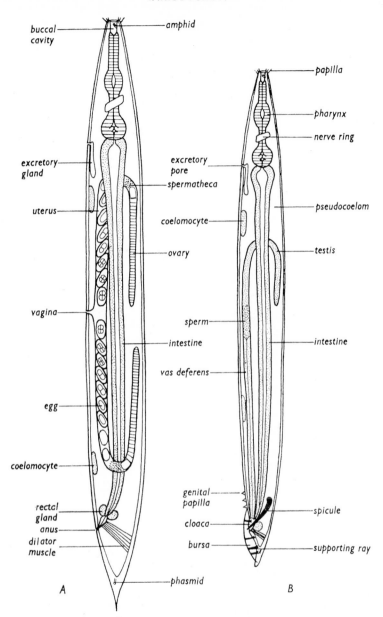

Fig. 12-2. General morphology of a nematode (hypothetical). (*A*) Female. (*B*) Male. Lateral view. (Lee, The Physiology of Nematodes. Courtesy, W. H. Freeman & Co.)

and lipids. Collagen is present in the cuticle of *Ascaris*. RNA, ascorbic acid, adenosine triphosphate, hemoglobin, acid phosphatases, and other enzymes occur in the inner cortex of nematodes, suggesting that this complex membrane is not an inert covering, but is in a state of constant metabolic activity,[3] and is involved in the synthesis of proteins. At the present time we have insufficient information to make any broad generalizations concerning the nutritive function of the nematode cuticle (see Read[20]) but it is important in osmoregulation and ion regulation.

The hypodermis is sometimes called the epidermis. It lies beneath the basal layer of the cuticle, and is not sharply delimited from

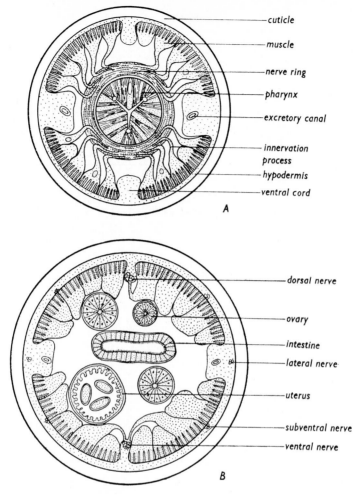

cuticle

muscle

nerve ring

pharynx

excretory canal

innervation process

hypodermis

ventral cord

A

dorsal nerve

ovary

intestine

lateral nerve

uterus

subventral nerve

ventral nerve

B

Fig. 12-3. Transverse sections of a nematode. (*A*) Pharyngeal region. (*B*) Middle region. (Lee, The Physiology of Nematodes. Courtesy, W. H. Freeman & Co.)

the body musculature to which it is attached. It is a thin layer with four longitudinal thickenings or **cords** containing the nuclei of the hypodermis. It may be cellular or syncytial. Once the juvenile forms have developed, the number of nuclei present in the hypodermis remains constant.

Muscular System

There are two kinds of muscles in nematodes, (1) somatic (unspecialized), which consist of a single layer lying next to the hypodermis; and (2) specialized, which have a variety of functions depending on their location (e.g., spicular muscles to protrude the spicules in males). The body wall muscles are placed longitudinally and are responsible for the snake-like movements of the worms. The heavily fibrous zone of each myofiber is attached to the hypodermis, but the other, less fibrous end of the muscle cell (seen clearly in cross section, Figs. 12–3B, 12–5) is connected to either the dorsal or ventral nerve cord from which it receives its motor stimulation.[9] Apparently the number of muscle cells remains constant during the growth of the worm. Between the muscle layer and the digestive tract is the body cavity known as a pseudocoelom, which functions as a hydrostatic skeleton.

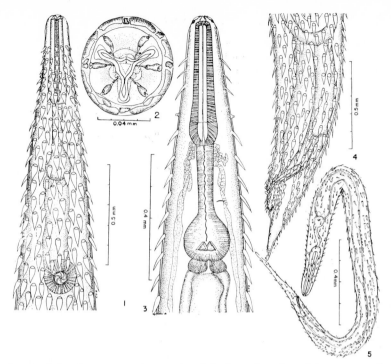

Fig. 12-4. *Paratractis hystrix*, a nematode parasite of the large intestine of the water turtle, *Podocnemis dumeriliana*, showing pronounced spiny body surface. (Sarmiento, courtesy of J. Parasitol.)
(1) Anterior portion of female, showing the arrangement of cuticular projections and the excretory pore. (2) En face view of the head. (3) Anterior portion, showing the esophagus and papillae. (4) Posterior end of female. Lateral view. (5) Larva bearing cuticular projections.

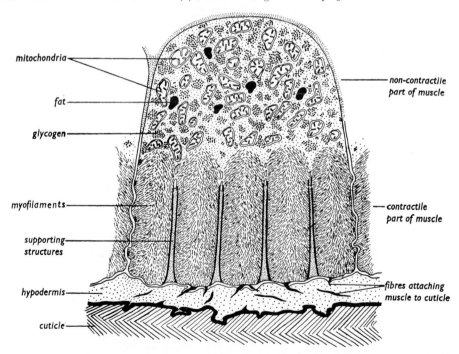

Fig. 12-5. Transverse section through a muscle of the body wall of *Nippostrongylus brasiliensis* towards the end of the muscle. Drawn from several electron micrographs. (Lee, The Physiology of Nematodes, University Reviews in Biology, Oliver and Boyd.)

In any muscle system some antagonising mechanism must be available. That it is largely hydrostatic in nematodes, based on the fluid contents of the pseudo-coelom, cannot be seriously questioned. The nematode body is cylindrical, the muscles are arranged longitudinally and the antagonization of the muscle system is dependent on the action of the internal pressure:volume. Thus, when the muscles contract, so as to straighten the body, they will elongate the contracted muscles of the opposite side of the body and there will be a concomitant increase in the internal pressure. However, the antagonizing effect is only required to act on antero-posteriorly arranged muscle units but the pressure generated will act in all directions, radially as well as longitudinally. If there is no method of compensating for, or overcoming, this, the body would not straighten but would simply tend to swell and become, or tend to become, spherical. The cuticle must, therefore, be able . . . to stretch antero-posteriorly but must not stretch radially, that is it must be anisometric.[12]

The above concept may be too rigid because many nematodes are able to pull back the anterior part of the body, thus foreshortening and widening the body.

Digestive System

The alimentary tract of nematodes is a fairly simple tube made of cells arranged in a single layer. The mouth leads to a buccal capsule (not always present), thence to a muscular esophagus that empties into the intestine. Minute projections called *microvilli* lining the inner surface of the intestine have been described for several species (see *Ascaris*, p. 300). The anus is located near the posterior tip of the worm, and a slight enlargement, the rectum, lies just anterior to the anus.

The food of nematodes consists of host tissues (*Nippostrongylus*), host blood or mucus (hookworms), or various secretions and host intestinal contents (*Ascaris*). Absorption of nutrients may take place through the body surface and the worm's gut. However, the cuticle of some species (e.g., *Ascaris lumbricoides*) seems to be relatively impermeable to glucose and amino acids.[5] Digestive enzymes have been demonstrated in the gut and

pharynx, and some worms may be capable of extracorporeal digestion. For a review of nutrition in intestinal helminths, see Read,[20] and Lee.[16]

Nervous System

The nervous system consists basically of a ring of nervous tissue around the esophagus (Fig. 12–2), and another nerve ring around the posterior region of the intestine. Longitudinal nerves, generally four main ones, connect the rings and extend to the extremities of the body.

Phasmids (Fig. 12–6) are small, paired, postanal organs associated with the nervous system in many parasitic nematodes. These structures have an obscure function, but frequently they are called chemoreceptors or olfactory receptors; they may have a flushing action. Phasmids are pouch-like and possess a minute canal leading to the outside. These glands are used as a basis of classification (e.g., *Aphasmidia*—without phasmids; *Phasmidia*—with phasmids). All nematodes possess a pair of similar organs, called **amphids,** in the head region. These organs are more conspicuous in free-living marine species than in parasites, and they are considered as chemoreceptors. Inglis has suggested that amphids act as mechano-stretch receptors. Figure 12–7 shows the position of amphids and head papillae near the mouth. The papillae are sensory and probably tactile in function. For a review of neuromuscular functions, see Debell.[9]

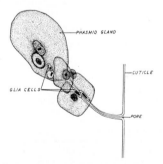

Fig. 12-6. A phasmid in a female *Spironoura*. The phasmid gland is a flushing (or "scent"?) gland, and the opening to the outside is a postanal lateral pore. (Redrawn from Chitwood and Chitwood, Introduction to Nematology.)

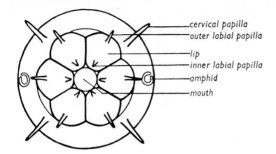

cervical papilla
outer labial papilla
lip
inner labial papilla
amphid
mouth

Fig. 12-7. En face view of a nematode head showing the positions of the mouth, lips, amphids and papillae. (Jones, Plant Nematology, MAFF Technical Bulletin No. 7. Courtesy of Her Majesty's Stationery Office.)

Excretory System

Considerable morphologic variation occurs among excretory systems. The class Secernentea possesses longitudinal ducts opening to the outside through a midventral pore that is associated with ventral gland cells called **renetts** and an excretory duct. The longitudinal ducts are often connected by a cross tube, giving the whole apparatus an "H" shape. The class Adenophorea usually has a single execretory gland that is cervico-ventral in position and opens by way of a short duct through a pore on the midventral line. Ammonia is the major end-product of nitrogen metabolism in the infective stages of the parasites, but urea and uric acid are frequently present. See Arthur and Sanborn[1] for a consideration of osmotic and ionic regulation in nematodes, and Rogers[22] for a review of nitrogenous components and their metabolism.

Respiration and Circulation

Nematodes do not have blood vessels, but body fluids that contain hemoglobin bathe the various organs. Hemoglobins may also occur in the tissues of the body wall. These pigments have an affinity for oxygen in some but perhaps not all nematodes. The requirement for oxygen varies greatly among worms, but even the most completely anaerobic worms may still require minute amounts of free oxygen.

The hemoglobin in body fluids may be a by-product of digestion, playing no role in respiration, but that in the body wall probably accepts oxygen from the surrounding medium. The common stomach worm of ruminants, *Haemonchus contortus*, and the cecal worm of fowl, *Heterakis gallinae*, utilize molecular oxygen readily under laboratory conditions, and presumably their tissue hemoglobin plays the same role in vivo.

In a general review of carbohydrate and energy metabolism of nematodes, Saz[26] stated the following:

All nematodes studied to date are capable of consuming oxygen upon *in vitro* incubation under appropriate conditions. The physiological role of this gas in the metabolism of some helminths is established, in others it is poorly understood, and in still others it is decidedly inhibitory to survival. It can be said that a relatively large group of parasites is similar to *Ascaris* in that there is no oxygen requirement for the normal physiological production of energy in the adult stage. . . . Regardless of the status of oxygen, however, substrates are not oxidized to completion by any of the helminths studied to date. Products accumulate, indicating either the absence or participation of rate-limited terminal respiratory pathways. . . . Terminal respiratory pathways, and particularly mechanisms for electron transport associated with such pathways, are still poorly understood.

Reproductive System

Nematodes are usually dioecious and commonly exhibit sexual dimorphism. This condition of one sex being of a different size, shape, or color from the other reaches an extreme in *Trichosomoides crassicauda*, a parasite of the urinary bladder of the rat. The male worm of this species is a tiny creature compared with the female, and he lives like a parasite in her uterus.

The female reproductive system consists of one or two coiled tubules uniting to form a vagina that opens through a vulva. The vulva is usually located on the anterior portion of the body. The distal ends of the tubes form the ovaries, the portions next to them are the oviducts, and the remainder are the uteri. Lipids tend to be abundant in the re-

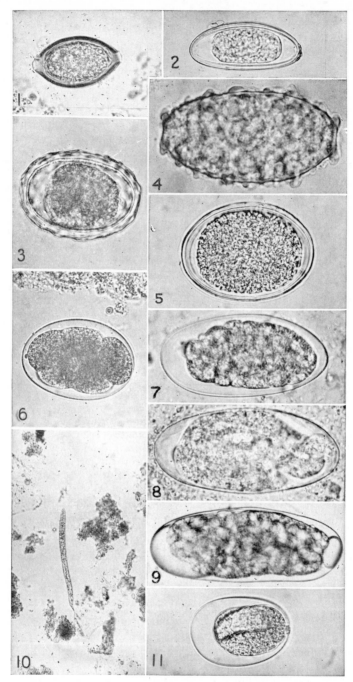

Fig. 12-8 Nematode eggs. (1) *Trichuris trichiura*. (2) *Enterobius vermicularis*. (3) *Ascaris lumbricoides*, fertilzed egg. (4). *Ascaris*, unfertilized eggs. (5) *Ascaris*, decorticated egg. (6) Hookworm. (7) *Trichostrongylus orientales*, immature egg. (8) *Trichostrongylus*, embryonated egg. (9) *Heterodera marioni*. (10) *Strongyloides stercoralis*, rhabditoform larva. (11) *Strongyloides* egg (rarely seen in stool). All figures × 500 except 10 (×75) (Hunter, Frye and Swartzwelder, A Manual of Tropical Medicine. Courtesy, W. B. Saunders Co.)

productive organs of male and female nematodes.

The chemistry of nematode eggs has received considerable study, but only a start has been made towards completely understanding the complicated processes involved in the formation and composition of the eggshell and the exact role that each layer plays. Eggs of Strongyloidea, Oxyuroidea and Ascaridoidea (Fig. 12–8) all have primary envelopes produced by the eggs themselves and consist of lipid coats, membranes and shells. Many variations among simple, complex, thin and thick layers are found. The egg membranes of Strongyloidea and the exterior eggshell of Oxyuroidea consist of proteins, whereas the interior eggshell of Oxyuroidea consists of chitin. Eggshells of Ascaridoidea are constructed of protein lamellae with chitin layers interposed, and are covered with an exterior membrane which consists of a quinone-tanned protein. Some ascarids have tertiary envelopes produced by the uterine wall.

Male worms possess reproductive organs that are likewise modifications of long coiled tubes. The worms usually have only one testis, which is the distal end of a tube that continues as the vas deferens and joins the lower end of the gut at the cloaca. Before this junction occurs, the vas deferens enlarges and forms a seminal vesicle, or storage sac, for sperm. The sperm of some species of worms resemble amebas. The terminal end of the male organs may be called the ejaculatory duct. Transfer of sperm to the female worm is aided by a pair of spicules (Fig. 12–1) in many species of roundworms. These long, hardened structures may be thrust out through the cloaca, and may serve the additional function of tactile sensory organs.

Another male structure, the **gubernaculum,** is a sclerotized thickening of the cuticle in some species of worms. This organ is formed from the spicule pouch; it lies on the dorsal side of the cloaca and probably helps to guide the spicules as they are thrust out. The **crura** is part of the gubernaculum. A larger, less well-developed organ, the **tele-**

mon, lies on the ventral and lateral walls of the cloaca. It develops as a thickening of the cuticle of the cloacal lining and helps to direct the spicules in copulation. The posterior end of some male species is flared and curved in such a manner as to suggest a hood. This structure is called the **copulatory bursa** (Figs. 13–11, 13–15); it helps the male worm to hold itself to the female during copulation. The walls of the bursa may be supported by finger-like rays. The number and arrangement of these rays serve as diagnostic features in species identification.

LIFE CYCLES

Life cycles take a variety of forms among the thousands of species of nematodes that infect animals. Early developmental stages within the egg are basically the same for all nematodes and are well represented in Figure 12–9 which illustrates the process in the egg of *Contracaecum aduncum,* a parasite of marine fish. One of the simplest life cycles is exhibited by *Trichuris (Trichocephalus) trichiura,* the common whipworm of man (described on p. 327), and by similar species in many other mammals. Eggs of whipworms pass out of the body with feces and develop into embryos within a few weeks. They are then infective to a new host and gain entrance by being ingested. Embryonated eggs may remain viable for many months if they remain in moist areas. When ingested, they pass to the cecum where they hatch and, in about four weeks, mature. As is characteristic of nematodes, the sexes are separate. Mating occurs as soon as the adult stage is reached, and not long thereafter the females start producing eggs. See Rogers and Sommerville[24] for a detailed review of chemical aspects of growth and development.

Molting

The shedding of the cuticle occurs by the deposition of a new cuticle and by **ecdysis** (shedding of the old). A new cuticle may be formed without the shedding of the old, which becomes the sheath of some juveniles (Fig. 12–10). The shedding, or exsheath-

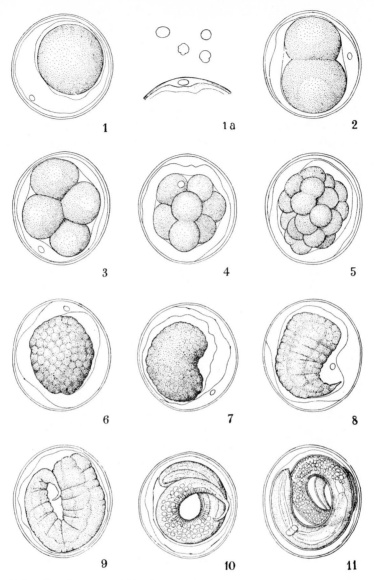

Fig. 12-9. Development within the egg of *Contracaecum aduncum*, a nematode of fish. (St. Markowski, courtesy of Bull. Acad. Polon. Sci.)

ment, is brought about by an **exsheathing fluid** that is released between the old and the new cuticles and is produced in response to a stimulus from the host. Digestion of the old cuticle is accomplished with the aid of the enzyme leucine aminopeptidase. Evidence indicates that an endocrine link connects the reception of the host stimulus and the release of the exsheathing fluid by the parasite.

The process of exsheathment in the nema-tode worm *Phocanema* (=*Porrocaecum*) in the muscles of codfish and the gut of seals is under neurosecretory control.[8] The neurosecretory cells occur in the dorsal and ventral ganglia of the worm. A study of exsheathment mechanisms of nematode larvae in vitro has shown that the infective larvae of *Dictyocaulus viviparus* demonstrate an absolute requirement for the enzyme pepsin. *Trichostrongylus colubriformis* shows a relative requirement for

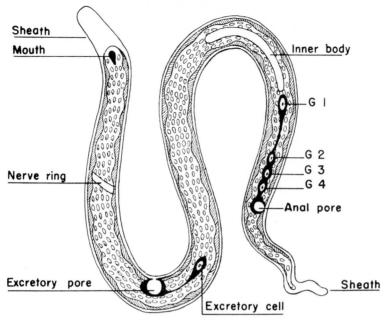

Fig. 12-10. A juvenile (microfilaria) nematode showing the sheath and other structures. (United States Navy Medical School Laboratory Guide.)

pepsin, whereas *Haemonchus contortus* infective larvae are indifferent to pepsin. Exsheathment in the latter species is rapid and complete in various salt solutions under an atmosphere of CO_2.[27] Four molts probably occur during the life cycles of nematodes.[23]

Infective Stage

The infective stage, whether a larva or egg, is a bridge between free-living and parasitic modes of life. The organism does not feed or grow, and thus is very different from other stages. It must be able to "recognize" the host. Its food reserves are carbohydrate or fat, and, although the need for oxygen varies greatly, oxygen is consumed when available at atmospheric pressure. The mechanism of osmoregulation of infective stages is unknown, but permeability of eggshells is known to be low.[23]

A nematode parasite that shows an unusually variable type of life cycle is the threadworm, *Strongyloides stercoralis*. Under optimal environmental conditions the adult worms live in the soil and carry on a nonparasitic existence. When environmental conditions are unfavorable, the juveniles be-

come infective to a new host (p. 282).

Many complex factors (*e.g.*, thickness of water film, enzymes, larval movement) influence the entry of a worm into the skin or the mucosa of a host, and our knowledge of the factors controlling the subsequent behavior of the worm is meager. Madsen[18] has suggested that the tissue phase or "dormant" period in the life cycle of a worm is a phenomenon that depends on the degree of resistance of the host rather than an innate behavioral pattern of the parasite.

Many filarial worms (see p. 312) make use of insect vectors. *Wuchereria bancrofti*, which causes **elephantiasis** in man, is one of the best known examples.

The development of various species of filarial nematodes to the infective stage in their arthropod hosts has been described as taking place in muscles, the fat body, the parenchyma, the Malpighian tubes and the hemocoel, but no one has reported this development as occurring in the nervous system, gonads, alimentary canal or salivary glands.[15] Wherever they develop, the juvenile worms (Fig. 12–11) are usually highly specific to the particular type of tissue selected. These

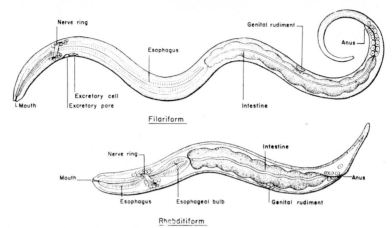

Fig. 12-11. Juveniles of intestinal roundworms. (United States Navy Medical School Laboratory Guide.)

juvenile worms often pass through two distinct stages. The first is known as a **rhabditiform** larva, the major characteristic of which is an esophagus that constricts posteriorly and then enlarges to form a bulb. The second stage is a narrower, longer worm, known as a **filariform** larva, with an extended esophagus that does not constrict or form a bulb.

Many parasitic roundworms, after entering an arthropod host, become surrounded by chitin or other special membranes, or become impregnated with a brown deposit forming a "capsule" around them. Studies have shown that this encapsulation or melanin deposit is a defense mechanism on the part of the host.[15] Sometimes this reaction occurs only when a parasite is introduced into an abnormal host.

Larva Migrans

Nematode larvae commonly migrate through an unnatural or inadequate host in which the larvae do not develop into adults. For example, larvae of the cat hookworm, *Ancylostoma brasiliensis*, may wander into the skin of man, causing "creeping eruption" characterized by visible tracks with red, painful, swollen advancing ends, often associated with intense itching. This kind of infection is called **cutaneous larva migrans.** The same kind of larval infection in internal organs, as occurs with larval *Toxocara canis* in man, is called **visceral larva migrans.**

Visceral larva migrans, as defined by Sprent,[29] is "a biological phenomenon to which all animals are subject ... [comprising] ... the invasion of, and migration through, any of the tissues of the animal body by nematode larvae, whether they be derived from nematodes which are natural parasites of the affected host, or from parasites which normally mature in other hosts." Beaver,[2] however, recommends the restriction of the definition "to include only the prolonged migration and long persistence of larvae whose behavior clearly reflects that which occurs in a normal intermediate or paratenic host." Examples in man include larval migrations of *Toxocara canis*, *Toxocara cati*, *Gnathostoma spinigerum*, *Sparganum* species, and probably certain species of the hookworm *Ancylostoma*. Beaver introduced the term **paratenesis** to "denote the passage of infective-stage larvae without essential development through a series of transport (paratenic) hosts to the final host with the transport host serving at the same time to maintain the infective-stage larvae from one season of transmission to the next." Much more work needs to be done on larval morphology, behavior, physiology and immunity.

GENERAL CONTROL

Control of some of the more important nematodes is mentioned in the discussions of

TABLE 12–1. Phylum Nematoda

Class	Order	Suborder	Superfamily
Secernentea (Phasmidia)	Rhabditida	Rhabditina	Rhabdiasoidea
			Rhabditoidea
	Strongylida	Strongylina	Syngamoidea
			Ancylostomatoidea
			Strongyloidea
		Trichostrongylina	Trichostrongyloidea
			Heligmosomatoidea
		Metastrongylina	Metastrongyloidea
			Protostrongyloidea
			Pseudalioidea
	Ascarida	Ascaridina	Ascaridoidea
			Heterocheiloidea
		Heterakina	Heterakoidea
			Subuluroidea
			Aspidoderoidea
			Cosmocercoidea
		Oxyurina	Oxyuroidea
			Syphacioidea
	Spirurida	Spirurina	Spiruroidea
			Thelazioidea
			Gnathostomatoidea
			Acuarioidea
			Physalopteroidea
		Camallanina	Camallanoidea
			Dracunculoidea
		Filariina	Filarioidea
			Onchocercoidea
	Tylenchida		
Adenophorea (Aphasmidia)	Trichinellida	Trichinellina	Trichinelloidea
	Dioctophymatida	Dioctophymatina	Dioctophymatoidea
			Eustrongyloidea

the various species in the following chapters. However, one of the newest ideas for control is the use of predaceous fungi. Some of these fungi can be grown in culture on synthetic media, but others must be grown in their natural hosts. The endozoic fungus, *Harposporium anguillulae*, for example, sends its mycelia throughout the tissues of nematodes of the genera *Rhabditis* and *Panagrellus*.

For epidemiology and control of some nematode infections of grazing animals, see Michel.[19] For general references to the nematodes, see Chitwood and Chitwood,[6] Hyman,[11] Lavoipierre,[15] Lee,[16] Rogers,[21]

Sasser and Jenkins,[25] Skrjabin,[28] Stoll,[30] Thorne,[31] and Wallace.[32]

CLASSIFICATION

Hyman[11] has presented strong evidence for grouping, as classes, the Rotifers, Gastrotricha, Kinorhyncha, Nematoda and Nematomorpha under the phylum Aschelminthes (a term meaning "worms with a cavity"). Other authorities prefer to consider each of these groups as a separate phylum. The above classification is taken from M. B. Chitwood.[7] Only orders and superfamilies containing parasitic representatives are listed.

REFERENCES

1. Arthur, E., and Sanborn, R. C.: Osmotic and ionic regulation in nematodes. *In* Florkin, M., and Scheer, B. T., (eds.): Chemical Zoology. Vol. III, pp. 429–464. New York, Academic Press, 1969.

2. Beaver, P. C.: The nature of visceral larva migrans. J. Parasitol., *55*: 3–12, 1969.

3. Bird, A. F.: Chemical composition of the nematode cuticle. Observations on individual layers and extracts from these layers in *Ascaris lumbricoides* cuticle. Exp. Parasitol, *6*: 383–403, 1957.

4. Bird, A. F., and Bird, J.: Skeletal structures and integument of Acanthocephala and Nematoda. *In* Florkin, M., and Scheer, B. T., (eds.): Chemical Zoology. Vol. III, pp. 253–288. New York, Academic Press, 1969.

5. Cavier, R., and Savel, J.: Le métabolisme protéique de l'*ascaris* du porc, Ascaris *lumbricoides* Linné, 1758, est-il ammoniotélique ou uréotilique? C. R. Soc. Biol., *238*:2448–2450, 1954.

6. Chitwood, B. G., and Chitwood, M. B.: An Introduction to Nematology. Rev. ed. Baltimore, B. G. Chitwood, 1950.

7. Chitwood, M. B.: The systematics and biology of some parasitic nematodes. *In* Florkin, M., and Scheer, B. T., (eds.): Chemical Zoology. Vol. III, pp. 223–244. New York, Academic Press, 1969.

8. Davey, K. G., and Kan, S. P.: Endocrine basis for ecdysis in a parasitic nematode. Nature, *214*:737–738, 1967.

9. Debell, J. T.: A long look at neuromuscular junctions in nematodes. Quart. Rev. Biol., *40*:233–251, 1965.

10. Eckert, J., and Schwarz, R.: Zur Struktur der Cuticula invasionfähiger Larven einiger Nematoden. Zeit. f. Parasitenk., *26*:116–142, 1965.

11. Hyman, L. H.: The Invertebrates: Acanthocephala, Aschelminthes, and Entoprocta, Vol. III. New York, McGraw-Hill, 1951.

12. Inglis, W. G.: The structure of the nematode cuticle. Proc. Zool. Soc. Lond. *143*:465–502, 1964.

13. Jamuar, M. P.: Electron microscope studies on the body wall of the nematode, *Nippostrongylus brasiliensis*. J. Parasitol., *52*:209–232, 1966.

14. Kan, S. P., and Davey, K. G.: Molting in a parasitic nematode, *Phocanemae decipiens*. III. The histochemistry of cuticle deposition and protein synthesis. Can. J. Zool., *46*:723–727, 1968.

15. Lavoipierre, M. M. S.: Studies on the host-parasite relationships of filarial nematodes and their arthropod hosts. II. The arthropod as a host to the nematode: a brief appraisal of our present knowledge, based on a study of the more important literature from 1878–1957. Ann. Trop. Med. Parasit., *52*:326–345, 1958.

16. Lee, D. L.: The Physiology of Nematodes. San Francisco, W. H. Freeman and Company, 1965.

17. ———: The structure and composition of the helminth cuticle. *In* Dawes, B. (ed.): Advances in Parasitology. Vol. 4, pp. 187–254. New York, Academic Press, 1966.

18. Madsen, H.: On the interaction between *Heterakis gallinarum, Ascaridia galli,* "Blackhead" and the chicken. J. Helminth, *36*:107–142, 1963.

19. Michel, J. F.: The epidemiology and control of some nematode infections of grazing animals. *In* Dawes, B. (ed.): Advances in Parasitology. pp. 211–282. New York, Academic Press, 1969.

20. Read, C. P: Nutrition of intestinal helminths. *In* Soulsby, E. J. L. (ed.): Biology of Parasites. pp. 101–126. New York, Academic Press, 1966.

21. Rogers, W. P.: The Nature of Parasitism. The Relationship of Some Metazoan Parasites to Their Hosts. New York, Academic Press, 1962.

22. ———: Nitrogenous components and their metabolism: Acanthocephala and Nematoda. *In* Florkin, M., and Scheer, B. T. (eds.): Chemical Zoology. Vol. III, pp. 379–428. New York, Academic Press, 1969.

23. Rogers, W, P., and Sommerville, R. I.: The infective stage of nematode parasites and its significance in parasitsm. *In* Dawes, B. (ed.): Advances in Parasitology. Vol. I, pp. 109–177. New York, Academic Press, 1963.

24. ———: Chemical aspects of growth and development. *In* Florkin, M., and Scheer, B. T. (eds): Chemical Zoology. Vol. III, pp. 465–499. New York, Academic Press, 1969.

25. Sasser, J. N., and Jenkins, W. R.: Nematology. Fundamentals and Recent Advances, With Emphasis on Plant Parasitic and Soil Forms. Chapel Hill, Univ. North Carolina Press, 1960.

26. Saz, H. J.: Carbohydrate and energy metabolism of nematodes and acanthocephala. *In* Florkin, M., and Scheer, B. T., (eds.): Chemical Zoology. Vol. III, pp. 329–360. New York, Academic Press, 1969.

27. Silverman, P. H.: Exsheathment mechanisms of some nematode infective larvae. J. Parasitol., *49*:50, 1963.

28. Skrjabin, K. I. (ed.): Essentials of Nematology. Vol. III. Trichostrongylids of Animals and Man. Israel Program for Scientific Transla-

tions (for N.S.F. and Dept. Agric., Wash., D.C.), 1960.

29. Sprent, J. F. A.: Visceral larva migrans. Aust. J. Sci., *25*:344–354, 1963.

30. Stoll, N. R.: Biology of nematodes parasitic in animals. J. Parasitol., *48*:830–838, 1962.

31. Thorne, G.: Principles of Nematology. New York, McGraw-Hill, 1961.

32. Wallace, H. R.: The bionomics of the free-living stages of zoo-parasitic and phyto-parasitic nematodes—a critical survey. Helminthol. Abst., *30*:1–22, 1961.

Chapter 13

Class Secernentea (=Phasmidia), Orders Tylenchida, Rhabditida, Strongylida

ORDER TYLENCHIDA

Nematodes of this order are parasites of plants. They are discussed on page 328.

ORDER RHABDITIDA

Superfamily Rhabdiasoidea

The evolution of a parasitic mode of life among roundworms might well have started with groups like the rhabditoids (see p. 566). Many of these worms live in decaying flesh, dung, decomposing plant material or similar substances in which transfer to the intestine of an animal or the tissues of a plant is relatively easy. Members of this large order possess head sense organs in the form of papillae, and the amphids are reduced to small pockets.

Strongyloides stercoralis. This intestinal parasite (Fig. 13–1) of man is worldwide in its distribution but is found primarily in warm countries. The incidence of infection is usually very low, but it may run as high as 25 per cent in favorable areas. Chimpanzees, dogs and cats may possibly be reservoir hosts, but some authorities believe the worms in these animals belong to different species. Related species occur in numerous other mammals. (For a study of the comparative morphology of several species of *Strongyloides* from various animals, see Little.[10]) Parasitic females of *S. stercoralis* are about 2 mm. long × 50 μ wide, while the free-living males average 0.7 mm. long × 45 μ wide. They are called **threadworms,** and the disease they produce is known as **strongyloidiasis.**

Strongyloides stercoralis shows considerable variability in its life cycle, apparently being able to adjust the mode of its development to changing demands of the environment. Under favorable conditions of moisture, temperature, and food availability, the adult worms live in the soil. This free-living period begins with an egg that develops into four rhabditiform stages, the last one forming an adult male or female. Under warm, moist, shaded conditions, however, a second type of life cycle occurs. The free-living adults mate and produce eggs that pass through two rhabditiform stages and a filariform stage. The female filariform larvae are infective to man. Apparently, males are not normally involved in human infection, since they have rarely been observed. However, Levin[9] has reported the presence of adult and larval males in human stools.

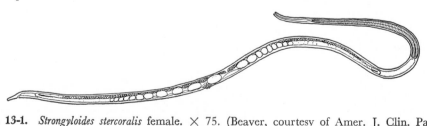

Fig. 13-1. *Strongyloides stercoralis* female. × 75. (Beaver, courtesy of Amer. J. Clin. Path.)

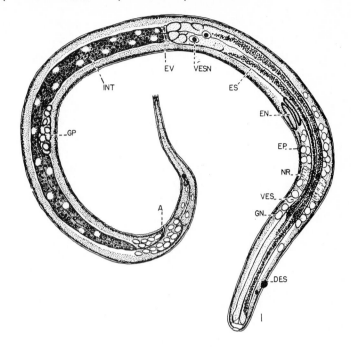

Fig. 13-2. *Strongyloides stercoralis* larva after 12 days in rat abdominal muscle (unstained). × 600. A, anus; DES, dorsal esophageal gland; EN, nucleus of excretory cell; EP, excretory pore; ES, esophagus; EV, esophago-intestinal gland; GN, ganglionic nucleus; GP, genital primordium; INT, intestine; NR, nerve ring; VES, ventral esophageal gland; VESN, nucleus of ventral esophageal gland. (Nichols, courtesy of J. Parasitol.)

The parasitic larvae may infect dogs and cats by being eaten, but to infect man they usually penetrate the skin, burrow through to blood vessels, and get to the lungs by way of the heart. Occasionally they are found in other tissues (Fig. 13–2). In the alveoli of the lungs, filariform larvae develop into the adolescent stage; they are coughed up and swallowed. These females, averaging 2.2 mm. × 40 μ in size, penetrate the lining of the small intestine, where they reproduce parthenogenetically and lay eggs that average 50 × 32 μ (Fig. 12–8). The duodenum is possibly the site of preference, but the worms may be found elsewhere in the intestine. The eggs hatch in the gut wall and the new juveniles migrate from the tissue to the lumen of the intestine and so find their way outside to soil.

A third type of life cycle of *Strongyloides stercoralis* involves **auto-infection.** Newly-hatched juveniles in the intestine may remain in the lumen of the gut, develop into filariform juveniles and burrow through the mucosa and into blood vessels. They are then carried to the lungs and make their way back to the intestine, as described above, where they

mature. Other filariform juveniles may crawl through the anus and re-enter the body by penetrating the skin of the perianal region. They also go to the lungs and from there to the intestine. Auto-infection, however, appears to be rare.

Symptoms may occur with the penetration of the juveniles through the skin—itching and swelling of the area are typical. Penetration of the lungs may cause coughing, a bronchial pneumonia and other pulmonary disorders. Worms in the intestinal mucosa sometimes cause diarrhea, abdominal pain, anemia, loss of weight, constipation or bloody feces. Allergic effects may also occur.

Superfamily Rhabditoidea

This group of nematodes is found on or in a wide variety of animals, primarily invertebrates, as well as on or in plants. The worms possess two esophageal bulbs.

Rhabditis coarctata. This species must be carried as ensheathed juveniles on the surface of dung beetles to fresh dung before they can mature. *R. ocypodis* is carried on the gills of crabs.

Neoaplectana glaseri. As with most

rhabditoids, this species is generally free-living, but its larva often penetrates and destroys the larvae and adults of certain beetles. *N. glaseri* has been used as a biological control agent against the Japanese beetle.

ORDER STRONGYLIDA

Superfamily Strongyloidea

The Greek term "strongylos" means round or compact; it is used in many scientific names other than for roundworms. However, the name "strongyle" has become a somewhat popular term to denote most roundworms of large domestic animals. Actually, strongyles belong primarily to members of the subfamilies Strongylinae and Cyathostominae. The name is confusing, since neither *Strongyloides stercoralis* nor *Rhabditis strongyloides* are strongyles. Strongyles are found in horses and other equids, and in elephants, rhinoceroses, rodents, swine, and other mammals. Species common to horses are *Strongylus equinus, S. edentatus,* and *Triodontophorus* species. The larger strongyles measure up to 4 cm. in length; they are stout, grayish worms found anchored to gut mucosa.

A ring of fence-like projections known as the **corona radiata** surrounds the mouth of these worms. There are no teeth or cutting plates. Male worms of this order have an expanded posterior end that is known as a **bursa,** which sets them apart from most of the other nematodes. This bell-shaped extension of the cuticle is supported by rays that exhibit a definite pattern. Females have a pointed tail. The mouth is not surrounded by the conspicuous lips that characterize ascaroids, and the esophagus does not have a bulb.

This group contains many important parasitic worms of domestic animals and man. Most of them suck blood by grasping a portion of the intestinal mucosa in their mouths; they keep the blood from clotting by secreting an anticoagulant. The host intestinal lumen is essentially anaerobic, but parasites obtain oxygen from the host red blood cells that they eat. The presence of fresh blood in worms gives some of them a bright red color.

Strongylus edentatus has a characteristic life cycle. Eggs reach the ground with host feces. In one or two days, the first-stage rhabditiform larva hatches from an egg and develops into the second-state rhabditiform larva. Within a few days the elongated, ensheathed, infective larva has developed. It cannot feed but may live for several months, sometimes throughout the winter. The host eats the infective larva along with vegetation. The larva penetrates the intestinal wall and goes to the connective tissue under the abdominal peritoneum, where it causes hemorrhagic nodules and increases in size for about three months. It then migrates back to the gut wall, where it produces other nodules, matures, and drops into the intestinal lumen.

Superfamily Syngamoidea

Syngamus trachea. The gapeworm, *Syngamus trachea* (Fig. 13–3), is a red nematode that lives in the tracheas of various species of poultry. It derives its common name from the fact that heavily infected birds apparently try to get rid of the annoyance caused by this worm by gaping, coughing, swallowing, or stretching the neck. Male worms average about 4 mm. in length and the females, about 17 mm. The male and female worms are permanently fused to unite the genital openings of each worm.

Eggs are coughed up by the host, swallowed, and find their way outside with feces. In moist soil the eggs become infective in a week or two, depending on the temperature, and may be eaten by a chicken or some other bird, or the eggs may hatch in the soil and the larvae eaten. The worm is versatile in its life history: the larvae may enter an earthworm, slug, or snail, or the egg may be eaten by one of these intermediate hosts. Thus, chickens or other birds can become infected by eating snails, slugs or earthworms.

Within the birds the larvae penetrate the gut wall to the circulatory system and are carried to the lungs, where they penetrate

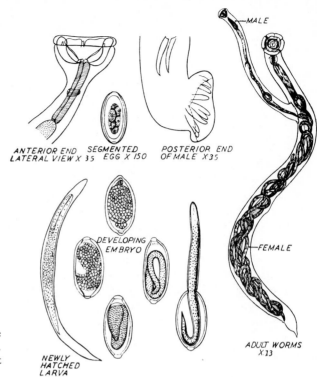

Fig. 13-3. *Syngamus trachea*, the gapeworm.   Redrawn from M. Neveu Lemaire. (Courtesy of Vigot Fréres, Paris.)

blood vessel walls and other tissues, enter lung cavities, and crawl to the trachea where they become established. Within two weeks they are mature; they live 23 to 92 days in chickens, and 48 to 126 days in turkeys. Birds may suffocate if the worms are present in large numbers. General cleanliness in bird management is extremely important, and care should be taken not to raise birds in areas contaminated from previous infections.

Stephanurus dentatus. This common kidney worm of pigs (Fig. 13–4) may also

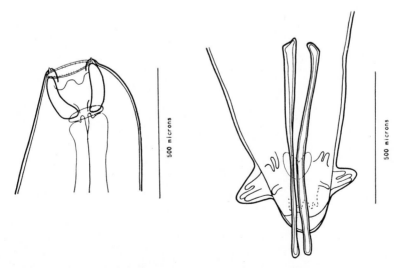

Fig. 13-4. *Stephanurus dentatus* anterior and posterior end of male. (Morgan and Hawkins, Veterinary Helminthology. Courtesy of Burgess Publishing Co.)

inhabit the liver, lungs, spleen, muscles, spinal canal and body cavities. Male worms are 30 to 45 mm. long. They appear black and white because some internal organs can be seen from the outside.

Female worms in kidneys produce oval eggs that average a little more than 100 μ in length; thus they can just be seen without a microscope. The eggs pass down the ureter, into the urinary bladder and to the outside with urine. A heavily infected pig may pass a million eggs a day.

In moist shaded soil, the eggs hatch into tiny, free-living larvae that grow and molt twice within the next three to five days. The young worms are then infective to pigs, which eat them with contaminated food. Larvae may also enter their hosts through cuts or sores on the skin, or they may even penetrate the unbroken skin. Within the body, larvae find their way via blood and lymph vessels to the liver, where they may remain for several months. Then the parasites make their way to the kidneys where they become mature. The entire life cycle usually requires six to eight months. Most larvae do not reach maturity.

Larvae and adults cause abscesses, hemorrhage, adhesions, loss of weight and death. The economic loss in the United States due to this worm alone has been estimated to be at least $72,000,000 annually. Treatment is ineffective. Good management practices help to keep down infection. Keeping the area around hogs dry and exposed to sunlight prevents the development of enormous numbers of eggs and larvae.

Oesophagostomum. The various species of *Oesophagostomum* have a common type of life cycle that can be represented by *O. dentatum* (Fig. 13–5), the nodular worm of swine. In this cosmopolitan species the males are about 9 mm. long, while the females average about 12 mm. long. The species name is based on the presence of a ring of toothlike projections bordering the mouth, and a ring of very short projections within the buccal capsule. The male has a pair of slender spicules about 1 mm. long and a tri-lobed bursa. The vulva of the female is posterior in position.

The life cycle begins with the fertilized eggs, which are about 70 to 40 μ in size. The eggs escape the host with feces and, in moist soil, they hatch in a day or two. After another few days or a week, the larvae have developed into the infective stage and are ready to be eaten by a hog. If the worms get into the large intestine they penetrate the intestinal wall, causing inflammation and the formation of nodules about 1 mm. in diameter—thus the common name of the worm. Eventually the juveniles leave the nodules, enter the lumen of the intestine, and develop into sexually mature worms. Symptoms of infection range from none to anemia, weakness, diarrhea, and emaciation.

Oesophagostomum columbianum, the sheep nodular worm, and *O. venulosum* are common nematodes of the intestines of farm mammals, especially sheep and goats. Eggs similar to those of hookworms usually leave the body of the host with feces and hatch in moist soil, but some eggs hatch within the body of the host. Adults are about the size of hookworms (12 mm. long); sexes, as usual, are separate. Soon after they mature, mating occurs and a new generation of eggs appears.

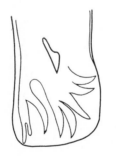

Fig. 13-5. *Oesophagostomum dentatum*, anterior and posterior end of male. (Morgan and Hawkins, Veterinary Helminthology. Courtesy of Burgess Publishing Co.)

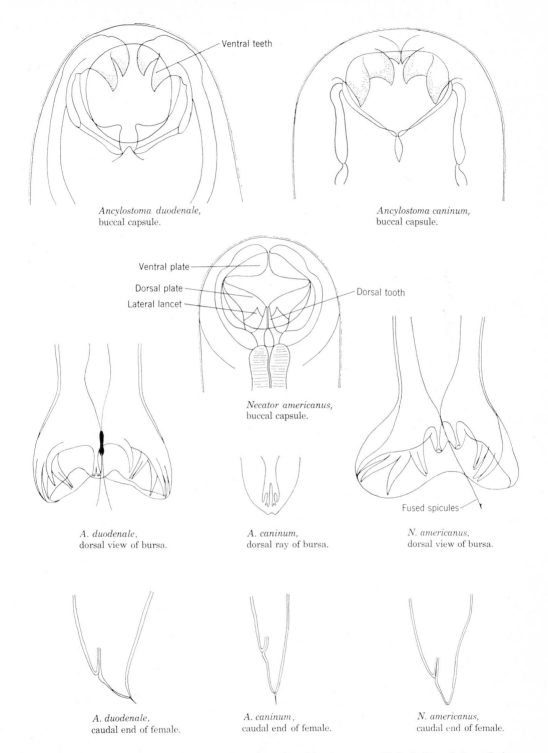

Ventral teeth

Ancylostoma duodenale,
buccal capsule.

Ancylostoma caninum,
buccal capsule.

Ventral plate
Dorsal plate
Lateral lancet

Dorsal tooth

Necator americanus,
buccal capsule.

Fused spicules

A. duodenale,
dorsal view of bursa.

A. caninum,
dorsal ray of bursa.

N. americanus,
dorsal view of bursa.

A. duodenale,
caudal end of female.

A. caninum,
caudal end of female.

N. americanus,
caudal end of female.

Fig. 13-6. Anterior and posterior ends of three species of hookworms. (Scheil, Parasitology Laboratory Manual. Courtesy of John Wiley & Sons).

Superfamily Ancylostomatoidea

Hookworms have a buccal capsule containing ventral teeth or cutting plates. The male bursa is normally conspicuous. The name "hookworm" is said to derive from the position of the anterior end, which is bent backward (dorsally). It is also purported to derive from the hook-like appearance of the bursal rays. Figure 13–6 shows diagnostic features of three common species.

Dogs and cats are hosts to four common species of hookworms: *Ancylostoma caninum* (in dogs), *A. tubaeforme* (in cats), *A. braziliense* (primarily in cats), and *Uncinaria stenocephala* (in dogs, foxes, wolves). The last two species may also occasionally occur in man.[4,5] The life histories of these worms are essentially similar to that of *A. duodenale* of man; the possibility of insects acting as paratenic hosts should not be overlooked. A fourth species,

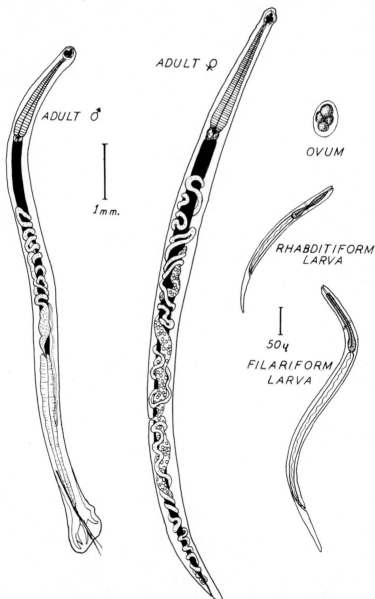

Fig. 13-7. The hookworm *Ancylostoma duodenale*.

Necator americanus of man, has been reported from dogs but it is rare in these hosts. *Uncinaria lucasi* is the worst enemy of fur seal pups on the Pribilof Islands. During the two-month stay of these young seals on the rookeries they die by the thousands from hookworm infection, covering the rookeries with their dead bodies.[13]

In general, hookworms are host specific, but the larval stages may penetrate the skin of "foreign" hosts. Thus the larvae of *Ancylostoma braziliense* sometimes enter and burrow in human skin. They cause intense itching, and the affliction is known as "creeping eruption." The major mechanical damage occurs in the host intestine, which bleeds because of minute lacerations produced by the jaws of the worms. The parasites suck blood and other fluids, and the serum, if not whole blood, appears to be needed for normal functioning of the worm's esophagus and for the incorporation of glucose into the parasite's glycogen.[14] Infection generally results in bloody diarrhea and anemia. The liver and kidneys may be damaged. General edema and hypoalbuminemia are common.[3] Death of the host may occur, but the most characteristic result is a draining of health and energy. About 630 million people have hookworms, and the estimated daily loss of blood due to this disease is the equivalent of the total volume of blood of 1,500,000 persons.[18] Dr. N. D. Stoll[18] has stated: "Now that malaria is being pushed back, hookworm remains *the* greatest *infection* of mankind in the moist tropics and subtropics. In my view, it outranks all other worm infections of man combined, with the possible exception of ascariasis, in its production of human misery, debility, and inefficiency in the tropical world."

The two most important hookworms of man are *Ancylostoma duodenale*, the Oriental species, and *Necator americanus*, the American hookworm. Modern travel has provided ample opportunity for both species to become worldwide in their distribution. Many other varities of hookworms occur in vertebrates, but they all are essentially similar in struc-

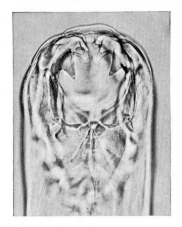

Fig. 13-8. En face of view of *Ancylostoma*. (Whitlock, Diagnosis of Veterinary Parasitisms. Lea & Febiger)

ture and in their life cycle. *Ancylostoma duodenale* will be discussed as a representative of the entire group.

Ancylostoma duodenale (Figs. 13–7, 13–8). This hookworm can be found in the small intestine of millions of people, chiefly in Europe, Africa and the Orient. It is the dominant species in the Mediterranean area, India, China and Japan. The male worm is 8 to 11 × 0.4 to 0.5 mm in size. The posterior end of its body is flared and forms a bursa supported by fleshy rays with a characteristic pattern. A pair of long spicules passes from the genital canal to the outside through the cloaca. A gubernaculum, sclerotized as are the spicules, is also used during copulation to help guide the spicules. The females average 10 to 13 mm. × 0.6 mm. in size. The posterior end of the body tapers to a rather blunt point. The vulva is located at a point about two-thirds the length of the body from the anterior end. Eggs are ovoidal, thin-shelled and measure 56 to 60 μ × 34 to 40 μ. When they are found during fecal examinations, they are usually already in the early stages of segmentation (Figs. 12–8, 13–9).

The life cycle (Fig. 13–9) starts with a fertilized egg that, by the time it reaches the soil, is well on its way to becoming a juvenile. The daily output of eggs from a single female worm is probably 10,000 to 20,000. Within

24 hours in moist, warm soil, rhabditiform larvae (Fig. 13–7) hatch from the eggs. Free oxygen is essential for hatching and for further development. Larvae grow rapidly, molt twice, and in about a week become nonfeeding, slender, filariform juveniles (Fig. 13–7, 13–10). This third larval stage is infective to man. The young worms crawl to a high point of dirt, vegetation, or bit of rock, so long as it is moist, and wait for a new host to come along. Excess water at this stage of the life cycle is injurious to the worms. Alternate wetting and drying is particularly harmful, hence frequent rains, with dry weather in between, tend to rid the soil of hookworm juveniles.

If a filariform juvenile comes in contact with the skin of a new host, it burrows into the skin and, if it gets deep enough, enters a blood or lymph vessel and is eventually carried to the lungs. Here it leaves the bloodstream, penetrates the lung tissue, and ar-

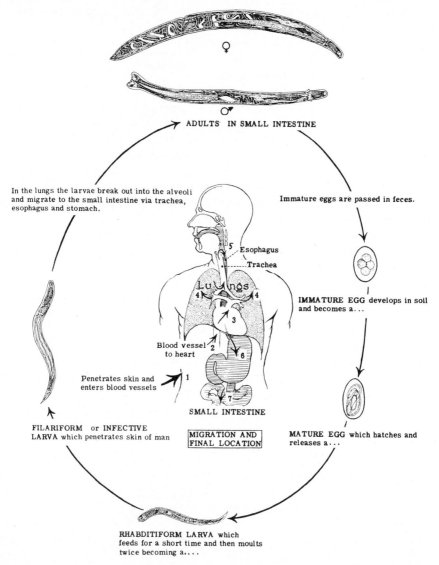

In the lungs the larvae break out into the alveoli and migrate to the small intestine via trachea, esophagus and stomach.

ADULTS IN SMALL INTESTINE

Immature eggs are passed in feces.

5 Esophagus

Trachea

Lungs

4

4

IMMATURE EGG develops in soil and becomes a . . .

3

Blood vessel 2
to heart

6

Penetrates skin and
enters blood vessels   1

7

SMALL INTESTINE

FILARIFORM or INFECTIVE
LARVA which penetrates skin of man

MIGRATION AND
FINAL LOCATION

MATURE EGG which hatches and
releases a . . .

RHABDITIFORM LARVA which
feeds for a short time and then moults
twice becoming a . . . .

Fig. 13-9. Life cycle of hookworms. (Courtesy of National Naval Medical Center, Bethesda, Maryland. Figures of the adult male and female are from Looss, 1905.)

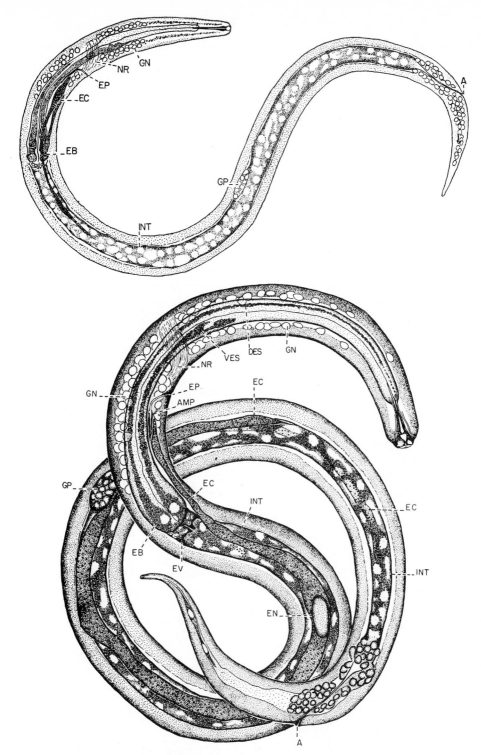

Fig. 13-10. Typical hookworm infective juveniles (unstained). (*Top*) *Necator americanus.* × 430. (*Bottom*) *Ancylostoma caninum.* × 700. A, anus; AMP, excretory ampulla; DES, dorsal esophageal gland; EB, esophageal bulb; EC, excretory column; EN, nucleus of excretory canal; EP, excretory pore; EV, esophago-intestinal valve; GN, ganglionic nucleus; GP, genital primordium; INT, intestine; NR, nerve ring; VES, ventral esophageal gland. (Nichols, courtesy of J. Parasitol.)

rives in the air spaces. From the lungs the worm passes up the trachea to the mouth cavity and is swallowed, and thus arrives in the small intestine. A final molt occurs and the young worm attaches itself to the wall of the intestine, begins to suck blood, and matures to the adult stage. About five weeks elapse between penetration of the skin and the production of new eggs. A patient suffering from severe hookworm disease may discharge with his feces six million eggs per day. A single hookworm may produce in its lifetime (about five years) up to 54 million eggs. Adult female worms may invade the intestinal epithelium and lay eggs. Only rhabditiform larvae have thus far been found in this locality, but if they develop into filariform larvae in the host the danger of auto-infection is obvious.

Experimental studies[12] with dogs have shown that, after oral administration of hookworm larvae, the parasites do not migrate to the lungs but develop directly in the tissues and lumen of the gut. The possibility of natural infections of young dogs by way of the mouth is, therefore, high. Also, a significant resistance to infection in groups of dogs and bitch pups follows vaccination with x-irradiated hookworm larvae.[11] Transmammary passage of larvae in dogs is common, but not in utero transmission.[19]

Symptoms of infection start with the "ground itch" occurring during the penetration of the skin by filariform larvae. If pyogenic bacteria enter with the larvae there may be itching, burning, erythema, edema and vesicle formation. "Creeping eruption" (cutaneous larva migrans, see p. 278) may occur if human skin is penetrated by larvae of other species of hookworms from animals or even by the species from man. Pulmonary inflammation and minute hemorrhages are associated with lung penetration. Anemia of the iron deficiency type is the principal host reaction to intestinal infection by adult worms. Other symptoms are pallor, fever, abdominal pains, diarrhea, listlessness, myocarditis and eosinophilia. Diagnosis is confirmed by finding hookworm eggs in stool

specimens. Control measures include proper sewage disposal and education of the people in endemic areas concerning sources of infection. It is possible that acquired immunity plays a major role in restraining hookworm infection in man.

Some autopsy material in Java has revealed, in the lining of the human small intestine, nodules containing not only adult specimens of *Ancylostoma duodenale* but eggs and larvae as well. This discovery suggests a method of auto-infection, but the extent of its prevalence is unknown.

Necator americanus. The same general life cycle occurs with *Necator americanus* and, from the medical point of view, presents essentially the same problems as does *Ancylostoma duodenale*. *A. ceylonicum* is a rare intestinal worm of man in southeast Asia. It was formerly believed to be the same species as *A. braziliense*.

Superfamily Trichostrongyloidea

Haemonchus contortus. The twisted stomach worm or "wireworm" (Fig. 13–11) is a worldwide inhabitant of the abomasum of sheep, goats, cattle and other ruminants. Females measure 18 to 30 mm. in length, whereas males are 10 to 20 mm. long. The common names arise from the fact that, in the female, the white ovaries are twisted around the red intestine. Males have a more uniform red color.

The life cycle begins with eggs, about 85 × 45 μ in size, which pass from the host to the ground with feces. Here they hatch and within four days they reach a sheathed infective stage. A new host acquires the parasite by eating infective larvae with grass or other food.

The effects on the host include anemia, digestive disturbances, loss of weight, often a swelling under the jaw, and susceptibility to other infectious agents.

Since details of the life history of *Haemonchus contortus* are relatively well understood, this parasite has been the subject of much study in immunology and host resistance. Free-living infective larvae in the host aboma-

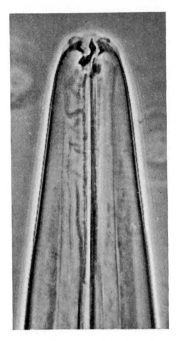
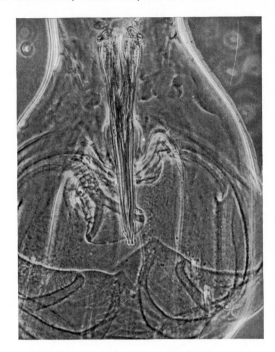

Fig. 13-11. (*A*) Lateral view of anterior end of *Haemonchus*. × 400. (*B*) Spicules and bursa of *Haemonchus*. Note barb on spicules and asymmetrical dorsal lobe. × 138. (Whitlock, Diagnosis of Veterinary Parasitisms. Lea & Febiger)

sum cannot become parasitic unless they are stimulated to begin the process of exsheathment. The chief factors that combine to act as a "physiologic trigger" are a temperature of about 38° C. and dissolved gaseous carbon dioxide at appropriate concentrations. The action of these factors may be influenced by reducing agents, hydrogen-ion concentration and the redox potential. The result of the stimulus is the release of an exsheathing fluid by the larvae. This fluid, probably a metabolite, not only brings about the molting process, but apparently also stimulates a reaction in the abomasum that can be called protective immunity. For details, see Rogers and Sommerville,[15] and Soulsby and Stewart.[17]

Ostertagia ostertagi (Fig. 13–12). Another stomach worm of ruminants is *Ostertagia ostertagi*, which lives in the abomasum and, rarely, in the intestine. Its common name is the brown stomach worm or the medium worm, and it is smaller than

Haemonchus contortus. Males average about 7 mm., and females about 9 mm. in length.

The life cycle of this worm is similar to that of *Haemonchus contortus:* the infective third-stage larvae appear on the fifth and sixth day after the eggs reach the ground. Symp-

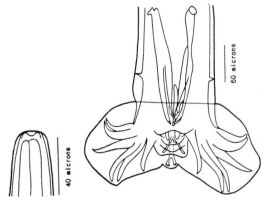

Fig. 13-12. *Ostertagia ostertagi*, anterior and posterior ends of male. (Morgan and Hawkins, Veterinary Helminthology. Courtesy of Burgess Publishing Co.)

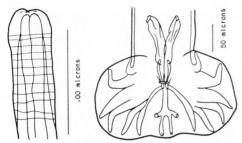

Fig. 13-13. *Cooperia punctata,* anterior and posterior ends of male. (Morgan and Hawkins, Veterinary Helminthology. Courtesy of Burgess Publishing Co.)

toms of infection include anemia, diarrhea and emaciation. (See Jarrett[8] for an account of pathologic and pathogenic aspects of parasitic gastritis in sheep and cattle.) Among other well known species of this genus is *Ostertagia circumcincta,* which occurs in sheep, goats, and many wild ruminants throughout the world.

Cooperia punctata. This small, reddish worm (Fig. 13–13) may be found in many countries in the small intestine of various ruminants, especially cattle. The males average a little over 5 mm. long, and the females about 6 mm. in length. The life cycle is not fully known but it probably resembles that of *Haemonchus contortus.* Symptoms of infection in calves include diarrhea, weakness, anemia, and emaciation.

Nippostrongylus brasiliensis. This small reddish worm, also called *N. muris,* is a common parasite of the small intestine of wild rats, but it is normally not found in laboratory rats. The worm has been used in various studies of the physiology of parasitic nematodes. The male is 3 to 4 mm. in length, whereas the female is 4 to 6 mm. long. In both sexes the head is somewhat enlarged. In the male the bursa is asymmetrical in shape and the spicules are long and slender. Eggs pass from the intestine of the host, and the young worms within them develop into the infective stage in one day under favorable environmental conditions. The life cycle is similar to that of the hookworm, *Ancylostoma duodenale* (p. 289), and the adult stage is reached in about a week. After entering the

host body, the larvae go first to the lymph glands where they are cleansed of their bacteria and other foreign material, then to the lungs.[7]

Nematodirus spathiger (Fig. 13–14). This is a common parasite of cattle, sheep, goats, and many wild ruminants; it inhabits the small intestine. The male ranges from 10 to 15 mm. in length and the female varies between 15 and 23 mm. Third-stage larvae are, as usual, the infective stage, and they are eaten by cattle or sheep while the hosts feed on grass or other vegetation. The symptoms of infection are not pronounced, but infected cattle probably exhibit the same symptoms as occur with other intestinal worms, although to a milder degree. Another species in ruminants is *N. helvetianus.*

Trichostrongylus. Seven or eight species of *Trichostrongylus* (e.g., *T. axei, T. colubriformis, T. vitrinus, T. capricola, T. orientalis*) have been reported from man. All of these worms are also found in domestic animals. *T. colubriformis,* for example, causes "black scours" in ruminants. It usually occurs in the small intestine. *T. axei* is found in the abomasum of ruminants and in the stomach and small intestine of horses. The others are also parasitic in ruminants. The world incidence of infection in man is probably not over one to two per cent. In some areas of the Orient, however, there is a ten to 20 per cent incidence. Local areas of Iran show infection rates as high as 60 per cent.

In general, the appearance and life cycle of these worms resemble hookworms. Trichostrongyles have a rather unpretentious anterior end, tapering to a blunted point without any accessory structures and without a prominent buccal capsule. Males possess a posterior copulatory bursa and paired copulatory spicules that are thick, short, irregular and brown in color. The life cycle is relatively simple. Adults (6 to 8 mm. long) live in the small intestine, and eggs leave the host with feces. The eggs are greenish in fresh stools, slightly longer and narrower than those of hookworms and are more pointed at one end.

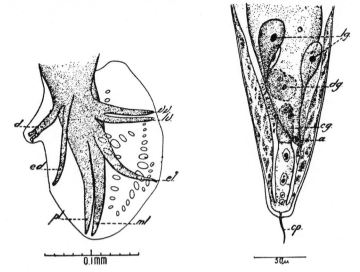

Fig. 13-14. *Nematodirus spathiger*, posterior ends of male and female; the latter shows the terminal process (cp). d, Dorsal ray in the distinct dorsal lobe; ed, externodorsal ray; pl, posterolateral and ml, mediolateral rays close together for most of their length; el, externolateral ray separate from the other laterals; lv, lateroventral and vv, ventroventral rays close together; lg lateroventral gland; dg, dorsal gland; cg, caudal gland; a, anus. (May, Proc. U. S. Nat. Museum)

In the soil there is a period of a few days before the larvae hatch and develop into the infective stage. During this free-living stage they are sheathed and are resistant to desiccation. The survival of free-living stages of *Trichostrongylus colubriformis* has been studied by Anderson,[2] who found that at −10° C. repeated freezing had no appreciable effect as compared with continuous frozen storage, but at −28° C. and −95° C., alternate freezing and thawing killed practically all the larvae after the second freezing. Infective larvae stored for 308 days at 4° C. were still able to infect a worm-free sheep.

Man becomes infected by swallowing the larvae, usually with contaminated food or drink, although the larvae may penetrate human skin. In the intestine of man the larvae mature in about 25 days, and the adult worm may live for eight years. Symptoms of infection are usually absent or mild.

Since the eggs are easily confused with those of hookworms, many cases of trichostrongyliasis have gone unreported. The world in-

Fig. 13-15. *Hyostrongylus rubidus*, anterior and posteror ends of male. (Morgan and Hawkins, Veterinary Helminthology. Courtesy of Burgess Publishing Co.)

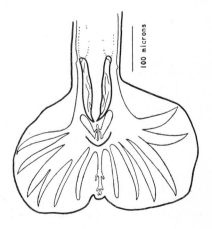

cidence of this disease in man is undoubtedly higher than usually believed. Probably more than 50,000,000 persons throughout the world harbor these worms. The incidence of infection in animals is extremely high.

Hyostrongylus rubidus. The red stomach worm of swine (Fig. 13–15) is worldwide in its distribution. The males are 4 to 7 mm. long and the females, 5 to 10 mm. long. Eggs in the soil hatch in 39 hours and the infective third-stage larvae develop in one week. Pigs become infected by eating these larvae. As with the other worms described above, *H. rubidus* sucks blood.

Dictyocaulus viviparus. This lungworm of cattle (Fig. 13–16) may be found in other large mammals such as moose, elk, bison, deer or pigs. Eggs are produced in the lungs of the host and often hatch before leaving the body. The larvae or eggs are coughed up and are usually swallowed and pass from the body with feces. Unhatched eggs are usually hatched at this time. It has been estimated that an infected animal can pass five million of these first-stage larvae a day. After about four days in moist soil, the larvae develop into the second and then the third stage. The latter is the infective stage that may be ingested by cattle or some other host. The larvae may be spread by clinging to spores of fungi that are discharged over the soil.

From the host intestine the larvae migrate by way of the lymphatics or bloodstream to the lungs, trachea or bronchi. Here they mature, and undergo a prepatent period of about a month. The adults are white, thread-like and 5 to 8 cm. long. They cause difficult breathing, coughing, foaming around the mouth, loss of appetite, diarrhea, fever, and even death. Five thousand larvae can kill a calf.

Superfamily Mestastrongyloidea

Metastrongyles. In this atypical group of the superfamily the buccal capsule is absent or rudimentary; bursa are reduced or absent; the vulva is situated near the anus; and the worms usually require an intermediate host. Most metastrongyles are called "lungworms," and they usually inhabit the respiratory or circulatory systems of carnivores, ungulates, rodents and primates throughout the world. For a bibliography of lung nematodes of mammals, see Forrester *et al.*[6]

Metastrongylus elongatus, M. salmi and *Choerostrongylus pudendotectus* are all thread-like, white worms that may reach 60 mm. in length, although some are considerably shorter. They live in air passages of the lungs of pigs, where they lay their eggs. These eggs are coughed up, swallowed, and passed from the host with feces. Various species of earthworms belonging to the genera *Helodrilus, Lumbricus* and *Diplocardia* eat the eggs, which hatch in the intestine. The larval parasitic worms leave the lumen of the gut and enter the walls of the esophagus, crop, gizzard and intestine.

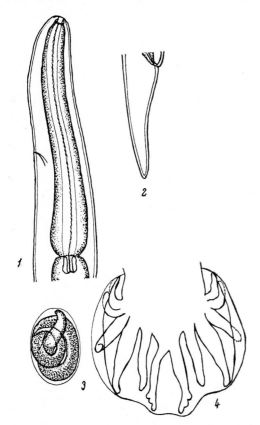

Fig. 13-16. *Dictyocaulus viviparus*, a lungworm of cattle and other mammals. (1) Anterior end. (2) Posterior end of female. (3) Egg with larva. (4) Bursa of male. (After Skrjabin *et al.*, 1954)

After a period of growth they enter the circulatory system of the annelid. In about a month they are infective to swine that eat the earthworms. Larvae are released into the pig intestine where they penetrate the walls and make their way via the lymphatic and blood vessels to the lungs. In three to four weeks after being eaten by pigs, the larvae have matured in the lungs and start producing eggs.

Infected pigs suffer from malnutrition, lung hemorrhage, difficult breathing, and coughing. The parasites may kill young pigs, and they may carry the virus of swine influenza.[16]

Angiostrongylus cantonensis. This is the rat lungworm, which may cause cerebral angiostrongylosis (eosinophilic meningoencephalitis) in man. The infection apparently occurs by eating raw prawns, land crabs, or, accidentally, small mollusks in uncooked vegetables. These invertebrates serve as intermediate hosts of the parasite, which occurs in western Asia, Australia, the islands of the Pacific, Madagascar, Ceylon and Sarawak.[1] The larvae have also been found in frogs, and have been experimentally established in pigs and calves.

Metastrongyles of the superfamily Protostrongyloidea live in the lungs of certain ruminants and deposit unsegmented ova; those of the superfamily Filaroidioidea live in the lungs of carnivores and monkeys and produce first-stage larvae. *Prostostrongylus rufescens*, the red lungworm, is a common species found in sheep and goats. Lungworm larvae often use a snail or slug as an intermediate host.

REFERENCES

1. Alicata, J. E.: Present status of *Angiostrongylus cantonensis* infection in man and animals in the tropics. J. Trop. Med. Hyg., *72*:53–63, 1969.
2. Andersen, F. L., Wang, G.-T., and Levine, N. D.: Effect of temperature on survival of the free-living stages of *Trichostrongylus colubriformis*. J. Parasitol., *52*:713–721, 1966.
3. Ball, P. A. J.: The relationship of host to parasite in human hookworm infection. *In* Taylor, A. E. R. (ed.): The Pathology of Parasitic Diseases. pp. 41–48. Oxford, Blackwell Sci. Pub., 1966.
4. Burrows, R. B.: Comparative morphology of *Ancylostoma tubaeforme* (Zeder, 1800) and *Ancylostoma caninum* (Ercolani, 1859). J. Parasitol., *48*:715–718, 1962.
5. Cameron, M. D.: The six hookworms of man, dog and cat: their modes of infection and treatment. Southwest Vet., *15*:292–295, 1962.
6. Forrester, D. J., Forrester, G. M., and Senger, C. M.: A contribution toward a bibliography on the lung nematodes of mammals. J. Helminth., Supp., 1966.
7. Gharib, H. M.: A preliminary note on the bacteriostatic properties of some helminths of animals. J. Helminth., *35*:225–232, 1961.
8. Jarrett, W. F. H.: Pathogenic and expulsive mechanisms in gastrointestinal nematodes. *In* Taylor, A. E. R (ed.): The Pathology of Parasitic Diseases. pp. 33–40. Oxford, Blackwell Sci. Pub., 1966.
9. Levin, M. B.: Infestation with *Strongyloides stercoralis*. J. Lab. Clin. Med., *28*:680–682, 1943.
10. Little, M. D.: Comparative morphology of six species of *Strongyloides* (Nematoda) and redefinition of the genus. J. Parasitol., *52*:69–84, 1966.
11. Miller, T. A.: Influence of age and sex on susceptibility of dogs to primary infection with *Ancylostoma caninum*. J. Parasitol., *51*:701–704, 1965
12. Nagahana, M. and Yoshido, Y.: Complete development and migratory route of *Ancylostoma duodenale* in young dogs. J. Parasitol., *51*:52, 1965.
13. Olsen, O. W.: A fur seal pup is born. Turtox News, *35*, 32, 1957.
14. Roche, M., and Layrisse, M.: The nature and causes of "hookworm anemia." Amer. J. Trop. Med. Hyg., *15*:1029–1102, 1966.
15. Rogers, W. P., and Sommerville, R. I.: The physiology of the second ecdysis of parasitic nematodes. Parasitology, *50*:329–348, 1960.
16. Shope, R. E.: The swine lungworm as a reservoir and intermediate host for swine influenza virus. III. Factors influencing transmission of the virus and the provocation of influenza. IV. The demonstration of masked swine influenza virus in lungworm larvae and swine under varied conditions. J. Exp. Med., *77*:111–126, 127–138, 1943.
17. Soulsby, E. J. L., and Stewart, D. F.: Serological studies of the self-cure reaction in sheep infected with *Haemonchus contortus*. Aust. J. Agricult. Res., *11*:595–603, 1960.
18. Stoll, N. R.: On endemic hookworm, where do we stand today? Exp. Parasitol., *12*:241–252, 1962.
19. Stone, W. M., and Girardeau, M.: Transmammary passage of *Ancylostoma caninum* larvae in dogs. J. Parasitol., *54*:426–429, 1968.

Chapter 14

Class Secernentea, Orders Ascarida, Spirurida

ORDER ASCARIDA

Three prominent lips around the mouth are important characteristics of this group of worms. Usually there is no pronounced posterior bulb in the esophagus, but there is, in many of the species of this order, an anterior muscular portion of the esophagus followed by a posterior **ventriculus**. The latter may be short, and an anterior **intestinal cecum** and/or a posterior **esophageal diverticulum** may arise from its posterior edge. The tails of both males and females come to a point, but the female tail is blunter. The tail of the male is usually coiled. Life cycles of the ascaridoids usually involve one or more intermediate hosts; that of *Ascaris lumbricoides* represents an unusual form of behavior. Members of this order are primarily parasites of vertebrates, and some species have considerable economic importance.

Superfamily Ascaridoidea

Toxocara and Toxascaris. *Toxocara canis* (Fig. 14–1) and *Toxascaris leonina* are common intestinal roundworms of dogs and cats and related carnivores. Adult male worms range in length from 7 to 9 cm., whereas females are 10 to 17 cm. long. *T. canis* is somewhat longer in average length than *T. leonina*. Neither species makes use of an intermediate host in its life cycle. *Toxascaris leonina* is more closely related to *Ascaris suum* of swine than it is to *Toxocara canis*.[8] The two are easily distinguished by examining their eggs (Fig. 14–2). *Toxascaris* eggs are colorless, have almost smooth shells and prominent vitelline membranes. Eggs of *Toxocara* are light brown, have a thicker, rougher, proteinaceous coat with less prominent vitelline membranes.

Infections are normally obtained by eating

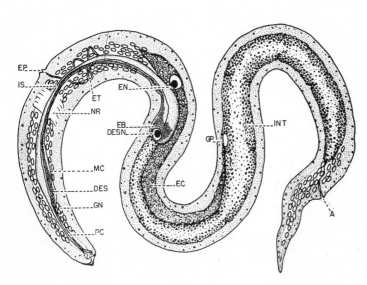

Fig. 14-1. Infective second-stage *Toxocara canis* larva pressed from egg. $\times$ 624. A, anus; DES, dorsal esophageal gland; DESN, nucleus of dorsal esophageal gland; EB, esophageal bulb; EC, excretory column; EN, nucleus of excretory cell; EP, excretory pore; ET, excretory tubule; GN, ganglionic nucleus; GP, genital primordium; INT, intestine; IS, isthmus; MC, metacorpus; NR, nerve ring; PC, procorpus. (Nichols, courtesy of J. Parasitol.)

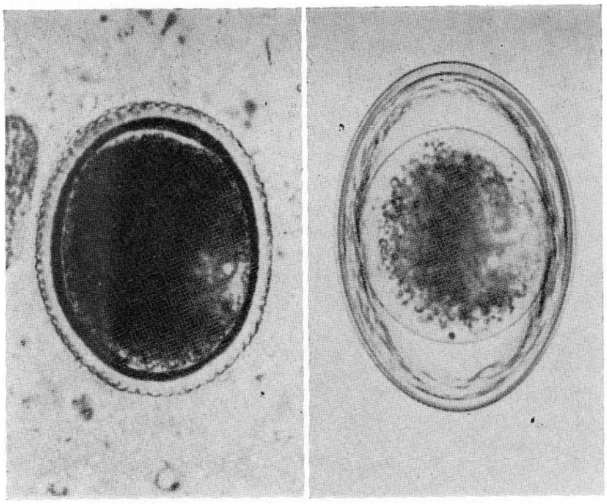

Fig. 14-2. Eggs of *Toxocara cati* (left) and *Toxascaris leonina* (right). *T. leonina* may be found in dogs as well as in cats. The one illustrated came from a tiger. It is 89 μ long. (Georgi, Parasitology for Veterinarians. Courtesy of W. B. Saunders Co.)

the eggs, which hatch and complete their first molt in the host duodenum. Both species have four larval stages in their life cycles. *Toxascaris* larvae enter the wall of the intestine, where they continue their development, then return to the lumen of the gut and there mature. *Toxocara* larvae penetrate the gut wall and, via the hepatic circulation, arrive in the liver, then migrate to the lungs, alveoli, trachea, mouth, esophagus, stomach and small intestine, where they mature. *Toxocara* can produce 200,000 eggs per day. Adults probably live for about four months.

Toxocara larvae are capable of living in a wide variety of abnormal hosts, including man—most frequently children. Children eat eggs, which hatch in the gut. In these hosts juvenile worms often do not follow normal patterns. They may migrate through the body producing visceral larva migrans, which usually causes mild symptoms but can be fatal. *T. canis* larvae may pass through the placenta of a mother and into the unborn fetus. Infection in dogs is especially harmful to puppies. The animals lose their appetites and their abdomens tend to enlarge.

Ascaris lumbricoides. When a man has "worms," he probably harbors the common intestinal roundworm, *Ascaris lumbricoides* (Fig. 14–3). This parasite is worldwide in its distribution, and in some localized areas of the Orient its incidence of infection is 100 per cent. The worm possesses typical nematode structures and is often used in biology laboratories as a representative of the Nematoda. A very closely related species, *A. suum*, lives in the intestines of pigs.[29]

Ascaris lumbricoides is the largest nematode of the human intestine. An adult female may reach 50 cm. in length (20 inches), but the average is about 27 cm. Characteristically, the males are smaller than the females. There may be only one worm in the intestine or there may be many. After medication, one man expelled 667 worms. An autopsy of another man revealed 1,488 worms, an astonishing burden of large parasites to be carried in the intestine. The authors have seen patients in the Orient so heavily parasitized with worms that a bowel movement produced nothing but a large mass of ascarids and tapeworms. It has been estimated that the glycogen consumption in one year of all the ascarids in all the people of China equals the carbohydrate value of 143,000 tons of rice.

Three prominent lips surround the mouth, each possessing a pair of minute papillae on its lateral margin (Fig. 14–4). The mouth cavity leads to a muscular esophagus that is attached to the midintestine. The latter organ extends almost to the posterior extremity of the body, where it empties into a short rectum. The rectum opens directly through the anus in the female and into the cloaca in the male.

The cuticle contains fibrous protein as well as small amounts of carbohydrates and lipids. The three basic cuticular layers (see p. 267)

Fig. 14-3.   *Ascaris lumbri-coides*, adults.   (Courtesy of Dow Chemical Co.)

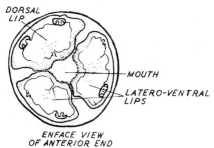

Fig. 14-4.   *Ascaris lumbricoides*, anterior end.

have been modified in *Ascaris* to form nine layers with an arrangement of spiral fibers. The outer cortex is not digested by pepsin or trypsin but can be digested by ficin and papain.   Evidence indicates that the cuticle is not permeable to amino acids and sugars. If this is correct, "we are led to the simple idea that the gut must play an important role in absorption."[25]   For a description of growth of the cuticle and fine structure of the body wall, see Watson.[30]

The ability of *Ascaris* to digest foodstuffs is not thoroughly understood.   Amylase, lipase, esterase and protease have been found in the intestine of this worm, but the exact role these enzymes play is still uncertain.   Glucose and fructose are rapidly absorbed and seem to be quickly metabolized.[27]   The inner lining of the intestine is covered with microvilli, which increase the surface area and possibly have a

secretory function.   There is evidence[7] that *Ascaris* in pigs ingest large numbers of cells that have been sloughed off from the host intestinal epithelium and lymphocytes that have found their way into the lumen.

Ascaris possesses two distinct types of hemoglobin: (1) that which comprises about two per cent of the hemolymph, and (2) that which occurs in the body wall of the worm. The functions of these hemoglobins are obscure.   Their respiratory function is questionable, since under anaerobic conditions the body-wall hemoglobin is deoxygenated in six hours but the hemolymph hemoglobin remains unchanged.   In general, the composition of the hemolymph is similar to the composition of the host intestinal fluids, so far as total ions, solids and ash are concerned. More specifically, 4.9 per cent of the hemolymph of *Ascaris* is composed of proteins, of which 2.8 per cent are albumens and 2.1 per cent are globulins.   Magnesium, calcium and potassium are kept at fairly regular concentrations in this fluid, but sodium passes freely through the cuticle.   The worm behaves like an osmometer.   Most of the lipids of *Ascaris* are deposited in the reproductive tissues, especially the ovaries and uterine eggs, and undoubtedly they are catabolized during embryonation.   Starving worms apparently are unable to utilize these lipids.

Males range from 15 to 32 cm. in length

(average 32 cm.), with a width of about 3 mm. The posterior end curves ventrally. Male genitalia form a long coiled tube situated in the posterior half of the body, consisting of testis, collecting tubule and ejaculatory duct, the latter opening into the cloaca. A pair of retractable spicules (Fig. 14–5) lies in the end of the ejaculatory duct. There is no gubernaculum, but there are numerous pre-anal and postanal papillae.

In females the vulva lies near the junction of the anterior and middle thirds of the body, and it leads to the vagina, branching to form paired genital tubules. Each tubule consists of a seminal receptacle, uterus, oviduct and ovary (Fig. 14–6). These tubules are more or less parallel with each other and they follow a tortuous course throughout the posterior two-thirds of the body. The uterus is wider than the other parts of the tubule, and it accommodates the eggs. A "genital girdle" surrounds the body at the level of the vulva, but it is not always evident. The girdle seems to be an indication of estrus, and appears when the sperm supply becomes exhausted.[3]

The life cycle of *Ascaris lumbricoides* starts with the fertilized egg within the body of the female. These eggs, 45 to 75 × 35 to 50 μ, are ovoidal with a thick transparent shell and an outer, coarsely mammillated, albuminous covering. They are not embryonated when voided with the feces of the host. Unfertilized eggs are longer, narrower, more elliptical, and usually possess an irregular albuminous coating. Each female may lay as many as 200,000 eggs per day. The eggs are laid in the small intestine of the host and they contain *refringent inclusion bodies*, which contribute directly to the formation of the vitelline membrane and possibly to the chitinous shell.[10] The shell of *Ascaris* eggs is highly resistant to desiccation and to chemicals. Eggs in ten per cent formalin have remained viable for weeks and have developed active larvae within them.

Growing and molting of the embryo occur within the egg but the larvae do not hatch in the soil. Full embryonation requires two to three weeks. When the eggs have developed to the second-stage juvenile (Fig. 14–7), they are infective to man, and when ingested they reach the small intestine where they hatch. Hatching is a complicated process that has been described as follows.[26] Dissolved host gaseous carbon dioxide at low redox potentials, undissociated carbonic acid and the pH are all involved in producing a stimulus that causes the eggs to produce a hatching fluid, which contains the enzymes esterase, chitinase, lipase, and protease. The esterase alters the vitelline membrane of the eggs, thus permitting the other enzymes to hydrolyze the hard shell. The inner membrane forms a bulge at this point and the larva escapes by stretching and finally bursting through this bulge. Occasionally the

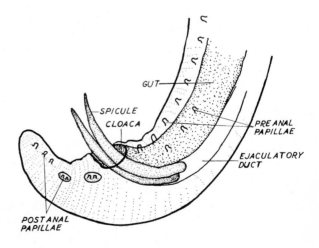

Fig. 14-5. *Ascaris lumbricoides*, posterior end of a male worm. The spicules emerge from the ventral surface of the worm.

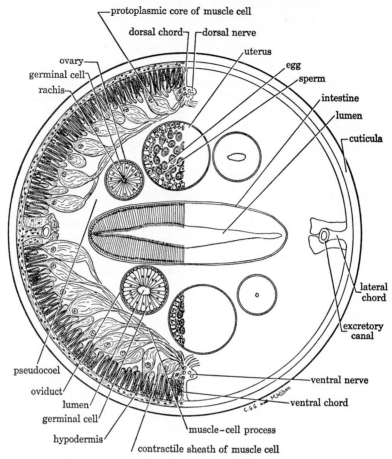

Fig. 14-6. *Ascaris lumbricoides*, cross section of the body through the region of the female gonads. (Brown, Selected Invertebrate Types. Courtesy of John Wiley & Sons)

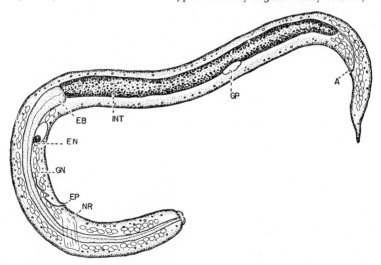

Fig. 14-7. *Ascaris lumbricoides*, second stage juvenile. × 684. A, anus; EB, esophageal bulb; EN, nucleus of excretory cell; EP, excretory pore; GN, ganglionic nucleus; GP, genital primordium; INT, intestine; NR, nerve ring. (Nichols, courtesy of J. Parasitol.)

entire shell is hydrolyzed, leaving only the thin vitelline membrane surrounding the larva, which then ruptures the membrane.

The larvae are now in that part of the body where one would expect them to remain and mature. Instead, they usually seem to possess a wanderlust, and they burrow through the intestinal wall to the lymphatics or blood vessels and make the same tour through the body as do hookworm larvae. They go through the heart to the lungs, where they spend a week or more, then migrate to air spaces, up the trachea to the mouth and are swallowed. Back in the small intestine where they started, they now remain and become mature. The tour through the body suggests that these ascarids have evolved from an ancestor related to the hookworm, and have simply eliminated the free-living larval stage in the soil.

The odds against any one egg reaching maturity are indicated by the vast numbers of eggs produced by one female. As mentioned previously, one worm may produce as many as 200,000 eggs per day. A worm probably does not live for more than a year, but if it should live that long and produce eggs at that rate, it would produce 73,000,000 eggs during its lifetime. If many worms are present at one time, we can easily understand why even an unconcentrated fecal smear almost always contains some eggs.

Symptoms of *Ascaris* infection vary with the phase of the infection. Immature worms leaving the intestine and migrating through the body can cause annoying or serious trouble, depending on where they go. Headache, muscular pain, coughing or fever are typical symptoms. If a young worm 5 to 7 cm. long should leave the intestine and wander up the alimentary canal to creep out of the nostrils, its appearance would be more embarrassing than serious. But if the worm should pierce the gut wall and get into the heart or brain, it could cause death. Larval worms that migrate into lung spaces and remain there for several days or more can produce pulmonary symptoms. Adult worms in the intestine frequently produce no symptoms, but

susceptible hosts or those with heavy infections may experience vague discomfort, no appetite, occasional intestinal pain, nausea, diarrhea, or constipation. Poorly nourished children in areas of heavy infection may reveal lumps of worms on rectal examination. Masses of worms are palpable through the abdominal wall. *A. lumbricoides* can produce an infection in experimental pigs, but the symptoms differ markedly from those produced by *A. suum*.[11]

Only a partial immunity to reinfection with *Ascaris lumbricoides* is acquired by human beings. Larvae of a superimposed experimental infection are surrounded by host cells that check the larval migration and finally destroy them. As with other nematode parasites, the different developmental stages of *Ascaris* release qualitatively different antigens. Evidence of circulating antibodies is seen by the formation of precipitates around larvae placed in immune serum. Studies of *A. suum* in experimental rats show that immunization of this rodent host is possible. At least 14 antigens from *A. suum* have been reported.

Ascaridia galli. This nematode (Fig. 14–8) is a common large parasite of the small intestine of various domestic and wild birds. It has a worldwide distribution. The males average about 50 mm. long and the females may be longer than 100 mm., although they usually measure about 90 mm. in length. These worms possess a pre-anal sucker. Eggs do not hatch in the soil, but when infective and eaten by a suitable host, they hatch in the intestine. During part of the growth period the juveniles burrow into the intestinal mucosa for a few days, but sometimes this burrowing does not occur. Occasionally young worms wander elsewhere in the host body, even getting into the oviduct and becoming enclosed within the egg shell.

Superfamily Heterocheiloidea

Certain members of the family Anisakidae whose larvae occur in fish have been known to cause tumors in the stomach of man. When a man eats uncooked, infected sea fish such as herring, he may suffer from anisakiasis, an

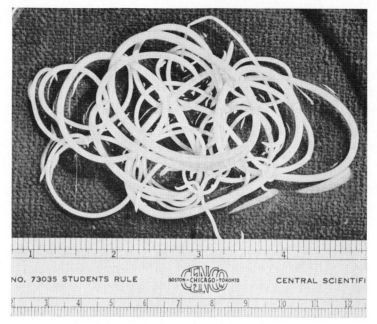

Fig. 14-8. The common chicken roundworm, *Ascaridia galli*. (Courtesy of Dow Chemical Co.)

intestinal eosinophilic granuloma (herring-worm disease). Cooking or freezing fish for 24 hours eliminates danger of infection. Probably, other nematode larvae or ova in animal food eaten by man are also significant as etiologic agents of this type of infection.[2,31] Larval stages of *Contracaecum* are common in marine fish.

Superfamily Heterakoidea

Heterakis. The remarkable association between the protozoan flagellate, *Histomonas meleagridis*, which causes blackhead disease in poultry, and the cecal nematode, *Heterakis gallinae* (Fig. 14–9), has already been mentioned (p. 52). The male worm averages about 10 mm. in length, while the female is usually about 13 mm. long. Eggs pass from the host with feces and are eaten by earthworms, which serve as vectors.[22] When infected worms are swallowed by chickens, turkeys, and presumably ducks, geese, guinea fowl, quail or other birds, they hatch within a few hours and the freed larvae burrow into the cecal mucosa for a day or two. They return to the cavities of the cecae and mature

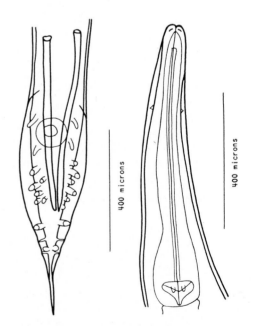

Fig. 14-9. *Heterakis gallinae*, posterior and anterior ends of a male. (Morgan and Hawkins, Veterinary Helminthology. Courtesy of Burgess Publishing Co.)

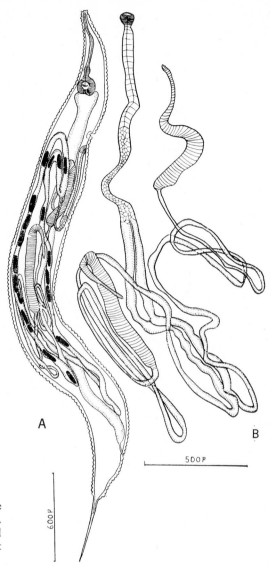

Fig. 14-10. *Pharyngodon mamillatus* from the rectum of a lizard showing. (*A*) A young female. (*B*) The dissected genital structures. (Chabaud and Golvan, courtesy of Arch. de l'Institut Past. du Maroc.)

in three to four weeks. The distribution of this worm is probably worldwide.

Although the presence of the adult nematode worms seems to have little effect on the host, the eggs of the worm carry the flagellate *Histomonas meleagridis*, which may cause serious disease, especially to turkeys. The problem of how and when the protozoan parasite enters the egg and when it leaves is still to be solved, but there is evidence that the larval worm as well as the egg may carry the flagellate.

Superfamily Oxyuroidea

Oxyuroids are characterized by an esophagus with a posterior bulb, males with one or two equal spicules, and both males and females with pointed tails (few exceptions in some males). All members of the group are parasitic, primarily in vertebrates, and their common name is *pinworm*.[19] There are no intermediate hosts. General anatomy of the group is represented in Figures 14–10, 14–11, which show *Pharyngodon mamillatus* from the rectum of the lizard, *Eumeces algeriensis*, and

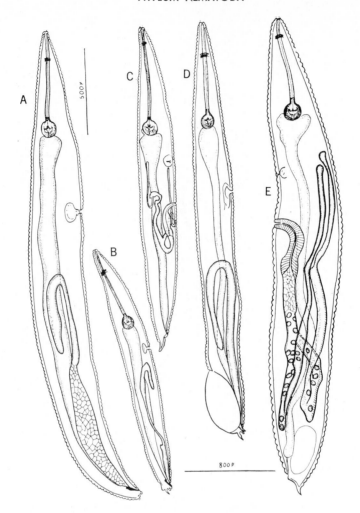

Fig. 14-11. *Thelandros bulbosus* from the cecum and rectum of a lizard. Lateral alae have been omitted. (*A, B, D*) Males; (*C, E*) females. (Chabaud and Golvan, courtesy of Arch. de l'Institut Past. du Maroc.)

Thelandros bulbosus from the cecum and rectum of the lizard, *Chalicides ocellatus polylepis*.

Oxyuris equi is the common pinworm in the cecum and colon of horses. Males average about 11 mm. long, whereas the females may be over 100 mm. in length (40 to 150 mm). The worm is definitely pathogenic and heavy infestations may cause considerable irritation to horses, which tend to rub the base of the tail region on fences or other objects. Sheep, rabbits, and other vertebrates may become infected with related species.

Superfamily Syphacioidea

Syphacia. This genus (Fig. 14–12), is found in the intestine of rats and mice, and an accidental infection has been reported from man. Persons handling rats and mice always run the risk of acquiring rodent parasites but, fortunately, these parasites are rarely pathogenic to man. *S. muris* appears to be the common species in rats and *S. obvelata*, the common species in mice. There is some evidence of possible cross infection between these hosts. The worms bear a superficial resemblance to *Enterobius vermicularis*, but the "tail" of both male and female is long and pointed. Males average about 1.3 mm. in length, and females about 4.5 mm. *Syphacia* eggs (125 × 35 μ) tend to be oval in outline but one side is flattened.

Enterobius vermicularis. The cosmopolitan pinworm or seatworm of man is *Enterobius vermicularis* (Fig. 14–13). The adult female parasite is usually most abundant in the cecum and appendix. It is a short whitish worm shaped like a narrow

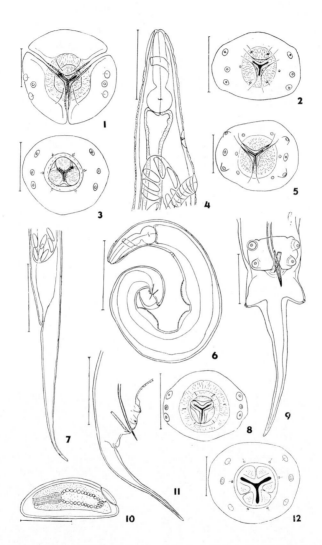

Fig. 14-12. *Syphacia*, several species from the intestines of various mammals showing especially the oral region. (Scales for Figures 1, 5, and 12 represent 25 μ; for 2 and 8, 30 μ; for 3, 9, and 10, 50 μ; for 4 and 6, 300 μ; for 7, 500 μ; for 11, 100 μ.)

(1) En face view of *S. thompsoni* from *Glaucomys sabrinus macrotis*, in Wisconsin. (2) En face view of *S. obvelata* from *Microtus* in Alaska. (3) En face view of *S. citelli*. (4) Anterior end of *S. arctica*, female. (5) En face view of *S. arctica*. (6) Male, *S. arctica*. (7) Tail of female, *S. arctica*. (8) En face view, *S. peromysci* from *Peromyscus maniculatus bairdii* in Wisconsin. (9) Ventral view, posterior extremity of *S. arctica*. (10) Egg of *S. arctica*. (11) Side view, posterior extremity of *S. arctica*. (12) En face view, *S. eutamii* from (type locality and host) *Eutamias minimus*, Grand Marias, Minn. (Tiner and Rausch, courtesy of Natural History Miscellanea)

Fig. 14-13. *Enterobius vermicularis.* (*Left*) Male, × 26, showing characteristic coiling of the hindbody. (*Right*) Internal structures of the female, × 26, are obscured by the mass of eggs. (Beaver, courtesy of Amer. J. Clin. Path.)

spindle. A pair of lateral cephalic alae ("wings") are situated at the anterior end. The mouth is surrounded by three lips or labia, and the mouth cavity leads to an esophagus with an extra or prebulbar swelling and a distinct bulb. Males range from 2 to 5 × 0.1 to 0.2 mm. in size, and each possesses a strongly curved posterior end. At this end there is a pair of small caudal alae that are supported anteriorly and posteriorly by pairs of papillae. The gubernaculum is lacking and, in contrast with *Ascaris* (p. 301), there is only one spicule. Females range from 8 to 13 × 0.3 to 0.5 mm. in size, each with a long

tapering tail. The anus lies at the junction of the middle and posterior thirds of the body, and the vulva is situated in front of the junction of the anterior and middle thirds of the body. Eggs (Fig. 14–14) are flattened on the ventral side and they measure 50 to 60 × 20 to 30 μ, have relatively thick shells, and are embryonated when laid.[17]

The eggs are usually not deposited in the host's intestine but they remain in the body of the female worm until she crawls through the host anus, usually at night. A person can sometimes feel the worms crawling in his rectum. The female worm contains

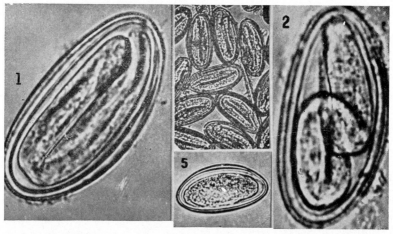

Fig. 14-14. Eggs of *Enterobius vermicularis.* (Original photomicrographs of Kouri.) The eggs are asymmetrical, double walled, colorless, and contain a tadpole-shaped or worm-shaped embryo. (*1, 2*) Egg with vermiform embryo. (The embryo emerges through the convex part, located in the cephalic pole of the egg— the adherence zone.) (*5*) Non-embryonated egg. (After Gradwohl and Kouri, Clinical Laboratory Methods and Diagnosis. Courtesy of C. V. Mosby Co.)

about 11,000 eggs, and the pressure of these eggs sometimes causes her to burst, scattering them on the body of the host or on the bedclothes. If the worm does not burst, she rapidly discharges her eggs, dies and becomes desiccated. Eggs are scattered about on clothes, bedclothes, hands and on the body of persons infected, and even in the dust of the room. In classrooms, up to 300 eggs have been found per square foot (30 cm. square) in dust. The eggs mature rapidly and are infective within a few hours.

In women and girls, a pinworm may crawl into the genital opening and cause inflammation and irritation. The person infected may experience intense itching in the anal region or only a mild tickling sensation. Children, especially, may become irritable and not sleep well. The general symptoms are anal or vaginal pruritus, sleeplessness, irritability, nausea, constipation or diarrhea. Since these symptoms are common to many kinds of intestinal parasites, they have little diagnostic value. Frequently there are no symptoms.

Man becomes infected by inhaling contaminated air, sucking fingers, or ingesting contaminated food or drink. When mature eggs are swallowed, they hatch in the small intestine and the larvae migrate to the cecum, appendix, colon or ileum where they mature in about a month. Male and female worms may become attached to the gut wall and produce inflammation. It has been estimated that about 209,000,000 people in the world are infected with this worm. Unlike most worms we have been describing, man is the only host for *Enterobius vermicularis*. Nonhuman forms occur in the chimpanzee, the lar gibbon and the marmoset. Children are more commonly infected than are adults, and the parasite is more prevalent in cooler countries than in hot climates where children wear fewer clothes.

Diagnosis is usually made by a perianal swab. Place cellophane tape, sticky side out, over the end of a tongue depressor or other suitable object, and then apply it to the perianal region. Then place the tape, sticky side down, on a microscope slide, taking care not to wrinkle it more than necessary. It is best to let a drop of toluene or N/10 sodium hydroxide flow under the tape to make the eggs more visible. The preparation is examined under low power of the microscope, switching to intermediate power for positive identification. Prevention includes frequent laundering of night clothes and bedding, keeping hands and fingernails clean, frequent bathing, and keeping rooms as dust-free as possible.

ORDER SPIRURIDA

Superfamily Spiruroidea

Both sexes of these slender worms of moderate size usually possess two lateral lips surrounding the mouth, but there may be four or six small labia. The buccal capsule is cuticularized and the esophagus does not have a bulb. In the female the vulva is usually located near the middle of the body. Most males have two unequal spicules.

These worms are parasites of the digestive tract, respiratory system, eyes, nasal cavities and sinus sacs of vertebrates. The life cycle involves one or two intermediate arthropod hosts (e.g., beetles, grasshoppers, flies, cockroaches, crustaceans). In a typical life cycle, thick-walled embryonated eggs from the final host are swallowed by the arthropod, which is subsequently eaten by a final host.

Gongylonema pulchrum. This worm (Fig. 14–15) is a parasite of sheep, goats,

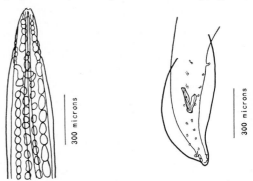

Fig. 14-15. *Gongylonema pulchrum*, anterior and posterior ends of male. (Morgan and Hawkins, Veterinary Helminthology. Courtesy of Burgess Publishing Co.)

pigs, horses, cattle and other mammals in which it lies embedded in the mucosa or sub-mucosa of the esophagus and oral cavity. It is worldwide in distribution. Males may reach 62 mm. in length and females may be as long as 145 mm. Eggs pass from the host in feces and are eaten by dung beetles or cock-roaches, in which the larval worms develop. About a month after ingestion, these larvae become encysted in the insect and are in-fective to a new vertebrate host that eats the insect. Man is occasionally infected.

Superfamily Thelazioidea

Thelazia. *Thelazia callipaeda*, the oriental eye worm, and *T. californiensis* are nematodes that are normally found in the eyes of verte-brate animals and occasionally the eyes of man. They are 5 to 20 mm. long and usually lie on the surface of the eyeball. When dis-turbed they tend to make their way under the eyelids or into the conjunctival sac. Charac-teristics of the genus include a mouth without definite lips, a short buccal cavity, usually numerous pre-anal papillae. Males may or may not have alae, and the spicules are usu-ally unequal. Members of the genus are ovoviviparous, and occur in the eyes, nasal chambers and mouths of mammals and birds, in the air sacs of birds, and in the intestines of fish.

Superfamily Gnathostomatoidea

Gnathostoma spinigerum. These short, fat worms are normally parasites of cats and of other animals, and only accidentally of man.[23] The worms are worldwide in distri-bution but are found mainly in the tropics. In the normal host they live in stomach tumors; thus eggs get to the outside with feces. The eggs must get to water where they hatch into free-swimming larvae that are eaten by small crustacea such as *Cyclops*, which in turn are eaten by invertebrates or vertebrates which are then eaten by cats. The parasites must develop, in turn, in each of these hosts to become infective to the next host. Adults are 3 to 5 cm. long and about 8 mm. wide. They possess distinct, enlarged, spiny heads,

and the spines continue down one-half of the body.

In man the worms are often found in an immature condition in skin tumors although they may become lodged in almost any other organ.

Superfamily Dracunculoidea

Nematodes belonging to this group do not possess definite lips, but six conspicuous labial papillae and eight external papillae are present. The esophagus generally has a muscular portion and a posterior, broader, glandular portion.

Dracunculus medinensis. The Guinea worm (Fig. 14–16) occurs in Africa, parts of Asia and rarely in South America. Areas of abundant rainfall are freer from the worm than are drier locations. The adult female measures 750 to 1200 mm. in length (up to 4 feet), with a diameter of about 1.25 mm. The male is much shorter, averaging only about 25 mm. in length. These worms live in the connective tissue of man and other vertebrates, especially just under the skin, and then can migrate from one site to another.

The life cycle starts with the development of young worms within the body of the female parasite. By the time the young are ready to emerge from the uterus, the female has produced a hole in the host's skin into which a portion of the worm's uterus projects. When the infected host skin comes in contact with water, as when bathing or washing clothes, myriads of young worms pass from the uterus into the water. The larvae are approximately 600 μ long. As they swim around, some of them may be eaten by the small freshwater crustacean, *Cyclops* (Fig. 16–8, p. 347). Within this crustacean vector the young Guinea worms migrate to the hemocoel, undergo one or two molts and become infective in ten to 20 days. The definitive hosts (man, dogs, cats, or various wild mammals) acquire the infection by in-gesting *Cyclops* in drinking water. In these mammals the larvae leave the *Cyclops* while it is being digested, and penetrate the host's intestinal wall. They migrate to connective

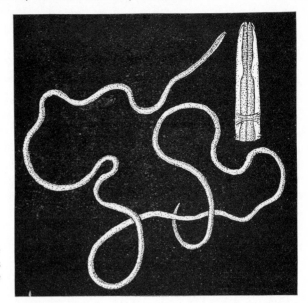

Fig. 14-16. *Dracunculus medinensis*, the Guinea worm. A female specimen is shown with an enlarged view of the anterior end.

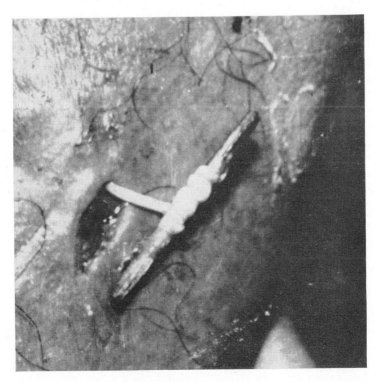

Fig. 14-17. The nematode, *Dracunculus medinensis*, being removed from the leg by slow winding on a matchstick. (Courtesy, Institute of Parasitology and Malariology, University of Teheran, School of Medicine)

tissue where they mature in about a year. After fertilization the males die and are absorbed by the host. When the females become gravid they migrate to the skin which they pierce, thus providing the locus for the development of a small ulcer.

Symptoms are absent until a skin sore begins to form. Then there may be nausea, diarrhea, giddiness, skin rash, itching or asthma. The sore is usually on feet, legs or arms but occasionally on other parts of the body, the location corresponding to the parts that most often get wet. At first a reddish pimple develops, then enlarges and forms a blister that eventually breaks. The sore is usually not more than 5 mm. in diameter, but it can get considerably larger, and sometimes it becomes secondarily infected with bacteria.

Medical treatment is of limited value. Surgical removal of the worms is the best procedure. Some Africans and Asians remove the worms by slowly winding them on small sticks (Fig. 14–17), but this feat must be done gradually, a few centimeters a day, or the worm will break and cause severe inflammation. Control measures center around keeping people with the sores from contaminating wells, laundry and bathing waters, and encouraging *Cyclops*-eating fish to become established in streams or ponds used by the people.

Superfamily Onchocercoidea

These long, thin, tapering worms have no lips around the mouth, and the esophagus does not possess a bulb. The life cycle involves a blood-sucking insect.[16] Adults are **filarial** worms and they produce **microfilariae.** Some of the species are important, pathogenic parasites of man. See Kessel[21] for a discussion of the ecology of filariasis, and Kagan[20] for immunologic methods for diagnosis of filariasis. Nelson[24] has summarized information on the role of animals in the transmission of filarial infections to man.

Splendidofilaria fallisensis. This parasite inhabits the subcutaneous tissues of domestic and wild ducks in North America. A careful study has been made of the life cycle of this parasite in white Pekin ducklings[1] and the results of this study will serve as an introduction to the Filarioidea.

Microfilariae in the blood of ducklings are taken into the body of a black fly (*Simulium rugglesi* and *Eusimulium anatinum*). In the hemocoel of the fly, the young worms (Figs.

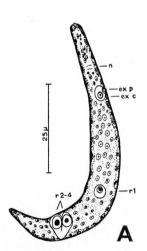

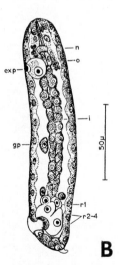

Fig. 14-18. Developmental stages of *Splendidofilaria fallisensis*. (*A*) Early larva. (*B*) Advanced first-stage larva with well-formed esophagus and intestine, lateral view.

ex c, Excretory cell; *exp*, excretory pore; *gp*, genital primordium; *i*, intestine; *n*, nerve ring; *o*, esophagus; *r* 2–4, rectal cells ("genital cells" of most authors); *r* 1, rectal lumen. (Anderson, courtesy of Can. J. Zool.)

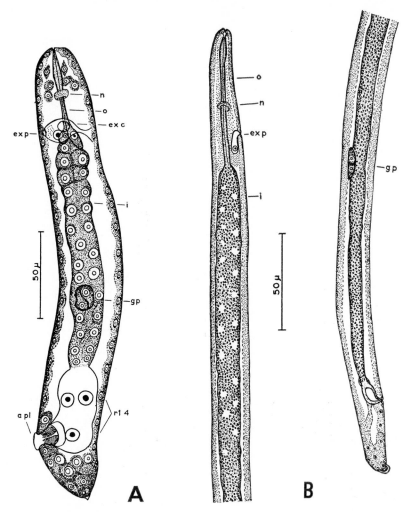

Fig. 14-19. Developmental stages of *Splendidofilaria fallisensis*. (*A*) Newly emerged second-stage larva, lateral view. (*B*) Third-stage larva from head of fly, lateral view, anterior and posterior ends.

a pl Anal plug ("nail-like structure" of Brug); *ex c*, excretory cell; *ex p*, excretory pore; *gp*, genital primordium; *i*, intestine; *n*, nerve ring; *o*, esophagus; *r* 1–4, rectal cells ("genital cells"). (Anderson, courtesy of Can. J. Zool.)

14–18, 14–19) develop into the infective third-stage larvae in one to two weeks, depending on the temperature. When the infected fly bites a duck, the worms are introduced into the blood and they migrate to subcutaneous tissues, where they mature. Mating occurs, and the females produce young which, about a month from the time the bird became infected, can be found in the peripheral blood of the host. These minute young worms exhibit diurnal periodicity, appearing in the peripheral blood only during the daytime.

Wuchereria and Brugia. Worms belonging to these genera live in the lymphatic vessels of their hosts in tropical countries and, like many other nematodes, give birth to young (Fig. 14–20). During development, a transparent sheath forms around the embryo before birth and the young or microfilariae are born well developed beyond the egg stage. A mosquito vector is involved in the life cycle.

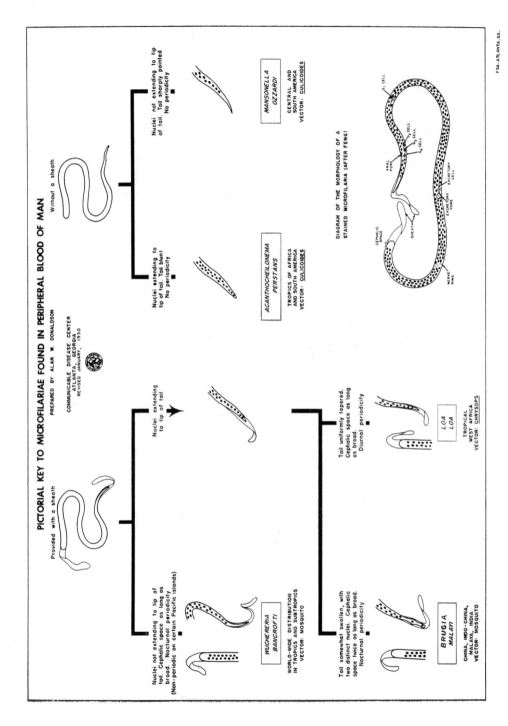

Fig. 14-20. Microfilariae found in peripheral blood of man. (Federal Security Agency U.S. Pub. Health Service.)

The genus *Wuchereria* contains only the species *bancrofti*, which lives in man, whereas the genus *Brugia* contains the species *malayi*, found in man and other mammals, and several other species found only in nonhuman mammals. Infection of man with *Brugia malayi* apparently does not occur outside of Asia, where it is found in rural areas.

Adult female worms average 82 mm. long and 0.25 mm. wide—similar in appearance to a short piece of fine thread. The males average 40 mm. in length by 0.1 mm. in width. The head is only slightly larger than the tapering posterior end, and the mouth is a simple hole without lips or other organs. The mouth leads directly into the esophagus, without a buccal cavity, and the esophagus does not possess the bulges and constrictions so characteristic of some of the other nematodes. The vulva opens to the outside of the body in the midregion of the esophagus. The male possesses two spicules of unequal length, and a gubernaculum.

The types of behavior of microfilariae in the peripheral blood of vertebrates are outlined as follows[9]: (1) periodic occurrence (e.g., *Wuchereria bancrofti*), in which a pronounced peak in microfilaria count may occur at some point in each 24-hour period; (2) continuous occurrence (e.g., *Brugia malayi*) in which microfilariae may be found in appreciable numbers throughout each 24-hour period, although a consistent minor peak in the microfilaria count may occur (subperiodic). *Wuchereria bancrofti* occurs as two biologically different forms: (1) the periodic form, which shows markedly nocturnal periodicity and occurs in the humid tropical zone throughout the world; (2) the diurnally subperiodic form, which is restricted to Polynesia and is transmitted by day-biting mosquitoes. In both these forms, the sequence of development from the infective larva to the adult worm in man is unknown.

Microfilarial periodicity depends on alternate (1) accumulation of the microfilariae in the lung capillaries, usually by day (active phase), and (2) approximately even distribution throughout all the circulating blood, usually by night (negative phase). This periodicity is oriented to the established circadian cycle of the host (sleeping and waking habits) instead of day and night as such. The period of a circadian cycle or rhythm approximates that of the earth's rotation, i.e., 24 hours. These rhythms are ubiquitous in living organisms. The microfilariae have their own endogenous rhythm, which is synchronized with the cycle of the host.[13] The periodicity appears to be related to the difference in oxygen tension between venous and arterial blood during day and night.[14,15] Microfilariae of *Wuchereria bancrofti* (Fig. 14–21) disappear during the day, probably because they accumulate in the small ar-

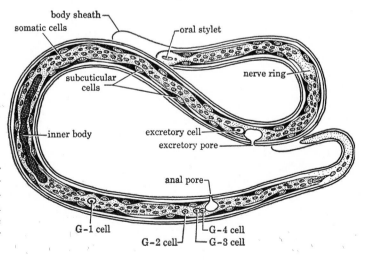

body sheath
somatic cells
oral stylet
subcuticular cells
nerve ring
inner body
excretory cell
excretory pore
anal pore
G-1 cell
G-2 cell
G-3 cell
G-4 cell

Fig. 14-21. *Wuchereria bancrofti*, microfilaria juvenile. G refers to a genital cell. (Redrawn from Fülleborn, from Brown, Selected Invertebrate Types. Courtesy of John Wiley & Sons, Inc.)

terioles of the lungs. If the host breathes oxygen at night, when the microfilariae are normally in the peripheral blood, the parasites accumulate in the lungs.[14] The circadian rhythm of body temperature probably plays a part in regulating this periodicity.

Like *Wuchereria bancrofti*, *Brugia malayi* has two forms: (1) a periodic form, which shows a marked nocturnal periodicity, and (2) a subperiodic form, in which a minor nocturnal peak occurs, but the larvae are present in peripheral blood continuously through each 24-hour period.[9]

Whatever the cause of the migration of young worms to the peripheral circulation, if they are there when man is bitten by a mosquito (*Culex, Anopheles, Aëdes, Psorophora, Mansonia*), they are taken into the stomach of the insect where they begin a period of growth and migration that usually requires about two weeks. During this time the microfilariae molt a few times and become elongated, infective, filariform larvae. They may migrate to various parts of the mosquito's body during their development, but most of them eventually reach the mouth parts where they are in a position to penetrate the skin or wound when the insect bites another man (or the original host). Apparently, larvae are dropped on the surface of the skin by the mosquito.

Unlike many other parasite life cycles that require an intermediate host, the filariae do not multiply in the mosquito. Development is intracellular within the muscle fibers, but tissue reaction is rarely observed. A light infection probably does the mosquito no harm, but a heavy infection is injurious and greatly increases mosquito death rate. As Edeson and Wilson[9] state: "Differences in the parasites may be just as important as differences in their vectors. Generalization can be most misleading. Each host-parasite relationship demands separate study."

The details of penetration of the larvae of *Wuchereria bancrofti* through human skin are still somewhat obscure. Experiments have been made by placing the infective larvae on intact skin, on scarified skin and on areas that have been pierced with a needle to simulate a mosquito bite. In each of these experiments the worm larvae failed to get through the skin. Apparently the mosquito is needed, indicating an absolute dependency on the intermediate host. See page 406 for a description of mosquitoes and their life cycles.

Worm larvae travel from the point of entry into human skin to the lymphatics, where they mature to adults. Of several manifestations of filarial infection, the best known begins with lymphangitis of the limbs or trunk, acute inflammation of the scrotum or its contents, and lymph node enlargement. The blocking of lymph vessels by masses of adult worms and growth of host tissue at the blocked area cause the swelling of extremities so characteristic of advanced cases of this type of filariasis. When the disease has caused the enlargement of such organs as the scrotum, breast, or legs, it is called **elephantiasis** (Fig. 14–22). Elephantiasis lesions tend to be irreversible.

Symptoms range from none to fever, tenderness of infected parts, eosinophilia, inflammation, and transient swelling. Anxiety caused by fear is of considerable importance. Thick blood smears are preferable in examinations for the microfilariae. Microfilaraemia, however, cannot appear at least for several months after the first bite of an infected mosquito.

Prevention involves the use of appropriate insecticides, mosquito control measures and moving from areas of infection. However, *Wuchereria bancrofti* live for ten years, making control by mosquito eradication difficult. Control by giving an entire population small amounts of the drug diethylcarbamazine has been suggested.[13] This type of procedure has proven to be partially successful against malaria in Brazil where chloroquine was added to cooking salt in local experimental communities.

Brugia malayi is the cause of widespread filarial disease of man in the Far East from Korea south through China, Philippines, Malaya, Indonesia and New Guinea, as well as Ceylon and India. It is transmitted by mosquitoes of the genera *Mansonia* and *Anopheles*, and occurs also in monkeys and cats.

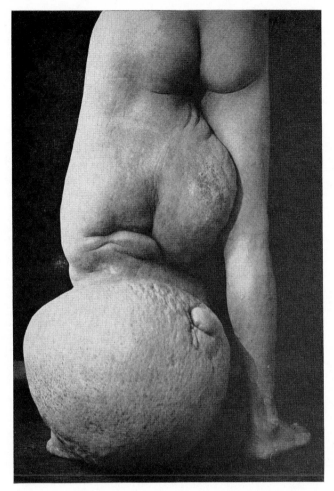

Fig. 14-22. Elephantiasis of the left leg. (Courtesy of the Mayo Clinic)

Clinical features include elephantiasis typically involving the feet and legs. For further details see Buckley[5] and Wilson.[32]

Closely related to *Brugia malayi* is *B. pahangi*, which is found principally in carnivores, but also in a wide range of other kinds of mammals, including primates. It has been reported from Malaya and East Pakistan, and has been transmitted to man experimentally. The mosquito vectors are species of *Mansonia*, *Aëdes* and *Armigeres*. The microfilaria, *B. patei*, has been found in dogs and cats in East Africa. For details of *B. pahangi*, see Buckley,[5] and Schacher.[28]

Onchocerca volvulus. This subcutaneous parasite (Fig. 14–23) causes the disease known as **onchocerciasis** in man in Africa and tropical America. The adult worms are usually found in fibrous skin tumors that appear as nodules almost anywhere on the body, but rarely on the lower parts of the legs. The life cycle starts with the fertilized egg within the body of the female. This zygote develops into a microfilaria (Fig. 14–24) that escapes the egg membranes and leaves the mother worm. Many thousands of microfilariae, 250 to 300 μ long, gather in the nodules, wander in connective tissues and find their way into superficial lymphatic vessels. Apparently, they do not get into blood vessels.

The insect vector is the black fly, *Simulium*, several species of which may bite man and so

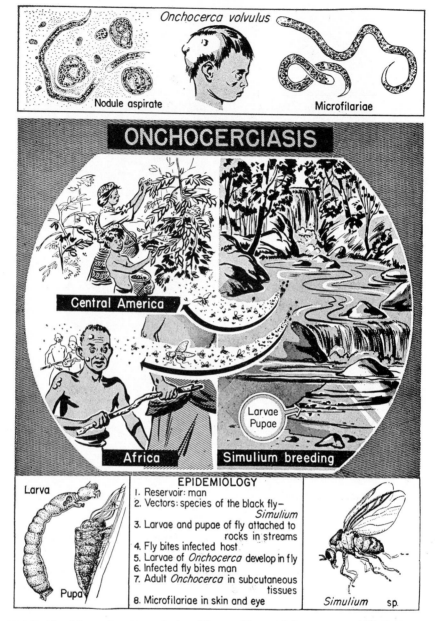

Fig. 14-23. Epidemiology of onchocerciasis. (Hunter, Frye and Swartzwelder, A Manual of Tropical Medicine. Courtesy of W. B. Saunders Co.)

pick up the microfilariae. Within the fly the young worms migrate to the thoracic muscles and develop into the rhabditiform, then the filariform stage in about a week. Filariform larvae migrate to other tissues, especially the proboscis, and when the fly bites again the filariae get back into human skin or subcu-

taneous connective tissue. Within a year mature worms appear in nodules, and the cycle is complete.

Adult males are 20 to 50 mm. long and the females may reach 700 mm. (almost 28 inches)[18] but they are usually shorter. The worms occur in a tangled mass in the nodules. A light infection consists of a single nodule, but occasionally over 100 nodules appear on one person. Usually these nodules contain degenerating worms, whereas healthy adult worms are more often found free in subcutaneous tissues.

The infection usually causes an itching dermatitis, but the original nodule is commonly painless. As the nodules increase in number and age they may fill with pus. They gradually become fibrous and eventually are calcified. Sometimes skin involvement is not accompanied by nodular formation. A serious infection occurs when the worms get into the eye. There is no satisfactory drug treatment, but surgical removal of the nodules usually has good results. Control consists of destroying black flies directly with insecticides and by eliminating their breeding places (see Burch[6]).

Several species of *Onchocerca* occur in gorillas, cows, deer, water buffalos, donkeys, camels and African antelopes.

Loa loa. This parasite (Fig. 14–25) is the

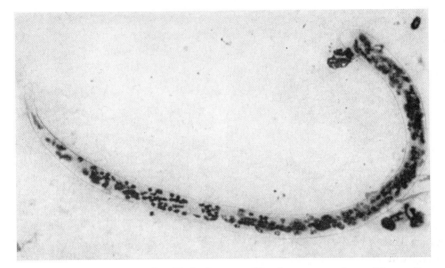

Fig. 14-24. *Onchocerca volvulus*, microfilaria from scarification preparation. (Photomicrograph by Zane Price, in Markell and Voge, Diagnostic Medical Parasitology. Courtesy of W. B. Saunders Co.)

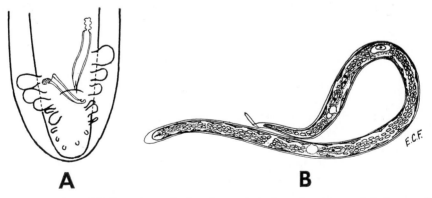

A   **B**

Fig. 14-25. *Loa loa.* (*A*) Posterior end of male, ventral view, showing caudal alae, papillae and copulatory spicules. × 180. (From Faust, after Yorke and Maplestone, Nematode Parasites of Vertebrates. Courtesy of J. and A. Churchill, Ltd.)

(*B*) Microfilaria of *Loa loa.* × 666. (From Faust, after Fülleborn, Arch. Schiffs Tropen-Hygiene; courtesy of Johann Ambrosius Barth.)

eye worm of Africa. The adult parasite averages about 50 mm. in length, and it lives in
the connective tissues of various parts of the
body of man, gorillas, and monkeys. Female
worms deposit sheathed embryos (microfilariae) that migrate to the bloodstream and
are taken up by the bite of a horse fly or
mango fly (*Chrysops*), which belongs to the
family Tabanidae. The fly feeds during all
daylight hours, but its biting peak is during
the middle of the day. Microfilariae are
ingested from a blood pool charged with an
anticoagulant from the fly. Apparently
little damage, if any, is done to the insect.[12]
The parasites become infective filariform larvae within the insect in about ten days and
they are transferred to man by the bite of the
fly. The worms probably require the best
part of a year to become mature, and they
wander in connective tissue, periodically
crawling over the eye just under the corneal
conjunctiva.

Monkeys may possibly serve as host
reservoirs for this worm. The parasite can be
transmitted by practically all species of
Chrysops, but the exact pathway of larval
migration either in the fly or in the mammalian host has not yet been determined. As
with many other larvae of nematodes, the
microfilariae of *Loa loa* are periodic in their
cycle. During the day the larvae remain in
the peripheral circulation where they can be
picked up easily by the blood-sucking fly,
whereas at night they are concentrated in lung
capillaries. How this periodicity is controlled
is still a mystery, but it does not completely
depend on such factors as oxygen tension,
CO_2 concentration or exercise, although
some physiologic factors are involved.

The infection is usually painless, but
allergic responses may occur. The worms
can be removed from the eye fairly easily.

Dirofilaria immitis. The "heartworm"
lives in the pulmonary artery and heart of
dogs and in some wild carnivores such as the
fox or wolf. It is transmitted by mosquitoes,
and is worldwide in distribution. The first
larval stage develops in the hemocoel of the
insect abdomen, and the second stage de-

velops in the hemocoel of the abdomen and
thorax. The mosquito seems not to be
affected by the parasite. The adult male
worm is about 14 cm. long, whereas the female
is about 27 cm. in length. See Winter[33] for
details of pathogenicity.

Immature *Dirofilaria* of various species may
infect man as conjunctival or subcutaneous
parasites. One species from the conjunctivae
has been named *D. conjunctivae*, a relatively
small filaria in which the caudal papillae of
the males exhibit a distinct asymmetry in
number and distribution. Occasionally adult
Dirofilaria have been reported from man,[4] an
abnormal host, but there is no evidence that
viable microfilariae are produced by female
Dirofilaria in man.

REFERENCES

1. Anderson, R. C.: The life cycle and seasonal transmission of *Ornithofilaria fallisensis* Anderson, a parasite of domestic and wild ducks. Can. J. Zool., *34*: 485–525, 1956.
2. Asami, K., *et al.*: Two cases of stomach granuloma caused by *Anisakis*-like larval nematodes in Japan. Amer. J. Trop. Med. Hyg., *14*:119–123, 1965.
3. Beaver, P. C., and Little, M. D.: The genital girdle in relation to estrus in *Ascaris lumbricoides*. J. Parasitol., *49* (Sect. 2, Supp.):46, 1963.
4. Beaver, P. C., and Orihel, T. C.: Human infection with filariae of animals in the United States. Amer. J. Trop. Med. Hyg., *14*:1010–1029, 1965.
5. Buckley, J. J. C.: Occult filarial infections of animal origin as a cause of tropical pulmonary eosinophilia. E. Afr. Med. J., *35*:493–500, 1958.
6. Burch, T. A.: The ecology of onchocerciasis. *In* May, J. M. (ed.): Studies in Disease Ecology. New York, Hafner, 1961.
7. Davey, K. G.: The food of *Ascaris*. Can. J. Zool., *42*:1160–1161, 1964.
8. Douglas, J. R., and Baker, N. F.: Some host-parasite relationships of canine helminths. *In* McCauley, J. E. (ed.): Host-Parasite Relationships. Proceedings of the Twenty-Sixth Annual Biology Colloquium, April 23–24, 1965. pp. 97–115. Corvalis, Oregon, Oregon State University Press, 1965.
9. Edeson, J. F. B., and Wilson, T.: The epidemiology of filariasis due to *Wuchereria bancrofti* and *Brugia malayi*. Ann. Rev. Entomol. *9*: 245–268, 1964.

10. Foor, W. E.: Electron microscopic studies of the refringent inclusion bodies in oocytes and fertilized eggs of *Ascaris lumbricoides*. J. Parasitol., *51* (Sect. 2, Supp.):49, 1965.

11. Galvin, T. J.: Development of human and pig *Ascaris* in the pig and rabbit. J. Parasitol., *54*:1068–1991, 1968.

12. Gordon, R. M.: The host-parasite relationships in filariasis. Trans. Roy. Soc. Trop. Med. Hyg., *49*:496–507, 1955.

13. Hawking, F.: Advances in filariasis especially concerning periodicity of microfilariae. Trans. Roy. Soc. Trop. Med. Hyg. *59*(1):9–25, 1965.

14. Hawking, F., and Gammage, K.: The periodic migration of microfilariae of *Brugia malayi* and its response to various stimuli. Amer. J. Trop. Med. Hyg., *17*(5):724–729, 1968.

15. Hawking, F., Pattanayak, S., and Sharma, H.: The periodicity of microfilariae. XI. The effect of body temperature and other stimuli upon the cycles of *Wuchereria bancrofti*, *Brugia malayi*, *B. ceylonensis* and *Dirofilaria repens*. Trans. Roy. Soc. Trop. Med. Hyg., *60*:497–513, 1966.

16. Hawking, F., and Worms, M.: Transmission of filaroid nematodes. Ann. Rev. Entomol., *6*:413–429, 1961.

17. Hulínská, D.: The development of the female *Enterobius vermicularis* and the morphogenesis of its sexual organ. Folia Parasitol., *15*:15–27, 1968.

18. Hyman, L. H.: The Invertebrates: Acanthocephala Aschelminthes and Entoprocta. Vol. III. New York, McGraw-Hill, 1951.

19. Inglis, W. G.: The oxyurid parasites (nematoda) of primates. Proc. Zool. Soc. London, *136*:103–122, 1961.

20. Kagan, I. G.: A review of immunologic methods for the diagnosis of filariasis. J. Parasitol., *49*:773–798, 1963.

21. Kessel, J. F.: The ecology of filariasis. *In* May, J. M. (ed.): Studies in Disease Ecology. New York, Hafner, 1961.

22. Lund, E. E., Wehr, E., and Ellis, D.: Role of earthworms in transmission of *Heterakis* and *Histomonas* to turkeys and chickens. J. Parasitol., *49* (Sect. 2):50, 1963.

23. Miyazaki, I.: On the genus *Gnathostoma* and human gnathostomiasis, with special reference to Japan. Exp. Parasitol., *9*:338–370, 1960.

24. Nelson, G. S.: Filarial infections as zoonoses. J. Helminth., *39*:229–250, 1965.

25. Read, C. P.: Nutrition of intestinal helminths. *In* Soulsby, E. J. L. (ed.): Biology of Parasites. pp. 101–126. New York, Academic Press, 1966.

26. Rogers, W. P., and Summerville, R. I.: The infective stage of nematode parasites and its significance in parasitism. *In* Davies, B. (ed.): Advances in Parasitology. Vol. 1, pp. 109–177. New York, Academic Press, 1963.

27. Sanhueza, P., *et al.*: Absorption of carbohydrates by intestine of *Ascaris lumbricoides in vitro*. Nature, *219*:1062–1063, 1968.

28. Schacher, J. F.: Morphology of the microfilaria of *Brugia pahangi* and of the larval stages in the mosquito. J. Parasitol., *48*:679–692, 1962.

29. Schwartz, B.: Experimental infection of pigs with *Ascaris suum*. Amer. J. Vet. Res., *20*:7–13, 1959.

30. Watson, B. D.: The fine structure of the body wall and the growth of the cuticle in the adult nematode *Ascaris lumbricoides*. Quart. J. Micr. Sci., *106*:83–91, 1965.

31. Williams, H. H.: Roundworms in fishes and so-called "herring-worm disease." Brit. Med. J., *1*:964–967, 1965.

32. Wilson, T.: Filariasis in Malaya. A general review. Trans. Roy. Soc. Trop. Med. Hyg., *55*:107–129, 1961.

33. Winter, H.: The pathology of canine dirofilariasis. Amer. J. Vet. Res., *20*:360–371, 1959.

Chapter 15

Class Adenophorea (=Aphasmidia), Orders Anoplida, Trichinellida, Dioctophymatida; Nematodes Parasitic in Plants

ORDER ANOPLIDA*

Superfamily Mermithoidea*

These slim, smooth nematodes range in size from a few millimeters to 50 cm. in length. They are free-living in the soil or in water as adults but are parasitic during the larval stages, especially in insects but also in other invertebrates. There is no buccal capsule, and the long esophagus (sometimes half the body length) proceeds directly from the mouth opening. The unusual intestine consists of two or more rows of enlarged cells filled with food reserves.

Agamermis decaudata. This species (Fig. 15–1) is parasitic in the body cavity of grasshoppers. Young worms, 5 to 6 mm. long in moist soil, enter grasshopper nymphs, but in so doing the worms usually leave half of

* This order and superfamily was not included in the table on page 279 because the list as prepared by Dr. MayBell Chitwood was not intended to include all the taxa.

their bodies outside. There is a node or breaking point at about the middle of the body, and at this point the two portions of the worm part company, the anterior half taking up its residence in the insect and the posterior half disintegrating in the soil. The necessary organs for continuing life and for reproduction accompany the anterior half, thus all is well. After remaining in grasshoppers for one to several months, the worms emerge through the body wall and lie in the soil during the winter. The next spring they mature, mate, produce eggs that hatch and liberate more young which enter other grasshoppers, and so the cycle is complete.

The burden of worms that a grasshopper may carry is enormous, as can be seen in the illustration, and considerable damage to the viscera may result in death. Some insects become infected by eating mermithid eggs. These parasites may infect crustacea, spiders and snails, in addition to insects.

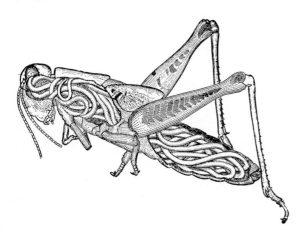

Fig. 15-1. Grasshopper infected with juvenile *Agamermis decaudata*. (Hyman, The Invertebrates, courtesy of McGraw-Hill Book Co.)

ORDER TRICHINELLIDA

Superfamily Trichinelloidea

The anterior portion of the body of these worms is filamentous, sometimes markedly so, while the posterior part is wider and often considerably shorter. The mouth is not surrounded by lips, the buccal capsule is small or rudimentary, the esophagus lacks a bulb and is a slender tube lying in a row of large cells called **stichocytes.** These stichocytes are collectively known as the **stichosome.** The anus is terminal in both the male and female. Males usually possess a single ensheathed spicule, but some of them lack a copulatory apparatus. The vulva is normally located near the junction of the two portions of the body. Most females are oviparous and the life cyle is usually direct, but a few species make use of an intermediate host.

Capillaria annulata. The cropworm (Fig. 15–2) of poultry uses an earthworm in its life cycle. Eggs are deposited in the mucosa of the crop and are freed when this layer sloughs off and is carried down the intestine. The eggs become embryonated in a few weeks and are eaten by earthworms. Poultry become infected by eating the infected earthworms. The larval nematodes are liberated in the bird's crop, penetrate the mucosa, and mature to adults that soon mate, and a new batch of eggs is produced. Adult males average about 15 mm. in length, whereas the females average about 40 mm. Since heavily infected

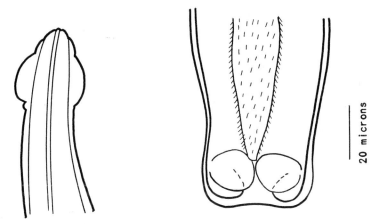

Fig. 15-2. *Capillaria annulata.* Anterior and posterior ends of the male. (Morgan and Hawkins, Veterinary Helminthology, courtesy of Burgess Publishing Co.)

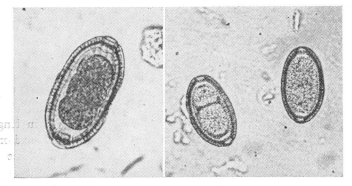

Fig. 15-3. *Capillaria hepatica.* Photomicrographs of the eggs of the parasite obtained from lesions produced by this parasite in the liver of a rat. (Gradwohl, Kouri, Clinical Laboratory Methods and Diagnosis, courtesy of C. V. Mosby Co.)

birds do not eat well, they lose weight, become thin and often die.

Capillaria hepatica is a liver-worm that infects rodents, ground squirrels, rabbits, beavers, monkeys and other animals, and occasionally man. The worm sometimes causes hepatic cirrhosis, but probably it is not especially pathogenic. The brown, pitted, thick-shelled eggs (Fig. 15–3) (about 50 μ in diameter) are discharged in the liver but do not develop further unless freed from the liver and exposed to air. A new host acquires the infection by eating decomposed infected liver or the feces of an animal that has eaten infected liver. In the new host the eggs hatch in the intestine and the young larvae burrow through the intestinal wall and make their way to the liver, where they mature. Adult males are about 4 mm. long and 100 μ wide, whereas the females are twice as wide and about 10 mm. long.

Trichinella spiralis. The trichina worm, *Trichinella spiralis* (Figs. 15–4, 15–5), causes the widespread disease *trichinosis*. There has been a significant decline in the prevalence of this disease over the past ten years in both man and hogs. At least 30,000,000 people in the United States, however, are infected with the larvae of this worm. Some estimates are as high as 60,000,000. It is especially prevalent in Europe and the United States and, curiously enough, rare or absent in most of the Orient and Australia.

Adult female worms are 3 to 4 mm. long × 60 μ wide. Adult males average about 1.5 mm. long × 45 μ wide and possess two large conical posterior papillae. The worms live in the intestinal mucosa of the duodenum and jejunum of man, pigs, wild boars, rabbits, walruses, rats, beavers, raccoons, skunks, seals, bears, polar bears, ermine, wolves, wolverines, lynx and many other mammals. The incidence of infection in arctic marine mammals is not high. Experimental infections of adults have been established in birds. The parasites show typical nematode structures, and they differ from many roundworms in that the females do not lay eggs.

Young worms develop within the slender

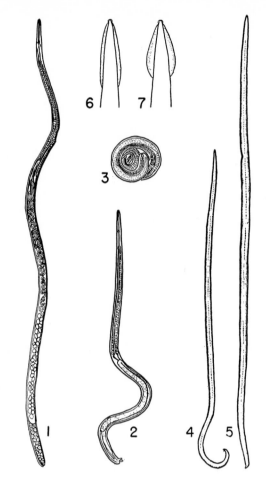

Fig. 15-4. *Trichinella spiralis* adults compared with *Ascaris* and *Toxocara*. *1–3. Trichinella spiralis. 1,* Adult female, × 67. *2,* Adult male, × 67. *3.* Infective stage larva, × 97. *4* and *5. Ascaris lumbricoides. 4.* Male with typical flexion of posterior end, × ⅜. *5.* Female, × ⅜. *6* and *7. Toxocara* spp. Anterior end showing cervical alae. *6. T. canis,* × 7½. *7. T. cati (mystax),* × 7½. (Beaver, courtesy of Amer. J. Clin. Path.)

female, and molting occurs during intrauterine development, resulting in the formation of second stage larvae.[1] The larvae average about 100 μ in length when they emerge from the vulva, and one adult female may produce 1500 in three to four weeks. One theory suggests "that because only two molts occur during the life cycle of *T. spiralis,* the so-called adults are actually third stage neotenic larvae."[17] These young worms

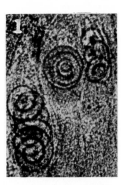
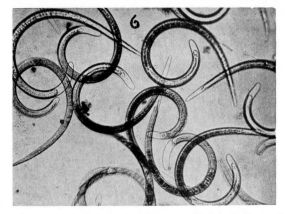

Fig. 15-5. *Trichinella spiralis.* Experimental infestation of white rat. (Original photomicrographs [of Kouri] of material supplied by Dr. G. Bachman, Puerto Rico.) *Left.* Larvae in diaphragm muscle. *Right.* Larvae freed by artificial digestion of muscle. (Gradwohl and Kouri, Clinical Laboratory Methods and Diagnosis, courtesy of C. V. Mosby Co.)

burrow through host tissues to the venous blood vessels, and connective tissue, thence to all parts of the body. They may leave the vessels and enter any organ, and they have been recovered (in hogs) from the stomach wall, testes, liver, brain, lungs, intestinal wall, pancreas, wall and contents of the urinary bladder, heart and spinal cord (Fig. 15–5). They may even pass through the placenta and into a fetus or from a mother to her young while suckling. The vast majority of the larvae, however, enter striated muscles, especially the diaphragm, intercostals, extraocular, lingual, larynx, pectoral, biceps, deltoid, gluteus and gastrocnemius. Here they grow and molt and, unless eaten by another host, they gradually become encapsulated and eventually calcified. In the diaphragm of one human case, 1000 larvae per gram of muscle were found. They may live for six months or as long as 30 years.

The life cycle can continue only if the infected muscle is eaten by another host (Fig. 15–6). There are a variety of infection chains, such as rat to rat, pig to pig, pig to rat. Man usually becomes infected by eating uncooked pork. In the small intestine of man the larvae are liberated and grow to sexual maturity in two days. There is a two-way chemical attraction between males and females.[2] The males probably die soon after

copulation, and the viviparous females live for about a month, producing young most of the time. Adult worms are essentially harmless.

Symptoms of infection in man usually appear during the second week after he consumes poorly cooked infected pork. These symptoms include headache, hemorrhages under the skin, fever, difficult breathing, edema, soreness of the infected muscles, and eosinophilia. There is no specific drug against trichinosis. Prevention consists of cooking meat (especially pork) thoroughly to kill the larvae. Quick freezing (lowering the temperature immediately to $-35°$ C.) apparently kills most of the larvae, and ionizing radiation (x-rays, gamma rays, high energy electrons) also seems to be effective. Meat scraps fed to hogs should be sterilized, and rats and mice should be eliminated.

Intradermal skin tests aid in diagnosis. One serologic test consists of placing living *Trichinella* juveniles in the blood serum from a person suspected of harboring the parasite. If antibodies are present, a precipitate forms around the worms, especially at body openings. For details, see Kagan.[5] Eleven antigens have been reported,[19] but antigens from adult worms are qualitatively different from those from larvae.[13] Partial immunity to trichinosis has been gained in experimental

LIFE CYCLE OF TRICHINELLA SPIRALIS

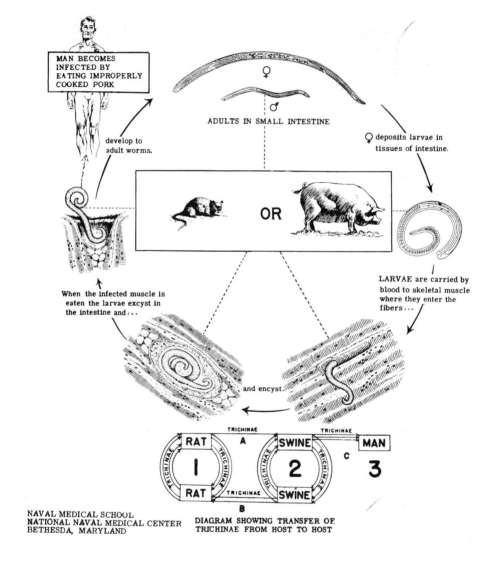

Fig. 15-6.   Life cycle of *Trichinella spiralis*.   (Courtesy of National Naval Medical Center, Bethesda, Maryland.)

animals by such devices as intraperitoneal injections of living larvae, heat-killed larvae or dried and powdered larvae, and subcutaneous injections of serum from heavily infected rabbits. Acquired immunity begins with reactions to specific antibodies. Tissue damage to the host leads to nonspecific allergic inflammation, and may be followed by expulsion of adult worms.[9]

Trichuris trichiura. This worm is one of the most common helminth parasites of man, especially in tropical and subtropical regions, and apparently the same species lives in pigs and monkeys. Similar species may be found in many other animals. *Trichuris trichiura* eggs found in the gut contents of the frozen body of an Inca girl in Chile, buried at an elevation of 17,658 feet at least 450 years ago,

indicate the presence of this parasite in South America before the Spanish conquest, and raise an interesting question concerning the possible New World origin of *Trichuris*. It is worldwide in its distribution.

Trichuris trichiura (Figs. 15–7, 15–8) is commonly known as the "whipworm" because of the characteristic shape of its body. The anterior three-fifths is a capillary tubule that contains the esophagus; the posterior two-fifths are more fleshy and contain the intestine and sex organs. The esophagus has a reduced musculature and is associated with a row of large secretory cells, the stichocytes.

Male worms have a ventrally coiled caudal extremity and measure 30 to 45 × 0.6 mm. A single spicule protrudes through the retract-

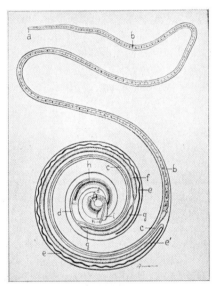

Fig. 15-7. Male *Trichuris trichiura.* (After Guiart, modified by Kouri in "Lecciones de Parasitologia y Medicina Tropical." Courtesy of Editorial Profilaxis S.A., Havana, Cuba.) *a, b, c, d.* Digestive tract: *a,* mouth in the anterior end of the parasite; *b,* esophagus occupying the entire length of the thin portion of the parasite and formed by a fine duct which passes through a single layer of cells; *c,* intestine; *d,* cloaca. *e, e', f, g, h, i.* Male genitalia: *e,* testicle; *e',* vas deferens; *f,* seminal vesicle; *g,* ejaculatory duct; *h,* spicule; *i,* sheath or thorny prepuce. (Gradwohl and Kouri, Clinical Laboratory Methods and Diagnosis, courtesy of C. V. Mosby Co.)

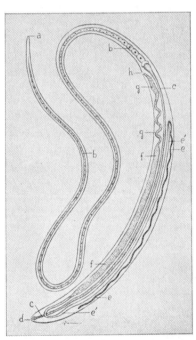

Fig. 15-8. Female *Trichuris trichiura.* (After Guiart, modified by Kouri in "Lecciones de Parasitologia y Medicina Tropical." Courtesy of Editorial Profilaxis S.A., Havana, Cuba.) *a, b, d, d.* Digestive tract: *a,* mouth in the anterior end of parasite; *b,* esophagus in the thin portion, formed by a fine duct which passes through a single layer of cells; *c,* intestine; *d,* anus. *e, e', f, g, h.* Female genitalia, simple; *e,* ovary; *e',* oviduct; *f,* uterus filled with eggs; *g,* long and sinuous vagina; *h,* vulva. (Gradwohl and Kouri, Clinical Laboratory Methods and Diagnosis, courtesy of C. V. Mosby Co.)

able sheath at the posterior end of the body. The sheath possesses a bulbous end and has numerous recurved spines. The female measure 35 to 50 × 0.7 mm. The vulva lies at the anterior extremity of the fleshy portion of the body. Eggs are barrel-shaped, brown in color, with an outer and an inner shell and transparent polar prominences. Eggs measure 50 to 54 × 23 μ.

The life cycle begins with eggs that are deposited in the cecum, appendix, ileum, colon or rectum, where the adults live. The eggs pass from the host body to the soil with feces and they develop embryos within a few weeks. These eggs may remain viable outside of the host for many months if they lie in moist areas, but development of the larvae within them may be delayed by dryness or cold. When a favorable weather change occurs, the bulk of the accumulated organisms continues development, thus making possible a massive infection. Embryonated eggs are infective to a new host, and they gain entrance to this host by being eaten. Obviously the more chances there are for fecal contamination of food and water, the greater the incidence of infection. When ingested, eggs are passed to the host cecum, where they hatch and where the young larvae burrow into the intestinal wall. After a few days they leave the intestinal wall and may go to some other part of the intestine to mature. In a few months, mature worms are ready to mate and produce eggs, thus completing the cycle. (See also Fig. 12–8(1), p. 274.) Four molts are required from egg to adult.

In a light infection, symptoms are usually absent. Rarely there may be digestive disorders and anemia, toxic disturbances, intestinal obstruction or even perforation of the gut wall. Sometimes there are symptoms that are similar to those of hookworm infection. Whipworms are avid blood suckers.[3,10] A heavy infection may cause inflammation, eosinophilia, hemorrhage, anemia, diarrhea, blood and mucus in feces, dyspnea, nausea, loss of weight, abdominal pain, fever and prolapse of the rectum. Secondary bacterial infection may occur. Diagnosis is confirmed by finding the eggs in feces. Sanitary sewage disposal and personal cleanliness are obvious important preventive measures.

ORDER DIOCTOPHYMATIDA

Superfamily Dioctophymatoidea

These moderate to long worms each possess a mouth lacking lips but surrounded by papillae. There is no bulb on the esophagus. Males possess a bell-shaped bursa without rays. Females lay eggs with thick, pitted shells and end-plugs. Birds and mammals are the primary hosts. Some of the worms may use fish as intermediate hosts. *Eustrongylides*, *Hystrichis* and *Dioctophyma* are genera of particular interest to parasitologists. The latter will serve to illustrate the order.

Dioctophyma renale. Since the males range in length from 140 to 450 mm. and the females range from 200 to 1,000 mm. in length, the common name of the parasite is the giant kidney worm. These reddish worms may be found in the kidneys of dogs, wolves, raccoons, minks, weasels, otters, foxes, martens, seals, and other mammals in the United States and the Orient. Apparently the parasites occupy the right kidney much more frequently than the left one. Destruction of the kidney may be complete.

Although the life cycle has been described as involving two intermediate hosts, a leech-like oligochaete worm and the black bullhead, *Ameiurus melas*, experimental studies by Karmanova[7] have shown that a fish may be a reservoir host, but is not necessary for completion of the cycle. Dogs are infected with larvae of *Dioctophyma renale* that hatch from eggs ingested by branchiobdellid worms.

NEMATODES PARASITIC IN PLANTS

Most of the important plant nematodes belong to the ORDER TYLENCHIDA, commonly included only in a course in plant parasitic diseases. They are considered here at the end of the nematode discussion because the ORDER DORYLAIMIDA also contains parasites that live in or on plants. In this brief presentation only the genera *Xiphinema* and *Trichodorus* have been selected from the order Dorylaimida. All the other genera belong to the order Tylenchida.

The nematode parasites of plants are often called "eelworms" (Fig. 15–9). They do enormous damage to cultivated plants and cause the loss of many millions of dollars a year to farmers. This loss is due to the direct damage done by the worms and by other organisms that enter the damaged areas.

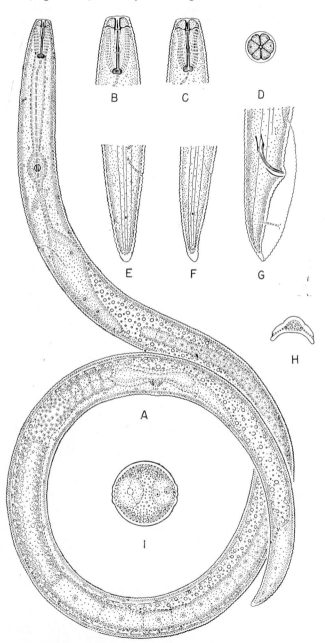

Fig. 15-9. *Pratylenchus vulvus.* Structure of root-lesion nematode. *A.* Adult female. *B* and *C.* Heads of females. *D.* Face view of female. *E* and *F.* Female tails. *G.* Male tail. *H.* Diagrammatic cross-section through male tail. *I.* Cross-section through female body. Greatly magnified. (Maggenti and Allen, courtesy of Calif. Agriculture, University of California, Division of Agricultural Sciences.)

Nematodes may even transmit plant viruses. Fanleaf of grapes, for example, is a worldwide virus disease spread by the dagger nematode, *Xiphinema index*.

In structure, plant nematodes are much like the familiar ascarid worms in vertebrate digestive tracts, except they are smaller. The smallest are only about 400 μ long, while few reach a length much over 3 to 4 mm. One of the most characteristic structures in these worms is a stylet, or spear, in the anterior end. This structure apparently helps the parasites to penetrate plant tissues. Life cycles consist of eggs, larvae and adults, but the details vary considerably depending on the species of nematode, species of plant host, and temperature and other environmental factors. In general, the female worm lays eggs that hatch either in the soil or in the host plant. If the host plants are not available, the eggs frequently will not hatch but will remain dormant for years. Even larval stages of some forms can remain alive for a surprisingly long time. The larvae of the wheat nematode *Anguina tritici*, for example, can live in galls for 20 years. In any case, once nematode larvae get into a plant, they begin to feed on plant tissues. Probably all crop and ornamental plants are attacked by nematode parasites.

Some nematodes are ectoparasitic on parts of plants that appear above the ground. *Aphelenchoides besseyi*, for example, feeds on leaves or on the developing buds. This parasite may also be endoparasitic, feeding on tissues within the stems and leaves of the strawberry plant. *A. ritzema-bosi* may even be ectoparasitic on some parts of gooseberry or blackberry plants and endoparasitic on other parts. Apparently the host plant determines the nature of the parasitism. Gall formation is one of the responses to endoparasitic activity. Galls may be formed in stem tissues or in a leaf or even in flowering tissues. *Anguina* spp. will produce galls in various flowering and other parts of plants. *Ditylenchus dipsaci* is a stem and bulb nematode that lives in the stems of wheat, alfalfa, potato, and leaves of onion, daffodil, garlic and other plants. Intercellular lamellae of the host plant break down and the tissues become loose and spongy. Secondary infection may be produced by bacteria or fungi. *D. dipsaci* may live for as long as 21 years in the dry state.

Underground parts of plants may be

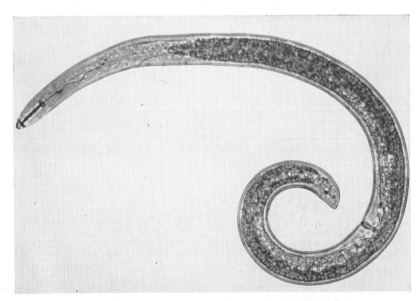

Fig. 15-10. Mature female of *Helicotylenchus dihystera*—a plant pathogenic nematode. (Courtesy of W. T. Mai, Cornell University.)

attacked by either ectoparasites or endo-parasites or by both. The ectoparasitic worms simply lay their eggs on or near roots and when young worms hatch they start to feed on the tender parts of roots, such as the tip. *Trichodorus* and *Xiphinema* are two genera whose species behave in this manner. Plant roots swell and the plant becomes stunted or dies. These worms ordinarily do not actually enter plant tissues. Many other species do, however, penetrate into the underground parts of plants. *Pratylenchus* (Fig. 15–11), *Radopholus*, *Ditylenchus* and *Helicotylenchus* (Fig. 15–10) are genera whose species get into roots and usually lay their eggs inside the plant.

The biochemical aspects of the method of injury to potato stems by the nematode *Ditylenchus destructor* have been analyzed by Myuge.[12] He kept the worms for one day in water and then analyzed the water for enzymes and other substances. He found that the water contained amylase and proteo-lytic enzymes that contained sulfhydril groups and that were especially active in a weakly-acid medium. Apparently the nature of necrosis caused by the worms consists of protein coagulation due to ammonia poison-ing, consequent dehydration and protein oxidation. Starch hydrolysis leads to dis-ruption of natural osmosis followed by a dehydration of affected tuber tissues. An increase in size of plant cells bordering on the focus of invasion also occurs.

Hypotheses have been advanced suggesting that enzymes in nematode saliva induce changes in cell division and morphology. Sayre, . . . for example, suggests that *Meloidogyne* larvae secrete a proteolytic enzyme which makes available essential amino-acids for nematode growth and at the same time releases indolacetic acid in root tissues. Indolacetic acid might also be released by proteolytic activity on structural protein. The splitting of peptide bonds on the protein chain would release a number of different amino-acids including tryptophane which is a precursor of indolacetic acid. Thus a high level of indolacetic acid would accumulate in the tissues around the nematode and induce cellular changes and gall formation. This is an attractive hy-pothesis for there is evidence that nematodes can secrete proteolytic enzymes, that amino-acids do accumulate in *Meloidogyne* galls, and, as Krusberg . . . has shown, free tryptophane occurs in lucerne galled by *Ditylenchus dipsaci* but not in healthy plants. Krusberg suggests that tryptophane could easily be converted to indolacetic acid by nematode or plant enzymes and so induce the galling reaction. Further weight is added to the hypothesis by the fact that plant growth hormones induce symptoms in plants similar to those associated with nematode invasion and that indole compounds

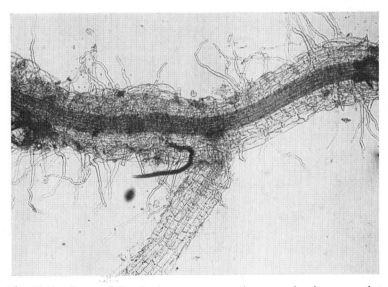

Fig. 15-11. Larva of *Pratylenchus penetrans* entering an orchard grass rootlet. (Courtesy of W. T. Mai, Cornell University.)

are present in the galled tissues of *Abelmoschus esculentus* infested with *Meloidogyne javanica*. . . . Hypotheses have also been advanced to explain other plant reactions to nematode infestation and, as Dropkin . . . has pointed out, histo-chemical methods and enzyme analysis will help to give the necessary information. It is along such lines that future work on histopathology will move.[22]

Some plant nematodes are host specific while others seem to have little preference for host plants. Some groups (*Aphelenchoides* and *Ditylenchus*) possess representatives that are even able to live and reproduce on fungi as well as on higher plants. Plants, like animals, can be resistant to parasites, and various strains resistant to certain nematodes have been developed.

Family Heteroderidae

This important family of destructive plant parasites consists of many species belonging to three genera that can be identified according to the following key.[20]

1. Female, a white saccate body with thin cuticle, eggs ejected into a mucoid mass about the posterior end; frequently gall-forming
   Meloidogyne Goeldi, 1887

Female, a tough, white saccate body, retaining most or all of eggs; not gall-forming

2. Body becoming a brown cyst at maturity, vulva near terminus
   Heterodera Schmidt, 1871

Body remaining white, vulva slightly posterior to middle of body
   Meloidodera Chitwood *et al.*, 1956

Meloidogyne is one of the many genera that typically forms galls, but in some species galls do not appear. The galls have given rise to the name "root-knot" nematodes for the group. Thousands of plants in a small area may be attacked, resulting in wilting or death. Damage is done to many farm and garden plants including potatoes, peas, cereals, tomatoes, sugar beets, beans, clover, water-melon and decorative plants. Unfortu-nately, the presence of the worm parasites is not usually known until they occur in such large numbers that serious damage is well under way. *Meloidogyne marioni* (Fig. 15–12) is an example of one species that causes great damage among cultivated plants.

Eggs usually range from 38 to 90 μ in length, depending on the species. The first larval molt occurs within the egg, which

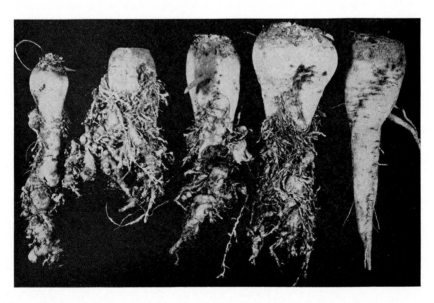

Fig. 15-12. *Meloidogyne.* Root-knot nematode injury to sugar beets. Beet on right from soil treated with a nematocide. (Thomason and Lear, courtesy of Calif. Agriculture, University of California, Division of Agricultural Sciences.)

hatches into worm-like larvae about 0.5 mm. long, with rounded heads and pointed tails. If these preparasitic larvae are in the soil, they crawl about without feeding until they find a suitable plant, and then they penetrate large roots, rhizomes or tubers. These second-stage larvae use a piercing organ, or stylet, to penetrate the roots and to make their way to softer tissues near the center of the root. Here the worms molt three times, gradually increasing in size and in sexual dimorphism. The host reacts by forming giant cells on which the larvae feed. Hypertrophy of host tissue cells and the giant cells start the formation of the gall, which is associated with disruption of the xylem. The whole infested area becomes distorted with the ball or knot of diseased tissue within which may be found all stages of the parasite, from eggs to adults. Enlarged females become spherical or pear-shaped and may reproduce parthenogenetically. They may live in a gall or in other tissues. Young roots with heavy infestations may be covered with female worms that are attached to the host. At the posterior end of each worm is an egg mass in a gelatinous egg-sac. Thus, larval worms may be produced either within a plant, and migrate to other tissues, or may hatch in soil and attack other plants.

Heterodera is a genus of cyst formers. The female cuticle of these worms is transformed into a tough, brown, resistant cyst (Fig. 15–13) within which the eggs develop into larvae. Eggs average 46 × 100 μ. The first molt of the developing worm occurs within the egg. Larvae are liberated from the cyst and penetrate plant roots in which, after a series of molts, they mature. Some females swell and become lemon-shaped. The head and neck region remain embedded in host tissue while the rounded body, which becomes the brown cyst, is on the outside. The cyst "hatches" in one to several years and the larvae enter tender plant roots. Infected plants die.

The potato root eelworm, or golden nematode, *Heterodera rostochiensis;* the sugar beet eelworm, *H. schachtii;* the cereal root eelworm,

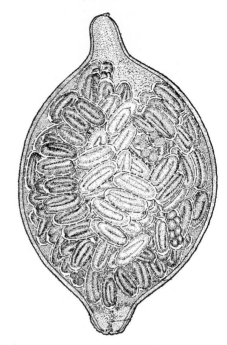

Fig. 15-13. *Heterodera.* Sugar-beet nematode cyst containing eggs   Maggenti and Allen, courtesy of Calif. Agriculture, University of California, Division of Agricultural Sciences.)

H. major; and the pea eelworm, *H. gottingiana;* are examples of plant nematodes of economic importance. The potato root eelworm causes widespread damage and is a typical example of the group. Males average a little over 1 mm. in length and are slender and worm-like whereas the females become shortened and flask-shaped. There is a head-like projection and a sac-like body which is the "cyst." These cysts average about 0.6 mm. in diameter. Both males and females gradually become extruded to the root surface, and when development is complete the males usually leave the host plant and live an independent life in the soil. The mature sac-like females often remain attached to the plant by their head ends (Fig. 15–14). Mating occurs at this stage and the sac becomes filled with eggs. The females die and the sac turns dark in color, becomes tough and resistant, and is now a cyst, often containing several hundred eggs. Detached cysts remain in the soil ready to open under the right condi-

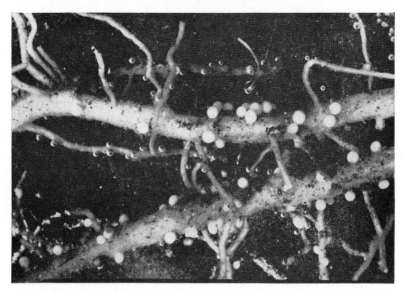

Fig. 15-14.   Immature females of the golden nematode *Heterodera rostochiensis* attached to potato roots. (Courtesy of W. T. Mai, Cornell University.)

tions, liberating the infective larvae which are about 450 μ long.   Hatching depends on many factors, but apparently the presence of other potato plants plays a role.   There seems to be a diffusate from the roots of near-by potato plants which stimulates hatching.   The life cycle normally requires five to seven weeks and a single generation is produced each year.

The golden nematode can take up to 120,000 roentgens or more of radiation without being killed.   The lethal dose for man is about 650 roentgens.   The nematode can thus survive after being irradiated with a dose strong enough to kill 180 men.   Some other plant nematodes can withstand doses higher than 600,000 roentgens.

Soil nematodes face many dangers besides desiccation and inability to find food or a proper host.   Numerous species of soil fungi capture nematodes by producing adhesive "buttons" to which nematodes adhere. Others form mycelial rings in which the worms become enmeshed.   In either case the fungi invade the captured worms and digest them, a process that takes about 48 hours.

Soil fumigants increase crop yield enormously by killing the nematodes, but, of course, their effectiveness is limited to those parasites that are in the soil.

For a detailed account of nematodes parasitic in plants, see Christie,[4] Krusberg,[8] Thorne,[20] and Webster.[23]   For a comparison of plant-parasitic nematodes with others, see Peters.[16]

IMMUNITY

Nematodes do not undergo multiplication in host tissue, so the opportunities to produce a wide variety of antigens is not as great as with many other parasites.   Resistance is, of course, due to many factors in addition to the production of antibodies.   Active acquired immunity to *Ascaris suum* infection in pigs and rodents is due primarily to antibodies with secondary cellular cooperation.[18]

In general, mature hosts acquire good immunity against their gastrointestinal roundworm parasites whereas young hosts do not. Delay in the production of antibodies by young hosts is probably an important factor. In any case, the types of reaction depend on the nature of antigen and degree of parasitic infection.   As with other parasites, tissue-dwelling forms evoke a greater response than lumen forms.   Some antigens are shared between larvae and adult worms and some are not. Even with one species of worm in one host there may be several antigen-antibody systems in operation.

One commonly observed reaction to the host's immune response is the inhibition of the development of a worm. The system is reversible since removal of the inhibiting mechanism (e.g., removal of worms to another host) results in continuation of development. The origin of the antigens is still somewhat obscure but it is clear that the exsheathing fluid is an important source in developing worms.

Only a few of the many species of nematodes being studied for antigen-antibody reactions or other types of host resistance can be mentioned here. *Onchocerca volvulus*, which causes onchocerciasis in man, evokes active immunization by the production of antibodies that react with other than the specific antigen (heterophilic). *Dirofilaria uniformis* induces host production of several types of antibodies. *Trichinella spiralis* evokes two homocytotropic antibodies in host serum. Diagnosis of *Trichinella* in man and animals can be made by using a soluble antigen fluorescent antibody test.

There is considerable variation in the nature of the response of worms to antibodies. With *Nematospiroides dubius* in mice, immunity seems to be aimed at removing early parasitic stages. Immunity is shown by reduced worm burdens, altered distribution of larval and adult worms, stunting of adult worms and more intense cellular reactions to encysted larvae.[15] With *Nippostrongylus brasiliensis* in rats, the most noticeable changes are resorption of spermatozoa, a drop in egg production, changes in the cytology of the worm intestinal cells, and the appearance of many large droplets, possibly lipid, in the cytoplasm of various worm organs.[11] With pinworm of mice, *Syphacia obvelata*, on the other hand, striking age resistance is developed, but no resistance of a more specific nature or immunity. This situation is probably due to the lack of contact between host issues and the worms.[14]

One technique for determining the presence of immune bodies in host serum is to place a drop or two of the serum on a slide, add some of the parasites (e.g., nematode larvae), protect them with a sealed cover glass and incubate. Immune serum is indicated by the formation of precipitins in any of the physiologic openings of the worms. In larval nematodes these openings would be the buccal cavity, anal pore and excretory pore.

The esophagi of hookworms contain a substance that has a proteolytic action. It is possible to dissect the esophagi from these worms, grind them in saline with a tissue grinder, and extract the enzyme. When this enzyme from the dog hookworm, *Ancylostoma caninum*, is injected into dogs, these animals become at least partially immune to a challenge infection with hookworm. This immunity is indicated by the fact that a second infection is light, the worms being fewer in number and smaller than in a first infection. The experiment indicates that antibodies acting as antienzymes are formed in hosts as a response to substances that possess enzymatic activity in the secretions and excretions of worms. Immunity results from the inhibition of enzymatic activity in the parasite.[21]

REFERENCES

1. Berntzen, A. K.: Comparative growth and development of *Trichinella spiralis in vitro* and *in vivo* with a redescription of the life cycle. Exp. Parasitol., *16*:74–106, 1965.
2. Bonner, T. P., and Etges, F. J.: Chemically mediated sexual attraction in *Trichinella spiralis*. Exper. Parasitol., *21*:53–60, 1967.
3. Burrows, R. B., and Lillis, W. G.: The whipworm as a blood sucker. J. Parasitol., *50*:675–680, 1964.
4. Christie, J. R.: Plant Nematodes. Gainesville, Univ. Florida Agric. Expt. Sta., 1959.
5. Kagan, I. G.: Trichinosis: A review of biologic, serologic and immunologic aspects. J. Infect. Dis., *107*:65–93, 1960.
6. ——: A review of immunologic methods for the diagnosis of filariasis. J. Parasitol., *49*: 733–798, 1963.
7. Karmanova, E. M.: The life cycle of the nematode *Dioctophyme renale* (Goeze, 1782), a parasite in the kidneys of carnivora and of man. Doklady Akad. Nauk SSSR, *132*:1219-1220, 1960.

8. Krusberg, L. R.: Biology of plant-parasitic nematodes. J. Parasitol., *48*:826–829, 1962.

9. Larsh, J.: The present understanding of the mechanism of immunity to *Trichinella spiralis*. Amer. J. Trop. Med. Hyg., *16*:123–131, 1967.

10. Layrisse, M., Aparcedo, L., Martinez-Torres, C., and Roche, M.: Blood loss due to infection with *Trichuris trichiura*. Am. J. Trop. Med. Hyg., *16*:613–615, 1967.

11. Lee, D. L.: Changes in adult *Nippostrongylus brasiliensis* during the development of immunity to this nematode in rats. Parasitology, *59*:29–39, 1969.

12. Myuge, S. G.: Trophical characteristics of potato stem nematodes. Abs. J. Referat. Zh. Biol., No. 1, 832. Orig. Pub: Izv. AN SSSR, Ser. biol. No. 3, 357–359, 1958.

13. Oliver-Gonzalez, J.: Seminar on immunity to parasitic helminths. III. Serological studies on stage specificity in *Trichinella spiralis* Exper. Parasit., *13*:13–17, 1963.

14. Panter, H. C.: Studies on host-parasite relationships: *Syphacia obvelata* in the mouse. J. Parasitol., *55*:74–78, 1969.

15. ———: Host-parasite relationships of *Nematospiroides dubius* in the mouse. J. Parasitol., *55*:33–37, 1969.

16. Peters, B. G.: Plant-parasitic nematodes compared with others. Nematologia *10*:1–4, 1964.

17. Shanta, C. S., and Meerovitch, E.: The life cycle of *Trichinella spiralis*. II. The muscle phase of development and its possible evolution. Can. J. Zool., *45*:1261–1267, 1967.

18. Taffs, L. F.: Immunological studies on experimental infection of pigs with *Ascaris suum* Goeze, 1782. VI. The histopathology of the liver and lung. J. Helminth., *47*:157–172, 1968.

19. Tanner, C. E.: Immunochemical study of the antigens of *Trichinella spiralis* larvae. II. Some physiochemical properties of these antigens. Exp. Parasitol., *14*:337–345, 1963.

20. Thorne, G.: Principles of Nematology. New York, McGraw-Hill, 1961.

21. Thorson, R. E.: The stimulation of acquired immunity in dogs by injections of extracts of the esophagus of adult hookworms. J. Parasitol., *42*:501–504, 1956.

22. Wallace, H. R.: The Biology of Plant Parasitic Nematodes. London, Edward Arnold Ltd., 1963.

23. Webster, J. M.: The host-parasite relationships of plant-parasitic nematodes. *In* Dawes, B. (ed.): Advances in Parasitology. vol. 7, pp. 1–40. New York, Academic Press, 1969.

Section VI
PHYLUM ARTHROPODA

Chapter 16

Introduction; The Crustacea

INTRODUCTION

Members of this phylum possess an exoskeleton with jointed appendages. The body is divided into segments that are variously grouped into functional units such as head, thorax and abdomen; the digestive tract is complete and the circulatory system forms a hemocoel which is the body cavity. The coelom is reduced. Respiration is accomplished by tracheae, gills, book lungs or body surface. Malpighian tubules serve as excretory organs in most species. The brain is dorsal whereas the main nerve cord is ventral in position. Paired ganglia usually occur in each somite, and eyes are simple or compound. Sexes are separate.

Some crickets puncture stems of raspberry plants and lay their eggs within the pith. Are these crickets more parasitic than those grasshoppers and crickets that simply chew the stems? No more forceful way to emphasize the difficulty of defining "parasite" can be found than to select a textbook on entomology and to read about the multitudinous methods by which insects have solved the problem of obtaining food. Among the most destructive plant pests are the chinchbugs, plantlice, mealybugs and other Hemiptera whose sucking mouth parts, extraordinarily rapid rate of reproduction, and general behavior combine to effect a devastating invasion of plants the world over. But are they not simply plant feeders instead of parasites? The ovipositor of the female of many leaf hoppers is adapted for lacerating plant tissues, and eggs are deposited in longitudinal rows on the stems or under the leaf sheath. Certainly this habit is as much a parasitic one as are the habits of lice.

Some general references are Borror and DeLong,[3] Chapman,[6] Clausen,[7] Furman,[13] Herms,[14] and Imms.[17]

Respiration

Both endoparasitic insects and free-living aquatic insects must be able to extract dissolved oxygen from the surrounding liquid or semiliquid medium, or to retain a connection with an atmospheric supply. Some of the most common devices employed by aquatic larvae, however, are not known to occur among those of endoparasitic habits. In parasitic forms with a closed tracheal system, respiratory exchange occurs through a thin cuticle that covers a voluminous tracheal supply. Some early instars of parasitic insects are either atracheate or they have a tracheal system filled with fluid, requiring oxygen diffusion directly into the hemolymph. The paired, richly tracheated tail filaments of *Cryptochaetum* larvae (small flies, endoparasites of coccid hemipterans) and the blood filled caudal vesicle of braconid larvae (Hymenoptera) play important roles in respiration. In a few parasitic Hymenoptera and in most endoparasitic Diptera larvae, a connection exists between the open tracheal system of the parasite and the atmosphere. Examples are to be found among the Encyrtidae and minute Hymenoptera that frequently attack such insects as mealybugs and scales (Coccidae). The structure and behavior of the primary larval parasite are correlated with a respiratory modification of the egg shell—the aeroscopic plate or band.[18] This structure is a strip of modified chorion acting as an air-channel, and the larval spiracles are applied to the inner surface of the strip. The ovarian egg consists of two ovoid bodies connected by a narrow tube, and the entire contents of the egg remain in the posterior body when oviposition is completed. Larvae may possess spiracles, but all larvae may obtain oxygen from the hemolymph of the host by diffusion. Some larvae remain partly enclosed posteriorly by the shell so that the spiracles may gain contact with the air-bearing structures on the egg proper. Some parasitic larvae (conopid flies in adult bees and wasps) become attached to a tracheal trunk of their host. Other larvae (e.g., *Cryptochaetum*) perforate the body wall or trachea and thus place their spiracles in direct contact with the atmosphere. Evaginations of the proctodaeum, spine-like processes on or adjacent to the posterior spiracles, ribbon-like extensions of cocoons of ectoparasites, all illustrate the wide variety of ways in which parasitic insects have solved the problem of obtaining oxygen. A word of caution is needed here. Clausen[8] has pointed out, "In most instances the adaptations are assumed to have that function in the absence of any other apparent purpose, and though the assumption may be logical, yet there still remains the necessity for experimental work."

Reproduction

Most insects are **oviparous** but a few species reproduce by other ways than simply by laying eggs. The more common and interesting of these ways are described briefly below.

Viviparous insects produce larvae or nymphs instead of laying eggs. Sometimes the phenomenon is little more than a retention of the eggs until they hatch in the reproductive tract, but frequently it involves an elaborate modification of the morphology and physiology of the parent, and in such instances it may be called "pseudoplacental viviparity." Viviparity occurs in scattered representatives of many orders, but it is particularly common among the parthenogenetic Aphididae (see

pages 378), the Strepsiptera (see page 393) and the Pupipara (see page 421).

Parthenogenesis occurs when eggs undergo full development without having been fertilized. The phenomenon exists widely throughout the phylum Arthropoda, and it may appear in one or a few species of otherwise bisexual genera. Parthenogenesis is common among the Lepidoptera, Hymenoptera and Amphididae. It may be facultative when it co-exists with bisexual reproduction, or obligatory when males are absent or rare. Parthenogenetically developing insect eggs may appear combined with viviparity or with paedogenesis.

Paedogenesis is the phenomenon of reproduction by immature individuals of a species. It is essentially parthenogenesis combined with neoteny. For example, the paedogenetic larvae or pupae of some insects (e.g., the beetle *Micromalthus debilis*, the gall midges *Miastor* and *Oligarces*, and the polyctenid bug *Hesperoctenes*) start giving birth before molting.

Polyembryony is the development of two or more embryos from a single egg, and it is common in certain parasitic Hymenoptera (see page 382), in a few Strepsiptera (page 393), and much more rarely in other orders of insects. Sometimes, as in the case of the chalcid wasps, *Copidosoma* and *Litomastix*, one parasitic egg may give rise to as many as 1,000 larvae, thereby achieving a high reproductive potential. Field studies, however, indicate that females of polyembryonic species tend to produce fewer eggs than do related nonpolyembryonic species. Also, polyembryonic species are not as effective as parasites when compared with other kinds of insects.

Functional **hermaphroditism** seems to have been developed only twice among insects. The scale insect, *Icerya purchasi*, produces normal males and hemaphrodites but no females. The outer cells of the hermaphrodite gonad form ovarioles, and the inner cells form sperm. The eggs may develop parthenogenetically, or they may be fertilized by sperm from the same individual, or more rarely they may be fertilized by sperm from normal males with which the hermaphrodite mates. Each phorid fly, *Termitostroma*, possesses a pair of ovaries and one testis. Probably, in this fly cross-fertilization is the rule, but self-fertilization can occur.

Ticks exhibit many varieties of reproductive behavior. The sexes may be about equal in size or extreme sexual dimorphism may occur, with the male as a parasite on the female (e.g., in *Aponomma*). In some species of ticks the female is free from the male entirely and thus always reproduces parthenogenetically; in most cases, however, copulation takes place. This act is intriguing in these animals because the male opens the female genital aperture with his mouth parts, deposits a bundle of his sperm within, and soon dies.

Parasitism by immature stages rather than by the adult occurs widely among the entomophagous insects, especially the Hymenoptera. Insect eggs are either laid on the body of the host (frequently the larval form of another insect), or in the body, or sometimes in the egg of the host. The newly hatched parasite may confine its activity to the surface of its host or it may eat its way into the host. The problem of nourishment does not worry the parasite because the invader is either sitting on or surrounded by a vast mountain of food. Some parasitic wasps drill a hole through the bark of trees and deposit their eggs in the burrows of their host. The eggs hatch and wasp larvae seek out their hosts.

In at least one instance, a larval fly has developed a special organ, like that of its host, to enable it to get out of the host's cocoon. The fly, *Systropus conopoides*, parasitizes the caterpillar of *Sibine nonaerensis*, remaining with its host until the cocoon is formed. When the fly has finished its preliminary development, it emerges from the cocoon by drilling a hole by means of a special spine on its head, thus imitating the host pupa. This type of parallelism is called **homeopraxy.**

CLASSIFICATION OF SOME PARASITIC ARTHROPODS

Phylum Arthropoda
  Class Isopoda
    Suborder Flabellifera
    Suborder Gnathiidea
    Suborder Epicaridea
  Class Copepoda
  Class Branchiura
  Class Amphipoda
  Class Cirripedia
  Class Insecta
    Order Mallophaga
      Suborder Amblycera
      Suborder Ischnocera
    Order Anoplura
    Order Hemiptera
      Suborder Homoptera
      Suborder Heteroptera
    Order Hymenoptera
      Superfamily Ichneumonoidea
      Superfamily Cynipoidea
      Superfamily Chalcidoidea
      Superfamily Proctotrupoidea
      Superfamily Scolioidea
      Superfamily Apoidea
    Order Coleoptera
    Order Strepsiptera
    Order Lepidoptera
    Order Siphonaptera
    Order Diptera
      Family Culicidae
      Family Simuliidae
      Family Tabanidae
      Family Psychodidae
      Family Oestridae
      Family Tachinidae
      Family Calliphoridae
      Family Sarcophagidae
      Family Muscidae
      Family Hippoboscidae
  Class Arachnoidea (=Arachnida)
    Order Acarina
      Suborder Mesostigmata
      Suborder Ixodides
        Family Argasidae
        Family Ixodidae
      Suborder Trombidiformes
      Suborder Sarcoptiformes

CRUSTACEA

Class Isopoda (Pillbugs, Wood Lice)

These crustaceans possess bodies that are usually flattened dorsoventrally, without a carapace, with sessile eyes, and with a short abdomen whose segments are often partly fused. Each of the seven characteristically free segments of the thorax bears a pair of legs. Parasitic species are abundant and they tend to favor crustaceans and fish as hosts. Most parasitic species have been found in ocean water, although a few freshwater forms have been described.

Suborder Flabellifera

Family Cymothoidae. In this family, a whole range of types exists, with gradations from actively swimming predatory species to parasites whose adult stages are permanently fixed to the host and incapable of locomotion. The free-swimming species may be exemplified by *Cirolana borealis*, which possesses powerful biting jaws. The cirolanid isopods

are mainly scavengers rather than predators, but they sometimes attack a cod caught on a hook or otherwise at a disadvantage, and viciously gnaw their way into the body so that the fish is soon literally nothing but skin and bones. They have even been known to attack humans.

Cymothoa and *Nerocila* represent the genera of isopods that, as adults, cling to the gills or skin of their fish hosts by means of strong hook-claws (Figs. 16–1, 16–2). *Nerocila* and *Anilocra* are always external, i.e., on skin or fins. *Cymothoa* and *Livoneca* occur in gill chambers or mantles.   *Ceratothoa* parasitizes flying fish,

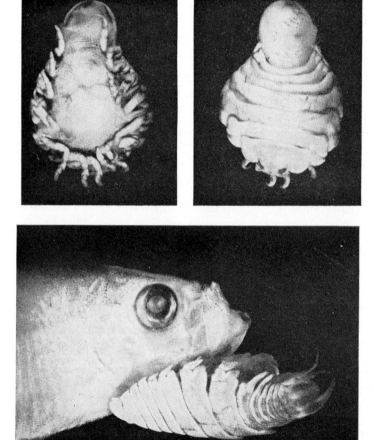

Fig. 16-1. Parasitic isopods. *Top.* Ventral and dorsal views of female *Riggia paranensis*. *Bottom.* A marine sardine (*Clupea*) with an adult female *Nerocila orbignyi* clinging to its head. (Szidat, courtesy of Archiv. f. Hydrobiologie)

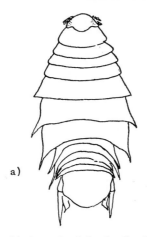
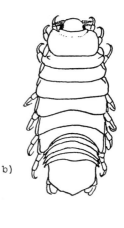

a)   b)

Fig. 16-2. Parasitic isopods of the family Cymothoidae from freshwater South American streams. *a. Nerocila orbignyi*, a sea form from the mouth of the La Plata river. *b. Livoneca symmetrica* from freshwater in British Guiana. (Szidat, courtesy of Archiv. f. Hydrobiologie)

often clinging to the tongue and sometimes almost preventing the unfortunate fish from closing its mouth. Another genus, *Ichthyoxenos*, occurs within cavities of its host skin, and each cavity encloses a male and a female parasite.

Livoneca convexa presumably begins its life cycle as a free-living planktonic individual.[20] Parasitic males enter gill chambers of the host fish (*Chloroscombrus orqueta*), whereas females, probably commensal, may enter the oral cavity of the same fish. It is not known, however, whether females enter the mouth directly or move there from the gill chamber. Only the adult male shows evidence of causing direct damage. Other flabelliferan isopods (families Corallanidae and Excorallanidae) also are parasites of fish.

Sexual dimorphism is the rule among parasitic isopods, and the phenomenon of **protandrous hermaphroditism** occurs among the flabelliferans. This phenomenon starts when the parasite first attaches itself as a functional male to a host. Later, the male becomes a female, develops a brood pouch and produces eggs. The presence of a female on a fish inhibits further development of the male toward the female phase. The nature of this influence is unknown, but it has been proven experimentally with *Anilocra physodes*, an external parasite that can be removed from the host. Since both male and female are temporary parasites, sexual dimorphism cannot be attributed to the parasitic mode of life. The protandric male is not easily recognized, and a young sexually immature female has probably often been mistaken for the male. Any given species of cymothoid usually possesses a wide range of favored hosts, and it may also be found free in the plankton. Host specificity, therefore, is often not marked.

Family Aegidae. Members of this family have piercing and sucking mouth parts enabling them to suck the blood of their hosts. The strong hooked claws on the anterior pairs of legs help the isopods to cling closely to the skin of their victim, but they are still able to leave the host and swim about in the water. After an ample meal the digestive canal of *Aega* becomes distended into a large bag of semisolid blood, and this mass, when removed and dried, is the "Peter's Stone" of old Icelandic folklore, to which magical and medicinal virtues were attributed. *Aega spongiophila* lives, not on fish as do other species of the genus, but within a sponge.

Suborder Gnathiidea

Family Gnathiidae. Another family of isopods, also containing only temporary parasites, are the Gnathiidae. Unlike those that are parasitic as adults, the gnathiids are

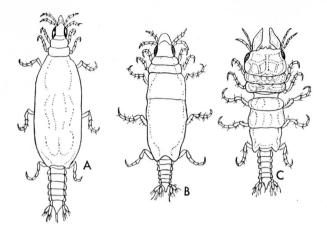

Fig. 16-3. *Gnathia maxillaris.* Sars. *A.* Larva. *B.* Adult female. *C.* Adult male (Sars). (Baer, Ecology of Animal Parasites, courtesy of The University of Illinois Press)

parasitic as larvae. These larvae are temporary dwellers on fish, feeding exclusively on blood. Adult gnathiids may be found in plankton or in mud dredged from the bottom. Morphologic differences among the larvae, adult males and adult females are so pronounced that each type was originally described as a distinct genus. *Praniza, Anceus* and *Gnathia* (Fig. 16–3) are representative genera. Adults of this family do not feed, possessing neither a mouth nor a gut, but the larvae attach themselves to the skin and gills of fish and gorge themselves with blood. After three larval phases, each phase separated by a molt, the adult stage is reached. The male gnathiid uses its powerful mandibles for digging a burrow into the mud at intertidal levels, and into this burrow the male and female withdraw to end their lives in private.

Suborder Epicaridea

Contrary to the habits of the cymothoids and the gnathiids, which are temporary parasites at either the adult or the larval stages, the suborder Epicaridea consists of isopods parasitic in both larval and adult stages. This suborder is divided into two superfamilies, the Bopyrina and the Cryptoniscina, both of which are parasitic on crustacea.

Epicarids are particularly interesting because their life cycle involves two hosts, and because sexes are determined epigamically (i.e., after fertilization), depending on hostal

environmental factors. See Baer[2] for details.* The first epicarid larva resembles a small isopod, and possesses piercing and sucking mouth parts, and claw-like appendages with which it attaches itself to the surface of free-swimming copepods. This kind of larva is called an **epicaridium,** and while it remains upon its copepod host, it undergoes six successive molts and changes progressively into two distinct larval stages known as the **microniscium** and the **cryptoniscium** stages (Fig. 16–4). Upon reaching the latter stage of development, the parasite leaves its copepod host, proceeds to the sandy or muddy sea-bottom, and there seeks a crustacean (e.g., crab or shrimp), into whose branchial chamber or brood pouch it enters. Within this second host the parasitic cryptoniscium stage develops in one of two directions according to whether it belongs to the superfamily Bopyrina or to the superfamily Cryptoniscina.

If the cryptoniscus larva is a bopyrine it molts and thereby loses most of its pleopods, and is now known as a **bopyridium.** This larval stage initially always develops into a female; but successive larvae, either attached to the same host or to the female parasite, all become males. The question as to whether sex in parasitic isopods really depends upon the environment was studied by a series of

* Subsequent experiments seem to cast a question on whether the process of sexual determination is uniform throughout the suborder.

experiments conducted by Reverberi and Pitotti[23] and by Reinhard.[22] *Stegophryxus hyptius*, a bopyrid ectoparasite on the abdomen of the hermit crab, *Pagurus longicarpus*, seeks the definitive host as a sexually undifferentiated and sexually undetermined cryptoniscus larva. Larvae that settle directly on the host develop into females, and those that attach themselves to a female bopyrid develop into males. Presumptive females, if removed from the host at an early stage and placed in the brood pouch of a female *Stegophryxus*, will change into males. Reverberi and Pitotti have successfully transformed the smaller of a pair of bopyrid females into a male by rearing the females *in vivo*, without their host crab. Sexual differentiation, therefore, may depend upon environmental factors, at least for the

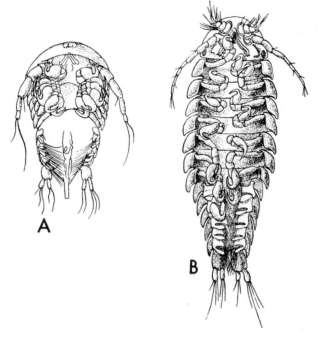

Fig. 16-4. *A. Cepon elegans*, microniscus larva. *B. Portunion kossmanni*, cryptoniscus larva. (Giard and Bonnier in Baer, Ecology of Animal Parasites, courtesy of University of Illinois Press)

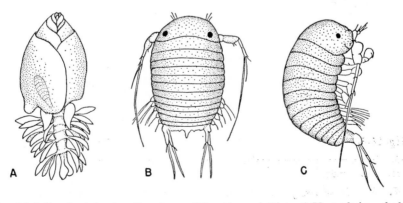

Fig. 16-5. Adult female and epicardium larva of *Stegophryxus hyptius*. *A*. Ventral view of adult female. The dwarf male, although not visible externally, is shown within the brood pouch by a dotted outline to indicate its position and relative size. × 5. *B*. The epicaridium or first larvae stage shown in dorsal view. × 120. *C*. Lateral view of epicaridium larvae. × 120. (Reinhard, courtesy of Biological Bulletin)

forms tested. Extra females or males on a host soon disappear, leaving only a single adult female paired with a single functional male (Fig. 16–5). Several pairs of the same or different species, however, have been found on one host, and complemental (nonfunctional?) males are not infrequent.

Bopyrus squillarum lives in the gill cavity of the prawn, Leander serratus, and it causes a large swelling on one side of the host carapace. The parasite has a flat and distorted body, with an enormous mass of eggs on the ventral side of the female. The mouth parts form a short beak with which the parasite sucks the blood of its host. The bopyrid isopods, especially those of the family Entoniscidae, frequently are the cause of a complete or partial atrophy of host gonads.

Most authorities believe that the entoniscids, while appearing to be internal parasites, are actually external, being surrounded by an invagination of the external chitinous covering of the host, and attached to its surface by thread-like chitinous tubes.[1] For example, Pinnotherion vermiforme (Figs. 16–6, 16–7) occurs inside of the thorax and abdomen of the pea crab, Pinnotheres pisum—itself a parasite of the clam, Mytilus edulis. The male isopods have been found free in the body of the host, and hence they have been called true internal parasites, but they may have accidentally broken through the thin chitinous membrane from their normal position among the pleopods of the female. The female entoniscid undergoes a pronounced morphologic change, and the surface of its pleopods

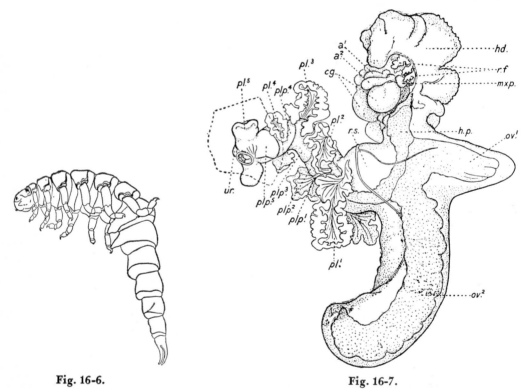

Fig. 16-6.   **Fig. 16-7.**

Fig. 16-6. *Pinnotherion vermiforme*, male, showing characteristic ventral curvature. (Atkins, courtesy of Proc. Zool. Soc. London)

Fig. 16-7. *Pinnotherion vermiforme*, female. *a.*[1] Vestige of right antennule. *a.*[2] Vestige of right antenna. *cg.* Cephalogaster. *hd.* Hood-region of the brood-chamber. *h.p.* Hepatopancreas. *mxp.* Maxilliped. *ov*[1], *ov.*[2] Anterior and posterior ventral processes of the gonad. *pl.*[1-5] Right pleural lamellae; *plp.*[1-5] Pleopods. *r.f.* Respiratory (?) folds. *r.s.* Position of the "receptaculum seminis." *ur.* Last abdominal somite. × 6.3. (Atkins, courtesy of Proc. Zool. Soc. London)

becomes corrugated and filled with numerous blood lacunae which serve in a respiratory function. In the same female the oostegites form a brood pouch, part of which appears as a hood around the head. Epicaridian larvae are expelled from this pouch through a secondary opening.

Probably all the parasitic isopods of the superfamily Cryptoniscina are protandrous hermaphrodites. The males become mature in the cryptoniscan stage. The life history of this group starts in the same manner as that of the bopyrids, and it progresses similarly to the cryptoniscus larval stage. From here on the life history is markedly different. The larvae, after entering the branchial chamber or brood pouch of crabs, become protandrous hermaphrodites of a delayed type since the males and females are morphologically distinct. An excess production of eggs causes the gravid female to undergo morphologic and presumably physiologic degeneration. Common among the degenerative changes is an asymmetrical disappearance of appendages from one side of the body. More drastic, however, is an atrophy of some internal organs of the female, leaving little more than a bulky sac packed with eggs.

Class Copepoda

A fine plankton net towed through almost any natural body of fresh or salt water soon collects multitudes of organisms consisting chiefly of minute crustaceans. The most abundant kinds of crustaceans in such hauls are usually copepods, whose bodies furnish the major basic food supply for all larger aquatic animals. The best-known freshwater member of the class Copepoda is *Cyclops* (order Cyclopoida, Fig. 16–8) which serves

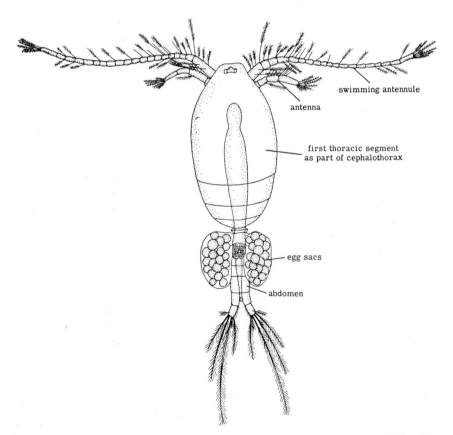

Fig. 16-8. *Cyclops.* (Bullough, Practical Invertebrate Anatomy, courtesy of Macmillan)

as the intermediate host for several helminth parasites of man and of other vertebrates. *Cyclops* is often called a "water flea."

Most copepods are free-living, but many are parasitic and, like so many other parasitic groups, they range in habits from the most casual and temporary contact with a host to a rigid and permanent attachment as adults or larvae within the host body. Most groups of aquatic animals (especially marine) may serve as hosts for at least one species of copepod during part of its life cycle. Echinoderms, annelids, ascidians, mollusks, arthropods, fish and whales are universally invaded. The parasite may appear practically identical with free-living species, or it may be so modified as an adult, that were it not for the larval stages, one would hardly be able to determine even the phylum to which the copepod belongs (Fig. 16–11). Morphologic simplification in copepods may sometimes be correlated with the distribution of their hosts in time, but such simplification seems to provide little assistance in the determination of the affinities of the parasites themselves.

Diagnostic features of copepods are as follows: free-living or parasitic, without compound eyes or carapace; with biramous or uniramous palpas (or none) on the mandibles; typically six pairs of thoracic limbs, but none on the abdomen (i.e., posterior to the genital apertures); first pair of thoracic (invariably cephalo-thoracic) limbs always uniramous, sixth pair often uniramous, other pairs usually biramous; suctorial proboscis, common in parasitic species, formed by the upper and lower lips enclosing specialized mandibles; two pairs of antennae, one pair of mandibles, two pairs of maxillae and one pair of maxillipeds; head often greatly modified not only because of the specialized mouth parts, but because cephalic appendages (usually the second antennae) have become transformed into grappling organs.

The sexes in copepods are nearly always separate. The male as a rule is much smaller than the female, whose pair of ovisacs ("egg string") is usually conspicuous. Some groups (e.g., Harpacticoida) have only one sac. The female of *Chondracanthus merluccii* is 12,000 times larger than the male, but this sexual dimorphism is extreme. Hermaphroditism is rare. Most parasitic species appear to have several breeding seasons each year. In some instances (especially in all of the Chondracanthidae and some of the Lernaeopodidae), the males are sessile on the females. In such cases the male loses the power of locomotion upon reaching maturity. In still other species, one act of copulation is apparently sufficient to fertilize the female for life, and the male is then not found in persistent association with the female. The embryos remain attached to the external opening of the oviducts, enclosed in the ovisac until they hatch, a process often requiring several weeks. Nauplii larvae are, in the first stage, equipped

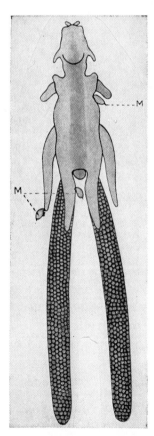

Fig. 16-9. *Chondracanthus merluccii,* female, dorsal view, with attached males, × 8. (Scott and Scott, The British Parasitic Copepoda, The Ray Society, courtesy of B. H. Blackwell Ltd.)

with three pairs of appendages and one or more pairs of caudal setae (balancers), and two distal antennulary setae. The three appendages represent the first and second antennae and the mandibles of the adult (Figs. 16–12, 16–25). Many parasitic forms hatch at a later stage, with more appendages.

Copepod larvae normally pass through six nauplius stages (the last three often called "metanauplius"), and five copepodid stages, but abbreviated development may take place, especially in parasitic species.

Order Cyclopoida

Family Chondracanthidae. *Chondracanthus* (Fig. 16–9) is common on gills and in the mouths of various fish. The female body of *C. gibbosus* within the gill cavity of the marine angler or fishing "frog" (*Lophius piscatorius*) is extended into paired lobes, giving it a curious irregular shape; and its appendages are greatly reduced. The mouth is flanked by sickle-shaped jaws, and the whole body may reach a length of about 2.5 cm. At the posterior end of the female body, just at the point where the egg masses are attached, close inspection will reveal a minute, maggot-like object clinging by means of hook-like antennae. This is the male that is attached, like a secondary parasite, to its enormously larger mate.

Cucumaricola notabilis is a peculiar copepod which forms amorphous cysts within the coelom of the holothurian, *Cucumaria frauenfeldi*. Fertilized eggs and nauplius larvae, as well as adults, are contained within the cysts. There are two copepodid stages, the first an active swimming form and the second a quiescent form. Although this species is totally unlike any copepod known to parasitize either echinoderms or other types of hosts, and although it is of uncertain systematic position, being assigned[21] tentatively with the Chondracanthidae, it is illustrated in Figures 16–10 to 16–13 because of the interesting extreme amorphous condition of the female copepod.

Order Notodelphyoida

Family Notodelphyidea. This family consists of copepods that almost universally inhabit the body cavities of tunicates. The group is a well-defined natural unit because of the preservation of fundamental characteristics in spite of a high degree of adaptive radiation. The family is characterized by the occurrence of a prehensilly modified articulated hook as the terminal member of the second antennae, and the development of a brood sac enclosed within the body.[16] The life cycle probably involves five naupliar stages and six copepodid (subadult) stages. The second copepodid stage is the infective one, but the sequence of developmental events and the possible metamorphoses in the life histories are almost entirely unknown, and

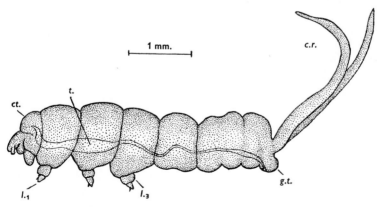

Fig. 16-10. *Cucumaricola notabilis,* a copepod parasite of the holothurian, *Cucumaria.* Mature male parasite, side view showing position of left reproductive organs. *cr.* Caudal rami. *ct.* Cephalothorax. *g.t.* Genital tubercle. *1.* Trunk appendages. *t.* Testis. (Paterson, courtesy of Parasitology)

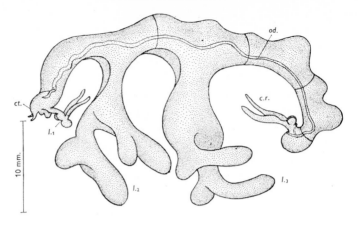

Fig. 16-11. *Cucumaricola notabilis.* Mature female, side view, with oviduct seen through translucent body wall. *c.r.* Caudal rami. *ct.* Cephalothorax. *1.* Trunk appendages. *od.* Oviduct. (Paterson, courtesy of Parasitology)

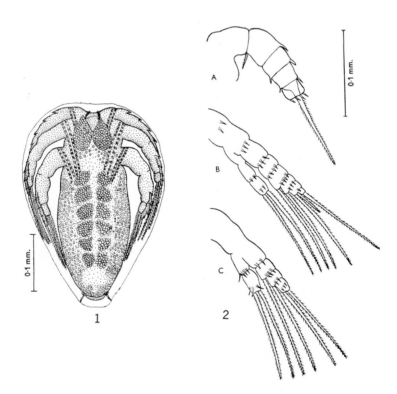

Fig. 16-12. *Cucumaricola notabilis.* *1.* Ventral view of nauplius larva seen through the egg membrane. *2.* Appendages of nauplius larva. *A.* Attennule. *B.* Antenna. *C.* Mandible. (Paterson, courtesy of Parasitology)

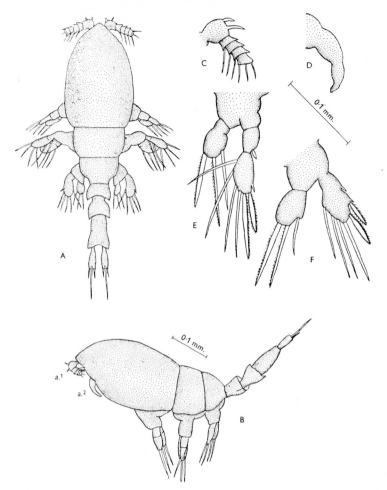

Fig. 16-13. *Cucumaricola notabilis.* First copepodid larva. *A.* Dorsal view. *B.* Side view. *a*[1], *C.* Antennule. *a*[2], *D.* Antenna. *E.* Second trunk appendage. *F.* Third trunk appendage. (Paterson, courtesy of Parasitology)

the mode of penetration has not been ascertained. Species whose appendages are the most reduced come from the more specialized habitats such as those found in the common cloaca of the systems of zooids of a compound ascidian. In another such form, there is an extreme degree of parasitic adaptation in the formation of a cyst within the ventral blood vessel of representatives of two genera of solitary tunicates.

Other symbiotic organisms commonly associated with notodelphyids in ascidians are: epizootic ciliates, sporozoa, hydroids, flatworms, nemerteans, nematodes, polychaetes, pinnotherid crabs, and, most commonly, other symbiotic copepods and amphipods. Illg[16] has described North American representatives of this copepod family, and concerning the interrelationships among the symbiotic organisms in tunicates, he notes that "the possibilities of complex cycles of nutritive relationships form one of the more obvious ecological corollaries of the biotic complex of ascidicolous organisms."

Xenocoeloma (order not definitely known) represents an extreme type of modification involving host tissue. This genus is parasitic on marine annelids, and the adult parasite is reduced to the hermaphroditic gonads plus some muscles. The mass is enclosed in a cylindrical outgrowth of the host epithelium which contains a gut-like prolongation of the host coelom (Fig. 16–14). The life history is incompletely known, but the nauplius that emerges from the eggs resembles that of the monstrillids in the absence of digestive organs. The manner of access to the host has, unfortunately, never been observed.

Order Caligoida

Family Caligidae. The Caligidae is possibly the most widely distributed family of copepods parasitic on fish, and it probably contains the largest number of species (e.g., *Caligus* and *Lepeophtheirus*). The family Chondracanthidae, however, contains a greater number of genera. *Caligus* (Fig. 16–15) is ectoparasitic, chiefly on fish, has a suctorial mouth tube, and retains the power of swimming in the adult stage. This genus possesses two semicircular structures on the frontal margin of the cephalic shield that are probably sensory rather than attachment devices. *Caligus* does not always remain on one host.

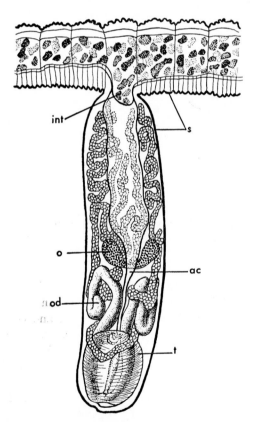

Fig. 16-14. *Xenocoeloma brumpti C* and *M*, section of the parasite and the annelid host. *ac.* Axial cavity formed by the host's coelom. *int.* Host's gut. *od.* Oviduct. *o.* Ovary. *s.* Skin of both the host and the parasite. *t.* Testis (Caullery and Mesnil in Baer, Ecology of Animal Parasites, courtesy of University of Illinois Press)

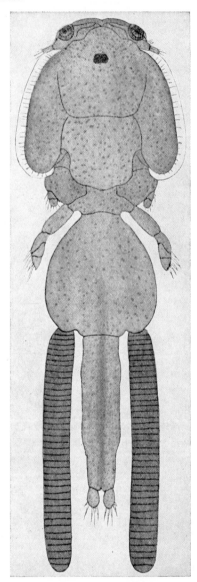

Fig. 16-15. *Caligus pelamydis* female, dorsal view. × 22.5. (Scott and Scott, The British Parasitic Copepoda, The Ray Society, courtesy of B. H. Blackwell Ltd.)

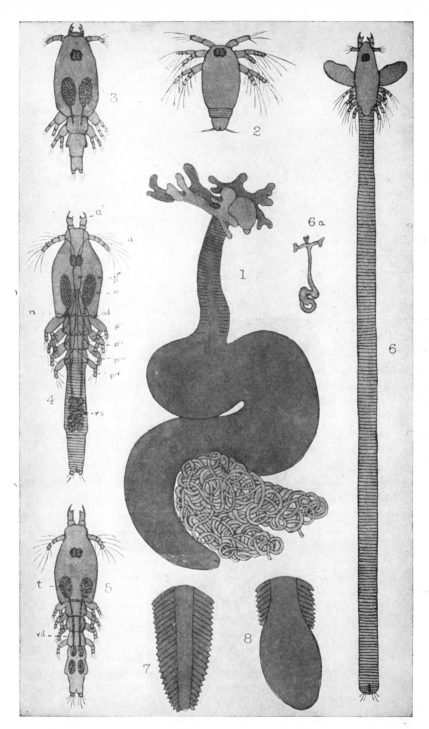

Fig. 16-16. *Lernaeocera branchialis*. *1.* Mature female side view, × 4. *2.* Newly hatched nauplius, × 51. *3.* Very young female, unfertilized, dorsal view, × 51½. *4.* Fertilized female, after leaving gills of flounder, dorsal view, × 27½. *5.* Mature male, dorsal view, × 28½. *6.* Fertilized female, "Pennell stage," just after attachment to gills of a Whiting, dorsal view, × 15½. *6a.* A later stage than *6*, side view, natural size. *7.* Apex of the gill-ray of a flounder, × 18. *8.* Apex of the gill-ray of a flounder, showing malformation caused by young parasite, × 18. (Scott and Scott, The British Copepoda, The Ray Society, courtesy of B. H. Blackwell Ltd.)

353

Dinemoura producta (similar in general appearance to *Caligus*), common on the great basking shark, has the same structures as does *Cyclops* but with many modifications related to a parasitic habit. Respiration in copepods is accomplished by a diffusion of gases through the body surface. Obviously the surface-volume ratio is of vital significance. *Dinemoura* is about 2 cm. long, whereas *Cyclops* is only about 2 mm. long. One, therefore, has a mass about 1000 times greater than that of the other, but if its shape were identical with that of *Cyclops*, the external surface of *Dinemoura* would be increased only about 100 times. Even taking into account the relatively more sendentary existence of the parasitic species, there would not be sufficient surface for respiration. These considerations probably explain the existence, in *Dinemoura*, of a number of flattened lobes projecting from the dorsal surface of the thoracic segments. The head of *Dinemoura* has a median eye and short second antennae armed with strong hooks for clinging to the skin of the host. The mouth parts are adapted for piercing and sucking. The appendages used for swimming and the caudal rami with bristles are essentially similar in basic plan to those in *Cyclops*. The entire head is shaped like a shallow cup whose edges fit closely against the host skin, much like the fit of a limpet on a rock. This modification aids greatly in preventing dislodgement by the pressures of water currents and waves. Trailing behind the body of the female is a pair of egg strings that may be as much as four times as long as the body.

Family Lernaeopodidae. *Lernaeocera*, often called the "gill maggot" (Fig. 16–16), is commonly seen attached to the gills of marine cod, and its bloated, S-shaped, blood-red body looks much more like a worm than it does a crustacean. The larval nauplius stage (unsegmented three pairs of legs) proves its kinship with the arthropods. After a free-living period of growth, including several molts, the nauplius is transformed into a "cyclops" or "chalimus" stage when it becomes parasitic on the gills of certain flatfish (e.g., *Pleuronectes*) by means of suctorial

mouth parts, accompanied by a reduction of its limbs. The chalimus attaches itself to gills of the host, first by the second antennae, then by a chitinous secretion that is extruded as a laterally flattened thread extending into the gill tissue. This stage is followed by a resumption of the power of movement when the parasite leaves its host and for a time lives a free-swimming, adult, sexually mature existence very similar to that of *Cyclops*. At this time fertilization of the eggs takes place, and the development of the male ceases. The female, however, now seeks a new host, generally a member of the cod family (e.g. *Gadus merlangus*), and again the gills are particularly attractive to the parasite. The anterior end of the female is buried into the host tissue, and it becomes curiously modified to form a branched anchor resembling short roots. The genital somites become greatly enlarged and vermiform, and the egg mass appears as a cluster of tightly coiled filaments. In proporticn, the appendages of the thorax become minute and nonfunctioning.

The salmon gill maggot, *Salmincola salmonea*, is a common and widespread copepod parasite of the Atlantic salmon (*Salmo salar*). The mature female parasite, when seen attached to a gill filament of its host, "somewhat resembles a gymnast hanging from a vertical bar."[12] Attachment to the gill-filament is effected by means of a secreted "bulla" which is applied to the gill surface. A thin sheet of living gill tissue partly covers the bulla. A pair of prehensile maxillipeds lies behind an oral cone, and from the sides of the cephalon the maxillary "arms" (second maxillae) extend and converge on the bulla to which they are permanently fixed. The thorax bears no appendages. The paired egg sacs measure 4 to 11 mm. in length.

The life history of this copepod has been described by Friend[12] and will be used here to exemplify the life history of the group. Figure 16–17 presents a diagram of the life-history, and Figures 16–18, 16–19 and 16–20 illustrate details of external anatomy of various developmental stages of the related genus *Achtheres*. During the first river phase of the fish, and the first sea phase, totalling from two plus years to seven plus years, the salmon is not attacked by gill maggots, although it acquires sea-lice. At the beginning of the second river phase the sea-lice are lost and the maiden fish acquires larval gill-maggots which swim from beneath the gill covers of other and older fish. In the fall the host re-enters the sea, and during its sojourn (up to two years) in marine waters its copepods thrive and grow but do not breed. When the salmon re-enter the rivers their copepod parasites begin to breed.

The first copepodid larvae of *Salmincola salmonea* may live free for up to six days, but when they come in contact with the gills of a salmon they attach themselves "by means of a button and thread," and molt to become the second copepodid larvae. The next molt is followed by the sexually mature male or the first stage female, both of which may move freely over the gill surface, or they may attach themselves by means of chelate appendages. During this stage copulation takes place and the male disappears. Meanwhile the frontal gland in the female has been elaborating the attachment bulla which is planted on the gill tissue in its final position. The female then molts and the distal ends of each second maxilla are forced as plugs into corresponding sockets in the bulla, and permanent attachment is achieved. The female then grows to mature size. Development to this stage requires about five or six months. One female may produce two or more generations of larvae.

Order Monstrilloida

Family Monstrillidae. These copepods are free-living as adults, but parasitic during the intermediate stages in the blood vessels of various polychaete worms, or in the body cavity of prosobranch snails. The nauplius of *Monstrilla* hatches without a mouth or gut, and it burrows into the body of its host, discards its chitinous exoskeleton and loses its limbs. By the time it reaches the host body cavity it consists only of a naked mass of

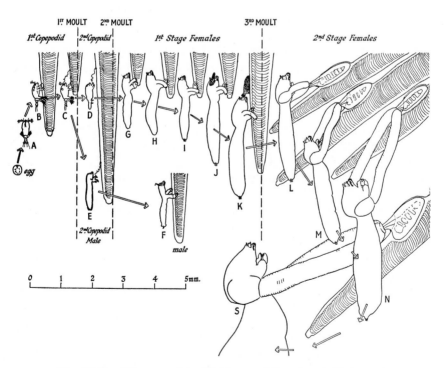

Fig. 16-17. Diagram of the life-history of the salmon gill maggot.
(Friend, courtesy of Trans. Royal Society of Edinburgh)

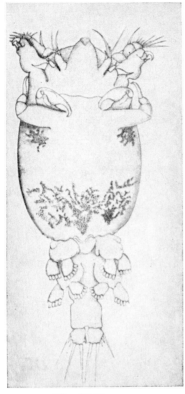
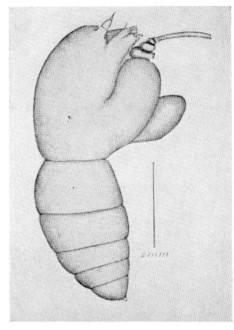

Fig. 16-18.   Fig. 16-19.

Fig. 16-18. The free-swimming larva, first copepodid stage of *Achtheres*. Ventral view of first copepodid larva, showing appendages. (Wilson, courtesy of Proc. U.S. Nat. Museum)

Fig. 16-19. The second copepodid stage of *Achtheres*. Side view of male larva. (Wilson, courtesy of Proc. U.S. Nat. Museum)

embryonic cells, which then become surrounded by a thin cuticle, and which develop a pair of long flexible processes that apparently represent antennae. The antennae absorb nutrients. The food thus taken lasts the entire remaining life cycle, throughout which there are no functional mouth parts. The mass of parasitic cells within the host gradually develops the organs of the adult, which bores its way outside by means of rows of hook-like spines surrounding the pointed posterior end of the sac. Upon reaching the surface, the enclosing membrane bursts and the adult parasite is free, appearing similar to *Cyclops*. The animal thus passes its whole life cycle without a gut, and between the two free-swimming stages, one at each end of its

life, it exists as a bag of parasitic cells absorbing nourishment from the internal fluids of its host (see Fig. 16–21 of *Haemocera*).

Further details about parasitic copepods may be found in works by Dudley,[11] Scott and Scott,[24] Wilson,[26] and Baer.[2]

Class Branchiura

All modern European literature and most of the more up-to-date American references agree in referring this group, morphologically well-differentiated from copepods, to a systematic category equivalent to that of copepods.

Branchiurids are crustacea that are temporarily parasitic on fish. They superficially resemble copepods, but differ from the latter in the possession of compound eyes, lateral

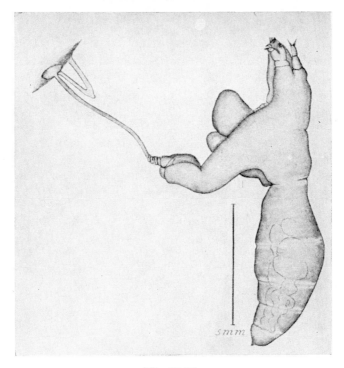

Fig. 16-20.

Fig. 16-20.   Young adult of *Achtheres ambloplitis*. Side view of sexually mature male, attached to spine of gill arch. (Wilson, courtesy of Proc. U.S. Nat. Museum)

carapace-like head-lobes fused to the sides of the first thoracic somite, an opening for genital products between the fourth pair of thoracic limbs, and a proximal overhang of some of the thoracic exopodites. Other diagnostic features include an unsegmented, limbless, bilobed abdomen, and four pairs of biramous thoracic limbs.

Branchiurids are commonly called "carp lice," or "fish lice," and they are to be found on both fresh- and salt-water fish. The females deposit their eggs on stones, bits of wood and other objects, and the larvae resemble adults. The most common genus, *Argulus* (Fig. 16–22), has a pair of suckers situated on the second maxillae, and a poison spine in front of the proboscis. It clings to the skin and gills, and appears to have little host preference. See Bowen and Putz[4] for reference to pathogenicity.

Class Amphipoda

This class is characterized as follows: body often laterally compressed, absence of a cara-pace, abdomen flexed ventrally between the third and fourth somites, telson usually dis-tinct, habitat usually marine. The most commonly seen (but not most common) species are those called "sand fleas." Al-though many amphipods are associated in a casual manner with marine invertebrates, few species are parasitic. Some members of the family Gammaridae have suctorial mouth parts and they lead a semiparasitic life. A number of forms burrow into other animals in order to obtain food. *Phronima sedentaria* forms a "case" from a salp, *Pyrosoma*, in which it and the young live, but it does not feed on the salp. Its food is the soft-bodied prey that comes within reach of the entrance.

The class Amphipoda may be divided into

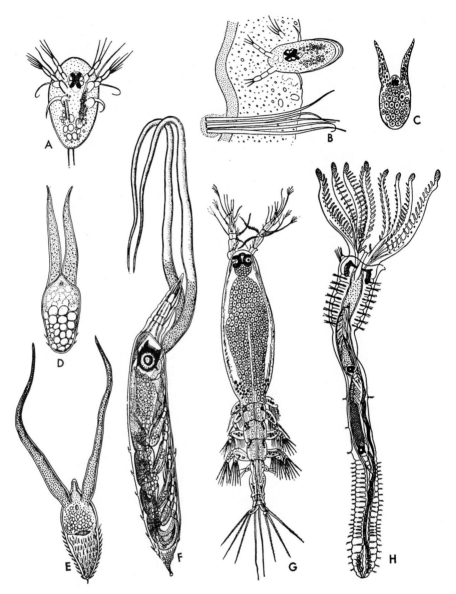

Fig. 16-21. *Haemocera danae* (Clap). *A.* Nauplius. *B.* Nauplius in the act of penetrating into the body of its host. *C–E.* Successive larval stages showing the development of the appendages and also of the spinous sheath enclosing the larva. *F.* Fully formed copepodid. *G.* Adult female copepod devoid of a mouth. *H.* Annelid with two copepodid larvae in its coelomic cavity. (Malaquin in Baer, Ecology of Animal Parasites, courtesy of University of Illinois Press)

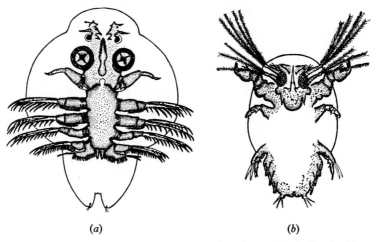

(a)   (b)

Fig. 16-22. *Argulus. a.* Ventral view of the female. *b.* Newly hatched larva. (Cameron, Parasites and Parasitism, courtesy of John Wiley and Son)

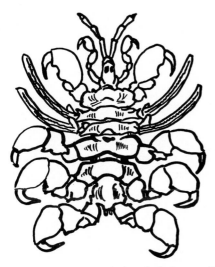

Fig. 16-23. *Cyamus ceti*, the whale louse. (Bate and Westwood, A History of British Sessile-eyed Crustacea, courtesy of John Von Voorst)

the following suborders, of which the first is the largest: Gammaridea, Hyperiidea, Ingolfiellidea and Caprellidea.

Family Cyamidae

Parasitic Caprellideans. Five genera of amphipods living on various Cetacea belong to this family. The semiparasites have dorsoventrally flattened bodies, reduced abdomens, and claws on their legs; they cling to the skins of their hosts. *Cyamus*, the whale "louse"

(Fig. 16–23), is the best known, and it has a wide geographic range. Only a few reports have been made on cyamids of the small, toothed cetaceans commonly known as dolphins and porpoises, and additional collections from these mammals are needed. *Syncyamus* (Fig. 16–24) occurs in the blow-hole and angle of the jaw of a dolphin from Panama Bay. Cyamid mouth parts are not adapted for sucking blood; instead, the animals probably feed on mucus, bacterial and algal growth, *etc.* on skin.

The amphipods are unique among crustacean parasites in being unable to swim at any period of their life history. The young settle down near their parents, and masses of individuals of all sizes may be seen clinging closely together on the skin of their host, often intermingled with barnacles. Other genera are *Paracyamus*, *Platycyamus*, and *Isocyamus*. For further details see Bowman,[5] Margolis,[19] and Leung.[15]

Class Cirripedia (Barnacles)

Barnacles are among the most highly modified crustaceans. They abound in marine waters attached to rocks, pilings, shells of other animals, or on almost any other solid surface that provides a firm purchase. Like so many other sessile animals, barnacles, with few exceptions, are hermaph-

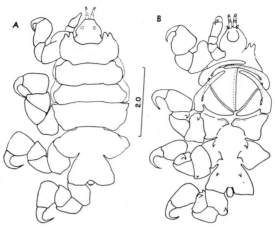

Fig. 16-24. *Syncyamus pseudorcae.* *A.* Female, dorsal. *B.* Female, ventral.
(Bowman, courtesy of Bull. Marine Sci. of the Gulf and Caribbean)

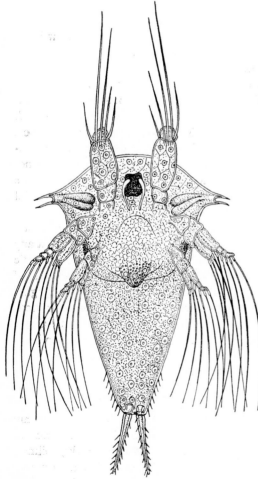

Fig. 16-25. *Sacculina carcini,* nauplius larva.
(Delage in Baer, Ecology of Animal Parasites,
courtesy of University of Illinois Press)

roditic, and their bodies are enclosed in a calcareous shell composed of plates which commonly overlap or are fused to furnish a formidable bulwark against pounding waves or grazing predators. Charles Darwin's two volumes[9] on barnacles are classic standard reference works.

Cirripedia are crustaceans without compound eyes in the adult, with a carapace (except in rare instances) as a mantle over the trunk, with typically six pairs of biramous thoracic limbs, and usually a mandibular palp that is never biramous. A free-swimming nauplius larva (Fig. 16–25) hatches from the cirriped eggs, and, after a feeding period of a few days, it molts several times and changes into a **cypris** larva (resembling an ostracod) with a bivalve shell. The shell valves, unlike those of an ostracod, are united along both the dorsal and ventral surfaces (Fig. 16–26). The larval barnacles possess the typical crustacean features of chitinous, jointed, two-branched appendages heavily fringed with bristles. After a week or more, the cypris larva settles to the bottom on a solid object and becomes attached by the antennae at the head end with the aid of a cement gland. By a series of remarkable changes the adult stage is reached and the animal is permanently situated, "standing on its head," spending its life in reproduction and in sweeping food and water into its shell

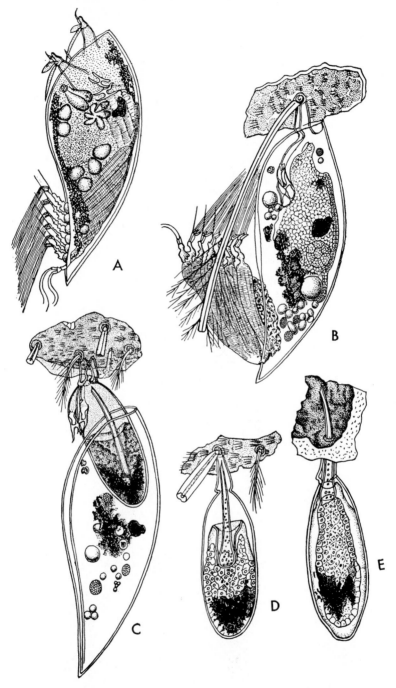

Fig. 16-26. *Sacculina carcini* Thomps. *A.* Cypris larva. *B.* Cypris attached to its host by its antennae and shedding its locomotory appendages. *C.* Kentrogon larva. *D–E.* Stages of penetration of the kentrogon into the crab. (Delage in Baer, The Ecology of Animal Parasites, courtesy of University of Illinois Press)

by means of its lacy appendages, which reach out like a delicate casting net.

The species of barnacles range in habits from completely free-living to casual commensalism to extreme pathogenic parasitism. Many forms illustrate phoresis in which host specificity is pronounced, as in those species which are attached only to gorgonian corals, decapods, starfish, sharks, or whales.

Examples of commensal cirripeds are: *Chelonibia*, which has developed a branched system of roots that penetrate into manatees and into the bone of the plastron of marine turtles; *Coronula, Tubicinella, Cryptolepas*, and *Xenobalanus*, whose calcareous shell plates grow into the skin of whales; and *Anelasma squalicola* (Fig. 16–27), which becomes partly buried in the skin of sharks. In the latter species a pear-shaped portion is embellished with branched, root-like appendages that apparently secrete an enzymatic substance that dissolves the surrounding muscle tissues of the host. *Rhizolepas annellidicola* is another parasitic barnacle which is anchored to its host (the annelid, *Laetmonica producta*) by a system of foot-like appendages. Two other examples are *Platylepas* on turtles, manatees and sea snakes, and *Alepas* on various medusae.

Order Rhizocephala

The most extremely modified parasitic cirripeds belong to this order, whose members live on or within the bodies of other crustacea, mostly decapods. The larval stages appear

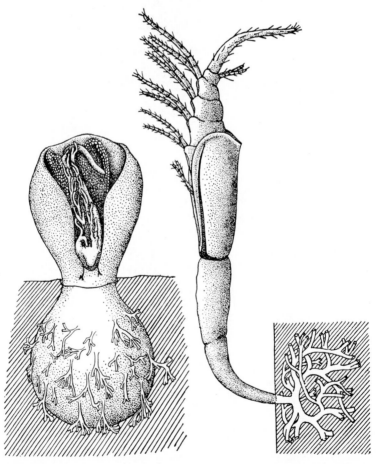

Fig. 16-27. *Anelasma squalicola* (left) partly buried in the skin of a shark; and *Rhizolepas annellidicola* (right) attached to an annelid host. (Baer, Ecology of Animal Parasites, courtesy of University of Illinois Press.)

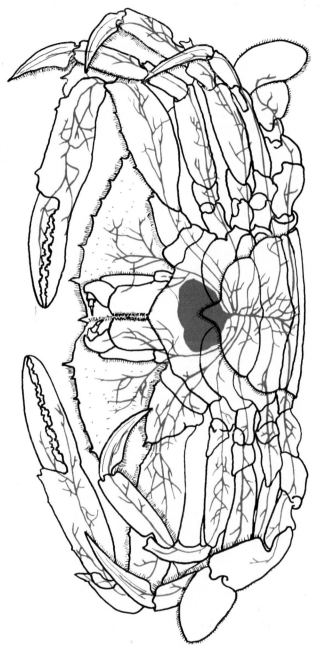

Fig. 16-28.   The root-like branches of *Sacculina* permeating the body of its crab host.

identical with those of free-living barnacles but there is no trace of a mouth or gut in the parasites. The adult rhizocephalid possesses neither appendages nor segmentation, and it is anchored to the host by a stalk from which roots proceed into the host tissues. The best-known Rhizocephala are *Sacculina carcini* and *Triangulus munidae*. The remarkable life history of *Sacculina* starts with a nauplius possessing the characteristic frontal horns of cirriped nauplii.[10] The cypris larva of *Sacculina* is active after dark and it attaches itself by its antennae to a crab at the base of a bristle. Soon after attachment the whole trunk with its muscles and appendages is thrown off and a new cuticle is formed under the old one. The body of the parasite becomes an amorphous cellular mass within the two valves of the cypris. From these cells the **kentrogon** larva (a mass of embryonic cells) is developed, and by means of a short dart-like tube that pierces the integument of the host, the kentrogon flows into the body of the crab. The parasitic embryonic cells are carried by the host circulation to the central surface of the gut where they begin to multiply and spread out in all directions as root-like branches. A rounded cellular mass (the future **sacculina externa**), containing the rudiments of genital organs and a ganglion, now appears at the base of the root system, and the parasite migrates to a spot on the ventral side of the crab near the single diverticulum of the gut. At this point the parasite makes a hole through the host integument, and the external portion of the barnacle is thrust through this hole, and it grows into the hermaphroditic adult. The **sacculina interna** continues to grow and it may permeate almost the entire body of the host (Fig. 16–28). A "sacculinized" crab may live with its barnacle parasite for at least two years, and it may become infected with two or more individuals. The crab, however, ceases to molt and its metabolism, particularly that relating to sexual development and activity, is profoundly modified. (See page 473 for a discussion of the effects of this parasite on the sex determination of the hosts.)

Thompsonia, parasitic on crabs, hermit crabs, and other decapods, becomes so extremely modified as to resemble a fungus in both appearance and function. Like *Sacculina*, the parasitic rootlets are diffused throughout the host tissue, but in *Thompsonia* these rootlets give off sacs which attain an external position at a molt of the host. The sacs contain nothing but ova which ripen (probably by parthenogenesis) into cypris larvae.

Rhizocephala have been reported from fresh water. *Sacculina gregaria* in the crab, *Eriocheir japonicus*, travels with its host from brackish water to fresh water, but this parasite is able to reproduce only when the crab returns to brackish water to incubate its eggs and shed its larvae. Rhizocephala apparently are tolerant of a variety of different hosts, thus demonstrating a lack of close host specificity. For further details of parasitic barnacles, see Baer,[2] and Smith.[25]

REFERENCES

1. Atkins, D.: *Pinnotherion vermiforme* Giard and Bonnier, an entoniscid infecting *Pinnotheres pisum*. Proc. Zool. Soc. Lond., Part II, 319–363, 1933.
2. Baer, J. G.: Ecology of Animal Parasites. Urbana, Univ. Illinois Press, 1951.
3. Borror, D. J., and DeLong, D. M.: An Introduction to the Study of Insects. ed. 3. New York, Holt, Rinehart and Winston, 1970.
4. Bowen, J. T., and Putz, R. E.: Parasites of freshwater fish: IV, Miscellaneous 3. Parasitic copepod *Argulus*. Fish Disease Leaflet No. 3. Bur. Sport Fisheries and Wildlife. Washington, D.C., 1966.
5. Bowman, T. E.: A new genus and species of whale-louse (Amphipoda: Cyamidae) from the false killer whale. Bull. Mar. Sci. Gulf Caribbean, 5:315–320, 1955.
6. Chapman, R. F.: The Insects, Structure and Function. Amer. New York, Elsevier Publ. Co., 1969.
7. Clausen, C. P.: Entomophagus Insects. New York, McGraw-Hill, 1940. 688 pp.
8. ———: Respiratory adaptations in the immature stages of parasitic insects. Arthropoda, *1*:197–224, 1950.
9. Darwin, C.: A Monograph on the Sub-class Cirripedia, with Figures of all the Species. Vol. 1 (1851), Vol. 2 (1854). London, Ray Society.

10. Day, J. H.: The life-history of *Sacculina*. Quart. J. Micros. Sci., *77*:549–583, 1935.

11. Dudley, P. L.: Development and systematics of some Pacific marine symbiotic copepods. A study of the biology of the Notodelphyidae, associates of Ascidians. Seattle, Univ. Washington Press, 1966.

12. Friend, G. F.: The life-history and ecology of the salmon gill-maggot *Salmincola salmonea* (L) (Copepod Crustacea). Trans. Roy. Soc. Edinburgh, *60*:503–541, 1941.

13. Furman, D. P.: Manual of Medical Entomology. Palo Alto, California, The National Press, 1961.

14. Herms, W. B.: Medical Entomology. ed. 5. (rev. by M. T. James). New York, Macmillan, 1961.

15. Leung, Y.: An illustrated key to the species of whale-lice (Amphipoda, Cyamidae), ectoparasites of Cetacea, with a guide to the literature. Crustaceana, *12*:279–290, 1967.

16. Illg, P.: North American copepods of the family Notodelphyidae. Proc. U.S. Nat. Mus., *107*:463–649, 1958.

17. Imms, A. D.: A General Textbook of Entomology. London, Methuen & Co. Ltd., 1957.

18. Maple, J. D.: The eggs and first instar larvae of Encyrtidae and their morphological adaptation for respiration. Univ. Calif. Pub. Ent., *8*:25–122, 1947.

19. Margolis, L.: Notes on the morphology, taxonomy, and synonymy of several species of whale-lice (Cyamidae: Amphipoda). J. Fish. Res. Bd. Canada, *12*:121–133, 1955.

20. Menzies, R. J., Bowman, T. E., and Alverson, F. G.: Studies of the biology of the fish parasite *Livoneca convexa* Richardson (Crustacea, Isopoda, Cymothoidae). Wasman J. Biol., *13*:277–295, 1955.

21. Paterson, N. F.: External features and life cycle of *Cucumaricola notabilis* nov. gen. et sp., a copepod parasite of the holothurian, *Cucumaria*. Parasitology, *48*:269–290, 1958.

22. Reinhard, E. G.: Experiments on the determination and differentiation of sex in the bopyrid *Stegophryxus hyptius* Thompson. Biol. Bull., *96*:17–31, 1949.

23. Reverberi, G., and Pitotti, M.: Il Ciclo Biologico e la Determinazione Fenotipica del Sesso di *Ione Thoracica* Montagu. Bopiride Barassita di *Callianassa laticauda* Otto. Pubb. Staz. Zool. Napoli, *19*:111–184, 1942.

24. Scott, T., and Scott, A.: The British Parasitic Copepoda. Vol. I. Copepoda Parasitic on Fishes. V. II. Plates (72 plates). The Ray Society. London, Dulau & Co., 1913.

25. Smith, G.: Rhizocephala. Fauna und Flora des Golfes von Neapel., *29*:1–122, 1906.

26. Wilson, C. B.: (Numerous papers on North American copepods) Proc. U.S. Nat. Museum, *25* through *64*, 1902–1924.

Chapter 17

Class Insecta I

ORDER ORTHOPTERA

Family Blattidae

Cockroaches belong to this family. They are flattened insects whose head is turned so far under that the mouth parts are directed backwards. The antennae are long and slender and the wings, when present, are brownish and semitransparent. They possess chewing mouth parts.

Cockroaches are worldwide in distribution and comprise about 2,260 species belonging to approximately 250 genera. The more important species from the parasitologist's point of view are the American cockroach (*Periplaneta americana*), 30 to 40 mm. in length; the German cockroach (*Blatella germanica*), 14 to 16 mm. long; the Oriental cockroach (*Blatta orientalis*), about 25 mm. long; and the Australian roach (*Periplaneta australasiae*), similar to *P. americana* but lighter in color. A very large species (*Blaberus cranifer*) is found in Cuba. It may reach 65 mm. in length by 35 mm. in width, or even larger.

The life history begins with eggs that are deposited in leathery packets (**oothecae**), which sometimes protrude from the genital aperture of the female. Eggs hatch into nymphs that gradually metamorphose into adults. Scent glands in the adult give these insects a characteristic odor.

Roaches often frequent filthy places and do not have discriminatory tastes in food. They readily eat starchy and sugary materials as well as sputum and feces. They have been charged with transmitting viruses, bacteria, protozoa, tapeworms and thornyheaded worms. Probably they are not important vectors from the medical or veterinary point of view, but certainly they may be accidental carriers of parasitic spores, cysts, eggs or even motile stages.

LICE

In 1842 Henry Denny, working on lice, wrote, "In the progress of this work, however, the author has had to contend with repeated rebukes from his friends for entering upon the illustration of a tribe of insects whose very name was sufficient to create feelings of disgust." Such feelings for lice have inspired the application of the terms "cooties," "crabs," and "vermin" for these tiny insect pests of the skin. "Plant lice," "book lice," "bark lice," "dust lice," and "louse flies" are also insects, but not closely related to true parasitic lice. Also, the crustacean "fish lice" and "wood lice" add to the confusion of common names. Most beginning students of parasitology hear a great deal about lice, but see only a few preserved specimens. A careful examination of a bird's feathers, or a quick combing of a mammal's fur with a fine-toothed comb will generally reward the searcher with numerous lice as well as with other arthropod ectoparasites. When present in large numbers lice may cause an intolerable itch, or they may be the vectors of serious disease.

All lice are parasitic on the surfaces of birds and mammals. In temperate climates lice are most numerous during February and March, and least numerous between June and August. Reasons for this kind of seasonal variation in numbers are not well understood. The lice are divided into the order Anoplura, which contains species whose mouth parts are adapted for sucking the blood and tissue fluids of mammals—hence often called "sucking lice"; and the order Mallophaga, containing the lice whose mouth parts are adapted for chewing epithelial structures on the skins of their hosts. The chewing habit has given rise to the term "biting lice" for the Mallophaga, but they do not actually bite. The classification used here is that devised by Hopkins.[5]

The Mallophaga have broad heads, at least as broad as the thorax, whereas the Anoplura have heads that are narrower than the thorax (compare Figs. 17–2 and 17–4). Another readily distinguishable feature is the presence in the Mallophaga and the absence in the Anoplura of pigmented, heavily sclerotized mandibles.

All lice are wingless, have dorsoventrally flattened bodies, short antennae with three to five segments, and reduced or absent eyes. The thorax, indistinctly segmented, bears one pair of spiracles. The short legs possess tarsi whose claws are used for grasping feathers or hair. The abdomen, always without cerci, generally bears six pairs of spiracles. Unlike most arthropod parasites, all lice spend their

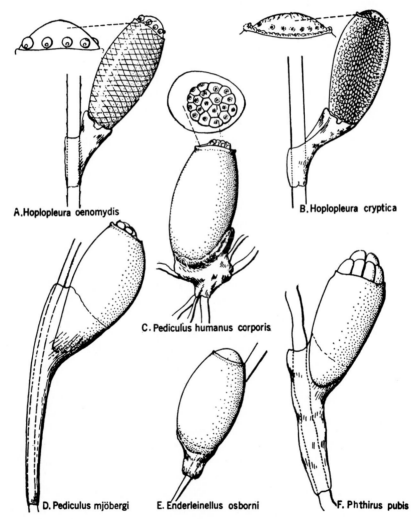

A. Hoplopleura oenomydis
B. Hoplopleura cryptica
C. Pediculus humanus corporis
D. Pediculus mjöbergi
E. Enderleinellus osborni
F. Phthirus pubis

Fig. 17-1.   Eggs of lice.   (Ferris, The Sucking Lice.   Memoirs of the Pacific Coast Entomol. Soc., Courtesy of Calif. Acad. Sci., San Francisco)

entire lives on the bodies of their hosts, and infestation from one host to another is accomplished by direct contact. For this reason host specificity is relatively well marked (see p. 536). Lice eggs, known as "nits," are attached by the female to the feathers or hairs of the host (Fig. 17–1).

An interesting method of emergence of the nymph from the egg of the pigeon chewing louse, *Columbicola columbae*, was described by Rothschild and Clay.[11] The egg cap (operculum) is provided with small pores through which air may pass to the inside. The nymph within the egg, when ready to leave, swallows air which passes through the digestive tract and is forced to the bottom of the egg. The air pressure thus accumulated gradually thrusts the nymph up against the operculum which is pushed open by the head of the nymph. The human louse nymph, *Pediculus humanus*, is said to emerge in the same manner, but other lice and the bedbug, *Cimex lectularius*, and certain fleas open their eggs either by the use of hatching spines on the head, or by pressure caused by an enlarged head distended with amniotic fluid that has been swallowed.

Both Anoplura and Mallophaga possess a relatively simple life cycle with gradual metamorphosis. The first nymph, which develops within the egg, is structurally similar to the adult, differing chiefly in its smaller size, absence of color and underdeveloped sex organs. Four nymphal instars, each separated by a molt (ecdysis), precede the adult stage. In *Columbicola* each nymphal stage lasts six or seven days, and between each molt the body becomes successively larger and darker in color. After the third and final molt the nymph becomes either a male or a female sexually mature adult. In many species the female, which is almost always larger than the male, far outnumbers the latter. Sometimes the male of a species is rare, or has never been found. Although parthenogenesis is known to occur in at least one species of mammalian louse, the rarity or apparent absence of males may be attributed to their death immediately after mating.

Many lice, particularly those whose diet is chiefly keratin, possess within their bodies symbiotic intracellular bacteria which presumably aid in the digestion of food. The bacteria pass from louse to louse by way of the eggs. Similar bacteria are found in ticks, mites, bedbugs, and in some blood-sucking Diptera. The bacteria possibly play a role in host specificity, being unable to cope with "strange" blood if the louse gets on the wrong host.

Order Mallophaga

This order contains nearly 3000 known species of lice, primarily parasitic on birds, but also on dogs, cats, horses, cattle, goats, sheep, and other mammals, including marsupials. The Mallophaga have not been convicted as effective carriers of disease-producing microorganisms. Feather lice range in size from the minute males of *Goniocotes*, which scarcely reach 1 mm. in length, to the large hawk-infesting *Laemobothrion*, which attains a length up to about 10 mm. The mouth parts, adapted for chewing epithelial materials, may have severe effects on the hosts. Literature on the diet and digestive processes of members of this order has been reviewed by Waterhouse.[12] He notes that lice eat the protecting sheaths of growing feathers, feather-fiber, down, skin-scurf, scabs, blood, their own eggs and cast skins, and probably also mucus and sebaceous matter. The lice on birds apparently can digest keratin (see note above on symbiotic bacteria). Some species on mammals ingest hair, and it is possible that these species chew off pieces from the sides in somewhat the same manner as does a beaver when it fells a tree. However, mammalian Mallophaga probably prefer a diet resembling that of many of the species parasitic on birds (e.g. epidermal scales, skin-scurf, wool-wax). Although they possess chewing mouth parts, some species of Mallophaga regularly feed on the freshly-drawn blood of their hosts. For example, *Menacanthus stramineus*, the body-louse of poultry, actively feeds on its host's blood, but other species prefer blood clots.

It should also be noted that some species of Mallophaga parasitic on birds possess piercing mouth parts, and that they feed, as do the Anoplura, on the blood and tissue fluids of their hosts. One genus, *Piagetiella*, has selected the throat-pouches of pelicans and cormorants as a preferred place to live.

It is possible that certain birds attempt to rid themselves of lice by picking up ants and placing them on their feathers or by perching on an ant hill with wings outstretched. It is presumed that ants eat lice, but this presumption has been questioned. Another interesting idea is that the birds eat ants, and the resulting formic acid in the bird discourages the lice. We do know that practically all mammals that eat ants are nearly free from lice. Scratching by the host undoubtedly helps to rid the animal of fleas, but the effectiveness of this habit against lice is questionable. However, Eichler counted 20,000 lice on the skin of one dog which, due to a defect, could not scratch itself. Damage to feathers caused by lice is probably very light, but it is difficult to separate damage caused by lice from that caused by feather mites. Although the Mallophaga are not effective carriers of human disease, one species (*Trichodectes canis*) acts as the intermediate host for a dog tapeworm, and the genus *Dennyus* on swifts serves as the intermediate host of a filaria (*Filaria cypseli*).

The order Mallophaga is divided into the three suborders Amblycera, Ischnocera, and Rhynchophthirina. The latter is sometimes considered as a separate order, and it is represented by only one species, *Haematomyzus elephantis*, parasitic on African and Indian elephants. The other two suborders (Fig. 17–2) will be described briefly.

Fig. 17-2. The two main types of feather lice. *Left*, Amblycera. *Right*, Ischnocera (Rothschild and Clay, Fleas, Flukes and Cuckoos, courtesy of Wm. Collins Sons and Co. Ltd.)

Suborder Amblycera

Species of lice belonging to this suborder are parasitic on both birds and mammals, and they include the feather, shaft and body lice of poultry, and a number of pests on guinea pigs and other rodents. They may be distinguished by the antennae, which are almost always composed of four segments, the third being stalked (Fig. 17–3). The antennae always lie in a ventrolateral groove on each side, and they may or may not project beyond the side of the head. The maxillary palps, sometimes absent, may also project beyond the sides of the head in preserved specimens, and may, therefore, be mistaken for antennae. The mandible chewing motion is in a hori-

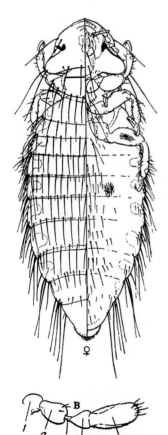

Fig. 17-3. *Menopon gallinae,* female. Note that the antenna (B) of this species, unlike that of most Amblycera, has five segments. (Lapage, Veterinary Parasitology, courtesy of Oliver and Boyd Ltd.)

zontal plane. The ninth and tenth segments of the abdomen are fused.

Examples of the suborder Amblycera are as follows:

Menopon gallinae is often called the shaft louse of fowl. The female is about 2 mm. long and it lays its eggs in clusters on the feathers (Fig. 17–3).

Menacanthus stramineus (Eomenacanthus) is the "chicken body louse" of fowl and turkeys. The female is 3.3 mm. long and it prefers the skin to feathers. It may seriously injure adult fowl. One chicken may be infested with more than 6,000 lice of several species (Fig. 17–4).

Heterodoxus contains species which occur on dogs (in warm countries only, not in Europe), kangaroos and wallabies.

Suborder Ischnocera

Lice belonging to this suborder are commonly found on cattle, equines, goats, sheep, dogs, cats, and also on fowl and other birds. The lice can be distinguished by their easily visible filiform antennae composed of three to five segments (Fig. 17–2). There are no maxillary palps. The first and second, and the ninth and tenth segments of the abdomen are fused, and the eleventh segment may not be visible.

Examples of the suborder Ischnocera are as follows.

Bovicola (Damalinia) is found on cattle, sheep, goats, deer, other two-toed ruminants and horses. *B. bovis (Trichodectes bovis)* is called the "cattle biting louse." It is the most widely distributed louse on cattle in Britain, and is also common in North America. The lice in winter are found at the base of the tail, on the shoulders and along the back, unless there is a very heavy infestation, in which case they may be found all over the host body. The irritation caused by these active pests may be severe, and the infested cattle often try to rid themselves of the lice by biting the skin and rubbing themselves against tree trunks, fence posts and rocks.

Trichodectes contains species found on dogs, martins, weasels, badgers, skunks, and other small mammals. The female is almost 2 mm. long, and it may serve as the vector of the larval stage of a dog cestode (Fig. 17–5).

Cuclotogaster (= *Lipeurus*) *heterographus* is found on the skin and feathers of the head and neck of fowl, partridge and other birds. It is often called the fowl head louse, and it may be seriously injurious. The female is about 2.45 mm. long, and its eggs are laid singly on the feathers (Fig. 17–6).

Order Anoplura (=Siphunculata)

About 300 species of sucking lice have been described, and they are all parasitic on mammals. In addition to their characteristic piercing mouth parts, the lice can be distin-

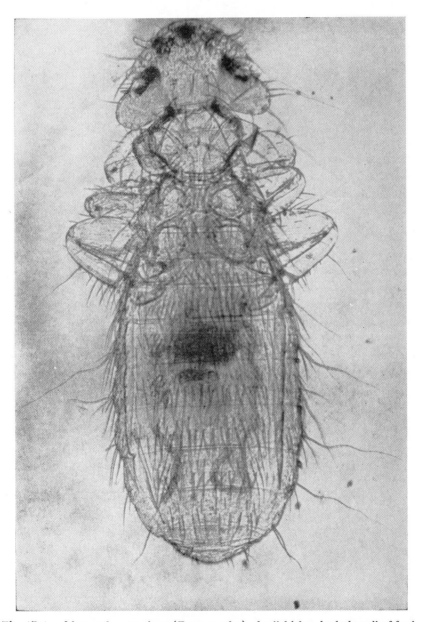

Fig. 17-4. *Menacanthus stramineus (Eomenacanthus)*, the "chicken body louse" of fowl. (Lapage, Veterinary Parasitology, courtesy of Oliver and Boyd Ltd.)

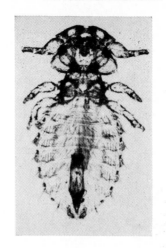

Fig. 17-5. *Trichodectes canis.* Female, *left;* male, *right.* (Georgi, Parasitology for Veterinarians, courtesy of W. B. Saunders)

guished by the small, fused thorax and by their antennae, which are usually composed of five segments and are always visible. The eyes are reduced or absent, and the third pair of legs, often broad and flattened, is usually the largest. Each tarsus has only one claw. The single pair of thoracic spiracles opens on the dorsal side of the mesothorax. The six pairs of abdominal spiracles are arranged as those of the Mallophaga. Only seven of the nine abdominal segments are visible externally. The mouth parts, adatped for piercing and sucking, are retracted into the head when not in use, thereby obviating the need for a proboscis.

The family Linognathidae contains *Linognathus vituli*, the bluish-black, long nosed cattle louse. Its first pair of legs is much smaller than the others. The family Neolinognathidae contains only two species, parasitic only on elephant-shrews in East and South Africa. The genus is *Neolinognathus*. The family Haematopinidae contains *Haematopinus suis*, the largest of all lice to be found on farm stock. The female is 4 to 6 mm. long, and the male 3.4 to 4.75 mm. long. It is the only species of louse that infects the domestic pig (Fig. 17–7).

Family Pediculidae

Human lice. This family contains the human head louse and the body louse of the genus *Pediculus*, and the human pubic or crablouse belonging to the genus *Phthirus* (Fig. 17–8). *Pediculus humanus* occurs in two forms, the head louse, *P. humanus capitis*, and the body louse, *P. humanus humanus* (= *P. humanus corporis*). The two look very much alike. A typical head louse and a typical body louse are rather easily distinguished, but they overlap in appearances and movements. Head lice average approximately 2.4 mm. long and are slightly smaller than body lice. Body lice, however, are seldom if ever found on the head, whereas head lice may be found on the body. The lice are adapted for clinging to hairs, but the body lice have found refuge in the clothing with which man has compensated for his nudity. Head lice and body lice can interbreed and produce fertile offspring which may possess characters intermediate between the two parents.

Copulation between a male and female human louse begins when the male crawls underneath the female, approaching her from behind. When the tips of their abdomens unite the female rears up to a vertical position, lifting the male with her. The two lice then return to a horizontal position and remain united for 30 minutes to several hours. The female may lay up to 300 eggs during her life which, under optimal conditions, lasts about a month. The oval eggs, or nits, are laid singly, and they measure about 0.8 by 0.3 mm. At 30° C. the eggs hatch in eight or nine days and the young nymph is 1 mm. long.

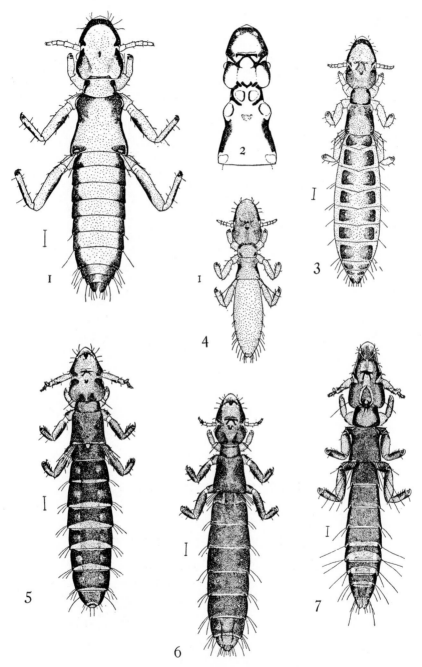

Fig. 17-6. *Cuclotogaster* (*Lipeurus*), various species.
(Kellogg, courtesy of Leland Stanford Jr. Univ. Publ.)

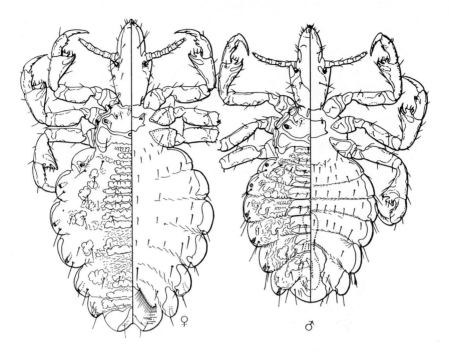

Fig. 17-7. *Haematopinus suis.* (Ferris, The Sucking Lice, Memoirs of The Pacific Coast Entomological
Society, courtesy of Calif. Acad. Sci., San Francisco)

LICE COMMONLY FOUND ON MAN

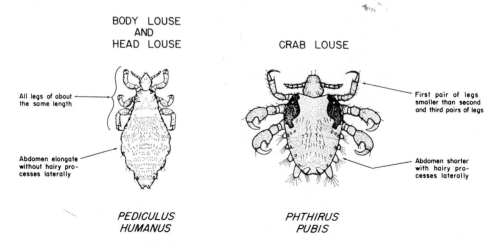

Fig. 17-8. Lice commonly found on man.

Adult lice can suck up as much as 1 mg. of blood at a time, but probably they prefer to take smaller quantities at frequent intervals. At 30° C. they can survive only about three days of starvation. Experiments have shown that lice can move at a rate of 9 inches (23 cm.) per minute. They prefer a temperature of 29° to 30° C., and they avoid, when possible, any change in humidity. Their immediate past experience conditions the response to environmental changes—hence different individuals often exhibit different responses. Movement toward dark areas is chiefly a response to directed light received by the horizontally placed eyes.

Phthirus spp. possesses a wide thorax that constitutes most of the body: and the coxae are far apart at the margins of the thorax. *P. pubis* (Fig. 17–8) frequents the pubic hairs and perianal regions of man, but it may wander to other parts of the body, including the head. It is smaller and much broader than the other two human lice, and its legs have the appearance of being attached to the edge of the somewhat flattened body. The forelegs are slender with long fine claws, whereas the middle and hind legs are thick, with thick claws. The adult seems to be unable to survive for longer than a day when removed from its host, and its total span of life is about one month. Although transmission from person to person is accomplished by close contact, it is a grave injustice to assume that sexual contact is the only mode of transfer.

The Effect of Sucking Lice on Their Hosts. *Haematopinus suis* probably transmits the virus of swine fever, and lice in general probably carry tularemia and leishmaniasis. *Polyplax spinulosa* of black and brown rats transmits the rickettsia that causes typhus. Nevertheless, the sucking lice are not known to be of any great importance as vectors of domestic and wild animal diseases. They may be called host irritants, and because they often increase rapidly in numbers they frequently are the cause of scratching, restlessness, biting, loss of sleep, and interruption of feeding. During the biting process the saliva

injected into the wound prevents the host's blood from coagulating as it is sucked through the slim mouth parts. It is the reaction of the host cells to the louse saliva that causes the symptoms of irritation. Lice seem never to be satisfied for long with one meal, but go on feeding all the time. Blood exuding and clotting at the sites of the bites may form a fertile location for the growth of bacteria, or an effective attractant for flies whose larvae may cause serious malady (see p. 422).

In man, *Pediculus humanus* can transmit impetigo, trachoma and cholera by simple mechanical contamination, and this louse is the normal vector of exanthematous typhus caused by *Rickettsia prowazeki*, of trench fever caused by *R. quintana*, and of louse-borne relapsing fever caused by *Spirochaeta recurrentis*. The rickettsia that are ingested by the louse with its blood meal multiply in its gut cells, which eventually burst, thus liberating large numbers of these parasitic bodies into the louse feces. Man acquires typhus by rubbing infected louse feces into an abrasion of the skin. The infected louse dies in about ten days after acquiring the cellular parasites because of its ruptured gut cell. Hans Zinsser's book, *Rats, Lice and History*,[14] gives a vivid picture of the manner in which lice and typhus fever have influenced the history of mankind.

Trench fever became widely prevalent during World War I and has since disappeared. The spirochaetes of relapsing fever penetrate the louse gut and multiply within the fluid of the body cavity. They remain in this fluid until the louse host is crushed, at which time the spirochaetes escape and may enter a human host through an abrasion of the skin. Obviously the habit of "popping" lice with the finger nails, and the disgusting practice of biting lice to kill them is seriously hazardous if relapsing fever is prevalent.

Head lice are frequently a serious problem among urban children, but much more rare among adults. Body lice are largely confined to those adults and children who do not change their underwear frequently. Conditions of crowding, especially during sleeping hours, are particularly conducive to the spread of lice. For these reasons such groups as destitute people, refugees, vagrants, prisoners, and armies provide fertile fields for lice populations. A chronicler contemporary of the well-known church dignitary St. Thomas à Becket has written that on the morning after the churchman was murdered his vestments were removed, and the haircloth underwear was so infested with lice

that they "boiled over" as the cloth was stripped from the cold body.

Primary defense against human lice involves personal cleanliness of body and clothing. It is well to remember that even in civilized communities there are always some chronically lousy individuals whose feeble efforts at personal cleanliness are ineffectual, or who have ceased to worry about their condition and who passively offer their parasites to every passerby.

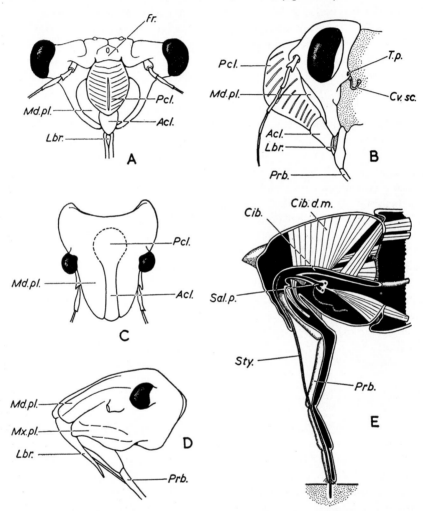

Fig. 17-9. Head and mouth parts of *Hemiptera*. *A, B.* Frontal and lateral views of *Magicicada septendecim* (after Snodgrass, 1935). *C, D.* The same of a Pentatomid, *Euschistus variolarius* (after Snodgrass, 1935). *E.* Section of head of *Graphosoma italicum* in feeding position (after Weber, 1930).

Acl. Anteclypeus. *cib.* Cibarium. *cib.d.m.* Cibarial dilator muscles. *Cv.sc.* Cervical sclerae. *Fr.* Frons. *Lbr.* Labrum. *Md.pl.* Mandibular plate. *Mx.pl.* Maxillary plate. *Pcl.* Postclypeus. *Prb.* Proboscis (labium). *Sal.p.* Salivary pump. *Sty.* Stylets. (Imms, A General Textbook of Entomology, courtesy of E. P. Dutton and Co.)

ORDER HEMIPTERA (BUGS)

Plant bugs and their relatives are predominantly phytophagous (plant feeders), having piercing and suctorial mouth parts. A considerable number of species show a strong tendency toward entomophagy, and in a number of instances plant feeding has been abandoned entirely and the predatory role is obligatory. Here, again, the delicate line separating a predator from a parasite defies discovery. The bugs typically possess two pairs of wings, the anterior pair being most often of a harder consistency than the posterior pair. The slender, segmented beak constitutes the most easily recognizable feature of bugs. Palpi are atrophied, and the labium is in the form of a dorsally-grooved sheath receiving two pairs of bristle-like stylets (modified mandibles and maxillae) (Fig. 17–9). Metamorphosis is incomplete.

Probably no other group of insects is so directly concerned with the welfare of man because of the vast amount of injury to plants brought about directly and indirectly by thousands of species of bugs. Chinch-bugs, leaf-hoppers, plant lice, cotton stainers, white flies, tea blight bugs, scale insects, and mealy bugs are among the most destructive.

Suborder Homoptera

The extraordinarily rapid rate of reproduction occurring in many members of the Homoptera constitutes a cardinal factor bearing upon the devastation for which these insects are responsible. Only a few of the parasites of plants will be mentioned because the subject of plant pathology, especially as it involves the insects, is so vast that it requires many volumes, and a special course of study, to do it justice. For this reason, and because of precedent, the animal parasites of plants have not been used in the traditional course in "Parasitology." Nevertheless, these parasites are excellent examples of the state of parasitism; they are easy to obtain, and they lend themselves readily to a study of basic principles.

Family Psyllidae

This is a family of "jumping plant lice" which frequently become serious parasites of fruit trees (pears, apples) and other plants. Nymphs and sometimes also adults of the genus *Psylla* (Fig. 17–10) damage the blossoms and stunt the shoots, and they often produce gall-like malformations. Nymphs of many species secrete large amounts of a white, waxy substance. The potato or tomato psyllid, *Paratrioza cockerelli*, transmits a virus disease to potatoes, tomatoes, peppers, and eggplants.

Family Aleyroididae

These minute "white flies" resemble tiny moths whose wings are dusted with a mealy, white, powdery wax (Fig. 17–11). The adults are usually observed while they are actively feeding on leaves, but the young, except the first instar, are sessile, and they look like scales. A conspicuous and characteristic organ, present in both nymphal and adult stages, is the vasiform orifice that opens on the dorsal surface of the last abdominal segment. Within the orifice is a tongue-shaped

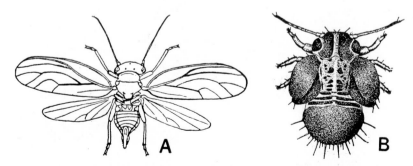

Fig. 17-10. *A. Psylla mali* (after Carpenter). *B. Psylla pyricola*, nymph in last instar (after Stingerland). All enlarged. (Imms, A General Textbook of Entomology, courtesy of E. P. Dutton and Co.)

organ upon which honeydew collects from the anus. White flies are abundant throughout the world, especially in the tropics and subtropics. They are parasitic on citrus trees, tomatoes, cucumbers and other plants. Damage results from the process of sucking sap from leaves.

Family Aphididae

Aphids or plant lice (Fig. 17–12) belong to this family. These familiar soft-bodied insects are frequently found in large numbers, consisting of individuals in all stages of development, sucking the sap from young shoots or leaves of plants. Diagnostic morphologic features of these insects are the cornicles or "honey tubes" situated at the posterior end of the abdomen. The cornicles are the channels from glands that secrete a waxy fluid that acts as a protection against predaceous enemies. Honeydew is emitted through the anus. The antennae are relatively long; the front wings are considerably larger than the hind wings, and when at rest they are generally held vertically above the body.

The life cycle of many aphids is unusual and complex, and the physiologic mechanisms underlying host-alternation, parthenogenesis and polymorphism are not clearly understood. Undoubtedly, seasonal changes in the physiology of the growing host plant are at least partly responsible for the complexities of the cycle. Aphids commonly cause a curling or wilting of the food plant, and they frequently serve as vectors of a number of important virus and fungus plant diseases. For details see Kennedy and Stroyan.[8]

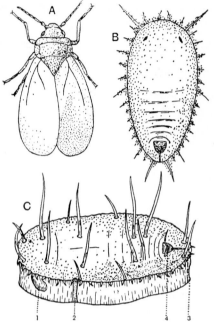

Fig. 17-11. The white fly *Trialeurodes vaporariorum*. *A.* Imago × 50. *B.* Larva in first instar × 150. *C.* Puparium × 65.

1. Adult eye. *2.* Thoracic breathing fold. *3.* Caudal breathing fold. *4.* Vasiform orifice. (Lloyd, Ann. App. Biol. 9, in Imms, A General Textbook of Entomology, courtesy of E. P. Dutton and Co.)

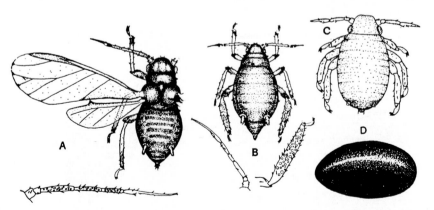

Fig. 17-12. *Aphis fabae. A.* Male. *B.* Oosporous female. *C.* Fundatrix, first instar. *D.* Egg. The antennae are also shown under higher magnification, together with the tarsus in *B.* (From original drawings by J. Davidson in Imms, A General Textbook of Entomology, courtesy of E. P. Dutton and Co.)

Superfamily Coccoidea

These are the familiar "mealybugs" (e.g. *Pseudococcus*) and "scale insects," all of which are characterized by the more or less degenerate, wingless females with obscure segmentation and often atrophied legs and antennae. The males have a single pair of wings or, rarely, they are wingless, and, lacking mouth parts, they do not feed. The males look like small gnats with a style-like process at the end of the abdomen (Fig. 17-13).

Like the aphids, many coccoidea secrete honeydew which renders them attractive to ants; and many secrete a powdery or clear wax or lac. These secretions have been used by man for centuries in the preparation of shellac and candle wax. The life histories of this family have not been studied extensively, but they include parthenogenesis, oviparity, and viviparity. The first instar nymphs are provided with functional legs, and their mobility ensures dispersal of the species.

A great many species of the coccoidea are injurious parasites of cultivated and other

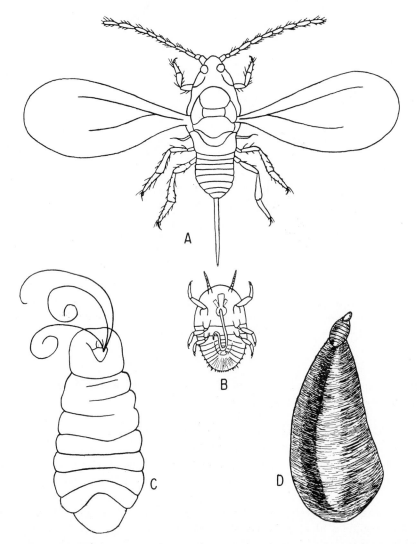

Fig. 17-13. The oystershell scale, *Lepidosaphes ulmi* (Linn.). *A.* Adult male. *B.* Newly hatched young, or crawler. *C.* Adult female. *D.* Scale of female. (Borror and Delong, An Introduction to the Study of Insects, courtesy of Rinehart and Co.)

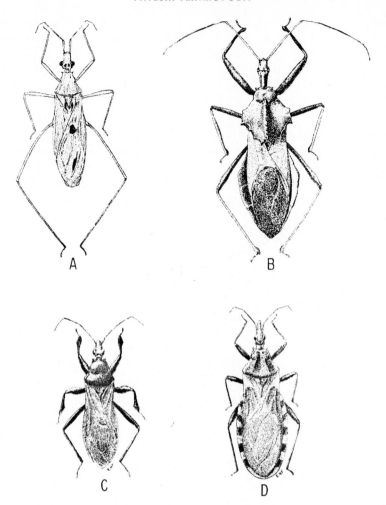

Fig. 17-14. Assassin bugs. *A. Narvesus carolinensis* Stal. × 2. *B.* The wheel bug, *Arilus cristatus* (Linn.), × 1½. *C. Melanolestes picipes* (Herrich-Schaeffer), × 2. *D.* A blood-sucking conenose, *Triatoma sanguisuga* (LeConte), × 2. (Courtesy of Froeschner and the American Midland Naturalist)

plants. Some species prefer only one kind of host (e.g., *Cryptococcus fagi* on *Fagus sylvaticus*), whereas others appear to have little host preference. *Lepidosaphes ulmi*, for example, is known to infest about 130 widely separated species of plants. The San José scale [*Aspidiotus* (= *Quadraspidiotus*) *perniciosus*] on deciduous fruits, and the red scales (*Aonidiella aurantii* and *Chrysomphalus aonidum*) on citrus trees, are fine examples of coccoidea whose feeding habits and general biology result in serious damage to their hosts. Some mealybugs are vectors of virus diseases.[4]

Suborder Heteroptera

Family Reduviidae

The 3000 or so described species of these bugs exhibit a wide range of variation in form. The subfamily Triatominae contains bugs of medium size to large, and usually of a black or brownish color, but sometimes with bright red or yellow markings (Fig. 17–14). The head is characteristically long, narrow, and with a neck-like portion immediately behind the eyes. The rostrum (beak) is three-segmented and bent sharply back under the head when at rest, and the antennae are filiform and are inserted on the sides of the head between the eyes and tip of the snout. The abdomen is often widened at its middle, exposing the margins of the segments lateral to the folded wings. The wings of Heteroptera are generally placed, when at rest, so as to reveal a conspicuous triangular area immediately posterior to the thorax. The reduviids commonly live on the blood of other insects, but a few triatomids ("assassin bugs") attack higher animals, including man. *Rhodnius prolixus* and species of *Triatoma* and of *Panstrongylus* are the natural vectors for *Trypanosoma cruzi*, the causative agent of Chagas' disease, a frequently-fatal form of human trypanosomiasis (see p. 31). Many species of assassin bugs will inflict a painful bite if carelessly handled, and severe allergic symptoms occasionally occur. The bugs are regularly found in nests or burrows of rodents or other host animals. For details of these bloodsucking bugs, see Buxton.[2]

Family Cimicidae

The bedbugs are parasitic on birds and mammals. Characteristic morphologic features (Fig. 17–15) are an oval, flattened body about 4 mm. long without wings, and covered with many hair-like spines. The compound eyes are conspicuous, but ocelli are absent; the rostrum lies in a ventral groove, and the

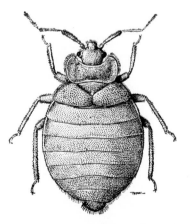

Fig. 17-15. Male bedbug. (The Bed-Bug, courtesy of British Museum Natural History Economic Series No. 5, 1954)

tarsi are three-jointed. The thorax consists of three segments: an anterior large prothorax extending forward on either side of the head; a mesothorax whose visible portion is a small triangle; and a metathorax hidden from view dorsally by two small pads which are rudimentary forewings.

Members of the genus *Cimex* are associated with birds, bats, man and other mammals, and have a worldwide distribution. In North and Central America *Haematosiphon inodorus* is a parasite of poultry.

The two most common bedbugs attacking man that are found the world over are *Cimex lectularius* of temperate climates, and *C. hemipterus* of tropical countries. Both species, colored dark mahogany or bright chestnut, hide in cracks, crevices, or under rugs by day, and emerge at night to feed on human blood. The sucking beak contains two pairs of stylets designed for piercing the skin. After puncturing the host skin the bug pumps saliva into the wound through the salivary tube, and then sucks up a mixture of the saliva and host blood. The saliva probably prevents coagulation of blood, and it is responsible for the irritating effects of the bite. Nymphs also feed on blood, and when the bug has finished feeding it usually pauses a moment, then runs away as fast as possible. The bugs are able to endure long fasts, and some have been kept alive without food for over a year. *C. lectularius* is found naturally also on chickens, rabbits and bats.

Bedbugs are generally considered as loathsome creatures because of their nocturnal, stealthy attacks and because of their peculiar pungent odor. One of the earliest references in English to these pests is that of Humphrey Lloyd who translated the following statement made by Pope John XXI in the sixteenth century: "Small stynkinge wormes which live in paper and wod, called Cimices."

Before the female bedbug can lay fertile eggs, she must not only mate but she must feed. The pearly white, slightly curved, operculated eggs are normally laid on rough surfaces like those found in crevices and behind wallpaper. A quick-drying cement fastens each egg (about 1 mm. long) securely to the surface of the material on which it is deposited. Upon hatching after six to ten days, the nymphs, about the size of a pinhead, appear very similar to the adult, and immediately seek shelter. There are five molts before the adult stage is reached. For details on the ecology of bedbugs, see Johnson.[7]

Bedbugs and Disease. Bedbugs are not responsible for the transmission, as natural vectors, of disease. Mechanical transmission of infections by means of contaminated feet of the insects may take place, and leptospirosis is known to be transmitted from guinea pig to guinea pig by the bite of bedbugs; but, in spite of some conflicting accounts and persistent suspicion, no clear evidence has been presented for bedbugs being more than harmless pests of mankind. McKenny-Huges and Johnson,[10] however, have called attention to the fact that "in infested areas it is often possible to pick out children from buggy homes by their pasty faces, listless appearance and general lack of energy. It can be argued that the house in which bugs are tolerated will also be the home of malnutrition, dirt and other causes of physical infirmity. Such causes cannot be held solely responsible, and sleepless nights with constant irritation due to the injection of the minute doses of bedbug saliva into the blood are likely to contribute largely to the ill-health of children and even of certain adults. Some fortunate people are not affected by the bites of bedbugs; others gain immunity after repeated biting; whilst others, less fortunate, are always susceptible."

ORDER HYMENOPTERA (WASPS AND ANTS)

Hymenoptera, of which there are probably about 100,000 described species, exhibit an enormous variation in behavior. The phe-

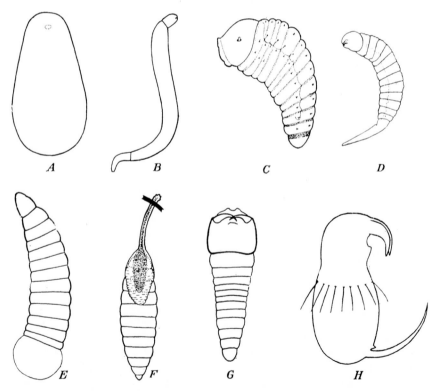

Fig. 17-16. First instar larval forms of some parasitic Hymenoptera. *A.* Sacciform. *B.* Asexual. *C.* Hymenopteriform. *D.* Caudate. *E.* Vesiculate. *F.* Encyrtiform. *G.* Mandibulate. *H.* Teleaform. (Clausen, Entomophagous Insects, courtesy of McGraw-Hill Book Co.)

nomenon of a complex social organization among ants, bees and wasps has captured the attention of everyone who has even a mild interest in the kaleidoscopic life of the insect world. The highly evolved condition which parasitism has reached in the order has been independently acquired among species belonging to diverse superfamilies. Even in the suborder Symphyta, which includes all of the more primitive members of the order, some species are parasites, as larvae, of plants.

The order consists chiefly of ants, bees, wasps, sawflies, and ichneumon flies. Parasitism is prominently displayed by members of the order, and insects are favored as hosts. Diagnostic features of the order include two pairs of membranous wings interlocked by means of hooklets; mouth parts adapted for biting, sucking or lapping; an abdomen usually constricted at its base and its first segment fused with the metathorax; an ovipositor modified for piercing, sawing or cutting; larvae often legless; pupae commonly in cocoons; metamorphosis complete.

The phenomenon of **polyembryony** (production of two or more embryos from a single egg) occurs in some of the parasitic Hymenoptera. During this process an egg, laid in the egg of a host insect, produces polar bodies which, instead of being discarded, become centers of cytoplasm surrounded by a membrane called a **trophamnion**. The whole embryonic mass of parasitic tissue divides to form a number of morulae, each of which becomes an embryo. In this manner a brood of 2 to about 3,000 parasites may be produced within a single host. Here the cost of the lives of hundreds or thousands of larval parasites is the life of one host—a bargain indeed!

Typical parasitic hymenopteran insects are entomophagous (insect-eating) and parasitic only during the larval stages. There is a great variety in the manner and place of oviposition by the female, ranging from the common internal placement of the egg within the body of the host to its deposition on foliage or in plant tissue far removed from the animal host. Upon hatching, the parasitic

larvae (Figs. 17–16, 17–17) feed upon the host, which is usually itself a larva or an egg and is thereby destroyed. The fact that a given larva of some species may be either ecto- or endoparasitic, according to the species of the host on which it feeds, indicates that the nature of the host is partly responsible for the behavior of the larva. The larvae of *Onchophanes lanceolator*, for example, is ectoparasitic on the caterpillars of *Cacoecia sorbiana*, but endoparasitic in caterpillars of other genera of Lepidoptera. Another wasp larva may be found attached to spiders. Insect parasites may pupate inside the host, on its outer surface, or even entirely apart from it. Details of the processes of host selection, site selection for oviposition, and placement of the egg on or in the host body vary immensely, but one example is outlined to emphasize the complexity of the habits of parasitic Hymenoptera. This example is *Tiphia*, described on page 390.

Some of the larger or more important groups of parasitic Hymenoptera are described and illustrated in the following discussions. Examples have been selected more or less at random to indicate the diversity of the parasitic mode of life as adopted by these insects. The system of classification and the factual information are taken largely from Imms.[6] All of the examples belong to the suborder Apocrita, which includes the vast majority of the Hymenoptera, and which is characterized by a deeply constricted abdomen between the first abdominal segment (propodeum) and the second, and by larvae without legs. A series of numerous superfamilies is recognized, and six of these are mentioned.

Venom. Parasitic Hymenoptera paralyze and sometimes kill their hosts prior to oviposition, but some species apparently never inject venom into the host. The host reaction to the venom may be immediate or it may be delayed for several minutes. The venom of the wasp *Microbracon hebetor* diluted to 1 part in 200,000,000 parts of the host hemolymph is sufficient to cause permanent paralysis. The site of action in the host appears to be the neuromuscular junction.

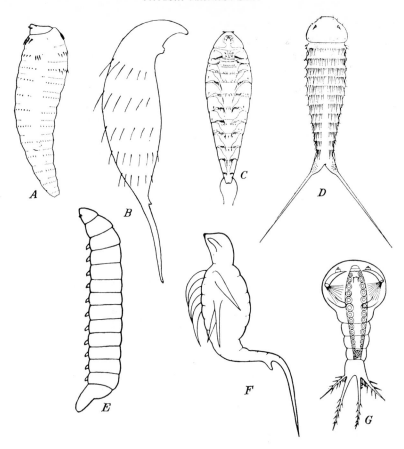

Fig. 17-17. First instar larval forms of some parasitic Hymenoptera. *A*. Microtype. *B*. Mymariform. *C*. Planidium. *D*. Agriotypiform. *E*. Polypodeiform. *F*. Eucoiliform. *G*. Cyclopiform. (Clausen, Entomophagous Insects, courtesy of McGraw-Hill Book Co.)

Superfamily Ichneumonoidea

Ichneumon flies (Fig. 17–18) are predominantly parasites of Lepidoptera, but many of them use as hosts the Hymenoptera, Coleoptera, Diptera, Arachnida, and a few other arthropods. Most species of the parasites are probably seldom restricted to any individual species of host, and in their behavior the ichneumons are among the most highly evolved of all solitary insects. Practically all are external parasites of the immature stages of their hosts. The presence of a caudal tail on the newly-hatched ichneumon larva of many species is a prominent feature. The head in the young larva is large and often strongly sclerotized, but older larvae (third instar) generally become maggot-like with reduced heads. Fully fed, mature larvae construct silken cocoons. Adults commonly display remarkably long ovipositors, that of *Megarrhyssa* reaching up to fifteen cm. (six inches) in length.

Braconid wasps (Fig. 17–19), like the ichneumonids, select a wide variety of hosts, but the Lepidoptera are the most commonly parasitized. The braconids are closely similar in both structure and habits to the ichneumonids. The well-studied *Apanteles glomeratus* is a common parasite of the larva of *Pieris* (Lepidoptera) which may support 150 parasites. Mature larval parasites (or parasitoids) gnaw their way through the skin of their host and then produce a mass of sulphur-yellow cocoons irregularly heaped together.

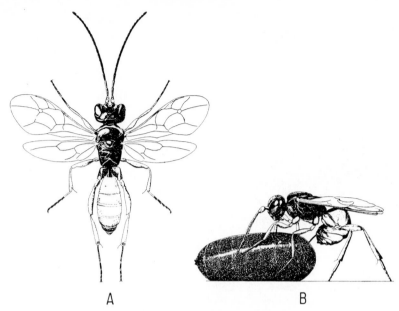

A   B

Fig. 17-18. A hyperparasitic ichneumon, *Phygadeuon subfuscus* Cresson (Gelinae). *A*. Adult male. *B*. Female ovipositing in puparium of host. The host of this ichneumon is a tachinid fly, *Aplomyiopsis epilachnae* Aldrich, which is parasitic on the Mexican bean beetle, *Epilachna varivestis* Mulsant. (Courtesy of U.S.D.A.)

Fig. 17-19. A female braconid (*Lysiphlebus testaceipes*) ovipositing in an aphid. (Clausen, Entomophagous Insects, courtesy of McGraw-Hill Book Co.)

Superfamily Cynipoidea

Gall wasps (Fig. 17–20) are small, often minute, darkly colored insects that lay eggs in plant tissues, mainly on oaks, thereby producing swollen galls around the developing larvae. Males are rare, parthenogenesis common, and the pedicel-equipped eggs hatch into legless larvae which do not spin a cocoon. The parasites gain shelter and nutrition from the galls, or are inquilines in them. The forms of galls produced by these insects are almost endless and they may occur in all parts of the plant, from the roots to the flowers. An explanation of the phenomenon of gall-formation has not yet been made to the satisfaction of botanists or entomologists. All that can be said at the present time is that galls are produced as the result of reactions of the cambrium and other meristematic plant tissues in response to the stimuli induced by the presence of the living larval parasite. In addition to the insects that actually produce a gall, the latter is frequently the abode of Diptera, Coleoptera, Lepidoptera and other Hymenoptera.

Superfamily Chalcidoidea

Chalcid wasps are small insects with elbowed antennae, and they probably comprise the largest group in the order Hymen-

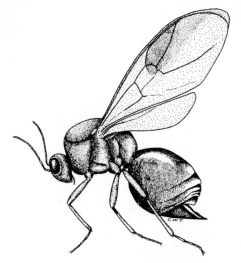

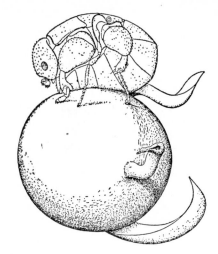

Fig. 17-20. A gall wasp. *Diplolepis rosae* (Linn.). This species develops in the mossy rose gall. (Borror and DeLong, An Introduction to the Study of Insects, courtesy of Rinehart & Co.)

Fig. 17-21. Male *Blastophaga psenes* fertilizing the female, the latter within a galled flower of the fig. (Imms, A General Textbook of Entomology, courtesy of E. P. Dutton and Co.)

optera. The bulk of the species are either parasites or hyperparasites of other insects, and they are of great economic value in the biologic control of insect pests (see p. 493). Some larvae infest the seeds or stems of plants, others are hyperparasites of other insects which, in their turn, destroy insect species harmful to man. The orders most frequently parasitized are Lepidoptera, Homoptera, Diptera and Coleoptera. Enormous numbers of eggs and larvae of Lepidoptera succumb to attacks by chalcid wasps but the pupal stage is rarely invaded. The Coccoidae (Hemiptera), often called "scale insects" or "mealy bugs," are the most universally attacked of any group of insects. For a general account of the biology of the group see Clausen.[3]

Family Agaonidae

Among the most remarkable chalcids are the "fig insects" which maintain a symbiotic relationship with various species of *Ficus*. The best known species is *Blastophaga psenes* (Fig. 17–21), which is a symbiont of the fruit of *Ficus carica*. In the words of Imms:

The eggs of this Chalcid are laid in the ovaries of the caprifig and give rise to galls therein. The male imago emerges first and, on finding a gall containing a female, commences to gnaw a hole through the wall of the ovary and fertilizes the female while the latter is still *in situ*. The female leaves the receptacle through the opening at its apex, and laden with adherent pollen, flies to a neighboring fruit. If the latter be in the right condition she seeks the opening and gains admission into the interior of the receptacle, where she commences oviposition. Should the caprifig, from which she has emerged, be suspended in a tree of the Smyrna variety, she enters a fruit of the latter, but subsequently discovers that she has selected a wrong host, since the flowers are of such a shape that they do not permit oviposition within them. After wandering about for a while, she usually crawls out of the receptacle and incidentally pollinates the flowers. Most of the males die without ever leaving the receptacles in which their development took place. *B. psenes*, while in the fig, sometimes becomes the host of another wasp (*Philotrypesis*) which may be called a "cleptoparasite."[9]

The solitary wasp, *Specoidea*, is also an example of cleptoparasitism. The female wasp provides its nest with paralyzed prey, lays its eggs, and the larval parasites feed on the stolen prey.

Other Families

Hardly a single order of insects is immune

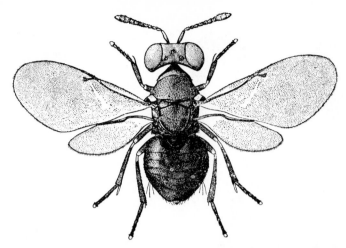

Fig. 17-22.   A typical chalcid, *Blastothrix sericea*, female, magnified.
(Imms, A General Textbook of Entomology, courtesy of E. P. Dutton and Co.)

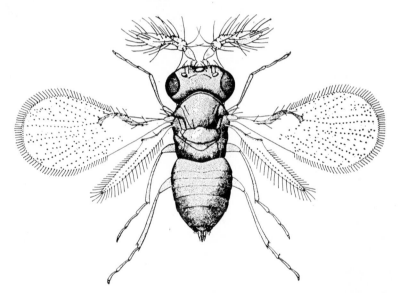

Fig. 17-23.   The adult male of *Trichogramma minutum*.
(Clausen, Entomophagous Insects, courtesy of McGraw-Hill Book Co.)

from attack by members of Encyrtidae. *Blastothrix* (Fig. 17–22) has been studied extensively.   In the family Trichogrammatidae, the genus *Trichogramma* (Fig. 17–23), which usually parasitizes the eggs of Lepidoptera (as many as 20 individuals may develop within one egg) has been used extensively in connection with biologic control. *T. minutum* alone has been found to attack more than 150 host species representing seven orders of insects.   The family Pteromalidae is one of the most common, and some of the species in this family are external gregarious parasites of lepidopterous and coleopterous larvae and pupae (Fig. 17–24).   Among the Aphelinidae, *Eretmocerus serius* has been used effectively for the biologic control of "white flies" (Hemiptera).   This parasite (Fig. 17–25) has the unusual habit of being ectophagous during the first and a portion of the second stage, and then entering the body of its host when the latter becomes a pupa.

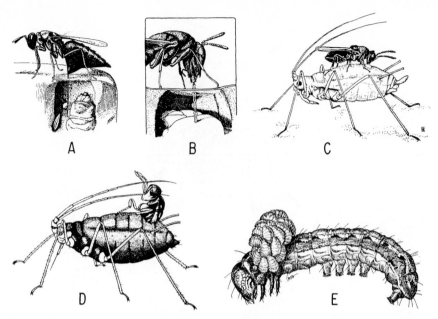

Fig. 17-24. Feeding and emerging chalcids. *A. Habrocytus* (Pteromalidae) ovipositing. *B. Habrocytus* feeding at the tube made by her ovipositor. *C. Zahropalus inquisitor* (Howard) (Encyrtidae) feeding at a oviposition puncture made in the abdomen of an aphid. *D.* Adult of *Aphelinus jucundus* Gahan (Aphelinidae) emerging from an aphid. *E.* A colony of *Euplectrus* larvae (Eulophidae) feeding on a caterpillar. (*A* and *B*, courtesy of Fulton and the Entomological Society of America; *C* and *D* courtesy of Griswold and Entomological Society of America; *E*, courtesy of U.S.D.A.)

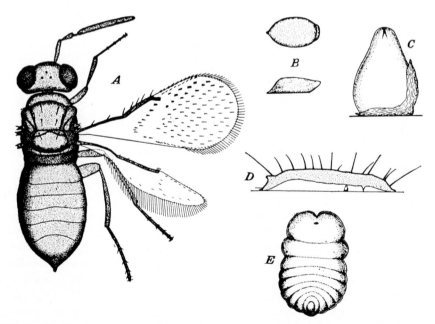

Fig. 17-25. Developmental stages of *Eretmocerus serius*. *A.* The adult male. *B.* The egg, dorsal and lateral views. *C.* The first-instar larva. *D.* Host larva with first-instar larva beneath the body. *E.* The mature larva. (Clausen, Entomophagous Insects, courtesy of McGraw-Hill Book Co.)

Marietta carnesi (Fig. 17–26) is a tiny aphelinid wasp, a little over a millimeter long, which parasitizes another wasp, *Comperiella bifasciata*, which is parasitic in a scale insect. The first wasp is thus a hyperparasite on the primary parasite of the scale insect.

Superfamily Proctotrupoidea (Serphoidea)

All the members of this superfamily are parasitic, some on eggs of other insects, some on larvae and pupae; some are hyperparasites and a few are inquilines. The majority of species develop cocoons, but in the aphid-infesting forms the body of the host protects the pupa. *Platygaster dryomyiae* (Fig. 17–27) exhibits polyembryony. *Mantibaria manticida* (*Riela*) exhibits an exceptionally advanced type of parasitism. It is an interesting case of an adult hymenopteran insect that feeds on the blood of its host. Development takes place within the egg mass of the praying

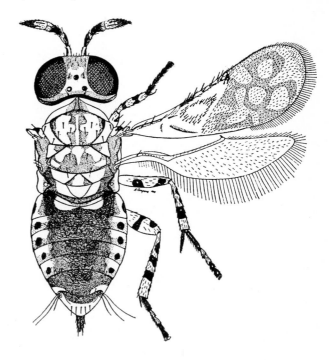

Fig. 17-26. *Marietta carnesi*, female. A hyperparasite of the primary parasite of a scale insect. (Compere, courtesy of Univ. Calif. Publ. Entomol.)

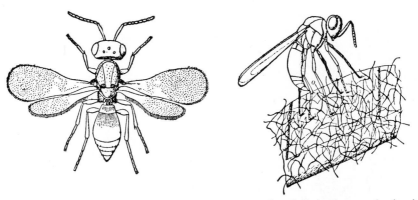

Fig. 17-27. *Platygaster dryomyiae*, female. On the right the female is in the act of oviposition in an egg of *Dryomyia* (Cecidomyidae). (Imms, General Textbook of Entomology, courtesy of E. P. Dutton and Co.)

mantis, *Mantis religiosa*. After the adult parasites settle down on the adults of the host, the female parasite gnaws through the chitinous veins at the base of the wings of the mantis. Here the parasites cast off their wings and lead ectoparasitic lives at about the only spot where the parasite can escape being brushed off by the formidable front legs of the mantis. If the mantis is a female and has commenced oviposition, the adult parasite migrates to the genital region to lay its eggs in the host ootheca while the latter is being formed. Parasites that settle upon the male mantis perish with their hosts.

Superfamily Scolioidea

Scolioid wasps belong to an extensive group including some of the most primitive members of the stinging forms, as well as some of the largest of the Hymenoptera. Many of these wasps are parasitic, usually on coleopterous larvae. The wasps are generally large, hairy, and dark in color. Their bodies are often marked with spots or bands of yellow or red, and their wings frequently display a metallic iridescence.

Family Tiphidae

This is the dominant family of hymenopteran parasites attacking scarabaeid beetle grubs in the soil. They are solitary wasps which develop as external parasites, usually upon the final larval instar of the host. The genus *Tiphia* (Fig. 17–28) is the most cosmopolitan and the most common of the family, and the adult female commences its parasitic activities by burrowing into the soil to gain access to the cell occupied by the grub. The following account is taken from Clausen,[3] and it describes the behavior of *T. popilliavora*. As soon as the female wasp finds a grub:

she first crawls over the dorsum of the body from the rear, then curls the abdomen down and around the side, and inserts the sting in the center of the thorax, usually between the first two segments. This stinging is repeated until the grub is quiescent. She then turns to the abdomen and commences an extensive kneading of the ventral surface with the mandibles, beginning with the first segment and continuing for its length. When this is complete, she grasps the lateral margin with the mandibles and coils the body transversely over the dorsum and to the ventral surface, forming almost a complete circle. The tip of the abdomen is applied to the groove between the fifth and sixth abdominal segments, near the margin, and is rhythmically moved backward and forward for several minutes, thus broadening the groove and possibly rasping away a portion of the integument to permit of more ready penetration by the larva. In the course of this preparation, any egg or young larva that may be present as a result of an earlier oviposition is rubbed off or broken. The egg is finally extruded and is firmly attached by a mucilaginous material. It lies transversely in the groove, with the anterior pole directed toward the median ventral line of the host body. The wasp may then quit the body or remain for a period of feeding. . . . The grub recovers from effects of the sting in twenty to forty minutes.

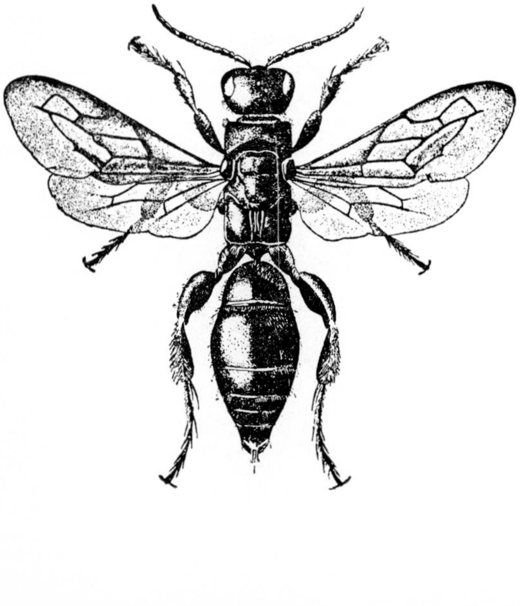

Fig. 17-28. *Tiphia transversa*, female vespoid wasp. (Imms, A General Textbook of Entomology, courtesy of E. P. Dutton and Co.)

Superfamily Apoidea

Social and solitary bees are best known through the voluminous literature on the honeybee, one of the best understood of all insects. Of the numerous parasitic genera of the family Apidae the largest is probably *Nomada* whose members are similar to wasps in appearance, and are usually black and yellow in color, with almost bare bodies. Species of this genus are usually parasitic on different species of *Andrena*, a common solitary bee of the Holarctic region, and on other genera of bees.

ORDER COLEOPTERA
(BEETLES)

Over a quarter of a million species of beetles have been described, and a great many more are undoubtedly yet to be found by man. The order contains a relatively small number of parasitic forms, most of which are ectoparasites. Probably the most distinctive feature of beetles is the elytron or front wing. Most beetles have four wings, but the front pair is thickened, leathery or hard and brittle, and it serves as a protective sheath over the longer and membranous hind wings which are the only ones normally used for flight. The elytra generally meet in a straight line down the middle of the back.

Family Staphylinidae

This family comprises the rove beetles. Adults are recognized by their slender bodies, short elytra, and by their habit of elevating the abdomen when disturbed. Only a few species of this family are true parasites, the best known being in the genera *Coprochara*, *Aleochara*, and *Baryodma*. The larvae of these beetles live within the puparia of Diptera, but they are obligate external parasites upon the pupae.

Family Rhipiphoridae

This family comprises a few hundred species, all parasitic as larvae in the larvae of the Hymenoptera, cockroaches and a few other insects. The adult beetles are conspicuous with their streamlined bodies, comb-

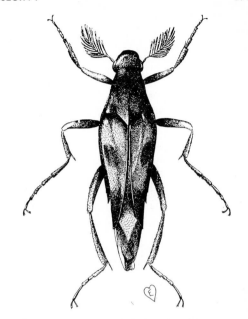

Fig. 17-29. The adult female of *Macrosiagon pectinatus*, a rhipiphorid beetle. (Clausen, Entomophagous Insects, courtesy of McGraw-Hill Book Co.)

like or fibrous antennae of the males, and varied color patterns (Fig. 17–29). The triungulinids (first instar larvae, Fig. 17–30) of all species are equipped with a caudal sucker, and they are able to assume an erect position with their legs free for grabbing on to passing insects or other objects. Each larval instar is markedly different from the others, a condition known as hypermetamorphosis. *Metoecus paradoxus*, common in Europe, is a parasite of the larvae of wasps (*Vespa* spp.). Only a portion of the first stage is passed internally, just beneath the skin, and the second larval stage of the parasite is found as a collar encircling the neck of the host.

Family Meloidae

The members of this very large family are called blister beetles or oil beetles. The long-legged, soft-bodied adults have a strongly deflected head and often reduced elytra, or no wings at all. The larvae, as in the case of the family Rhipiphordae described above, exhibit hypermetamorphosis (Fig. 17-31). Although the adult meloids are entirely plant

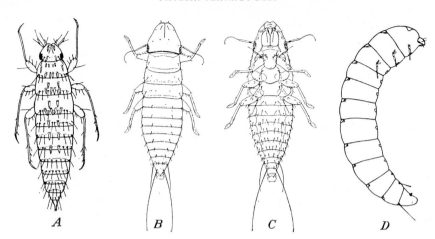

Fig. 17-30. First instar larvae of the Rhipiphoridae. *A. Rhipidius denisi* Chob. (after Chobaut). *B* and *C. Macrosiagon flabellatum F.*, dorsal and ventral views, before feeding. *D.* The same at the completion of the endoparasitic phase, immediately before the first molt, showing the extreme distention of the body and the wide separation of the segmental plates. (Grandi in Clausen, Entophagous Insects, courtesy of McGraw-Hill Book Co.)

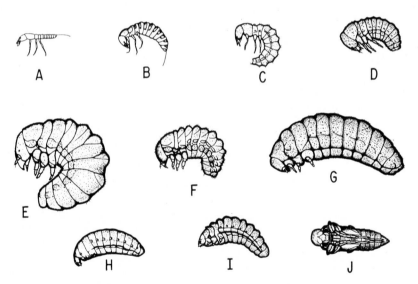

Fig. 17-31. Larval and pupal instars of the black blister beetle, *Epicauta pennsylvanica* (DeGeer), showing hypermetamorphosis. *A.* Newly hatched first instar, or triungulin. *B.* Fully fed first instar. *C.* Second instar. *D.* Third instar. *E.* Fourth instar. *F.* Newly molted fifth instar. *G.* Gorged fifth instar. *H.* Sixth instar (coarctate larva of pseudopupa). *I.* Seventh instar. *J.* Pupa. (Courtesy of Horsfall and the Arkansas Agricultural Experiment Station)

feeders, the larvae are parasitic or predaceous; the parasitic species prefer honey cells of various bees of the families Megachilidae and Andrenidae as their transient homes. In the majority of cases the relationship between the beetle larva and the bee larva is not one of true parasitism. The parasite generally consumes the egg of the host, then remains to feed upon the food mass stored in the cell by the parent bee. Many species of these beetles are predators on the egg pods of locusts. An example of a species that attacks bees is *Tricrania sanguinipennis*. Another example is *Sitaris muralis*, which attacks bees of the genus *Anthophora*. Eggs of the parasite are deposited near the nest of the bees in August, and the newly hatched triungulinids hibernate until the following spring when the more successful individuals attach themselves to the hairy bodies of the bees. *Anthophora* constructs cells in the ground, and when the female bee deposits an egg in a honey-filled cell, a triungulid slips off her body, alights on the egg, and thereby becomes imprisoned in the sealed-up cell where it eats the egg and honey, molts several times and becomes an adult plant-feeding beetle.

ORDER STREPSIPTERA (TWISTED-WINGED PARASITES OR STYLOPIDS)

Stylopids are small, endoparasitic insects that exhibit a striking sexual dimorphism in the adult stage. The males have branched antennae, large protruding eyes, club-like structures (halteres) instead of fore wings, and large hind wings. The females are wingless and normally they remain in the host enclosed in a puparium which protrudes slightly from the body of the host. A few females leave the host and have a larviform structure with a terminal gonopore. Only about 300 species are known, and these species have commonly been placed within the order Coleoptera. All species, so far as is known, complete their larval development within the body of the host in a manner similar to that of the beetle *Rhipidius* in cockroaches (see p. 392). The first instar larvae of stylopids bear a striking resemblance to those of the Rhipiphoridae (Coleoptera, see p. 391). The minute adult males are not encountered by collectors nearly as often as are the females, and many species have consequently been described on the basis of one sex only. In common parlance, insects harboring these parasites are said to be "stylopized."

Although a few records of Orthoptera and Hemiptera as hosts to the Strepsiptera are known, the preferred hosts are members of the Homoptera (chiefly Auchenorrhyncha) and of the Hymenoptera (chiefly Vespoidea, Sphecoidea and Apoidea). The following account of the biology and habits of these parasites is based upon *Xenos vesparum*, a parasite of wasps and bees (see Imms[6] and Clausen[3]). The male lives only a few hours after emerging from the host, but the female remains permanently endoparasitic with its cephalothorax protruding through the body wall of the wasp or bee. Copulation takes place by the male alighting on the host and inserting the aedeagus (penis plus lateral structures) into the aperture of the brood canal of the female. Larvae hatch within the body of the

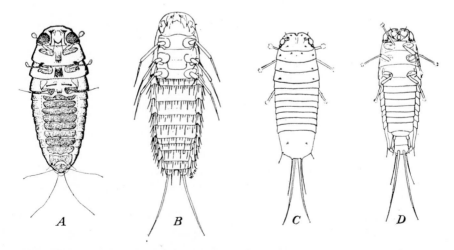

A   *B*   *C*   *D*

Fig. 17-32. First instar larvae of the Strepsiptera.
(Clausen, Entomophagous Insects, courtesy of McGraw-Hill Book Co.)

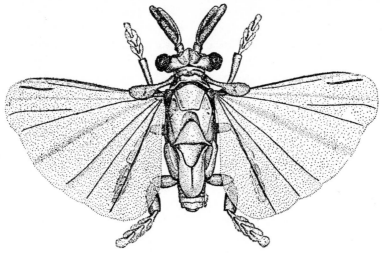

Fig. 17-33. *Stylops shannoni*, male, enlarged.
(Imms, General Textbook of Entomology, courtesy of E. P. Dutton and Co.)

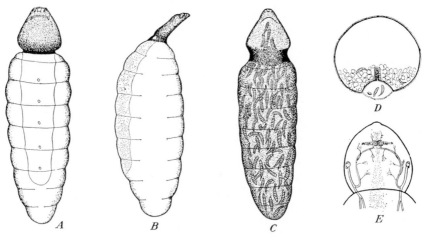

Fig. 17-34. Adult females of the Strepsiptera, with details. *A. Stylops melittae* Kirby, showing the genital openings on the cephalothorax and the brood chamber. *B.* The same, lateral view. *C.* A gravid female of *Halictophagus curtisii* Dale, showing fully developed triungulinids distributed throughout the body. *D.* A diagrammatic cross section through the fourth abdominal segment of *S. melittae*, showing the outer portion of the genital canal and several triungulinids in the brood chamber. *E.* Cephalothorax of *Xenos vesparum* Rossi, dorsal view, showing the single pair of spiracles, the anterior commissure and the longitudinal tracheal trunks, which divide in the first thoracic segment. (Redrawn after Nassanov, 1892, 1893 in Clausen, Entomophagous Insects, courtesy of McGraw-Hill Book Co.)

female and issue in large numbers, sometimes several thousand, through the genital canals into the brood pouch. Larvae eventually emerge through the brood canal, and they remain as active creatures upon the body of the host until opportunity is afforded for escape. The first instar larvae are known as triungulinids (Fig. 17–32). The larvae probably leave the first or "maternal" host when the latter are closely associated with others on flowers or in the nest. By simple attachment to adult wasps or bees the larval parasites are transported to larval hosts, within which they speedily burrow through the body wall and take up lodgings in body spaces between the organs, pushing them out of position. Ab-

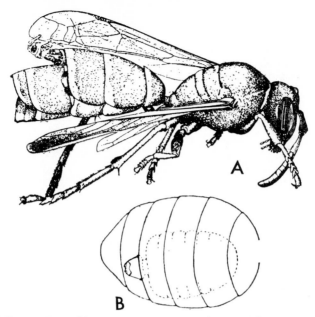

Fig. 17-35. *A. Polistes gallicus*, with a male *Xenos vesparum* (a stylopid) hatching from the puparium. *B.* Abdomen of *Andrena vaga*, with adult female of *Stylops* sp. (Baer, Ecology of Animal Parasites, courtesy University of Illinois Press)

sorption of nutrients from the host blood results in growth of the parasite, and after the seventh instar the parasitic larva works its way outward and protrudes from the body wall of the host which is by now in the pupal stage of development. The male parasite now undergoes pupation, and soon the winged insect is liberated (Figs. 17–33, 17–35). The female (Figs. 17–34, 17–35) becomes a white grub-like creature remaining within the host.

The effects of stylopids on their hosts vary considerably, depending on numbers of parasites within one host, length of time which the parasites spend within the host, seasonal and other considerations. Generally, however, marked changes in the growth and development of the host occur, including pronounced alterations of the secondary sexual characters.[13] Much confusion has arisen from the description of "new" species of stylopized wasps and bees. For further details, see Bohart.[1]

REFERENCES

1. Bohart, R. M.: A revision of the Strepsiptera with special reference to the species of North America. Univ. Calif. Pub. Ent., 7:91–160, 1941.
2. Buxton, P. A.: The biology of a bloodsucking bug, *Rhodnius prolixus*. Trans. Entomol. Soc. Lond., 78:227–236, 1930.
3. Clausen, C. P.: Entomophagous Insects. New York, McGraw-Hill, 1940.
4. Ferris, G. F.: An Atlas of the Scale Insects of North America. 6 vols. Stanford, California, 1937–1953.
5. Hopkins, G. H. E.: The host-associations of the lice of mammals. Proc. Zool. Soc. Lond., 119:387–604, 1949.
6. Imms, A. D.: A General Textbook of Entomology. London, Methuen & Co. Ltd., 1957.
7. Johnson, C. G.: The ecology of the bedbug, *Cimex lectularius* L. in Britain. J. Hyg., 41:345–461, 1942.
8. Kennedy, J. S., and Stroyan, H. L. G.: Biology of aphids. Ann. Rev. Entomol., 4:139–160, 1959.
9. Kuttamathiathu, J. J.: The biology of *Philotrypesis caricae* (L.), parasite of *Blastophaga psenes* (L.) (Chalcidoidea: Parasitic Hymenoptera). Proc. XVth Internat. Congr. Zool., London, p. 662–664, 1958.
10. McKenny-Hughes, A. W., and Johnson, C. G.: The Bedbug. Its Habits and Life History and How to Deal With It. Brit. Mus. Nat. Hist. Econ. Ser., 5, London, 1954.
11. Rothschild, M., and Clay, T.: Fleas, Flukes

and Cuckoos. A Study of Bird Parasites. London, Collins, 1952.

12. Waterhouse, D. F.: Studies on the digestion of wool by insects. IX. Some features of digestion in chewing lice (Mallophaga) from bird and mammalian hosts. Aust. J. Biol. Sci. (Melbourne), 6:257–275, 1953.

13. Wheeler, W. M.: The effects of parasitic and other kinds of castration in insects. J. Exp. Zool., 8:377–438, 1910.

14. Zinsser, H.: Rats, Lice and History; being a study in biography, which after twelve preliminary chapters indispensable for the preparation of the lay reader, deals with the life history of typhus fever. Boston, Little, Brown, 1935.

Chapter 18

Class Insecta II

ORDER LEPIDOPTERA (MOTHS AND BUTTERFLIES)

The vast majority of butterflies and moths are entirely free-living, feeding on plant materials, but in one family, the Epipyropidae, comprising minute moths, the entomophagous habit reaches its highest development among all Lepidoptera. Some of these moths are obligate external parasites attacking principally Homoptera of the family Fulgoridae (leaf-eaters). All species of these parasites are limited to a single host during larval development. Instances of true parasitism by members of other lepidopterous families are exceedingly rare.

In a typical example of a parasitic moth, the female deposits numbers of eggs upon the foliage of the food plant of the host. Young larvae wander about in search of a host, and when it is found, they attach themselves to the host body with their heads directed caudally, and feed upon its secretions. The mandibles of the parasite become embedded in an aperture in the integument of the host abdomen, but the effects of parasitism upon the host appear to be slight. The family Cyclotornidae also includes parasites; the first instar larvae of *Cyclotorna* (in Australasia) parasitize Homoptera, whereas the second instar larvae live in ants' nests.

Eye-frequenting moths occur in Africa and Southeast Asia. These insects feed from lachrymal secretions of cattle, water buffaloes, sheep, antelopes, pigs, horses, elephants, and other mammals, possibly including man. The moths may be vectors of infectious organisms causing eye disease.

ORDER SIPHONAPTERA (FLEAS)

Over 1,300 species of fleas have been described, and undoubtedly there are many more not yet discovered by man. The adults only are parasitic, and they, like the lice, are restricted to birds and mammals on whose blood they feed. Fleas have bodies that are laterally compressed, and the first segment, the coxa, of each leg is large, aiding in the jumping ability for which fleas are justly famous. Most of us may recall how difficult it is to catch a flea. This difficulty is the result of the small size of the flea, its laterally compressed and slippery body, the backward directed spines, and the strong, active legs tipped with claws. These characteristics enable fleas to jump from one host to another and to move easily between hairs and feathers. The compact structure of a flea is adapted to a forward movement only, and to change even the field of vision requires an alteration in the position of the entire insect.

Probably the most convenient place to look for a flea is on the body of a domestic dog or cat. Both of these hosts may be attacked by the dog flea, *Ctenocephalides canis*, the cat flea, *C. felis*, the human flea, *Pulex irritans*, the poultry flea, *Echidnophaga gallinacea*, or occasionally by other species of fleas. Many wild animals are infested with *Pulex simulans*. For a comprehensive work on fleas of western North America, see Hubbard.[6]

Structure

Conspicuous features of the external anatomy of fleas are shown in Figures 18–1 to 18–5. The antennae of males are nearly

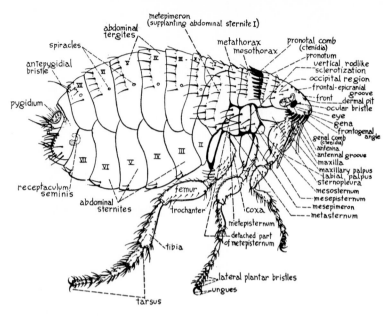

Fig. 18-1 *Ctenocephalides felis*, the cat flea, adult female. (Dept. of Health
Education and Welfare, Public Health Service, Atlanta, Ga.)

always longer than those of females, and
during copulation the male takes up a position
underneath the female and holds her firmly
with his antennae from below. The spines
along the mouth edge serve to prevent hairs
or feathers of the host from interfering with
the action of the mouth parts of the flea
during feeding. The body of the male has
an upward tilt posteriorly, whereas the female
body is evenly rounded terminally.

Fleas are truly "encased in a suit of armour,"
and each segment of the thorax, the prothorax, mesothorax and metathorax, may be
regarded as a membranous ring of adjoining
plates. The notum of the prothorax is often
armed with a row (comb) of heavy, pigmented spines (one row on each side) called
the **pronotal ctenidium** or **pronotal comb.**
The spines help the flea to move through
the feathers or hairs and they also help to
protect underlying parts.

The abdomen consists of ten segments, and
each segment, like those of the thorax, has a
dorsal sclerite called the **tergum,** and a
ventral sclerite called the **sternum,** except
that the first segment lacks a sternum. On
the abdomen these plates overlap, permitting
considerable flexibility inside the abdomen,
and when a flea has become engorged with a
blood meal the plates appear as islands
separated by broad, bare bands of skin. The
shape and bristle pattern of the seventh abdominal segment in female fleas has considerable taxonomic importance.

Tergum nine in the male is greatly modified
to form a clasping apparatus used during
copulation with a female. The clasper is of
primary value in the identification of males.
The exceedingly complicated genital structures of the male are illustrated in Figure 18–4.
The ninth segment of both male and female
fleas has on its tergum a dorsal sensory plate
called the **pygidium** (sometimes called the
sensilium) covered with pits, bristles and
hairs (Fig. 18–5). The whole structure suggests a tiny pin cushion, and it possibly
functions in the detection of air currents.
Such a function would assist the flea in finding a host which may be moving about.

The **spermatheca** (Fig. 18–6) is taxonomically the most important genital structure of
the female flea. It consists of a wide **head**

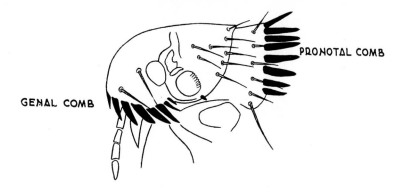

CTENOCEPHALIDES CANIS

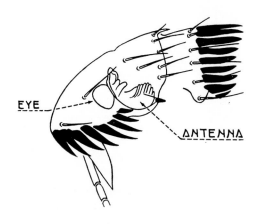

CTENOCEPHALIDES FELIS

Fig. 18-2. Head of *Ctenocephalides canis*, the dog flea, and of *C. felis*, the cat flea. (Capriles, courtesy of Puerto Rico J. Pub. Health Trop. Med.)

or **reservoir**, and a terminal, long sausage-shaped **tail** or **appendix**.

The alimentary tract is the internal organ system of particular importance in disease transmission. The mouth leads to a thick-walled pharynx equipped with pumping muscles, thence to a narrow gullet or esophagus which enters a pear-shaped **proventriculus** or **gizzard**. The gizzard is provided internally with a series of spines that project backward in front of the entrance to the stomach. These spines help to crush the blood cells of the host. Between the gizzard and the stomach is a valve which prevents the food in the stomach from being vomited during the process of digestion. So effective is the valve that pressure applied from behind sufficient to burst the stomach wall does not force food back into the esophagus. At the posterior end of the large stomach are situated four tubular glands which function as kidneys. The short intestine is equipped distally with six small, oval, rectal glands. A pair of salivary glands lies on each side of the stomach, and they are connected by a common duct that leads to the pharynx. Muscular attachments at the expanded end of the salivary duct constitute a salivary pump.

During the process of biting and feeding, the piercing mouth parts enter the host skin

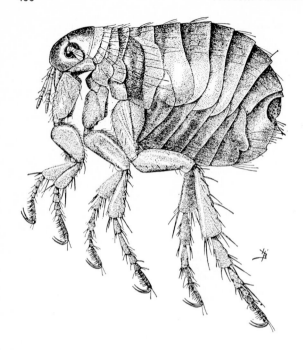

Fig. 18-3. *Pulex irritans*, the human flea, adult female, greatly enlarged. (Bishop Farmers Bulletin, 683.)

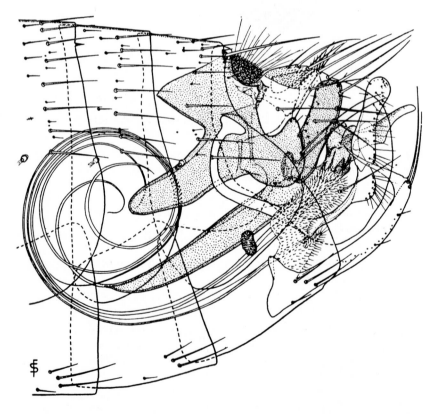

Fig. 18-4. *Ceratophyllus gallinae*, male. Clasper, coiled penis, sensilium and other features of the posterior end of the abdomen. (Lapage, Veterinary Parasitology, courtesy of Oliver and Boyd, Ltd.)

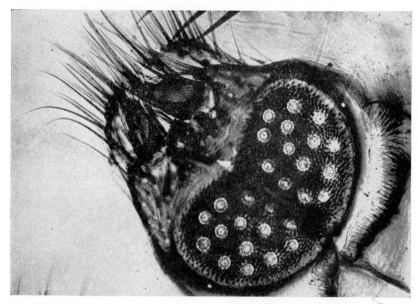

Fig. 18-5. The pygidium (sensilium) of a flea. × 435. (Rothschild and Clay, Fleas, Flukes and Cuckoos, courtesy of Wm. Collins Sons and Co.)

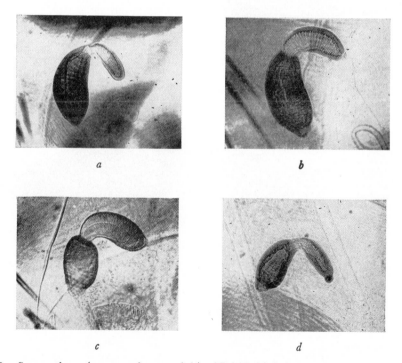

Fig. 18-6. Spermatheca (receptaculum seminis) of British bird fleas and mammal fleas. *a. Ceratophyllus arei* (from carrion crow). *b. Ceratophyllus borealis* (from rock pipet). *c. Monopsyllus sciurorum* (from red squirrel, for comparison with *a* and *b*). *d. Ceratophyllus columbae* (from rock dove). (Rothschild and Clay, Fleas, Flukes and Cuckoos, courtesy of Wm. Collins Sons and Co.)

and the flea thrusts its head downward, elevating the abdomen and hind legs. The rostrum parts of the palpi lie flat on the skin of the host, apparently serving to check any tendency to warping on the part of the slender piercer, and possibly also to assist in levering up the mouth parts at the end of the meal. The mouth parts are withdrawn with a sudden jerk. When a flea bites, the salivary pump pours out a stream of saliva that reaches the host blood vessels via the groove on the inner surface of the laciniae. At the same time the pharyngeal pump works to draw up the host blood, mixed with saliva, and forces it into the esophagus and stomach where it is digested.

Life History and Habits

During their life cycles fleas pass through a complete metamorphosis from egg to larva to pupa to adult (Fig. 18–7). The egg is relatively large, smooth, oval and translucent. If a flea-infested house cat is permitted to sleep on a black cloth during the night, the flea eggs may easily be seen scattered over the cloth the next morning. Mingled with the

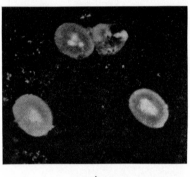

A

B

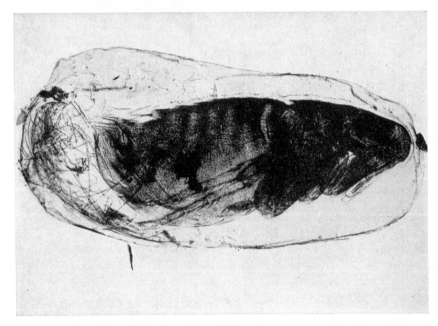

C

Fig. 18-7. Life cycle of a flea. *A.* Eggs, × 29. *B.* Larvae, × 19. *C.* Pupa within cocoon, × 37. (Rothschild and Clay, Fleas, Flukes and Cuckoos, courtesy of Wm. Collins Sons and Co.)

eggs are bits of dried blood and hairs. The human flea, *Pulex irritans* (Fig. 18–3), may lay well over 400 eggs during her lifetime. In two to ten days the eggs hatch into eyeless and legless but very active larvae. The strongly sclerotized head bears two short antennae, and the 13 other body segments (three in the thorax, ten in the abdomen) bear numerous bristles.

Anal struts consisting of a pair of blunt, hooked processes distinguish flea larvae from those of dipterous insects. Under favorable conditions the larvae may reach their third stage in about two weeks, but development may be delayed for six months or more. The larvae feed on various organic debris to be found in the host's nest, crevices in floors and under rugs. Blood derived from the excreta of adult fleas is sometimes a necessary part of the larval diet. The larvae of bird fleas thrive on broken-down sheaths of feathers and on epidermal scales of young birds.

The larva molts twice before reaching the third stage, when it spins a cocoon within which it pupates. The larval stage may last from one to 24 weeks. The pupae may live for a week up to a year, depending on the species and such environmental factors as temperature and moisture. The entire life history of a flea may be as short as 18 days, or it may last for many months. The fully-formed adult flea may lie quiescent for an indefinite period of time before it begins an active existence. In a pamphlet published by the British Museum of Natural History[2] one may find the following interesting observations.

> ... flea's cocoons are extraordinarily sensitive to mechanical disturbance of any kind. Persons entering a long-deserted house sometimes have cause to complain of hordes of fleas appearing 'suddenly' after a short time. It is probable that in such cases fleas resting in the cocoon, beneath floors, in cracks, etc., have come out in response to the vibrations caused by people moving in their proximity. Bird fleas show the same phenomenon. A migrant (Wheatear) returning to an old nest (in a rabbit hole) in the spring, and shot immediately afterwards, has been found to harbour fleas of a rare and local

species breeding there. A simple experiment will illustrate what has probably taken place. If a cocoon showing dark against the light be lifted gently on a card which is tapped sharply, the enclosed flea comes out, in most cases, at once. Mammals and birds, then, revisiting old haunts are extremely liable to catch fleas merely by brushing against 'resting' cocoons. Some species go even further towards meeting their hosts halfway—*e.g.*, in the spring, the species (*Ceratophyllus styx*) found on the sand martin leaves the cocoon and congregates in swarms round the entrance of the burrows, where on a fine April morning one may see them waiting for the arrival of visitors. Such habits are plainly of great value to a species, or group of species, and may account for sporadic outbreaks of fleas, after long intervals. Of course, where hosts are numerous and the conditions of temperature and moisture are suitable the 'resting' stage will be shorter.

Bat fleas tend to crowd onto females before the bats migrate to summer colonies. Some fleas (especially small mammal fleas) have a predilection for male hosts. In contrast, the human flea is more numerous on women than on men. The explanation for this type of behavior is undoubtedly related to activities of one sex that favor dissemination of fleas, and to the influence of host sex hormones or other hormones that are circulating in the blood.

Some fleas require a blood meal before they will copulate. Many males die soon after mating, and the female lives long enough to lay a quantity of eggs that are fertile only if the female has previously had a meal of blood. Although seasonal changes in climate affect the numbers of eggs laid and the duration of larval stages, bird fleas have a more sharply defined breeding season than do mammal fleas. The moisture requirements of the larvae are of cardinal importance in determining both survival and transfer to new hosts.

The hormone cycles of the host play an important part in the regulation of the flea breeding cycle and numbers. Minute amounts of hyrdocortisone injected into a rabbit will induce maturation of fleas feeding upon its blood. Rothschild[10] has vividly described the manner in which the breeding of a rabbit flea depends upon the sexual cycle of its host.

Some fleas (e.g., the oriental rat flea, *Xenopsylla cheopis*) have regular seasons of rarity and abundance, depending on the climate. The activity and length of life of an adult flea appear to be little affected by low atmospheric humidity, which readily kills the larva, but it is very sensitive to temperature changes. When disturbed, adult fleas frequently sham death and lie with their feet held tightly to the body, in which condition they are easily blown about. A human flea has been known to jump to a height of over 18 cm. (7 inches) and a horizontal distance of 33 cm. (13 inches).

Fleas and Human Disease

Bubonic plague, one of the most serious bacterial diseases of mankind, has in former years wiped out many hundreds of thousands of people during a single epidemic. The disease probably retarded western civilization by 200 years. Although human plague is now largely under control, it is an ever present danger because the bacterial causative organism (*Pasteurella pestis*) is transmitted through fleas from rodents to man. Plague was introduced to man in North America in the year 1900, and since that time over 500 persons have contracted the disease, and 321 of these persons have died of it.[6] About 60 different kinds of fleas are associated with the house mouse, black rat and brown rat, and of these fleas at least eight are known to be able to carry the plague bacteria, which lives in the blood of its host. Rats are the principal culprits in the transfer of infected fleas to man, and the oriental rat flea, *Xenopsylla cheopis*, is the most important transfer host. *Nosopsyllus fasciatus* is also of worldwide importance as a carrier of plague bacteria. The bacilli in an infected flea so congest its gizzard and stomach that blood sucked from a mammalian host fails to pass into the stomach, and when the sucking efforts cease, the blood flows back into the wound, thereby infecting a new host.

Xenopsylla cheopis and *Nosopsyllus fasciatus* are also transmitting agents for a nonepidemic typhus of man called "murine typhus." The flea-borne disease is caused by a rickettsia, and it normally occurs in rats. The etiologic agent of epidemic typhus can also be picked up by the oriental rat flea, and the flea might play a small role in the transmission of this disease to man, but the louse is pre-eminent as a vector of typhus fever (see p. 375).

Other diseases known to be associated with fleas include tularemia (caused by *Pasteurella tularensis*) in man, rabbits and rodents; salmonellosis in man (caused by *Salmonella enteritidis*); myxomatosis of wild and domestic rabbits (caused by a virus); anemia and dermatitis. In addition to the above, several parasitic worms infect fleas. The cysticercoid stages of several tapeworms (*Dipylidium caninum* of dogs and cats, *D. sexcoronatum*, *Hymenolepis diminuta* of rodents and man) develop in larvae of several species of fleas. The larva of a filarial worm (probably *Dipetalonema reconditum*) to be found in subcutaneous tissues of dogs, develops in dog and cat fleas. The larva of the chigoe flea (genus *Tunga*) infests the skin of birds and mammals, causing intense itching and ulceration. *Tunga* and *Neotunga* are remarkable in that the female burrows beneath the skin of its host and passes the entire adult life as an internal parasite. For a review of fleas and associated disease see Jellison.[8]

Fleas themselves are also parasitized by gregarines and other Protozoa in the mid-gut, by a hymenopterous insect larva (*Bairamlia fuscipes*) in the flea larva, by mites that live in bird nests and that destroy both flea larvae and pupae, and by a chalcid fly.

ORDER DIPTERA (FLIES)

Injurious insects cost the United States approximately four billion dollars a year. This sum includes losses due to such activities of insects as damage to livestock, crops, buildings, and stored products, and to various control measures. Much of this tremendous loss is caused directly or indirectly by parasitic insects and a significant amount of the damage is done by flies, gnats, mosquitoes and other members of the order Diptera. Of the many orders of insects, this one possesses the greatest number of parasitic forms. Although some flies are parasitic as adults, most of the parasitic species have free-living adults whereas their larval stages live in invertebrate or vertebrate hosts. Since so many flies deposit their eggs in decaying organic matter, it is not surprising that some of them have developed the habit of depositing eggs in fetid wounds of other animals. The developing larvae, or maggots, then have an opportunity

to penetrate deeper into the bodies of their hosts. For an account of feeding habits of blood-sucking arthropods, see Weitz.[12]

Structure

Insects belonging to the order Diptera are characterized by the possession of only one pair of wings, hence the name **diptera,** which means "two wings." Immediately behind the wings occurs a pair of club-shaped organs called "halteres." These organs are considered to be vestigial wings. The three body divisions, head, thorax, and abdomen, which

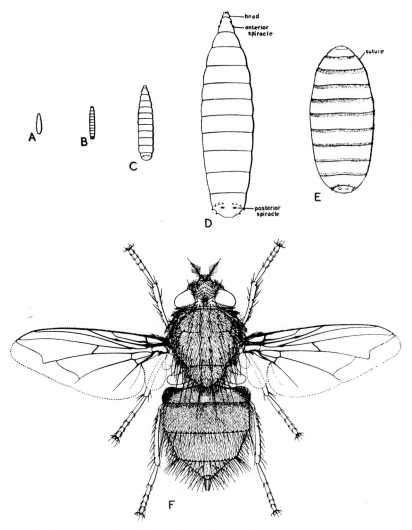

Fig. 18-8. *Calliphora vicina.* The blue bottle fly showing the characteristic appearance of *A,* the egg; *B,* the first stage larva; *C,* the second stage larva; *D,* the third stage larva; *E,* the puparium; *F,* the adult or imago. (Smart, Insects of Medical Importance, courtesy of British Museum, London)

characterize insects, are especially marked in the Diptera. The adult head usually possesses prominent eyes and a pair of antennae (Fig. 18–8). The thorax supports legs and wings. The wings are membranous and may be smooth, hairy or, as in the mosquitoes, may possess scales. Venation of the wings is an important diagnostic feature (Fig. 18–9). The abdomen is usually distinctly segmented and bears the genitalia at its posterior end. Some of the major morphologic details for flies are illustrated in Figure 18–9. The life cycles of Diptera usually involve the egg, larva, pupa, and adult. The egg may hatch within the body of the female, but most dipterous insects lay eggs. Larvae of flies are called maggots and, like other larvae, they undergo a series of molts while growing.

Families of Diptera

Only families containing the more important parasites of man or of animals will be

discussed. The student may consult any modern text on entomology for a more complete presentation.[5,7] The following families are considered here: Culicidae, Simuliidae, Tabanidae, Oestridae, Tachinidae (= Larvaevoridae), Calliphoridae, Sarcophagidae, Muscidae, Hippoboscidae and Psychodidae. All of these families, except the last, are illustrated in Figure 18–11.

Family Culicidae

Mosquitoes transmit the causative agents of such serious diseases as human and avian malaria, dengue fever, yellow fever, fowl pox, elephantiasis, and other forms of filariasis. Thus it is not surprising that, on the basis of weight alone, the literature on mosquitoes is staggering.

At least 2,000 species of mosquitoes occur in the world and they may be found in almost every country. These insects (Fig. 18–12) are characterized by slender bodies

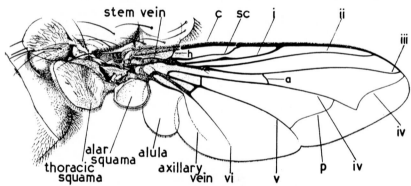

Fig. 18-9. Right wing of *Calliphora vicina*. *a*, Cross vein; *c*, costa; *sc*, subcosta; *p*, posterior cross vein. Longitudinal veins numbered by Roman numerals. (Smart, Insects of Medical Importance, courtesy of British Museum, London)

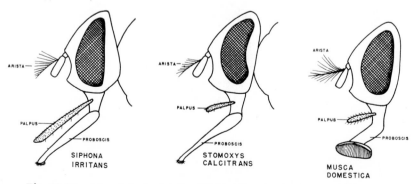

Fig. 18-10. Morphologic details of fly heads. (Federal Security Agency)

and long legs. The mouth parts of adult females form a blood-sucking proboscis (Figs. 18–15, 18–16). The antennae of the male tend to be bushier and thus more prominent than those of the female. Mosquitoes in general may be distinguished from similar flies by their wing venation. Of special value is the presence of two bifurcated veins toward the apex of each wing. These divided veins are separated by a single vein (Fig. 18–12). The veins are covered with scales.

Recognition of the various genera and species of mosquitoes is of such great importance to the medical parasitologist that space is given here to pictorial keys and other illustrations. These keys are self-explanatory and should aid in the identification of the common forms, especially those found in the United States.

The life cycle of mosquitoes involves the usual egg, larva, pupa and adult stages; the larvae and pupae live in water. Eggs, about

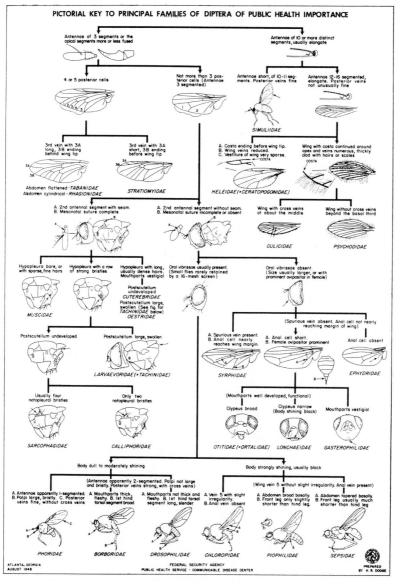

Fig. 18-11. Pictorial key to principal families of diptera of public health importance. (Federal Security Agency)

a millimeter in length, are laid by members of the genus *Culex* in groups on the surface of water. These eggs are glued together in the form of tiny rafts. Members of the genus *Anopheles* deposit their eggs singly without gluing them together and they remain afloat by virtue of lateral air chambers (see illustration in key, Fig. 18–13). Upon hatching, the larvae swim about and search for food (e.g., minute algae). They are sometimes called "wrigglers" because of the way they move through the water. Larvae come to the surface periodically and obtain air through breather tubes or siphons, a habit which leads to their destruction at the hands of zealous antimalarial workers. The young insects grow and molt four times and after the last molt they become pupae. A pupa is shaped like a comma with a very large head equipped with siphons. The pupae are so active they are called "tumblers." They do not feed, but during the few days of their existence the adult structures develop. When they molt the pupae come to the surface of the water, thus permitting the winged adults to take up their aerial and terrestrial life. After mating, the male soon dies but the female of some species lives during the winter

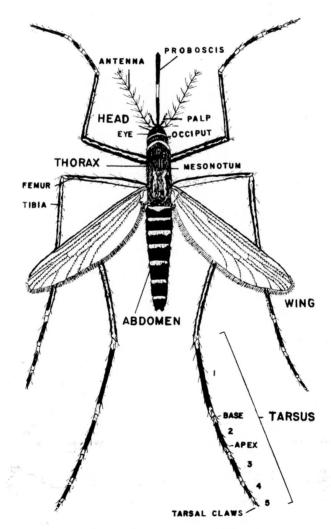

Fig. 18-12. *Culex tarsalis*, adult. (Owen, The Mosquitoes of Wyoming, courtesy of Univ. Wyoming Pub.)

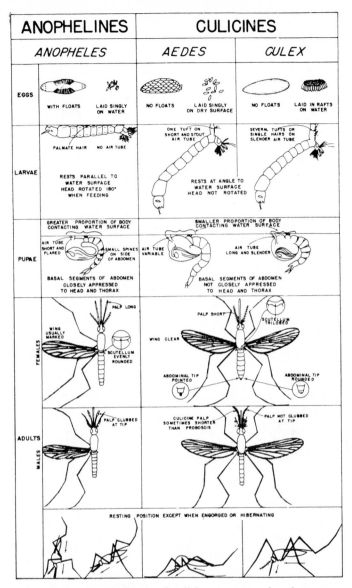

Fig. 18-13. Anophelines and culicines, comparative characters. (United States Public Health Service)

in a dormant condition in protected places like sheds, barns and abandoned houses. The sperm within her body remain viable, ready to fertilize the eggs that appear in the spring.

Mosquitoes are probably best known for transmitting *Plasmodium*, the causative organism of malaria. The life cycle of this protozoan parasite is described in Chapter 4, p. 88. Man is usually more interested in what this parasite does to himself than he is in how the

parasite affects the mosquito. The protozoa undergo greater changes in the mosquito than they do in man, suggesting that the mosquito reacts more violently than does man to plasmodial infection. Mosquitoes, however, are probably better adapted to the parasite than is man, but we know relatively little about the adjustments a mosquito has to make. It should be noted that even some anopheline mosquitoes, which are the normal vectors of *Plasmodium*, do not become infected even after a meal of parasitized human blood. This fact is not surprising, because some people are also refractory to infection, but our knowledge of parasite-host relations between *Plasmodium* and mosquitoes is limited.

At least five species of mosquitoes are known to transmit monkey malaria. Four of these are found in Malaya and one in India.[4] The species most commonly found in the mangrove forests of coastal Malaya is *Anopheles hackeri*. Much has to be learned, however, about the biology of transfer of malaria to monkeys and apes.

Enemies of mosquitoes are legion. These insects may be parasitized by other sporozoa (microsporidia, coccidia) as well as by gregarines (*Lankesteria*, *Caulleryella*), flagellates (*Herpetomonas*, *Crithidia*), ciliates (*Lambornella*, *Glaucoma*), trematodes (*Agamodistomum*, *Pneu-*

moncoeces larvae), filaria, mites (Hydrachnidae) and blood-sucking midges. As if these internal and external parasites were not enough, mosquitoes have many larger natural enemies to plague them. Larval stages are attacked by water beetles, by voracious dragonfly larvae and by salamanders, frogs, and fish. Adult mosquitoes are eaten by wasps and predacious flies, spiders, lizards, birds, and bats.

Destruction of immature mosquitoes can be accomplished by adding oil to the water in which they live. This technique prevents them from obtaining oxygen because they cannot attach to the surface layer of oil with their breathing tubes. The chief destructive effect, however, is due to the volatile toxic substances that occur in most oils used (e.g., kerosene, diesel fuel). Various larvicides, such as copper sulphate, Paris green or

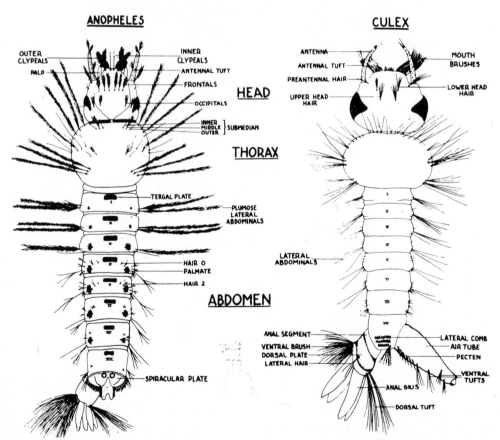

Fig. 18-14. Comparative characters of *Anopheles* and *Culex* larvae. (United States Public Health Service)

naphthalene, may also be added to the water to kill larvae and pupae. Students should remember that these chemicals may also kill desirable organisms. Insect sprays containing pyrethrum, lindane, or any of a number of other chemicals are effective against adult and immature mosquitoes.

Elimination of breeding places is a highly recommended method of mosquito control. Swampy areas may be drained and filled, and such containers as cisterns or small vessels emptied or covered.

Family Simuliidae

The family Simuliidae is worldwide in distribution and it includes small insects called blackflies or buffalo gnats. The immature stages of these flies are aquatic, normally preferring running water. Both larvae and pupae are attached to rocks or to plants under water. Adults are hump-backed (Fig. 18–17), hence the name "buffalo gnat." They sometimes occur in enormous swarms, especially near water; and they cause considerable annoyance to livestock and man by their vicious bites. Some species will gather in tremendous numbers on domestic mammals and on birds. The flies may cause such an annoyance to poultry that the birds may leave their nests. Aggressive female insects sometimes get under the wings of young birds and suck so much blood that the birds die. Danger from the gnats is not only due to loss of blood and possibly bacterial infection of bitten areas, but also to the toxic saliva of the insect and to parasites which the fly might transmit.

The major parasite of man that simuliid flies carry is the roundworm, *Onchocerca volvulus*, which infects man in some of the warmer countries such as Africa, Central America, and Mexico. The flies may also transmit tularemia and the protozoan parasite, *Leucocytozoon*. The latter organism is a blood inhabitant of birds, especially of turkeys and ducks. The name indicates that this parasite inhabits white blood cells. There is evidence, however, that the host cells are erythrocytes which are quickly altered and superficially resemble leukocytes (p. 102). *Leucocytozoon simondi* is found in the peripheral blood of

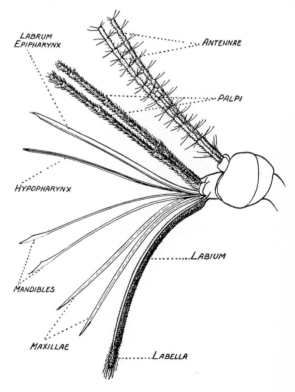

Fig. 18-15. Mouth parts of female mosquito (*Anopheles quadrimaculatus*) dissected out of the sheath. Note that the mandibles have finely serrated, blade-like lips and that the maxillae have pointed chitinous lips with fewer and coarser serrations than the mandibles. The proboscis is made up of all parts labeled except the antennae and palpi. (Fox, Insects and Disease of Man, courtesy of the Blakiston Company)

ducks, while *L. smithi* is a similar parasite of turkeys. Common insect vectors are *Simulium rugglesi* and *Eusimulium anatinum*. The flies attach themselves to the necks of the birds and

feed for two or three minutes before flying away. The use of insecticides and other control measures against the fly are only partially effective.

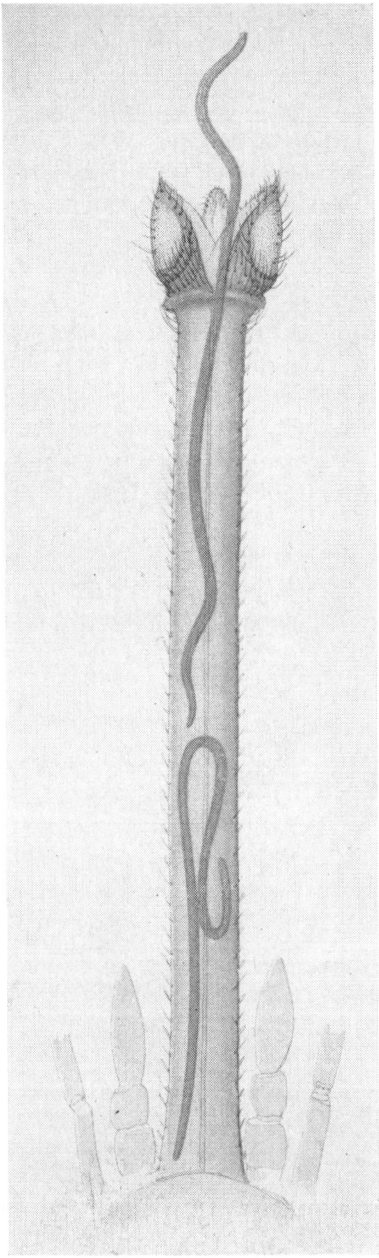

Family Tabanidae

Tabanidae is another family of flies whose females suck blood. These insects are considerably larger than the Simuliidae and are known as horse flies, deer flies, or clegs (Fig. 18–18). They are stout flies, each with a broadly triangular head, most of which is composed of eyes. Immature stages are typically aquatic or to be found in wet earth, a habit which accounts for the presence of the adults near freshwater or marshy areas. Many species, however, may fly several miles from their breeding places. As with many other groups of flies, the males feed on plant juices and pollen. Sexes may be distinguished by the distance between the eyes, as seen from above. Male eyes are close together or **holoptic,** whereas the female eyes are distinctly separated or **dichoptic.** Genera important to the parasitologist are *Tabanus* and *Chrysops.*

Various species of *Tabanus* transmit the flagellate blood parasite, *Trypanosoma evansi,* which causes the disease **surra** in horses, cattle, dogs, elephants and in other animals (p. 36). Members of this genus of flies also carry anthrax and tularemia to man and animals. Tularemia is also transmitted by *Chrysops* which, in addition, may carry the filarial worm, *Loa loa* (p. 319). Different species of the genus *Chrysops* feed at different times of the day. *C. silacea* and *C. dimidiata* do most of their biting during the morning and then later in the afternoon; *C. langi* and *C. centurionis* do the bulk of their biting in the evening at sunset.

Family Psychodidae

The subfamily with which we are specially interested is the Phlebotominae or **sandflies,** which have piercing mouth parts and are vectors of leishmaniasis (p. 38). A few of the important species which transmit *Leishmania* to man and other vertebrates are *Phlebotomus*

Fig. 18-16. Mature larva of *Wuchereria bancrofti* escaping from the proboscis of the mosquito *Culex quinquefasciatus* (Francis, courtesy of United States Public Health Service Hygienic Laboratory Bull. No. 117)

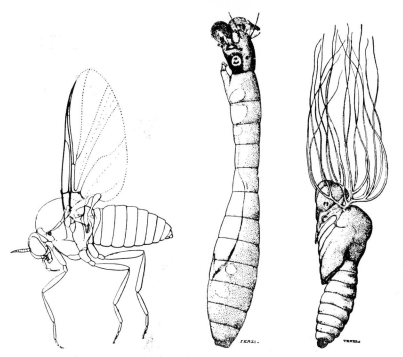

Fig. 18-17. *Simulium*, a black fly. Left figure, a side view of adult. Middle, a larva; right, a pupa. (Smart, Insects of Medical Importance, courtesy of British Museum)

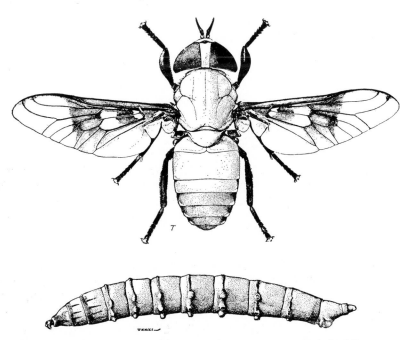

Fig. 18-18. *Tabanus latipes* with its larva. (Smart, Insects of Medical Importance, courtesy of British Museum)

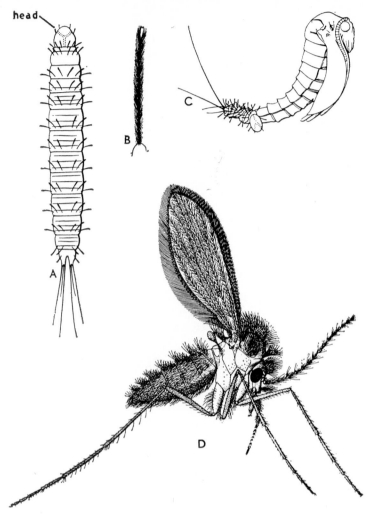

Fig. 18-19. Larva, pupa and adult female of *Phlebotomus papatasii*. (Smart, Insects of Medical Importance, courtesy of British Museum)

argentipes, P. papatasii, P. perniciosus, P. sergenti, P. chinensis, P. longipalpis and *P. intermedius*. The genus of the New World species has recently been changed to *Lutzomya*.

Sandflies (Fig. 18–19) may be found during the day in protected moist places such as caves, burrows or other holes in the ground, sheds, crevices or in loose piles of humid material. Eggs are laid in the soil or in other material that is shady and moist, and the larvae develop in three to four days. After feeding for about two weeks they change into the pupal stage, which lasts another week.

The entire life cycle takes one to three months, depending on the temperature. The female sucks blood at night, especially at times when there is no wind, since these insects are extremely weak fliers.

Family Oestridae

Warble flies are so called because their larvae usually lie under the skin of their hosts and cause a swelling known as a "warble." The flies have been placed in the family Oestridae by some workers and in the family Hypodermatidae by others.

The best-known members of this group are the ox warble, *Hypoderma bovis* (Fig. 18–20), a stout hairy fly that looks a little like a bee, and a similar species, *H. lineatum*, the heelfly. The adult heelfly is hairy, black and yellow, and several times larger than a housefly. In the spring each female fly lays hundreds of eggs on the leg or body hairs of cattle. In a few days the eggs hatch into tiny, white spiny larvae that crawl down the hairs and burrow through the skin. For several months the larvae wander through the intestines, liver, heart, muscles, or other organs, apparently enjoying the protection and warmth of their host. Some of them spend many weeks in the connective tissues of the esophagus. Apparently the first instar larvae migrate to the vertebral region. Here they produce small swellings or warbles in the skin. They make a tiny hole in the skin for air and they increase in size to about 25 mm. in length. During this period they turn dark brown. In the

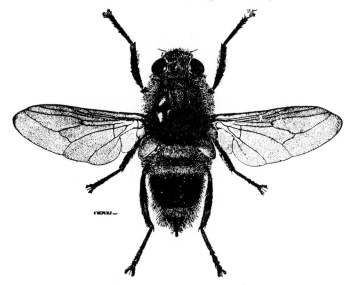

Fig. 18-20. *Hypoderma bovis*, the ox warble, adult. (Smart, Insects of Medical Importance, courtesy of British Museum)

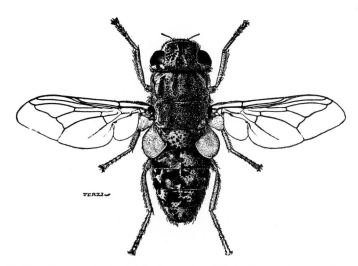

Fig. 18-21. *Oestrus ovis*, the adult sheep nostril fly. (Smart, Insects of Medical Importance, courtesy of British Museum)

spring or summer, the grubs, or maggots, emerge through the holes and drop to the ground and pupate. They crawl under loose soil or trash, and after two to seven weeks, depending on the temperature, the adult flies emerge.

One to three hundred millions of dollars are lost annually in the United States because of damage to hides and irritation and annoyance to cattle. The animals "get off their feed," lose weight, and are generally unthrifty. Milk production diminishes. The holes in the skin make the hide unfit for high quality leather.

Dermatobia hominis, a somewhat similar fly though not so hairy, may occur as a maggot in the skin of man in Central and South America. This warblefly is about 12 mm. in length. The nonbiting females attach their eggs to the bodies of mosquitoes (*Janthinsoma lutzi*), other flying insects, or even ticks. The eggs thus get a free ride to a new host. When the mosquitoes alight to feed on man or on other warm-blooded animals, the warblefly eggs hatch and the larvae penetrate the skin of their new host.

Oestrus ovis (Fig. 18–21) is a fly which, instead of laying eggs, deposits its larvae or maggots in the nostrils of sheep, goats and a few other mammals such as deer. The maggots crawl into the nasal passages, and, during the next few weeks, they molt twice. When fully grown and about 25 mm. long, they come out of the nose, drop to the ground and pupate. From the pupa the adult fly emerges. It is also called a gadfly or nasalfly. This fly may attack shepherds or other persons who have recently eaten goat's milk products and whose breath smells of such food as cheese. The fly deposits its larvae in the shepherd's nostrils, on his lips or even on his eyes, and sometimes it causes considerable damage.

The names **botfly** and **warblefly** are used somewhat interchangeably. Thus *Gasterophilus intestinalis* (Fig. 18–22) may be called a horse bot or warblefly. This and related genera are usually placed in a separate family, Gasterophilidae. The adult of this botfly also looks a little like a bee. It attaches its eggs to hairs on the legs or bodies of horses. As soon as the eggs become infective, the larvae pop out when they are brushed by the moist, warm lips of the horse. These larvae are spiny and they become attached to the horse's lips or tongue and thus are easily swallowed. They may, however, burrow through the mucous membranes of the mouth and make their way to the stomach through various tissues. The larvae attach themselves to the lining of the stomach, sometimes in such large numbers that this membrane is almost covered with them. After some months of maturing, the maggots pass from the host's body with feces, drop to the ground and pupate. Adult flies emerge from the pupae and when they start laying eggs the life cycle is complete.

Family Tachinidae

The family Tachinidae contains many species of flies, all of which are parasitic on insects or other arthropods (e.g., terrestrial isopods). Some of the tachinids look like houseflies and some appear more like bees or wasps, but most of them possess long bristles on the tips of their abdomens. A tachinid normally deposits its eggs on the body of its host (e.g., larvae of butterflies or beetles), and larvae burrow inside to eat. The larva literally eats itself out of house and home, killing its host. When mature, the fly crawls out of the host and pupates on the ground. Some tachinids lay their eggs on foliage and when the larvae emerge they crawl on and into a host, or they are infective after being swallowed by the host. For comments on the evolution of the tachinids, see page 571.

Some flies parasitize grasshoppers. Both nymphal and adult grasshoppers (*Chortophaga viridifasciata*) in Tennessee have been found to be inhabited by the tachinid fly, *Ceracia dentata*, and by the sarcophagid, *Blaesoxipha hunteri*.[11] The parasites occur as larval stages either free in the hemocoel of the host or, as they get older, attached to tracheal trunks, an air sac or abdominal spiracles. As if the difficulties in reaching maturity for *C. dentata* are not enough, these parasitic flies may themselves be parasitized (hyperparasitism) by a hymenopteran (*Brachymeria tegularis*), which may occur within the puparia of the flies.

Family Calliphoridae

Blowflies may look somewhat like houseflies but are often a little larger and frequently are metallic green or blue in color. They have the habit of laying their eggs on organic masses such as dead animals, excrement, open sores or exposed cooked or uncooked food. The eggs hatch into maggots which proceed to eat the material around them. Occasionally, infested meat is cooked for the dinner

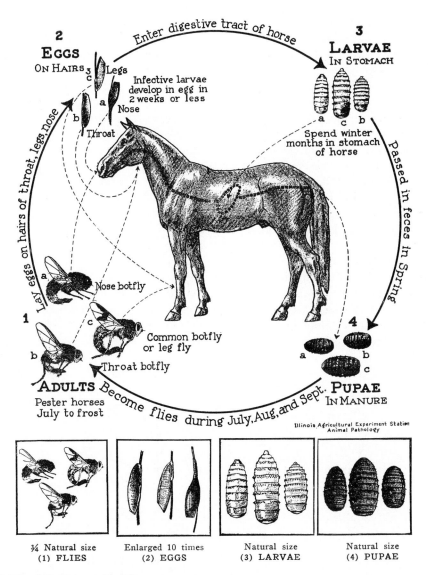

Fig. 18-22.   Life history of botflies.
1. Flies pester horses from July to frost, laying their eggs on the hairs of the nose, throat, and legs.
2. In the eggs larvae develop in two weeks or less and enter the digestive tract.   *3.* Larvae spend the winter months in the stomach, where they are a drain on the animal; in the spring they pass out in the manure.   *4.* In the ground pupae develop from the larvae. During July to September these pupae become flies, which start the life cycle all over again.   (Thorp and Graham, courtesy of University of Illinois College Agriculture)

table, but the infestation passes unnoticed until, upon carving the meat, the dead maggots fall out upon the plate, much to the disgust and distaste of the diners.

The genus *Calliphora* includes the bluebottle flies whose larvae are commonly found in decaying meat. Eggs are also laid in exposed foul wounds of man and animals. The maggots eat only the putrid tissues in the wound. Occasionally live *Calliphora* maggots are swallowed in cooked meat, and rarely they pass through the body. The emergence of these maggots from the intestine accounts for some of the reported cases of myiasis in man. The term **myiasis** refers to the presence of parasitic maggots in or on man or animals (see p. 422). *Phaenicia sericata* and *Phormia regina* are two species whose larvae have been used by physicians to clean wounds and to treat such diseases as osteomyelitis in man.

Lucilia bufonivora lays its eggs on the back of a toad or other amphibian.[1] The eggs hatch within a day, apparently due to the chemical nature of the skin glands. The tiny larvae crawl to the head region and on the surface of the eye. By blinking its eyes, the toad carries the maggots to the lacrimal ducts, which lead to the nasal cavity. Here the larvae molt and form the second stage which actively feeds on host tissues and eventually destroys the cartilaginous nasal septum. When ma-

ture, the larvae drop out of this enlarged nasal cavity and pupate in the ground.

Chrysomya bezziana is a green blowfly which seems to have a preference for wounds and body spaces of man and of wild and domestic animals. The larvae (Fig. 18–23) may be found in the nose, eyes, ears, alimentary canal, urinary passages, genital organs or tiny cuts or sores in the skin. The eggs may even be laid in the pierced ear lobes of women. At first the small maggots are not noticed, but when they grow to a centimeter or more in length and begin to burrow through the skin near the wound, they may produce a tumor of considerable size, especially if many worms are present. This species is found in the Philippines, India, and Africa. Maggots of the clusterfly, *Pollenia rudis*, crawl from eggs in the soil to earthworms (*Allolobophora chlorotica*) and enter the openings of the male sex ducts. After a dormant period of eight months the larvae feed and grow and destroy much of their host's tissues. After several molts the mature larvae leave the earthworm host and pupate in the ground.

Cochliomyia hominivorax is a bluish blowfly about twice the size of a housefly. It lays its eggs in neat rows on the edges of wounds of mammals, especially on fresh cuts. The eggs hatch in a day or less into maggots which feed on living flesh. These maggots are called

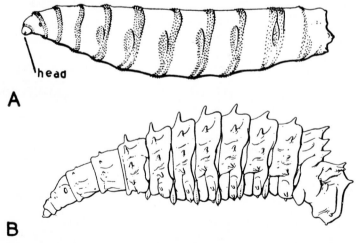

Fig. 18-23. *Chrysomya bezziana.* *A.* The mature third stage larva, × 8. *B.* The mature third stage larva of *C. albiceps*, × 8. (Smart, Insects of Medical Importance, courtesy of British Museum)

"screwworms." The wound attended by screwworms has a foul odor. In about a week the maggots drop to the ground and pupate in the soil. In another week, if the weather is warm, adults emerge. In still another week or less new eggs, about 200 in a batch, are deposited on the edge of a cut, sore, navel of new born animals, or on some other abrasion. This screwworm is an obligatory parasite of warm-blooded animals and is responsible for 20 million dollars worth of damage annually to United States livestock. Untreated animals usually die. General control involves doing all that is possible to prevent cuts and scratches on livestock. A new method of control consists of releasing laboratory reared sterilized male flies in an infected area. Females do not mate a second time, so those which happen to mate with sterile males cannot reproduce. This trick played on unsuspecting females has resulted in a significant reduction in screwworm infection in relatively isolated areas.

Other blowflies, principally *Phormia regina*, the black blowfly, *Cochliomyia macellaria*, the secondary screwworm fly, and *Phaenicia sericata*, the greenbottle fly, possess larvae that are called fleeceworms. They attack the hair of sheep and goats and also enter wounds. Eggs are usually deposited on soiled fleece or on old or new sores. In addition to injuring the quality of fleece and aggravating

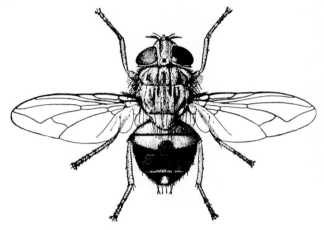

Fig. 18-24. *Cordylobia anthropophaga*, the tumbufly, female, × 4. (Smart, Insects of Medical Importance, courtesy of British Museum)

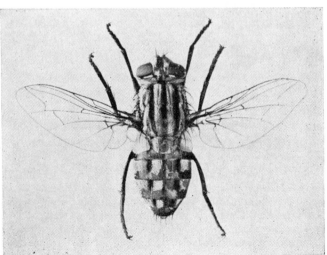

Fig. 18-25. *Sarcophaga haemorrhoidalis*, a fleshfly, female, (Mönnig, Veterinary Helminthology and Entomology, courtesy of Baillière, Tindall and Cox)

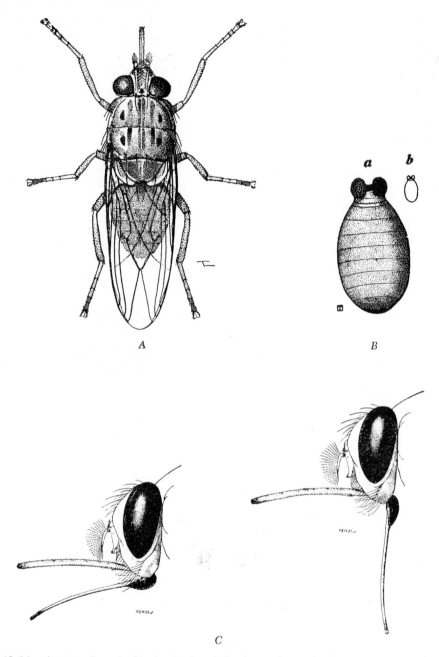

A   *B*

C

Fig. 18-26. A tsetse fly. *A. Glossina longipennis* in the resting attitude assumed by the living fly. Note the position of the wings. × 4 (after Austen). *B.* Puparium of tsetse fly. *a,* enlarged; *b,* natural size. The posterior end is toward the top of page. (After Austen from Castellani and Chambers.) *C.* Head of species of *Glossina* showing how the proboscis with its bulb-shaped base is lowered from the palpi for the act of feeding. The palpi are not inserted into the wound but remain horizontal as the piercing proboscis is sunk into the skin (Castellani and Chambers. Smart, Insects of Medical Importance, courtesy of British Museum)

wounds, they may carry the germs of anthrax, plague, undulant fever, tularemia and enteric diseases.

The tumbufly of Africa, *Cordylobia anthropophaga* (Fig. 18–24), measures about 10 mm. in length and is yellowish brown in color. Eggs are deposited on the ground where they hatch. When the appropriate host (man and other mammals) lies on the ground, larvae burrow through the skin into their new home. As the larva grows, it produces a boil-like swelling in the skin, which has a small hole at the top of the boil. Through the hole the maggot breathes and discharges its waste. When mature, the maggot emerges through the hole, drops to the ground and pupates.

Family Sarcophagidae

The family Sarcophagidae contains the fleshflies. Fleshflies are rather large, gray insects with longitudinal black stripes on the dorsal surface of the thorax (Fig. 18–25. Notice the checkered appearance of the abdomen). Like the blowflies, the fleshflies are scavengers and the larvae may be found in wounds, carrion, sores or body cavities of man and animals. Unlike the blowflies, most of the fleshflies do not lay eggs but are larviparous. The genus *Sarcophaga* is worldwide in distribution. The genus *Wohlfahrtia* contains species that parasitize man and animals, thus causing myiasis in these hosts. A larva of *Wohlfahrtia magnifica* has even been found in the cavity of an infected tooth.

Family Muscidae

The family Muscidae is well known for furnishing man with some of his most serious insect parasites. They include the tsetse fly, *Glossina*[3] (Fig. 18–26), carrier of the causative organisms of African sleeping sickness (p. 420, 35); the stablefly, *Stomoxys calcitrans*, which can transmit (mechanically) the causative organism of Oriental sore (p. 38). The common housefly, *Musca domestica*, also attacks livestock. This is not a biting species but it may act as a mechanical carrier of disease such as dysentery, cholera, anthrax, typhoid

fever, and yaws. It may also transmit worms. The roundworm, *Habronema muscae*, is a parasite in the intestine of horses. Larvae of the worm leave the body of a horse in feces and enter the bodies of fly larvae. When the flies mature the worms have reached their final larval stage, and they mature in the horse when flies are swallowed.

Pupipara

Pupipara are peculiar looking flies which are ectoparasitic on birds and mammals. They are blood-sucking and are called louse flies, bat flies, tick flies, or keds. They usually are wingless, flattened diptera that can be found on the bodies of animals or in bird nests or bat roosts. Some of them possess and retain wings; others shed them after reaching their hosts. Wild animals are attacked as well as horses, sheep, and other domestic forms. That ectoparasitic flies are found on bats is not surprising when one considers the probability that ancestral flies deposited their eggs in bat droppings. From this start the association undoubtedly became progressively closer until today there are such forms as the completely wingless pupiparous species, *Nycteribia biarticulata*, which at first glance looks more like a louse than it does a fly. The pupipara belong to several families.

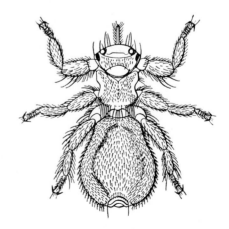

Fig. 18-27. *Melophagus ovinus*, the sheep ked. × 4. (Smart, Insects of Medical Importance, courtesy of British Museum)

Family Hippoboscidae

The family Hippoboscidae includes the more common species of Pupipara. An important member of the family is the sheep ked or "sheep tick," *Melophagus ovinus*, a wingless, leathery flattened insect (Fig. 18–27) which is hardly recognizable as a fly. It is about 6 mm. long and, like other pupipara, does not lay eggs but gives birth to larvae already advanced to almost the pupal stage.

Pseudolynchia canariensis, another hippoboscid, is a winged louse fly that lives on pigeons. In addition to causing annoyance to the birds, this ectoparasite carries an endo-parasite of birds, the malarial-like blood parasite, *Haemoproteus*.

Myiasis

Myiasis, as defined by Zumpt[13] is "the infestation of live human and vertebrate animals with dipterous larvae, which, at least for a certain period, feed on the host's dead or living tissue, liquid body-substances, or ingested food." (See Fig. 18–28.) At least 187 species of myiasis-producing larvae are recorded in the world literature. They belong primarily to the families Calliphoridae and Oestridae, but are also found among the

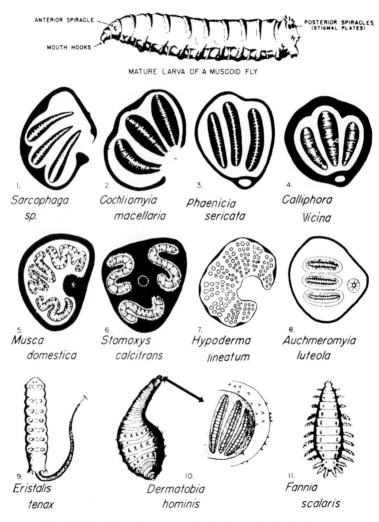

ANTERIOR SPIRACLE   POSTERIOR SPIRACLES (STIGMAL PLATES)

MOUTH HOOKS

MATURE LARVA OF A MUSCOID FLY

1. Sarcophaga sp.
2. Cochliomyia macellaria
3. Phaenicia sericata
4. Calliphora Vicina
5. Musca domestica
6. Stomoxys calcitrans
7. Hypoderma lineatum
8. Auchmeromyia luteola
9. Eristalis tenax
10. Dermatobia hominis
11. Fannia scalaris

Fig. 18-28. Key characters of myiasis-producing fly larvae. Figures 1 to 8 are stigmal plates. (Federal Security Agency)

Muscidae, Chloropidae and Gasterophilidae. Several examples have already been given. Those in the intestine produce intestinal myiasis, whereas those in the genital organs or urinary tract produce urinogenital myiasis. Some maggots occur only in wounds and sores, especially foul and suppurating sores, whereas others actually get into living flesh, producing boils or weals in the skin or, as we have seen, preferring the chambers of the nose, mouth, eyes, or ears.

Intestinal myiasis is usually caused accidentally by contaminated food or drink. Fly eggs or maggots may occur on over-ripe fruit, raw meats, cheese, salads, dried fish, dirty water, or milk. Urinogenital myiasis is usually also accidental. Flies may lay their eggs on the exposed and unclean pubic area. Some of the genera of flies involved in accidental myiasis are *Calliphora, Musca, Fannia, Stomoxys, Drosophila, Sepsis,* and *Tipula*. Flies which deposit their eggs in or near foul wounds and whose larvae thus have an opportunity to invade the wound belong mainly to the genera, *Sarcophaga, Phormia, Cochliomyia* or *Chrysomyia*.

Identification of many of these maggots is difficult. One of the common diagnostic characters is the nature of the posterior spiracles. Figure 18–28 illustrates these spiracles as they appear in the larvae of some myiasis-producing flies.

Invertebrates have their share of parasitic fly larvae. Some Hymenoptera may contain maggots which actually attach their spiracular openings to the host trachea in order to breathe. The snail, *Helicella vergata*, may become so infested with fly maggots that the parasites kill their host and eat practically all of the dead tissues.

One of the few studies[9] of the biochemical effect of myiasis on the host involved the white-footed mouse, *Peromyscus leucopus*, and the botfly, *Cuterebra angustifrons*. Development of the larva in the mouse was accompanied by progressive reductions in the albumin-to-globulin ratio of the host. These reductions were correlated with the number of larvae present. Ingestion of albumin by the botfly was proposed as a partial explanation for the reductions.

REFERENCES

1. Baer, J. G.: The Ecology of Animal Parasites. Urbana, Ill., Univ. of Illinois Press, 1951.
2. British Museum of Natural History: Fleas as a Menace to Man and Domestic Animals. Their Life-History, Habits and Control. ed. 6. London, British Museum of Natural History, Econ. Ser. 3, 1949.
3. Buxton, P. A.: The natural history of tsetse flies. An account of the biology of the genus *Glossina* (Diptera). Mem. London Sch. Hyg. Trop. Med. No. 10. H. K. Kewis & Co., 1955.
4. Cheong, W. H., Omar, A. H. B., and Warren, M.: The known vectors of simian malaria in Malaya today. Med. J. Malaya, *20*:327–329, 1966.
5. Herms, W. B.: Medical Entomology. ed. 5, rev. by M. T. James. New York, Macmillan, 1961.
6. Hubbard, C. A.: Fleas of Western North America. Their Relation to the Public Health. Ames, Iowa, Iowa State College Press, 1947.
7. Imms, A. D.: A General Textbook of Entomology. ed. 9, rev. by O. W. Richards and R. G. Davies. London, Methuen & Co., 1957.
8. Jellison, W. L.: Fleas and disease. Ann. Rev. Entomol., *4*:389–414, 1959.
9. Payne, J. A., Dunaway, P. B., Martin, G. D., and Story, J. D.: Effects of *Cuterebra angustifrons* on plasma proteins of *Peromyscus leucopus*. J. Parasitol., *6*:1004–1008, 1965.
10. Rothschild, M.: Fleas. Sci. Amer., Dec., 44–53, 1965.
11. St. Amand, E., and Cloyd, W. J.: Parasitism of the grasshopper, *Chortophaga viridifasciata* (Degeer) (Orthoptera: Locustidae), by Dipterous larvae. J. Parasitol., *40*, 83–87, 1954.
12. Weitz, B.: Feeding habits of bloodsucking arthropods. Exp. Parasitol., *9*:63–82, 1960.
13. Zumpt, F.: Myiasis in Man and Animals in the Old World. Washington, D.C., Butterworth, 1965.

Chapter 19

Class Arachnoidea (= Arachnida)

ORDER ACARINA (MITES AND TICKS)

Acarina abound almost everywhere, but except for ticks, which are large enough to be easily recognized, they are little known to biologists in general. A handful of soil is likely to contain several to thousands of specimens. All ticks, however, and certain mites, are parasitic on or in terrestrial vertebrates during at least one stage in their life cycles. They may be the direct causative agents of disease, or they may transmit pathogenic microorganisms, or they may serve as reservoirs of infection. Hundreds of species live on or in plants, often causing serious damage. The feathers of birds and the hairs and skin of mammals are favorite habitats, while internal organs of both vertebrates and invertebrates are frequently invaded by mites. Fourteen pounds of ticks were once removed from the skin of one horse in three days, and as many ticks were still left on the suffering animal. Freshwater ponds, streams, rivers, and the oceans all have their mite faunas. The Acarina are not small enough to be handled like protozoa, yet they are too small to be treated like insects, and they are not soft-bodied enough to be studied like worms. For these reasons they require special techniques for their collection and preparation for study.

Most acarologists, entomolgists and zoologists consider the Acarina to be an order of the class Arachnoidea. The group, however, is sometimes listed as a subclass, or even as a separate class, of arthropods. The mites and ticks are readily separated from other arachnids by the possession of a distinct gnathosoma (an anterior capitulum bearing mouth parts) and by the absence of a clearly recognizable division between the cephalothorax and abdomen. The phylogeny of the Acarina is obscure, and most authorities on the group consider it to be polyphyletic in origin.[6]

Morphology

As in other arthropods, the tegument consists of an outer cuticle and a single layer of epithelial cells that secrete it. The cuticle may be membranous or leathery and sometimes has hard plates or shields. Special structures such as glands, setae and sensory organs are derived from tegumentary cells. Figure 19–1 illustrates details of the external anatomy.

Mouth parts consist of a pair of chelicerae that usually terminate in small pincers (chelae) possessing a dorsal, fixed digit and a ventral, movable digit; and of a pair of pedipalps (or palpi), usually consisting of four to six segments, sometimes modified as a thumb and claw. The majority of Acarina possess three pairs of walking legs as larvae and four pairs as adults. These legs are usually divided into six segments which terminate in well-developed claws. Of primary importance to the systematics of the group are the anatomic features of the respiratory system. The suborders are established largely upon the basis of the numbers and location of the stomata (spiracles or openings of trachea).

Many of the larger mites rather closely resemble the ticks, but mites are usually small forms in which the hypostome is hidden and unarmed, whereas ticks are larger forms in which the hypostome is exposed and armed with teeth or hooks. Two or more simple

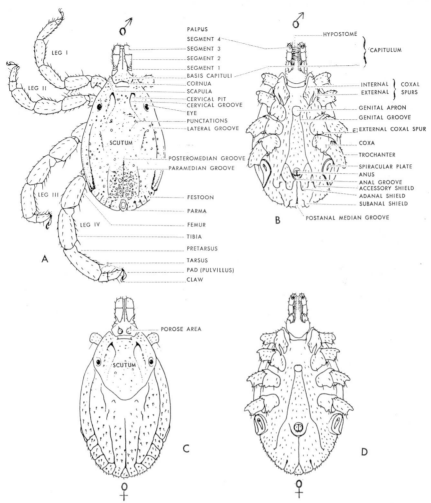

Fig. 19-1. Key morphologic characters. Hypothetical male and female ticks, family Ixodidae. *A* and *B*. Male, dorsal and ventral views. *C* and *D*. Female, dorsal and ventral views. (Hoogstraal, *African Ixodoidea*, U.S. Naval Medical Res. Unit No. 3, Cairo.)

eyes may be present, but they never occur on the capitulum. The foregut is subdivided into a buccal cavity, a pharynx or pumping organ and an esophagus. The midgut is a thin-walled stomach, and the hindgut terminates in a sacculate rectum. Although there is a single-chambered heart in some forms of ticks, in most Acarina the circulatory system consists only of colorless blood without a heart.

Ticks and some mites may be differentiated as follows:

Ticks	*Mites*
1. Body clothed with short hairs or bare.	1. Body clothed with long hairs.
2. Hypostome exposed and possessing teeth.	2. Hypostome hidden and unarmed.
3. Larger forms, all macroscopic.	3. Usually small forms, many microscopic.
4. Body texture leathery in appearance.	4. Body texture usually membranous in appearance.

For an introduction to acarology see Baker and Wharton.[6]

Ticks

Suborder Ixodides

Families Argasidae and Ixodidae. About 800 different species of ticks occur in the world. They belong to two families, the Argasidae, which have soft, tough bulbous bodies that obscure the mouth parts and most of the legs; and the Ixodidae, or hard ticks, which have a dorsal chitinous plate (**scutum**) and a prominent capitulum. Although they are "hard," the females of this family can enlarge greatly while feeding.

The life cycle begins with eggs in masses deposited by the female on the ground, in cracks and crevices in houses, or in nests and burrows of animals. Favorable conditions include moisture, abundant vegetation and numerous hosts. Female ticks may lay as many as 30,000 eggs. The six-legged larvae, often called "seed ticks" (Fig. 19–2), which hatch from the eggs must find a host and begin feeding on blood. After molting, the larvae become nymphs, which have eight legs but no genital opening. Nymphs also molt after feeding, and become adults. The female

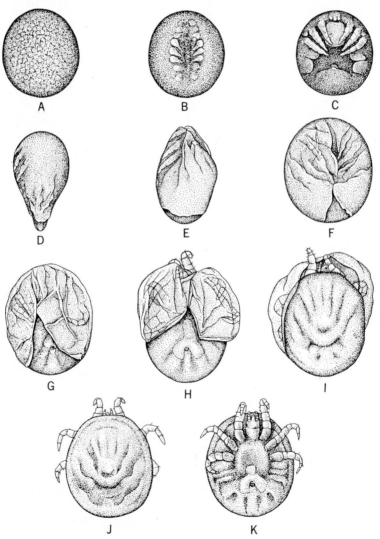

Fig. 19-2. *Ornithodoros moubata* egg and larva. (Hoogstraal, *African Ixodoidea*, United States Naval Medical Research Unit No. 3, Cairo.)

increases greatly in size after feeding, often reaching four times its original adult length, and increasing from 1 to 450 mg. in weight. The male, however, is enclosed in a nonelastic tegument that prevents much increase in size during feeding. Ordinarily the parasites climb on to a host and feed between molts (Fig. 19-3), and drop to the ground and molt or deposit eggs. For a comprehensive account of feeding in ectoparasitic Acari with special reference to ticks, see Arthur.[2] The complete life cycle may require a few weeks to two years, and it varies considerably among the numerous species of ticks. See Figure 19-4 for a key to genera.

Both sexes are blood suckers. It has been estimated that as many as 200 pounds of blood may be withdrawn from a large host animal by ticks in one season. After a meal of blood the female tick is ready for copulation and oviposition. The spermatophores are introduced into the vagina by the capitulum of the male. Hard ticks usually lay from 2,000 to 8,000 eggs after a pre-ovipositional period of three to 24 days (e.g., for Dermacentor variabilis). The time of oviposition varies from two to six weeks. Soft ticks lay fewer eggs (100 to 200) in several batches following successive blood meals.

Adult ticks may withstand starvation for several years, and, as Rothschild and Clay[11] have said, they "are the great exponents of the gentle art of waiting." In addition to waiting for food, both sexes may wait many months for a mate, and may remain together in copulation for over a week. Most ticks are intermittent parasites of mammals, birds and reptiles, usually spending most of their lives on the ground where they seek the shade. They often crawl up onto bushes or other vegetation and wait for any suitable host to come along. The majority of the ixodids (Fig. 19-4) in the immature stages parasitize different hosts from those preferred during the adult stage. *Dermacentor andersoni* (*D. venustus*), for example, may be found as nymphs and larvae on rodents, and as adults on sheep and man. On the other hand, the larvae of *Boophilus annulatus* become attached to cattle and remain on these hosts until the adult tick is ready to lay eggs.

Tick-borne Diseases

The several disorders and diseases of vertebrates traceable to ticks may be divided into two groups as follows (see Arthur[1]):

A. *Local inflammatory and traumatic damage at at the site of attachment* may be a mild inflammation and itching, or it may be far more serious, such as the invasion of the auditory canal by the spinose ear tick, *Otobius megnini*, causing edema, hemorrhage, thickening of the stratum corneum and partial deafness. The sharply toothed chelicerae cut an opening into the epidermis, then the recurved teeth of the hypostome serve as effective anchoring organs.

Fig. 19-3. *Acanthodactylus* lizard infested with nymphs of *Hyalomma*. (Hoogstraal and Kaiser, courtesy of Ann. Entom. Soc. Amer.)

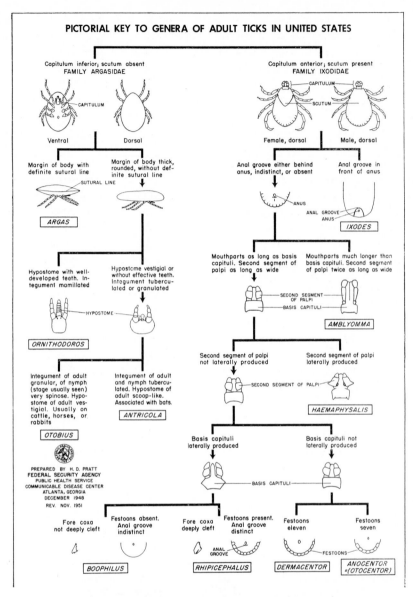

Fig. 19-4.   (Courtesy of Federal Security Agency.)

Such an invasion of the auditory canal is common in cattle. Salivary fluid from the tick prevents coagulation of the host blood as it is sucked through a tubular stylet within the mouth cavity. Adults of this tick do not feed at all. The genus *Amblyomma* may be found on man, other mammals, birds and reptiles (Fig. 19–5). *A. maculatum* sometimes occurs in enormous numbers on such ground-feeding birds as meadow larks, and it is also a serious parasite of livestock, attacking the inner surface of the outer ear.

B. *Systemic damage* may result in tick paralysis or in a less severe form of sensitization reaction to a toxic substance secreted by the salivary glands of the tick. Tick paralysis is commonly found in domestic animals and occasionally in man, especially children. About 12 ixodid ticks have been implicated (e.g., *Dermacentor andersoni*). The toxic sub-

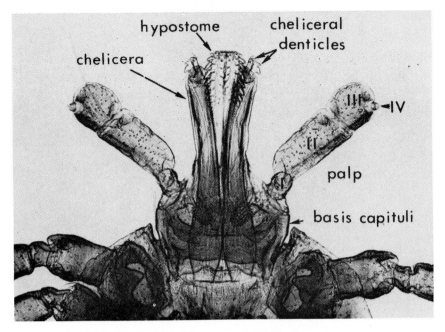

Fig. 19-5. Capitulum of *Amblyomma* sp. (Georgi, Parasitology for Veterinarians, courtesy of W. B. Saunders.)

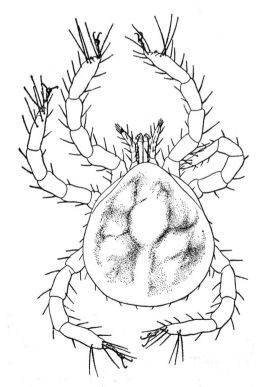

Fig. 19-6. *Argas americanus.* Six-legged stage. (Hassall, United States Department of Agriculture.)

stance, possibly elaborated by the tick ovaries or ova, causes a progressive ascending flaccid motor paralysis, elevation of temperature, impairment of respiration, speech and swallowing, and occasionally death due to respiratory or cardiac paralysis. Treatment of man involves complete removal of the ticks.

In addition, ticks frequently serve as vectors of diseases caused by viruses, rickettsias, bacteria, and protozoa.[13] Among the viral diseases transmitted by ticks are tick fever of sheep and lymphocytic choriomeningitis of rodents. Rabies of many mammals in the U.S.S.R. has been reported as being transmitted by ticks, possibly by *Argas persicus*.

Relapsing fever of vertebrates, caused by spirochaetes (*Borellia*) has been described from many areas over the world. Transovarial passage of *Borellia* occurs through several generations of ticks. Tick-borne strains of *Borellia* are commonly transmitted to birds by the genus *Argas* (Fig. 19–6), and to mammals by the genus *Ornithodoros* (Fig. 19–7). These ticks (*Ornithodoros*) possess a remarkable ability to survive under conditions

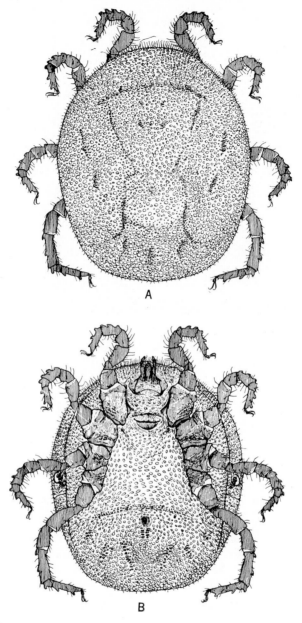

Fig. 19-7. *Ornithodoros moubata*, dorsal and ventral views. (Hoogstraal, *African Ixodoidea*, United States Naval Medical Research Unit No. 3, Cairo.)

of starvation and low humidity. A thin layer of wax in the epicuticle greatly reduces water loss, and the ability to close the spiracles and to extract water from moist air assists the tick in its regulation of water balance. Relapsing fever in man is characterized by recurrent febrile paroxysms, and three or four such attacks recur at intervals of about a week until immunity is established. The mortality rate is about 4 per cent. Except for the relatively few instances when ticks invade human residences, human relapsing fever results from man's intrusion upon the natural habitat of the tick and its host. At least 13 different species of *Ornithodoros* are known to be vectors of various species of the spirochaete, *Borellia*, throughout the world. Relapsing fever may also be transmitted by mites.

Texas cattle fever (red water, splenic fever, Mexican fever) is widely distributed among cattle in Europe, Africa, the Philippines and North, Central and South America. Theobald Smith was the first worker experimentally to incriminate ticks as vectors of disease. He and Kilbourne[12] described the causative agent, the protozoan *Babesia bigemina*, and they discovered the surprising fact that the parasites are carried from the infected mother tick to her offspring through the eggs. The parasites multiply within the red blood cells of cattle and within tissues of the tick. *Boophilus annulatus* in North and South America and in Europe, *Rhipicephalus appendiculatus* in tropical Africa, and other species of ticks are common vectors. The protozoan genus *Babesia* causes other diseases commonly known as "piroplasmosis" in domestic animals.

Tularemia is caused by a plague-like bacterium, *Pasteurella tularensis*, and its symptoms are similar to those of typhoid fever. It is found in mammals, birds and man, but is most common in rabbits and ground squirrels. The disease is characterized by an ulcer localized at the site of inoculation, and it is transmitted through the bite or fecal contamination of *Dermacentor andersoni* (Fig. 19-8), *D. variabilis*, and *Haemaphysalis leporispalustris*. The bacterium may pass from one generation of tick to another through the eggs. The disease may be acquired by man by simply handling diseased animals or their infected ticks. It is seldom fatal. Other tick-borne bacterial diseases of mammals are anthrax, erysipeloid and brucellosis.

American (Rocky Mountain) spotted fever, caused by *Rickettsia rickettsi*, is an acute febrile disease of wild rodents transmissible to man and laboratory animals. It is uncommon in the Rocky Mountain regions of Canada and the United States, but more common in eastern United States, and in various parts of Central and South America. The disease is transmitted by several species of ticks, especially *Dermacentor andersoni* which during its immature stages is a parasite of rodents, but as an adult is found on larger mammals including man. After an incubation period of five to ten days, a chill and rise in tempera-

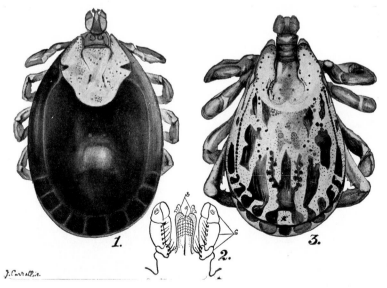

Fig. 19-8. *Dermacentor andersoni*. *1.* Dorsal view of female. *2.* Head showing hypostome (*a*), chelicerae (*b*) and palp (*c*). *3.* Dorsal view of male. Enlarged. (Stitt's Diagnostics and Treatment of Tropical Diseases, courtesy of the Blakiston Division, McGraw-Hill Book Co.)

ture is followed by a rash over the face and trunk. Male, female, and immature ticks may infest man.

Other tick-borne rickettsial diseases are: heartwater disease of sheep, goats and cattle; hemoglobinuric fever of domestic animals; and anaplasmosis of domestic animals. Although dogs frequently are infested with ticks, cats rarely have them. The brown dog tick, *Rhipicephalus sanguineus* (Fig. 19–9), is one of the most common. Other ticks to be found on dogs include *Dermacentor andersoni*, *D. variabilis*, *Amblyomma maculatum* and *Ixodes scapularis*. For a review of ticks in relation to human rickettsial diseases, see Hoogstraal.[8] For ticks and protozoan diseases, see Arthur.[3]

Mites

Mites abound in soil, humus, stored food, marine and fresh water, and as parasites of

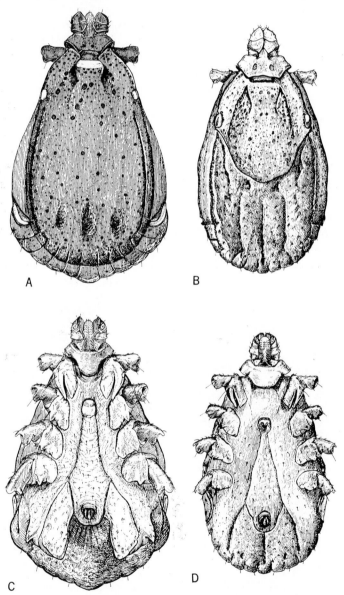

Fig. 19-9. *Rhipicephalus sanguineus sanguineus.* The male is on the left, female on the right. (Hoogstraal, *African Ixodoidea*, United States Naval Medical Research Unit No. 3, Cairo.)

plants and animals. Mites vary in size from less than 0.5 to about 2.0 mm. As a group they are free-living, but the suborder Mesostigmata contains the largest number of parasitic genera whose members are mostly lymph feeders as larvae or nymphs, and blood feeders (haematophagous) as adults. The life cycle from egg to larva (six-legged) to nymph (eight-legged) to adult is completed in from about eight days to more than four weeks. Mites feed on decaying organic matter or on the tissues of dead or living organisms. Some

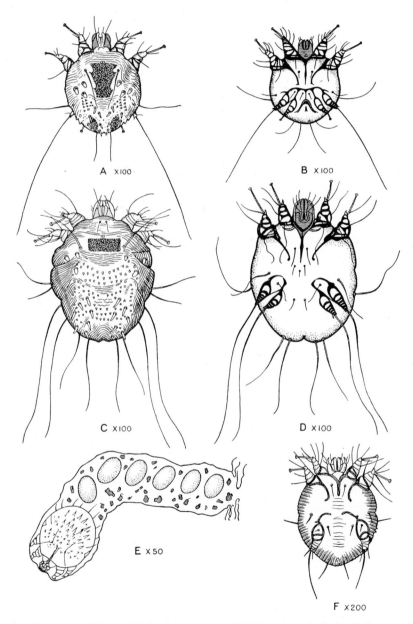

A ×100   B ×100

C ×100   D ×100

E ×50   F ×200

Fig. 19-10. *Sarcoptes scabiei. A.* Male, dorsal view. *B.* Male, ventral view. *C.* Female, dorsal view. *D.* Female, ventral view. *E.* Female with eggs in burrow. *F.* Hexapod larva, ventral view. (*A* to *D,* redrawn from Munro; *E,* redrawn from Banks; *F,* redrawn from Blanchard. Belding, Textbook of Clinical Parasitology, courtesy of Appleton-Century-Crofts.)

species use insects as a means of transportation as well as a source of food. Mites are most numerous in temperate zones. Parasitic species generally exhibit highly specialized structures such as enormous clawlike processes on the first pair of legs, used for grasping hairs of the host. Other species are adapted for such habitats as under the scales of snakes (the snake mite, *Ophionyssus natricis*); on the skin and feathers of birds where one genus, *Harpyrynchus*, lives in the feather follicles which become enlarged to form tumors; or within the lungs of snakes (family Entonyssidae). A family of mites allied to the latter (family Pneumonyssidae) contains free-living and parasitic forms, and one species may be found encapsulated in the bronchi or lungs of some Old World monkeys. Other mites may be found in the nasal passages of sea lions, in breathing tubes of domestic fowl, and in numerous other locations on or within the host. For a general account, see Hughes.[9]

Nonburrowing species, called **psoroptic** mites, pierce the skin, cause inflammation, exudations, itching and scab formation. For these reasons they are known as scab mites, and they cause **scabies.** A second group, the **sarcoptic** mites, consists of species which burrow into the skin and cause sarcoptic mange. The word **mange** is used rather loosely to include the effects of both types of mites. An infestation of mites is termed **acariasis,** and when the infestation is in the skin, producing channels in which eggs are deposited, the term **sarcoptic acariasis** is used (e.g., in human scabies caused by *Sarcoptes scabiei*). When mites deposit their eggs at the base of the host hairs or on the skin, producing scabs, the term **psoroptic acariasis** is used (e.g., in sheep scab caused by *Psoroptes communis* var. *ovis*).

Bird mites may be beneficial as well as harmful. A group (Cheyletidae) of "delousing" mites on birds prey on feather mites (Analgesidae) and possibly also on the eggs of feather lice (Mallophaga), and thereby destroy these ectoparasites and relieve the bird hosts of at least a portion of their misery.

Invertebrates may also be infested with these parasites. For example, mites (*Unionicola*) have been found on mussels (*Anodonta*). Other species infest amphipods.

Family Sarcoptidae (Itch and Mange Mites)

Numerous species of mites occasionally infest man, but *Sarcoptes scabiei* is responsible for human scabies. This species is so similar to forms of the same genus that infest dogs, cats, rabbits, foxes, pigs, horses and cattle that all of them are commonly considered biologic races (see p. 549) of one species. *S. scabiei* (Fig. 19–10) lives in cutaneous burrows where the gravid female deposits one to a few oval eggs daily for four to five weeks. The larvae, which have six legs, frequently produce lateral tunnels. Cutaneous lesions begin to develop in a few days after initial infestation, but the characteristic intense itching does not start until a month or so. The microscopic fecal pellets from the parasites are responsible for vesiculation and associated pruritus. Infestations similar to those just described for man are common on domestic animals. Scaly-leg mites on poultry cause lifting of the scales and a swollen condition of the shank with deformity and encrustation.

Family Psoroptidae (Scab Mites)

These mites, which cause psoroptic acariasis, differ from the itch mites in possessing long, slender legs, all four pairs of which extend beyond the margin of the elongate body (Fig. 19–11). The best known of these mites is *Psoroptes equi* var. *ovis* of sheep. Other varieties of this species infest cattle, horses and goats, and one species causes ear canker in rabbits. The mites are to be found on the surface of the body among the scabs at the base of the hairs, and generally in areas most thickly covered with hair. *Otodectes cynotis*, which closely resembles *Psoroptes*, is responsible for a common ear infestation of dogs, foxes and cats.

Family Listrophoridae

The myocoptic mange mite, *Myocoptes musculinus* (Fig. 19–12), is a widespread hair-clasping parasite of white or brown laboratory mice. It sometimes causes considerable trouble, especially when mice are kept for long-term experiments in crowded cages.

Oribatid mites (Fig. 19–13) commonly serve as intermediate hosts of tapeworms belonging to the order Cyclophyllidea. The worms *Moniezia expansa* of sheep, *Bertiella studeri* of primates, *Cittotaenia ctenoides* and *C. denticulata* of rabbits, *Anoplocephala perfoliata*, *A. magna*, and *Paranoplocephala mamillana* of equines, *Moniezia benedeni* of ruminants, *Thysaniezia giardi* of sheep and goats, and others are well-known examples.

Family Trombiculidae (Chiggers and Harvest Mites)

Larval stages (chiggers) of the families Trombidiidae and Trombiculidae are para-

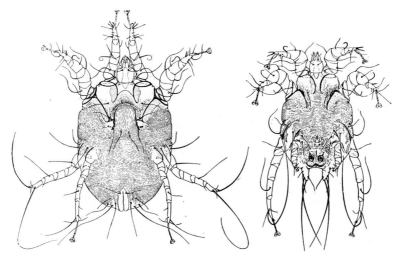

Fig. 19-11. *Psoroptes equi.* *Left.* Ventral view of female. *Right.* Ventral view of male. (Baker, *et al.*, A Manual of Parasitic Mites of Medical or Economic Importance, courtesy of National Pest Control Assoc. Inc.)

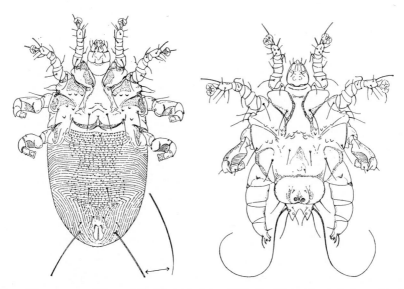

Fig. 19-12. *Myocoptes musculinus.* *Left.* Ventral view of female. *Right.* Ventral view of male. (Baker, *et al.*, A Manual of Parasitic Mites of Medical or Economic Inportance, courtesy of National Pest Control Assoc. Inc.)

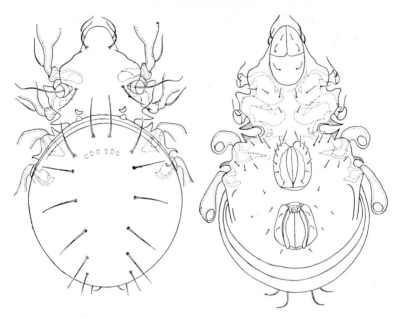

Fig. 19-13. An oribatid mite, *Metabelba papillipes*, female. *Left*. Dorsal view. *Right*, Ventral view. Legs removed. (Baker, *et al.*, A Manual of Parasitic Mites of Medical or Economic Importance, courtesy of National Pest Control Assoc. Inc.)

sitic, whereas the postlarval stages are predators, usually on soil arthropods. The larval trombidiids are parasitic on arthropods, and the trombiculid chiggers are parasites of a great many kinds of vertebrates, even fish, but especially of mammals (for details, see Wharton and Fuller[14]). The larvae of the family Trombiculidae frequently cause an intense, intolerable itch on the skin of man and animals. These mites are bright red or orange in color (hence the common name "red bug"), just visible to the naked eye, and they live as adults in grassy and bushy terrain frequented by domesticated animals or wild rodents. The six-legged larvae (Fig. 19–14) are parasites which, contrary to popular belief, do not burrow into the skin. The adult body is divided into an anterior portion (cephalothorax), bearing the mouth parts and the two anterior pairs of legs, and a larger posterior portion bearing the two posterior pairs of legs. The eggs are deposited singly or in small clusters on moist ground, most frequently in damp places well-covered with vegetation, and in six days the fully developed larva hatches. The larva attaches itself to a

vertebrate host and remains on its skin for a few days to a month. Saliva from the parasite dissolves a minute spot of the skin, and a tubular structure is formed. This structure, called a stylostome, is almost as long as the body of the mite, and it becomes filled with semi-digested tissue debris on which the parasite feeds. *Eutrombicula* and *Trombicula* are the most common genera.

Scrub typhus (mite typhus) in the Far East is transmitted by the larvae of species of the genus *Trombicula*, and is caused by *Rickettsia tsutsugamushi*. Various rodents serve as reservoirs of the disease, which in man may result in a high mortality rate.[4] On epidemiologic grounds the mites are suspected of transmitting epidemic hemorrhagic fever, a virus disease in Korea, Manchuria and Siberia.

Among the false spider mites (Tenuipalpidae), there are a number of species that are economically important. (e.g., species of the genus *Brevipalpus* on citrus and ornamentals in many parts of the world), which commonly feed on the lower surfaces of leaves. Some species feed on bark, and others on flower

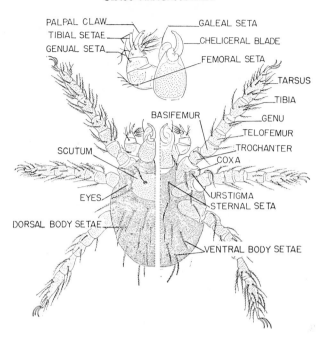

PALPAL CLAW
TIBIAL SETAE
GENUAL SETA
GALEAL SETA
CHELICERAL BLADE
FEMORAL SETA
TARSUS
TIBIA
BASIFEMUR
GENU
TELOFEMUR
TROCHANTER
SCUTUM
COXA
URSTIGMA
STERNAL SETA
EYES
DORSAL BODY SETAE
VENTRAL BODY SETAE

Fig. 19-14. *Trombicula alfreddugesi.* Dorsal and ventral views of larva to illustrate external morphologic characters. (Baker, *et al.*, A Manual of Parasitic Mites of Medical or Economic Importance, courtesy of National Pest Control Assoc. Inc.)

heads; a few form plant galls. The needle-like cheliceral stylets of these mites pierce the epidermis, the chlorophyll of the plant host is lost, and the plant tissues acquire a silvery appearance that later becomes rusty (for details of these mites see Pritchard and Baker[10]).

Family Pyemotidae (Predaceous Mites)

The family contains the North American "grain itch" mite, *Pyemotes ventricosus*, which feeds on larvae of insects that infest wheat and other grains, and also on the larvae of other insects, many of which are harmful (e.g., the cotton-boll weevil). These soft-bodied mites frequently swarm over the surface of the human body and burrow superficially into the skin, but unlike the Sarcoptidae, they soon leave the skin.

Family Dermanyssidae (Gamasid Mites: Parasitoidea)

The bloodsucking mites are parasites of reptiles, birds, and mammals, and they frequent nests and burrows. The "red mite," "roost mite" or "chicken mite" (*Dermanyssus gallinae*) of poultry, and the "tropical rat mite," *Ornithonyssus bacoti*[6] (= *Bdellonyssus* or *Liponyssus*) (Fig. 19-15) are examples of animal pests that are also injurious to man. The latter species is concerned with the transmission of certain rickettsial and virus diseases of man. *Dermanyssus gallinae* is a common parasite of pigeons, chickens, sparrows and other birds. Roost mites may be so abundant as to cause severe anemia and even death to young birds. After feeding they leave the host and hide in crevices.[7]

Nasal mites (Rhinonyssidae) belonging to the genus *Sternostoma* are primarily parasitic in the nasal cavities of birds. Some species, such as *S. tracheacolum*, the canary lung mite, is an internal parasite of the trachea, air sacs, and bronchi of the lungs, but not of the nasal cavities.

Families Acaridae and Glycyphagidae (Food Mites)

"Grocer's itch," "copra itch," "miller's itch," "cottonseed itch," "barley itch" are

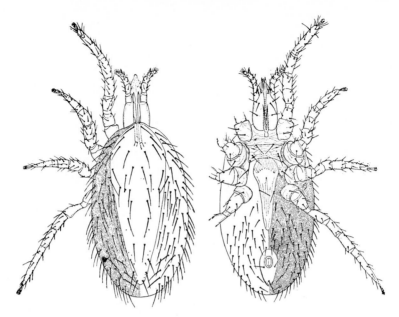

Fig. 19-15. *Ornithonyssus bacoti. Left.* Dorsal view of female. *Right.* Ventral view of female. (Baker, *et al.*, A Manual of Parasitic Mites of Medical or Economic Importance, courtesy of National Pest Control Assoc. Inc.)

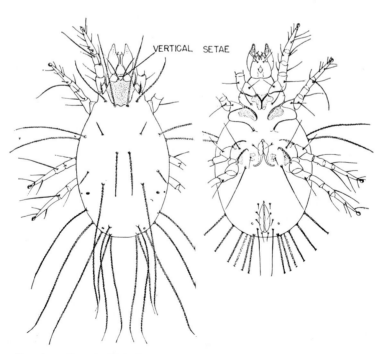

Fig. 19-16. *Tyrophagus lintneri. Left.* Dorsal view of female. *Right.* Ventral view of female. (Baker, *et al.*, A Manual of Parasitic Mites of Medical or Economic Importance, courtesy of National Pest Control Assoc. Inc.)

some of the common names given to human infestations of very tiny mites belonging to the family Acaridae (= Tyroglyphidae) or to the family Glycyphagidae. These mites (Fig. 19–16) feed upon and develop rapidly in such foods as grain, cheese, dried meat, dried fruit, insect collections, stored seeds, as well as such materials as stuffing in furniture. Literally millions of them may appear in a few days. Persons handling infected stored food products may be attacked temporarily by the mites and experience severe dermatitis. It has been found that contact with the living mites is unnecessary to produce the symptoms because they appear readily after infested material is rubbed on the skin, or after dust from it is blown on the skin, or even after the dust is inhaled. Hence the bodies or excretions of the mites are toxic, although allergy also plays a part in the production of symptoms.

Fig. 19-17. *Demodex canis.* Ventral view of female. (Baker, *et al.*, A Manual of Parasitic Mites of Medical or Economic Importance, courtesy of National Pest Control Assoc. Inc.)

Family Demodicidae (Follicular Mites)

A worm-like mite that infests hair follicles and sebaceous glands of various mammals is *Demodex folliculorum*. It has an elongated abdomen and stumpy legs, and may sometimes be found in clusters of 200 or more. Although infestation of human skin is common, particularly on the scalp and face, it is generally harmless, but in dogs the mites (Fig. 19–17) produce a severe type of mange, and in hogs and cattle they may be the cause of skin tubercles.

For details on parasitic mites of medical and economic importance, see Baker *et al.*,[5] and Traub.[13]

The Treatment of Tick and Mite Infestations

Specific acaricides are characterized by the lack of, or low, toxicity to insects and to mammals, but virtually nothing is known about the primary mode of action of these compounds. There appears to be a relatively high susceptibility of the larval stage to the acaricides. Most studies of this problem have dealt primarily with phytophagous mites, but the usual recommendations for control of parasitic mites and ticks suggest chlorinated hydrocarbon insecticides rather than the more specific acaricides. Ticks in South Africa and Australia have become highly resistant to DDT, BHC, toxaphene and other chemicals.

REFERENCES

1 Arthur, D. R.: Ticks and Disease. New York, Harper & Row, 1962.
2. ———: Feeding in ectoparasitic acari with special reference to ticks. *In*, Dawes, B.: Advances in Parasitology. Vol. 3, pp. 249–298. New York, Academic Press, 1965.
3. ———: The ecology of ticks with reference to the transmission of Protozoa *In* Soulsby, E J. L. (ed.): Biology of Parasites. Emphasis on Veterinary Parasites. pp. 61–84. New York, Academic Press, 1966.
4. Audy, J. R.: The role of mite vectors in the natural history of scrub typhus. Proc. Tenth Internat. Congr. Entom., *3:* 639–649, 1956 (1958).
5. Baker, E., *et al.*: A Manual of Parasitic Mites of Medical or Economic Importance. New York National Pest Control Assoc., Inc., 1956.
6. Baker, E. W., and Wharton, G. W.: An Introduction to Acarology. New York, Macmillan, 1952.
7. Hollander, W. F.: Acarids of domestic

pigeons. Trans. Amer. Micros. Soc., *75*:461–480, 1956.

8. Hoogstraal, H.: Ticks in relation to human diseases caused by *Rickettsia* species. Ann. Rev. Entomol., *12*:377–420, 1967.

9. Hughes, T. E.: Mites or the Acari. Bristol, England, The Athlone Press, 1959.

10. Pritchard, A. E., and Baker, E. W : The false spider mites (Acarina: Tenuipalpidae). Univ. Calif. Publ. Entomol., *14*:175–274, 1958.

11. Rothschild, M., and Clay, T.: Fleas, Flukes & Cuckoos. London, Wm. Collins and Sons, 1952.

12. Smith, T., and Kilbourne, F. L.: Investigations into the Nature, Causation, and Prevention of Texas or Southern Cattle Fever. Washington, D.C., Bureau Anim. Indust., 1893.

13. Traub, R.: Some considerations of mites and ticks as vectors of human disease. *In* Maramorosch, K. (ed.): Biological Transmission of Disease Agents, pp. 123–134. New York, Academic Press, 1962.

14. Wharton, G. W., and Fuller, H. S.: A manual of the chiggers. Mem. Ent. Soc. Wash., *4*:1–173, 1952.

Section VII
MISCELLANEOUS PHYLA

Chapter 20

Miscellaneous Phyla

PHYLUM PORIFERA

Sponges are rarely parasitic. One species, *Cliona celata*, bores into the shells of mollusks. This species may reduce oyster shell substance by as much as 40 per cent by this tunneling. The activity of the sponge may cause a break through the inner shell layer. These breaks are repaired if the water is warm (above $7°$ C.), but if the water is cold the hole remains unrepaired. The adductor and hinge muscles may also be damaged.

PHYLUM COELENTERATA

Peachia sp. and *Edwardsia* sp. are sea anemones whose larval stages become parasitic (inquinilism) on the surface of the body or in the gastrovascular cavity of other anemones (medusae) or in ctenophores. The parasites maintain their position by sucker-like adaptations of their mouths, and they feed on particles of food carried by a current created by ciliary action of the host.

Hydrichthys is a hydroid that lives as a colony on the bodies of fish and, occasionally, on crustaceans. The parasite has lost its tentacles and each polyp feeds on the blood and tissues of the host, which has been injured by the root-like outgrowths, or stolons, sent into the host flesh by the parasite.[18]

The minute anthomedusan, *Zanclea costata*, with abortive tentacles, becomes attached to the nudibranch, *Phyllirrhoe*, and appears to be parasitic on the snail. As the latter matures, however, it utilizes the medusa as a pelagic vehicle, and later discards it. Thus the nudibranch may be considered to be the parasite.[14]

The planula of some of the Narcomedusae become attached to the medusae of other hydroids. They develop in a normal manner and bud off young medusae.

PHYLUM CTENOPHORA

Gastrodes parasiticum (Fig. 20–1) is a ctenophore whose young stage is parasitic (commensal) in the mantle of the tunicate, *Salpa fusiformis*. The ctenophore is a flattened organism about 1 mm. in diameter, round in outline, with a concave oral surface and an arched aboral surface. The larvae of these ctenophores are of the planula type, thus suggesting a relation to the Coelenterata. The larval *Gastrodes* develops characteristic ctenophore rows of ciliated "combs," statocysts, tentacles and gastrovascular system. The parasite leaves its host, matures as a free-living organism, and produces planula larvae that enter new hosts.[7]

MESOZOA

The Mesozoa include two groups of highly specialized, cellular endoparasites of marine invertebrates. The phylogenetic origins of

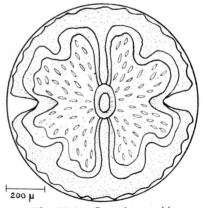

200 μ

Fig. 20-1. *Gastrodes parasiticum*, a parasitic ctenophore of a tunicate.

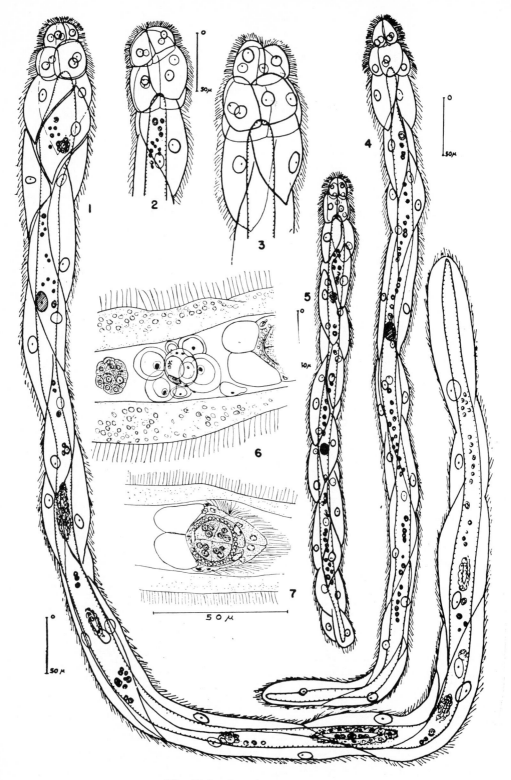

Fig. 20-2. *Legend on opposite page.*

these groups, the Dicyemida and Orthonectida, are obscure; some authorities consider them to be distinct and separate phyla.[11] Clearly, the mesozoans are incompletely known and deserve a great deal more attention. The adults of both groups are of two types: the **nematogens,** which bear vermiform embryos, and the **rhombogens,** which produce free-swimming ciliated larvae. For details of the anatomy, life history and systematic relationships, see review papers by Caullery,[2] Grasse,[4] Hyman,[7] McConnaughey,[15][16][17] and Stunkard.[23]

Dicyemida

Benthic cephalopods in many parts of the world contain vermiform dicyemid parasites. With few exceptions, they are found attached to the renal appendages of the host by means of a modified anterior end (calotte), hanging into a urine-filled renal coelom. Their simple structure consists of a single external layer of ciliated cells surrounding one or more axial reproductive cells. None of the tissue and organ systems common in multicellular organisms is present.

The life cycle involves asexual and sexual generations. Nematogens in immature cephalopods are characterized by vermiform embryos that are contained in the internal reproductive cell. A transition to the rhombogen phase occurs in mature cephalopods. This stage is characterized by the presence of nonciliated hermaphrodites (**infusorigens**) that give rise to free-swimming infusoriform larvae.

The infusoriforms grow within the parent rhombogen. When mature they are released, pass with the urine and escape from the host. The development outside the host is unknown, as is the manner of infection of new hosts by larval nematogens.

Dicyema sullivani (Fig. 20–2). This parasite will serve as an example of the group. It is a parasite of *Octopus bimaculoides* from southern California and Baja California, Mexico. The adult stages (nematogens and rhombogens) are slender, averaging .75 to 1.5 mm. long. Vermiform embryos at the time of emergence measure about 100 μ; the lengths of the infusoriforms are 40 to 48 μ.

A generalized classification of the dicyemids is as follows:

Family DICYEMIDAE

Genus *Dicyema* (e.g., *D. sullivani* from the kidneys of *Octopus bimaculoides*)

Dicyemennea (e.g., *D. brevicephaloides* from the branchial heart coela of *Rossia pacifica*)

Pleodicyema (e.g., *P. delamarei* from the kidneys of *Bathypolypus sponsalis*)

Pseudicyema (e.g., *P. truncatum* from the kidneys of the cuttlefish, *Sepia*)

The following two genera are sometimes placed in a separate family.

Family CONOCYEMIDAE (=HETEROCYEMIDAE)

Genus *Conocyema* (e.g., *C. polymorpha* from the kidneys of *Octopus vulgaris*)

Microcyema (e.g., *M. vespa* from the kidneys of *Sepia officinalis*)

Fig. 20-2. *Dicyema sullivani*, a mesozoan parasite of *Octopus bimaculatus*. All figures made with the aid of a camera lucida.

1. Nematogen containing vermiform larvae in various stages of development. The axial cell nucleus and two small accessory nuclei present in the axial cell are hatched to distinguish them from the somatic cell nuclei.

2. Anterior end of a nematogen in side view, showing the calotte (polar cap) and parapolar cells.

3. Anterior end of a nematogen in side view.

4, 5. Very young emerged nematogens prior to the formation of vermiform larvae.

6. An infusorigen from a rhombogen showing the free nucleus, one spermatogonium and one primary spermatocyte in its central cell, plus four fully formed sperm cells. Developing egg cells are on its periphery, the largest with a sperm attached to its edge. The paranucleus lies nearby in the axial cell of the parent rhombogen. The trunk cells of the rhombogen show the typical protein reserve granules.

7. Infusoriform, frontal optical section.

Note the spiral twist of the calotte, and the four axoblasts already present in the axial cell. One of the larvae has 31 somatic cells, the other 33. (McConnaughey, courtesy of J. Parasitol.)

Orthonectida

The orthonectids form a rare group of parasites found only in marine organisms. They occur in members of five invertebrate phyla: ophiuroid echinoderms; acoel, rabdocoel and polyclad turbellarians; gastropod and bivalve mollusks; polychaete annelids and nemerteans. Knowledge of the life cycle is more complete for orthonectids than for the dicyemids. Since most of the work has been done on *Rhopalura ophiocomae*, this will serve as an example.[11] Free swimming adults, upon release from plasmodia in host ophiuroids, are typically minute (0.1 to 0.8 mm.), either elongate or oval in shape. Externally the somatoderm is marked off into rings by circular grooves. The reproductive cells are confined to the interior, much the same as in the dicyemids, but form a mass composed of several hundred cells. Ciliated larvae develop, are released from the female, and enter the genital clefts of new host brittlestars. Here they disintegrate, liberating germinal cells that penetrate the host cells. Multiplying asexually, they form plasmodia that spread throughout the tissues and coelomic spaces, eventually giving rise to more males and females.

Caullery's[2] classification will be followed pending further investigation of the group. Some authorities feel that the similarities between the rhopalurids and the pelmatosphaerids are superficial, perhaps indicating that these two groups are diphyletic and not closely related.

Family RHOPALURIDAE

    Genus *Rhopalura* (e.g., *R. ophiocomae* from the ophiuroid *Ophiocoma neglecta*)

    Ciliocincta (e.g., *C. sabellariae* from the polychaete *Sabellaria cementarium*)

    Stoecharthrum (e.g., *S. giardi* from *Scoloplos mulleri*)

Family PELMATOSPHAERIDAE

    Genus *Pelmatosphaera* (e.g., *P. polycirri* from the annelid *Polycirrus haematoides*)

PHYLUM NEMERTEA (RHYNCHOCOELA)

Almost all nemerteans (ribbon worms or proboscis worms) are free-living, bottom-dwelling, marine animals whose soft bodies are equipped with a conspicuous proboscis. A few commensal and parasitic species have been described. *Nemertopsis actinophila* is a slender form that lives beneath the pedal disk of sea anemones. *Carcinonemertes* (9 to 70 mm. long) may be found on the gills and on egg masses of crabs. Some species of *Tetrastemma* live in the branchial cavity of tunicates. *Gononemertes parasitica* is a commensal species found on crustaceans.[9] *Malacobdella* and *Uchidaia* (Fig. 20-3) have been found in the mantle cavity of clams.[10]

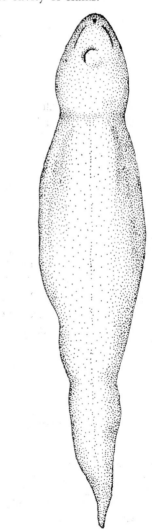

Fig. 20-3. *Uchidaia parasita*, a nemertean from the mantle cavity of a clam. (Redrawn from Iwata, 1966)

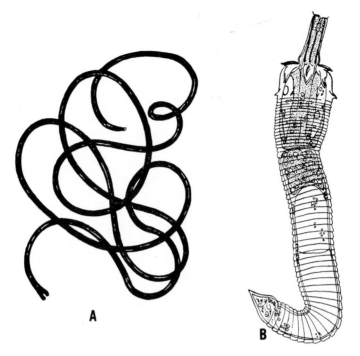

Fig. 20-4. *A. Gordius* sp. An adult male "horsehair worm," × 1.5. (Original) *B.* Larva of *Gordius aquaticus*. (Baer, Ecology of Animal Parasites, courtesy of Univ. Illinois Press.)

PHYLUM ROTIFERA

Rotifers are dioecious, microscopic animals that are abundant in fresh water, brackish water and moist soil. A few species live in the ocean, and a few are parasites. A ciliary organ (corona) is located at the anterior end of the body; the pharynx is provided with movable jaws (mastax); and typical flame-bulb protonephridia are present. Male rotifers are generally much smaller than females, and are rarely seen.

Seison moves about like a leech on the gills of the crustacean, *Nebalia*. *Pleurotrocha* may be found on colonial vorticellids (Protozoa), *Daphnia*, *Cyclops*, insect larvae and *Hydra*. Various species of *Proales* invade the heliozoan, *Acanthocystis*, pond snails, the tips of filaments of the alga, *Vaucheria*, and *Volvox*. *Albertia* is a worm-like, transparent rotifer that lives in the coelom of earthworms and other annelids.

In parasitic rotifers the head, corona and mastax tend to be reduced. For details see Hyman[8] and Rees.[20]

PHYLUM NEMATOMORPHA

Occasionally when one looks in farmyard water-troughs he sees a long (up to 1 m.), narrow, dark-brown worm. Because of a common notion that horse hairs become transformed into these worms, they are popularly called "horsehair worms," "hair worms" or "hair snakes" (Fig. 20-4). Horse hairs, of course, have nothing to do with the worms. In the United States there are two common varieties, *Gordius robustus* and *Paragordius varius*. In Europe the common species is *Gordius aquaticus*.

Adults and larvae possess a degenerate, nonfunctional digestive tract. Food is absorbed through the tegument. The anterior end of the adult is translucent with a dark ring behind the clear area. The posterior portion of the male ends in two broad branches, and the posterior portion of the female is usually bluntly tapered, but it may end in three broad branches.

Each female lays one million or more eggs

in strings in water or in moist soil. Within each egg an embryo develops that possesses an armed proboscis and head region that is separated from the rest of the body by a septum (Fig. 20–4). The proboscis becomes a perforating organ that enables the larva to emerge from the egg. Soon after escaping, the young worm usually penetrates an arthropod, but it may form a cyst and be eaten. Arthropods serving as hosts include beetles, cockroaches, grasshoppers, crickets, centipedes and millipedes. If an appropriate host becomes infected, the larval worm makes its way to the body cavity or to fat bodies, where it starts to mature to the adult stage. If eaten in the cyst form, the cyst wall is digested, releasing the worm. A cyst may remain viable in water for two months. If eaten by an inappropriate host (e.g., some insects, snails, certain fish), the cyst wall is digested but the parasite secretes a new cyst covering and is able to withstand passage through the intestine and thus gets back to soil or water. In this manner the worm may be transported long distances. If eaten again, this time by a suitable host, it can reach the adult stage. Just before reaching maturity in the insect, the worm escapes, possibly with the aid of a digestive enzyme. Worms do not leave the insect unless they can enter water. During their stay in the insect they may bring about castration of their host. In water the worms copulate and soon afterwards egg laying commences. The adults die in the fall of the year after all eggs have been laid. See Hyman[8] for details.

Most hair worms belong to the class Gordioidea and inhabit fresh water. A few belong to the class Nectonema and are pelagic in the ocean. Larvae in the latter forms enter crabs or hermit crabs.

PHYLUM ANNELIDA

Commensalism among annelids is relatively common but parasitism is rare. It is sometimes difficult, however, to distinguish between the two types of relationships. Leech-like oligochaetes (*Branchiobdella* and *Bdellobrillus*) are found attached to the ex-

posed exterior surface or gills of crayfish and other crustacea, and are undoubtedly commensals. Some species appear to be parasitic.[6] *Ichthyotomus sanguinarius*, on the other hand, is a bloodsucking parasite attached to the fins of the eel, *Myrus vulgaris*, and to other fish. The attachment organ involves two protrusible stylets that articulate with one another. The worm reaches 10 mm. in length and is dorsoventrally flattened. It is a neotenic larva, becoming sexually mature when only 2 mm. long. *Histriobdella* is another neotenic polychaete found in the branchial chambers of the European and Norwegian lobsters. This annelid is probably a commensal. *Stratiodrilus*, another genus that becomes sexually mature while in the young stage, is found on freshwater crayfish in Australia, Madagascar and South America. One species of oligochaete, *Friderica parasitica*, lives on another oligochaete, an earthworm. This parasite possesses a centrally located sucker, and the posterior half of its body is flattened. *Schmardaella lutzi* is an oligochaete that lives in the ureters of South American tree frogs. A transparent oligochaete, *Pelmatodrilus planariformis*, is a flattened species whose host is an earthworm in Jamaica. The parasite lives on the surface of the worm.

Myzostomids are polychaetes that often inhabit crinoids. The worms are so modified in shape that they are almost unrecognizable as annelids. They are dorsoventrally flattened, protandrous hermaphrodites, becoming first males, then females. Some species burrow into their hosts and form cysts within which a male and a female are sometimes found. *Myzostomum pulvinar* occurs in the intestine of a crinoid, while *Protomyzostomum polynephris* lives in the coelomic cavity of the ophiuroid, *Gorgonocephalus*. *P. nephris* feeds on the genital glands of the ophiuroid and causes partial castration of the host. Some mysoztomids are so firmly attached to the surface of crinoids that they leave permanent scars.

Some of the polychaete parasites are lumbrinerid-like worms belonging to the superfamily Eunicea. Some invade other members of the same superfamily and they sometimes grow so large that they are tremendous in proportion to their hosts. All of these lumbrinerid-like parasitic polychaetes belong to the family Arabellidae or at least show affinities to this family. The parasites have been found in echiuroids (*Bonellia*), and in members of the following polychaete families: Eunicidae, Onuphidae, Syllidae and Terebellidae. The parasites usually invade the body cavity or vascular system, normally while the host worms are in early developmental stages. Such parasites include *Drilonereis benedicti* in *Onuphis magna*, *Oligognathus parasiticus* in *Bonellia viridis*, and *Haematocleptes terebelloides* in *Terebellides stroemii*. One species, *Haplosyllis cephalata*, attaches itself by its pharynx to the cirri of an eunicid worm. This method of attachment by the pharynx is also seen with the ectoparasitic *Parasitosyllis*, which lives on other polychaetes and on nemerteans.

Leeches

Leeches belong to the class Hirudinea and may be found in the sea in fresh water or on land. Their general anatomy is illustrated in Figure 20–5. They range in length from 1 to 20 cm., but most are 2 to 5 cm. long. The distinction between a micropredator and a parasite is often blurred, so leeches are sometimes not included in a discussion of parasitism. Some of them are predaceous, feeding on worms, insect larvae, snails and other invertebrates, which they usually swallow whole, whereas others suck blood, feeding on invertebrates or vertebrates. Their hosts are mainly marine or freshwater fish, but they also attack amphibians, turtles, snakes, and terrestrial animals including man. Because of their habits, leeches may be called predaceous or bloodsucking parasites.

Some leeches are on their hosts only during periods of feeding, whereas others (e.g., many fish leeches) leave their hosts only for breeding. Some take a step further and attach their cocoons to the host, and a few (some of the rhynchobdellids) become sedentary and never leave their hosts.[1,13] When a leech is

CLASS HIRUDINEA

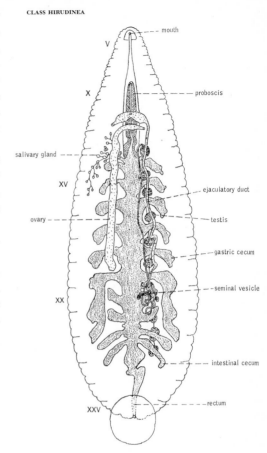

Fig. 20-5. The leech *Glossiphonia complanata*, ventral view. (After Harding and Moore) (From Pennack, R. W., Fresh-water Invertebrates of the United States. Copyright 1953, The Ronald Press Company, New York.)

on an aquatic host, the surrounding water makes firm sucker attachment possible. Terrestrial leeches use their urine to provide the necessary fluid. This procedure occurs in tropical land leeches (Haemadipsidae), which attack both mammals and birds. Leeches may get part way into the host's body. The genus *Theromyzon*, for example, may be found on the nasal membranes of shore and water birds.

Hirudo. This genus contains the medicinal leech that has been used to clean wounds of man. It normally sucks blood from the skin and occasionally from the nasal membranes or lining of the mouth cavity. *Hirudo* can ingest two to five times its own weight during one feeding period, and 200 days may be required for the digestion of one large meal. Since the leech is able to fast for a year or more, a meal every six months would appear to be ample.

Several species of aquatic leeches produce hirudiniasis in man. *Limnatis nilotica* infests the upper respiratory and digestive tracts and may cause severe injury. It is acquired through drinking water, and has been reported from Asia, the Middle East, southern Europe and Africa.

PHYLUM PENTASTOMIDA (TONGUE WORMS)

In 1969, J. T. Self[21] reviewed the biologic relationships of the Pentastomida and he stated that these bloodsucking endoparasites "constitute an aberrant group of helminths which, like onycophora, have lost most affinities to other major classes. Superficially acarin resemblances are thought to reflect larval adaptations only. Modern studies suggest that the pentastomes share both arthropod and annelid characters but that one cannot justify assigning them to either phylum. They are, therefore, assigned the status of an independent phylum with the belief that they share a common ancestry with the Annelida and Arthropoda."

Of the two orders, the Cephalobaenida, whose life cycles involve insects, fish, amphibia and reptiles as secondary hosts, are the more primitive. The order Porocephalida generally require mammals as secondary, and reptiles as definitive hosts.

The pentastome cuticle, like that of arthropods, contains chitin, and papillae (e.g., two at the posterior end of the Cephalobaenida) are the only sense organs. The mouth is framed by a chitinous structure, and the digestive tube is relatively simple, ending with an anus usually in the last segment. Muscle tissues are striated, similar to those of arthropods, but locomotion is sluggish, although the parasites can migrate through host tissues. The single ovary in females may extend from one end of the abdomen to the other, and in the male the single testis occupies one-half to one-third of the body cavity.

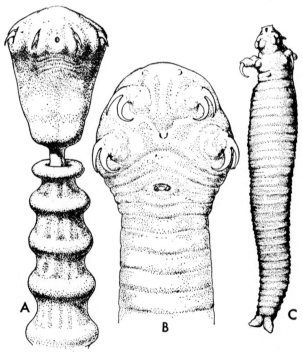

Fig. 20-6. *A.* Anterior extremity of *Armillifer annulatus*. *B.* Head of *Leiperia gracilis*. *C.* Entire specimen of *Raillietiella mabuiae*. (Baer, Ecology of Animal Parasites, courtesy of University of Illinois Press)

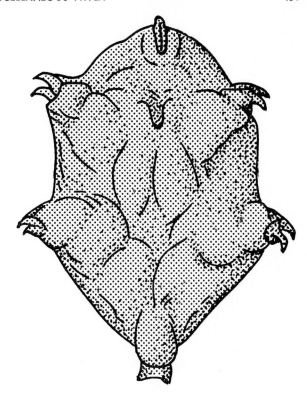

Fig. 20-7. Newly hatched larva of *Porocephalus crotali*, ventral view. (Modified from Penn, courtesy of J. Parasitol.)

The pentastomids are legless and worm-like, but near the mouth are two pairs of hollow, curved, retractile hooklets that are rudimentary appendages (Fig. 20–6). Sharp hooks on the anterior end are anchoring devices. The immature stages (Fig. 20–7) are mite-like in appearance, with two or three pairs of legs. Adult pentastomids feed on blood and mucosal cells in the mouth, esophagus or respiratory passages of their hosts. In abnormal hosts, such as humans, there is extensive phagocyte infiltration as well as fibrosis.

Relatively little is known of life histories and general biology of these worms, although some life cycles in the genus *Linguatula* have been fairly well understood for many years. (See Nicoli and Nicoli[19] for a general review.) In most genera, such as *Armillifer* (Fig. 20–8), *Sambonia, Raillietiella, Porocephalus, Linguatula* and *Megadrepanoides*, a life cycle involving two vertebrate hosts would be expected because the adult parasites normally occur in carnivorous lizards or snakes. These hosts regularly feed on smaller reptiles, amphibians and mammals in which larval pentastomids have been found. For example, adults of *Porocephalus crotali* live in the lung cavities of crotaline snakes, especially of those species of rattlesnakes (*Crotalus*) that range along the western parts of North, Central and South America.[3] The parasite eggs, which reach the outside through the snake sputum, are readily ingested by muskrats, opossums, bats, armadillos, raccoons, and other mammals, in which they hatch in the small intestine. Larvae migrate to the viscera (e.g. lungs and liver), where they become encapsulated in the tissues and where, in about three months, they develop into nymphs. If the nymph-infected tissue of the mammal is ingested by an appropriate snake, the nymphs migrate up the esophagus into the tracheae and lungs, where they become adult pentastomids. Nymphs of *Armillifer* have been found in man in the Congo. The cockroach, *Periplaneta americana*, is the intermediate host for the pentastomid, *Raillietiella hemidactyli*, which

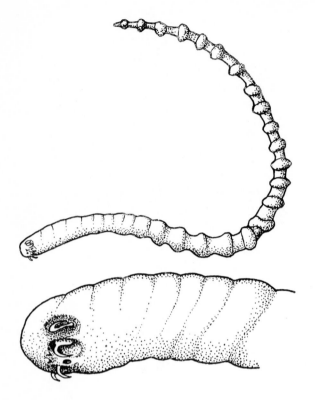

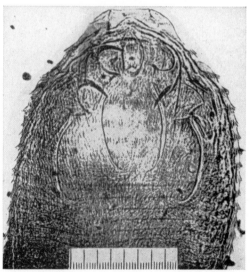

Fig. 20-8. *Armillifer* sp. from the lung of a python in Indonesia. *Top*. Entire specimen. *Bottom*. Enlarged view of head. The preserved worm was 72 mm. long.

lives in the lungs of lizards.[12] Undoubtedly a thorough search would reveal other examples of insects serving as intermediate hosts of these parasites. For a discussion of host-parasite relationships, see Self and Kuntz.[22] For a study of the biologic relationships of the group and a bibliography on the Pentastomida, see Self.[21]

Linguatula serrata (Fig. 20–9) is one of the best-known species of this group. The adult female is tongue-shaped, 100 to 130 mm. long and up to 10 mm. wide, whereas the male is about 20 mm. long and 3 to 4 mm. wide. Eggs containing embryos with rudimentary legs are deposited in the nasal passages and frontal sinuses of mammals (commonly dogs) and are discharged in nasal secretions. Upon reaching water or moist vegetation, embryonation is completed, and if the eggs are ingested by the intermediate host (e.g., cattle, goats, sheep, rabbits, rats, man), hatching occurs in the digestive tract, and the larvae migrate through the intestinal wall and become lodged in the liver,

Fig. 20-9. *Linguatula serrata*. Head end. Each division of the scale represents 10 μ. (Sprehn, courtesy of Parasitologische Schriftenr.)

mesenteric nodes and other viscera. After a number of molts, requiring a period of five to six months, the nymph stage is attained.

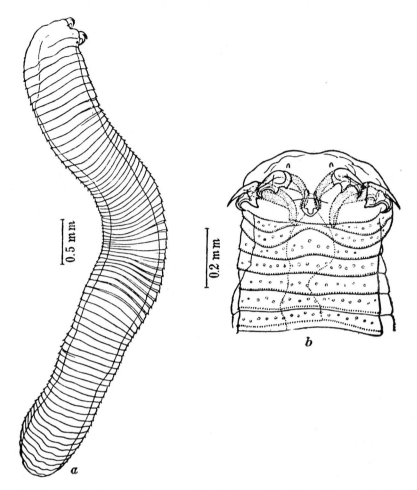

Fig. 20-10. *Pentastomum solaris.* *a.* Entire worm, lateral view. *b.* Anterior end, ventral view. From the
lungs of a crocodile. (Tubangui and Masilungan, courtesy of Philippine J. Sci.)

Nymphs lie encapsulated with the host tissue, and they are the infective stage for carnivorous animals that feed on the herbivorous intermediate hosts. Although the general belief is that nymphs are digested out of their capsules and immediately migrate to the nasal passages of the definitive host, there is evidence[5] that the nymphs quickly leave their cysts after the death of the intermediate host, and cling to the mucous membrane of the mouth of the carnivorous host. *L. rhinaria* is a common parasite of livestock, and it is a facultative parasite of man in both adult and nymphal stages. This species also occurs in rats, dogs, and in such experimental animals as guinea pigs. In the genera *Linguatula*, *Pentastomum* (Fig. 20–10), *Sebekia* and *Leiperia*, each hook is double in both pairs.

PHYLUM MOLLUSCA

Relatively few mollusks become parasites. As with other groups, it is sometimes difficult to decide whether a mollusk is a commensal or a parasite. The clam, *Entovalva mirabilis*, is an internal parasite of a holothurian. Other species of the same genus of clams may also be found in these hosts. Some commensal clams live on or in sea urchins, sponges, sipunculids, crustaceans, polychaetes, and other invertebrates. These mollusks belong to several genera (e.g., *Modiolaria*, *Montacuta* and *Lepton*) and they may injure their hosts, but the relationships are not well understood.

Glochidia are larval members of the clam family Unionidae. They belong to several genera and species, and are parasites of fish. The young freshwater clams are discharged in vast numbers from the mother clam and become attached to the gills and body surface of various fish. These glochidia (Fig. 20–11) possess two small valves that aid in attachment with a pincer-like action. The host epidermis grows over the parasite. Later, host tissue beaks down, and within the chamber thus formed each young clam metamorphoses. It then breaks out of the chamber, drops to the sand and matures into an adult clam.

Specific glochidia occur on fish inhabiting the same biotope as that of the adult clam, but the segregation is ecologic because, experimentally, the glochidia can be induced to fix themselves to many different kinds of living supports, and, in nature, the speed of the fish and level at which it swims influences the fixation of the larval clam.

Parasitic snails are either prosobranchiate or opisthobranchiate gastropods. The relationships of the pyramidellids (e.g., *Odostomia*, Fig. 20–12, and *Turbonilla*) to their hosts are similar to those of a micropredator to its macroprey. In this group the radula has been lost, and feeding is by means of an elongate proboscis armed with a piercing stylet and esophageal pump. These small snails are adapted for tapping the body fluids of many hosts, mainly annelids and mollusks. *O. eulimoides* (family Pyramidellidae) attacks oysters, causing malformations of the edge of the shell of the host. The parasite then forms small pockets in the shell and penetrates toward the adductor muscle. The margin of the shell becomes thickened, and the adductor muscle becomes covered with shell deposits. Sand gets into the host and may cause suffocation and death. As many as seven parasitic snails have been found in one oyster. Some species of snails (*Epitonium* and *Opalia*) are micropredators or semiparasites that feed on coelenterates.

Ectoparasites like *Stylifer*, *Thyca* (Fig. 20–12), and *Eulima* are found securely attached to or embedded in the surface of echinoderms, especially asteroids and crinoids. A permanent hole is made in the body wall of the host so that the proboscis can be inserted into host tissue.

Endoparasites such as *Asterophila*, *Entoconcha*, *Enteroxenos* and *Gasterosiphon* have lost nearly all the organ systems commonly associated with typical gastropods. Only the presence of a veliger larval stage indicates their molluscan ancestry. *Entocolax* (Fig. 20–13), a worm-like parasite, penetrates the body wall of holothurians. It becomes attached to the lining of the body wall and hangs freely in the body cavity, sweeping up host fluids through a proboscis into a greatly

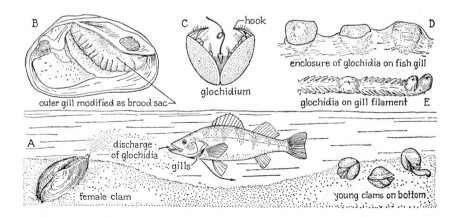

Fig. 20-11.   Life cycle of glochidia.   (Storer, General Zoology, courtesy of McGraw-Hill)

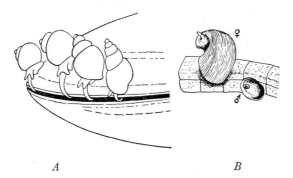

Fig. 20-12. *A. Odostomia scalaris* with proboscides inserted between the valves of a clam. *B.* Male and female specimens of *Thyca stellasteris* on the arm of a crinoid. (Baer, Ecology of Animal Parasites, courtesy of University of Illinois Press)

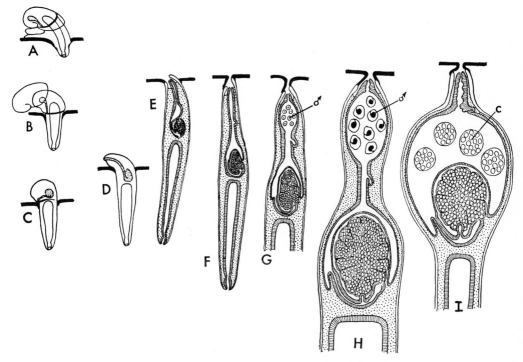

Fig. 20-13. Supposed mode of entrance of *Entocolax* into its host. (Baer, Ecology of Animal Parasites, courtesy of University of Illinois Press)

reduced intestine. A siphon allows release of reproductive products to the outside. *Entocolax mirabilis* feeds on its host by inserting its proboscis-like esophagus into a blood vessel and taking blood into its stomach. The parasite is about 8 cm. long.[24]

REFERENCES

1. Baer, J. G.: Ecology of Animal Parasites. Urbana, Ill., Univ. Illinois Press, 1951.

2. Caullery, M.: Classe des Orthonectides (Orthonectida Giard, 1877). *In* Grassé, P.-P. (ed.): Traité de Zoologie: Anatomie, Systématique, Biologie. Tome IV. Fasc. I. Plathelminthes, Mésozoaires, Acanthocéphales, Némertiens. Paris, Masson et Cie, 1961.

3. Esslinger, J. H.: Development of *Porocephalus crotali* (Humboldt, 1808) (Pentastomida) in experimental intermediate hosts. J. Parasitol., 48:452–458, 1962.

4. Grassé, P.-P.: Classe des Dicyemides. *In*

Traité de Zoologie: Anatomie, Systématique, Biologie. Tome IV. Fasc. I. Plathelminthes, Mésozoaires, Acanthocéphales, Némertiens. Paris, Masson et Cie, 1961.

5. Hobmaier, A., and Hobmaier, M.: On the life cycle of *Linguatula rhinaria*. Amer. J. Trop. Med., *20*:199–210, 1940.

6. Holt, P. C.: The branchiobdellida: epizootic annelids. The Biologist, *50*:79–94, 1968.

7. Hyman, L. H.: Phylum Mesozoa. *In* The Invertebrates: Protozoa through Ctenophora. Vol. I. New York, McGraw-Hill, 1940.

8. ———: The Invertebrates: Acanthocephala, Aschelminthes, and Entoprocta The Pseudocoelomate Bilateria. Vol. III. New York, McGraw-Hill, 1951.

9. ———: The Invertebrates: Smaller Coelomate Groups. Vol. V. New York, McGraw-Hill, 1959.

10. Iwata, F.: *Uchidaia parasita* nov. gen. et nov. sp., a new parasitic nemertean from Japan with peculiar morphological characters. Zool. Anz., *178*:122–136, 1966.

11. Kozloff, E. N.: Morphology of the orthonectid *Rhoplaura ophiocomae*. J. Parasitol., *55*: 171–195, 1969.

12. Lavoipierre, M. M. J., and Lavoipierre, M.: The American cockroach, *Periplanata americana* as an intermediate host of the pentastomid, *Raillietiella hemidactyli*. Med. J. Malaya, *20*(1):72, 1965.

13. Mann, K. H.: Leeches (Hirudinea). Their structure, physiology, ecology and embryology. With an appendix on the systematics of marine leeches, by E. W. Knight-Jones. New York, Pergamon Press, 1962.

14. Martin, R., and Brinckmann, A.: Zum

Brutparasitismus von *Phyllırrhoe bucephala* Per. and Les. (Gastropoda, Nudibranchia) auf der Meduse *Zanclea costata* Gegenb. (Hydrozoa, Anthomedusae). Publ. Staz. Zool. Napoli, *33*:206–223, 1963.

15. McConnaughey, B. H.: The rhombogen phase of *Dicyema sullivani*, McConnaughey. J. Parasitol., *46*(5):608–610, 1960.

16. ———: The Mesozoa. *In* Dougherty, E. C., *et al.* (eds.): The Lower Metazoa, Comparative Biology and Phylogeny. Berkeley, Univ. of California Press, 1963.

17 ———: The Mesozoa. *In* Florkin, M., and Scheer, B. T. (eds.): Chemical Zoology. Vol. II. Porifera, Coenlenterata, and Platyhelminthes. New York, Academic Press, 1968.

18. Miyashita, Y.: Occurrence of a new *Hydrichthys* in the Pacific Coast of Japan. Annot. Zool. Jap., *20*:151–153, 1941.

19. Nicoli, R. M., and Nicholi, J.: Biologie des Pentastomides. Ann. Parasitol. Hum. Comp., *41*:255–277, 1966.

20. Rees, B.: *Albertia vermicularis* (Rotifera) parasitic in the earthworm *Allolobophora caliginosa*. Parasitology, *50*:61–66, 1960.

21. Self, J. T.: Biological relationships of the Pentastomida; a bibliography on the Pentastomida. Exper. Parasitol., *24*:63–119, 1969.

22. Self, J. T., and Kuntz, R.: Host-parasite relations in some Pentastomida. J. Parasitol., *53*:202–206, 1967.

23. Stunkard, H. W.: The life history and systematic relations of the Mesozoa. Quart. Rev. Biol., *29*:230–244, 1954.

24. Tikasingh, E. S., and Pratt, I.: The classification of endoparasitic gastropods. Syst. Zool., *10*, 65–69, 1961.

Section VIII
PARASITE-HOST RESPONSES

Chapter 21

Parasite-Host Responses

INTRODUCTION

Prosperous parasitism requires the evolution of certain essentials for maintenance of a parasite within the body of a host. (1) Entry into the host must be comparatively easy. Many parasites accomplish this all-important first step passively by being swallowed as a cyst, egg or larval stage. Active entry usually requires special organs, such as histolytic glands. (2) A parasite must be able to remain within the particular organ or organs to which it has become adapted. For the larger parasites this requirement is often met in part by the development of mechanical devices, such as suckers or hooks. (3) The basic physiologic processes common to all living things must be allowed to function. The parasite must, therefore, find within its new habitat all the necessary factors for such processes as respiration, assimilation, growth, and reproduction. (4) The parasite must be able to counteract the defense mechanisms of its host and the adverse consequences of numerous other results of a hostile environment. Some of these deterrents to successful parasitism are: host enzymes, foreign proteins, hormones, toxic substances, phagocytes, antibodies, and changes in osmotic and pH conditions of the surrounding medium. The parasite finally achieves success when the host becomes tolerant, or even cooperative.

In general, we find a trend of modification, a sort of direction of evolution that varies in character among the groups of parasites. Nematodes, for example, tend to increase in size. To be impressed with this trend, one has but to compare the sizes of free-living nematodes that are found in soil or water with the average size of the parasitic forms. Trematodes, on the other hand, have not responded to parasitism with such a pronounced increase in size, but they have undergone a greater increase of complexity.

When a parasite inhabits two or more different types of hosts in its life cycle, the effect of the host on the parasite is different in each host, and when one host harbors several different kinds of parasites, the effect of the host generally differs with each type of parasite.

Adaptation of a parasite to its host, with the concomitant adjustment of the host, are the chief factors in successful parasitism. This adjustment evolves slowly, and it may require millions of years. Man has probably been parasitized by the malarial parasite, *Plasmodium*, since the beginnings of the human race, but the parasite has not yet become completely adapted to man. Its present degree of success is maintained partly by enormous reproductive output to meet the fantastic odds against the completion of a normal cycle by any one zygote. That it has a long way to go before becoming well-adapted to its host is probably indicated by the severity of the host reaction. After all these thousands of years, there are still about 200,000,000 cases of malaria annually in the world, with about 1 per cent mortality. A well-adapted parasite causes little or no harm to its host—obviously, the internal parasites remain alive only so long as their hosts live.

One of the many kinds of interesting parasite-host relationships is demonstrated by a host and its insect parasitoid (parasitic only in the larval stages). Salt[54] has shown that, far from being a purely passive victim, such a host is often able to impress its mark upon the insect that destoys it. The host influences the size, form, rate of development, and behavior of its parasite, and probably its fecundity, longevity and vigor. "There can be no doubt that the host may bequeath to its parasite an important and sometimes striking legacy of morphological, physiological, and behavioristic characteristics."

Failure of the host to adjust properly to its parasites often results in weakness, incapacities or disease, but failure of a parasite to adjust to its host is even more critical. The parasite has the problem of maintaining its basic organization, yet retaining enough flexibility to adapt itself to changing conditions in new surroundings. This adjustment is a delicate process, a critical balance that is in a state of continuous flux. No wonder that parasites are often not suited to life in a host other than the one to which they have become accustomed. A thorough understanding of the physiologic aspects of parasite-host relations cannot be realized until a careful study is made of the biochemical natures of the specific parasite and particular host under consideration. Generalizations are hazardous.

The over-all result of parasitism, so far as the parasite is concerned, has been described as a life of ease without much effort. As was mentioned in an earlier chapter, parasites obtain "free board and lodging" with abundant food, and perhaps some freedom from competition and from predators. One should not, however, consider the life of a parasite to be easy, comfortable and without labor. There is constant danger: host antagonism must be met, and adaptations must be made. All organ systems are involved. The general effect on internal parasites is an increase in the structure and function of those systems associated with reproduction and attachment, and a simultaneous simplification of the sensory and locomotor systems.

NUTRITION: EFFECTS ON HOST

Important concepts arising from studies on nutritional aspects of parasitism were summarized by Geiman[21] as follows: "(a) the concept that inception and severity of the specific infection and disease depend upon a particular nutritional deficiency and the specific nutritional requirements of the parasite; (b) the concept that *synergism* (where nutritional deficiency of the host results in an increase of severity of infection) is dominant over antagonism (where a nutritional deficiency results in decreased frequency of severity of infection); and (c) the concept that pathogenesis following an upset in dynamic equilibrium between the host and parasite may result from changes in the metabolism of either the host or parasite from either exogenous or endogenous factors."

Loss of food due to the nutrient needs of parasites may have undesirable consequences for the host, but it is usually not a serious problem. Many parasites even make partial compensation for this loss. They may incompletely oxidize their food so that they excrete partially oxidized substances. These substances are frequently utilized by the host. Thus the host ingests the food, digests it, "feeds" some of it to parasites which derive energy from partial breakdown of the digested material, and which excrete the rest to be used completely by the host. As is discussed briefly in Chapter 22, page 498, the rumen of cattle and of other ruminants contains many ciliates. It is reasonable to suppose that the host benefits not only by digesting the vast numbers of protozoa that die, but also by utilizing some of the intermediate breakdown products of cellulose (p. 500).

The popular notion that a person with worms is constantly hungry is not borne out by fact. Intestinal worms may eat the food that their host swallows, but two or three worms consume a negligible fraction of any one host meal. Malnutrition, however, is a common associate of parasites in general. This condition is due to many factors other than actual robbery of food by parasites. The symptom is often spoken of as "off its feed" and is common with worm infections in domestic animals. A few parasites will usually do no harm, nor do they normally produce observable effects on the host. When an animal suffers malnutrition due to parasitic infection, it is usually the result of a long-standing disease or of an unusually heavy infection, or both.

The malarial parasite, *Plasmodium*, and the roundworm, *Ascaris*, are good examples of organisms that may cause malnutrition in man. This condition is due to a reduction of effective protein supplies in the host by vomiting, loss of appetite for food, diarrhea or liver damage; and to increased protein requirements, due to feeding of the parasite, protein destruction and the building up of antibodies from a protein base.[47] Protein deficiency impairs normal metabolism of the liver, hence the effect of many parasites on the host is reflected in liver abnormalities such as fatty degeneration. In addition to malaria, kala-azar and schistosomiasis are diseases involving this type of liver impairment in man.

NUTRITION: EFFECTS ON PARASITE

In the numerous studies that have been made on the effect of the host's diet on the well-being of the parasite, there exist many different opinions, but one might conclude that a deficiency in the host diet has the effect of rendering the host more vulnerable to parasitic infections. Whatever the cause, when certain essential substances are reduced or omitted, the overall resistance of the host to the parasite decreases. This observation applies particularly to intestinal parasites and it implies that the intestinal environment becomes more favorable to the parasite during periods of malnutrition. It would seem that a healthy host is generally more free from parasites than is an unhealthy one. Although a moderately unhealthy host often harbors increased numbers of parasites, a definitely ill host is often in too poor a condition even to support its parasites.

The kind of nutritive material ingested by larval wasps, *Trichogramma*, affects the development of wings in the male parasite. Those wasps that feed on moth eggs (the normal host) generally develop into winged adults, whereas those that feed on alderfly eggs (abnormal hosts) develop almost exclusively into wingless adults.

In regions of South America where human malnutrition is prevalent, chronic malaria is common, whereas in other regions where malnutrition is virtually absent, malaria is present but rarely in a chronic form. A series of experiments on the diet relationships of canaries and various species of *Plasmodium* have been conducted to help solve this problem.[4] Most of the birds on poor diets had more severe primary attacks of malaria than did the controls. In general, these primary attacks were characterized by greater numbers of parasites, higher parasite peaks (the greatest number of parasites per 10,000 erythrocytes counted in a single day during an infection), longer patient periods, and more severe symptoms. The mortality rate was highest and relapses occurred only in birds on deficient diets. Immunity was greatly reduced or entirely absent in a number of nutritionally deficient birds.

Experiments of the sort just described may be misleading in several ways. If the omission of certain substances in the host diet does not appear to affect a parasite, the negative response is not necessarily an indication that the parasite does not require that substance. The parasite may not depend on its host's diet as a source of that substance. The suggestion has been made that the average *Ascaridia galli*, a roundworm living in the intestine of chickens, does not need vitamin A, B-complex, or D. This suggestion was based on the fact that alteration of the host diet does not seem to affect the nematode. It would be safer to conclude only that the nematode might not depend upon its host for these vitamins.

A diet consisting largely of milk or milk sugar has an adverse effect on intestinal helminth and protozoan fauna. There is no evidence that the milk directly injures the worms, and probably the effect is due to an alteration of the environment. A diet rich in lactose changes the intestinal emptying time and alters such reactions as oxidation-reduction potential, pH and the synthesis of vitamins by intestinal microorganisms. As yet we are not familiar enough with the biochemistry of parasites to understand fully the effects of these changes on intestinal parasites. Certainly we cannot assume that the nature of the metabolism of vitamins or of any other substance is necessarily the same in the parasite as in the host.

The types of carbohydrates in the host diet have a gross effect on the size of the rat tapeworm, *Hymenolepis diminuta*.[49] Tapeworms in rats that receive starch as the sole carbohydrate are much larger than those in rats that receive only dextrose or sucrose. The reasons for this difference are obscure but they may be associated with differences in the availability of carbohydrate to the parasite, with the effects of different carbohydrates on the bacterial flora of the gut (and, therefore, an indirect effect on the worms), or with different effects on the physiology of rat and worm. Starvation of a rat causes a reduction in the weight of its tapeworms and, curiously enough, increases the ability of the parasite to utilize glucose. Starvation of a hamster causes a reduction in weight of its tapeworms, but does not increase the worm's ability to utilize glucose. Tapeworms of the dogfish shark also lose weight when their host is starved, and this loss can be prevented by the oral administration of starch. This effect of starvation on the weight of intestinal tapeworms would hardly justify starving a man in an effort to rid him of worms. The nature and quantity of the host dietary carbohydrate also has a definite effect on the growth and egg production of many worm parasites.

HOST CELL REACTIONS

The following short review of certain cell reactions to parasites is adapted from Soulsby.[57] The account is limited primarily to the cell types that are closely associated with immune responses.

A characteristic feature of almost all parasitic infections is a marked cellular response. Abnormal multiplication of cells (hyperplasia) is a common response and is seen in the bile duct during fascioliasis, in the urinary bladder during schistosomiasis, and in the gastrointestinal mucosae during various helminth and protozoan infections. The host cells may lose their function or even acquire new functions (metaplasia). Frequently, an adult or larval parasite is trapped in host tissue composed of lymphoreticular elements whose growth has been initiated by the accumulation of lymphocytes, plasma cells and macrophages. This type of reaction by the host is obviously a form of local immunity. In cutaneous leishmaniasis, the macrophages in which the leishmania multiply undergo hyperplasia until the area is infiltrated with lymphocytes and plasma cells. Proliferation of the macrophages ceases and eventually the cutaneous lesion subsides.

There is evidence[31] that plasmodia may produce a toxin that profoundly affects the basic metabolic processes of host tissue cells by inhibiting respiration and oxidative phosphorylation. These events begin during the erythrocytic phase of the cycle.

Blood Cells

Neutrophil accumulations are characteristic during the initial stages of parasitic infection. The role they play is unclear but they are associated with inflammation and tissue destruction and are particularly abundant around dying or dead parasites. The neutrophil response is usually replaced by one of lymphoid cells and macrophages. An interesting phenomenon is the frequently observed adhesion of leukocytes to the surface of parasites, particularly helminth larvae (e.g., *Ascaris suum* and *Nector americanus*). Immune serum is apparently involved in some instances, but the entire problem deserves much more study.

Eosinophilia is another characteristic phenomenon of many, but by no means all, parasitic infections. Usually the increase in numbers of these cells is not associated with the early acute inflammatory stages of infection, but with a later stage of inflammation. The reason for the accumulation of eosinophils is obscure. Most theories involve histamine, but one theory suggests that it is due to antigen-antibody complexes.

Cysticerci of the tapeworm, *Taenia crassiceps*, were injected into mice intraperitoneally.[19] Leukocytosis developed in peritioneal fluid and blood, and the number of eosinophils rose from less than 100 per ml. to 95,000 per ml. Monocytes and lymphocytes increased in numbers intraperitoneally. Leukocytosis and eosinophilia were most pronounced when the mice were beginning to recover from the cysticerci, and least when recovery was overcome. Mice that were most successful in overcoming cysticerci were those that developed high eosinophilia. Female mice that were fed worm eggs developed peripheral eosinopenia during the first two weeks, then changed to eosinophilia without leukocytosis.

Some of the pathology described above is due to direct destruction of host tissue by mechanical or chemical means, and it should be emphasized again that this type of phenomenon must not be confused with true reaction of the host.

The loss of red blood cells resulting in anemia is a common effect of parasitism. One of the best-known examples is anemia due to the presence of hookworms in the intestine. This is the result of the removal of blood due to the feeding habits of the worm, and it can be simulated by simply withdrawing blood at intervals from the patient. In a healthy host the condition can usually be remedied by iron therapy. In man the erythrocytic count may drop below 2,000,000 per ml. (from about 5,000,000) and the hemoglobin may drop below 30 per cent of normal. This drop results in, ". . . a severe pallor, extreme languor and indisposition to play or work, popularly interpreted as laziness; a flabbiness and tenderness of the muscles; breathlessness after slight exertion; enlargement and palpitation of the heart, with weak and irregular pulse; edema, making the face puffy and the abdomen 'potbellied'; a fishlike stare in the eyes; reduced perspiration; more-or-less irregular fever, and heartburn, flatulence and abdominal discomfort. The appetite is capricious, and

frequently there is an abnormal craving for coarse 'scratchy' substances such as soil, chalk, and wood."[5] In addition to these reactions, the removal from the body of the protein and iron may be particularly serious during pregnancy.

Anemia may also be due to the failure of the blood-forming organs to function normally or to receive appropriate constituents for the formation of normal blood cells. The fish tapeworm, *Dibothriocephalus latus*, apparently causes this type of pernicious anemia in its human host.

Protozoan parasites may also produce an anemia in their host. The malarial organism, *Plasmodium*, the sleeping sickness parasite, *Trypanosoma*, the causative agent of cattle piroplasmosis, *Babesia* and various coccidia, all evoke this response. Red cells may be destroyed by occupancy, but where no mechanical destruction is possible the effect is probably due in part to lytic substances. It is best, however, to think of the entire body of the host as reacting to parasites in the blood. This approach emphasizes the complexity of the problem and indicates the involvement of other organs, such as the spleen, and other processes such as production of acidosis, anoxia, phagocytosis, vascular endothelial changes, temperature effects, and antibody formation.

HOST TISSUE REACTIONS

Direct destruction of host tissue by mechanical or chemical activities of parasites obviously is related to host reactions. If a parasite produces a chemical substance that destroys host tissue, the destruction is not a host reaction; but if, in addition, inflammation occurs, the host is reacting to the parasite. Local tissue reaction in normal infections may not be directed against the parasite in order to exterminate it. There is evidence to support the opinion that the local tissue reaction of the host is really directed towards the repair of damage caused by the parasite burrowing through the host and may be without effect on the parasite.[10] Mechanical

464 PARASITE-HOST RESPONSES

or chemical destruction of host tissues by parasites, however, is probably related to immune reactions of the host. The first place to look for localized reactions is, of course, at the spot of attachment or lodging. One has only to pick off a firmly attached parasite to see at least mild inflammation of host tissue. The inflammatory reaction may progress to a swelling or to tissue degeneration, or cancerous-type growths may occur.

Mechanical pressure of parasites may set up inflammation, cause the formation of connective tissue, rupture host tissues, produce hemorrhages, or initiate a variety of other reactions. Pressure effects may be the direct or indirect result of single parasites in a small space, as occurs when a young ascarid crowds into a minute blood vessel. Single, but greatly enlarged, parasites, such as bladder worms in various organs, may bring about similar effects. Large numbers of parasites such as adult nematode worms in lymph vessels may evoke serious mechanical damage.

The tendency of a host to wall off a parasite, to separate itself from a foreign body from which it is otherwise unable to get away, is a general, widespread reaction. The larval stage (acanthella) of the acanthocephalan, *Moniliformis dubius*, within the hemocoel of a cockroach, *Periplaneta americana*, becomes surrounded by a capsule formed from closely packed vesicles. The vesicles arise from long tubular protrusions that extend from the surface membranes of hemocytes.[34]

Various species of marine clams were infected with cercariae of the trematode *Himasthla quissetensis* by Cheng et al.[6] Except in *Tapes philippinarum*, where metacercariae were found in tissues, they were enveloped by a noncellular, parasite-secreted inner cyst wall. In *T. philippinarum*, only the metacercariae found in the testes were enveloped by such a wall. Most of the metacercariae were also surrounded by an outer wall comprised of host leukocytes and either myofibers or connective tissue fibers, apparently depending on the type of fibers available in host tissue immediately surrounding the parasites.

The leukocytes were believed to be attracted by the inner wall.

Defense reactions of insects to metazoan parasites have been studied and reviewed by Salt.[55] He reported that blood cells are always involved and they act by forming a cellular capsule from which a connective tissue envelope usually develops and which often contains melanin. Little evidence exists that other tissues of insects are involved in defense reactions. The defense reaction is elicited primarily when the parasite is in an unusual host. In their usual hosts, some parasites do not elicit a defense reaction, and others are able to endure it in a dormant state or to resist it. The type and intensity of reaction varies with physical and physiologic factors of both parasite and host. Annelids, mollusks, crustacea, acarina and larval echinoderms encapsulate metazoan parasites. Blood cells, coelomic corpuscles and fibroblasts are involved and the reaction is essentially the same as that which occurs in insects.

Host reaction to eggs of the tapeworm, *Taenia mustelae*, has been reported in detail.[18] When two field mice were fed worm eggs, the mice succumbed within the first two weeks from excessive liver damage and internal hemorrhage. In other mice, cestodes began to appear after eight days. The young tapeworms increased in size and then ruptured the liver capsule. Many of the worms migrated elsewhere. Some appeared subcutaneously, producing hemorrhage, some were found in the thymus, and others caused a severe hemorrhage into the peritoneal cavity. By the seventeenth day of development, the mice had produced isolation membranes around some of the larvae.

When *Trichinella spiralis* larvae occur in muscles, calcium salts may be produced by the host and deposited in parasite tissues, thus calcifying the annoying organisms. If the response by the host is not adequate, and the concentration of parasites is sufficiently high, the host may, of course, die. The biochemical aspects of these relationships are complex and not clearly understood, but certainly some adjustment to a parasite must be made or the host is in danger. Trichinosis may be so severe in swine that death of the host occurs as early as seven to ten days after infection. The numbers of infected hogs that die, however, are extremely few. Even the numbers of infected hogs that show symptoms are far less than one per cent.

The heterophyid fluke, *Cryptocotyle lingua*, in the gut of terns and dogs, may cause a pronounced reaction on the part of the host.[66] The worms live between the villi and in the lumen of the intestine. "They denude the epithelium, produce a copius exudation of mucus, cause pressure atrophy, necrosis, a sloughing of tissue, hyperemia, infiltration of eosinophils and plasma cells, and hyperplasia of the fixed tissue elements." The injury is mainly mechanical and if the host does not die, a "self-cure" seems to take place. This result suggests an immunity reaction. From the medical point of view one can be glib about such pathogenic effects of parasites on their hosts. From the basic biochemical point of view, however, little is known of the true relationship between the physiologic activities of the parasite and the reaction of host tissues. It is easy to observe that a certain organ is damaged, but it is extremely difficult to say with certainty what reaction or reactions caused the damage.

Severe thickening of blood vessels of any infected organ obviously may interfere with normal functioning of that organ. Blood flow may be impeded when heart worms are present in large numbers in dogs. Worms in the brain may cause mental disturbances in man. Liver function may become abnormal if this organ is heavily infected with flukes. Other metabolic disturbances caused by parasites include biochemical changes in the host serum and physiologic disturbances (e.g., increase in fat). *Sacculina* is probably toxic to its host as indicated by the results of injection of parasite extracts into crabs: paralysis and death usually occur. When *Sacculina* grows into internal organs, the crab ceases to molt. This reaction may be associated with the increase of the lipid content of the hemolymph in the infected crab.

WEIGHT GAIN

Increase in the size or weight of the host may accompany infection with certain parasites. This phenomenon occurs with invertebrates and vertebrates. Females of the crustacean *Gammarus zaddachi* infected with plerocercoids of the tapeworm, *Diplocotyle*, are significantly larger than noninfected individuals. Infected males remain about normal size.[59] Larvae of the beetle, *Tribolium*, and other larvae grow more rapidly and larger when heavily infected with the microsporidian, *Nosema*. They die, however, before reaching maturity. A yellow oil was extracted from the protozoan spores. Injection of this oil into larvae of the silkworm, *Hyalophora cecropia*, produced similar activity. The same results also occur when the host is injected with normal molting hormone. Apparently the parasites can substitute for the host endocrine gland.[16] Shell diameters of snails infected with larvae of the blood fluke, *Schistosoma mansoni*, were consistently larger than those of controls in the first several weeks of an infection experiment.[45] Later, the infected snails showed stunted growth as well as physiologic castration due to the infection.

Some species of carpenter ants infected with metacercariae of the fluke, *Brachylecithum mosquensis*, are more obese than noninfected ants and, unlike the latter, they do not conceal themselves but crawl on exposed surfaces where they are easily found by birds that are the next hosts of the fluke. This behavior seems to be a remarkable example of an animal that sacrifices its life for its parasites. The "sacrifice," of course, is induced by the parasite.

Weight gain in parasitized mammals seems to be more lasting than in invertebrates. Young female white mice subdermally infected with small numbers of growing scoleces of spirometrid sparganum gain weight more rapidly than their controls.[37] Infected male mice show a less consistent and less pronounced weight gain. The infected mice seem to be in good health. Infection with this sparganum stimulates growth rates of mice and hamsters. An insulin-like agent that promotes glucose utilization of isolated rat epididymal adipose tissue *in vitro* has been extracted and concentrated from these parasites. It is suggested that secretion of this agent results in increased permeability to glucose not only in its own tissue, but also in that of the host.[22] Weight gain in rats and mice occurs when they are infected with *Trypanosoma lewisi* or *T. duttoni*.[27,28]

A possible clue to an explanation of animal host weight gain is found in plants. The principal growth regulatory substance of higher plants is indoleacetic acid (IAA). Many plant pathogens produce IAA if tryptophan is present in the culture fluid. Numerous microbial parasites of plants produce either *in vivo* or *in vitro* substances that can be identified as IAA or other secretions that are growth-promoting (e.g., gibberellins produced by the pathogen *Gibberella fujikuroi*). Possibly, gall-producing insects also liberate growth-stimulating substances. Protozoa in animal hosts may also be involved in this type of relationship.[16]

HOST AGE

Age resistance to parasites is common. Usually, the older the host, the greater the resistance. When older hosts are infected with many different species of parasites, however, the hosts often have larger numbers of any one species than do younger hosts. Part of the explanation for these facts is associated with immunity reactions, but other factors undoubtedly play a role. A change in diet may be significant. For example, a decrease in carbohydrate with a corresponding increase in protein intake will, in some cases, protect the host against intestinal parasites. Often this change in diet occurs in man. There is also evidence that older vertebrate hosts have more duodenal goblet cells than do younger hosts, and that these cells are associated with resistance in older animals. The mucus contained in these cells apparently has an inhibitory effect on some parasites. In some animals, however, susceptibility to parasites increases as the host becomes older.

The sex of the host may also influence age resistance. Resistance against *Ancylostoma caninum* in dogs, for example, commences at an earlier age in the bitch, and these hosts are significantly less susceptible to primary infection than are adult males.[35] The reasons for such differences vary and cannot always be attributed to host hormones. Differences in feeding behavior may be one important factor. A few studies have been made on the effect of flatworm parasites on the health of fish. In one such study,[20] it was found that, in general, young fish were more seriously affected than older fish, and this difference was usually observed when the fish were crowded. For a further discussion of age resistance, see p. 487.

Wijers[65] demonstrated the effect of host age on infection rate. He fed flies (*Glossina palpalis*) of various ages on a monkey infected with *Trypanosoma gambiense*. In flies newly emerged from the pupal case, the infection rate was 21 per cent; in two-day old flies, it was four per cent; in three-day old flies, one per cent, and in older flies, no infection occurred. Probably, softer membranes in younger flies permitted easier migration of the trypanosomes. (See also Chapter 22.)

Host preference may be remarkably selective. Some mosquitoes are apparently able to distinguish between the sex and age of the host. Filarial parasites of man tend to show lower infection rates and parasite densities in females and older people. The more efficient vectors thus may be those that prefer males to females and the younger age groups to older age groups.[38]

PHYSICAL ADAPTATIONS OF PARASITES

Parasites have developed a fascinating variety of organs for maintaining themselves within or on their hosts. It would require an entire book to describe in detail all the different behavior techniques and the hooks, suckers, holdfasts, spines, teeth, adhesive secretion, scales, lips, papillae, setae, alae, flagella, pseudopodia and other structures employed for this purpose.

Modification of insect and arachnid bodies for maintaining themselves on their hosts are especially noteworthy. Lice, fleas, and bedbugs are wingless. To maintain their positions, however, they must be able to cling to host hairs or feathers or scales, so usually they have developed strong, large claws. Some ticks, mites and lice have claws shaped in such a manner that they closely fit the hairs of their hosts.

Some parasites require pronounced modifications only in the immature stages. Warble and botflies find it necessary to provide some means of anchoring their eggs to the hairs of their hosts. Eggs of the warblefly of cattle are attached by means of a stalk near the base of the hair. When the larvae emerge they burrow into the skin. The eggs of the botfly of horses are attached to the outer ends of the hair, without stalks. This position of attachment is undoubtedly related to the necessity of rapid transit of the larva from the egg to the mouth of the horse. Transition has to occur during the moment the host brushes its hair with its lips.

Adaptations by loss of organs or of functions are significant features of the lives of many parasites. This loss has a major effect on the parasite, and it probably renders the parasite more host-specific. There is a gutless parasitic snail (*Thyonicola serrata*) that lives in a sea cucumber. An even more drastic loss of organs occurs in *Sacculina*, described on previous pages of this book. This barnacle has become plant-like in general appearance (Fig. 16–28, p. 363), and the adult has lost its alimentary canal, its appendages, and its segmentation. The body of the parasite is a sac-like structure from which branched processes extend throughout the host crab. These processes serve as anchoring devices as well as feeding organs.

The size of the host may affect the proportions of parts of the bodies of its insect parasites. Thus, in *Trichogramma*, one of the chalcid wasps that parasitizes moths, modifications of the wings and antennal bristles are a response to the size of the host. Since modification of certain parts of insect parasites or even a sufficient discrepancy in size between males and females might prevent copulation, the host could be responsible for a reproductive barrier between male and female parasites.

PHYSIOLOGIC ADAPTATIONS OF PARASITES

Physiologic processes of the parasite may differ markedly from those of the host, and profound changes may take place in a parasite as it adjusts to living within its environment. Glycolytic enzymes of parasites and of hosts, for example, may differ in various ways: (*1*) in kinetic properties, e.g. affinity for substrates; (*2*) in substrate specificity; (*3*) in immunochemical properties; and (*4*) in reaction with chemotherapeutic agents.[48] Loss of physiologic characters is as important as loss of morphologic characters. The worm, *Ascaris*, for example, has lost cytochromes and the cytochrome-oxidase system, and many enzymes of the tricarboxylic acid cycle. This worm has also lost the mechanism for the metabolism of arginine phosphate. Obviously, any loss of this nature must be accompanied by the loss of the need for the end-products of these reactions or by the development of a method of obtaining the needed substances from the host or from other associated organisms.

We know relatively little about the physiologic processes that take place in an insect parasite as it seeks its host. The olfactory sense plays a role, but there appears to be a distinct difference in the manner in which a host attracts a male insect and the manner in which it attracts a female insect of the same species. Of course, the difference may lie with the insects and not with the host. Some hymenoptera females, which are able to deposit either fertilized or unfertilized eggs "at will" in a host larva, apparently are induced, by unknown means, to deposit fertilized eggs in the larger hosts. In this direct manner the host may have an effect on the sex ratio of its parasites. Age is related to size, and we have already noted that the effect of parasites on the host is conditioned by the age of the host, but the age of the host also influences the size of some of its parasites. The tapeworm *Hymenolepis diminuta*, for example, varies in size with the age of the rat in which it lives, becoming larger in older rats. The basis for the difference in worm size, however, is difficult to establish since rats of different ages differ themselves in size and in many other developmental characteristics. See page 487 for a consideration of host age in relation to parasite numbers.

One of the requirements for a parasitic mode of life is an ability to adjust to the oxygen variations that occur within a host. Although most, if not all parasites, use oxygen as an hydrogen acceptor when it is available, other substances are widely used. The ability of a normally anaerobic parasite to oxidize sugar completely in the presence of oxygen is known as the **Pasteur effect.** This effect has been demonstrated for many worms. In general, most worms show little Pasteur effect but the nematodes, *Eustrongylides ignotus* and *Litomosoides carinii*, show marked effect. As oxygen requirements are studied it becomes increasingly evident that there are probably no truly obligate anaerobe animal parasites.

The QO_2 of some digenetic trematodes studied by Vernberg[62] varied with the species and with the stage of development, but in general the respiration rate was found to be directly proportional with the oxygen tension down to a partial pressure of three per cent or lower. Below this level the parasite was able to regulate its rate of respiration.

When the host's normal reactions are weakened or otherwise altered, its pathogenic parasites may increase in numbers. Such a cause-and-effect relationship has been induced experimentally by placing mice under the influence of alcohol. The development of the malarial parasite (*Plasmodium berghei*) in the blood of these mice was subsequently advanced. Some people like to justify drinking by claiming that the alcohol kills any "bugs" that may be present in the body.

The experiment with mice would seem to indicate just the reverse.

The strong reciprocal influence between a host and its parasites is clearly illustrated when a population of parasites (e.g., protozoa) is passed repeatedly through experimental hosts. Such passage may, on the one hand, enhance the virulence of the parasite and kill its hosts, or, if the host is an "unnatural one," the virulence of the parasite may be reduced. For other examples of reciprocal influences see the discussion on general ecology of parasitism, especially size relationships, host age and parasite numbers (p. 487), movements and hibernation (p. 486), parasitocoenosis (p. 494) and symbiotic cleaning (p. 502).

Sporozoites of the coccidian, *Eimeria necatrix*, introduced into chickens are taken up or engulfed by macrophages that carry the sporozoites into the intestinal epithelium.[61] Such transport of parasites by macrophages is common, but in this case the parasites are unharmed, and they are liberated by the degeneration of macrophage cells—a sort of taxi service to the preferred site of infection, but upon reaching its destination the taxi falls apart to discharge its passengers. The parasite is obviously protected against the normal destructive processes of the macrophages. It is illogical to think that the host protects the parasite, so one must assume that the parasite produces some substance that protects itself.

Further examples of physiologic adaptations of parasites occur throughout the other chapters of this book. See in particular discussions of the more important parasitic diseases of man.

SURROUNDING SUBSTANCES

The chemistry of the surrounding medium has a profound influence on a parasite. Obviously, if the chemical composition of the surrounding medium is antagonistic, the parasite is apt to die. It is surprising, however, how resistant some parasites can be to changes in their chemical environment. Blood was drawn from the hearts of chickens infected with *Plasmodium gallinaceum* and then it was defibrinated and placed in feeding tubes about 7.5 to 10 cm. long by 1.5 cm. in diameter.[14] The tubes were covered with membranes prepared from gut sausage casing, which were relaxed in water and alcohol. Mosquitoes (*Aëdes aegypti*) were allowed to feed upon the blood in these tubes for 30 to 45 minutes, and the tubes were agitated at about five-minute intervals to prevent the interference of sedimentation with the number of gametocytes ingested by the insects. In some of the blood cell suspensions, various concentrations of saline were substituted for serum. Within a range of 0.7 per cent to 1.15 per cent saline, no significant difference occurred in the intensity of infection. The intensity was slightly lowered at 0.85 per cent but was materially lowered at 1.3 per cent. The osmotic pressure of the host blood obviously affected the behavior of the parasite. The mean infection rate determined by this method of *in vitro* feeding was similar to that induced by bird feeding. Again we must use caution in interpreting the results. As Eyles[14] has pointed out, when considering the relationship between a mosquito and the plasmodium it carries, the vertebrate host cannot be ignored. There are factors in the blood of the vertebrate host that influence the subsequent infection in the mosquito.

The dog hookworm, *Ancylostoma caninum*, requires an active factor in dog serum for the utilization of glucose. This factor is also present in the sera of animals resistant to the hookworm, but other adverse factors occur in these host animals. Cat serum is devoid of the adverse factors and thus can tolerate the dog hookworm.[15]

Why is an intestinal parasite not digested by the enzymes of its host? This question is frequently asked by students. A common answer is that the living parasite produces "antienzymes." Undoubtedly, part of the answer to the question lies in those processes that prevent an animal from digesting its own intestinal mucosa. There is, however, little real evidence for the existence of antienzymes in parasites. Certainly the nature of the

outer covering plays an important role in preventing worms from attack by host enzymes. This cuticle, or tegument, is probably composed of a polymerized mucoprotein, the molecules of which may be protected from proteolytic enzymes by a shell of glycogen molecules. The cuticular cortex of living nematodes is not digested by pepsin or trypsin, although the under layers are readily dissolved by these enzymes. When a worm dies it is quickly digested, indicating that death is accompanied by changes in the cortex which permit penetration of the digestive enzymes. In the case of protozoa, we have to fall back on the postulation that some kind of "inhibitor" produced by the parasites comprises part of the answer.

The effects of host alimentary canal secretions on its parasites have received little attention. If a parasite produces a substance that stimulates the production of host intestinal secretions, these secretions may have an important regulating effect on the parasite. We know that certain materials (such as fatty acids) produced as end-products of helminth metabolism may increase host intestinal secretions, but their effects, in turn, on the parasite are yet to be understood. The problem is one that deserves careful study and that undoubtedly has considerable significance in the parasite-host relationship. We know little about the variations of secretory activity from one site in the intestine to another. Bile salts secreted by the liver are resorbed chiefly in the ileum, but apparently not at all in the duodenum. Thus parasites such as *Giardia*, *Ascaris* and *Trichostrongylus* find in the duodenum a markedly different biochemical environment than obtains in other parts of the gut.

The stomach of sheep apparently produces a substance that causes parasitic larval nematodes to secrete a fluid (hormone?) which, in turn, causes the worms to molt.[30] Also, a host stimulus appears to induce eggs of *Ascaris* and *Ascaridia* to secrete a fluid ("hatching fluid") that causes the eggs to hatch.

There is clear evidence that the behavior of

a parasite within its host is influenced by the presence of another parasite. For example, the roundworm, *Ascaridia*, in chicks thrives better when the chick is also infected with another roundworm, *Heterakis*, than when alone. Also, *Heterakis* found in chicks infected with the protozoan, *Histomonas meleagridis*, are smaller than those from chicks free from the protozoans.[52] (See p. 495 on concurrent infections.)

Such studies emphasize again the necessity for studying the host and its parasites as a biologic unit. This unit not only involves the host as the environment of the parasites, but it involves the external environment of the host as well. The internal environment of the host obviously includes its parasites. Parasitologists and ecologists must labor together for many years before most of the details of the environmental factors that influence parasitism are readily understood (see Chapter 22).

HORMONES

Only a beginning has been made on an understanding of the relationships involved in the effect of various hormones of the host on parasites of the digestive tract. As might be expected, some of the information is conflicting. In midly hyperthyroid chickens, for example, the roundworm, *Ascaridia galli*, attains significantly greater lengths than do worms from normal and mildly hypothyroid birds. But the reverse condition has been reported for the worm, *Heterakis gallinae*, which attains a greater length in mildly hypothyroid hosts. Possibly the two worms behave differently toward the hormone thyroxin, and possibly different research techniques led to different results. We cannot assume that because one species of intestinal worm is found to react in a certain manner to a hormone, another species of worm in essentially the same environment will react in a similar manner. Many experiments will have to be made under varying and controlled conditions before we can draw general conclusions.

The effect of thyroxin on the cysticercoids of the tapeworm, *Hymenolepis nana*, of mice throws some light on this problem. If mice are given 3 mg. of thyroid extract by mouth daily for a month prior to infection, a slightly higher percentage of these cysticercoids will develop than is usual in normal or hypothyroid animals. It would seem that, in general, a decrease in thyroxin production on the part of the host would favor at least those parasites in its intestine that rely heavily on carbohydrates as a source of food. Thyroxin has a stimulating effect on the absorption in the intestine of carbohydrates and fats. It follows that hypothyroidism slows the rate of absorption of these foods and thus makes more of them available to helminths or to any other organisms that might use them.

A remarkable effect of host hormones on parasites was suggested by Szidat.[60] He found a trematode (*Genarchella genarchella*) in snails of the species *Littoridina australis* living in a lagoon near Buenos Aires. The unusual aspect about this fluke was that its entire life cycle could be completed within the snail. Thus, the first intermediate host has become the only essential one, while the true final host, the fish, *Salminus maxillosus*, has become unnecessary. The cercarial stage is suppressed and there is no second intermediate host. The rediae produce metacercariae that mature and lay numerous eggs containing miracidia within the rediae. Szidat suggested that when the original fish host became adapted to fresh water in a former epoch, it produced excess hormones from the thyroid gland and from the hypophysis. The action of these hormones might have been responsible for shortening the life cycle of the parasite.

Even protozoa of invertebrates apparently are influenced by host hormones. Our knowledge of the hormones of invertebrates is limited, but we have evidence of the presence of sex hormones, growth hormones, neurohormones, and perhaps many others. Since hormones are nonspecific, we might look for the same types of reactions in parasites of invertebrates as those we observe in parasites of vertebrates. We know, for example, that glandular changes in termites and in wood-roaches have a direct bearing on the life cycles of their flagellates (p. 474), but we know little more than that. The field is wide open for research.

Sexual differences between hosts and concurrent differences among their parasites have been studied rarely. One such study of lice on the meadow vole and on the deer mouse showed that the rate of infestation of voles was significantly lower in females than in males, but that no such difference was apparent in the mice. Also, the average numbers of lice and the increase in infestation rate with host age were significantly greater in male voles than in females.[8] Obviously the male vole provides a more favorable environment for the growth and development of lice, but the reasons for such differences are not clear. Stahl[58] found that only in young mice, during the third to tenth week after experimental infection with *Aspiculuris tetraptera*, was there a heavier burden of worms in male hosts. The male rat is more susceptible to infection with the nematode, *Nematospiroides dubius*, than is the female, but hormones are not necessarily involved.[11]

During the breeding season, female rabbits have a much greater infection with two nematodes, *Trichostrongylus retortaeformis* and *Graphidium strigosum*, than do males. In females whose ovaries have been removed, the numbers of both nematodes are the same as in males. If the fallopian tubes are removed during the breeding season, the numbers of worms in the females are intermediate between those in normal females and in males. Reproductive hormones of the host apparently influence the host-parasite relationships.[12] The magnitude of parasitic infections may vary considerably not only between the sexes but at different times of the year, and the two variables may interact, as shown above. Interpretations based on a single study of one sex at one time of the year may thus be grossly misleading.

Hymenolepis diminuta, the common tapeworm of rats, has also been the subject of much experimentation on the effects of hormones of the host on the parasite. An elaborate series of experiments on the growth of this worm in normal and in castrated male and female rats was performed by Addis.[1] The rats were maintained on complete diets and on diets with certain deficiencies. Some of the animals on each type of diet were given sex hormones while others were not. Adult, immature, pregnant and nonpregnant rats were used. Growth of the tapeworms was normal in male rats that were kept on a vitamin-deficient diet, but in female rats, also kept on a vitamin-deficient diet, the tapeworms were stunted in growth. Castration of male rats, whether kept on a vitamin-deficient diet or not, stunted the growth of the tapeworms. Normal growth of the worms could be restored by injecting or feeding testosterone or progesterone. When female rats were castrated (spayed), there was no effect on the worms, but when the female rats were kept on a vitamin-deficient diet the growth of the tapeworms was stunted in both castrated and noncastrated rats. Injection of male sex hormone into castrated female rats on a vitamin-deficient diet caused a slight improvement in growth of the worms. Injection of male sex hormones into normal females on a vitamin-deficient diet did not improve growth of the tapeworms. The only definite conclusion that Addis presented was that the worms depended on testosterone or progesterone for normal growth in male rats. Results like these are hazardous to interpret without a great deal more corroborative research.

Castration of the rat host causes a decline in the rate of synthesis of glycogen from sodium pyruvate and glucose in *Hymenolepis diminuta*. Host fasting, on the other hand, accelerates the rate of synthesis of glycogen from glucose but not from pyruvate. The significance of these types of reactions may be argued, but they serve to point out an area of challenging and productive research.[9]

When the testes are removed from a mammal, one of the results is a deposition of fat in various tissues of the body. If a castrated rat harbors the tapeworm, *Hymeno-*

lepis diminuta, fat is also deposited in the tissues of the worm. This occurrence is difficult to explain but it might be related to a blocking effect on protein synthesis. Additional effects of castration of the host on this tapeworm are: lowered ability to become established in the gut, reduced transamination (the reversible transfer of amino groups in amino acids), and lowered carbohydrate synthesis as noted above. These effects indicate a general reduction in the level of metabolism.

Another indication of the relationship between the sex hormones of the host and the life of the parasite is seen in *Schistosoma mansoni* in mice. These blood flukes are especially interesting because they are bisexual. A group of castrated male albino mice was infected with the flukes, while a group of uncastrated mice was similarly infected as controls. Nine weeks later all the mice were sacrificed and a count was made of the adult parasites found. There was a significant reduction in the number of male *Schistosoma* only in the castrated hosts, indicating a beneficial effect of male sex hormones of the host on the male parasites. In another study of worms in mice it was noted that male mice harbored twice as many pin worms as did female mice. Whether this difference was due to sex hormones or to other factors is unknown.

Several studies have indicated that the resistance of male hosts (e.g., rats,[11] frogs[26]) to helminths is increased after the experimental administration of the female hormone estradiol. Quite a different kind of relationship from those mentioned above has been reported for the mosquito, *Anopheles quadrimaculatus*, and the nematode larvae of *Dirofilaria immitis*.[67] If the mosquito is decapitated immediately after it has fed on an infected dog, the normal metamorphosis and growth of the nematode takes place. Thus the "interruption of the gonadotropic hormone cycle in these mosquitoes, as observed by the failure of their ovarian development, does not interfere with the development of the parasite."

INFLUENCE ON SEXUAL PROCESSES OF HOSTS

The effects of parasitic infection on the reproductive structures, function and behavior of the host has received little attention. The evidence indicates, however, that such effects are widespread. They may be due to mechanical or biochemical factors or both, but the specific nature of the biochemical responses are not yet clarified.

A study of 15 bulls, 8 rams and 30 ewes that had been discarded because of low sexual potency or sterility provides an example of a general sexual response to parasitism. The animals were found to be infected with various worm parasites. They were all given anthelminthics, and after "dehelminthization," sexual normalcy, for the most part, was restored. One cannot assume from this report, however, that any animal with a heavy burden of worm parasites suffers from sexual deficiencies.

Egg-laying capacities in snails infected with larval trematodes are commonly reduced and growth is often retarded. In a study of the snail, *Biomphalaria pfeifferi*, infected with *Schistosoma mansoni* larvae, however, snails of all ages experienced a temporary acceleration in growth rates but suffered shortened life spans. The more intense the infection of individual snails, the greater was the increase in growth rate. Although some eggs were produced by snails infected before maturity, complete sterility occurred in snails infected after maturity. Eggs laid by infected snails had a higher sterility rate and a lower hatching rate than eggs laid by noninfected snails. In another study[46] of snails infected with *Schistosoma mansoni* in the laboratory, extensive migration of large numbers of cercariae, intense tissue reactions, and degenerating trapped cercariae were apparently important factors in causing suppression of fecundity and inducing accelerated shell growth, followed by stunting of shell growth and a high mortality rate. It has been suggested that larval trematodes alter the metabolic temperature response of snail hosts.[29] This may be one of the factors responsible for the effects described above.

The shrimp, *Leander serrifer*, is parasitized by either of two species of bopyrid isopods, *Bopyrus squillarum* and *Diplophyrus jordani*.[68] These parasites obtain their food by sucking blood. Breeding characteristics of the shrimp are normal so long as the parasites remain on the host. If the parasites are removed or fall off, however, the breeding characters disappear completely from the pleopods at the next molting. But if the parasites occupy the branchial chamber of the shrimp, the breeding characters never develop. Thus, under appropriate conditions, the parasites are apparently able to inhibit normal sexual behavior of the host, as well as to ensure normal breeding characteristics.

Certain hermit crabs parasitized by rhizocephalids (e.g., *Sacculina*, Fig. 16–28, p. 363) show some degree of sex reversal. Males may be feminized as indicated by altered appendages. The parasitized female crabs seem to be more resistant, since they do not take on male characteristics. Possibly the parasites produce a toxin that inhibits normal development of male characters. Certainly the gonads of a host are sometimes altered enough to prevent the production of normal gametes. *Sacculina* in a male crab may die before the host testes have completely degenerated. The crab is often able to recover, but may then produce both sperm and ova, thus becoming hermaphroditic.[50]

Parasitic castration may be caused by two different sorts of animal associations. On the one hand, the parasites may live within the gonads, causing direct castration by a more or less complete atrophy of these organs. On the other hand, an indirect alteration of gonads and of the secondary sexual characters may take place, leading to complete castration or to the production of individuals more or less intersexual in appearance. (See Reinhard[50] and Hartnoll.[23])

Direct destruction of the gonads occurs when the warblefly, *Cuterebra emasculator*, infests the chipmunk, *Tamias listeri*.[3] Many trematode larvae destroy the hermaphroditic or unisexual glands of the snails they infect. The trematode, *Distomum megastomum*, may obliterate the gonads of the crab, *Portunus depurator*. Tapeworm larvae may bring about castration of fish gonads.[64] Some mermithid nematodes parasitize female midges (chironomids) and bring about a certain amount of sex reversal. Male characters may appear in the insects, and in extreme cases the ovaries are replaced by testes.

A holotrich ciliate lives in the testes of starfish (*Asterias rubens*) in the Plymouth area of the United States Atlantic coast. The parasites are especially abundant in well-fed medium-sized starfish, in which they occur in up to 28 per cent of the males during March to April. The effect of these protozoans on their hosts is a breakdown of all the germinal tissue in the testes. Most of the starfish suffer complete castration. Marine copepods may also become castrated by a parasitic dinoflagellate.

Parasitic worms may become castrated by their own parasitic protozoa. To be a parasite and a host at the same time apparently presents fearful problems. The tapeworm, *Catenotaenia dentritica*, which lives in a squirrel, is parasitized by a haplosporidian, *Urosporidium charletyi*, which brings about castration of its worm host. The common sheep tapeworm, *Moniezia expansa*, is sometimes so parasitized by the microsporidian, *Nosema helminthorum*, that not only the gonads but all other tissues of the worm are infected. Hyperparasitism of worms leading to possible castration has also been noted with Microsporida in the fish acanthocephalid, *Echinorhynchus*; with the flagellate, *Giardia muris*, in the fluke, *Echinostoma revolutum*, from a mouse; and in several species of Microsporida in other trematodes and cestodes.

Indirect parasitic castration is much more common than direct destruction of the gonads. The parasitic isopods of the suborder Epicaridea are all parasitic on prawns and crabs, and they commonly cause a degeneration of the gonads of their hosts. The isopod, *Liriopsis pygmaea*, parasitic on the barnacle, *Peltogaster curvatus* (itself attached to hermit crabs), feeds during its young stages on the host juices. The epicarid apparently diverts food from the ovary, which consequently degenerates.

Among the many other examples of a crustacean host being castrated or sterilized by a parasitic crustacean is the hermit crab, *Pagurus longicarpus*. This crab may be burdened with two species of isopods, a bopyrid, *Stegophryxus hyptius* (Fig. 16–5, p. 345), and an entoniscid, *Paguritherium alatum*. The bopyrid lives attached to the abdomen of its host whereas the entoniscid lives within the hemocoel. Usually the crab is not parasitized by both of these organisms at the same time. *Stegophryxus* seems to have little effect on its host but *Paguritherium* brings about complete castration of the female crab, and practically complete atrophy of the gonads of the male crab.[51]

Strepsipteran insects (Figs. 17–30, 17–34, pp. 392, 394) are small peculiar forms whose larvae are endoparasitic in various members of the insect orders Hymenoptera and Homoptera. One result of this parasitism is atrophied ovaries in the female host and sterility in the male. External genitalia and secondary sexual characteristics may also be altered. Plague-carrying fleas on rodents may be castrated by a nematode, *Heterotylenchus pavlovskii*.[25]

Castration of vertebrate hosts by parasites apparently rarely occurs. There may be sexual differences, however, in reactions to worm parasites. Female hamsters are more resistant to the roundworm, *Nippostrongylus brasiliensis*, than are male hamsters, but this sexual difference is absent in rats, which are hosts to the same worms.

Several theories have been developed to explain the effects of parasitism on the sexual development of the host. These theories include the production of toxic substances (mentioned previously), interference with general nutrition, interference with fat metabolism, hormone changes, or mechanical destruction of gonad tissues. A clue to the specific causes of parasitic castration may be found in a study of morphogenic effects produced by the gonads through hormonal influences. If the eye stalks of the prawn

(*Leander serratus*) are removed, not only does a reaction in the chromatophores occur, but also a rapid development of the ovary, and eggs are produced out of season. Obviously an endocrine gland in the stalk directly or indirectly acts on the growth of the ovary. If a parasite modifies the hormonal balance of its host, the genital glands and secondary sexual characters of that host clearly may become altered.

A suggestion has been made that the words "parasitic castration" are inappropriate when discussing the effect on invertebrates, because secondary sexual characteristics in such hosts are apparently independent of the gonads. The implied effect of such castration on these characters, therefore, is nil.[2] It would seem appropriate, however, to use the phrase to indicate destruction or alteration of gonad tissues by parasites without reference to any further effect on the structure or behavior of the host.

INFLUENCES ON SEXUAL PROCESSES OF PARASITES

A relation between the sexual cycles of an animal and its intestinal organisms has been noted in the wood-feeding roach, *Cryptocercus*.[43] This insect is host to symbiotic intestinal protozoa. If part of the brain (pars intercerebralis) is removed from the roach before the beginning of the molting period, the sexual cycles of the insect are entirely suppressed. If the portion of the brain is removed during the molting period, those cycles which have not completed gametogenesis are usually blocked. In roaches whose sexual cycles have been thus blocked the symbiotic intestinal protozoa usually degenerate. Control of molting is apparently more directly a function of the hormone **ecdysone,** produced by the prothoracic gland. During each molting period of the cockroach the intestinal flagellates undergo sexual cycles, but normally not at other times. If, however, ecdysone is injected into adult insects that lack prothoracic glands and also into intermolt nymphal hosts, gametogenesis is induced in the flagellates.

The flagellate, *Barbulanympha*, is one of many that lives in the gut of *Cryptocercus*, and normally it begins its gametogenesis from 25 to 30 days before the molting of the host. If this parasite is transferred to a defaunated nymph 15 days before the insect is ready to molt, the flagellate will not start gametogenesis, and it will die. If *Trichonympha*, another roach parasite, is transferred in like manner to another defaunated host, the flagellate will begin its sexual cycle at its usual time which, in this case, is only five to six days before molting of the host begins. These results indicate again that the host is responsible for initiating the sexual cycles of its parasitic flagellates. There is also evidence that not only does the host start the cycles, but that it controls these cycles during their courses of development.[7]

Maturation and egg-laying in the rabbit flea is a response to a rise in levels of corticosteroids and estrogens in the blood of the rabbit. Fleas pair only on the newborn young that are one to eight days old. Ovarian regression and resorption of developing oocytes of the flea is controlled by the levels of progestins in the blood of the doe. This relation is a remarkable example of dependence of a parasite on the hormones of the host, and it suggests that the flea might be used as an indicator of the level of certain hormones in the blood of its host.[53]

The sex of some parasitic isopods is apparently determined by the host. In *Ione thoracica*, the first bopyridium larva to become attached to a crab is always a female, whereas all the rest are always males. Just what causes the parasites to develop into one sex at one time and the other sex at another time is still a mystery. (See p. 341 for a more detailed discussion of parasitic isopods.)

STRESS AND PARASITISM

Stress causes many reactions, including an alteration in the hormonal balance of a host. This imbalance may decrease inflammation of tissues and weaken other resisting mechanisms. Stress is thus undoubtedly a factor in parasitism because anything that lowers host resistance is apt to favor the establishment of parasites. The sensitive periods of parasite development are particularly subject to the effects of stress on the host.

Cortisone is used experimentally to determine its influence on the relation between host and parasite. Several investigators have reported increased susceptibility of mice to intestinal worms as a result of injecting cortisone. Mice that are usually resistant to infection with the roundworm, *Strongyloides ratti*, can become infected if given daily doses of cortisone. This response might be anticipated when one remembers that stress is followed by the production of adrenocorticotrophic hormone (ACTH) from the pituitary gland, and consequent release of adrenal glucocorticoids, which diminish normal inflammatory responses of the host. In one study,[32] cortisone at a dose of 75 mg. per Kg. of body weight significantly increased susceptibility of worm-free mice to infection with the pinworm, *Aspiculuris tetraptera*. Cortisone also has a profound effect on resistance of experimental mice to the tapeworm, *Taenia taeniaeformis*. The drug makes refractory hosts highly susceptible, but it must be given to the laboratory animals before the twelfth day of infection.[44] The effect of cortisone may differ according to the sex of the host. Experimentally-infected female mice and naturally-infected female jackrabbits are more susceptible to infection with the larval tapeworm *Taenia multiceps* than males.[13] Cortisone elevates the infection rate in male mice but not in females. The drug alters the distribution pattern of the parasite, but this alteration is similar in both sexes of host. Cortisone is most effective when it is injected between 1 to 14 days after mice have been infected with the parasites.

Even ectoparasites may benefit from host stress, or at least from the injection of the host with ACTH or cortisone. In an experiment with lambs, daily injections of ACTH for one month broke down resistance which had developed against keds (Fig. 18–27, p. 421). Similar results occurred after cortisone injection. It was concluded that physiologic or environmental stress, such as pregnancy or undernourishment, can affect the basic annual ked population cycle.[39] As one might expect, all experiments do not lead to the same conclusions. Working with the rat, Villarejos found that cortisone did not increase susceptibility to amebiasis.[63] Many more studies of the effect of cortisone should be made.

Population density, sexual cycles, changes in the blood picture, antibody formation, metabolism of basic foodstuffs, tissue damage and many other body activities and reactions are intimately related to stress, and, therefore, may influence the parasite burden. Trichomoniasis may be predominately due to emotional stress.[33,36] Acute amebiasis in kittens is probably due more to psycho-physical responses to stress than to infection with the parasite *per se*.[24] Parasites may aggravate pathogenic conditions due to nonparasitic factors. Injurious effects of partial starvation and low environmental temperature on mice, for example, may be more serious if the animals are infected with *Trypanosoma duttoni*,[56] and infectious bronchitis in chickens may be more serious if ascarids are also present.

Experimental work on stress and its relation to the flagellate, *Trichomonas*, in the cecum of ground squirrels has shown that there is a direct correlation between the numbers of these protozoan parasites and the degree of stress to which the hosts are subjected.[40,41] Stress factors being studied include hunger, temperature, mechanical irritation, crowding, darkness, light and fighting. Curiously enough, caging the squirrels, without subjecting them to additional stress seems to have as much influence on the flagellate numbers as do most of the planned stress factors. Fighting among the squirrels was especially effective in promoting an increase in numbers of cecal flagellates. Murie (personal communication) has reported a significant increase in the numbers of two unidentified cecal protozoa in mice in cold-exposed and high density groups. Cold stress, in the form of reduced temperature for two successive nights, doubled the numbers of cecal amebas in ground squirrels compared with the ameba count in field controls.[42]

REFERENCES

1. Addis, J. C.: Experiments on the relations between sex hormones and the growth of tapeworms (*Hymenolepis diminuta*) in rats. J. Parasitol., 32:574–580, 1946.
2. Baer, J. G.: Ecology of Animal Parasites. Urbana, Ill., Univ. of Illinois Press, 1951.
3. Bennett, G. F.: Studies on *Cuterebra emasculator* Fitch 1856 (Diptera: Cuterebridae) and a discussion of the status of the genus *Cephenemyia* Ltr. 1818. Can. J. Zool., 33:75–98, 1955.
4. Brooke, M. M.: Effect of dietary changes upon avian malaria. Amer. J. Hyg., 41(1): 81–108, 1945.
5. Chandler, A. C., and Read, C. R.: Introduction to Parasitology. ed. 10. New York, John Wiley & Sons, 1961.
6. Cheng, T. C., Shuster, O. N., Jr., and Anderson, A. H.: A comparative study of the susceptibility and response of eight species of marine pelecypods to the trematode *Himasthla quissetensis*. Trans. Amer. Micr. Soc., 85:284–295, 1966.
7. Cleveland, L. R., and Nutting, W. L.: Suppression of sexual cycles and death of the protozoa of *Cryptocercus* resulting from change of hosts during molting period. J. Exper. Zool., 130(3):485–514, 1955.
8. Cook, E. F., and Beer, J. R.: A study of louse populations on the meadow vole and deer mouse. Ecology, 39:645–659, 1958.
9. Daugherty, J. W.: The effect of host castration and fasting on the rate of glycogenesis in *Hymenolepis diminuta*. J. Parasitol., 42(1):17–20, 1956.
10. Dawes, B.: Death of *Fasciola hepatica* L. weakened by X-irradiation. Nature, 200: 602–603, 1963.
11. Dobson, C.: Certain aspects of the host-parasite relationship of *Nematospiroides dubius* (Baylis) II. The effect of sex on experimental infections in the rat (an abnormal host). Parasitology, 51:499–510, 1961.
12. Dunsmore, J. D. Influence of host reproduction on numbers of trichostrongylid nematodes in the European rabbit, *Oryctlogus cuniculus* (L). J. Parasitol., 52:1129–1133, 1966.
13. Esch, G. W.: Some effects of cortisone and sex on the biology of coenuriasis in laboratory mice and jackrabbits. Parasitology, 57:175–179, 1967.
14. Eyles, D. E.: Studies on *P. gallinaceum*. II. Factors in the blood of vertebrate host in-

fluencing mosquito infection. Amer. Jour. Hyg., *55*(2):276–296, 1952.

15. Fernando, M. A., and Wong, H. A.: Metabolism of hookworms. III. The effects of normal sera on the carbohydrate metabolism of adult female *Ancylostoma caninum*. Exper. Parasitol., *17*:69–79, 1965.

16. Fisher, F. M.: Production of host endocrine substances by parasites. *In* Cheng, T. C. (ed.): Some biochemical and immunological aspects of host-parasite relationships. Ann. N.Y. Acad. Sci., *113*:63–73, 1963.

17. Foster, A. O.: Parasitological speculations and patterns. J. Parasitol., *46*:1–9, 1960.

18. Freeman, R. S.: Life history studies on *Taenia mustelae* Gmelin, 1790 and the taxonomy of certain taenioid cestodes from Mustelidae. Can. J. Zool., *34*:219–242, 1956.

19. ———: Studies on responses of intermediate hosts to infection with *Taenia crassiceps* (Zeder, 1800) (Cestoda). Can. J. Zool., *42*:367–385, 1964.

20. ———: Flatworm problems in fish. Can. Fish. Cult., No. 32, pp. 11–18, May, 1964.

21. Geiman, Q. M.: Comparative physiology: mutualism, symbiosis, and parasitism. Ann. Rev. Physiol., *26*:75–108, 1964.

22. Harlow, D. R., Mertz, W., and Mueller, J. F.: Insulin-like activity from the sparganum of *Spirometra mansonoides*. J. Parasitol., *53*:449–454, 1967.

23. Hartnoll, R. G.: Parasitic castration of *Macropodia longirostris* (Fabricius) by a sacculinid. Crustaceana, *4*:295–300, 1962.

24. Josephine, M. A.: Experimental studies on *Entamoeba histolytica* in kittens. Amer. J. Trop. Med. Hyg., 7:158–164, 1958.

25. Kurochkin, Y. V.: *Heterotylenchus pavlovskii*, sp. n., a nematode castrating plague-carrying fleas. Doklady Akad. Nauk USSR (Transl. by AIBS, Washington, D.C.), *135*:952–954, 1961.

26. Lees, E., and Bass, L.: Sex hormones as a possible factor influencing the level of parasitization in frogs. Nature, *188*:1207–1208, 1960.

27. Lincicome, D. R., Rossan, R. N., and Jones, W. C.: Growth of rats infected with *Trypanosoma lewisi*. Exper. Parasitol., *14*:54–65, 1963.

28. Lincicome, D. R., and Shepperson, J.: Increased rate of growth of mice infected with *Trypanosoma duttoni*. J. Parasitol., *49*:31–34, 1963.

29. Lukina, A. P.: Vilyanie gel'mintoyov na vosproizvoditel' nuyu sposobnost organisma. [Effect of helminths on the reproductive capacity of the host.] Veterinariya *12*:32–35, 1958. Referat Zhur Biol., 1959, No. 84846.

30. Madsen, H.: On the interaction between *Heterakis gallinarum, Ascaridia galli*, "blackhead" and the chicken. J. Helminth., *36*: 107–142, 1962.

31. Maegraith, B. G.: Pathogenic processes in malaria. *In* Taylor, A. E. R. (ed.): The Pathology of Parasitic Diseases. pp. 15–32. Oxford, Blackwell Sci. Pub., 1966.

32. Mathies, A. W., Jr.: Certain aspects of the host-parasite relationship of *Aspiculuris tetraptera*, a mouse pinworm. III. Effect of cortisone. J. Parasitol., *48*:244–248, 1962.

33. McEwen, D. C.: Common factors in *Trichomonas* vaginitis. Gynaecologia, *149* (Supp.):63–69, 1963.

34. Mercer, E. H., and Nicholas, W. L.: The ultrastructure of the capsule of the larval stages of *Moniliformis dubius* (Acanthocephala) in the cockroach *Periplanata americana*. Parasitology, *57*:169–174, 1967.

35. Miller, T. A.: Influence of age and sex on susceptibility of dogs to primary infection with *Ancylostoma caninum*. J. Parasitol., *51*:701–704, 1965.

36. Moore, S. F., and Simpson, J. W.: The emotional component in *Trichomonas* vaginitis. Amer. J. Obstet. Gynec., *68*:947, 1956.

37. Mueller, J. F.: Parasite-induced weight gain in mice. *In* Cheng, T. C. (ed.): Some biochemical and immunological aspects of host-parasite relationships. Ann. N.Y. Acad. Sci., *113*:217–233, 1963.

38. Muirhead-Thomson, R. C.: Factors determining the true reservoir of infection of *Plasmodium falciparum* and *Wuchereria bancrofti* in a West African village. Trans. Roy. Soc. Trop. Med. Hyg., *48*:208–225, 1954.

39. Nelson, W. A.: Development in sheep of resistance to the ked *Melophogus ovinus* (L). II. Effects of adrenocorticotrophic hormone and cortisone. Exper. Parasitol., *12*:45–51, 1962.

40. Noble, G. A.: Stress and parasitism. I. A preliminary investigation of the effects of stress on ground squirrels and their parasites. Exper. Parasitol., *11*:63–67, 1961.

41. ———: Stress and parasitism. II. Effect of crowding and fighting among ground squirrels on their coccidia and trichomonads. Exper. Parasitol., *12*: 368–371, 1962.

42. ———: Stress and parasitism. III. Reduced night temperature and the effect on pinworms of ground squirrels. Exper. Parasitol., *18*: 61–62, 1966.

43. Nutting, W. L., and Cleveland, L. R.: Effects of glandular extirpations on *Cryptocercus* and the sexual cycles of its protozoa. J. Exper. Zool., *137*(1):13–38, 1958.

44. Oliver, L.: Studies on natural resistance to *Taenia taeniaeformis* in mice. II. The effect of cortisone. J. Parasitol, *48*:759–762, 1962.

45. Pan, C-T.: Generalized and focal tissue responses in the snail, *Australorbis glabratus*, infected with *Schistosoma mansoni*. *In* Cheng, T. C. (ed.): Some biochemical and immunological aspects of host-parasite relationships. Ann. N.Y. Acad. Sci., *113*:475–485, 1963.

46. ———: Studies on the host-parasite relationship between *Schistosoma mansoni* and the snail *Australorbis glabratus*. Amer. J. Trop. Med. Hyg., *14*:931–976, 1965.

47. Platt, B. S.: Protein malnutrition. *In* British Postgraduate Medical Federation: Lectures on the Scientific Basis of Medicine. Vol. 4, Chap. IX. London, Athlone Press, 1956.

48. Read, C. P.: The carbohydrate metabolism of worms. *In* Martin, A. W. (ed.): Comparative Physiology of Carbohydrate Metabolism in Heterothermic Animals. Seattle, Univ. of Washington Press, 1961.

49. **Read**, C. P., and Rothman, A. H. The role of carbohydrates in the biology of cestodes. I. The effect of dietary carbohydrate quality on the size of *Hymenolepis diminuta*. Exper. Parasitol. *6(1)*:1–7, 1957.

50. Reinhard, E. G.: Parasitic castration of crustacea. Exper. Parasitol., *5*:79–107, 1956.

51. Reinhard, E. G., and Buckeridge, Sister, F. W.: The effect of parasitism by an entoniscid on the secondary sex characters of *Pagurus longicarpus*. J. Parasitol., *36*:131–138, 1950.

52. Rogers, W. P., and Summerville, R. I.: Chemical aspects of growth and development. *In* Florkin, M., and Scheer, B. T. (eds.): Chemical Zoology. Vol. III, pp. 465–499. New York, Academic Press, 1969.

53. Rothschild, M.: Fleas. Sci. Amer., *213*:44–53, 1965.

54. Salt, G.: The effects of hosts upon their insect parasites. Biol. Rev., *14*:239–264, 1941.

55. ———: The defense reactions of insects to metazoan parasites. Parasitology, *53*:527–642, 1963.

56. Sheppe, W. A., and Adams, J. R.: The pathogenic effect of *Trypanosoma duttoni* in hosts under stress conditions. J. Parasitol., *43*:55–59, 1957.

57. Soulsby, E. J. L.: Lymphocyte, macrophage, and other cell reactions to parasites. *In* Immunologic Aspects of Parasitic Infections. Pan American Health Org., Sci. Publ. No. 150, pp. 66–90, 1967.

58. Stahl, W.: Influence of age and sex on the susceptibility of albino mice to infection with *Aspiculuris tetraptera*. J. Parasitol., *47*:939–941, 1961.

59. Stark, G. T. C.: *Diplocotyle* (Eucestoda), a parasite of *Gammarus zaddachi* in the estuary of the Yorkshire Esk. Britain. Parasitology, *55*: 415–420, 1965.

60. Szidat, L.: Über den Entwicklunszyklus mit progentischen Larvenstadien (Cercariaeen) von *Genarchella genarchella* Travassos 1928 (Trematoda, Hemiuridae). Tropen & Parasit., *7*:132–153, 1956.

61. Van Doornick, W. M., and Becker, E. R.: Penetration and invasion of the intestinal mucosa of the chicken by the sporozoites of *Eimeria necatrix*. J. Protozool., *3*, Supp., 2, 1956.

62. Vernberg, W. B.: Respiration of digenetic trematodes. Ann. N.Y. Acad. Sci., *113*:261–271, 1963.

63. Villarejos, V. M.: Cortisone and experimental amebiasis in the rat. J. Parasitol., *48*: 194, 1962.

64. Wardle, R. A., and McLeod, J. A.: The Zoology of Tapeworms. Minneapolis, Univ. of Minnesota Press, 1952.

65. Wijers, D. J. B.: Factors that may influence the infection rate of *Glossina palpalis* with *Trypanosoma gambiense*. I. The age of the fly at the time of the infected feed. Ann. Trop. Med. Parasit., *52*:385–390, 1958.

66. Willey, C. H., and Stunkard, H. W.: Studies on pathology and resistance in terns and dogs infected with the heterophyid trematode *Cryptocotyle lingua*. Trans. Amer. Micr. Soc., *61*:236–253, 1942.

67. Yoeli, M., Upmanis, R. S., and Most, H.: Studies on filariasis. II. The relation between hormonal activities of the adult mosquito and the growth of *Dirofilaria immitis*. Exper. Parasitol., *12*:125–127, 1962.

68. Yoshida, M.: On the breeding character of the shrimp, *Leander serrifer*, parasitized by bopyrids. Annot. Zool. Japonenses, *25*: 362–365, 1952.

Section IX

ECOLOGY OF PARASITES

Chapter 22

General Considerations

INTRODUCTION

The concepts of ecology in the study of parasites frequently have been ignored, but they are essential to the understanding of parasitism. Ecology is the basis for much of the discussion of such problems as invasion of the host, reactions of host and of parasites, chemistry of parasitism, parasite-host specificity and evolution of parasites and their hosts. In this chapter we shall emphasize the general principles of ecology through a consideration of parasite communities and their immediate environments.

Early works on the ecology of parasites dealt with the epidemiology of human disease in the tropics. More recent works have focused the attention of parasitologists on the necessity of an ecologic approach to all studies in parasitology if a broad and accurate understanding of parasitism is to be attained. May,[48,49] for example, has said that disease is "that alteration of human tissues that jeopardizes their survival *in a given environment*." Academician E. N. Pavlovskii[56] has emphasized the "doctrine of nidality" which may be stated as follows (from Audy[2]): "A disease itself tends to have a natural habitat in the same way as a species: many diseases, and especially the zoonoses, have natural habitats in well-defined ecosystems where pathogens, vectors, and natural hosts form associations or biocenoses within which the pathogen circulates; therefore, a landscape is an epidemiological factor because its characteristics are those of the local ecosystem (hence Pavlovskii's term 'landscape epidemiology')." Audy himself is concerned with the diseases of particular populations of hosts, and his approach "is to consider a host-species A (and its own specially modified environment) as the habitat and food supply of a large array of parasitic and commensal organisms, a number of which may cause disease in individuals of A in suitable circumstances. The whole assemblage makes up a parasite-pattern characteristic of A, and differing from that of species B largely because B and A (*i*) have genetically different constitutions, (*ii*) usually occupy different niches, so that each is exposed to a different milieu and to different 'occupational' hazards, and (*iii*) may belong to different biocenoses." Such an approach makes clear Audy's statement that "symptomatic disease due to an infective agent is a special response which the agent may rarely produce."[1] Ecologic field studies that involve groups of hosts and their parasites become much more profitable than the same studies that ignore the parasites. The biologist who is not a parasitologist is not accustomed to using parasites as "ecologic labels" or "biologic tags" that provide an abundance of information about the habits and habitats of their hosts—information waiting to be tapped (Audy[3]).

See Baer[4] for a general discussion of the ecology of animal parasites. A study by Kates[36] on the ecologic aspects of helminth transmission in domestic animals is recommended reading for a general background, and for a review of the life cycles of numerous animal parasites of veterinary importance. For these parasites Kates discusses the effects of temperature, moisture, oxygen, and the importance of protection from freezing, desiccation and direct sunlight. He also discusses the influences of climate and weather.

POPULATIONS AND COMMUNITIES

A biotic **community** is a unit of various organisms loosely held together by the interdependence of its members. A **population** is a smaller, more intimately associated group (usually composed of members of one species) within the community. Whereas each individual has its own characteristics, communities and populations have additional characteristics as the result of the aggregate. The more important characteristics of a population as distinct from those of an individual are: density, birth rate, death rate, age distribution, dispersion and form. A group of parasites, such as microfilariae in blood, trichina in muscle, or amebas in the gut, may behave as a population whose cohesion largely depends upon mechanical rather than chemical factors. The community concept, however, has not been studied in detail by many parasitologists. One might consider the parasites in an organ of the host as a community quite apart from the host—the latter being the external environment of the community. But since the host is a living organism as intimately associated with its parasites as the latter are with themselves, the host and its parasites should be studied as a community of organisms.

Thus the student should develop a community concept when studying parasitology. When a parasite is studied by itself, apart from its environment, only a part—and often a small part—of its total biology can be understood. The community principle in ecologic theory emphasizes the orderly manner in which diverse organisms usually live together. An ecologic complex is formed by the parasite, the vector, the host and various features of the host's environment. But this complex is far more than the sum of its parts. It is something new and forever changing.

THE ECOLOGIC NICHE

The "habitat" is the place where an organism lives, such as the intestine, but the "ecologic niche" is the organism's position or status within the habitat, and it results from the organism's structural adaptations, physiologic responses, and specific behavior. In other words, the niche of a parasite is what it *does* in its habitat. The niche depends on where the organism lives *and* on what it does. The niche of a parasite cannot be described accurately until one knows details of the mutual interactions between abiotic and biotic environments, numbers of individuals, and the effects of interaction of the parasites among themselves and with their hosts. If two species with the same requirements find themselves in the same niche, competition will tend to force one of them to be eliminated. Species that are phylogenetically closely related normally do not occupy the same niche. When the usual sigmoid growth curve is plotted for two populations and both curves are steeper when they are separate than when they are interacting, they are competing. Recall, for example, that *Giardia*, *Ascaris*, *Trichinella*, *Strongyloides*, *Necator*, *Hymenolepis* and *Taenia* all occur in the small intestine of man. These parasites do not occupy exactly the same ecologic niche because a niche, as we have already stated, is not only the space occupied, but the parasite's place in that space—involving food, period of activity, and other behavioral factors. All of these parasites, when crowded in one portion of an intestine, do occupy the same habitat, and this situation is possible because of the absence of enemies and the presence of an abundance of food.

The occupied nest of a bird is an excellent subject for the study of various niches of a small community. Numerous populations of arthropods—some of them parasites of birds, others free-living—are usually found in such nests. These populations are not stable because the various species of organisms have different requirements for food, temperature, humidity and light. Obviously, then, the populations change with changes in the seasons. Within the nest there may be levels at which some species are more abundant than others. Out of a total of 3469 arthropod inhabitants of a great tit's nest, 490 were found in the inner lining, 2277 in the middle layers, and 702 in the outer layers. In a flycatcher's nest the relative positions were found to be the reverse, over half of the total of 1568 arthropods being located in the inner lining. Such differences in nests help to explain the differences among the parasitic arthropods on the bodies of birds.

In 1934 Gause[27] first demonstrated experimentally that the rule of one species to a niche is true in a high proportion of cases. This rule is known as **Gause's principle.** We do not know how great an overlap must exist before one species of parasite, or even a free-living form, pushes out the other. Schad[67] has shown that, although there are ten or more species of colon-inhabiting oxyuroid nematodes in the European tortoise, there are striking differences in linear and radial distribution of the worms as well as differences in feeding habits, and, probably, in responses to seasonal changes and age of the host.

When two species compete for the same source of energy, the energy (food) will be distributed between the two according to what has been termed the **differential equation of competition.** As the population of each species grows, however, there is a change in the equation, and the available energy becomes unequally distributed because one species always multiplies more rapidly than the other. According to Gause, owing to its advantages, mainly a greater multiplication potential, one of the species in a mixed population drives out the other entirely. This generalization was based on a study of populations of free-living protozoa, but the principle also applies to populations of parasites. When the host competes for the same food as that eaten by its parasites, the problem obviously becomes more complicated.

An **ecotone** is a transition between two or more communities. In this junction area the populations of two communities overlap and the area contains some forms peculiar to it—forms not found elsewhere. Often both the number of species and total numbers of individual animals are greater in the ecotone. This tendency for increased variety and density in the ecotone is called the **edge effect.** Such effects are clearly demonstrated in tidal areas along marine shores. One might predict that if the population of a vertebrate species (A) overlaps that of a closely related species (B), and if (A) and (B) each has a distinct fauna of parasites, the ecotone would contain hosts with more varieties of parasites than typical for either (A) or (B). But such studies on the ecology of parasitism have not attracted the attention of many workers.

LIMITING FACTORS

Limits on growth, development and distribution may be imposed by practically all environmental agencies both inside and outside the bodies of parasites and their hosts. The ecologic minimum and maximum (e.g., essential chemicals and food) help to establish the character of the niche, and the range between these extremes represents the limits of tolerance. The productivity of a parasitic species, whether measured in terms of the reproductive rate or of energy, may be rich or poor as compared with an adjacent species because of limiting ecologic factors. Of particular significance in the molding of such closely knit animal societies as insect colonies and parasite-host associations are accumulations of excretory and secretory products, and the metabolism of food.

There is an infinite variety of limiting or controlling factors whose actions depend on the needs of the organism, but these needs themselves vary constantly with the individual and with time and place. Thompson[76] has warned us against too bland an assumption that distribution of an organism is chiefly determined by the action of one or two limiting factors. "The simple truth is that the natural control of organisms is primarily due, not to any complex cosmic mechanisms or regulatory factors, but rather to the intrinsic limitations of the organisms themselves . . . An excessive reverence for the idea of adaptation and the assumption that organisms are indefinitely plastic and capable of fitting themselves into almost any situation at relatively short notice, has led to some rather inaccurate statements on the subject."

TEMPERATURE AND CLIMATE

Climatic factors are often ignored in a study of parasites unless they are directly related to the control of diseases of economic importance, such as the various plant and animal diseases transmitted by insects. The temperature at which a parasite grows is of major significance. Many species of protozoa and helminths (e.g., *Hymenolepis nana*) have been maintained experimentally at abnormally high temperatures and they have consequently changed their morphology as well as their physiologic reactions to a marked degree. Helminths sometimes go through more than a 60° C. change in temperature from host to host during the normal course of their life cycles. A study[69] of the effects of temperature on the cestode *Schistocephalus* suggests that two enzyme systems may be present, each of which responds differently to temperature changes. One enzyme system may control somatic growth in plerocercoid larvae, with a peak efficiency near 23° C., while the other system may control maturation in adult worms, with a peak efficiency near 40° C.

Temperature affects both the reaction of the host salamander *Triturus v. viridescens* to the parasite *Trypanosoma diemyctyli* and the response of the parasite to the host. Infections of *T. diemyctyli* occur in adult salamanders, and a critical point of change in the nature of the infection from a pathogen to a non-pathogen is at about 20° C. The infection is pathogenic at lower temperatures only, and at higher temperatures the metabolic rate of the host is great enough to reduce the numbers of trypanosomes, probably through the production of antibodies.[7] Here we have an example of an ecologic factor (temperature) exerting a direct influence on the equilibrium that exists between a host and its parasites.

The third-stage larva of the cattle stomach nematode, *Haemonchus contortus*, is the infective stage for grazing animals that become infected by eating the larvae with grass. The main factors concerned with the vertical migration of the larvae up the leaves are temperature and humidity. Most larvae are in the grass blades during early morning and evening. The time of the morning maximum becomes progressively earlier while passing from winter to summer, and the time of the evening maximum becomes progressively later. The reverse is true during the second half of the year. A low humidity, accompanied by either low or high temperature, inhibits vertical migration, and the largest number of climbing larvae are to be found during rainy seasons.[61] Free larvae of this worm and other livestock nematodes can develop experimentally in cultivation over a temperature range of about 10 to 35° C. or more, but the optimum is between 20 and 30° C. Larval mortality is high above 40° C., but some larvae may survive for short periods at 50° C. or more. See Kates[36] for a review of temperature, climate and other ecologic aspects of helminth transmission in domestic animals.

Plerocercoid larvae in the stomach of *Gymnodactylus* (a lizard) migrate well if the host is kept at room temperature, but at 37° C. migration does not take place. Relatively few experiments have dealt with the effects of high temperatures on developing parasites, but Voge[83] found that adverse effects on *Hymenolepis diminuta* are greatest when exposure occurs during the sensitive period of maximal larval growth and development (three to five days). Major indicators of sensitivity to high temperature (38.5° to 40° C. for 24 hours) are failure of scolex-withdrawal, and inhibition of infectivity for the mammalian host. Such exposures to high temperature have little or no effect when applied at other times during development. The nature of the temperature effects is not well understood, but it may be related to dehydration.

Trypanosoma rotatorium is a highly polymorphic parasite in the blood of frogs. Studies have been carried out to determine how temperature affects the density of the flagellates in peripheral blood, and to determine the internal temperatures of bullfrogs in their natural environment. The peripheral parasitemia of the parasite in *Rana catesbeiana* was found to be markedly affected by temperature.

"Over the long term, high temperatures are always coincident with high peripheral parasitemia and vice versa; over the short term, increases in temperature bring about a corresponding increase in parasite level, and vice versa. . . . It is proposed that the control of peripheral parasitemia is due to changes in the level of metabolic activity of the host."[6]

A careful study of the ecology of the ciliate, *Urceolaria*, has shown that population changes in the parasite are not influenced by changes in the host population, but that the changes (fluctuations) are largely initiated by fluctuations in the surrounding populations of bacteria, which fluctuations in turn are influenced by rainfall.[62] Temperature was found to be a secondary factor only. These ciliates tolerate marked chemical changes in the water in which they live and they feed chiefly on bacteria that are on or near the freshwater triclad turbellarian flatworms that serve as hosts.

Many parasites are said to be not only more abundant in the tropics but less "adapted" to colder climates. Such conclusions often result from considering each facet of information separately rather than studying and appreciating the full complex of ecologic and epidemiologic implications. Otto[54] has sharpened the focus of parasitologists on ecologic problems with relation to amebiasis by stating that:

The question of whether amebiasis is better provided for survival and transmission in the tropics or in cooler climates involves the consideration of several different ecological factors. It seems to me that there has been a too easy assumption that the organism is *per se* adapted for the tropics and again that it is commonly transmitted in the water supply. In general, both the infection rate and the prevalence of disease is highest in the tropics. But what does this signify with reference to the ecology of the parasite and the epidemiology of the disease? . . . How do we reconcile the tropical distribution of this infection with the indications that the transfer stage, the cyst, is best equipped for survival and distribution in the colder climates?

The recognized and recorded water-borne epidemics of amebic dysentery that have resulted in the highest percentage of infection have all occurred in northern communities. Quite obviously there are significant gaps in our knowledge of even this well-known parasite, and the gaps involve taxonomic considerations, metabolism of the ameba, and environmental requirements and influences.

A number of protozoan parasites have been preserved by freezing, and after thawing were found to be in good condition and able to grow and reproduce in a normal manner. Levine and Andersen[46] listed the more important papers on this subject, and reported their own success at slowly freezing cultivation forms of the flagellate *Tritrichomonas foetus* and keeping them for 2048 days at $-95°$ C. in the presence of 10 per cent glycerol. Eleven per cent of the parasites were alive after the 5.6 years, and new cultures were readily initiated with them.

In addition to the strictly climatic influences, other physical factors that play a role in parasite ecology are **radiation** (relatively not important for internal parasites, but important for hosts), **biogenic salts,** and **atmospheric gases.** Oxygen is one of the most common limiting factors. A liter of air contains about 210 ml. of oxygen but a liter of water contains no more than 10 ml. at sea level, and the solubility of oxygen is increased by low temperatures and decreased by high salinities. Parasites and their hosts do not always vary in the same direction with environmental variations. This fact was demonstrated during an investigation of the relation between the range of snails (*Bithynia tentaculata, Dreissensia polymorpha*) and their ciliate parasites in fresh and in brackish waters.[59] As the salinity of the water increased, the ciliate, *Conchophthirus acuminatus*, decreased in abundance whereas the ciliate, *Hypocomagalma dreissenae*, increased in numbers. Thus the host snail is adapted to live in a wide range of salinities but its ciliate parasites vary markedly in this ability. Other important physical factors of the environment are **currents** and **pressures** (especially important for hosts, e.g., water currents for snails, insects, fish; wind currents for insects, birds, plants; barometric pressure, probably not of direct importance); **soil** (recall that soil is not only a factor of the environment and organism, but is produced by them as well); **transparency** of water and air (important in the penetration of light).

Seasonal Variations

The results of a few investigations of seasonal variations of parasites are presented below, but final generalizations on the subject cannot be made at the present time. The winter's accumulation of worm eggs may all hatch at once upon the advent of warm weather in June or July. A period of drought may also be responsible for a sudden massive invasion of worm larvae. On the other hand, adverse conditions, such as the drying of ponds or winter ice, generally mean fewer or less vigorous parasites.

In considering seasonal variations of ecto-parasitic arthropods, it is well to remember that flea larvae probably have difficulty in developing in soil that is too moist, but that ticks and mites pass all but the egg stage largely on the host, or, if molting on the ground, are capable of extended locomotion to favorable environments. Hence a very dry ground or a very wet ground would be limiting factors for fleas but not for the other ectoparasites. A cumulative flea index on the California ground squirrel, *Citellus beecheyi beecheyi*, showed that the index reached an annual maximum during August, September and October.[65] These months represent the hot, dry season before the fall rains occur, with an associated decline in the mean temperature. The most abundant species of flea was *Echidnophaga gallinacea*, which, as might be expected, is found most commonly in localities where the hot, dry seasons are the longest.

A seasonal variation in the incidence and development of cestodes in fish of temperate climates has frequently been reported.[34] A study of a seasonal cycle of the tapeworm, *Proteocephalus stizostethi*, disclosed that in the yellow pike-perch, *Stizostedion vitreum*, from Lake Erie, viable embryonated eggs occur only in June.[15] The fish are free from the worm in late summer, and new infections are obtained early in the fall, and the worms mature during the following summer. During a year's survey of the protozoa of some marine fish at Plymouth, England, we found evidence of seasonal variations of Myxosporida of the dragonette (*Callionymus lyra*), the heavy infections being more abundant in the winter, but in the same fish hosts the heavy haemogregarine infections were more abundant during the summer.[51]

A seasonal increase of host sexual activity possibly is correlated with a seasonal decrease in the extent of parasitism in some animals, but no clear evidence for this bit of speculation is available. It is possible that unfavorabe weather may result in decreased sexual activity of an animal, with resultant increase of parasites. Many snails are more heavily infected with trematodes during the fall than during the remainder of the year. Synchronous development of life cycles between host and parasite are well known among insects, but such relationships have not been observed as frequently among other host-parasite associates. The trematode, *Schistosoma mansoni*, may stop its development in the snail, *Australorbis glabratus*, when these hosts go into estivation in natural habitats that are subjected to annual drought. Thus climatic conditions adverse for the snail induce a resting stage (diapause) during larval development of its trematode parasites.[5]

Movements and Hibernation

Dogiel[19] has concluded that hibernation of freshwater fish causes a number of definite changes in their parasites.

It provides an excellent example of the complexity of factors influencing the dynamics of the parasite fauna. The influence of hibernation can more properly be considered as being the sum of the influence of several factors. Involving cessation of feeding, it bars the path of infestation for intestinal parasites. In effect it brings about a lowering of intensity or even disappearance of some intestinal helminths. The drop in water temperature retards sexual and asexual reproduction and leads to a reduction in numbers of ectoparasites (the ciliates, monogeneans). The loss of mobility and the overcrowding of fish in its winter hollows exposes it to infestation with other ectoparasites (leeches, carp lice) which are not affected by low temperature. Finally, changes in the physiological conditions of the gut, brought about by inanition, lead to destrobilation of some cestodes. All these factors contribute to the sum total of the influence of hibernation.

Hibernation of experimental hamsters may either prevent or retard the development of *Trichinella spiralis*.[13] "Forty-eight and seventy-two hours of hibernation at 5° C. gave complete protection to 4 hamsters receiving 200 *T. spiralis* larvae when onset of hibernation was within the first thirty-six hours." Adult helminths in hibernating ground squirrels appear to be much more sensitive to temperature changes than are the protozoan parasites.

HOST SIZE, AGE, AND PARASITE NUMBERS
Host Size

Little attention has been paid to the question of size of parasite in relation to size of its host, although parasitologists generally expect to find bigger parasites in bigger hosts. For example, among the parasitic arthropods within the species of any genus, the larger species of parasites tend to be found on the larger hosts. Often when, in the literature, one finds the statement that the size of the parasite is greater in a given kind or size of host than it is in another, there is no notation of the **numbers** of parasites present. The average size of large numbers of a given species of parasite crowded in a host organ is generally smaller than the average of a few parasites in the same organ.

The **crowding effect** is a reduction in size of individual parasites inversely proportional to the number of parasites in a given infection. This effect has been noted by a large number of parasitologists, but there is still some question as to its cause. It is easy to say that one or two worms in a host organ have more room in which to grow, but that 100 worms in the same organ are crowded for space and have to compete for limited supplies of food. Another logical answer is that harmful metabolic by-products of the worms retard their growth. But without the support of experimental evidence, such answers are only generalized speculations.

Crowding of hookworms may reduce the number of parasite eggs produced, and crowding of mermithid nematodes may result in the elimination of female parasites. Studies on the tapeworm, *Hymenolepis diminuta*, however, have provided the most clues to the causes of reductions in size of parasites. This worm apparently is independent of protein in the host diet, and possibly cestodes in general have no fat requirement. *H. diminuta* appears to have a definite, though small, free oxygen requirement, and it is this free oxygen, which presumably enters the gut by diffusion from the surrounding tissues, that may be a limiting factor for the cestodes. Roberts[63] found that *H. diminuta* reaches its maximal size when five or fewer worms are present in one host (rat), and that this maximum is "apparently governed by an interaction of host size, strain, diet and health, and the genetic capacity of the worm." Roberts studied the chemical embryology of *H. diminuta* and showed that crowding markedly affected the carbohydrate and lipid content of developing worms. The carbohydrate content in young worms from small populations rose much higher than in worms from denser populations. In young and in adult populations, there was a pronounced decrease in the total protein and phospholipid content as crowding increased. The crowding effect appears, therefore, to have mainly a nutritional basis. These results substantiate those of Read[60] and others. In addition to competition for host dietary carbohydrates, there may be some other way in which the environment is altered, thereby helping to reduce the size of crowded worms. One of these ways might be a "change in the predominant microbial flora, with consequent changes in pH, oxidation-reduction potential, etc."[63a]

Host Age

The relations between the age of the host and the kinds and numbers of its parasites vary considerably according to the circumstances and to the host group under consideration. As a broad generalization, however, and with many exceptions, older animals have larger numbers of parasites than do younger animals of the same species. This statement is not in conflict with the fact that immature laboratory animals are often more susceptible to experimental infection than are adults. Young mice, for example, are much more easily infected with *Trypanosoma cruzi* than are adult mice. But young mice have fewer numbers of normal mice parasites than do adult mice. In addition, parasites successfully introduced into the bodies of young animals are often not permanent in the experimental hosts. Age resistance may be the result simply of an extension and heightening of natural resistance. Immunity factors undoubtedly play a role in susceptibility, and in young hosts the antibody response is slower than in adults. Dogiel[17] has reviewed many aspects of this problem and he stated that: "It appears that age immunity usually develops among mammals and birds, but not among the lower vertebrates and the invertebrates."

The relationships between the age of the host and the kinds and numbers of its parasites may be separated into: (*a*) those factors concerned with the results of parasitism (i.e., the changes within the host due to the presence of parasites), and (*b*) those relationships concerned with the normal conditions of host anatomy and physiology encountered by the parasite upon its first contact with the host. The differences between these two groups of factors, however, are not always clear. If a parasite prefers a younger host we often do not know, for example, whether the preference is due to the absence of immunity on the part of the younger host or to the presence of a mechanical barrier, such as a thick skin, in the older host. The results of a few studies on this problem are presented below, but many more investigations are needed. Chapter 21 also contains information pertinent to this problem.

A clear demonstration of the age factor was made by Wijers[84] in his study of the infection rate of *Glossina palpalis* with *Trypanosoma gambiense*. He found that, after feeding flies of various ages on infected monkeys, there was a 21 per cent infection rate in newly emerged flies, a four per cent rate in two-day-old flies, a one per cent rate in three-day-old flies, and no infection in older flies. The peritrophic membrane of the fly is probably responsible for these differences. The trypanosomes must get through or around the membrane, which is a chitinous tube lying free in the midgut. The membrane is soft and discontinuous in new flies, but becomes more rigid within a day or two.

Many species of miracidia favor young snails over older hosts. Certain species rarely or never penetrate snails over two or three months old, but in other species the age limitation does not appear to exist. This preference for young snail hosts has been demonstrated with the Schistosomatidae and Spirorchiidae, and with *Paragonimus kellicotti* and *Clinostomum marginatum*. Preference by miracidia for young snails does not necessarily mean that young snails are more heavily infected with trematodes. On the contrary, older snails are generally more heavily infected, and they are often killed by their trematode parasites.

Fish hosts have been used extensively as a basis for the study of the effects of host age on parasitism. The average number of worms per freshwater fish usually shows a regular arithmetic increase with age and size of fish. Such worms as the tapeworm, *Dibothriocephalus*, the nematodes, *Camallanus*, *Raphidascaris*, and the flukes, *Tetraonchus*, *Dactylogyrus*, *Azygia*, increase both in numbers and effects on their fish hosts with the latter's age. A few parasites are apparently not affected by host age; others, such as the myxosporidian, *Henneguya oviperda*, reach their maximal numbers in three-year-old fish, then decline in intensity. Studies by Gorbunova,[29] and Dogiel and Petrushevéskii[18] of fish in Russia confirm the generalization that the numbers of parasitic species in most fish increase regularly with the age of the host. The pike

(*Esox lucius*), the roach (*Leuciscus rutilus*), and the salmon received particular attention during these studies.

No age resistance to *Trichinella spiralis* in mice was noted by Larsh and Hendricks.[42] In young mice, however, a significant majority of adult worms was located in the posterior one-half of the small intestine, whereas the reverse was true in old mice. This difference in localization was attributed to the difference in intestinal emptying time. When mice of different ages were experimentally infected with the pinworm, *Aspiculuris tetraptera*, the female host became resistant to infection at the time of first estrus, whereas the male mice gradually developed resistance as they advanced in age.[47] This sexual difference in age response may be related to the level of gonadal hormones.

In many fish the parasite infestations increase up to a certain age of host maturity, then decrease. We have found that when average numbers of parasites (the nematode, *Spirocamallanus*) were used as a criteron of infection in the goby, *Gillichthys mirabilis*, the most heavily infected fish were always the older ones. If, however, the mode or median was used, the numbers of worms in the different age groups of hosts were roughly the same. A study of 100 wild rabbits (in an enclosure in Australila) showed that there was no relation to age and sex as to the level of infection with *Trichostrongylus retortaeformis*, but another nematode, *Graphidium strigosum*, was found in larger numbers in older and female hosts.[21]

A marked host age difference has been noted in the rate of infection of dogs by *Toxocara* and *Toxascaris*.[68] Almost 92 per cent of dogs of less than six months were infected with both species, but the rate dropped to 26.2 per cent in older dogs. This age difference was accounted for entirely by infections with *Toxocara canis* whose incidence was 75.6 per cent in younger dogs and 7.1 per cent in older dogs. *Toxascaris leonina* occurred at approximately the same rate in younger and older dogs.

Age resistance may be associated with the phenomenon of premunition, which is a state of resistance established against an infection after an acute infection by the same pathogen has become chronic, and which lasts for as long as the infecting organisms remain in the body. This phenomenon is probably related to the normal immune reactions of the host body.

Density of Populations

The experimental science of **population dynamics** has kindled the interest and imagination of biologists all over the world. In recognizing that the environment of each individual animal includes the population of which the animal is a part, the following two basic questions arise. How are populations controlled in nature? How can we define the relative "fitness" of an individual or of a species? If the demands of life are met, arguments on definitions of "fitness" and "success" are pointless, but one fact seems clear—an increase in population size does not necessarily mean that the individual or the species is fit; it usually means that the pressure of environmental resistance has decreased, or just that a normal cycle of overpopulation has occurred. Among the problems to be solved before answering the above questions are calculations of intrinsic fecundity in relation to rate of natural (actual) increase, dispersive ability, and the ability to search. All of these problems involve the numbers of individuals in a given area, hence the focus of population ecology is the density of a population and the processes that control this density.

The migrations within the host of such parasites as larval nematodes and *Plasmodium* are well known, but they have not often been considered in relation to the ecology of the total parasite-host complex. The effects of dispersal of any kind depend upon the status of growth form of the population, and on the rate of dispersal. If the population is well above or well below the "carrying capacity" of the host, dispersal may have more pronounced effects. The phenomenon of dominance among the several parasites of, for example, the blood of a vertebrate, must occur, but little attention has been given to it.

See page 511 for some accounts of parasite migrations in relation to the distribution of their hosts.

Density-Dependence. Much has been written about density-dependent factors in the parasite-host complex, but unfortunately the concepts of these factors have been defined in conflicting and confusing ways. Let us assume the existence of a population of wild pheasants infected with *Coccidia*. Let us also assume that this flock of pheasants has remained fairly constant in adult numbers (density) for several years, and that each year about one-fourth of them have died of coccidiosis. Now suppose that the size of the flock increases because of more favorable weather conditions resulting in more plant food. We would expect that, with an increase in numbers of birds (hosts), there would be a corresponding increase in numbers of parasites simply because, with more hosts to invade, there is more room for parasite reproduction and growth. We might then argue that because the population density of pheasants within the flock tends to be constant, the chances of any one bird dying of coccidiosis would be the same, the relative numbers of parasites per host remaining the same. It would seem logical to say that the density of the parasites depends upon the density of the host—more hosts, more parasites. But in this case the relationship is not simply a matter of numbers. The *effects* of the parasite on the host change with changes in numbers of parasites within each host, and the important factor is the effect on each *individual* pheasant of the population rather than on the whole host population as a unit. In cases where changing host densities do regulate the parasite population there is a significant time lag between the increase in host numbers and increase in their parasites.

The usual concept of density-dependence includes "all density-relationships in which the mean effect per individual of the population is higher at high densities than at low densities."[73] The more pheasants there are in a flock of a given size (area distribution), the less chance any one young bird has of

surviving to maturity if the flock is infected with coccidia. The effect of a limited food supply is density-dependent because the limitation operates more severely against a population at high densities than at low densities, *per individual of the population*. To put this principle in the words of Smith,[71] "the relative efficiency of both entomophagous insects and disease is greater when the host insect is abundant than when it is scarce, this being the effect of population density on the host-finding capacity of the parasite." According to Nicholson,[50] "the action of the controlling factor must be governed by the density of the population controlled."

DeBach and Smith[16] tested the above principle experimentally by making the parasites (*Mormoniella vitripennis*, a chalcid wasp) search for a density of 40 hosts (puparia of the house fly, *Musca domestica*) through three quarts of barley for 40 hours. The densities of parasites varied from 1 to 300. One of the factors in the host-parasite complex that acted as an automatic control was the effect of host population densities on the rate of increase of the parasite. Although this rate decreased with increments in host density, the higher the host density in relation to the density of the parasite, the greater was the *total* increase of parasite numbers. As a result, however, of competition and overlapping in searching for the host, causing a reduced fecundity in the parasite, the *next generation* of parasites actually decreased in numbers. Thus increases of parasites above a certain optimum will result in a subsequent actual decrease in total numbers of parasites. "In short, increases in host density act to increase numbers of parasites in the next generation, although not in direct proportion to host density, whereas increases in parasite density, beyond a certain point in relation to host density, result in a decrease in the total numbers of parasites in the next generation." These phenomena were previously observed by Flanders[25] in a similar experiment utilizing the chalcid, *Trichogramma* sp., and the eggs of the moth, *Sitotroga cerealella*. It was then pointed out that the limitations revealed by such experimental conditions do not prevail when multiple generations occur under natural conditions. Population oscillations of both host and parasite can occur in constant climatic conditions.

An overwhelming number of such theoretical considerations of the role of parasites in the natural control of insects have been made. Varley[80] demonstrated that the knapweed gallfly, *Urophora jaceana*, is density-dependent because it kills a much greater percentage as well as a larger total number of its hosts when the population of the host is high. Various workers have demonstrated that the number of eggs deposited increases with host density. The inherent reproductive capacity of a parasite is limited, but there is some dispute as to whether the number of eggs deposited by a parasitic insect ever attains this limitation. While working with chalcid wasps, Varley and Edwards[82] found that when the host population density is high, the calculated area of discovery falls to very low values and the number of hosts attacked by each parasite is independent of host density. If density is to be correctly measured, a determination of the type of distribution, size and permanence of species groups and degree of aggregation must be made.

As numbers of individuals increase, competition for space, food and other essentials obviously intensifies. The pteromalid wasp, *Neocatolaccus mamezophagus*, and the braconid wasp, *Heterospilus prosopidus*, attack full-grown larvae of the azuki bean weevil, *Callosobruchus chinensis*, and thus occur as parasites in the same ecologic habitat. In mixed populations of these three species the progeny of both wasps and of the host population increases in numbers rapidly at first and then maintains a nearly constant value. In comparing these findings with those resulting from a study of only single parasite species, Utida[79] concluded that the parasitization efficiency of each species of wasps is lowered when together they attack the same host population. *Neocatolaccus* demonstrated a high efficiency of parasitism at higher host densities, but *Heterospilus* exhibited more efficiency of parasitism at lower host densities. Efficiency

in these studies was measured in terms of progeny number. Under natural conditions, however, as pointed out by Smith,[70] the combined action of two or more competing parasites may give better control of a host than any one parasite acting alone.

For further details, see Burnett,[9] Odum,[53] Solomon,[72,74] Thompson,[77] and Varley.[81]

PERIODICITY

Periodicity of behavior is an intriguing phenomenon in the life cycles of many parasites, as well as of animals and plants in general. Reproductive cycles are common to all animals, and have been studied extensively, but the periodic and predictable appearance of large numbers of offspring never fails to excite the wonder of anyone who discovers it anew.

Regular fluctuations of numbers of amebic cysts in the feces of infected animals or of man is an example of periodicity in parasitic protozoa. A ten to 14-day period occurs between the appearance of maximal numbers of *Entamoeba coli* cysts in human hosts. *E. histolytica* also has its cycle of cyst formation, and laboratory technologists soon learn that a single examination for parasitic amebas may lead to inaccurate conclusions, since on one day a stool examination may disclose no amebic cysts whereas on the next day many cysts may appear.

The coccidian, *Isospora lacazii*, in the English house sparrow undergoes a diurnal periodicity in oocyst production (see p. 84 for the life cycle of a coccidian). The numbers of oocysts eliminated by the birds reach a peak daily between 3 P.M. and 8 P.M. and they occur in much fewer numbers preceding and following these hours. A logical assumption would be that this type of occurrence is associated with enhanced opportunity for the oocysts to be taken up by another host. Perhaps crowding together of birds in preparation for the night has a bearing on the problem.

Malarial organisms offer one of the best-known examples of cyclical behavior among protozoan parasites. The causative agent in benign tertian malaria is a typical example. *Plasmodium vivax* has a minimal prepatent period of eight days. What the parasite is doing in the body during these eight days immediately after the bite of a mosquito is discussed on page 92. A "clinical periodicity" of 48 hours follows this period. An event that occurs every 48 hours falls on every third calendar day; thus the name "tertiary malaria" is given to the disease caused by *P. vivax*. Every 48 hours many red blood cells of a man infected with this type of parasite rupture, owing to the completion of a parasite division process called schizogony. The rupture of millions of these cells, all approximately at the same time, with the liberation of their contents into the blood stream produces the chills and fever characteristic of malaria. The same type of cycle occurs in bird plasmodia. *P. gallinaceum* in ducks has a cycle of 36 hours, and it often starts in such a manner that rupture of red blood cells occurs at noon and at midnight alternately. Theories on the causes of circadian rhythms in vertebrates generally include the participation of endocrine glands. Hormones may well be involved in the regulation of malarial periodicity.

Periodicity among the multicellular parasites is as common as that among the protozoa. Cercariae, representing one of the larval stages in the life cycle of flukes, probably emerge from snails at times related to the light of day. Under laboratory conditions some cercariae leave their snail hosts only during the early evening. The student should never assume that behavior in the laboratory is exactly similar to behavior under normal conditions in the native habitat of an organism, but periodic activity related to certain daylight hours is so widespread among parasites that one is probably safe in concluding that the laboratory behavior of these cercariae is a reflection of normal periodicity.

Microfilarial worms in man are involved in one of the most interesting of all examples of periodicity among parasites. Adults of *Wuchereria bancrofti*, which causes "elephantiasis," live in blood and lymphatic vessels deep in the body where females produce living young. These larval worms, or microfilariae, may remain in the deeper vessels or they may come to the body surface between the hours of 10 P.M. and 4 A.M. The fact that the worm is transmitted by the bite of a mosquito leads to the conclusion that there is some relation between this behavior of the young worms and the visits of mosquitoes while the host is sleeping. The conclusion is problematical. In some of the islands of the Pacific the worms come to peripheral blood vessels during the day instead of the night, but in these areas the mosquito vector is a day-biting instead of a night-biting variety! In other places (e.g., the Philippines), both types of behavior can be found. In areas where the nocturnal periodicity exists, a man may change his working hours from daytime to nighttime and sleep during the day. If he is infected with *W. bancrofti*, the microfilariae respond to this change by remaining deep in the body at night and coming to the surface during the day. Some theories on the causes of this kind of periodicity are presented on page 315.

A different kind of periodicity has been described[10] for caryophyllaeid cestodes in the white sucker, *Catostomus commersoni*, in the Iowa River. These parasites exhibit a seasonal distribution that varies in time with the species of worm. All of the five species of cestodes employ oligochaete annelid worms as intermediate hosts. The periodicity of adult appearance in the fish host is probably controlled by the short life span of the adult parasites, whether or not embryonated eggs wait until after winter before hatching, and by the duration of annelid infection by the procercoid larval stages. The presence of one of the worm species (*Glaridacris*) appears to inhibit the establishment of procercoids of other species.

BIOLOGIC CONTROL

Mechanical, chemical, cultural, legislative and other methods of controlling agricultural pests have been used with varying degrees of success for centuries, but biologic control (i.e., the manipulation of biotic balance) offers the most hopeful promise of permanent success. Biologic control is applied ecology, and the agricultural entomologist must create an environment unfavorable for the occurrence of the arthropod at pest densities. One of the most unfavorable aspects of the environment for the increase of any animal is the presence of relatively large numbers of predators or parasites. Biologic control, then, involves the introduction and encouragement of predators and parasites that destroy insects that are destructive to agriculture. The objective of biologic control is to change the biotic equilibrium position from a population level at which the pest insect causes economic loss to a level at which its destructive action is negligible.

As we have already indicated, there is a tendency, in any single species, for its population density to fluctuate near an average level. Among the causes of such fluctuations is the interaction of a species with its natural enemies. One of the most interesting and economically important groups of such natural enemies of insects is the parasitic Hymenoptera sometimes designated as **parasitoids.** Although the terms "parasite" and "parasitoid" are frequently used as synonyms, not every parasite is a parasitoid. The latter is parasitic only as a larva, and it destroys its host so that it functions more like a predator, except that it destroys only a single host individual instead of several. It is also of large size in comparison with its host, which usually belongs in the same taxonomic group (i.e., Insecta).

Wasps and wasp-like insects (Hymenoptera) are commonly employed as agents of biologic control. The wasp, *Trichogramma minutum*, a parasite of the European corn borer and of the codling moth, is reared by the millions to help eliminate these insects as pests of important agricultural crops. A common sequence of events in the parasite-host relation is first the selection of a host, then the stinging of the host to render it immobile, then the laying of one or more eggs by means of the ovipositor which is jabbed into the host body

so that the eggs can be laid therein. The larvae that hatch within the host body, feed on the body juices and tissues, and eventually kill the host. Intensive studies of *Trichogramma* and of other parasites have revealed much important information on the ecology of parasitism, especially on population phenomena. The sugar cane borer (*Diatraea saccharalis*) has been partly controlled first by *T. minutum* in several countries, and later by the fly, *Lixophaga diatraeae*, in Cuba. The California black scale (*Saissetia oleae*) on citrus has been partly controlled in California by the chalcid wasp, *Metaphycus helvolus*, imported from Africa in 1937. The coconut moth (*Levuana iridescens*) in Fiji is controlled with a tachinid fly (*Ptychomyia remota*) with striking results. The Oriental moth (*Cnidocampa flavescens*) is controlled with the tachinid fly (*Chaetexorista javana*) in Massachusetts with fine results except during seasons following exceptionally cold winters. For details of these and of many other efforts at biologic control of populations of harmful insects, see Clausen.[14]

In a population of insects subject to parasitization by larval wasps (or subject to a contagious disease), the probability of survival of an individual is inversely correlated with the population density. The rate of increase of the parasitoid depends upon its inherent reproductive capacity and upon the number of hosts it can readily find; the population level of the host obviously depends not upon the host density but upon the *searching capacity* of the adult female wasp. Doutt[20] has described three distinct processes of selection which operate to restrict the host list of parasitic hymenoptera. (1) **Ecologic selection,** which implies that for a host and parasite to meet they must be geographically, seasonally and ecologically coincident. A combination of random and nonrandom searching appears to be most common. (2) **Psychologic selection,** which involves the highly developed nervous system and responses of the adult female parasite. The sense of smell has commonly been reported as primarily important in host detection. Ullyett[78] has found that if a cocoon of the South African brown-tailed moth (*Euproctis terminalis*), parasitized by the pupa of the ichneumonoid, *Pimpla bicolor*, is broken open in the forest, both the pupa and the hands and arm of the observer are covered in a few minutes by a swarm of the parasitic females. (3) **Physiologic selection,** which involves the physiologic compatibility of the wasp larva and its host.

The searching capacity of the adult female parasite, therefore, involves the finding of a host habitat and the finding of a host, but success depends also upon host acceptance and host suitability. Many insects, for example, have the ability to discriminate between parasitized and healthy hosts.[26,66] We know very little about the chemical factors that are at the basis of the selection of a host by a parasite. Few species of parasitoids limit their attack to a single host species, but normally a parasite will attack only a fraction of the species on which development is actually possible. Many host species never attacked in nature can be demonstrated experimentally to be suitable for parasitoid larval development.

The **Hopkins host selection principle** states that a given insect species that is capable of breeding in two or more hosts will normally continue to select for its offspring the particular host species on which its own life cycle was passed. This interesting behavior implies, therefore, that the host predetermines the selection by the ovipositing female of the same nutrient medium for her progeny as she herself enjoyed during her own development. The result of many generations of such selection tendencies is the development of special physiologic strains (see p. 549) of parasites, each with its own host preference.

Insects may be used to control other kinds of destructive or harmful invertebrates. For example, some species of snails are readily killed by larvae of dipterous flies, and ticks can be destroyed by insect larvae. From a study of a chalcid wasp parasite of whitefly larvae, Burnett[9] observed that the maintenance of the host-parasite system is a form of host protection because it ensures rapid decline in abundance of the parasite population.

During the course of a long series of studies of Malayan snails infected with larval trematodes, K. J. Lie and his associates found numerous incidences of double infections exhibiting interspecific larval antagonism within snail hosts. In a summary[44] of these studies, the authors pointed out the theoretical possibility of using the antagonistic behavior as a new form of biologic control of flukes pathogenic to man.

Two types of antagonism have been observed: (1) *Direct antagonism* is exerted only by dominant rediae that actively prey on other species of rediae, sporocysts, or even cercariae. Echinostome rediae appear to be facultatively predatory, probably preferentially predatory on larvae of other species if these happen to be accessible in the snail. (2) *Indirect antagonism* is exerted by rediae or sporocysts and, at least sometimes, by both trematode species on each other. Its mechanism is unknown, but it might involve competition for nutrients, production of inhibitory substances, or humoral and cellular responses of the host . . . patterns of interaction vary greatly between species, and *Echinostoma malayanum* rediae will consume whole live sporocysts and cercariae of *S. spindale*, whereas the relatively powerful rediae of *Paryphostomum segregatum* apparently cannot breach the tough walls of living *Schistosoma mansoni* daughter sporocysts. They can, however, ingest whole cercariae.

THE PARASITE-MIX

The combined populations of organisms, both flora and fauna, that live together in a host organ, or in the entire host, or in the host population are known as the **parasite-mix.**[52] A more technical appellation is **parasitocoenosis.** For example, the intestine of a vertebrate animal may contain large numbers of different kinds of bacteria, yeasts, protozoa and worms. All of these organisms constitute the parasite-mix of the intestine. When we recall that there is a continual change in the numbers, developmental stages and physiologic activity of these populations, we begin to appreciate the influence that one species has upon the other in the same community of parasites. A detailed knowledge of the parasite-mix is essential to the understanding of pathogenesis, clinical symptoms and nonsymptomatic carriers. A heavy infection with one species of pathogenic parasite is usually not accompanied by a heavy infection with another species of pathogenic parasite. This relationship has often been cited for commensal forms.

Concurrent Infection

There is clear evidence that the behavior of a parasite within its host is influenced by the presence of another species of parasite. For example, the roundworm, *Ascaridia*, in chicks thrives better when the chick is also infected with another roundworm, *Heterakis*, than when alone. Also, *Heterakis* in chicks infected with the protozoan, *Histomonas meleagridis*, are smaller than those from chicks free from the protozoan.[46]

Young white mice infected with the nematode, *Nippostrongylus brasiliensis*, just prior to infection with *Hymenolepis nana* var. *fraterna*, exhibit a marked resistance to the tapeworm.[41] Helminths may be capable of increasing or of decreasing host resistance to microorganisms. Viruses and helminths, and protozoa and helminths frequently are said to be mutually antagonistic. Arguments for such antagonisms, however, are mostly theoretical, and much experimental research is needed to produce actual evidence for the existence of these phenomena. Kilham and Olivier[37] reported that encephalomyocarditis (EMC) virus alone produced only mild pathogenic symptoms in laboratory rats, but when the rodents were given a combined infection with the nematode, *Trichinella spiralis*, the rats "experienced a high incidence of cropping and death while control rats remained free of disease." The cause of this phenomenon is unknown.

One such experiment[32] demonstrated that the fluke, *Diplodiscus*, is partly responsible for the absence of opalinids from the intestine of the green frog, *Acris gryllus*. In this host the tadpoles are heavily infected with the protozoa but lightly infected with the worm. The situation is reversed in adult frogs. If flukes are experimentally introduced into the rectum of opalinid-infected frogs, the opalinids either completely disappear or they become reduced in number.

Another study of the influence of a prior infection on a subsequent infection demonstrated that in mice the introduction of the roundworm, *Ancylostoma caninum*, 24 to 48 hours before an experimental infection with the nematode, *Trichinella spiralis*, caused a significant reduction in the number of adult *Trichinella* normally expected to be found in the small intestine. Of considerable interest is the fact that, if the hookworms were administered 12, 96, 144, or 192 hours prior to the *T. spiralis* infection, no such interference with the trichina occurred. The reduction in numbers of nematodes took place primarily in the anterior portions of the small intestine, and it may have been the result of nonspecific inflammation engendered by the hookworms.[30]

An interesting example of the influence of concurrent infection on the distribution of two helminths in the gut of their host was described by Chappell. He found that "when the cestode *Proteocephalus filicollis* occurred concurrently with the acanthocephalan *Neoechinorhynchus rutili*, in natural infections of the three-spined stickleback [*Gasterosteus aculeatus*], the distribution of each species in the gut was significantly different from when these species occurred alone. In concurrent infections partial spatial separation of the two species was observed in which both adults and plerocercoids of *P. filicollis* attached more frequently in the anterior intestine while individuals of *N. rutili* attached more frequently in the rectum. In single species infections individuals of both species were distributed more widely throughout the gut. The data are thought to exemplify competitive exclusion and are possibly the result of the adverse effect of each species on the environment of the other."[11]

Concurrent infections may be of a quite different nature than those examples men-

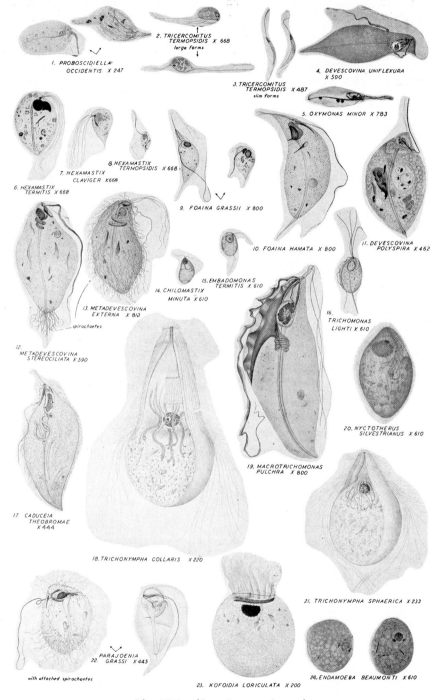

1. *PROBOSCIDIELLA OCCIDENTIS* X 247

2. *TRICERCOMITUS TERMOPSIDIS* X 668
large forms

3. *TRICERCOMITUS TERMOPSIDIS* X 487
slim forms

4. *DEVESCOVINA UNIFLEXURA* X 590

5. *OXYMONAS MINOR* X 783

6. *HEXAMASTIX TERMITIS* X 668

7. *HEXAMASTIX CLAVIGER* X 668

8. *HEXAMASTIX TERMOPSIDIS* X 668

9. *FOAINA GRASSII* X 800

10. *FOAINA HAMATA* X 800

11. *DEVESCOVINA POLYSPIRA* X 462

12. *METADEVESCOVINA STEREOCILIATA* X 590

13. *METADEVESCOVINA EXTERNA* X 813
spirochaetes

14. *CHILOMASTIX MINUTA* X 610

15. *EMBADOMONAS TERMITIS* X 610

16. *TRICHOMONAS LIGHTI* X 610

17. *CADUCEIA THEOBROMAE* X 444

18. *TRICHONYMPHA COLLARIS* X 220

19. *MACROTRICHOMONAS PULCHRA* X 800

20. *NYCTOTHERUS SILVESTRIANUS* X 610

21. *TRICHONYMPHA SPHAERICA* X 233

22. *PARAJOENIA GRASSI* X 445
with attached spirochaetes

23. *KOFOIDIA LORICULATA* X 200

24. *ENDAMOEBA BEAUMONTI* X 610

Fig. 22-1. (*Legend on opposite page*)

tioned above. If larval nematodes (*Toxocara canis*) are given to guinea pigs one week to one month prior to infection with *Entamoeba histolytica*, the presence of the nematode in the liver of the host does not affect development of cecal amebiasis, but it does apparently *increase* the number of positive amebic cultures (*in vitro*) obtainable from the liver in animals with cecal lesions caused by the ameba.[39] Some parasites of the intestine evidently account for the appearance of exacerbation of some bacterial diseases, and vice versa. "*Hymenolepis diminuta* developing in the presence of *Moniliformis dubius* are lighter, shorter, have a lower average weight: length ratio, and are limited to the posterior part of the intra-intestinal range they occupy in single infections. *M. dubius* in the concurrent infections are lighter, possibly shorter, have a lower average weight:length ratio, and tend to attach further anterior than in single infections." The similarities between these effects of concurrent infections and those of crowding suggest that the former are due to competition, possibly for carbohydrates.[33] If albino rats are experimentally infected with *Balantidium* plus ascarid larvae, the protozoa become much more pathogenic than when the worm larvae are absent.[8]

A human visitor to the tropics may return to temperate climates to find himself a walking zoo—the delight of any class in parasitology. It is only from extensive ecologic studies that the entire influence of the parasite-mix on the health and activities of the host can be ascertained. During such studies the entire life history of each parasitic species should be considered because each species

has a biotype (e.g., an organ of the host), but the parasite species may pass through and be specific for several markedly different biotypes during the course of its life cycle.

MUTUALISTIC INTESTINAL PROTOZOA AND BACTERIA

Insects

A classic example of mutualism is the cooperation between termites and their intestinal protozoan fauna. The hindgut of many termites is so crowded with protozoa (Fig. 22–1) that the protozoan bodies are often distorted from being pressed on all sides. The protozoa in nymphs of *Zootermopsis* constitute one-seventh to one-third of the total weight of host and parasites. Most of the larger flagellates in termites belong to the order Hypermastigida (see p. 51). The termite families Mastotermitidae, Kalotermitidae, Hodotermitidae and Rhinotermitidae are normally host to an abundance of flagellates. Significantly, most members of the largest and most specialized of termite families, the Termitidae, do not feed on wood, but the few members that do so feed possess in their intestines bacteria which presumably elaborate enzymes that digest wood. The first nymphal instar acquires its protozoa by proctodeal feeding at the anus of an infected termite, and during each molting period the protozoa are temporarily lost—to be regained as originally acquired. Only the adult, asexual, nondividing forms ordinarily occur in termites. Encystment of the protozoa does not take place.

If human beings possessed cellulose-digesting enzymes, we would not have so critical a

Fig. 22-1. Representative protozoa from termites. 1 from *Kalotermes occidentis*, Lower California. 2 and 3 from *Termopsis angusticollis*, California. 4 from *Kalotermes perezi*, Costa Rica. 5 from *Kalotermes minor*, California. 6 from *Kalotermes flavicollis*, California. 7 from *Kalotermes marginipennis*, Costa Rica. 8 from *Termopsis angusticollis*, California. 9 from *Kalotermes flavicollis*, Europe. 10 from *Glyptotermes parvulus*, Uganda. 11 from *Kalotermes occidentis*, Lower California. 12 from *Glyptotermes parvulus*, Uganda. 13 from *Mastotermes darwiniensis*, Australia. 14, 15, and 16 from *Amitermes beaumonti*, Canal Zone. 17 from *Kalotermes jeannelanus*, East Africa. 18 from *Termopsis angusticollis*, California. 19 from *Glyptotermes* sp., Uganda. 20 from *Amitermes silvestrianus*, Canal Zone. 21 from *Termopsis angusticollis*, California. 22 from *Kalotermes* (*Neotermes*) *connexus*, Hawaii. 23 from *Kalotermes simplicicornis*, California. 24 from *Amitermes beaumonti*, Canal Zone. (1 and 11 from Lewis; 5 from Cross; 23 from Light; all others from Kirby; 14, 15, 16, 20, 24 courtesy of Parasitology, Cambridge; all others courtesy Univ. Calif. Pub. Zool.)

world food shortage problem as we have today because people could survive quite well,
although perhaps not tastefully, on straw,
sawdust, wood shavings, bark and dead
branches. In a pinch, we could eat paper and
old furniture. Neither do the termites and
wood-eating roaches themselves elaborate
cellulose-digesting enzymes (cellulases), but
their protozoan and, probably, bacterial
guests, pay for the intestinal housing facilities
by digesting the food eaten by the host.

Termites can be defaunated by incubating
them at 36° C. for 24 hours or by submitting
them to two atmospheres of oxygen pressure
for four hours. After the protozoa are dead
the insects will continue eating wood and
other materials, but they will starve to death
because they cannot digest their food. If they
are soon refaunated they will survive.

Physiologic experiments dealing with *Cryptocercus* and its flagellates have not been so
extensive as those with termites, but in all
probability the roach is also dependent almost
entirely on the fermentation of cellulose by its
intestinal protozoa. One marked difference
between roach and termite protozoa is the
habit of encystment of many of the protozoa
in the roach during molting of the insect. The
roach retains its infection throughout its life.
Another difference is that encystment is accompanied by a sexual process, and the stimulus for the initiation of a sexual cycle in the
flagellate is extrinsic, attributable to the molting hormones of the host (see Chapter 21,
p. 474). Molting and the sexual cycles never
occur separately. Although the sexual cycles
of the flagellates are initiated by host hormones, the follow-up is directly influenced by
the rise in pH (7.2 to 7.4) of the gut fluids
during the molting period.

The role of extracellular symbiots (bacteria, spirochaetes) on the surfaces of protozoa in termites and in roaches remains
obscure. A great diversity in numbers, kinds
and distribution of both extracellular and
intracellular symbiots occurs in the flagellates of termites,[38] but experimental evidence
concerning their nature is lacking. The intracellular bacteria in parasitic flagellates may

well play a part in the digestion of cellulose.
Indeed, some workers[28] believe that the bacteria are responsible for all cellulose digestion.

Ruminants

The placid cud-chewing habit of cattle is
directly related to the digestion of cellulose,
and to the masses of symbiotic ciliates and
bacteria to be found in the first stomach of
these animals. In sheep and goats the holotrich ciliates (superficially resembling such
forms as *Paramecium*) occur in numbers between 160,000 and 200,000 per milliliter of
rumen content. The three most common
species in sheep and goats (also to be found in
cattle) are *Isotricha prostoma, Isotricha intestinalis*
and *Dasytricha ruminantium*. Hungate[35] found
an average number of 7000 *Diplodinium* per
milliliter of rumen contents in cows fed
timothy hay and concentrate, with a range
between 800 and 30,000! Considerable differences and fluctuations occur among the
various ruminants, but some protozoan species
are common to many kinds of hosts, even to
those widely separated geographically. The
microfauna of ruminants consists principally
of holotrichous ciliates belonging to the suborder Trichostomina, and spirotrichous ciliates belonging to the suborder Oligotricha,
particularly the family Ophryoscolecidae
(Fig. 22–2). The young ruminant acquires
its infection by ingesting infected saliva. See
Rogers[64] for an account of the metabolism
of ruminants.

The relationships among the bacteria,
protozoa and other microorganisms in
ruminant stomachs have been reviewed by
Hungate. Cattle ciliates contain a cellulase,
but the bacteria of the rumen also possess
cellulolytic properties. In addition, the possibility that intracellular symbiotic bacteria
are concerned with cellulose and starch digestion should not be overlooked. Hungate[35]
has said, "The faculty for an obligately
anaerobic existence is so widespread among
bacteria and so relatively uncommon in
protozoa that utilization of anaerobic intracellular bacteria for synthesis of essential
building materials (e.g., high-energy phos-

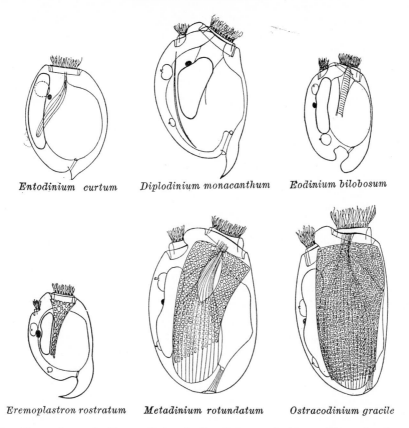

Entodinium curtum   Diplodinium monacanthum   Eodinium bilobosum

Eremoplastron rostratum   Metadinium rotundatum   Ostracodinium gracile

Fig. 22-2. Illustrations of ciliates of the family Ophryoscolecidae. These drawings illustrate sample genera and species from the stomach of a wild bull (*Bos gaurus*) in Mysore, India. (Kofoid and Christenson, courtesy of Univ. Calif. Pub. Zool.)

phate) might account for the ability of the protozoa to survive under anaerobic conditions. According to this viewpoint the fermentation products could well be those of the intracellular bacteria." The existence of intracellular symbiotic bacteria in rumen ciliates, however, has not been conclusively established, although there are bodies in the cytoplasm that *look* like bacteria. There is clear evidence,[31] however, that species of *Entodinium* and other genera actively ingest bacteria. The rumen protozoa are obviously suitable material for experimental investigations of the problem of "mutualism within mutualists."

Three important genera of ciliates in cattle are: *Diplodinium, Entodinium* and *Epidinium*. The role played by these organisms is less well understood than that played by termite

protozoa. It is more difficult to defaunate cattle than termites, but it can be done by starvation for several days followed by the administration of a copper sulfate solution. When this is done, cellulose and hemicellulose continue to be digested at the same rate as in untreated cattle. *Diplodinium*, however, can digest cellulose. This ciliate is probably mutualistic. Other genera, including *Isotricha, Dasytricha* and *Bütschlia*, readily absorb soluble carbohydrates of the host stomach, and convert the carbohydrates into starch, thereby helping to keep the sugars from being immediately fermented by the bacteria.

In addition to cellulose, the ciliates probably feed on bacteria and on other protozoa or various types of organic debris. *Diplodinium* possesses the enzyme *cellulase* or *cellobiase*, and it may store large amounts of glycogen.

Both *Entodinium* and *Epidinium* digest starch, and thus they, too, help to level out the fermentation process that occurs rapidly by action of the bacteria.

The question as to the amount of protein supplied the host in the form of protozoa can be answered by using Hungate's estimates for the bovine of 3000 *Diplodinium*, 3000 *Isotricha*, 5000 *Dasytricha* and 5000 *Entodinium* per milliliter. Calculation shows that about 66 gm. of protein are supplied the host in the form of protozoa each day. The exact contribution would fluctuate from day to day as numbers and kinds of ciliates vary. Oxford[55] has suggested that if the rumen ciliates excrete ammonia, there may actually be wastage of protein in the rumen through the conversion of plant protein into protozoan protein. If this wastage occurs, the protozoa may be called "food robbers," but more detailed knowledge is needed before this problem can be solved. The reserve starch storage in the protozoa does not appear to be a major mechanism by which carbohydrates are supplied to the host, but the starch is a reserve source of energy for fermentation. In this manner the polysaccharides in the protozoa are indirectly of value to the host. The average contribution to the host of both fermentation products and protein by the protozoa has been estimated to be about one-fifth of the total requirement. We know that the ciliates and the bacteria in a cow stomach synthesize B-complex vitamins, enzymes, amino acids and proteins. These substances are used directly by the host as food.

Turning now from ciliates to amebas we find clear evidence that bacteria, both pathogenic and nonpathogenic, play a role in the pathogenicity of *Entamoeba histolytica*.[58] Amebas from ameba-trypanosome cultures were unable to establish themselves in germ-free guinea pigs without bacteria, but the addition of either of two species of bacteria (*Aerobacter aerogenes* or *Escherichia coli*) not only permitted the ameba to become established, but promoted invasion of the tissues. One may conclude that "bacteria are involved essentially in the etiology of intestinal amebiasis and that

synergism of ameba and bacteria is a prerequisite to development of the disease."

Reflect for a moment on the tremendous churn of activity that is housed in an intestine. Bacteria, yeasts, spirochaetes, protozoa and worms may all be struggling for nutrients and space at the same time. By-products of metabolism are being poured into the mass continuously. Physiochemical conditions (such as pH and gases) are constantly being altered. The bacteria synthesize vitamins and other essentials for the protozoa and worms. No wonder that evolutionary processes among intestinal organisms and their hosts have resulted in relationships that bear the character both of coexistence and of antagonism.[75]

FOOD CHAINS

Food chains may be divided conveniently into the following three types. (1) The predator chain, which, starting from a plant base, goes from smaller to larger animals. (2) The parasite chain, which goes from larger to smaller organisms. (3) The saprophytic or saprozoic chain, which goes from dead organic matter into microorganisms and into a few larger forms such as fungi.

Potential energy is lost at each level of food transfer—that is, at each step in the chain. Photosynthetic plants fix only about two to three per cent of the energy of the sunlight that falls upon them, and when these plants are eaten by herbivorous animals, about ten per cent of the potential energy of the plants is used. When the herbivorous animals are eaten by small carnivores, perhaps 12 or 15 per cent of the potential energy in their bodies is used. A fourth step in the chain would be the devouring of small carnivores by large ones, and the percentage of potential food energy obtained would be slightly more than at the previous step in the chain.

In distinguishing between a predator and a parasite the matter of size of food is important. Although in the parasite chain organisms at successive levels (steps) are smaller and smaller instead of generally larger and larger,

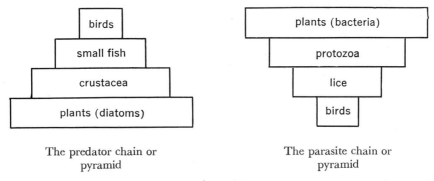

The predator chain or pyramid

The parasite chain or pyramid

Fig. 22-3.

as in the predator chain, there is no fundamental difference between the two. A predator usually disposes of its prey all at once, whereas a parasite, if it disposes of its host or part of its host, does so a little at a time—usually at a slow enough rate so as not to interfere seriously with the host's well-being. Both predator and parasite chains are limited to a few steps or links as shown in Figure 22–3. One significant feature in the evolution of parasitism is the parasitization of adjacent links in the predator food chain.

The parasite pyramid of food is upside down because most parasites are smaller and more numerous than are their hosts. If, instead of a pyramid of numbers we use a pyramid of mass or "biomass," we find that the parasite biomass and the predator biomass are the same because the total *weight* of individuals at successive levels or steps is plotted. The biomass pyramid indicates the over-all effect of food chain relations for the ecologic group as a whole. For a detailed discussion of this subject, see Elton[23] and Odum.[53] Since the number of organisms that can be supported depends on the number of food organisms being replaced as fast as some are eaten (rather than on the amount of food organisms present at any one time), parasites bathed in a continuous supply of nutrients furnished by the host are in a highly favored position. This situation helps to explain the presence of enormous numbers of parasites sometimes to be found in a single host.

ROLE OF METAZOAN PARASITES IN THE TRANSMISSION OF MICROBIAL INFECTIONS

Bacterial or viral infections often accompany general lowering of vitality which brings about reduced resistance to microbe-caused diseases. This reduction in resistance may result from heavy or long-standing infections with animal parasites. Introduction of infective bacteria or viruses is always a possibility when larger parasites injure host tissues.

Ectoparasites, such as sucking insects, mites and ticks, are notorious offenders in this regard. They can bite or pierce the skin or mucous membranes and thus destroy the host's first line of defense against microbial infection. Larval flesh flies have been known to eat away part of the mucous membranes of human nasal passageways, which thereafter become infected with bacteria, resulting in death of the host. Such flies may develop in any body opening.

The role of helminths in the transmission of viruses, bacteria and protozoa has been recognized or suspected for some 40 years, but although experimental infections have often been demonstrated, proven infections in nature have been relatively rare.[57] Examples of bacterial infections through the intervention of skin-penetrating nematodes are numerous. Laboratory experiments have shown, for example, that hookworm larvae contaminated with anthrax bacilli can infect guinea pigs with anthrax. Experiments of this nature that involve viruses have been few. Swine influenza virus can be transmitted by lungworms of swine, and the nematode, *Trichinella*, encysted in muscles of guinea pigs infected with lymphocytic choriomeningitis virus, has been known to transmit the virus to another host.

In a review of this subject, Woodruff[85] has emphasized the importance of understanding how helminths, particularly migrating larvae, can either transmit or aggravate infection with microorganisms. He suggested "that consideration of the possible part played by helminths in transmitting bacterial, viral and some protozoal infections would be well worth while and might clear up some of the many problems concerning their epidemiology. In particular, further studies should be carried out on transmission by *Toxocara* of poliomyelitis and toxoplasmosis; of the role of *Ascaris* larvae in disseminating intestinal micro-organisms, and of the part played by helminths in the development of the carrier state in cholera. This latter study should also be expanded to include *Salmonella* infections and carriers."

SYMBIOTIC CLEANING

Many shrimps, small fish and other organisms regularly move over the willing bodies of such free-swimming animals as fish and turtles to pick off food particles, damaged and necrotic tissues, copepods, isopods, bacteria, fungi and other bits of material that may become irritating to the "host." Limbaugh[45] and Eibl-Eibesfeldt[22] have described some of these relationships, which have been recognized for many years but have been considered as merely scientific curiosities (Fig. 22–4). The widespread occurrence of this phenomenon, especially in the marine environment, and its relation to the biology of ectoparasites and their hosts demands at least a brief consideration in any general treatment of the subject of symbiosis.

The cleaner fish represent at least eight families (e.g., Lebridae) and 26 or more species. Most of them do their cleaning only in the juvenile stages. These fish, for the most part, are so colored as to contrast with the environment. The cleaner shrimps also are usually brightly colored, often red, and they display themselves to their hosts in a

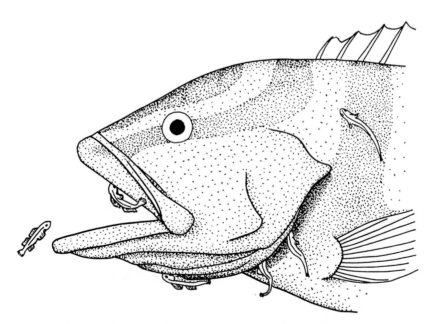

Fig. 22-4. Cleaner fish (*Elacatinus oceanops*) on the head of *Epinephelius*. (Modified from Eibl-Eibesfeldt, courtesy of Zeit. Tierpsychologie)

conspicuous manner. The shrimps characteristically wait at the entrance of their dark crevasses or holes with their bright antennae extended into the sunlit water, and they whip the antennae back and forth while the body is swaying. This behavior, plus the color, apparently attracts the fish, which need little enticement to seek relief from a burden of ectoparasites. Among the better known cleaner shrimps are: *Periclimenes yucatanicus, Hippolysmata californica,* and *Stenopus hispidus* (the common boxer shrimp). At least one crab has been seen to clean fish.

The host fish often opens its mouth, thereby inviting the shrimps to enter and pick its teeth or gill cavities. Occasionally the shrimp is allowed by the fish to make minor incisions in order to obtain subcutaneous parasites. Fish apparently come to particular spots that are called "cleaning stations," and there they may line up or crowd around waiting their turn to be cleaned. Parasites are removed rapidly as the shrimp flits over the body of its host. If the shrimp encounters a wound it will spend considerable time picking away the injured flesh. If all the known cleaners are removed from a given area, such as a portion of a reef in the tropics, the numbers of fish soon are reduced and those that remain tend to accumulate parasites and skin diseases.

THE IMMEDIATE ENVIRONMENT OF THE HOST

A review of much of the discussion on the preceding pages of this chapter can be made in the form of a case history by selecting an ecologic study of one kind of host and its parasites, and by noting the interplay of as many environmental factors as possible. Such a study was made by Fischthal over a period of years on parasites of northwestern Wisconsin fish, and summarized by him in 1953.[24] The following account is taken largely from his paper, and confined to his observations on lake habitats.

Physical Factors

Form of Basin. High productivity of aquatic organisms tends to favor a high percentage and intensity of parasitism in fish.

Greater biologic productivity is favored in a lake where the photosynthetic zone is closely superimposed over the decomposition zone. If the lake basin edge is steep, much of the essential decomposition material is removed to deep water where it is inaccessible, and as one consequence, the mollusks, turbellarians, annelids, insects and crustaceans that may be utilized as intermediate hosts for trematodes or cestodes find living conditions untenable.

Shore Line and Changes in Water Level. Other factors being equal, the longer the shore line, the greater the biologic productivity. Hence where there is an irregular shore line with resulting bays, coves and shallow water, there is more superposition of photosynthetic and decomposition zones and a greater diversity of bottom and margin conditions. These features favor an abundance of plankton and larger plants and animals, including parasites in all stages of development.

Water Movements. Movements of lake water may assist the dissemination of parasites by spreading larval stages to new hosts, or they may have an adverse effect by moving free-living larval parasites or parasitized intermediate hosts out of reach of the next host in the life cycle.

Wave Action. The action of waves tends to denude exposed shores and to reduce the incidence of parasitism except in those hosts that are capable of burrowing.

Temperature. Cold, deep water is not conducive to a high productivity. This limiting factor tends to result in a concentration of the fauna in the upper, warm layers, therefore aiding in an increased superposition of hosts and parasites. The metabolic rates of poikilothermic hosts are greatly increased with high temperatures. Fish generally harbor more parasites in the summer and early fall than in winter and early spring seasons. With a higher metabolic rate more food is required, and with an increase in food intake a fish acquires a higher degree and variety of parasitism.

Light. Increased light results in increased development of green plants and increased biologic productivity in general. The movement of many planktonic organisms is affected by light. Diurnal movements of various aquatic organisms, including fish, is well known. Light affects the emergence of cercariae from snails, some preferring darkness, others light, and still others favoring the sunrise or sunset periods. When heavy snow blankets ice on a lake, a prolonged shutting-out of light may occur, resulting in a winter-kill which destroys free-swimming larval stages and their hosts.

Chemical Factors

Dissolved Oxygen. In lakes a stratification of oxygen is closely correlated with a thermal stratification. The deeper waters are deficient in oxygen, thereby limiting the kinds and numbers of animals that can live there. A light winter-kill in lakes, resulting from a depletion of oxygen under ice, decreases the fish population. During summer months in plant-choked lakes and ponds, there may be oxygen exhaustion at night and oxygen replacement during the day, resulting in a partial kill of fish, plankton and associated parasites. Thus parasitism may be increased or decreased by the presence or absence of a sufficient quantity of oxygen.

Carbonates. Soft waters (i.e., with little carbonates) are usually unproductive, whereas hard waters are normally very productive. Mollusks require carbonates as shell building materials. Crustacea are reduced in numbers as the carbonates decrease. When these invertebrates disappear, the larvae of trematodes and acanthocephalans cannot find their intermediate hosts, and the fish experience a freedom from heavy infections. Leeches are generally much more abundant in hard water than in soft water. In contrast with an alkaline humus substrate, an acid humus substrate tends to poison most leeches, thereby reducing the opportunities for transfer of such protozoan parasites as haemogregarines and trypanosomes.

Hydrogen Ion Concentration (pH). Mollusks are scarce or absent in excessively acid lakes (in part soft water and bog lakes) be-cause the acid either dissolves or corrodes the carbonate shells, thereby reducing the trematode population. Intermediate hosts of cestodes and acanthocephalans are frequently reduced in the plankton of acid waters.

Pollution. It is common practice in the maintenance of commercial fish ponds to add a small amount of manure or of commercial fertilizer to increase the general biologic productivity. The natural drainage of manure into lakes increases the growth of algae, plankton and bottom organisms, thereby favoring the productivity of parasites. Excessive pollution with sewage or with industrial wastes is usually detrimental to all aquatic organisms.

Biologic Factors

Plankton. As we have already noted, plankton plays an essential part in the life cycles of many fish parasites, especially as it provides intermediate hosts. Plankton also serves as food for fish larvae and young.

Benthos. The bottom fauna includes snails, clams, aquatic insects, annelids, turbellarians and crustacea—all of which may serve as intermediate hosts for various worms of fish.

Aquatic Plants. We have already emphasized the importance of a close association between the photosynthetic zone and the decomposition zone for the maintenance of fish and their parasites. Trematodes in fish are particularly abundant in weedy bays and shorelines because of the greater superposition, in these habitats, of fish, snails and stages in the life cycles of the parasites.

Food Chain. Abundant evidence leads to the conclusion that fishes are most heavily parasitized during the seasons when they ingest the largest quantities and kinds of food. The fry of many fishes start to accumulate parasites as soon as they begin feeding on plankton, and if, when they become adults, they change their feeding habits, the nature of their parasitic faunas also changes. Many larval worms in fish owe their sanctuary to the fact that their fish hosts are eaten by birds in which the larvae reach sexual maturity.

Fish Populations. Different species of fish as well as individuals of one species are often in competition for food. This food may include invertebrates with larval worms specific to one of the species of fish, but not to others. Overlapping of two fish populations (the ecotone) may enhance parasitism if the parasite involved reaches maturity in both species, or if one host serves as the intermediate carrier for the parasite that matures in the other host species. The parasitic fauna of a fish species inhabiting inshore, shallow waters may differ considerably from the parasitic fauna in the same species of fish inhabiting offshore, deeper waters. Fish that migrate from one part of a lake to another tend to acquire more parasites than those that remain all of their lives in one general locality. Lakes with the greatest variety of habitats tend to have the greatest variety and incidence of fish and parasites.

QUESTIONS TO BE ASKED BY ANY INVESTIGATOR OF THE ECOLOGY OF PARASITES

When a student of parasitology finds a parasite in a host he should ask himself many questions concerning the parasite, concerning other parasites in the same host, concerning the host, and concerning the environment. Some of those questions have been asked in the preceding pages of this chapter, and many of them have been implied. A number of such questions are listed below, and they constitute a summary of important problems encountered in any consideration of the ecology of parasitism:

1. How does the presence of a population of one species of parasite influence the growth and development of another species in the same host organ or in another organ of the same host?
2. Is the host dependent in any manner upon its parasites?
3. What effect does a seasonal change have upon the numbers and kinds of parasites for each host?
4. Are the parasites more tolerant or less tolerant than the host to changes in the environment (e.g., salinity, temperature)?
5. Which environmental factors, both inside and

outside the host, encourage the maximal degree of pathogenicity?
6. What are the chemical and behavioral aspects of the ecologic niche of the parasite and of its host?
7. When two or more species of parasites compete for the same energy (food), what factors operate to favor one over the other?
8. Do the limiting factors of the host environment also directly limit the growth and development of the parasites?
9. How do population changes of the host influence populations of its parasites?
10. What are the size and age relationships between hosts and parasites?
11. What changes in numbers, kinds, and development of parasites are the result of migration of the host?
12. How are populations of parasites controlled in nature?
13. Do the parasites control the populations of their host?
14. What is the relationship between the food of the host and the kinds, number, and development of its parasites?
15. Does the proximity and availability of related host species influence the growth, development and distribution of the parasites?
16. How does the adult parasite differ from its developmental stages in response to the environment?
17. Is the parasite specific to its host or does it "enjoy" a wide range of host species?
18. How widely is the parasite distributed and what barriers prevent its further distribution?
19. What has been the evolutionary history of the parasite?
20. What is the effect of the sex of the host on the growth and pathogenicity of the parasite?

REFERENCES

1. Audy, J. R.: A biological approach to medical geography. Brit. Med. J., *1*:960–962, 1954.
2. ———: The localization of disease with special reference to the zoonoses. Trans. Roy. Soc. Trop. Med. Hyg., *52*:308–334, 1958.
3. ———: Parasites as ecological labels in vertebrate ecology. *In* Purchon, R. D. (ed.): Proceedings of Centenary and Bicentenary Congress of Biology. pp. 123–127. Singapore, Univ. of Malaya Press, 1960.
4. Baer, J. G.: Ecology of Animal Parasites. Urbana, Ill., Univ. of Illinois Press, 1951.
5. Barbosa, F. S., and Barbosa, I.: Dormancy during the larval stages of the trematode *Schistosoma mansoni* in snails estivating on the soil of dry natural habitats. Ecology, *39*: 763–764, 1958.

6. Bardsley, J. E., and Harmsen, R.: The trypanosomes of Ranidae. I. The effects of temperature and diurnal periodicity on the peripheral parasitaemia in the bullfrog (*Rana catesbeiana* Shaw). Can. J. Zool., *47*:283–288, 1969.

7. Barrow, J. H., Jr.: The biology of *Trypanosoma diemyctyli*, Tobey. III. Factors influencing the cycle of *Trypanosoma diemyctyli* in the vertebrate host. *Triturus v. viridescens*. J. Protozool., *5*:161–170, 1958.

8. Bogdanovich, V. V.: Balantidiaz s Soputstvuyushchei Glistnoi Invaziei (v Usloviyakh Eksperimenta). Med. Parazitol. i Parazitar. Bolezni, *31*:711–715, 1962.

9. Burnett, T.: Experimental host-parasite populations. Ann. Rev. Entomol., *4*:235–250, 1959.

10. Calentine, R. L., and Fredrickson, L. H.: Periodicity of caryophyllaeid cestodes in the white sucker, *Catostomus commersoni* (Lacépède). Iowa State J. Sci., *39*:243–250, 1965.

11. Chappell, L. H.: Competitive exclusion between two intestinal parasites of the three-spined stickleback, *Gasterosteus aculeatus* L. J. Parasitol., *55*:775–778, 1969.

12. Chitwood, M. B.: Intraspecific variation in parasitic nematodes. Systematic Zool., *6*: 19–23, 1957.

13. Chute, R. M.: Infections of *Trichinella spiralis* in hibernating hamsters. J. Parasitol., *47*: 25–29, 1961.

14. Clausen, C. P.: Biological control of insect pests. Ann. Rev. Entomol., *3*: 291–310, 1958.

15. Connor, R. S.: A study of the seasonal cycle of a proteocephalan cestode, *Proteocephalus stizostethi* Hunter and Bangham, found in the yellow pikeperch, *Stixostedion vitreum vitreum* (Mitchill). J. Parasitol., *39*:621–624, 1953.

16. DeBach, P., and Smith, H. S.: Effects of parasite population density on rate of change of host and parasite populations. Ecology, *28*:290–298, 1947.

17. Dogiel, V. A.: General Parasitology. Revised and enlarged by Yu. I. Polyanski and E. M. Kheisin, transl. by Z. Kabata. Edinburgh, Oliver & Boyd, 1964.

18. Dogiel, V. A., and Petrushevéskii, G. K.: Parazitofauna ryb Nevskoí Guby (Fish parasites of the Neva Bay). Trudy Lennini gradskogo Obshchestva Estestvoispytatele- (Trav. Soc. Naturalistes Lenningrad), *62*: 366–434, 1933. (In Russian with German summary.)

19. Dogiel, V. A., Petrushevskii, G. K., and Polyanski, Yu. I.: Parasitology of Fishes. Transl. by Z. Kabata. Edinburgh, Oliver & Boyd, 1961.

20. Doutt, R. L.: The biology of parasitic Hymenoptera. Ann. Rev. Entomol., *4*:161–182, 1959.

21. Dudzinski, M. L., and Myktowycz, R.: Relationship between sex and age of rabbits, *Oryctolagus cuniculus* (L.) and infection with nematodes *Trichostrongylus retortaeformis* and *Graphidium strigosum*. J. Parasitol., *49*:55–59, 1963.

22. Eibl-Eibesfeldt, I.: Über Symbiosen, Parasitismus und andere besondere zwischenartliche Beziehungen tropischer Meeresfische. Zeit. Tierpsychologie, *12*:203–219, 1955.

23. Elton, C.: Animal Ecology. London, Sidgwick and Jackson, 1927.

24. Fischthal, J. H.: Parasites of Northwest Wisconsin fishes. IV. Summary and limnological relationships. Trans. Wisconsin Acad. Sci., Arts & Let., *42*:83–108, 1953.

25. Flanders, S. E.: Effect of host density on parasitism. J. Econ. Ent., *28*:898–900, 1935.

26. ———: Mass culture of California red scale and its golden chalcid parasites. Hilgardia, *21*:1–42, 1951.

27. Gause, G. F.: The Struggle for Existence. Baltimore, Williams & Williams, 1934.

28. Ghidini, G. M.: Ricerche sull'attività cellulosolitica della flora e fauna intestinale di *Reticulitermes lucifugus* Rossi. Boll. Soc. Ital. Biol. Sper., *15*:220–221, 1940.

29. Gorbunova, M.: [Variations of the Parasitic Fauna of the Pike and the Roach With their Age.] *In* Dogiel, V. A.: [Problems of Ecological Parasitology.] Annals Leningrad State Univ., No. 7, 1–194, 1936. (In Russian.)

30. Goulson, H. T.: Studies on the influence of a prior infection with *Ancylostoma caninum* on the establishment and maintenance of *Trichinella spiralis* in mice. J. Elisha Mitchell Sci. Soc., *74*:14–23, 1958.

31. Gutierrez, J., and Davis, R. E.: Bacterial ingestion by the rumen ciliates *Entodinium* and *Diplodinium*. J. Protozool., *6*:222–226, 1959.

32. Hazard, F. O.: The absence of opalinids from the adult green frog, *Rana clamitans*. J. Parasitol., *27*:513–516, 1941.

33. Holmes, J. C.: Effects of concurrent infections on *Hymenolepis diminuta* (Cestoda) and *Moniliformis dubius* (Acanthocephala). I. General effects and comparison with crowding. J. Parasitol, *47*:209–216, 1961.

34. Hopkins, C. A.: Seasonal variations in the incidence and development of the cestode *Proteocephalus filicollis* (Rud 1810) in *Gasterosteus aculeatus* (L. 1766). Parasitology, *49*:529–542, 1959.

35. Hungate, R. E.: The Rumen and Its Microbes. New York, Academic Press, 1966.

36. Kates, K. C.: Ecological aspects of helminth transmission in domestic animals. Amer. Zoologist, *5*:95–130, 1965.

37. Kilham, L., and Olivier, L.: The promoting effect of trichinosis on encephalomyocarditis (EMC) virus infection in rats. Amer. J. Trop. Med. Hyg., *10*: 879–884, 1961.

38. Kirby, H., Jr.: Relationships between certain protozoa and other animals. *In* Calkins, G. N., and Summers, F. M. (eds.): Protozoa in Biological Research. pp. 890–1008. New York, Columbia University Press, 1941.

39. Krupp, I. M.: Amebic invasion of the liver of guinea pigs infected with the larvae of a nematode, *Toxocara canis.* Exper. Parasitol., *5*:421–426, 1956.

40. ———: Effects of crowding and of super-infection on habitat selection and egg production in *Ancylostoma caninum.* J. Parasitol., *47*: 957–961, 1961.

41. Larsh, J. E., Jr., and Donaldson, A. W.: The effect of concurrent infection with *Nippostrongylus* on the development of *Hymenolepis* in mice. J. Parasitol., *30*:18–20, 1944.

42. Larsh, J. E., Jr., and Hendricks, J. R.: The probable explanation for the difference in the localization of adult *Trichinella spiralis* in young and old mice. J. Parasitol., *35*:101–106, 1949.

43. Levine, N. D., and Andersen, F. L.: Frozen storage of *Tritrichomonas foetus* for 5.6 years. J. Protozool., *13*:199–202, 1966.

44. Lie, K. J., *et al.*: Implications for trematode control of interspecific larval antagonism within snail hosts. Trans. Roy. Soc. Trop. Med. Hyg., *62*:299–319, 1968.

45. Limbaugh, C.: Cleaning symbiosis. Sci. Amer., *205*:42–49, 1961.

46. Madsen, H.: On the interaction between *Heterakis gallinarum, Ascaridia galli,* "blackhead" and the chicken. J. Helminthol., *36*:197–142, 1962.

47. Mathies, A. W., Jr.: Certain aspects of the host-parasite relationship of *Aspiculuris tetraptera,* a mouse pinworm. I. Host specificity and age resistance. Exper. Parasitol., *8*:31–38, 1959.

48. May, J. M.: The Ecology of Human Diseases. New York, MD Publications, 1958.

49. May, J. M. (ed.): Studies in Disease Ecology. New York, Hafner Pub. Co., 1961.

50. Nicholson, A. J.: The balance of animal populations. J. Animal Ecol. (Supp.), *2*:132–178, 1933.

51. Noble, E. R.: Seasonal variations in host-parasite relations between fish and their protozoa. J. Mar. Biol. Assoc. U.K., *36*:143–155, 1957.

52. ———: Fishes and their parasite-mix as objects for ecological studies. Ecology, *41*: 593–596, 1960.

53. Odum, E. P.: Ecology. New York, Holt, Rinehart & Winston, 1963.

54. Otto, G. F.: Some reflections on the ecology of parasitism. J. Parasitol., *44*:1–27, 1958.

55. Oxford, A. E.: The conversion of certain soluble sugars to a glucosan by holotrich ciliates in the rumen of sheep. J. Gen. Microbiol., *5*:83–90, 1951.

56. Pavlovskii, E. N.: The ecological parasitology. J. Gen. Biol., *6*:65–92, 1945. (Russian with English summary.)

57. Philip, C. B.: Helminths as carriers of microbial disease agents of man and animals. *In* Maramorosch, K. (ed.): Biological Transmission of Disease Agents. pp. 159–169. New York, Academic Press, 1962.

58. Phillips, B. P., *et al.*: Studies on the ameba-bacteria relationship in amebiasis. Comparative results of the intracaecal inoculation of germfree monocontaminated, and conventional guinea pigs with *Entamoeba histolytica.* Amer. J. Trop. Med. Hyg., *4*:674–692, 1955.

59. Raabe, Z.: Investigations on the parasitofauna of freshwater molluscs in the brackish waters. Acta Parasitol. Polonica, *4*:375–406, 1956.

60. Read, C. P.: The role of carbohydrates in the biology of cestodes. VIII. Some conclusions and hypotheses. Exper. Parasit., *8*:365–382, 1959.

61. Rees, G.: Observations on the vertical migrations of the third-stage larva of *Haemonchus contortus* (Rud.) on experimental plots of *Lolium perenne* S 24, in relation to meteorological and micrometeorological factors. Parasitology, *40*:127–143, 1950.

62. Reynoldson, T. B.: Factors influencing population fluctuations of *Urceolaria mitra* (Peritricha) epizoic on freshwater triclads. J. Animal Ecol., *24*:57–83, 1955.

63. Roberts, L. S.: The influence of population density on patterns and physiology of growth in *Hymenolepis diminuta* (Cestoda: Cyclophyllidea) in the definitive host. Exper. Parasitol., *11*:332–371, 1961.

63a.———: Developmental physiology of cestodes. I. Host dietary carbohydrate and the "crowding effect" in *Hymenolepis diminuta.* Exper. Parasit., *18*:305–310, 1966.

64. Rogers, T. A.: The metabolism of ruminants. Sci. Amer., *198*:34–38, 1958.

65. Ryckman, R. E., Lindt, C. C., Ames, C. T., and Lee, R. D.: Seasonal incidence of fleas on the California ground squirrel in Orange County, California. J. Econ. Ent., *47*:1070–1074, 1954.

66. Salt, G.: Experimental studies in insect parasitism. IV. The effect of superparasitism on populations of *Trichogramma evanescens*. J. Exper. Biol., *13*:363–375, 1936.

67. Schad, G. A.: Niche diversification in a parasitic species flock. Nature, *198*:404–407, 1963.

68. Schantz, P. M., and Biagi, F. F.: Coexistence of *Toxocara* and *Toxascaris* in dogs in Mexico City. J. Parasitol., *54*:185–186, 1968.

69. Sinha, D. P., and Hopkins, C. A.: Studies on *Schistocephalus solidus*. 4. The effect of temperature on growth and maturation *in vitro*. Parasitology, *57*:555–566, 1967.

70. Smith, H. S.: Multiple parasitism: its relation to the biological control of insect pests. Bull. Ent. Res., *20*: 141–149, 1929.

71. ———: The role of biotic factors in the determination of population densities. J. Econ. Ent., *28*:873–893, 1935.

72. Solomon, M. E.: The natural control of animal populations. J. Animal Ecol., *18*:1–35, 1949.

73. ———: Dynamics of insect population. Ann. Rev. Entomol., 2:121–142, 1957.

74. ———: Meaning of density-dependence and related terms in population dynamics. Nature, *181*:1778–1781, 1958.

75. Stefanski, Vitol'd.: Biotsenoticheskie Otnosheniya Mezhdu Paraziticheskoi Faunoi i Bakterial'noi Floroi Pishchevaritel'nogo Trakta. (Biocoenotic relations between parasitic fauna and bacterial flora of the digestive tract.) Zool. Zhur., *34*, 992–999, 1955; Referat. Zhur., Biol., 1956, No. 63975, 1955.

76. Thompson, W. R.: On natural control. Parasitol., *21*:269–281, 1929.

77. ———: The fundamental theory of natural and biological control. Ann. Rev. Entomol., *1*:379–402, 1956.

78. Ullyett, G. C.: Some aspects of parasitism in field populations of *Plutella maculipennis* Curt. J. Entomol. Soc. S. Africa., *6*:65–80, 1943.

79. Utida, S.: Effect of host density upon the population growth of interacting two species of parasites. Experimental studies on synparasitism. Second report. Oyo-Kontyu, *9*:102–107, 1953. (In Japanese with English summary.)

80. Varley, G. C.: The natural control of population balance in the knapweed gall-fly (*Urophora jaceana*). J. Animal Ecol., *16*:139–187, 1947.

81. ———: Meaning of density-dependence and related terms in population dynamics. Nature, *181*:1778–1781, 1958

82. Varley, G. C., and Edwards, R. L.: The bearing of parasite behaviour on the dynamics of insect host and parasite populations. J. Animal Ecol., *26*:469–475, 1957.

83. Voge, M.: Sensitivity of developing *Hymenolepis diminuta* larvae to high temperature stress. J. Parasitol, *45*: 175–181, 1959.

84. Wijers, D. J. B.: Factors that may influence the infection rate of *Glossina palpalis* with *Trypanosoma gambiense*. I. The age of the fly at the time of the infective feed. Ann. Trop. Med. Parasitol., *52*:385–390, 1958.

85. Woodruff, A. W.: Helminths as vehicles and synergists of microbial infections. Trans. Roy. Soc. Trop. Med. Hyg., *62*:446–452, 1968.

Chapter 23

Distribution and Zoogeography

INTRODUCTION

Biogeography is a title under which is gathered a vast amount of fact and speculation on problems associated with the distribution of plants and animals in space. This field of inquiry impinges upon many others that are considered in this book, especially host specificity (p. 522) and evolution of parasitism (p. 545). Conclusions relating to the biologic aspects of geography may invite warm agreement and support, or, conversely, equally warm and entertaining disagreement. Thus, the value of a certain parasitic genus as an indicator of the distribution and evolution of its host may on the one hand be supported by conclusions drawn from biogeography, or, on the other hand, be exposed to most zealous criticism from the same source. In addition to the obvious practical importance of knowledge about the distribution of parasites, such as that of hookworms or of the insect vectors of trypanosomes, the purely philosophical aspects of biology and geography are not only important, but generally are more interesting. The spread of natural populations of parasites is a specialized aspect of biogeography of particular consequence for the parasitologist.

Present distribution of a species depends on: (1) the age of the species—the older it is the more time it has had in which to disperse; (2) the possibilities it has had for dispersal in the past; (3) the present opportunities it has for dispersal—this factor varies with its ability, if a parasite, to live apart from a host, and with the extent to which the host is bound to a particular habitat. Physical, chemical, mechanical and biologic agencies have operated from the remote beginnings of the first para-

sites to encroach upon the freedom of plants and animals. These factors may produce their effects on adult parasites directly, on larval parasites, on the availability of intermediate hosts, or on definitive hosts. The exact mode of operation, past and present, is often obscure.

Zoologic investigations of parasites and their hosts must be based on large numbers of faunal lists or distributional records. Such records for parasites are scanty, but those that exist suffice for a beginning of some generalizations concerning the historical and present relationships among hosts and parasites. When considering the historical record the student should remember that in all sciences built on history, much is hypothetical.

In this chapter we are again concerned with the environmental factors that combine to create the setting that determines the numbers and kinds of parasites to be found associated with one or more hosts. When members of the same species of host become separated in space, do their parasites remain the same? What are the environmental factors outside the host and inside the host that determine the distribution of the parasite? Complete answers to these and related questions must await further ecologic studies, but a good start has been made and we shall select examples from the growing literature, and formulate some tentative conclusions.

DISTRIBUTION AND CLIMATE

The distribution of parasites and of their hosts is directly and indirectly governed in a large measure by the climate. Climate varies according to latitude, longitude, altitude and season of the year; and it is the result of infinite and changing combinations of temperature, rain, wind, water currents, land and water masses, mountain ranges and vegetation. Plants that serve as food for animals grow only where there are appropriate conditions of temperature, moisture and soil.

Temperature is the most important single extrinsic factor that influences the existence of parasites. Large areas of water situated around small areas of land tend to equalize temperatures, whereas large areas of land, especially those remote from large bodies of water, tend to retain the sun's heat during the day, and in temperate regions, to lose the heat at night—thereby helping to engender intense cold during the winter months. In warm areas such as North Africa the hot, dry climate is responsible for torrid days and cool nights. In hot, wet climates (tropics), there is relatively constant warm temperature with a high humidity and often with few or no air currents (i.e., doldrums). Obviously the chances for survival and dispersal of such parasites as larval hookworms outside the body, and free swimming miracidia and cercariae depend directly upon temperature and moisture. Cysts and spores and invertebrate hosts may also be killed by unsuitable temperature and moisture conditions.

MICRODISTRIBUTION

The phrase "distribution of animals" usually suggests a spread over geographic areas, but it should also connote, for parasites, the spread within or on one host. When one host organ is considered, or even one cell, the term **microdistribution** is particularly appropriate.

The intense pace of competition for space and food forces parasites into almost every kind of available host tissue. Once within the tissue or space, the invader "selects" the best possible location where freedom for feeding and reproduction is at its maximum, within the limits imposed by the metabolism of the parasite and by the physiologic responses of the host. The presence of other pioneer parasites, which have won a head

start in the race for host tissue or space, renders even more complex the chemical adjustments that have to be made. At any one time, from a phylogenetic point of view, the several species of parasites together occupying one host appear to have become adjusted to a state of reposed rivalry, each respecting the other's territory, but ever ready to take immediate aggressive advantage of any weakness. This teleologic simile suggests some form of communication among the contestants, but, on the contrary, each parasite builds a selfish empire for its own species alone, and blindly pushes the borders of its realm.

Although "communication" is an inappropriate term to apply to most parasites, they do respond to each other. Male hookworms in dogs migrate toward the females, especially when the females are near the duodenum and the males are nearer to the ileocecal valve. Apparently the females produce a messenger substance that travels in the direction of fecal flow and that acts as an attractant to the males.[15]

Among the most interesting examples of microdistribution are the lice on birds and on man. Most bird groups have five or six species of lice, often many more.[2] For example, on the Tinamidae (tinamou group in South America), 12 species of lice belonging to eight genera and three families have been recorded from one species of bird host (*Crypturellus obsoletus*), while 15 species of lice belonging to 12 genera and three families were recorded from another host (*Tinamus major*). There is a general correlation between size and shape of the lice and size of feathers. Lice on the smaller feathers of the head and neck, where they are out of reach of the bird bill, tend to be broad, with larger mandibles and head, while lice on the longer, broader feathers of the back and wings are flattened and elongate.

A delicate regulation in the microdistribution of parasites within one host body is constantly operative. Parasites, to be sure, often are adapted to one organ, or to a part of one organ, or to one kind of cell, but the metabolic balance is easily upset and one species of parasite may overrun its usual boundaries. For example, a nematode of the intestine of a vertebrate host may occasionally become so abundant in numbers that it spreads into the stomach, gallbladder, liver and coelom. *Leishmania donovani* normally invades large endothelial cells of blood vessels and lymphatics, as well as a few monocytes of the blood, but it may also parasitize erythrocytes in the liver, bone marrow and spleen, especially in young children in advanced stages of leishmaniasis. In such a situation, this usurpation of space often discourages other kinds of parasites, or even eliminates them altogether.

Distribution of parasites within a host is governed by the same basic forces that control distribution of the hosts. Temperature, moisture, mechanical barriers, chemistry of surrounding medium, food supplies and other ecologic factors are always operative, as well as phylogenetic relationships, which may determine the degree of host specificity. In the fish called the "mud-sucker" (*Gillichthys mirabilis*), two closely related nematodes live in the mesenteries, but one of these worms is also occasionally found in the intestine. A physiologic difference between the two worms must be the explanation, because both worms appear to possess equal opportunity and equipment for penetrating the bile duct or intestinal wall.

Distribution of many kinds of parasites in a vertebrate body is initially determined by the course of the circulatory system—a natural distributing network for food, oxygen, metabolic products and parasites. Larval hookworms, *Ascaris*, *Wuchereria* and others, are each carried to all parts of the body by the blood and lymph, and each species of parasite is finally delivered to an organ according to its predilection.

No sharp distinction exists between microdistribution and any other kind; hence, before we move to the wider aspects of zoogeographic distribution, we shall consider one group of parasites, the lice, as an illustration of the complex relations between parasites in or on a host, and the distribution of the host. The distribution of lice, in the overwhelming majority of instances, is governed by the phylogeny of the hosts. This situation is in contrast with that of fleas, in which ecologic factors are of paramount importance. On the human body there is a high correlation between the amount of hair and the rate of infestation by lice. Girls generally have more lice than do boys. Evidence for such statements comes from studies of hair from shaven heads among troops, in prisons and orphanages. The more clothing people wear the more lice they tend to possess. Living habits of people affect their lice populations. If men live close together in ships, tents or barracks, the spread of lice is fostered. The temperature and humidity of the space between man's skin and his clothing—that is, the living space for lice—remains remarkably constant in different countries and in different seasons of the year. Therefore, since man stabilizes the climatic conditions on his surface, a wide geographic distribution of his lice is to be expected. From dry climates in Sahara and Iraq to the constant humid equatorial conditions of Ceylon, Congo and Tahiti, to the temperate lands of Europe and America, man's lice are readily found. Local absence of human lice is generally due to social or to hygienic habits of the people.

HOST MIGRATIONS

Migratory animals provide us with unique opportunities for studying the effects of changes in external environments. Few studies have been made on the parasites of mammals and birds in relation to migrations of the hosts, but migratory fish and their parasites have attracted the attentions of numerous parasitologists.

The circumpolar distribution of *Neoechinorhynchus rutili*, an acanthocephalan parasite, provides incontestable evidence of practically continuous geographic distribution of one species of parasite in freshwater fish of two continents, North America and Eurasia.[19] The adaptations of *N. rutili* to fish hosts are so flexible that the worms are found in seven families in Europe and eight families in North America, and in the two continents the following host families are parasitized in common: Salmonidae, Cyprinidae, Esocidae, Gasterosteidae and Percidae. The only obvious way a parasite that occurs so often in fresh waters can become so widely scattered over the world is through the utilization of a great diversity of hosts representing a diversity of habitats. Of the 40 or more species of fish that have been listed in the literature as hosts for *N. rutili*, only four (*Esox lucius*, the pike, *Gasterosteus aculeatus*, the threespine stickleback, *Pungitius pungitius*, the ninespine stickleback, and *Salvelinus alpinus*, the Arctic charr) are common to the two continents. Of particular importance are the wandering hosts such as salmon, trout, stickleback and charr, whose migratory habits often involve passage between salt and fresh water.

Migrations of thousands of miles are common among many fish, (e.g., tuna, salmon, eel), but few detailed investigations have been made that tell us whether the fish keep the same parasites during the whole course of migration. The larval eel lives for three years as a marine pelagic fish, and, according to Dogiel,[4] is completely free from parasites during this period. Its feeding habits are a mystery—nothing having been found in the digestive tract at this stage in its life cycle. A postlarval, marine stage of one year's duration is followed by a period of migration up a river where the fish becomes a bottom feeder and gains parasites. The first freshwater parasites in the young eel (about 70 mm. long) are those not requiring an intermediate host (e.g., *Myxidium giardi*, *Trichodina*, *Gyrodactylus*, some trematode larvae and *Acanthocephalus anguillae*). In later life the parasitic fauna of female eels is of a freshwater variety, whereas that of the male fish includes both freshwater and marine species. During the last months of their life the eels are said to be free from parasites.

Parasites of salmon change as the fish migrate from fresh to salt water and back again. Parasites from young freshwater forms show almost no host specificity and are also to be found in other freshwater fish in the same locality. The American west coast salmon is parasitized by the fluke-vector (*Nanophyetus*) of the canine salmon poisoning disease. The causative agent is a rickettsia. Fish that are heavily parasitized in fresh water migrate to the sea, and when they return two or three years later to spawn, they are practically free from the flukes. But the salmon again acquire thousands of metacercariae within a few weeks after entering fresh water.

When fish migrate to a new environment and become isolated there, or when they are introduced by man into new regions, their original parasite faunas become reduced in numbers. To put this conclusion in more general terms, the process of acclimatization leads to impoverishment of an animal's original parasites. A few relics of the past often remain, but, given time, the hosts tend to acquire new species of parasites not found in the original habitat. A host generally has a larger variety of parasites, particularly parasites peculiar to it, in the habitat where it has lived the longest. For example, the freshwater fish, *Lota lota*, has numerous parasites characteristic of other freshwater fish, but *L. lota* is a member of the Gadidae, a family consisting almost exclusively of marine species, and it also has a number of marine parasites that are reminders of its past.

Birds also sometimes migrate for thousands of miles, and carry their parasites with them. The state and extent of parasitism is directly related to the physiology of the host, and the physiology of migrating birds changes during their migrations. For example, some birds spend their summers in Alaska and their winters in the South Pacific. Do these birds possess the same kinds and numbers of parasites at both locations? Probably not, but practically no studies of this nature have been made with birds.

Migratory mammals (except for man) and their parasites offer to the parasitologist an almost untouched field for important basic research. Migratory whales and porpoises may lose their helminth parasites when they reach different environments.[3] One wonders what happens to the parasites of arctic land mammals when the hosts migrate to more temperate climates during the winter.[1]

Human migrations, especially to and from the tropics, have provided us with a great deal of information on the spread of disease and on the nature of immune reactions. "The migrations of populations have contributed largely to the development of animal parasites in new localities. Evidence favors the view that yellow fever, dengue, estivo-autumnal malaria, broad fish tapeworm infection, the hookworm infection produced by *Necator americanus*, Manson's blood-fluke infection, Bancroft's and other types of filariasis and dracunculosis were brought to the Western Hemisphere by the white colonists and their slaves imported from Africa, as were typhus fever, leprosy, smallpox, measles, mumps, syphilis, frambesia, and probably influenza. . . . Wherever climate, necessary intermediate hosts, and customs of the population were favorable, these diseases became established in the new soil."[5]

Trypanosomiasis in man and animals has received some intensive study from the point of view of migrating hosts. It is well known that the elimination of an intermediate host leads to the disappearance of a parasite from any given territory. Mammalian trypanosomes of the *vivax*, *congolense* and *brucei* groups are normally restricted to a tropical zone of Africa coinciding with the area of distribution of the tsetse fly, their transport host. *Trypanosoma vivax* is a striking exception to the above rule. In Africa, this species is transmitted to cattle by mechanical contamination of the proboscis of tsetse flies, but it is also found in West Indies, South America and Mauritius where it was introduced in the last century with infected cattle. The non-African strains are morphologically indistinguishable from the African ones, but the former are transmitted by horse flies (Tabanidae) in which the parasites cannot develop, and they are transferred mechanically, as in tsetse flies, by the proboscis. This substitution of one vector for another has enabled the

parasite to become widely distributed to distant lands.

The disease known as surra in domestic animals is caused by *Trypanosoma evansi* which is phylogenetically related to *T. brucei*, but it occurs only outside the area of distribution of tsetse flies. Its range includes the Palaearctic, Ethiopian, Madagascar, oriental and neotropical zoogeographic regions. Surra is also transmitted mechanically by the contaminated proboscis of horse flies. There is evidence from laboratory cross infection experiments that *T. evansi* originated from *T. brucei* in Africa. The disease nagana, caused by *T. brucei*, could originally have been contracted by camels which were brought into the "tsetse belt" of Africa; then with a combination of transfer of mode of infection to the mouth parts of horse flies, and migrations of camels, the new disease, surra, could have been extended far beyond the geographic boundaries of its ancestral disease nagana.

There is apparently a direct correlation between the relative scarcity of parasites and the ability of the host to adapt itself to widely different environments, as indicated by studies on numbers of parasites in widely dispersed hosts in comparison with parasites in hosts confined to one or two comparatively small areas. A wide geographic distribution of a host may be partly possible because of a relatively high resistance to parasitism. Much more work on this and related problems must be done, however, before convincing evidence and clear conclusions can be obtained.

DISTRIBUTION OF ARTHROPODS BY COMMERCIAL VEHICLES

In prehistoric times, when man's means of transportation was confined to his legs, the carriage of food stuffs and other articles and the driving of livestock sufficed slowly to transport insects and other arthropods, as well as other human parasites, from one location to another. The introduction of carts and canoes greatly increased the opportunities for man unwittingly to carry insects away from their natural habitats. Even when dugout canoes afforded the only means of transoceanic transportation, the spread of such insects as mosquitoes among the South Pacific islands by Polynesian voyagers was greatly facilitated. Thus certain mosquitoes of the *scutellaris* group of *Aëdes* commonly breed in beached canoes, and their eggs are resistant to drying. The use of sailing vessels and, later, steamships provided a means for numerous insect introductions to countries all over the world. Flies and cockroaches breed in galleys and quarters, various beetles infest stored foods, and mosquitoes commonly breed in many kinds of containers holding water. As the result of such transportation, *Culex fatigans* has become cosmopolitan, and *Aëdes aegypti* virtually pantropical.

The development of aviation, however, has been the most alarming and serious encouragement of accidental insect introduction on a global scale. Among the insects most commonly carried by aircraft are: Diptera (mosquitoes and flies), Hemiptera (bugs), Lepidoptera (butterflies and moths), Coleoptera (beetles), and Hymenoptera (ants, bees, wasps). Other insects often to be found in planes are cockroaches, lacewings, earwigs and termites. The hazards associated with air transportation are relatively greater than those with sea transportation, not only because of the much shorter time required for flight, but also because of the character of the international airports, which are usually situated in rural or semi-rural districts. The docks for ships, on the other hand, are located most frequently in the heart of heavily built up urban areas—presenting a limited choice of mosquito larval habitats.[7]

The accidental importation of *Anopheles gambiae* into Brazil from West Africa, about 40 years ago, led to a disastrous outbreak of malaria, causing intense suffering with more than 300,000 cases of the disease and 16,000 deaths. The same mosquito was introduced into Upper Egypt during World War II, and it initiated a serious malaria epidemic involving 170,000 cases with 11,889 deaths during 1942 to 1944.[22] Vigorously prosecuted campaigns, at great cost, ultimately eradicated *A. gambiae* from both Brazil and Upper Egypt.

A serious threat facing Southeast Asia today is the possible introduction by international air transportation of yellow fever from Africa or South America. New Zealand and the multitude of tropical Pacific islands south of 20°12′ S and east of 170° E. lack anopheline mosquitoes altogether. Hence there is a possibility of the introduction of malaria and yellow fever by airplanes carrying mosquitoes.

Aspects of insect quarantine were reviewed by Lee,[8] who stated, "The concept which lately has been described as insect quarantine simply implies considerations of the prevention of entry of noxious insects into areas where they are not known to occur, whether such areas be geographically or politically limited. Initially of course the emphasis is on the prevention of such entry into countries whether they be continents, major geographical units of continents, or islands large or small. Despite this initial emphasis, problems of prevention of spread of noxious insects within geographically or politically limited areas also arise and are generally considered within the field of insect quarantine."

Most countries now include provisions in their quarantine laws to guard against the special health hazards associated with air transportation. The International Sanitary Convention for Aerial Navigation of 1933/44, and the World Health Organization have made recommendations urging recognition of the importance of implementing insect quarantines, and incorporating detailed advice concerning spraying equipment, insecticides and disinsection techniques. As yet, however, there is little international uniformity in the interpretation of existing recommendations. Aircraft disinsection is generally considered to be a safe, simple and speedy safeguard, but it should not be regarded as affording complete protection.

DISTRIBUTION WITHIN RESTRICTED AREAS

A comparison of the parasites of the coastal cod and the winter cod (two subspecies) in the White Sea discloses differences in the respective faunas that may be attributed to differences in habits and habitats. The coastal cod feeds on the bottom and is infected with the fluke, *Podocotyle atomon*, obtained from its crustacean food, and with the ciliate, *Trichodina cottidarum*. The fish is heavily infected with intestinal stages of the nematode, *Contracaecum aduncum*, but lightly infected with the trematode, *Hemiurus levinseni*. On the other hand, the winter cod, which feeds primarily on plankton, is only lightly infected with *Contracaecum*, but every fish harbors *Hemiurus*.

The same species of fish in different parts of the White Sea sometimes possesses different parasites, and these differences appear to be related to hydrologic factors. On the other hand, the same manner of life in distantly separated (taxonomically) fish may lead to the acquisition of the same parasites. For example, the flounder, *Pleuronectes flessus*, and the wolf fish, *Anarrhichas lupus*, both feed on the sea bottom on the same animals, and they have nine species of parasites in common, but many of these parasites are not specific for these fish.

Infection of the flounder, *Pseudopleuronectes americanus*, with trematodes is heavier in inshore waters than in offshore waters, and close to shore the infection is heavier in fish that are taken in deeper water adjacent to open sea than in fish taken near shoals. Larger flounders have heavier infections than do smaller fish, and there is an absence of marked seasonal variations in the former. In casting about for an explanation for the differences between inshore and offshore fish, one should remember that near the shore there are usually many more kinds of other animals and plants associated with the variety of shoreline habitats (see p. 503). Many of these other animals may serve as intermediate hosts for parasites.

Lake Mogilny, situated on the island of Kildin in the Barents Sea, has fresh or brackish water down to a depth of about 5 meters, but below that level the water becomes heavily contaminated with hydrogen sulfide. Codfish are found in the lake, and they and other marine animals can live only in the layers between 5 and 12 meters deep. These animals can be considered descendants from marine forms that lived there when the waters of the lake were in communication with the surrounding sea. An examination of the parasites of these relic cod shows that the parasitic fauna becomes impoverished as compared with that of the same species of hosts living under normal marine conditions. Parasites normal to cod include the fluke, *Echinorhynchus gadi*, and the copepods, *Caligus curtus*, *Clavella uncinata*, *Clavella brevicollis*, and *Lernaeocera branchialis*. These parasites are absent in cod from Lake Mogilny. Also absent are several myxosporidia commonly found in marine cod. When intermediate hosts are involved, as with *Lernaeocera*, the absence of the parasite in the lake can easily be explained, but for the others there is no such ready explanation.[4]

CONCOMITANT STUDY OF HOSTS AND PARASITES IN DIFFERENT PARTS OF THE WORLD

A number of writers have contributed to a general principle which may be stated as follows: The systematics and phylogenetic ages of hosts can often be determined directly from the systematics and degrees of organization of their permanent parasites, and, conversely, the systematics and ages of parasites may be determined directly from the phylogenetic and taxonomic relationships of their hosts (see Rules of Affinity, p. 547).

Probably the first scientist to use parasites as indicators of the relationships and geographic distribution of hosts was von Ihering,[21] who based his conclusions upon a study of helminths. The concomitant and comparative study of hosts and parasites in different parts of the world has been labeled as the "von Ihering method." Many years ago, the well-known English helminthologist, H. A. Baylis, pointed out that von Ihering's facts were both inadequate and inaccurate. Von Ihering thought, for example, that the occurrence of the nematode, *Dioctophyme renale*, in wild Canidae in Europe and South America necessarily indicated that it had existed in their Upper Miocene ancestors. Actually this parasite has been recorded from other carnivores, and from the horse, pig, orangutan and man. Baylis was doubtful of relying on the von Ihering method when it was applied to helminths because the habits (particularly as regards food) and environment of the hosts have played a far more important part in determining their helminth fauna than have their phylogenetic relationships.

We should remind ourselves that when the relationships of the parasites are confused (e.g., the biting lice), and when cases of recent acquisition, divergent evolution, convergent evolution and discontinuous distribution occur, it is impossible to use parasites as infallible guides to the origins of hosts. Baylis has aptly warned us that "although the attempt to draw conclusions as to the relationships of animals from their helminth parasites may sometimes yield interesting results, it is fraught with so many pitfalls that it should be made with greatest caution." Mayr[12] has also issued a warning to those who would place too much emphasis on the importance of using parasites as a guide to host phylogeny. He says:

> We are dealing here with something very basic, with the whole principle of phylogeny, with the principle of this study of parallel phylogeny and we must be awfully sure of these tools we use, that we do not misuse them, and we must, at all times, allow for an occasional transfer of parasites, and we must allow for different rates of evolution, and we must realize . . . that the comparative anatomy is something more reliable. Two birds can exchange their parasites, nothing prevents this, but I have not yet seen two birds exchanging their heads, their wings or their legs. These have come down from its ancestors and not from from another bird that nested in a hole right next to it!

Keeping the limitations of the von Ihering method in mind, let us now turn to a more detailed consideration of the use of parasites as indicators of the evolutionary relationships of their hosts.

Parasites as Clues to Host Affinities and Evolution

The von Ihering method has been used in the study of the frog family Leptodactylidae, which is characteristic of (1) tropical and semitropical America (e.g., Patagonia), and (2) Australia and Tasmania. These frogs have been reported from nowhere else in the world. This situation can be explained on the basis of the existence of an original land bridge across Antarctica, or on the basis of convergent evolution. But all of these frogs have, in their intestines, opalinid parasites of the genus *Zelleriella* composed of very similar species. The presence of similar parasites could also be explained on the basis of convergent or parallel evolution, but for *both* host and parasite so to evolve may be too much to expect. Hence the first explanation above (the existence of a land bridge) has gained questionable support.[13] The concept of a former union of the great land masses of the world and their subsequent breaking apart with "continental drift" has not received much scientific support during the past 35 years. Recently, however, the question has become a live issue with renewed evidence for large vertical and horizontal movements of the earth's crust.[11]

A comparison of the trematodes of Australian frogs with their close relatives in frogs in Europe, America and Asia might lead one to support the view that the frogs of Australia originated in a hypothetical Palaearctic center in geologic times. But the study of the opalinids of these frogs, as noted above, suggests that the hosts originated in South America.

The entozoa of opossums of America and those of Australian marsupials are quite different. Australian marsupials have 14 genera of sclerostomes (e.g., *Strongylus*) but rarely pinworms. American opossums have no sclerostomes, but a peculiar pinworm (*Cruzia*) occurs commonly. Both host groups have a primitive tapeworm belonging to a cosmopolitan family, but the South American species is the more primitive. Hence the Australian marsupials appear not to be as closely related to the American opossum as has been supposed. These speculations are not inconsistent with the evidence presented above for the existence of a land bridge over the Antarctic. Fossil records suggest that a bridge connecting Patagonia and the Australian Region must have lasted longer than that between Patagonia and the Palaearctic Region.

Geologic studies furnish strong evidence for the existence during the Tertiary period of a wide band of water extending across what is now the Near East, Mediterranean, mid-Atlantic and Gulf of Mexico. This wide band is called the Tethys Sea. Arms of the Sea extended southward to areas now occupied

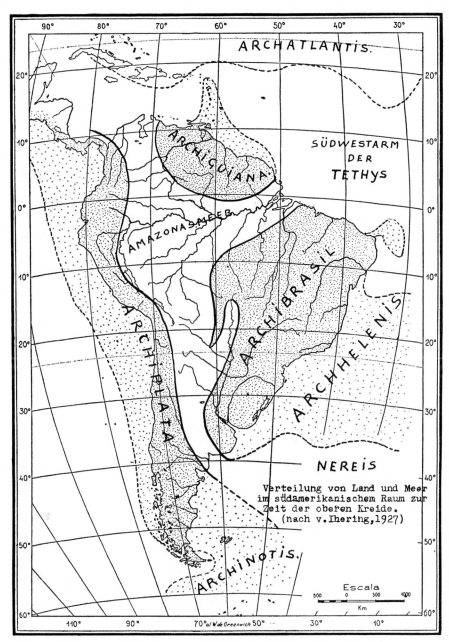

Fig. 23-1. Arms of the ancient Tethys Sea extending over the area now occupied by South America. (Szidat, Proc. XIV Internat. Congress on Zoology, Copenhagen)

by the Amazon and the La Plata-Parana-Paraguay river systems of South America (Fig. 23–1). Evidence for the former existence of these arms comes not only from geologic records, but from the flora and fauna of the present-day rivers and oceans. Among the animals that furnish this evidence are the trematodes and isopods that parasitize the fish. Lothar Szidat[16,17,18] of Argentina has made a detailed study of the zoogeographic implications of this whole area. Many of the freshwater fish parasites are characteristic of marine waters. Moreover, the nearest relatives of these freshwater parasites are sometimes to be found in the Caribbean and Mediterranean seas rather than in waters adjacent to the outlets of the rivers. Thus it is probable that, as the land masses were joined to form what we now call South America, and as the intervening waters changed from salt to fresh, due in part to the rising of the Andes and subsequent melting of snow, the marine organisms either died out or became adapted to a freshwater habitat. As a result of these changes, such fish as the Characinidae and Siluridae, now strictly freshwater inhabitants, possess parasites that changed more slowly than did their hosts (see p. 546), and that still exhibit marine features transported from their ancestral home. These parasites are relics of the Tethys Sea. Manter[10] pointed out that, of the 165 species of trematodes from marine fish of tropical Australia, 30 also occur elsewhere, and some of them have greatest affinities with those of the Caribbean. Dispersal must have occurred via the Tethys Sea millions of years ago.

The above considerations offer a ready explanation for the absence of freshwater cymothoid isopods on fish in the rivers of North America that flow into the Gulf of Mexico and in the rivers of Europe that flow into the Mediterranean—despite the fact that neighboring seas are rich in species of these parasitic isopods. Szidat's theory of evolution of certain digenetic trematodes from Tethys Sea fish is weakened by the fact that his fish hosts were all members of the family Anostomidae, whereas the Mediterranean hosts are the Mugilidae (mullets).

Many digenetic trematodes of marine fish are not widely distributed, but some significant geographic comparisons can be made.[9] For example, relatively strong similarities among trematodes occur between (1) the European Atlantic and the Mediterranean; (2) shallow waters at Tortugas, Florida and Bermuda; (3) shallow waters at Tortugas and the tropical American Pacific. Considerable dissimilarity occurs between (1) deeper waters and shallow waters at Tortugas; (2) shallow waters at Tortugas and the North Atlantic; (3) the Mediterranean and the Red Sea.

The nematodes offer poor material for discovering possible host affinities because of the existence of almost all possible gradations from free-living species to obligate parasites, and from strict host specificity to a very wide range of host tolerance, and because of the inadequacy of morphologic criteria to distinguish the species. Careful physiologic studies of nematode parasites are greatly needed. Evidence for the phylogenetic relationship between a South American bird, *Cariama cristata*, and the Eurasian bustard has been presented on the basis of an analysis of their helminths. Both birds are parasitized by the nematodes, *Subulura allodapa* and *S. suctoria*. In addition, both birds harbor species of the cestode genera *Chapmania* and *Idiogenes*. Physiologic segregation of tapeworms evidently occurred when vertebrates first split into present-day groups, resulting in a phylogenetic specificity. Hence we may conclude (a conclusion corroborated by taxonomic studies on birds) that *Cariama cristata* and the Eurasian bustard are related.

The effects of isolation of ectoparasites are more often reflected in morphologic changes than are those of endoparasites. The Mallophaga, or biting lice, exhibit a high degree of host specificity, and they may be used to elicit evidence of parallel evolution of hosts and parasites. Moreover, the lice probably were present on ancestral hosts, and evolved at a slower rate than did their hosts. Consequently, phylogenetic relationships of the hosts may be indicated by a comparison of their biting lice.

Distribution is governed largely by climatic and geographic factors in most groups of insects, but the biting lice generally spread over an enormous range in area and in climate, and geography is relatively unimportant. These lice thus are ideal parasites with which to estimate relationships among birds, because as the hosts have evolved, the lice have evolved, but at a slower rate. For example, there was long disagreement among ornithologists as to the systematic position of flamingos; some authorities placed them in the Ciconiiformes (storks) and others in the Anseriformes (ducks and geese). A study of flamingo Mallophaga shows that the lice correspond very closely to those of ducks and geese. Not one of the parasites suggests any close affinity between flamingos and storks. The Mallophaga of the three North African pelicans are not alike, and the lice of one of them (the Ethiopian pelican) are more like those on the Australian pelican and the South American pelican than those on the other two North African birds. Morphologically the lice of South American pelicans are sufficiently distinct as to be placed in a separate genus, and their hosts are placed in a separate subgenus. We have here a fine example of the effects of ecologic segregation of both hosts and their parasites.

A study of the Anopleura, or sucking lice, also presents interesting illustrations of the results of ecologic segregation. From information on present-day distribution of these lice we may conclude that they became isolated on their hosts long ago; yet, we find that the Australian marsupials do not possess sucking lice—a fact indicating that these hosts had already become isolated before the Anopleura arose as parasites.

Those mammals that are closely related to one another tend to have closely related or identical lice. The ground squirrels (*Citellus*) of North America are related, but they are different from those of Siberia. The lice, however, on these two groups of geographi-

cally separated squirrels appear to be identical. Because of the high degree of host specificity exhibited by lice, one can examine an unknown louse and tell, with little risk of error, from what kind of vertebrate it was taken, but unless the range of distribution of the group of hosts is known, one cannot tell from what part of the world the louse came. A conspicuous exception to this rule is found among the family Gyropidae, which occurs on a wide variety of hosts in South America but is not found elsewhere.

The distribution of parasites within the order Marsupialia conforms almost perfectly with the geographic distribution and antiquity of the hosts. For example, the biting lice of marsupials belong to the most primitive division of the Mallophaga, one family (Boopidae) infesting Australian hosts, and another (Trimenoponidae) infesting South American hosts.

The ostrich and the rhea both have sclerostome nematodes, indicating at least some measure of relationship between the two hosts. But sclerostomes occur in many kinds of grass-eating animals, such as horses, elephants, Australian marsupials, rhinoceroses, tapirs, and even in South American tortoises. Ten nematodes have been reported from the rhea, and five from the ostrich, but no species in common. Both birds are parasitized by one genus of louse (*Struthiolipeurus*), but rheas have, in addition, a second louse genus (*Meinertzhageniella*). Both birds have the mite, *Pterolichus bicaudatus*, the only species of parasite in common. The arthropod distribution and similarity weakly support evidence from feather structure and other anatomic features that the two orders of birds are related and not the relics of independent, unrelated stocks.

Space does not permit a review of all groups of mammals, but in general, before clear-cut, detailed lines of phylogenetic relationships can be established, much more research, particularly in the nature of careful comparisons of abundant collections of parasites as well as fossil mammals, must be made. The precise classification of parasites must be the basis of comparisons. For details of this problem, see especially Hopkins,[6] Vanzolini and Guimarães,[20] and Patterson.[14]

GENERAL RULES AND PRINCIPLES

1. Both the host and its environment determine the distribution of parasites.

2. Widely separated hosts may have the same species of parasites.

3. Distribution-pattern has one meaning for parasites, another for hosts.

4. The classification of the parasite is often of assistance when we presume to offer advice on obscure points about host classification.

5. When unrelated hosts live together and eat the same food they may possess some parasites in common.

6. The size and ecologic differentiation of the area in which the host lives is directly correlated with the diversification and distribution of parasites of that host. In general, the more diverse the environment, the more kinds of species of parasites exist in a given host.

7. In all questions concerning distribution of hosts and their parasites, the food factor of the host is of primary importance.

8. Parasite-host data may be used to suggest (a) genetic (phylogenetic) relationships among hosts, (b) places of origin and routes of dispersal of both hosts and parasites, and (c) ancient land connections between present and widely separated land masses.

9. Hosts that migrate for long distances tend to lose at least some of their parasites during the course of migration or soon after they arrive at the new location.

10. The process of acclimatization to a new geographic area leads to impoverishment of the host's original parasites, but a few relics of the past often remain.

11. Clues to the systematics and phylogenetic ages of hosts can often be obtained directly from the systematics and degrees of organization of their permanent parasites, and, conversely, clues to the systematics and ages of parasites may be obtained directly from the phylogenetic and taxonomic relationships of their hosts.

12. When a host is widely distributed (e.g., the clam, *Mytilus*, or the fish *Gasterosteus aculateus*) some of its parasites, such as protozoans, may accompany the host everywhere it wanders, but when intermediate hosts are involved, the distribution of the parasite may be closely restricted simply because the intermediate host has a narrow geographic range.

13. Never say that a species does not occur someplace simply because it has not (yet) been found there.

REFERENCES

1. Cameron, T. W. M.: Parasitology and the arctic. Trans. Roy. Soc. Canada, *51*, (ser. III): 1–10, 1957.
2. Clay, T.: Some problems in the evolution of a group of ectoparasites. Evolution, *3*:279–299, 1949.
3. Delyamure, S. L.: The Helminth Fauna of Marine Mammals in the Light of Their Ecology and Phylogeny. Moscow, Izdatelstvo Akademii Nauk, SSSR, 1955. (In Russian.)
4. Dogiel, V. A.: Problems of Ecological Parasitology. Ann. Leningrad State Univ. No. 7, 1–194, 1936. (In Russian.)
5. Faust, E. C., and Russell, P. F.: Clinical Parasitology. Philadelphia, Lea & Febiger, 1970.
6. Hopkins, G. H. E.: The host-associations of the lice of mammals. Proc. Zool. Soc. London, *119*:387–604, 1949.
7. Laird, M.: Insect introduction hazards affecting Singapore and neighbouring territories. Med. J. Malaya, *11*:40–62, 1956.
8. Lee, D. J.: The problems of insect quarantine. Proc. Linn. Soc. N.S.W., *76*:6–19, 1951.
9. Manter, H.: The zoogeography of trematodes of marine fishes. Exptl. Parasitol., *4*:62–86, 1955.
10. Manter, H. W.: Some aspects of the geographical distribution of parasites. J. Parasitol., *53*:1–9, 1967.
11. Maxwell, J. C.: Continental drift and a dynamic earth. Amer. Sci., *56*:35–51, 1968.
12. Mayr, E.: Evolutionary aspects of host specificity among parasites of vertebrates. *In* First Symposium on Host Specificity Among Parasites of Vertebrates. pp. 5–14. Inst. Zool., Univ. Neuchâtel, 1957.
13. Metcalf, M. M.: Parasites and the aid they give in problems of taxonomy, geographical distribution, and paleography. Smithsonian Misc. Collections 81, No. 8, 1929.
14. Patterson, B.: Mammalian phylogeny. *In* First Symposium on Host Specificity Among

Parasites of Vertebrates. pp. 15–49. Inst. Zool., Univ. Neuchâtel, 1957.

15. Roche, M.: Influence of male and female *Ancylostoma caninum* on each other's distribution in the intestine of the dog. Exper. Parasitol., *19*:327–331, 1966.

16. Szidat, L.: Beiträge zur Kenntnis der Reliktfauna des La Plata-Strom-systems. Arch. f. Hydrobiol., *51*:209–260, 1955.

17. ———: Über die Parasitenfauna von *Percichthys trucha* (Cuv. & Val.) Girard der Patagonischen Gewässer und die Beziehungen des Wirtsfisches und seiner Parasiten zur Paläarktischen Region. Arch. f. Hydrobiol., *51*:542–577, 1956.

18. ———: Versuch einer Zoogeographie des Süd-Atlantik mit Hilfe von Leitparasiten der Meeresfische. Parasitol. Schrift. Jena. No. 13, 98 pp. (Spanish summary.)

19. Van Cleave, H., and Lynch, J. E.: The circumpolar distribution of *Neoechinorhynchus rutili*, an acanthocephalan parasite of fresh water fishes. Trans. Amer. Micr. Soc., *69*: 156–171, 1950.

20. Vanzolini, P. E., and Guimarães, L. R.: Lice and the history of South American land mammals. Rev. Brasil. Ent., *3*:13–46, 1955.

21. von Ihering, H.: On the ancient relations between New Zealand and South America. Trans. Proc. New Z. Inst., *24*:431–445, 1891.

22. W. H. O.: Control of insect vectors in international air traffic. A survey of existing legislation. Int. Dig. Hlth. Legis., *6*:377–435, 1955.

Chapter 24

Parasite-Host Specificity

INTRODUCTION

Why are some parasites restricted to one species of host, whereas others flourish in a wide range of unrelated hosts? Why are some kinds of animals burdened with numerous parasites, while other kinds possess few parasites, if any? Seventeen species of fleas are found on swallows or martins and nowhere else, but swifts with similar nesting and feeding habits do not have a single flea restricted to themselves. Although most birds with large nests have many fleas, swan nests are free from these insects. The nematode, *Wuchereria bancrofti*, lives as an adult only in man, but *Trichostrongylus axei*, contrary to nematodes in general, is at home in a wide range and diversity of hosts. This species has been reported from the digestive tracts of the horse, ass, mule, sheep, cattle, goat, several species of wild ruminants, pig, and man, and experimentally in rabbits, hamsters and guinea pigs. Why do these differences exist?

The term "specificity," when applied to an animal or plant, refers to the things that make it distinct from other organisms. "Host specificity" refers to the restriction of a parasite to one or more host species. A high degree of host specificity may mean that the adaptations between host and parasite are so delicate and intermeshed that the parasite is unable successfully to survive in or on the body of another species of host. It may also mean, however, that external accidental, physical or environmental factors are responsible, and when these factors are removed the parasite is able to live in another host. The central problems in the study of specificity of parasite-host relationships concern mecha-

nisms that limit host selection and that control the ability of the parasite to invade the host and to survive within it.[51]

No parasite lacks host specificity. Obviously there are no parasites living in all kinds of vertebrate or invertebrate hosts. The term **monoxenous** indicates limitation to a single host, as occurs with adult *Wuchereria bancrofti*. **Oligoxenous** is used to describe parasites, like adult *Echinococcus granulosus*, which have a small host range. **Polyxenous** refers to the condition involving many suitable hosts, or relatively little host specificity, as occurs with *Fasciola hepatica*.

Problems of host specificity involve physiologic factors, and they can best be solved by the use of experimental methods in life history studies.

Strict host specificity exists between many parasite species and their hosts. Numerous flagellates are found only in the hindgut of termites. Many species of ciliates occur only in the cecum of horses. Diclidophoroidean trematode parasites on the gills of certain fish appear to be entirely specific to their particular species of host. Moreover, there is a definite site preference; *Diclidophora merlangi*, for example, occurs most frequently on the first gill arch of the cod, *Gadus merlangus*.[24] This exacting topographic relationship between parasite and host might be an important factor in the mechanism of host specificity, although it may be the result of variations in the flow of water over the different gills rather than of a choice exercised by the parasite.

Since many parasites employ more than one kind of host in their life cycles, the conditions determining the degree of host specificity are often markedly different in the several stages of a cycle. Indeed, the metabolic requirements of a parasite generally vary with its developmental changes, even within one host. In a study of the action of antimalarial drugs in mosquitoes infected with *Plasmodium gallinaceum* from chicks, the first indication of interference with parasite development usually occurred in the oocyst, but with one of the drugs used, the effect was first observed on the sporozoites.[52] The different physiologic demands of this parasite at particular stages in its development are demonstrated by marked changes in phosphatases and in nucleic acids in the parasite of the mosquito. Such studies again emphasize the necessity of using biochemical tools to disclose the basic nature of parasite-host specificity.

Whereas two strains of a given parasite might infect a final host with equal facility, they may not be equally infective to a given strain of intermediate host. In other words, a strain endemic in one area might be physiologically distinct from a strain endemic in another area, and this difference might be detected only on the basis of infectivity to an intermediate host. It is obvious, therefore, that in any consideration of specificity we must not confine our attentions to parasites, but must study their hosts with equal zeal. The host in many respects is equivalent to the environment surrounding free-living organisms, and the host, as an environment, is constantly developing specific responses and adaptations to its parasites (see Chapter 21).

ISOLATION OF PARASITE POPULATIONS

Before embarking on a more detailed discussion of the kinds, degrees and significances of host specificity, let us examine briefly some factors that are responsible for the isolation of parasite populations. These factors are basically of a genetic, biochemical, ecologic or physical nature.

1. When a species of host is divided into two or more population groups separated geographically in different environments, their respective parasite faunas are also different. This fact is a further indication of the influence of the environment of the host on its parasite-mix. Given sufficient time, a host population may become divided into non-interbreeding units, thereby forming new species of hosts and, consequently, changing the character of the parasites. Examples may be found in the abundant speciation among mites on isolated groups of lizards or on bats, which occupy very specialized biotopes. Pterygosomid mites (e.g., *Pterygosoma aculeatum* beneath the scales of lizards) undoubtedly originated on ancestral lizards before the hosts became segregated onto different continents.

A puzzling example of an obscure ecologic barrier is shown by the intestinal coccidian, *Eimeria mohavensis*, in the kangaroo rat *Dipodomys panamintinus*, and in *D. merriami*.[8] Both rats occupy the same geographic areas, and food contamination between the two seems to be inevitable. The two species of hosts presumably represent nonbreeding units originally derived from a common stock species. *D. panamintinus* is normally infected with the coccidian, but an examination of 200 *D. merriami* failed to reveal a single infected animal. Cross infection experiments in the laboratory, however, resulted in 90 per cent infection of *D. merriami* from the other rat. Obviously the ecologic barrier in nature, in this case, is not self-evident. Numerous experiments of a similar type have demonstrated a laboratory compatibility between certain other protozoan parasites (e.g., trypanosomes, amebas, trichomonad flagellates) and "unnatural" hosts. The effect of stress on the host and its possible relation to lowered resistance needs careful study. The stress factor might help to explain the results noted above.

2. A parasite may become transferred to a "foreign" host living in the same locality as the original host, and subsequently become isolated due to geographic separation of the two hosts. This type of ecologic segregation results from an association between two kinds of hosts close enough to permit their respective parasite populations to mingle before the host groups become isolated. The isolation

may, of course, be temporary. After isolation of a parasite upon a host takes place, both morphologic and physiologic changes may occur—given time and the operation of natural selection. If the separated host populations become reunited, the latter are sympatric species with the original ancestral parasites. The hosts may also exhibit secondary infections or infestations acquired during the period of separation.

Fleas from rabbits on Coronados Isles in the Gulf of California have become established on auklets, which are burrow-nesting birds. Likewise, puffins and shearwaters from the west coast of Britain have rabbit fleas on their bodies. On the Kerguelen Isles and Antipodes in the southern Atlantic the diving petrel, a gull, and a burrow-nesting parakeet have acquired a species of flea belonging to a group of marsupial fleas common to Australia.

3. The parasite may be unable to develop in any other host. This kind of isolation may be accomplished by one or more of the following six situations:

(1) *Absence of specific environmental conditions necessary for the growth and development of the parasite within other hosts.* This statement implies such environmental factors as food, oxygen, temperature, osmotic pressure, and water.

(2) *Resistance of the host.* Under this heading are included all activities by which the host defends itself against the presence of the parasite—immune reactions, age resistance, mechanical barriers, and others. Factors conditioning susceptibility and natural immunity of hosts are still partly obscure, but some of these are indicated below. For a discussion of immunity, see Chapter 21.

Ecologic aspects of parasitism can be studied effectively by introducing parasites experimentally into foreign hosts and then analyzing the reasons for failure or success in establishing a permanent parasite-host relationship. But the chief difficulty here is to establish adequate criteria for determining "success" of the parasite in the new host. Such criteria as volume of egg production, parasite size and number are extremely

variable. The student should remember that a parasite and its host represent a biologic system dependent for its maintenance on ecologic factors provided by *both* members of the association. If plerocercoid larvae of *Dibothriocephalus latus* are introduced into the stomachs of the lamprey, frog, toad, snake, and lizard, the typical larval migration takes place, although none of these animals is a normal host.[36] The experimental hosts are thus potential normal hosts for the fish tapeworm, and the absence of the worm as a natural parasite in these hosts is presumably due simply to the fact that fish are not a part of their natural diet. Morphologic features may prevent normal development of a parasite, as is shown by the fact that plerocercoids introduced into adult terrapins cannot migrate because of the rigidity of the stomach wall, whereas in the young turtle, the larvae migrate normally. Temperature alone may prevent or permit infection. Plerocercoids in the stomach of *Gymnodactylus* (a lizard) migrate if the experimental host is kept at room temperature, but not if the temperature rises to 37° C.

(3) *Inability of the parasite to enter the host.* Sometimes larvae are able experimentally to live and grow within a host, but the larvae are not equipped to penetrate the external surface. A lack of suitable means of transmission may also give an indication of narrow host specificity. *Trypanosoma equiperdum* in horses and mules is transferred normally only by sexual contact, but hypodermic injections of the parasites into the blood of laboratory rats, mice, rabbits and guinea pigs easily produced infections in these "foreign" hosts.

(4) *Presence of other parasites.* Whereas a given host may be infected with many species of parasites, only one or two of these species may be in a host body at the same time. At least 35 kinds of larval trematodes (21 strigeids, six plagiorchids, two schistosomes, one echinostome, one monostome, and four others) may be found in the freshwater snail, *Stagnicola emarginata*, but usually only two kinds (sometimes three or four) occur simultaneously in any one snail. Eleven species of

protozoa have been found in the marine fish, *Gadus merlangus*, but only two or three are generally observed in any single host. A related phenomenon is shown by the strigeid trematode cercariae (*Cotylurus flabelliformis*), which encyst in the same snail species that harbor the sporocysts. If sporocysts of *C. flabelliformis* have already lodged in the snail, the cercariae do not enter. If, however, sporocysts of other trematode species are present, their very presence appears to favor the penetration and encystment of cercariae of *C. flabelliformis*. The nature of these phenomena is not clearly understood, but it probably involves the entire defense mechanism of the host. A previous infection can affect the invasiveness of a parasite. (See the discussion of parasitocoenosis, p. 494.)

(5) *Resistance of the parasite.* The above four factors alone might suggest that the parasite would be successful if the host were willing, but the situation may be the reverse. For example, the parasite may simply be too large for the prospective host, or a physiologic incompatability may be due to biochemical processes in the parasite rather than in the host.

(6) *Genetic mutations.* Here we are apt to enter highly controversial territory, where changes claimed by some to be mutational are vigorously explained by others solely on the basis of selective adaptation. Only one example is mentioned, but others will be found on succeeding pages devoted to comparative host specificity. Probably the most celebrated reputed change in host specificity explainable as a mutation is that of the flagellate, *Trypanosoma rhodesiense*, which appeared rather suddenly in Rhodesia in 1909, supposedly arising from the morphologically identical *T. brucei* of domestic animals. Since then, the parasite has retained this infectivity for man as probably the only difference from *T. brucei*, which still stubbornly, and fortunately, refuses to parasitize human beings. One cannot, of course, be sure that this apparently sudden appearance of a new trypanosome of man was not the result of obscure epidemiologic factors, rather than of a genetic

mutation, for little was known about trypanosome epidemiology in the interior of Rhodesia a half century ago. Nevertheless, regardless of whether *T. rhodesiense* is a recent or an older species, the evidence seems strong that this example is an instance of a change in parasite-host specificity due to a mutation.

KINDS OF PARASITE-HOST SPECIFICITY

Host specificity is a function of physiologic specialization and evolutionary age. These two factors may not, of course, be mutually exclusive, but usually the older (phylogenetically) the parasite, the more specialized it becomes, and the more specialized its host becomes. This gradual increase in degree of specialization means that host and parasite become better and better adapted to each other and the parasite is less able to change physiologically (i.e., to mutate) enough to survive in a different kind of host. To put this generalization the other way around—if we select genera of parasites that are strictly confined to groups of hosts known to be closely related, we find that not only are the hosts usually specialized, but that the parasites are highly specialized. The student should remember, however, that as more intensive research discloses the existence of larger numbers of parasites, species once considered strictly host specific are often found not to be so.

We may divide parasite-host specificity into two broad categories. The first may be called **ecologic specificity,** wherein the parasite is capable of living in a foreign host but normally never reaches one because of an ecologic barrier. The second category may be called **physiologic specificity,** wherein a parasite is physiologically (genetically) compatible with its "normal" host, and is incapable of surviving in a foreign host because of physiologic incompatibility. When physiologic specificity involves a pattern of behavior, it may be called **ethologic specificity.** (For examples of the latter, see the discussion p. 538 under "The Search for a Partner.") Other examples are readily found among the digenetic trematodes. A convincing experiment demonstrating physiologic specificity was the one in which two common biting lice from fowl were reared in vitro.[4] Both *Lipeurus heterographus* and *Eomenacanthus stramineus* developed successfully under appropriate conditions of moisture and temperature, with chicken feathers and dried blood added for food. When feathers from the little green heron were substituted for chicken feathers, the *Lipeurus heterographus* lice died. This experiment indicated the presence of some chemical factor in feathers that is characteristic for a given group of birds and to which certain parasites are adversely sensitive. It is possible, however, that the heron feathers *lacked* something essential.

Physiologic compatibility often means that the parasite and its host have evolved together. This association over a period of many millions of years results in a **phylogenetic specificity,** which is often used to help solve problems of host taxonomy (see Chapter 25). Two related hosts, however, may possess closely related or identical parasites not because of the genetic relation between the hosts, but simply because the hosts have been feeding on the same food, which included larval stages and intermediate hosts of their common parasites. We should be reminded of the usual genetic variations in all species of parasites and hosts. Thus an individual member of host *A* might vary physiologically enough from the average to resist a normal parasite of *A*, and this individual variant might be so different from its siblings that it chemically resembles host species *B* enough to tolerate a parasite of host *B*. Such an extreme variant is not likely, but its possibility should not be overlooked.

Even when host specificity is phylogenetic, behavior may be important. "Thus the segments of *Taenia saginata* and *Taenia solium* in the faeces of man have a behaviour suited to the feeding habits of their respective hosts. The segments of *Taenia saginata* are active and move onto herbage where they are most likely to be eaten by cattle. On the other hand, the segments of *Taenia solium* are flaccid and remain in the faecal mass where they are more likely to be eaten by a pig."[34]

SPECIFICITY FACTORS RELATED TO INFECTION AND GROWTH

Under this heading we shall present some factors and processes that may affect host specificity, and that occur during each of three major stages, in turn, in the host-parasite association. This material is summarized from Rogers.[46]

The First Stage of the Association of the Parasite With the Host. The infective stage (spore, egg, larva) "is a 'resting' stage which requires factors from the host in order to resume development." The nature of these factors and their presence or absence often determine the degree of specificity before resumption of development commences. Those parasites with a direct life cycle (e.g., *Ascaris*) are generally more specific than those that employ intermediate hosts, and ecologic factors initially determine the range of hosts. A relatively high concentration of undissociated carbonic acid plus dissolved gaseous carbon dioxide appear to be essential for the exsheathment of larvae of the nematode, *Haemonchus contortus*. These and other requirements occur only in the rumen of ruminants, thus limiting the worm to these hosts (see p. 301 for details on the hatching of *Ascaris* eggs). The initial requirements for infection with other worms, however (e.g., *Ascaris lumbricoides*, *Toxocara mystax*, *Trichostrongylus axei*), may ensure that the early stages of development of the parasitic stages should take place in the appropriate host organ, but they are apparently not specific enough to limit the parasite to one species of host.[45] It has been suggested, without experimental evidence, that infective eggs or molting larvae require substances from the host for the initiation of parasite development.

The hatching factor (eclepic acid) excreted by roots of plants is essential for further development of the nematode plant parasite, *Heterodera rostochiensis*.[11] When hatching or excystment is induced by direct action of the host (as the action of digestive enzymes on cysts of larval tapeworms, or on metacercariae of *Opisthorchis sinensis*), host specificity is generally low.

Thus "the ecology and behaviour of the infective stage and of the host and the conditions that are necessary in the host for infection to take place affect specificity during the first stage of the association."

The Second Stage of the Association of the Parasite With the Host. The infective stage does not feed or grow, but as soon as infection has occurred the invading organism is truly parasitic, and must find necessary nutrients and be able to withstand any damaging physical or chemical factors in its environment. It is at this stage that the complex parasite-host relationships are often most "crucial in determining the range of hosts." For example, some tapeworms of elasmobranchs require the urea of the host gut fluid to help maintain normal osmotic function.[43] Dietary needs of many parasites (e.g., cestodes) are not found in the environments of free-living animals.[41]

The hormones of some hosts also probably influence their susceptibility to parasites (see p. 470). Natural or acquired resistance of a host (Chapter 21) obviously help to determine its receptiveness for a parasite. Rogers, however, has reminded us that, "as a rule we do not even know if the failure of a parasite to grow *in vivo* is due to lack of nutrients or to unfavorable chemical and physical features of the environment."

Young hosts often are more susceptible than are older hosts, but "in most hosts the development of age resistance, if it occurs, is a gradual process which does not become complete. It is reasonable to suppose, however, that the unfavourable features in the host that give rise to age resistance may often be similar to those which make an organism unsuitable as a host at any period during its life and so affect the range of hosts of parasites."

The Third Stage of the Association of the Parasite With the Host. The period of parasite reproduction is essential not only for maintenance of the species, but also for the production of infective agents. Host specifi-

city during this period may involve natural or acquired resistance, or it may be related to nutritional requirements of the parasite as well as to mechanical factors and changes in temperature. Such requirements may be greater during reproduction. Obviously, the cysts, spores, eggs or larvae of parasites must reach an environment where they can continue their development. The relationships between specificity and parasite reproduction are not well understood.

Each of the three stages of association described above involve some parasites of the vertebrate intestinal canal. Secretions that play an important role in the physiology of the intestine, such as bile and pancreatic juice, must also be involved in the success of parasites to begin and maintain their parasitic existence. In a study of the biochemistry of bile as a factor in determining host specificity, Smyth and Haslewood[49] emphasized that the chemical composition of bile varies from species to species, and they suggested "that specific bile salts may act as selective biochemical agents in host specificity by (a) stimulating eggs, cyst or spore hatching; (b) lysing or otherwise eliminating parasites in unsuitable hosts; (c) stimulating the metabolism in suitable hosts."

COMPARATIVE HOST SPECIFICITY

Now let us turn to the various groups of parasites, and inquire into the variations of host specificity which they display. Examples will be selected, more or less at random, to indicate the differences among these groups. Keeping in mind the many exceptions and many pitfalls when making generalizations, we can say that ectoparasites and their mammalian hosts display a rather consistent specificity due to parallel evolution. Such a conclusion is also true for monogenetic trematodes and cestodes of fish. In birds the problem is more difficult and unsettled, and there is much difference of opinion among ornithologists on bird taxonomy. Parasitologic evidence, however, especially from bird lice, is now arousing the interest of the ornithologists. We can also state that parasites with free-swimming stages able to enter the host skin tend to have a narrower range of host species than do those parasites that enter the host through the mouth. Also, in the more plastic groups of parasites, such as the copepods, the aquatic environment appears to be more favorable for the preservation of varieties than does the terrestrial environment. As a generalization, there is less host specificity when there are two intermediate hosts than when only one is employed. For example, the Pseudophyllidea (tapeworms) with two intermediate hosts (crustacea, fish) are less host-specific than are the Cyclophyllidea, with a single intermediate host.

Protozoa

The genetic complex of a parasite and of its host is obviously the basis for biochemical and physiologic patterns in determining the character of parasite-host specificity. The biochemical pattern embraces the production of metabolites that play a decisive role in parasite-host adjustments.

One kind of experimental manifestation of host specificity involves biochemical adjustments that favor adaptations to new hosts. If we select a sheep strain of *Trypanosoma vivax* that does not normally infect white rats, a small amount of sheep serum, when added to the trypanosome injection into rats, provides a congenial environment in which the parasites can develop.[7] After a period of adjustment, the parasites become adapted to living in the white rats without the aid of sheep serum, and, in fact, they produce a virulent infection that can be carried indefinitely from rat to rat by mechanical blood injections. If these parasites are transferred back into sheep, they lose their ability to infect white rats, but can regain this ability if again they are injected, with sheep serum, into rats. This evidence indicates an adaptation to a new environment rather than a mutational change.

Genetic changes may affect host specificity not only through the parasite but through its host. Very little is known about the genetics of vertebrate hosts as this factor affects their susceptibility to parasites. In mice the degree of resistance to *Plasmodium berghei* seems to be controlled through the agency of multiple genes. Several workers,[14] however, have been able to change the resistance of mosquito hosts to *Plasmodium* by selective breeding. To cite but one example, the infection rate of *Culex pipiens* for *P. elongatum* has been increased from 5 to 50 per cent in six generations by breeding from infected female mosquitoes.[32]

As we have already seen, cross infection experiments in the laboratory may result in the establishment of parasites in hosts never involved in nature. In a series of experiments, McGhee[27-31] found that, in a suitable red cell environment, the avian *Plasmodium lophurae* will invade and carry on at least part of its life cycle in the red cells of mice, rats, pigs, rabbits and man. Obviously, under these experimental conditions, some barriers to infection are removed, but the whole biochemical explanation must await further studies of this nature. In general, however, cells that have the highest potassium content are more susceptible to invasion by the "foreign" *Plasmodium*, and red cells from young mice or young rats are more easily infected than are cells from older animals. Infections may also be established in the reverse direction. Erythrocytes of duck and goose embryos can be infected with *P. berghei* from rodents.

Malarial parasites reach their highest degree of specialization in birds, and they probably originated among the reptiles during the Mesozoic period. A single species of bird may harbor several kinds of bird malarias, and avian *Plasmodium* species possess a wide latitude of host possibilities. *P. circumflexum*, however, appears to be restricted largely to robins. Among early mammals, malarial parasites were able to survive only in that mammalian stem leading to the modern primates. The malarial parasites of man are rather rigidly host-specific, although some of them are infective to anthropoid apes which themselves may harbor species of *Plasmodium* indistinguishable morphologically from those in man. The intracellular Sporozoa (e.g., *Plasmodium*, coccidia) tend, in general, to be much more host-specific than are those parasites (e.g., blood inhabiting flagellates, bile inhabiting Myxosporidia, intestinal amebas) that inhabit cavities and blood of their hosts. Manwell[26] concluded a review of parasite-host specificity, with special emphasis on blood protozoa, with this statement: "Thus it may be seen that, little as we know about the delicate interplay of factors making for mutual tolerance (or intolerance) of blood protozoa and their vertebrate hosts, we know even less about such relationships between the former and their vectors."

Each time a woodroach molts, most of its flagellates undergo sexual reproduction. Then follows the usual long process of fission (asexual reproduction) during the growth period of the host, and when it again molts its hormones induce the flagellates to reproduce sexually. On the other hand, the flagellate genera inhabiting the intestines of termites, *Zootermopsis*, e.g. do not undergo a sexual cycle, and the hosts must refaunate themselves after each molt. All protozoa from both roaches and termites can be removed from the insects without injury to the latter by increasing the oxygen tension. These conditions provide a prime opportunity to test the host-specific behavior of the parasites. Termite protozoa can live in and support the growth of nymphal and adult roaches until they molt (up to 221 days for nymphs), but roach protozoa can support termites for as long as one year, and can support a colony of *Zootermopsis* indefinitely.[35] The fact that both kinds of insects may be found in nature together in the same logs, and that reciprocal transfer of their flagellates is possible, lends evidence to support the belief that the termites and their protozoa are derived from a line of wood-eating roaches. There is a definite resistance to cross infection of these protozoans from one species of termite to another widely different species. The Protozoa are thus, in general, both morphologically and physiologically distinct.[9]

Ciliates inhabiting the stomachs of ruminants illustrate some interesting flexibility in host specificity. In general, the sheep and ox families, and to some extent the deer family, tend to harbor similar ciliate faunas, as shown by morphologic comparisons and by cross infection experiments. But ciliates from horse ceca do not infect cattle stomachs.

Certain types of bacteria can enhance the invasiveness and possibly the virulence of entamoebae. Experiments[17,37,38] have shown that without bacteria (at least with *Entamoeba histolytica* in guinea pigs), the host cannot even be infected (see p. 66). The role of the bacteria may be simply a physical one, providing a suitable microenvironment for the protozoans, but investigations on the role of bacteria or of other associates in cultivation with *E. histolytica* indicate that the answer is probably much more complex. The genus *Entamoeba* is widely distributed in vertebrates but its species appear to be fairly host-specific. The differences among the species are often not detectable morphologically, but physiologic species or at least biologic races are common. Much experimental work on cross infection studies needs to be done with the parasitic amebas.

Worms

Broadly speaking, parasitic worms with direct life cycles (e.g., *Ascaris*) are more host-specific than are worms with an indirect cycle (e.g., tapeworms). On the other hand, worms that have an indirect life history generally exhibit more specificity for their intermediate hosts than for their final hosts. Here, again, we may cautiously assume that the parasites are better adapted to their intermediate hosts, with which they have been associated for a longer period of time. The only really safe generalization to make is that among worms, as among the protozoa, there are wide ranges of host specificity. The ranges may be exemplified by some studies made on the worm parasites of moles and shrews.[40] One group of worms (e.g., the nematode, *Capillaria talpae*, and the tapeworm, *Choanotaenia filamentosa*) exhibits a narrow host specificity and is restricted to one species of mole, whereas another group (e.g., the trematode, *Panopistus pricei*) is less specific and freely invades several genera within the family Soricidae; but the nematode, *Parastrongyloides winchesi*, and other worms are widely distributed among the Talpidae and Soricidae within the order Insectivora. Finally, a few worms are unhampered with much restrictive host preference, and thereby able to live as parasites in widely differing systematic groups of hosts. An example of the latter is the acanthocephalan worm, *Polymorphus minutus*, which has been reported from freshwater fish, water birds and water shrews. See Cameron,[2] for a review of host specificity and evolution of helminthic parasites, and Schwabe and Kilejian,[47] for a discussion of chemical determinants of host specificity among flatworms.

Trematodes

Monogenetic trematodes are, as a group, markedly host-specific. This specificity is related to conditions for isolation that exist in the group, together with the habit of fastening eggs to the surface of the host, and with the several methods of attachment of adult worms (Fig. 6–4). When adhesive organs consist of hooks that are adapted to particular areas of the host surface (e.g., free edge of gill lamellae), the transfer of the adult worm from one host to another becomes difficult. When the parasites possess suckers, as do the Cyclocotylidae, and are thus able to move about in the gill chamber of a fish host, transfer to other hosts is more likely to take place. Examples of speciation of parasites in fairly close correspondence with speciation of their hosts are probably common throughout the whole group of monogenetic trematodes (e.g., *Hexabothrium* and *Erpocotyle* are found exclusively on elasmobranchs).

Free-swimming oncomiracidial larvae of the monogenean skin parasite *Entobdella soleae* respond to a specific substance secreted by the skin of the common sole (*Solea solea*).[21] This response is weak or absent when the parasites are experimentally exposed to other species of fish, or to detached pieces of skin from the other fish. The larvae are strongly attracted to isolated epidermis from *Solea solea*, and to agar jelly that has been in contact with the skin. Chemoreception probably is a major factor in host finding and host specificity among the monogenean trematodes in general.

Host specificity among monogenes is of a physiologic or ecologic nature, or both, but the specificity is pronounced and is of phylogenetic significance. The more primitive species of worms are generally to be found on the more primitive groups of fish, and those worms favoring freshwater fish are somewhat less host-specific than those on marine fish.[15] In spite of the high degree of host specificity at the species level, it is rather low at the generic level. When considering a higher taxonomic level, we often find a very wide range of host specificity. For example, the gyrodactyloids are distributed throughout teleosts, mollusks and amphibia, and the capsaloids are found in elasmobranchs, holocephalans, chondrosteins and teleosts. Here again, however, much work must be done, especially of an experimental nature, before the limits of host specificity can be ascertained.

One might expect the digenetic trematodes also to be highly host-specific because their parasite-host relationships are the result of a chain of complex adaptations between miracidia and their environment, sporocysts, rediae and snails, cercariae and their hosts, and, finally, between the adult worms and their vertebrate hosts. Too few complete life-history studies of digenetic trematodes have been made to warrant broad generalizations concerning host specificity of entire life cycles. Available evidence points strongly to a different degree of host specificity at each stage or level of the life cycle. Unfortunately, the possible varieties of intermediate as well as of final hosts have usually not been investigated. For these reasons the comments below, unless otherwise indicated, pertain only to adult trematodes.

Host specificity is not marked in all families of digenetic trematodes, but a study of collection records suggests that specificity prevails to a considerable extent, although closely related species may exhibit great differences in degree of specificity.[25] Thus *Schistosoma mansoni* is restricted to man and monkeys, whereas *S. japonicum* is a successful parasite of man, dogs, cats, pigs, cattle, horses, and others, and of the common laboratory animals. *S. haematobium* lives well in man, albino mice, hamsters, monkeys and baboons. It grows poorly in cats, albino rats, cotton rats, guinea pigs and goats; rabbits and dogs are refractory to infection. *S. incognitum* occurs in pigs and dogs in India, and it can be established easily in such laboratory animals as cats, sheep, goats, rabbits, guinea pigs, rats and mice. *S. spindale* is successful only in ungulates. Most of the species of blood flukes have no marked host specificity.

When distribution records of trematodes are analyzed, one normally finds that those genera that show a wide host tolerance are nonetheless limited to hosts that are related ecologically. For instance, all hosts from which about 50 genera of strigeids have been recorded are ecologically associated with water. Among the strigeid trematodes, families have been considered to be restricted to certain kinds of hosts. Members of the Diplostomatinae are parasites of birds, and the Alariinae are found exclusively in mammals. *Fibricola cratera*, a parasite of mammals, can easily be transferred to chicks, but whether avian hosts are infected in nature is not certain. The need for caution in formulating conclusions about host specificity is obvious.

The flukes, *Gorgodera amplicava*, in the frog, *Rana catesbeiana*, and *Gorgoderina attenuata* in *R. pipiens* are adapted to live in the urinary bladders of their hosts. To reach the bladder, larval stages must migrate through the body. It might be presumed that adult stages would be able to live in normal host organs that are compatible with the larval parasites, but experiments[13] in homotransplantion of adult worms showed that they failed to live in the new habitats. Frog bladder flukes implanted into the true urinary bladder of turtles remained normal for seven to ten days, and those transferred to the salamander, *Triturus v. viridescens*, were normal for 48 hours. This organ specificity may, therefore, be due to a resistance of the host against excystation of metacercariae or against postmetacercarial migration of juvenile trematodes. But the resistance does not operate against adult flukes artificially implanted into the urinary bladder in normally noninfective hosts.

Fasciola hepatica has become cosmopolitan in distribution and the adult may infect sheep, cattle, pigs, rodents, elephants, kangaroos and man. *Echinostoma revolutum* may infect various species of birds and mammals. These and other examples show that many digenetic trematodes tolerate a wide variety of hosts. Furthermore, a comparison of the morphology of one species taken from several different kinds of hosts demonstrates that the range of changes that occur have the same degree of magnitude which, under other conditions, would justify separation into one or more species. Specimens of *F. hepatica* taken from a cow could not be assigned, on the basis of morphology alone, to the same species as *F. hepatica* taken from a guinea pig.

A striking example of speciation related to parallel evolution involving digenetic trematodes and their hosts is the presence of five genera and ten species (all host-specific) of the Accacoeliidae in the intestine of the sunfish, *Mola mola*.

Several genera of invertebrates and vertebrates frequenting a lake district may be infected with the same genus or even species of trematode. Under experimental conditions, however, such unnatural hosts as chickens, ducks, rats, mice, and cats may successfully harbor the adults if fed sufficient numbers of metacercariae. Parasitologists often discover metacercariae or cercariae whose definitive hosts are unknown, but whose entire life cycle can be described because development can take place in a laboratory animal. A study of this nature began with metacercariae of *Cryptocotyle concavum* encysted in the skin of the stickleback fish, *Gasterosteus aculeatus*.[55]

Rediae and cercariae were obtained from the snail, *Amnicola longinqua*, but the identity of the normal definitive host (probably a bird frequenting the river area) could not be ascertained. Adult stages of this parasite, however, were readily recovered from the intestines of day-old chicks and ducklings 25 to 48 hours after feeding infected sticklebacks to the birds.

Experimental work with miracidia has often demonstrated a high degree of specificity for snails.[54] The miracidium of *Opisthorchis felineus*, for example, is attracted to the prosobranch snail, *Bithynia leachi*, but not to the closely related *B. tentaculata* that occurs in the same locality. Many such examples could be listed, but little is known about the nature of the attracting agent, which is possibly a water soluble skin secretion from the snail. Not all miracidia are host-specific, however.

As a final generalization, host specificity among digenetic trematodes is greater at the level of the intermediate host, especially the mollusk, than at the adult level. Thus, two distinct species of worms may live together as adults in the intestines of a bird or fish, but require different species of snails in which to complete their life cycles.

Cestodes

Tapeworms vary considerably in their range of host specificity, but adults tend to be more specific than do adults of most other groups of worms. Each order of bird and mammal possesses its own characteristic cestodes. Host specificity among these worms attains a high degree of perfection as the hosts become specialized. For instance, the cyclophyllids, with a very specialized type of internal anatomy, are found only in terrestrial vertebrates. Among the elasmobranchs are tetraphyllid tapeworms, which possess extravagant types of scolex structures. The sharks and rays each apparently harbor distinct species of cestodes. Snakes also possess distinct and characteristic species of tapeworms. *Dipylidium* and *Echinococcus* are found only in carnivores; *Moniezia*, *Thysanosoma* and *Stilesia*, only in ruminants, and so forth. But host specificity among cestodes reaches its highest development in birds. This relationship is perhaps best illustrated by water birds, such as grebes, loons, herons, ducks, flamingos and cormorants—birds that may occupy the same ponds or lagoons. Each bird species possesses its own tapeworm fauna. If we recall the above discussion about ecologic segregation, we recognize that among cestodes, contrary to the situation described for trematodes, host specificity is apparently more independent of ecologic segregation of their hosts, and more dependent on phylogenetic relationships. A quotation from Baer[1] summarizes the above statements on cestodes:

> . . . it is clear that both larval and adult tapeworms are associated with their hosts in a very intimate fashion. It is obvious that ecological segregation of the hosts originally produced isolation of the parasites in the different vertebrate groups. Yet, on the other hand, cestodes appear to be highly specialized from a physiological standpoint and to have become adapted to their hosts a very long time ago, as is shown by their present-day distribution. It is not possible, even experimentally, to break down this host specificity, as can be done for other parasites (trematodes). The data indicate that ecological specificity has here been replaced by phylogenetic specificity, a much more intimate type of association that arose thousands of centuries ago when cestodes first became parasitic in the ancestors of the species which today serve them as hosts.

Although a given species of adult tapeworm is limited to hosts belonging only to one class of vertebrate, species of worms of the same cestode genus may parasitize hosts belonging to different vertebrate classes. For well-known examples of this general rule we may cite the occasional presence of the dog tapeworm, *Dipylidium caninum*, and the rat tapeworm, *Hymenolepis diminuta*, in man.

Larval stages of cestodes frequently tolerate a much wider range of hosts than do the adults. *Hymenolepis diminuta* larvae have been reported from four different orders of insects and from myriapods. *H. gracilis* occurs in both copepods and ostracods. Larvae of the dwarf dog tapeworm, *Echinococcus*, have been found in many kinds of mammals. A study of these larval stages indicates that larval cestodes are ecologically segregated, but that there exist some forms in which specificity is independent of ecologic factors, and has resulted from physiologic adaptation. To cite one bit of evidence for this statement—when coracidia of *Dibothriocephalus latus* are fed to several species of freshwater copepods in one dish, some species of the latter are more favored hosts than others.

Opium-treated albino mice were experimentally much more readily infected with *Hymenolepis diminuta* than were those not so treated (intraperitoneally).[42] Opium slows the intestinal emptying time, and the effect of this process on the establishment of *H. diminuta* is pronounced. The fact that the parasite only rarely occurs in house mice suggests that the intestinal emptying time might be regarded as an explanation of apparent host specificity in this case.

The rostellum and suckers of the scolex probably play important roles in determining host specificity. "The more closely a scolex is adapted to the morphology of a particular type of mucosa, the narrower the host spectrum is likely to be. In this respect, the Pseudophyllidea may prove to be generally less host-specific than the Cyclophyllidea, although it is difficult to generalise."[48]

One possible explanation, on a physiologic basis, of the high degree of specificity shown by tapeworms is a dependence upon specific nitrogenous compounds secreted by the host intestine. The host specificity of tapeworms is undoubtedly related to specific biochemical characteristics of the worms, and to the chemical and physical properties of the environment within the host.

Nematodes

The study of host specificity among nematodes is particularly perplexing because of the wide variety of kinds of associations between these worms and other organisms, both plant and animal. An almost continuous series of associations from entirely free-living nematodes to obligatory parasites exists.

There is little evidence of parallel evolution of hosts and their nematode parasites. The most primitive genera and species of the parasites of vertebrates are found in mammals— more than two-thirds of the described strongyloids, usually considered the most primitive nematodes, are from mammals. There is little evidence of host specificity to support the common belief that nematodes from cavities or tissues of invertebrates are the oldest parasites of this phylum. Although in nematodes we do not find the phylogenetic specificity as exhibited by the cestodes, nor the ethologic specificity of trematodes, we do notice a broad host specificity in the more primitive species of hosts, becoming narrow in specialized species of hosts. Numerous plant nematodes vary widely in their tolerance of different kinds of hosts. During the course of evolution a "specificity by affinity of metabolism" becomes more and more pronounced, until the parasite is no longer able to adapt to a new host.

Brugia malayi can be successfully transmitted from man to forest and domestic animals by direct inoculation of infective larvae.[10] This observation is significant in a consideration of the question of reservoir hosts for this and other filarids. If we can experimentally transmit filarial worms from man to animal in the laboratory, can the parasites be transmitted in nature by mosquitoes or by *Chrysops* from animals to man? Probably so.

Most of the species of parasitic nematodes that are found in birds belong to genera that also occur in mammals, but in many groups of mammals the genera of nematodes are specific. Elephants harbor six distinctive genera and 20 or more species of strongyles, while six genera and about 18 species may be found in rhinoceroses. Horses and other equines harbor at least eight genera with more than 50 species of nematodes that are not found in any other group of animals.

Considerable work has been done on speciation of the hookworms, strongyloids and ascarids infecting man, and the results indicate that these parasites in vertebrates often form physiologic races (see p. 549). *Ascaris* in man and in pigs are almost identical in their morphology, but physiologic differences prevent successful cross infection under normal circumstances. A biochemical analysis of *A. lumbricoides* and *A. suum* discloses apparently identical constituents, but the carbohydrate fractions composed primarily of glycogen are antigenically distinct. Such studies are needed not only for the separation of *Ascaris* between man and pigs, but also for the separation of other species and races, such as the related *Toxocara* in cats and dogs. Specificity of nematodes whose larvae migrate through the body of the vertebrate host may be due primarily to the failure of larvae to complete somatic migration in a "foreign" host rather than to incompatibility between the adult worm and the host. Another kind of specificity in this group of worms is illustrated by the apparent preference by *Ascaridia galli* for male chicks over female chicks.

Sprent (personal communication) has emphasized that, among the ascaridoids, the larval behavior is a more sensitive indication of specific difference than is the adult morphology. The adult stage is the least likely to reveal specific differences, possibly because the environment of the vertebrate host intestine is more uniform than is the internal environment of the intermediate hosts. Thus, the migratory pathway, degree of growth and onset of various molts manifested by these larvae appear to be the most sensitive differentiating features.

Sprent[50] has made some significant observations on the changing specificity patterns during the life history of *Amplicaecum robertsi*, an ascaridoid from the carpet python. We quote from his summary:

> The snake is depicted as the apex of a food pyramid, whose base comprises a variety of animals ranging from earthworms to herbivorous animals. . . . The life history is thus regarded, not as a life cycle, but as a life pyramid; development proceeds according to a pattern of diminishing host-specificity. Host-specificity is wide at the base of the pyramid, so that second-stage larvae occur in a wide variety of paratenic hosts. Host-specificity narrows at the second moult which may occur in birds and mammals. It

narrows still further in the third stage because this larva, though it will survive in reptiles, birds and mammals, will not grow to a length at which it is capable of further development in the snake except in certain mammals. At the third moult, host-specificity shifts to certain reptiles but becomes eventually restrictive to the carpet snake, because this host alone appears to provide a suitable environment for maturation of the eggs.

In a study of patterns of evolution in nematodes, Inglis[19] concluded that, "In general parasitic nematodes are not host specific although they tend to be restricted to animals with similar feeding and ecological habits. The most spectacular example of a group which is apparently an exception to this is the pinworms of primates." See page 278 for a discussion of visceral larva migrans, a situation in which larval nematodes gain entrance to tissues, but do not reach adulthood because of antagonistic host responses. Poynter[39] has reviewed the problems of tissue reactions to nematode parasites.

Acanthocephala

The late H. J. Van Cleave has amply demonstrated that adult acanthocephalid worms show a relatively high degree of host specificity. For example, *Gracilisentis* and *Tanarhamphus* are found normally only in the gizzard shad. The Pacific pilotfish, *Kyphosus elegans*, and the closely related Atlantic pilotfish, *K. secatrix*, each harbors a distinct species of the acanthocephalan genus *Filisoma*. Likewise, *Moniliformis moniliformis* (probably a combination of several species) shows a narrow specificity in certain instances. Specificity in these worms is related to the nature of the life cycle in which no free-living stage has been reported, and in which an arthropod intermediate host is essential for all species.

Physiologic host specificity appears to be highly variable in this group of parasites. Acanthocephalans of carnivorous vertebrates cannot proceed from the arthropod to the final host unless the parasites pass first into an insectivorous host. Collection records indicate that there is little specificity among larval forms of at least some groups of Acanthocephala. A single species of *Centrorhynchus*, for example, has been reported as a larval parasite of lizards, snakes and frogs. Much experimental work on morphologic variation and physiologic host specificity under controlled conditions is needed. Even morphologic criteria are insufficient to establish exact systematics for these worms.[12]

Arthropods

Crustacea

The parasitic copepods are among the most diversified of all parasites, and almost the only ones that are found in the adult stage on both vertebrates and invertebrates. It has been suggested that copepods are probably the oldest parasites actually known, but they may share this position with prosobranch snails, which are exclusively parasitic in echinoderms, and with the group of annelids known as myzostomids, also exclusively in echinoderm hosts. Both prosobranchs and myzostomids have been recorded as fossils. Ecologic segregation of copepods on their hosts has often been accomplished with marked intimacy. Some of the species of blood feeders are restricted to one species of hosts (Fig. 16–16, p. 353). Although distinct host specificity among copepods occurs, further work of a statistical nature must be done with these crustaceans before we can formulate significant generalizations about them.

One group of parasites that has frequently been considered strictly host-specific is the parasitic isopods. Apparently each genus of many groups (e.g., entoniscids) is found on a particular host or group of hosts that appear characteristic for the parasite. For instance, *Danalia* and *Liriopsis* are found only on decapods. However, *Phryxus abdominalis* has been recovered from at least 20 species of shrimps belonging to two genera. Without a considerable amount of experimental research work on these and other forms, no definite conclusions can be made as to the host specificity of parasitic isopods in general. We can say that parasite-host relationships, in-

cluding specificity, among the isopods as well as among mites, barnacles and other arthropods, are essentially of an ecologic nature.

Hymenoptera

Insects as parasites are generally highly host-specific, but this specificity is predominantly ecologic, not physiologic. Evidence for this conclusion is gained when such activities as searching for the host habitat and for the host are eliminated experimentally, and the parasite is placed directly upon a "foreign" host. In such a situation the parasite very commonly goes ahead and lays its eggs on or in the new host with little or no hesitation.

The entomophagous Hymenoptera (see p. 382) are seldom monophagous (restricted to one kind of food), but they are far from indiscriminate in their attacks upon insect hosts. In nature these parasitic wasps have several potential species of hosts, and in the laboratory many more hosts may be discovered by experimental testing. Why then does the wasp normally select only one or two kinds of hosts in which to deposit its parasitic eggs? Several distinct processes of selection seem to occur, and are discussed under the heading of "Biologic Control" on page 492. The hymenopteran insect, *Aphelinus mali*, has been reported in the literature as attacking six or seven species of aphids, but current views suggest that most of these records are erroneous, and that most of the aphids may be synonyms of *Eriosoma lanigerum*, the woolly apple aphid.

The food of an insect host may affect the latter's suitability as a home for a parasite. For example, the hornworm, *Protoparce sexta*, when feeding on tomato, is a suitable host for the braconid wasp, *Apanteles congregatus*, but when this host is fed on dark-fired tobacco, the parasite dies before reaching maturity.

A sequence of ecologic processes leads to the attainment of host specificity of many Hymenoptera and of other entomophagous insects as well. Searching capacity and host specificity are correlated phenomena. The sequence is as follows: (*a*) the finding of the host habitat, (*b*) the finding (recognition) of the host, (c) acceptance by the host, and (d) host suitability for parasite reproduction. The quality of the host's environment rather than the qualities of the host itself appears to be the more important controlling factor in restricting the number of host species attacked.

Fleas

The habitat preference of a host may have a distinct effect on its arthropod parasites, and the latter may aid in host identification because of the degree of specificity attained. This relationship has been demonstrated[30] with the fleas of two closely related European wood mice, *Apodemus sylvaticus*, which lives in the fields of Normandy, and *A. flavicollis*, which lives in adjacent woods or areas of scrub growth. One might question the separation of these mice into two distinct species, and prefer to consider them as representing only two color phases or ecologic races of one species. When the mice fleas are scrutinized, however, they are found to belong to two distinct subspecies. *Ctenophthalmus agyrtes agyrtes* is restricted in Normandy to the woods in a district where it overlaps with the northwestern subspecies, *C. a. nobilis*, which lives in open country. Thus we have evidence from the host-preference pattern of fleas for the taxonomic separation of their mice hosts.

Among the reviews of the question of host specificity in fleas is that of Hopkins[18]:

Unlike many parasites, fleas (possibly rare exceptions) . . . pass their entire pre-adult life off the body of the host, their larvae being free-living feeders on organic dust, though the early stages usually take place in the host's dwelling. This means that the early stages of fleas, not being parasitic, are susceptible to the conditions, climatic and others, which govern the distribution of free-living animals to a far greater extent than animals which are parasitic in all their active stages. It also means that it is necessary for the newly-emerged flea to seek out a host, sometimes of one particular species, and that the period of starvation that the flea must undergo during this search can often be reduced and the search be prolonged (with better prospects of a successful conclusion) by the practice of polyhaemophagy (the ability to feed

on the blood of hosts other than the one normal to the flea in question). For these reasons it is common to find a flea on a host other than that (or those) to which it is normal, and such occurrences may range from the purely accidental presence of fleas on reptiles (from which they are probably unable to suck blood) through those in which a flea can obtain nourishment from the blood of a host on which it is extremely reluctant to feed, to instances in which a given species of flea has a number of hosts between which it shows little preference and on all of which the species can reproduce indefinitely.

Polyhaemophagy and promiscuity have been of considerable advantage to fleas, yet narrow specificity is indulged in by many species. When considered as a group, it is clear that the ecologic conditions in the nest (mammal or bird) are more important than is the host. For example, *Ceratophyllus garei* occurs in the nests of a great variety of birds if the nests are not too dry. *Pulex irritans* of man is sometimes abundant in pig-styes, and it is a true parasite of the badger. Hopkins states that, "The more promiscuous a flea is as regards the source of the blood on which it can mature its eggs, the more probable it becomes that random hopping will eventually result in the deposition of the eggs in an environment suitable for the development of the larvae, while the latter are not affected nearly so much by the question of whether the nest or burrow in which they find themselves was made by a rabbit or a bird, a squirrel or a mouse, as by the environmental conditions within it, particularly the relative humidity and temperature."

Lice

All lice are obligatory and permanent external parasites of birds and mammals. They cannot jump or fly, or even walk very well, and they spend their entire lives on the bodies of their hosts. For these reasons transfer from host to host is normally accomplished only when two host bodies are in close contact, as during copulation, feeding of young, or while standing together in herds. Obviously, then, there is no particular disadvantage to the parasite in being narrowly adjusted to one species of host. The death of the host inevitably means the death of the entire population of lice on its body. These limitations explain the intraspecific bounds of distribution, and the relatively extreme specificity in their host associations. It is of interest to note that there are no lice on bats, yet bats are heavily invaded by other parasites. Forty or 50 ectoparasites of several species are not uncommon on one bat. The reasons for the absence of lice on bats are unknown.

Authentic instances of lice distributions not explainable by host-phylogeny are rare, but interspecific transfers might occur during a struggle between prey and predator, and during the sharing of mud wallows, rubbing trees, roosting or perching spots, and as the result of the usurpation of a nest or burrow by an alien host. Phoresy (p. 5) may also result in interspecific transfer.

The fact that lice do not have even a resting stage of their life cycle off their host results in a correlation of their phylogeny with that of their hosts almost to the exclusion of other factors (see p. 570). Lice, therefore, almost always occur only on one host or on a small number of closely related hosts. An extreme degree of specificity occurs among lice on the Procaviidae (order Hyracoidea) where most of the subspecies of the host *Procava capensis*, have their own species or subspecies of louse (*Procavicola*). The chewing lice appear to be somewhat more host-specific than are the sucking lice. A few instances of incipient speciation of lice have been described; the best known is between *Pediculus humanus humanus*, the human body louse, and *P. humanus*, the human head louse.

The lice on primates all belong to the family Pediculidae, and no member of this family occurs on any other host. Both *Pediculus* and *Phthirus* (see p. 372) occur on man and higher apes, but not on monkeys. *Phthirus* includes species from the gorilla and chimpanzee, but since the records are from menagerie material it is not conclusively known that these apes are natural hosts of the crab louse. *Pediculus* has been recorded from the gibbon and from the chimpanzee. Apparently the orang is not infested with lice. Spider monkeys (*Ateles*) of tropical America are far removed anatomically from man and his ancestry, yet, curiously, the monkeys possess a species of *Pediculus* (sometimes separated as an independent genus, *Parapediculus*). It is possible that the lice were transferred from man to spider monkeys, but if such an event occurred it must have taken place in the remote past because considerable differences have been evolved between the parasites of man and those of monkeys. Although straggler lice frequently occur on birds, the specificity between feather lice and their hosts is relatively marked. For instance, the wing louse, *Lipeurus caponis*, occurs only on the wings, and the shaft louse, *Menopon gallinae*, is adapted to live within the shafts of large feathers.

Many biting lice exhibit an interesting mutualistic association with bacteria which, in nymphs and males, occurs in specialized myelocytes among the fat bodies. In adult females the bacteria accumulate in the ovary from whence they pass into the eggs and so are transmitted congenitally.[44] Lice from which the bacteria have been eliminated soon die. Because the bacteria occur in those lice that feed on blood, it has been suggested that one factor helping to determine host specificity is the inability of the bacteria to survive in a louse feeding on the blood of an abnormal host species. The causes of host specificity among lice are doubtless related to biochemical differences in the blood, skin, and plumage among the hosts.

Flies

To illustrate some details of the problems of parasite-host specificity among the Diptera, we have selected the Family Hippoboscidae of which all members are obligatory, bloodsucking ectoparasites of various orders of mammals and birds. These flies are flattened dorsoventrally, with a strong development of the sternal region of the thorax, which forms a smooth plate (Fig. 18–27). Some species hatch with normal wings which break off on reaching the host, others retain reduced wings and eyes, while one genus is completely without wings. Hipboscids occur on only five orders of mammals and on 18 out of 27 orders of birds. There are no species parasitic on both mammals and birds. Host-preference is more restricted in the flies of mammals than in those of birds, but the parasites show a wide diversity of behavior with all gradations from strict species-specificity to occurrences on host species of different orders. Hippoboscid host selection patterns are explained chiefly by ecologic factors; hence we find that, regardless of taxonomic affinities of the hosts, if they possess similar habits and habitats, they are likely to be burdened with similar parasitic flies. These considerations suggest that, when strict specificity exists, it is the result of geographic isolation, as in the case of flies on kangaroos. Conclusions similar to those expressed for Hippoboscids have been made for pupiparous Diptera of bats.

During the months of July and August at Fair Isle Observatory near Shetland and Orkney, five species of breeding passerine birds were trapped intensively and hippoboscid flies (*Ornithomyia fringillina*) were removed from each bird. Each fly was marked and then released on any one of the five species of hosts irrespective of the original source of the fly. A few marked flies were released without a host. After the lapse of an appropriate time the flies were again collected and 75 per cent of those recovered were found on the same individual birds on which they had been released, 18.5 per cent had moved to other birds of the same species, and 6.5 per cent had changed to a host species different from that on which they had been released. Of interest is the fact that male flies changed hosts more often than did females.[3] Experiments such as this one give us a better understanding of how parasites behave, but they do not provide answers to the question of *why* the parasites behave the way they do. What is behind the behavior? Is it phylogeny? serology? nutrition? chemotaxis? Or one might ask a negative question—why is there a lack of host specificity?

Although specificity of an insect parasite for a single species of host is rare, *Cryptochaetum iceryae*, a dipterous parasite, is restricted in the United States to the cottony cushion scale, *Icerya purchasi*, but in Australia it has been recorded on other species of the same genus of host. For details of this group of flies, see Theodor.[53]

Ticks and Mites

Although a few ticks and mites appear to be confined strictly to one host species, these groups of arachnids follow no hard and fast rule. The tick, *Dermacentor andersoni*, has a wide range of compatible hosts in all of its life stages. The more primitive trombiculid mites exhibit less host specificity than do those species which enjoy a more intimate association with their hosts. The latter group began phylogenetically as nest infesting species which developed first an ecologic type of host specificity, then as the intimacy became closer, a degree of physiologic host specificity was engendered. Ectoparasites such as mites and chiggers may appear to prefer one host to another, but this "preference" may be largely a matter of difference in the area of host skin exposed or available for infestation, or to differences in extent of host range or other host behavior.[33]

THE SEARCH FOR A PARTNER

A fruitful source of information on the nature of host specificity is a study of commensalism in which the resulting society of two or more individuals depends for its existence on the maintenance of highly specialized and precise "socially-adapted" behavior. In most of these relationships, the commensal makes an active search for its partner, and it is the nature of this search that now concerns us.

Laing[23] stated some principles (based on her work with parasitoidy among insects) that are applicable to all studies of symbiont behavior. She found, by the use of a choice-apparatus, that the chalcid wasp, *Alysia manducator*, is attracted by olfactory means to the environment where its host blow-fly larvae are to be found. This attraction is a result of a chemotaxis to some factor or factors in decomposing meat. The movements of the fly larvae in the meat evoke another stimulus that results in egg laying by the host. Laing concluded "that some parasites do first seek out a particular environment, in which they afterwards proceed to seek their hosts. The analysis of the process by which the parasite finds its host may be divided into two parts—the finding of the environment and the finding of the host in that environment—is, therefore, not merely a convenient theoretical division, but corresponds to an actual difference in the behavior of the parasite. . . . Not only, then, do some parasites find environments first and hosts later, they often use quite different senses for the perception of the two and make quite different movements to reach them. What those senses and movements are, however, will differ greatly with different parasites, and must be especially determined in each particular case."

Studies on the specificity of recognition of hosts by commensal polynoid annelids have shown that chemical attraction and recognition appear to be the usual mechanisms binding together such partners as the scaleworms, *Arctonoë fragilis*, with the starfish, *Evasterias troschelii*; *A. pulchra* with the sea cucumber, *Stichopus californicus*; *Hesperonoë adventor* with the echiuroid worm, *Urechis caupo*; *Polynoë scolependrina* with the terebellid, *Polymnia nebulosa*; and several others. An experimental technique devised by Davenport[5] consisted of a choice-apparatus or olfactometer in which "commensal worms were introduced into a Y-tube and were presented with a choice between streams from two aquaria. Material to be tested could be placed in either aquarium at random; similarly, connections with the aquaria were so arranged that streams to be tested could be introduced into either arm of the Y at random, thus making it possible for any consistent behavior resulting from uncontrolled inequalities in pressure or light to appear in the data from a large number of 'runs'. Such apparatus lends itself well to investigation of host

specificity in active forms which readily respond to streams of water carrying attractants."

All of the polynoid commensals so far investigated demonstrated strong positive responses to chemical stimulation by their hosts, and with few exceptions, this response was highly specific in spite of frequent close taxonomic affinity among hosts.[16] Davenport and his colleagues have demonstrated that specialized sorts of behavior may be induced by the presence of "host-factor." For example, the frequency of random turning in pinnotherid crabs is directly proportional to the concentration of "host-factor" in the crab's general environment.[6] This response obviously results in keeping the crabs in the vicinity of the clam host, like the ballet dancer who, while performing tight pirouettes, does not move very far. At higher concentrations the same chemical agent from the host has a directive influence on the crab, and induces the animal to move directly toward the clam.

A commensal sometimes is attracted to two unrelated hosts. In such a situation it seems likely that both hosts produce the same "attractant" (probably a metabolite), and that conditioning may not be immediately necessary.

Efforts have been directed toward correlating the *response* specificity of commensal worms on echinoderms with the known *host* specificity. Various categories of response specificity have been demonstrated, ranging from commensal populations that respond to their normal host alone (e.g., the polynoid worm, *Arctonoë fragilis* on the seastar, *Evasterias*) to commensals that appear to have no chemical discrimination, and that respond to many host animals. Moreover, all populations of a commensal species do not always behave in the same manner—some being much more specific than others. There are two populations of the polychaete worm, *Podarke pugettensis*, one a facultative commensal on several starfish (e.g., the web-star, *Patiria miniata*, and the mud-star, *Luidia foliolata*), the other free-living. The former population shows a strong tendency to respond positively to its host, but the other shows no such tendency. Experiments are needed to determine whether such differences are inherited or are conditioned.

The odor of a host is apparently a strongly determinative factor for mosquitoes in localizing the blood supplier.[12] Laboratory-bred *Anopheles atroparvus* reacted satisfactorily to airborne stimuli in an air-stream olfactometer, and the experiments pointed to the possibility of adaptation to the smell of a special type of host. Heat and moisture appear to be releasing stimuli for alighting on the host, but heat in addition has a directive influence. This thermotaxis is strongly activated by CO_2, which with other odors probably has an activating value only in the orientation process. Mosquito responses to chemicals are undoubtedly combined with visual responses in seeking and selecting a host. Preliminary experiments in the Congo suggest that, in searching for food, anophelines do not fly above 60 cm.

A problem related to the above is the explanation of preference for one host over another by parasites that may normally infect both. Worth[56] described an example of this problem involving cotton rats, and he said, "It would appear that for some reason the cotton rat is a favored host in the Everglades, being the carrier of more than five times as many individual ectoparasites as rice rats in the same environment despite a similar pattern of host infestations." The answer in this case lies in differences in the quality of host blood, microclimatic variations in the fur, differences in structure and texture of hair and skin, and grooming behavior. It is often difficult to detect such differences, and when they are detected, to evaluate their effects upon the parasites. To cite another example, the ciliate, *Trichodina*, is common on the gills of marine fish, but it is particularly abundant on benthic fish. Hence the preference is presumably associated with benthic life. The basic principle of variations among species, and variations among individuals within one species lies behind the obvious as well as the obscure patterns of parasite behavior.

GENERAL RULES AND PRINCIPLES

1. Parasite-host specificity is normally determined by ecologic factors or by physiologic factors, or both. The evolutionary age of the parasite-host combination is of fundamental importance.

2. In general, parasites with an indirect life cycle are less specific than those with a direct life cycle.

3. There is less host specificity when there are two intermediate hosts than when only one is employed.

4. Whereas two strains of a given parasite might infect a final host with equal facility, they may not be equally infective to a given strain of intermediate host.

5. The host is in many respects equivalent to the environment surrounding free-living organisms, but the host, as an environment, is constantly developing specific responses and adaptations to its parasites.

6. Specificity, once gained, may subsequently be lost.

7. When a species of host is divided into two or more population groups separated geographically in different environments, their respective parasite faunas normally exhibit differences.

8. A parasite may become transferred to a "foreign" host living in the same locality as the original host, and subsequently become isolated due to geographic separation of the two hosts.

9. After isolation of a parasite upon a host takes place, both morphologic and physiologic changes may occur.

10. Host specificity is fundamentally a function of physiologic specialization and of evolutionary age.

11. In comparing digenetic trematodes with cestodes, host specificity among trematodes is more pronounced in the larval stages, but the larval stages of cestodes frequently tolerate a much wider range of hosts than do the adult cestodes.

12. In nematodes, there is sometimes

greater specificity for the intermediate host than for the definitive host, but most nematodes are not markedly host-specific.

13. There is probably less host specificity among parasites than is generally assumed because numerous studies are continually disclosing new kinds of compatible hosts for species originally thought to be restricted to only one or a few kinds of hosts.

REFERENCES

1. Baer, J. G.: Ecology of Animal Parasites. Urbana, Ill., Univ. Illinois Press, 1951.
2. Cameron, T. W. M.: Host specificity and the evolution of helminthic parasites. *In* Dawes, B. (ed.): Advances in Parasitology. Vol. 2, pp. 1–34. New York, Academic Press, 1964.
3. Corbet, G. B.: The life-history and host-relations of a hippoboscid fly *Ornithomyia fringillina* Curtis. J. Animal Ecol., 25:403–420, 1956.
4. Crutchfield, C. M., and Hixon, H.: Food habits of several species of poultry lice with special reference to blood consumption. Florida Entomol., 26:63–66, 1943.
5. Davenport, D.: Specificity and behavior in symbioses. Quart. Rev. Biol., 30:29–46, 1955.
6. Davenport, D., Camougis, G., and Hickok, J. F.: Analysis of the behaviour of commensals in host-factor. 1. A hesioned polychaete and a pinnotherid crab. Anim. Behaviour, 8:209–218, 1960.
7. Desowitz, R. S., and Watson, J. J. C.: Studies on *Trypanosoma vivax*. IV. The maintenance of a strain in white rats without sheep-serum supplement. Ann. Trop. Med. Parasitol., 47:62–67, 1953.
8. Doran, D. J.: Coccidiosis in the kangaroo rats of California. Univ. Calif. Publ Zool., 59:31–60, 1953
9. Dropkin, V. H.: Host specificity relations of termite protozoa. Ecology, 22:200–202, 1941.
10. Edeson, J., and Wharton, R.: The experimental transmission of *Wuchereria malayi* from man to various animals in Malaya. Trans. Roy. Soc. Trop. Med. Hyg., 52:25–45, 1958.
11. Ellenby, C., and Gilbert, A. B.: Cardiotonic activity of the potato-root eelworm hatching factor. Nature, 180:1105–1106, 1957.
12. Golvan, Y. J.: La Spécificité Parasitaire Chez les Acanthocéphales. *In* First Symposium on Host Specificity Among Parasites of Vertebrates. pp. 244–254. Inst. Zool. Univ. Neuchâtel, 1957.
13. Goodchild, C. G.: Transplantation of gorgoderine trematodes into challenging habitats. Exper. Parasitol., 4:351–360, 1955.
14. Greenberg, J., and Trembley, H. L.: The apparent transfer of pyrimethamine-resistance from the BI strain of *Plasmodium gallinaceum* to the M strain. J. Parasitol, 40:667–672, 1954.
15. Hargis, W. J., Jr.: The host specificity of monogenetic trematodes. Exper. Parasitol., 6:610–625, 1957.
16. Hickok, J. F. and Davenport, D.: Further studies in the behavior of commensal polychaetes. Biol. Bull., 113:397–406, 1957.
17. Hoare, C. A., and Neal, R. A.: Host-parasite relations and pathogenesis in infections with *Entamoeba histolytica*, mechanisms of microbial pathogenecity. Symp. Soc. Gen. Microbiol., 5:230–241, 1955.
18. Hopkins, G. H. E.: Host-associations of siphonaptera. *In* First Symposium on Host Specificity Among Parasites of Vertebrates. pp. 64–87. Inst. Zool., Univ. Neuchâtel, 1957.
19. Inglis, W. G.: Patterns of evolution in parasitic nematodes. *In* Taylor, A. E. R. (ed.): Evolution of Parasites. pp. 79–124. Oxford, Blackwell Sci. Pub., 1965.
20. Jordan, K.: Where subspecies meet. Novit. Zool., 41:103–111, 1938.
21. Kearn, G. C.: Experiments on host-finding and host-specificity in the monogenean skin parasite *Entobdella soleae*. Parasitology, 57:585–605, 1967.
22. Laarman, J. J.: Host-seeking behavior of malaria mosquitoes. pp. 648–649. XVth Internat. Congr. Zool., Proceedings, 1959.
23. Laing, J.: Host-finding by insect parasites. 1. Observations on the finding of hosts by *Alysia manducator*, *Mormoniella vitripennis* and *Trichogramma evancescens*. J. Animal Ecol., 6:298–317, 1937.
24. Llewellyn, J.: The host-specificity, micro-ecology, adhesive attitudes, and comparative morphology of some trematode gill parasites. J. Mar. Biol. Assoc., 35:113–127, 1956.
25. Manter, H. W.: Host specificity and other host relationships among the digenetic trematodes of marine fishes. *In* First Symposium on Host Specificity Among Parasites of Vertebrates. pp. 185–198. Inst. Zool. Univ. Neuchâtel, 1957.
26. Manwell, R. D.: Factors making for host-parasite specificity, with special emphasis on the blood protozoa. *In* Cheng, T. (ed.): Some biochemical and immunological aspects of host-parasite relationships. Ann. N.Y. Acad. Sci., 113:332–342, 1963.
27. McGhee, R. B.: The ability of the avian malaria parasite, *Plasmodium lophurae*, to infect erythrocytes of distantly related species of animals. Amer. J. Hyg., 52:42–47, 1950.

28. ———: The influence of age of the animal upon the susceptibility of mammalian erythrocytes to infection by the avian malaria parasite *Plasmodium lophurae*. J. Inf. Dis., *92*:4–9, 1953.

29. ———: The infection by *Plasmodium lophurae* of duck erythrocytes in the chicken embryo. J. Exper. Med., *97*:773–782, 1953.

30. ———: The infection of duck and goose erythrocytes by the mammalian malaria parasite, *Plasmodium berghei*. J. Protozool., *1*: 145–148, 1954.

31. ———: Comparative susceptibility of various erythrocytes to four species of avian plasmodia. J. Inf. Dis., *100*:92–96, 1957.

32. Micks, D. W.: Investigations on the mosquito transmission of *Plasmodium elongatum* Huff, 1930. J. Nat. Malaria Doc., *8*:206–218, 1949.

33. Mohr, C. O.: Relation of ectoparasite load to host size and standard range. J. Parasitol., *47*:978–984, 1961.

34. Mönnig, H. O.: Measles in cattle and pigs: ways of infection. J. S. Afric. Vet. Med. Assoc., *12*:59–61, 1941.

35. Nutting, W. L.: Reciprocal protozoan transfaunations between the roach, *Cryptocercus*, and the termite, *Zootermopsis*. Biol. Bull., *110*: 83–90, 1956.

36. Pavlovskii, E.: [Conditions and factors affecting the formation of the host organism of a parasite in the process of evolution. Sketches of evolutionary parasitology, I.] Zool. Zhurnal, *25*:289–304, 1946. (In Russian with English summary.)

37. Phillips, B. P., Wolfe, P. A., and Bartgis, I. L.: Studies on the ameba-bacteria relationship in amebiasis. II. Some concepts on the etiology of the disease. Amer J. Trop. Med. Hyg., *7*: 392–399, 1958.

38. Phillips, B. P., *et al.*: Studies on the ameba-bacteria relationship in amebiasis. Comparative results of the intracecal inoculation of germfree, monocontaminated, and conventional guinea pigs with *Entamoeba histolytica*. Amer. J. Trop. Med. Hyg., *4*:675–692, 1955.

39. Poynter, D.: Some tissue reactions to the nematode parasites of animals. *In* Dawes, B. (ed.): Advances in Parasitology. Vol. 4, pp. 321–383. New York, Academic Press, 1966.

40. Prokopič, J.: The influence of oecological factors on the specificity of parasitic worms of insectivora. Folia Biol. (Prague), *3*:114–119, 1957.

41. Read, C. P.: Status of behavioral and physiological "resistance." Rice Inst. Pamphl., *45*: 36–54, 1958.

42. Read, C. P., and Voge, M.: The size attained by *Hymenolepis diminuta* in different host species. J. Parasitol., *40*:88–89, 1954.

43. Read, C. P., Douglas, L. T., and Simmons, J. E.: Urea and osmotic properties of tapeworms from elasmobranchs. Exper. Parasitol., *8*:58–75, 1959.

44. Ries, E.: Die Symbiose der Läuse und Federlinge. Z. Morph. Oekol. Tiere, (Berlin), *20*:233–367, 1931.

45. Rogers, W. P.: The physiology of infective processes of nematode parasites; the stimulus from the animal host. Proc. Roy. Soc. London, B, *152* (948): 367–386, 1960.

46. ———: The Nature of Parasitism. The Relationship of Some Metazoan Parasites to Their Hosts. New York, Academic Press, 1962.

47. Schwabe, C. W., and Kilejian, A.: Chemical aspects of the ecology of platyhelminthes. *In* Florkin, M., and Scheer, B. T. (eds.): Chemical Zoology. Vol. II, pp. 467–549. New York, Academic Press, 1968.

48. Smyth, J. D.: The Physiology of Cestodes. San Francisco, W. H. Freeman & Co., 1969.

49. Smyth, J. D., and Haslewood, G. A. D.: The biochemistry of bile as a factor in determining host specificity in intestinal parasites, with particular reference to *Echinococcus granulosus*. *In* Cheng, T. (ed.): Some biochemical and immunological aspects of host-parasite relationships. Ann. N.Y. Acad. Sci., *113*:234–260, 1963.

50. Sprent, J. F. A.: The life history and development of *Amplicaecum robertsi*, an ascaridoid nematode of the carpet python (*Morelia spilotes variegatus*). II. Growth and host specificity of larval stages in relation to the food chain. Parasitology, *53*:321–337, 1963.

51. Stefanski, W.: Quelles Conditions Exige le Parasite Pour s'Etablir dans son Hôte? Ann. Parasitol. Hum. Comp., *37*:661–672, 1962.

52. Terzian, L. A., Stahler, N., and Weathersby, A. B.: The action of antimalarial drugs in mosquitoes infected with *Plasmodium gallinaceum*. J. Inf. Dis., *84*, 47–55, 1949.

53. Theodor, O.: Parasitic adaptation and host-parasite specificity in the pupiparous Diptera. *In* First Symposium on Host Specificity Among Parasites of Vertebrates. pp. 50–63. Inst. Zool., Univ. Neuchâtel, 1957.

54. Vogel, H.: Der Entwicklungszyklus von *Opisthorchis felineus* (Riv.). Zoologica, *33*: 1–103, 1934.

55. Wootton, D. M.: The life history of *Cryptocotyle concavum* (Creplin, 1825) Fischoeder, 1903 (Trematoda: Heterophyidae). J. Parasitol., *43*:271–279, 1957.

56. Worth, C. B.: Observations on ectoparasites of some small mammals in Everglades National Park and Hillsborough County, Florida. J. Parasitol., *36*:326–335, 1950.

Section X
EVOLUTION OF PARASITISM

Chapter 25

Evolution of Parasitism

INTRODUCTION

All activities of animals are related, directly or indirectly, to their struggle for food, reproduction and protection. As Schiller long since declared, "the edifice of the world is only sustained by the impulses of hunger and love." From this struggle stems the wonderfully complex pattern of adjustments and changes that we call adaptations. It has been said that adaptations are the chief marvel of the living world, and their method of origin still the greatest problem of biology. Today we can approach only a little closer to the answers.

Current views relating to the nature of evolutionary mechanisms are based chiefly upon studies of highly specialized organisms, and since parasites are specialized, these views are particularly appropriate. Evolution depends primarily upon the occurrence of shifting gene frequencies in cross fertilizing organisms, followed by reproductive isolation.

Let us examine this statement on specialized evolution, and make sure that the technical terms are clearly understood. To begin with, mutations furnish material (i.e., variations) for the kind of natural selection that Charles Darwin described so well. Given, then, a certain frequency of mutations producing slight, haphazard changes, and given the selective action of the environment that preserves certain mutations and eliminates the others (e.g., genetic death), considerable structural and physiologic changes are possible within the time available for evolution. The factor of mutation plus the factor of recombinations of genes as the result of sexual reproduction produce the "shifting gene frequencies" referred to above. Gradual changes (structural and physiologic) within a species may proceed so far as to enable the systematist to distinguish a subspecies. Changes among individuals of the new subspecies may continue further, rendering them incapable of successful reproductive union with the other individuals of the species; hence, the changed individuals belong, by common definition, in a new species. These changes result in the development of reproductive isolation.

We must keep in mind that the organic environment is just as important as the inorganic environment in influencing the adaptations of a given species. The organic, or living, environment includes all forms of life with which a given individual or species comes into ecologic relation. Indeed, it is generally the organic environment that shows the more rapid and important alterations. There is a sharing of the host's body, and of ways of exploiting it, by viruses, bacteria, molds, fungi, protozoa, worms and arthropods. All of these organisms are part of the host's living environment. The dependence of adaptive trends on the organic environment includes, therefore, dependence upon internal and external parasites. This *biologic environment* is sometimes not fully appreciated by biologists. The combination of the host plus its parasite-mix evolves as a whole.

The entire life cycle of a parasite with all closely associated organisms, including the host, is a unit in evolutionary development. Thus, evolution of parasitologic systems (see Mattingly,[43] on the evolution of arthropod-borne vector systems) is as important as evolution of individual parasites. The statement just made assumes that parasitologic systems *do* evolve, and that there *is* an evolution of parasitism. We believe that this assumption is correct, but we must warn the student that all discussions of the evolution of parasites and parasitism are speculative.

Cameron[9] has stated that, "It is essential to remember that a parasite *must* have a host and its phylogeny and classification can only be interpreted in terms of the phylogeny and classification of the host. While it often happens that the host's phylogeny and classification are equally subjective, it is obvious that, in general, related hosts tend to have related parasites."

THE PACE OF PARASITE EVOLUTION

We have previously pointed out that many parasitic genera are found in widely separated groups (even phyla) of hosts, and obviously there is sometimes no correlation between the primitiveness of the host and the primitiveness of its parasites. The evolution of parasites, with clear exceptions, has not kept pace with the evolution of the hosts.

The mechanism by which parasites evolve more slowly than do their hosts may be explained as follows. Some members of a given species of host were able to survive more easily than others when the environment changed. Those more fortunate hosts were slightly different from those that were less suited (adapted) to the new environment, but both groups of hosts possessed the same parasites. During the sweep of time in a changing environment, all of the hosts gradually changed, but many of these changes involved external features, and the internal parasites were not subjected to the same degree or kinds of environmental influences as were their hosts. The parasite's environment, however, *did* eventually change, but the lag in time between the change of the host environment and of the parasite environment resulted in a difference in rate of evolution: the parasite evolved more slowly than did its host. Evidence for this conclusion is gained from the many examples of highly specialized hosts harboring some generalized parasites. For example, the phylogenetically ancient freshwater silurid and characinid fish of the Amazon and La Plata rivers harbor primitive marine trematodes and isopods. In general, however, the more specialized hosts possess the more specialized parasites (see "Rules of Affinity," p. 547).

During the evolution of bird lice the birds themselves underwent a rapid period of evolution, and by the end of the Eocene most of the modern families were established. Up to this time the lice must have been subjected to conditions of great evolutionary stimulus resulting in the division of the louse population into many partially isolated local populations, and by the end of the Eocene most of the present genera were probably also established. After this time the evolutionary pace of birds slowed down (Miocene birds can often be assigned to modern genera), and relatively few morphologic changes took place. Concurrently, there were reduced stimuli for changes in the bird lice, but in addition, the lice enjoyed a more constant environment than did their hosts, and they probably changed less rapidly than did the latter, as indicated by the existence today of many lice genera being restricted to one order of birds.

The complexity of these associations is illustrated further by the following examples. The bedbug, *Cimex lectularius*, which parasitizes man, is practically identical morphologically with the bug, *C. columbarius*, which parasitizes pigeons, and the two are fully cross-fertile in captivity. It is not known, however, if the two are reproductively isolated in nature. *Ascaris lumbricoides*, the common intestinal roundworm of man, is morphologically identical with *Ascaris* in pigs and probably in other hosts. But cross infection, although possible, is rare, indicating physiologic differences. Similarly, host-races are found with the tapeworm, *Hymenolepis nana*, of man and rodents, and with *Ancylostoma caninum*, the hookworm of dogs. Recall also the discussion of host specificity in Chapter 24. Without more information about each of these and many other examples, it is difficult or impossible to determine whether we are dealing with biologic races or with sibling species (closely related species living together in the same environment).

The interbreeding population of a parasite species may be considered as comprising all of the parasites of that species living on or in one host individual, especially if all developmental stages of the parasites occur on the host. The interbreeding population of a parasite species may also be considered to be all those parasites occurring on or in the entire host population, which may even be worldwide in distribution. Intermediate hosts have a different, and often wider, geographic range than do the final hosts. When intermediate hosts are involved, therefore, the effective interbreeding population of the parasite may be represented by a geographic distribution considerably wider than that of the final host alone. These considerations become exceedingly important when we study the evolution of parasites, because speciation in parasites, as in free-living animals, is the division of a single interbreeding population into two reproductively isolated ones. Thus, emphasis should be placed on the different gene pools that must be present if different species are to be maintained.[13]

When we consider organic change on the basis of genetic factors alone (mutations, random recombinations, random fixation or elimination), we realize that, in a group of a very few interbreeding individuals, the genetic constitution will change more quickly than it will in a large number of interbreeding individuals, chiefly because of sampling error. In a large population there are more varieties, so it takes a longer time to achieve a state of homogeneity. Absolutely maximal rates of phylogenetic evolution could occur, therefore, only in very small populations.

Although the last statement above is dogmatic, there is some contrary opinion. The above discussion assumes that parasites are conservative, but such conservatism may not be real, and parasites may be more evolutionarily progressive than their hosts. Jones[36] has presented evidence for this implied evolutionary opportunism of parasites:

First, instead of lagging behind their host's evolutionary advance, many parasites seem to have overtaken their hosts . . . the acquisition by ancient and primitive organisms (such as protozoa or flatworms) of a recent host (such as bird or mammal) is evidence of evolutionary opportunism . . . Second, parasite life cycles are extremely ingenious . . . it appears that only part of the parasite's life is spent in a nonstressful environment and that much of the life cycle involves highly precarious transitions, which require great adaptability of the successful parasite . . . Third, theoretical considerations of form, population structure, and environment of parasites do not necessarily support the view that parasites are conservative . . . Fourth, the effect of a stable, rich environment upon parasites in their definitive host may not be so conservative as the earlier discussion had suggested . . . perhaps the environmental factors which affect variation are neither positive nor negative in the enriched milieu, and the effect of such an environment upon evolution is actually not significant.

Therefore in conclusion it should be stated that while the question of the rate of the evolution of parasites is an intriguing one, there is at present no satisfactory answer to it. Continued research upon the morphology, taxonomy, life cycles, ecology, distribution, and evolution of parasites (and their hosts) may eventually make clear the way in which these animals evolved and the rate at which they are evolving.

RULES OF AFFINITY

Numerous writers have speculated about the value of parasites as clues to the evolution and affinities of hosts. Out of these speculations has come the word **parasitogenesis,** which refers to the evolution of relationships between the parasite and its host. When examining parasites for a clue to the phylogeny and relationships of hosts, one should consider all the parasites in a body, not just one species. Where evolutionary relationships have become obscure among hosts, a study of their parasites may clear the obscurity.

Several writers have proposed certain rules of affinity between parasites and the phylogeny of hosts. These rules are listed below, but they should be considered only as guides for study, for which *there are many exceptions.* Some parasitologists, for example, reject the view that there is any phylogenetic correlation between nematodes and their hosts. For more details, see Janiszewska,[35] and Szidat.[54]

The Fahrenholz Rule. Common ancestors of present-day parasites were themselves parasites of the common ancestors of present-day hosts. Degrees of relationships between modern parasites thus provide clues as to the parentage of modern hosts.

The Szidat Rule. The more specialized the host group, the more specialized are its parasites; and, conversely, the more primitive or more generalized the host, the less specialized are its parasites. Hence, among stable parasites, the degree of specialization may serve as a clue to the relative phylogenetic ages of the hosts.

The Eichler Rule. When a large taxonomic group (e.g., family) of hosts consisting of wide varieties of species is compared with an equivalent taxonomic group consisting of few representatives, the larger group has the greater diversity of parasitic fauna.

ADAPTATIONS AND PRE-ADAPTATIONS

The term "pre-adaptation" as first used by C. B. Davenport[17] suggests that the organism must have originally possessed characters that made it capable of living in a different environment. The term refers to the property in an animal or plant of suitability for some change in its habit or habitat. The concept of pre-adaptation has received some resistance from among biologists and psychologists, but the differences of opinion seem to be due to differences in definitions rather than to a general acceptance of the belief that adaptive mutations appear as chance mutations suitable for use in a situation before the situation arises.

Julian Hawes, late of the University of Exeter, England, once found some amebas in the intestine of a snake. The method of pseudopodial formation and the nuclear structure of the "parasites" were similar to the free-living *limax* amebas of the soil. The

amebas were tentatively considered as forms picked up with the snake's food, but when they were cultured in vitro they grew equally well in anaerobic and in aerobic media. If certain free-living soil amebas are able to grow and develop in a medium without free oxygen (that is, in a medium similar to the intestine of a vertebrate animal), the soil forms must have become pre-adapted to an intraintestinal life, at least so far as free oxygen requirements are concerned. If, also, they are pre-adapted to food conditions to be found in the vertebrate intestine, they are pre-adapted to the state of parasitism. These amebas give us a relatively simple picture of the role of pre-adaptations in preparing a free-living species for the parasitic habit. The important changes that fit a free-living form for the parasitic habit are physiologic, not the visible structural traits. The physiologic changes must, of course, stand the test of selection. Many species of parasites, such as some gregarines and tapeworms, possess hooks by means of which they attach themselves to the host. How did these complicated structures, which presumably were not present in free-living ancestral forms, ever evolve? Are the hooks essential for the parasitic life? The answer to the last question is, apparently, no, because we find that closely related species can get along very well without hooks. These groups of parasites probably were able to enter and live in their first hosts because of at least some physiologic compatibility. After they once became established, mutations affecting structural and physiologic changes that had adaptational advantages persisted.

Any genetic change that decreases the harm or increases the benefit of either parasite or host is adaptively advantageous and will be encouraged by selection. On the other hand, nonadaptive differentiation (indeterminate change or divergence) in a small population is called **genetic drift.** Hence, a geographic or host divergence may be linked with the nonadaptive divergence due to drift, which may result in the fixation or the elimination of genes entirely by chance with no relation to their selective value.

Can we expect a great amount of genetic change among parasites securely housed in the host? Have we not been accustomed to thinking that species change only if the environment changes? A changing environment, however, *selects* from among genetic varieties *already present* in the gene pool of the population. In the relatively stable environment of an adjusted parasite, the gene pool making up the parasite population gradually becomes larger because all of the nonharmful (including the beneficial) mutations tend to persist. Evidence for the truth of this statement may be obtained experimentally by transferring all the parasites from the intestine of one host to the intestine of another, but closely related, host. In the latter animal, natural selection operates to reduce the numbers of variations to those that can survive in the new environment.

Pre-adaptation, if it existed, enabled only a relatively few organisms to become symbiots. These first parasites were the ancestors of the immense varieties of present-day species.

BIOLOGIC (PHYSIOLOGIC) RACES

Let us assume that a mammal possesses a certain species of ectoparasite, *A*, that is adapted to live in the dorsal body hair. Let us also assume that, during the course of evolution, a group of these parasites begins to find the ventral part of the host's head more to its liking. This divergent group we will call *B*. We find that *A* and *B* are morphologically identical, but *B* becomes better adapted to live on the lower side of the host's head than does *A*. In other words, *A* and *B* are physiologically different, and each belongs to a different biologic race. This kind of divergent adaptation of separate groups within one species of parasite may progress to a state of full species differentiation, and all gradations from incipient physiologic subspecies to inter-sterile species are common. During the process of divergence, the appearance of visible morphologic differences lags behind the appearance of physiologic differences, and barriers to intercrossing are quickly estab-

lished by natural selection. The method by which such differentiation originates is not clear to biologists. Our greatest difficulty arises, of course, in the detection of minor, or even many major, changes of a physiologic nature.

Trypanosoma gambiense, T. rhodesiense and *T. brucei* are morphologically indistinguishable, but they differ in their host preference and in their effects upon the hosts. These flagellates are probably biologic races of one species. We have found that there is no way to distinguish clearly, on a morphologic basis, the forms of *Entamoeba* from sheep, pigs, goats and cattle.[46] These amebas probably belong to one species, but cross infection experiments will have to be made before proof can be established. Possibly, however, each group of amebas represents a true biologic species, but since they do not reproduce by sexual union, proof must be obtained by criteria less satisfactory than the interbreeding test. In the same manner the mange-mites of the species *Sarcoptes scabiei* are probably divided into biologic races, each adapted to a single host species—sheep, dogs, goats, camels, horses, rabbits, men, etc.

Hoare[26] has emphasized that "there is no fundamental difference between geographical races of free-living organisms separated spatially and biological races of parasites segregated by differences in hostal environment." There are numerous examples "of the existence among parasitic micro-organisms of intraspecific groups or 'types' in which the chief differential characters are represented by the pathological immunological manifestations produced by them in the host." Pathogenic trypanosomes, for example, can rapidly develop different "strains" that are immune to antibodies of the host.

The ichneumonid insect, *Nemeritis canescens*, normally parasitizes only the larvae of the meal moth, *Ephestia kuhniella*. Ichneumonids were reared experimentally on larvae of the wax moth, *Achronia grisella*. Whereas all adult females of the ichneumonid species possess a genetically-determined response to the odor of meal moths, those that were reared on the wax moth, or even brought into close contact with it immediately after emergence from the pupa, demonstrated an additional response (attraction) to the wax moth.[56] Later work showed that this larval conditioning depends on a tendency to be attracted by any olfactory stimulus characteristic of a favorable environment. Thorpe[55] concluded that "the theoretical importance of such a conditioning effect is that it will tend to split a population into groups attached to a particular host or food plant, and thus will of itself tend to prevent cross-breeding. It will, in other words, provide a non-hereditary barrier which may serve as the first stage in evolutionary divergence." As Huxley[31] put it, we have here "a beautiful case of the principle of organic selection . . . according to which modifications repeated for a number of generations may serve as the first step in evolutionary change, not by becoming impressed upon the germ-plasm, but by holding the strain in an environment where mutations tending in the same direction will be selected and incorporated into the constitution. The process simulates Lamarckism but actually consists in the replacement of modifications by mutations."

The causes of biologic race production are not clearly understood, as was pointed out above. If a species is subject to high mortality through the activities of a predator, we can speak of **predator-pressure**. Likewise we can use the term **competitor-pressure**. With a decrease in predator-pressure and a decrease in competitor-pressure there is a decrease in **selection-pressure**. That is, there are fewer agencies that tend to "kill off" the less favorably adapted individuals of a species. We are using Simpson's definition of the term "selection" as anything tending to produce systematic heritable change in populations between one generation and the next. We should now recall the phenomenon of continuous, more or less random mutations in all species. It follows that decreased selection-pressure is attended by increased variation. In the preface to the first edition of this book we stated that parasitism is "a life of

large income without work." The parasitic life has greatly reduced the danger from destruction by predators, and from interspecific competition for resources. From the arguments presented above, it follows that one would expect more variation among parasitic species as compared with nonparasitic species Among the parasitic protozoa we frequently find apparently distinct species with extremely small or no morphologic differences, with overlapping geographic or host distribution, and with a similarity in ecologic preferences—a puzzling situation indeed. Examples are illustrated by *Balantidium*, trypanosomes, *Plasmodium*, Myxosporida, *Hexamita*. Probably, the state of parasitism favors the persistence and the accumulation of genetic variations, which result in preservation of relatively large numbers of biologic races. These races are incipient species.

Nature, however, is full of checks and balances, and we must not forget that the parasite, even a mild one, carries on an assault against its host and that the host is a responsive environment. For these reasons, selection *does* take place and it limits the numbers of incipient species of parasites.

A clue to answers to such problems as those raised above lies in more careful studies of environmental changes within a host organ. The intestine, for example, contains a large community of populations forever growing, changing, dying. Fluctuating populations, streams of metabolic by-products, changes in enzyme concentrations, and, in general, the surge of chemical activity coincident with the processes of ingestion, assimilation, respiration and excretion of thousands of organisms crowded together do not support a stable or constant environment.

The student should be cautious when using such terms as "host-races," "biologic species" and "ecologic races," because of the obvious difficulty of distinguishing morphologically similar species (reproductively isolated) from morphologically similar but physiologically different races (reproductively compatible), and from phenotypically different populations that may or may not be reproductively isolated.

ORIGINS OF PARASITISM

On the preceding pages of this chapter we have dealt with mechanisms of speciation, using parasites and their hosts as examples. We could have substituted free-living forms, because whereas parasites often complicate the problem of interpretation, they do not present us with new, basic principles. We now turn to some questions on the origins of the parasitic habit. Assuming that the *first* parasites were derived from free-living species, how did they become parasites?

If we review the several theories on the origin of life, we recall that primitive free-living organisms gradually used up the available supply of complex organic substances and at the same time developed new enzymes that enabled them to synthesize those complex substances from available chemical precursors. Now, if the first virus-like microorganisms failed to develop these enzymes, and if they depended upon other primitive forms of life in whose bodies were to be found the only source of complete food, the virus-like organisms would, if they were to survive, become obligate parasites. When bacteria finally appeared, the viruses found a ready supply of hosts. To use the words of Dodson,[19] "The evolution of the bacteria made parasitism possible." On the other hand, the viruses have been called "degenerate" parasites.

Cytoplasmic components of a cell appear to be independent, in varying degrees, of the cell in which they are contained.[38] This independence supports the theory that these cytoplasmic components may have arisen through symbiotic association between microbial forms and colonies of "virus" or higher units. The cell nucleus, according to this view, is the descendent of the original virus colony with its surrounding envelope, and the cytoplasm consists essentially of the descendents of symbiotic organisms, together with products of the nucleus and substances resulting from their interaction. Such speculations on the role of symbiosis in evolution fires the imagination, but they lack experimental evidence to support them. Studies on DNA and RNA molecules may provide some evidence.

We may assume that at the time the first preparasite entered a host, the guest was either already resistant to host antibodies (that is, it was pre-adapted to life within the host), or it was able to change rapidly so as to become tolerant of the host. The host, on the other hand, had to be, or quickly became, tolerant of the guest. The changes had to be genetic if the relationship was to become permanent. In Chapter 21 of this book we mentioned several important problems that must be solved by any parasite. Problems of obtaining food, of leaving the host alive, and of regaining entrance are as important for continued success in parasitism as is the problem of combating host antibodies.

In many parasite-host relationships, the parasite is attracted to the host because of the presence of some kind of chemical compound elaborated by the host. Such compounds are presumably detected by the sensory apparatus of the parasite (see p. 538). One way in which a wide range of hosts might be secured by one species of parasite during its evolution is the fortuitous production by a "foreign" animal of a metabolite similar to the "attractant" produced by a host animal. Given sufficient adaptability on the part of the parasite, the foreign animal becomes another host. Davenport[18] felt certain that this kind of behavior takes place.

As we have already seen, the reproductive organs are the most conspicuous and important of all internal structures in parasitic worms. It has often been stated that animals that produce immense numbers of eggs must do so in order to compensate for the tremendous loss of eggs and young sustained by the species. This explanation is teleologic, but the *result* of the production of a large number of eggs is the assurance of survival of a few.

Secondary hosts could have been introduced into a life cycle because of the ability of an embryo stage or larva of the parasite to survive in many kinds of animals. The larva (e.g., of a cestode) would have been resistant to the antibodies of the new host, but it might have stimulated more antibody production so that, in order to survive, the larva would have had to protect itself by forming a cyst wall around itself. The new (transfer) host would, of course, have had to be part of the food supply of the original host to be of any value in completing the life cycle of the parasite.[63]

Most biologists agree that the intermediate host, particularly when it is an invertebrate, generally represents the ancestral single host. The assumption is made that at one time the parasite enjoyed the embrace of a single invertebrate host, but that infective stages, through the same evolutionary mechanisms as those that result in any parasitism, gradually became parasites in another host closely associated with the first one. A possible example is that of *Plasmodium*, the malarial parasite. In this case the mosquito is sometimes considered to be the "definitive" host harboring the sexual phases of the life cycle, and the vertebrate is the alternate secondary host. This evolutionary sequence would explain the more compatible adjustment and apparent lack of pathogenicity between *Plasmodium* and the mosquito. The significance of the degree of host reaction, however, is open to question. The mosquito vector is considered by some authors to represent a secondary involvement. Ball[4] has made a survey of instances in which the degree of pathogenicity would be a poor guide for determining the evolutionary time relationships between host and parasite. He has emphasized that a high degree of pathogenicity is not "*prima facie* evidence of recent and still imperfect development of the host-parasite relation." We must bear in mind that many factors beyond the mere length of association time play significant roles in determining the incidence and degree of pathogenicity of parasitic infections in new hosts.

The intriguing concept of "adaptation tolerance" has been proposed by Sprent.[50] The general idea is that, whereas new hosts react to the maximum of their potential against a new parasite, as parasites and hosts become adapted to each other, the latter gradually lose the ability of immunologic reactivity, so that they become tolerant, in the immunologic sense, of their parasites. Sprent suggested that this tolerance might have been achieved in two ways: (1) by the selection of parasites whose antigenic structure tended to approximate those of the hosts, so that the antigens of the parasites came to comprise the same immunologic determinants as the "self" components of the host, and thus became immunologically inactive (for example, the loss of those parasite enzymes that differ appreciably from enzymes present in the host); (2) by a tendency towards the obliteration of antibody patterns that correspond to antigens possessed by the parasite.

The adaptation-tolerance idea was stimulated by Burnet's[6] clonal selection hypothesis of acquired immunity. Burnet suggested that antibodies are produced by clones of lymphocytes, each clone representing a particular combining-pattern that corresponds to a particular antigenic determinant. He proposed that all possible antigenic patterns have their counterparts in the lymphocytes produced during the development of an individual. The net result is that, instead of the antigen directly influencing the synthesis of globulin, it merely selects an appropriate pattern and instigates the proliferation of a particular clone of cells. If, as a result of selection, parasites become associated with hosts that lack certain patterns, (i.e., those corresponding to parasite antigens), then the host is incapable of recognizing the parasite.

"PROGRESSIVE" AND "RETROGRESSIVE" EVOLUTION

Much of our discussion on the preceding pages has implied "progressive" evolution, that is, changes that in some way improve an organ or enhance a function. But evolution may also be "retrogressive." The development of the parasitic habit almost always involves some regression in evolution, largely because natural selection does not work against mutations that damage a useless character.

Whereas the loss of such structures as the intestine in the evolution of tapeworms (or at least in the evolution of the turbellarian-like ancestors of tapeworms) is clearly a result of "regression" in the usual anthropomorphic sense, from the worm's point of view it is "progressive," because the animal has evolved to a situation in which it no longer needs an intestine, and its physiologic energies can now be concentrated on the more important function of producing eggs.

When we say that an animal is highly specialized, we are saying that evolutionary changes during its phylogeny have been extended in complexity beyond those changes characteristic of animals that are less specialized. "Degenerate" changes are essentially a form of specialization, but obviously many structures and habits not of a degenerate nature are also specialized in parasites. Julian Huxley has defined "specialization" as improvement in efficiency of adaptation for a particular mode of life, and "progress" as an improvement in efficiency of living in general; but we are inclined to agree with J. B. S. Haldane who denies the existence of evolutionary *progress*, and calls it a mere anthropomorphism. The terms "degenerate," "regressive" and "retrogressive" are, of course, also anthropomorphic, and perhaps should not be used.

Organisms having a high degree of specialization do not give rise to new types. Only the relatively unspecialized group is able to give rise to new forms. Consequently, parasites as a whole are worthy examples of the inexorable march of evolution into blind alleys.

One category of parasites may be an exception to the generalization that parasitism is a comfortable shortcut to evolutionary blind alleys. Examples of insects that are parasitic as larvae (e.g., wasps that lay their eggs in the larvae of other insects) are well known. Recall, for a moment, the advantages of being a parasite, and compare the protection, ready food supply and shelter enjoyed by the parasite with the relatively greater struggle endured by the free-living animal. In other words, parasitism may not be as hard on the parasite as the free life is on the free-living animal. Recall also that the more specialized the parasite, the more irrevocable the habit. The benefits bestowed by parasitism are more marked in larval forms when growth of the parasite is rapid, but often the larva soon leaves its host and emerges as a free-living organism before there has been time to sacrifice its independence. Some entomologists believe that all the hymenoptera, as well as the Cyclorrhapha flies, including the house fly, have been descended from ancestors that were parasitic in their larval stages. The French zoologist Giard gave the name "placental parasite" to the mammalian fetus because it lives at the expense of its mother, the "host," and in all essential ways resembles a parasite that has adjusted to its environment with marked success. During the period of rapid growth of the fragile "parasite," it is far better cared for and protected than is its free-living mother. Obviously, the parasitic young is not a parasite of the host *species*, but only of the individual. Rothschild and Clay[49] have remarked that "a fundamental distinction can be drawn between the parasitic adult and the parasitic young, the full significance of which has not hitherto been fully appreciated. In the former, parasitism appears to lead to dependence and a loss of evolutionary potential, whereas in the immature stages, it may, on the contrary, prove to be a successful and progressive step." Thus, benefits can accrue to animals that are parasitic only in their larval stages.

On the following pages dealing with the origins of the several different groups of parasites, much of the discussion is no more than an intriguing exercise of the imagination. But new facts are daily coming to light and eventually the true pattern of parasite evolution will be revealed.

ORIGINS OF SPECIFIC GROUPS

Protozoa

Entozoic protozoa may have been derived from ectoparasites, as various authors have suggested. It does seem logical to suppose that free-living forms first became associated with hosts as casual commensals loosely attached to the skin or gills, then gradually fortified their position by moving into the mouth, gill chambers, anus, and other openings. But ectozoic forms are mostly primitive ciliates, and flagellates, and only a few genera such as *Trichodina* and *Hexamita* contain both ectozoic and entozoic species. Another logical guess as to origins of parasitic protozoa is that they were derived from species accidentally ingested by their future hosts. When we consider the large numbers of protozoa that are steadily ingested with the food of larger animals, and when we think of the nutritional benefits and the protection and moisture provided by the intestine, we appreciate the inescapable advantages for survival furnished by lodgings in a gut. Once established in the intestine, the parasite could migrate to all other parts of the body.

Since the parasitic habit among protozoa is not limited to exclusively parasitic groups but is scattered among orders containing free-living species, the parasitic habit probably arose frequently and independently from different groups of free-living ancestors. Sporadic and temporary invasion by free-living species into hosts may be comparable with the initial step in the origin of endoparasitism. For example, species of the Euglenida are sporadically found in tadpoles and in millipedes, and *Tetrahymena* is occasionally found in such sites as the digestive tracts of slugs, the coelom of sea urchins, the hemocoel of insects and the gills of the amphipod, *Gammarus pulex*.

When the complete life cycle of a protozoon is known, a critical examination of the various stages will often lead to information suggesting phylogenetic relationships with neighboring groups. Such a study of the opalinids (parasitic in the large intestine of tailless Amphibia), with particular attention to the infraciliature and mode of fission, has led to the removal of this group from the ciliates and the placement of it with the flagellates.

Baker[3] has reviewed the literature on the evolution of parasitic protozoa, and he has combined some current theories with some of his own to propose the following hypothesis, which we feel is the most logical of those postulated. The Sarcomastigophora, known as ameboflagellates, are the least-changed modern descendents of the original protozoa. From this assemblage of ameboflagellates, specialization probably began along three main directions. The first was characterized by a suppression of the ameboid phase, giving rise to the true flagellates; the second by suppression of the flagellate phase, giving rise to the amebae (Sarcodina); and the third by suppression of both ameboid and flagellar phases without completely losing either, giving rise to the Sporozoa. "The Ciliophora presumably arose from the flagellate line of development by an increase in the number of locomotor organelles and, subsequently, the development of a very complex pellicular and sub-pellicular morphology."

In 1935 Wenrich[62] made a comprehensive study of the hypothetical origin of parasitism among protozoa, and we shall select samples from his study. A number of investigations have shown that, when free-living protozoa first become entozoic, they do not necessarily undergo any marked morphologic modification. If protozoa intermediate in behavior and habitat between free-living and parasitic can be found, they should present a clue to an answer to the question of origin raised above. Such intermediate forms are common. Wenrich[61] described a holotrichous ciliate, *Amphileptus branchiarum*, found on the gills of tadpoles. The ciliate has a free-swimming stage that roams over the gills devouring ectozoic *Trichodina* or *Vorticella*. At other times, and more commonly, the "parasite" attaches itself to the tadpole gills by a thin membrane within which it gently rotates, pausing now and then to indulge its predacious tendencies to engulf masses of gill cells. The ciliate is, perhaps, in the process of changing over from a free-living predacious organism to a parasitic one. Other common species of *Amphileptus* are predacious.

Wenrich discovered some colorless euglenoids, belonging to the genus *Menoidium*, in the gut of the milliped, *Spirobolus marginatus*. He attempted to infect the millipeds by feeding both *Menoidium* sp. and *Euglena gracilis* to them, and he found that both (the former more successfully) were able to survive within the host intestine for a few days, but neither was able to become established as a permanent entozoic flagellate. Wenrich concluded that the *Menoidium* "displayed the facultative capacity of maintaining for a brief time, at least, an entozoic existence."

This situation led Wenrich to conclude that the host, representing a special and limited environment, has not had a marked directive influence on the evolution of its parasites; otherwise there should be more evidence of convergence in evolutionary trends among the parasites. It should be recalled, however, that hosts markedly influence physiologic changes in their parasites. Such changes, as we have already indicated, are the first to take place in evolutionary divergence and, indeed, are often the only significant modifications to occur. After all, the protozoa have had a longer time than any other animal phylum in which to evolve. They have invaded almost every possible ecologic niche, and some genera, such as *Hexamita*, are to be found in a wide variety of unrelated hosts without exhibiting significant morphologic changes.

Speculations on the evolution of the suborder Trypanosomatina have been made by Baker.[3] The promastigote body form (see p. 28) is generally considered to be the most primitive type of the family, and ancestral flagellates presumably were parasites of the gut of invertebrates, although some authors postulate a vertebrate origin. This ancestral promastigote probably led, on the one hand, to the genera *Leptomonas* and *Phytomonas*, and on the other hand, to the genera *Trypanosoma*, *Rhynchoidomonas* (including "*Herpetomonas*" as used by Wenyon), *Blastocrithidia* and *Crithidia*.

Methods of transmission offer clues as to the kinds of evolution experienced by the genus *Trypanosoma*. Thus the contaminative type of infection in *Trypanosoma cruzi* is evi-

dently the more primitive. Hoare[27] has suggested that the origin of inoculative transmission and its attendant form of parasite life cycle, as in *T. gambiense*, may be a secondary acquisition that originally developed in the hind-gut of the insect vector. Such trypanosomes may have been taken up by tsetse flies that began to transmit them mechanically to new vertebrate hosts, but when the flagellates adapted themselves to development in the proboscis and/or salivary glands, tsetse flies became their new obligatory transport hosts. Evidence for this hypothesis is presented by *T. vivax*, which develops only in the mouth parts of its insect vector, and by *T. congolense* (representing the next step in evolution), which develops in the mid-gut of the insect, and finally by the *Brucei*-group, which utilizes the mid-gut and then the salivary glands of the tsetse fly. A final bit of evidence for the described phylogenetic relations among trypanosomes is provided by the differences in susceptibility of their vectors to infection. Practically 100 per cent of triatomid bugs fed *T. cruzi* become infected, while fewer than one per cent of tsetse flies fed *T. brucei* become infected. The obvious conclusion is that *T. cruzi* and its bugs represent a much older and more stable association. Moreover, tsetse-borne trypanosomes easily lose the power to develop in the insect host and they may revert to mechanical transmission. Such a transformation is illustrated by *T. vivax* in cattle of South America, and by *T. evansi*, which presumably originated in Africa from *T. brucei* (see p. 36). A final step in the evolutionary series is *T. equiperdum*, which has become completely emancipated from an insect vector, and is transmitted directly from horse to horse by contact during the sexual act.

The evolution of the large and complex group of flagellates that inhabit the intestines of termites stemmed from the simple *Monocercomonas* or from a *Monocercomonas*-like form. This form has an uncomplicated parabasal body and axostyle, three free flagella, and an adherent or free-trailing flagellum. The Trichomonadidae have added a costa and an

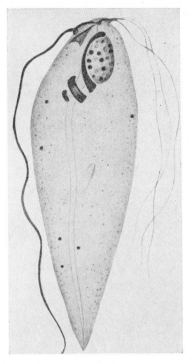

Fig. 25-1. *Devescovina cometoides*, from the termite *Cryptotermes dudleyi*. (Kirby, courtesy of Univ. Calif. Pub. Zool.)

undulating membrane in place of the recurrent flagellum (see Fig. 2–20, p. 46). *Devescovina* (Fig. 25–1) is similar to *Monocercomonas*, but it possesses a triangular cresta and its parabasal body is coiled around the axostyle. The Calonymphidae are derived from the Devescovinidae. For detailed discussion of these relationships, see Holland.[28]

Sarcodina have arisen from the Sarcomastigophora by a loss of the flagellar stage, and they appear to represent a polyphyletic group. Numerous examples of ameboid flagellates lend ample support to these conclusions. *Tetramitus* has ameboid stages but is usually classified as a flagellate. A line of evolution through this type of protozoon has lead to the ameba, *Vahlkampfia*. Another ameba, *Naegleria*, is strikingly similar to *Vahlkampfia* but unlike the latter it has flagellate stages. *Dientamoeba* has apparently arisen through another line involving the ameboid flagellate *Histomonas*.

Soil amebas (e.g., *Acanthamoeba*) may enter the noses of mice and monkeys, and migrate to the lungs or brain causing severe lesions. These amebas can be taken from the soil, grown in cultivation media, injected into laboratory animals and show immediate pathogenicity. Experiments like these prompt a re-evaluation of our concept that amebas have evolved from free-living forms slowly on their way to becoming parasites. Preadaptation to parasitism plays a significant role and undoubtedly the length of time it has taken for free-living organisms to travel the road to parasitism has varied widely.

Little attention has been given to the evolution of the subphyla Sporozoa and Cnidospora. A group of workers in Russia, headed by the late Cheissin, has been much interested in the taxonomy and phylogeny of these obligatory parasites. Cheissin[12] reminded us that they possess some structural characteristics suggestive of the flagellates (e.g., merozoites similar to promastigotes; flagellated microgametes similar to *Bodo;* sexual processes similar to those of phytomonads). However, an ameboid method of locomotion is common among these parasites. They probably arose from flagellates or ameboflagellates possessing life cycles similar to those of present-day phytomonads. The sporozoans and cnidosporans probably first became adapted to parasitism in the intestines of aquatic invertebrates, then moved to terrestrial and aquatic vertebrates. The gregarines today have retained the ancestral characteristic of inhabiting the intestinal lumen of invertebrates.

The primitive telosporean stock appears to have given rise to the gregarines, then to the coccidia (see Baker[3] for detailed speculations). The Haemosporina (including the malarial parasites and their close relatives) arose from the coccidia, but there is a controversy as to whether the ancestral group were coccidia of vertebrates or of invertebrates. In favor of the latter is the fact that the Haemosporina are less pathogenic to their invertebrate hosts than to their vertebrate hosts, and they are more host-specific to their invertebrate hosts.

In favor of the view that they have evolved from coccidian parasites of vertebrates is the tendency of coccidia to become tissue parasites of vertebrates, and the fact that malaria occurs in birds and reptiles that probably originated before the advent of blood-sucking flies. Finally, as Bray[5] pointed out, "if the haemosporidia (= Haemosporina) are coccidian by nature and originally insect parasites. . . . It is we who should be carrying the sporozoites or the sporogony stages and the mosquito whose gut should contain the exocoelomic schizogony stages and their haemocoelomic fluid which should suffer schizogony and gametogony stages." We are inclined to join those who believe that the Haemosporina have evolved from coccidian parasites of vertebrates.

The most comprehensive works on the phylogeny of the protozoa have been those concerned with the ciliates. The new approach to phylogenetic problems of the ciliates utilizes the subpellicularly located basal granules, or kinetosomes, which are intimately and indispensably associated with all external ciliary systems. This **infraciliature,** as it is called, is present even in the absence of external ciliature. The approach also focuses attention on the ontogeny or morphogenetic aspects of ciliate development. From an unknown zooflagellate ancestry the Gymnostomatida, a large order embracing a great variety of forms, is situated at the base of the ciliate evolutionary tree. Two examples of this order are *Amphileptus branchiarum*, common on the skin and gills of frog tadpoles, and *Chilodonella cyprini*, a skin and gill parasite of fish. The gymnostomes gave rise to the order Trichostomatida, in which is placed the well-known *Balantidium*. From the trichostomes arose the order Hymenostomatida, which stands at a major crossroads in ciliate evolution. A primitive member of this order is the highly pathogenic fish skin parasite, *Ichthyophthirius* (Fig. 5–4, p. 120). The hymenostomes gave rise to the Heterotrichida, but the former were also ancestors of other orders containing parasitic holotrichs. These orders include Peritrichida, the largest of all

ciliophoran orders, in which is placed *Urceolaria* and *Trichodina* (Fig. 5–7, p. 124), both to be found on or in many aquatic animals.

The subclass Spirotricha is characterized by a buccal ciliature composing the adoral zone of membranelles, the prominence of which is accentuated by an increasing loss of simple somatic ciliature. The first order, Heterotrichida, in this subclass contains *Licnophora* and other marine ectocommensals, and it probably gave rise, through the oligotrichs, to the order Entodiniomorphida. It is in the latter order that we find the highly specialized stomach commensals of herbivorous mammals. For details of ciliate systematics and evolution, see Corliss.[15,16]

The Flatworms

The first metazoa were almost certainly radially symmetrical animals, and speculations on the origins of the flatworms always run into the thorny question of an alteration from radial to bilateral symmetry. Hyman[32] reviewed the several theories of origin, and she presented convincing arguments to support the view that an ancestral planuloid form (resembling the ciliated-larva of coelenterates) gave rise on the one hand to coelenterates and on the other hand to the flatworms. This planuloid type of ancestor was probably a radially symmetrical and elongated organism without mouth or digestive tract, with a ciliated or flagellated epidermis, and with ingestion and digestion processes of the protozoan type. It was undoubtedly polarized with definite anterior and posterior ends. The planuloid theory is based upon the remarkable features of the primitive flatworms of the order Acoela (class Turbellaria). These small worms have the coelenterate characters of a ciliated epidermis, often syncytial, and often with basal muscle fibers. The interior of the worm is a solid mass of cells as in a planula larva, but there is a central, ventral mouth leading to a short pharynx. Gametes are differentiated from cells of the interior cell mass, the animals are hermaphroditic, and internal cross fertili-

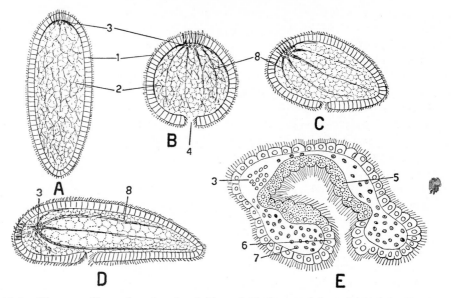

Fig. 25-2. Diagrams illustrating the planuloid-acoeloid theory of the origin of the Bilateria. *A.* Planula larva, mouthless with apical nerve center. *B.* Mouth formed, oral-aboral axis shortening. *C.* Body elongating in a sagittal (originally transverse) plane, nervous center shifting forward. *D.* Acoeloid stage, a bilateral creeping worm with anterior nerve center. *E.* Later stage, archenteron formed; actual embryonic stage of a polyclad (after Surface, 1907).

1, ectoderm; *2,* entodermal mass; *3,* nervous center; *4,* mouth; *5,* archenteron; *6,* stomodaeum; *7,* mesoderm; *8,* nerve cords. (Hyman, The Invertebrates, courtesy of McGraw-Hill Book Company)

zation is the rule. Obviously there is no difficulty in visualizing an evolution from planuloid ancestors to an acoeloid form (Fig. 25-2). Perhaps one objection to the hypothesis that all flatworms, including the Mesozoa, have been derived from a group of hypothetical, generalized, planula-like progenitors, may be based on the fact that there is no close comparison between the oncosphere and any known type of planula larva. Contrast, for example the figure of a pseudophyllidean oncosphere (coracidium) (Fig. 25-4, p. 560) with that of a planula. (See also papers by Stunkard.[51,52])

Major differences of opinion exist regarding the origins of the different groups within the phylum Platyhelminthes.[48] The order Rhabdocoelida (class Turbellaria) consists of small worms characterized by a mouth and unbranched digestive tract, protonephridia, oviducts, and nervous system usually with two main trunks. Most of these worms are free-living in salt, brackish, or fresh water,

but a few are parasitic in other turbellarians, mollusks, echinoderms and crustaceans. Within this order, the suborder Temnocephalida represents one of the few groups of parasitic rhabdocoels. Temnocephalids (Fig. 6-2, p. 131) are leech-like, almost devoid of cilia, and the anterior end is extended into two to 12 tentacles. The posterior end is equipped with one or two adhesive disks, the pharynx is barrel-shaped (doliiform), and the gonopore is single (common sex pore). Temnocephalids are ectocommensals on freshwater animals, chiefly crustaceans, but also on turtles and snails, frequently occurring within the branchial chamber. They do not derive nourishment from their hosts, but capture and devour insect larvae, rotifers and other small crustaceans. Another group of rhabdocoelids containing parasitic forms is the Dalyellioida within the suborder Lecithophora. Dalyellioids possess a doliiform type of pharynx, no proboscis, mouth generally at the anterior tip of the body, a single gonopore,

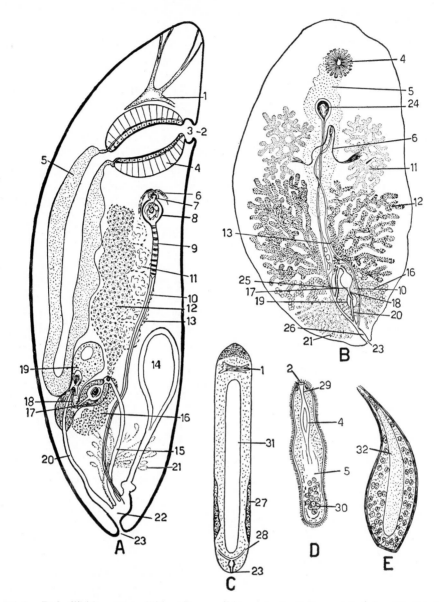

Fig. 25-3. Dalyellioida. Umagillidae, Fecampiidae. *A. Anoplodiera voluta* (after Westblad, 1930), from the gut of sea cucumbers, sagittal section. *B. Syndisyrinx* from the gut of California sea urchins, from life. *C–E. Fecampia* (after Caullery and Mesnil, 1903). *C.* Adult without mouth or pharynx. *D.* Juvenile stage with digestive system and eyes. *E. Fecampia* secreting the capsule containing many clusters of eggs and yolk cells.

1, brain; *2*, mouth; *3*, pharyngeal cavity; *4*, pharynx; *5*, intestine; *6*, sperm ducts; *7*, spermiducal vesicle; *8*, seminal vesicle; *9*, penis bulb; *10*, penis stylet; *11*, testis (only one shown in *A*); *12*, yolk gland (one shown in *A*); *13*, ejaculatory duct; *14*, uterus; *15*, vagina; *16*, ovary; *17*, seminal receptacle; *18*, insemination canal; *19*, copulatory bursa; *20*, bursal canal; *21*, cement glands; *22*, common antrum; *23*, common gonopore; *24*, long-stalked capsule in the uterus; *25*, ovovitelline duct; *26*, vagina; *27*, hermaphroditic gonad; *28*, gonoduct; *29*, buccal tube; *30*, yolk in intestine; *31*, space; *32*, worm inside capsule. (Hyman, The Invertebrates, courtesy of McGraw-Hill Book Company)

and no rhamnite tracts (Fig. 25–3). These parasites occur within the bodies of mollusks, echinoderms and other turbellarians.

As was suggested above, all the orders of Platyhelminthes have undoubtedly arisen either from the Acoela or from forms closely resembling them. The parasitic classes of flatworms probably arose from dalyellioid rhabdocoels. As Hyman[32] reminds us, it is noteworthy that these rhabdocoels occur primarily in mollusks and echinoderms. Mollusks are the chief intermediate hosts of flukes, and echinoderms are believed to be situated in the main line of evolution of vertebrates. The redial stage of trematodes resembles a rhabdocoel. Ancestral adult flukes presumably parasitized mollusks, and when vertebrates arrived, the flukes (by elaboration of adhesive organs already present in rhabdocoels, and by modification of the reproductive system in the direction of greater egg production) became adapted to verte-brate hosts, but retained their dependence upon molluscan hosts.

In a study of the evolution of parasitic platyhelminths, Llewellyn[41] constructed a tentative evolutionary scheme based on newly-hatched larvae (Fig. 25–4). The scheme assumed that the larvae may resemble the juvenile stages of the corresponding ancestral adults. A summary of Llewellyn's ideas is shown in Figure 25–5. He discussed all of the arguments for and against the theory that the Monogenea are more closely related to the cestodes than to the digenetic trematodes.[8] He subscribed to this theory, and based his conviction partly on the similarity between the "cercomere" of monogeneans and cestodes. He traced what he believes to be the evolution of endoparasitic strobilate cestodes from early monogenean ectoparasites of vertebrates. A comparsison[42] of the sulphur content of larval hooklets of various cestodes with the sulphur content of digenean

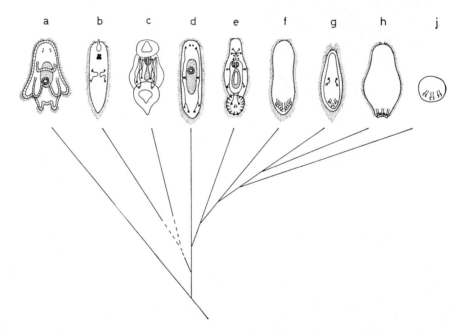

Fig. 25-4. Newly-hatched larvae of platyhelminths, arranged in a tentative evolutionary scheme. *a.* Muller's larva of a polyclad. *b.* Miracidium of a digenean. *c.* An aspidogastrean based on *Aspidogaster conchicola*, which has no cilia or eyes, although some aspidogastreans have these features. *d.* Juvenile of a rhabdocoel. *e.* Oncomiracidium of a monogenean. *f.* Decacanth (lycophore) of a gyrocotylidean. *g.* Coracidium of a pseudophyllidean. *h.* Larva of an amphilinidean (six hooks plus four supplementary sclerites). *j.* Hexacanth oncosphere of a cyclophyllidean. (Llewellyn, courtesy of Blackwell Sci. Pub.)

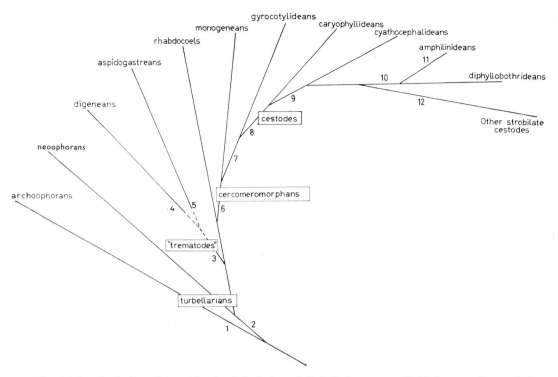

Fig. 25-5. Evolution of parasitic platyhelminths. *1.* Spiral cleavage, undivided ovary, flame cells? *2.* Irregular cleavage, germarium-vitellarium, flame cells present. *3.* Endoparasitism in mollusks. *4.* Polyembryonic larval multiplication in mollusk; pre-adult actively leaves mollusk and parasitizes vertebrates. *5.* No polyembryony; adults do not actively leave mollusk host. *6.* Ectoparasitism on vertebrates. *7.* Endoparasitism in intestine of vertebrates. *8.* Two-host life-cycle; six-hooked larva. *9.* Strobilization. *10.* Incorporation of a second intermediate host into life cycle. *11.* Progenetic development in second intermediate host, loss of definitive host, four supplementary larval hooks. *12.* Loss of tanning, closure of uterus, apolysis. (Llewellyn, courtesy of Blackwell Sci. Pub.)

spines and stylets provides some support to the concept that cestodes and monogeneans are more closely related to each other than either one is related to digeneans.

Let us now consider each helminth group in order.

Monogenea. One might imagine that the free-living ancestors of all trematodes began feeding on mucus and dead epithelial cells of their "hosts," particularly the mucus of the gills, and then gradually became adapted to live in deeper gill areas, then in the pharynx and finally in the intestine. Except for *Polystoma* (see below), however, there is no evidence for such an hypothesis for any of the Monogenea.

A better hypothesis is as follows: From a planuloid ancestor a rhabdocoel worm was developed. This worm gave rise to several forms, including the nemertians (Rhynchocoela), temnocephalans and dalyellioids. From the latter group a stock form, without asexual reproduction, with anterior mouth, paired testes, one or two ovaries, and with a copulatory bursa, gave rise to a form with a paired protonephridial system. This form then led to the Monogenea, the Aspidobothria, and the Digenea. The large majority of monogenetic trematodes are highly host-specific parasites of fish, particularly of the ancient group of cyprinoids, which suggests that this parasitic group is of great antiquity. The apparent lack of specificity in some genera (e.g., *Benedenia*, *Polystoma*) does not necessarily indicate recent origins. The genus *Polystoma*, occurring on the gills of tadpoles, in the bladder of amphibians and turtles, and in the mouth or nose of the latter, is of special interest. The larval parasites are found attached to the gills of tadpoles, and when host metamorphosis takes place the worms migrate to the alimentary canal and thence to the urinary bladder where they become adults. This unique life history suggests a method whereby ectoparasitic worms may have evolved into endoparasitic worms, but it does not support a supposition that entozoic Digenea were derived from ectozoic Monogenea. After all, the Digenea life history almost always involves a molluscan host, and original ancestral forms probably were free-living miracidia-like organisms.

The ancestral monogeneans may have been facultative skin parasites whose most direct descendents are the modern capsaloideans.[40] The early skin parasites gave rise to forms that sought the protection of gills and then acquired the habit of obligatory blood-feeding. The latter mode of nutrition led to symbiosis with bacteria and consequent acquisition of a genito-intestinal tract, and it also required a physiologic adaptation to the blood-borne antibodies of the host. These obligatory blood-feeding gill parasites, the ancestral polyopisthocotylineans, then radiated in development in correspondence with the radiation of their hosts. An ancient divergence from the Capsaloidea, characterized by viviparity, appears to have given rise to the Gyrodactyloidea.

Two possible explanations for the persistence of the ectoparasitic nature of the Monogenea have been suggested. First, their oxygen requirements may have prevented occupancy within the host, and second, being attached exclusively to aquatic hosts, there was no need to migrate inside to avoid desiccation. The latter explanation suggests that "getting inside" is the result of some sort of positive selection pressure—an unnecessary suggestion because present habitats of trematodes may simply reflect the original habitats adopted by ancestral forms.

Aspidobothria. Knowledge of a complete life cycle of an aspidobothrid is lacking, but

development is believed to be basically direct; that is, the young worm develops directly into an adult either in the same host or in another individual of the same or different species. The worms definitely do not belong to the monogenetic group because there is no posterior sucking disc or cuticular hooklets or anchors. The excretory pore or pores are posterior rather than anterior in position, and the intestinal tract is always rhabdocoelic in type, as in the turbellarians and in the digenetic gasterostomes. On the other hand, there is no evidence of an alternation of generations. The larva has no trace of ciliated epithelium.

The simplest type of host relationship among the aspidobothrids is that in which either a gastropod or a bivalve mollusk is the sole host, as in *Aspidogaster conchicola*. When such an infected mollusk is eaten by a fish, frog or reptile, the parasites sometimes demonstrate an ability to withstand digestion and to attach themselves to the wall of the vertebrate stomach or gut—suggesting the initiation of a second host as in the Digenea. A more complicated host relationship is that of *Stichocotyle nephropsis* in which encysted larvae are present in crabs and lobsters, while adults occur in biliary passages of elasmobranchs. The first-stage larva of the parasite may develop in a marine mollusk, thence to a crab or lobster. Thus we have a suggestion in this group of trematodes of how alternation of hosts may have evolved.[11]

Digenea. The Digenea probably have had a polyphyletic origin, with possibly only two primitive ancestors; the different groups converged in their evolution to similar-appearing forms today. When we recall that the elasmobranchs, an ancient group of fish, possess no characteristic genera of digenetic trematodes, and that American marsupials harbor trematodes belonging to species different from those in Australian marsupials, we must admit the relatively recent origin of the Digenea as compared with the Monogenea. But in any speculation of origins of the Digenea we must take into account the universal use of molluscan hosts, the universal employment of sexual reproduction, the habit of encystment of pre-adults, and the relatively greater host-specificity among the widely diversified larval stages. Wright[64] has suggested that speciation occurs most often as a result of parallel evolution between larval flukes and their molluscan hosts rather than final hosts.

The problem of the origin of the Digenea is made much more complex than that of the Monogenea because of the introduction, in the former, of intermediate hosts. The intermediate hosts probably were originally the final hosts containing the sexually mature adult parasites. A study of the germ-cycles in Digenea supports this theory. Further support is provided by the fact that sexually mature stages of Digenea are to be found in vertebrates of relatively recent origin. Cercariae may have originally attained maturity in a second host, but subsequently became modified into intermediate forms during the time the original definitive host was becoming a second intermediate host.

The evolution of digenetic trematodes may have begun with a life cycle containing a free-swimming planuloid larva (i.e., miracidium) that changed into a form resembling a cercaria or metacercaria. From this type of free-living life cycle the first host, in the form of a mollusk, was introduced. The free-swimming larva was eaten by the mollusk, or it became a commensal in the mantle and subsequently invaded the host tissues, and it evolved an elaborate process of asexual multiplication within its host. The next step in the evolutionary sequence probably came when the encysted cercariae (i.e. metacercariae) were ingested by a vertebrate host and matured in its intestine. The tail of the cercaria has been considered a reverse aquatic adaptation taking the place of cilia. (See Baer[1] and Chauhan.[11])

Initial phases of bladder formation are the same in the two major divisions of the Digenea, but the differences in the structure of the two bladder types have been recognized as of fundamental importance, and as representing stages in the evolution of the excretory system of the Digenea. A thin-walled bladder is probably the more primitive.[37]

Llewellyn[41] has reviewed these and other theories, and summarized his views as follows:

> Thus there are two distinct, and to some extent independent, phases of parasitism in the life-cycle of a digenean. If, as seems very probable, one of these phases was developed before the other, it is the molluscan phase which is likely to be the older (since it is to mollusc hosts that phylogenetic specificity is displayed), and it could well be that the first digeneans were parasitic when young and free-living when adults. The parasitism of vertebrates by the originally free-living adults could have developed later.
>
> Thus digeneans practise a kind of parasitism which is spread between two kinds of hosts which had, and have, no direct connection with each other (e.g. a predator/prey relationship as seen in cestodes) beyond that of co-existence in the same habitat.

Cestoidea. Speculation concerning the origins of tapeworms is particularly difficult because of the absence of free-living stages in their life cycles. Add to this deficiency the presence of ten-hooked and six-hooked embryos, holdfasts, proglottids, tetraradiate symmetry, and progenesis, and the difficulties are compounded indeed. The adult cestodes, as we have seen, are relatively highly host-specific, and they illustrate to a striking degree the morphologic specialization so typical of the parasitic mode of existence. These facts suggest an ancient origin.

The majority of helminthologists have believed that cestodes as well as trematodes have evolved directly and independently from rhabdocoele turbellarian ancestors. Probably the immediate ancestor of cestodes was a turbellarian without a gut (unless it was a primitive monogenean). The assumption that a tapeworm is "degenerate," having lost its gut during evolution as a parasite, is thus untenable. There is considerable difference of opinion as to the identity of which present-day cestode most nearly represents the original ancestral form. The Tetraphyllidea (scolex with four flexible, lappet-like suckers, adults mostly in sharks, freshwater fish, amphibians and reptiles) are often regarded as the most primitive group of tapeworms. Baer,[2] and others, have argued that the two-bothriate tapeworms are the most primitive of present-day forms, and have been derived from a prototrematode stock. The two-bothriate worms are basically similar to digenetic trematodes in many respects, and they apparently have given rise to the four-suckered and four-bothriate worms. Supporting evidence for this statement is derived from a study of *Haplobothrium*. Wardle and McLeod[59] have rejected this hypothesis on the grounds that proteocephalan genera (see p. 221), not two-bothriate genera, are the characteristic tapeworm parasites of primitive vertebrates. These authors tentatively suggest that the protocestode gave rise, through a process of delayed autotomy and secondary tetraradial symmetry, to the present tetra-fossate forms, and that the two-bothriate tapeworms are neotenic, persistent larval forms of the protocestode stock.

Since the most primitive hosts for adult cestodes are fish, we may expect to find the most primitive tapeworms in the cyclostomes, chimaerids and elasmobranchs. Cestodes have not been reported from cyclostomes or chimaerids, but in the latter archaic fish (e.g., the ratfish) is to be found the cestodarian, *Gyrocotyle* (Fig. 9–14, p. 219). Other genera inhabit the body cavity of ganoid and teleost fish, and, in one instance, the coelom of a turtle, but their relationships to the true tapeworms are obscure, and they are considered as an isolated, aberrant and pro-genetic group. A careful comparison of the structures and life histories of cestodes and cestodarians (see p. 218) has convinced some parasitologists that the latter group should be considered as a class separate from the former. Llewellyn[41] has reviewed the evidence for this conclusion, and stated that "there is much to be said then, for regarding the 'Cestodaria' as an artificial grouping of gyrocotylidean monogeneans and amphilinidean cestodes." The presence of vestiges of a haptor in all stages of amphilinids (see p. 218) suggests that the cestodarians are related to the mono-genetic trematodes; but the nature of the excretory system, presence of calcareous bodies, general biology, and other features, are more similar to the cestodes.[21] This group of highly specialized parasites probably represents the remains of a once flourishing assemblage of parasites from freshwater hosts. The original hosts migrated to the sea and became the ancestors of modern elasmo-branchs and chimaerids.

The freshwater ancestral hosts that gave rise to elasmobranchs have disappeared, but one freshwater derivation of these hosts has remained as an archaic, relatively unspecialized fish whose parasites should provide a fruitful source of speculation on evolution. This ganoid host is the bowfin, *Amia calva*, and it possesses a tapeworm, *Haplobothrium*, which is found nowhere else. The scolex of the worm (Fig. 25–6) suggests affinities with the tetrahynchid cestodes from elasmobranchs, but the internal anatomy is more similar to that of the pseudophyllids (see page 224). The body of the worm exhibits a curious secondary segmentation, and it often breaks into smaller pieces, the anterior end of each piece becoming differentiated into a secondary pseudophyllidean scolex. The life cycle of this unique tapeworm appears to involve two intermediate hosts, a copepod and a fish. Thus *Haplobothrium* possesses the morphologic features that may well have been characteristic of a primitive cestoid ancestor to both pseudophyllids and tetraphyllids.

Once the pseudophyllids were established in fish, an easy step to birds and mammals was possible because the latter group of hosts often feeds abundantly on infected fish. The bothriocephalids are to be found in teleost fish, but the related diphyllobothrids occur in birds and mammals.

The great majority of the genera of Cyclophyllidea, which include the common tapeworms of man, are to be found in birds. We have already pointed out that birds evolved very rapidly, and Baer[2] has suggested that the explosive evolution of birds is reflected in the rapid and diversified evolution of bird tapeworms. The evolution of mammals has progressed at a slower pace, and we

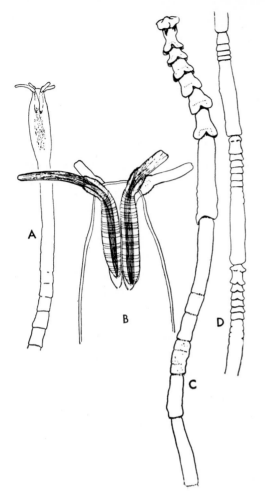

Fig. 25-6. *Haplobothrium globuliforme.* *A.* Scolex. *B.* Scolex greatly enlarged showing the four retractile tentacles. *C.* Portion of a strobila showing a pseudoscolex that resembles the scolex of pseudophyllidean species. *D.* Strobilia dividing secondarily into identical pieces (after Fuhrmann). (Baer, Ecology of Animal Parasites, courtesy of University of Illinois Press)

find comparatively few cestode genera in their intestines.

A common theory states that tapeworms were originally parasites of invertebrates that today are intermediate hosts. This theory is based on the fact that *Archigetes* (Fig. 10–6) is found in an invertebrate (the only such instance among tapeworms), and on such evidence as the use of insect hosts in the life cycle of *Hymenolepis nana.* The vertebrate host

becomes infected through ingestion of infected insects. When we recall that the direct type of life cycle is characteristic for cestodes, and is a secondary acquisition, we might consider the unusual use of insect hosts in this life cycle as a reversion to an original mode of life.

The theory presented in the above paragraph has been vigorously opposed by Baer,[2] who maintains that *Archigetes* is probably a neotenic procercoid and not an ancestral type of tapeworm. Moreover, the cestodarians are more closely related to the ancestral stock than are the cestodes, and the former group includes the amphilinids, which are coelomic parasites whose adults are possibly free-living.

A critical review by Stunkard[53] of the problem of *Archigetes* affinities contained the following statement: "... species of *Archigetes* from tubificid oligochaetes are progenetic larvae, morphologically indistinguishable from and probably identical with caryophyllaeid worms from fishes. These monozoic forms have sometimes been regarded as primitive, and the progenitors of merozoic cestodes, but the present information supports the idea that they are merely precocious progenetic larvae that have dropped the strobilate stage. It is possible that they represent an earlier phylogenetic stage."

Speciation of cestodes by isolation has probably occurred not only within groups of hosts that are phylogenetically related, but also within specialized hosts. Ancestors of the present-day genera of cestodes parasitic in both birds and mammals must probably be sought among the species that lived in Mesozoic reptiles.

Acanthocephala

In all probability the Acanthocephala originated as parasites of freshwater fish, with a single, invertebrate intermediate host. Potential transport hosts (i.e., paratenic hosts) have become ecologic intermediate hosts because immature worms are able to become re-encapsulated in vertebrate hosts that feed on arthropods. An appropriate final host may sometimes also become a potential intermediate host, and because of this adaptability there has been established a flexibility in parasite-host specificity. Acanthocephalans have become adapted to the bodies of reptiles, birds and mammals, and they undoubtedly represent an ancient group of parasitic worms.[2] The short, subglobular proboscis with a relatively small number of hooks seems to be the type, morphologically and phylogenetically, from which other modifications have been derived.[57]

Hyman[33] summed up the evidence on the possible origin of acanthocephalans as follows: ". . . comparisons do not furnish a decisive answer to the question whether the Acanthocephala are allied to the platyhelminths or to the aschelminths. The general structure is rather on the aschelminthic side, whereas the embryology presents more points of resemblance with the platyhelmintha." Cameron[9] called them a small group with elusive affinities and puzzling origin that are not even remotely related to the Nematoda and probably have no living relations. In 1967 Nicholas[45] reviewed the biology of the Acanthocephala and stated that little fresh information with a bearing on the phylogeny has come to light in recent years. A thorough review of the phylogeny of the group was presented by Golvan.[24]

Nematoda

The parasitic nematodes probably originated from more than one type of free-living ancestor. Parasitic and free-living species are remarkably alike. The chief differences are, in parasitic forms, an increase in complexity of genital organs and rate of reproduction, reduction of sense organs, and decrease in production of digestive enzymes. There is evidence[32] for the belief that the parasitic nematodes find their nearest free-living relatives among (1) the Kinorhyncha—microscopic marine nematode-like worms devoid of cilia, with a circlet of spines; and (2) the Priapulida, also marine, with cylindrical shape and warty appearance, a superficially

segmented trunk and terminal mouth and anus. A detailed review of the evolution of parasitic nematodes was made by Inglis.[34] Among many other good reviews are those of Cameron,[9] Chabaud,[10] Chitwood,[21] Dougherty,[20] and Osche.[47]

The parasitic nematodes in vertebrates form an increasing scale of complexity from those in elasmobranchs to those in mammals.[2] This distribution seems to suggest that these worms became abundant in vertebrates only after the hosts had taken to a terrestrial life, and that the parasites evolved in complexity with the evolution of their hosts. The larvae of these parasites often enter the lungs of their final hosts. It has been suggested that this site preference probably represents the vestige of an ancestral condition when the lungs, with their abundance of oxygen, were the only locations of infection. Hookworms, however, can develop directly in the intestine. A significant experiment has shown that inert particles injected under a frog's skin will follow the same route of migration as that pursued by larvae of *Rhabdias bufonis*, a parasite of the lungs.[23] The migratory habit of nematode larvae, as well as the penetration of host skin might, therefore, be due in part to mechanical causes, but such a conclusion is hazardous without further evidence. When intermediate hosts, usually arthropods, are involved, they almost invariably acquire infection by ingesting the larvae while the latter are still enclosed by the egg shell. One might regard the subsequent transference of the larva to the vertebrate host as simply an elaboration of the direct type of life history. The origin of larval migration might be related to the habit of ingesting infected secondary hosts—the final host becoming also an intermediate one.

Free-living phasmids are largely inhabitants of the soil, whereas the free-living aphasmids are more characteristically aquatic, and they include the great bulk of marine species. Life in the soil is peculiarly suited to the development of parasitism because of the opportunity and often necessity for a worm to seek the protective moisture of a host's body (plant or animal) when the soil becomes dry. Once a temporary refuge is gained, evolutionary processes resulting in stability of the relationship may occur; then parallel evolution of host and parasite logically follow. A thorough survey and census of species of nematodes needs to be carried out before we can formulate with assurance definite conclusions on evolutionary relationships, but some trends and tentative generalizations can be expressed.

The literature contains numerous papers dealing with speculations on the systematics and phylogeny among groups of parasitic nematodes. There are many conflicting views, often with little evidence to support them. Instead of attempting to present all of these views, we shall give only the ideas of one investigator (Inglis, 1965[34]), whose arguments and speculations appear to us to be as convincing, if not more so, as those of the others. Inglis's paper contains a review of all the pertinent works of other nematologists, with a critical analysis of their conclusions. He begins his own paper by stating in the introduction: "It is clear, however, that most nematodes are not markedly host specific, and as a consequence I will argue that the evolution of most groups of parasitic nematodes has been largely independent of the evolution of their hosts and has tended to occur in groups of hosts with similar ecological requirements and feeding habits. The hosts have simply acted as a substrate upon which the nematode parasites have speciated and evolved so that their evolution is largely a consequence of intra-host competition and not host evolution."

The following quotations, taken from the summary in Inglis's paper, contain the chief subjects and concepts that occupy the interests of those concerned with the evolution of nematode parasites.

1. All parasitic nematodes appear to have arisen from terrestrial free-living groups; the phasmidian parasites from the Rhabditina and the aphasmidian from the Dorylaimida. The terms phasmidia and aphasmidia are retained because of their value in delimiting these two major divisions and not because they indicate any major division of nematodes as a whole.

2. The phasmidian parasites have become parasites in animals on at least four major occasions giving rise to the groups recognized as the Drilonematoidea, Oxyuroidea, Cosmocercoidea and Strongylina.

3. The appearance of groups parasitic in the early aquatic vertebrates depended on the presence of an invertebrate which could act as an intermediate host and it is argued that the Ascaridoidea, Spiruroidea and Seuratoidea originated by way of a group which was parasitic in, or associated with, some terrestrial: aquatic invertebrate during the first three, or simply the third, larval stages. The Cosmocercoidea approach this condition in some cases and are considered to represent at least a model of the early conditions covering the transfer from terrestrial to aquatic hosts. The Seuratoidea are considered to represent the remaining members of the earliest groups of parasites to make this transfer to aquatic hosts.

4. The major division of the aquatic parasites is attributed to the development of a stomach in the host within which the Spiruroidea were restricted, while the Ascaridoidea remained largely in the intestine. Thus it is suggested that the infestation of vertebrates was prior to the appearance of jawed forms.

5. The significance of the third-stage larva in the phasmidian parasites is stressed as being the stage at which the parasites transfer from one environment to another, and it is shown that the life-cycle of the Tylenchida, parasitic in insects, is not fundamentally different from that of other phasmidian parasites.

6. The reduction of the length of time spent in the intermediate host in the life-cycles of many nematode groups is considered to be a result of selection pressures acting to reduce 'host-discordance' (i.e. the many differences in physiology etc. between the two, or more, animals in an indirect life-cycle). It is also pointed out that if a parasite utilizes more than one host at different stages in the life-cycle it must be plastic in its environmental requirements so that it is less likely to become tied to one host, or host group, and will speciate less easily. An indirect life-cycle also appears to be a pre-requisite to life in extra-gut localities within the host, as all such parasites have an indirect life-cycle.

[7.] Speciation in parasitic nematodes is considered to be wholly allopatric and not due to any evolutionary change in the host   This is suggested by establishing that niche diversification is common among parasitic nematodes and that intra-host competition can, and frequently does, occur. . . .

8. Trans-specific evolution is shown to be largely independent of the evolution of the hosts and to be due rather to the parasites speciating and evolving so as to utilize all available localities and niches within the host body. Any host restriction is almost wholly determined by the ecology and feeding habits of the hosts and, because most major host groups have similar food and ecological preferences it is possible to produce an apparent parallelism between the hosts and parasites This apparent parallelism, however, generally only applies when wide host and parasite groups are compared. . . .

10. As a consequence of the lack of host-parasite parallelism and the concept of intra-host competition it is pointed out that a group could appear which was so efficient a parasite that it could infest almost any host available. *Physaloptera* and *Capillaria* approach this condition and it is pointed out that the place and time of origin of both groups cannot be wholly established It is stressed that it cannot be established that *Capillaria* represents an archaic group of parasites, particularly as its absence from Madagascar suggests it may be of fairly recent origin.

Figure 25–7 illustrates the relationships among the major groups of phasmidian (class Secernentea) nematodes as postulated by Inglis.

The Arthropods

Conditions predisposing to parasitism among the arthropods, as well as among other animals, are: wanderlust (mites into cracks), saprophagous nutrition habits, sucking of plant juices, and, especially, the preference to live in crowded communities. As Rothschild and Clay[49] have said, the most favorable condition for the dawn and development of dependence is a social environment.

Various species of blood-sucking arthropods (e.g., some triatomids) obtain meals by tapping the blood-engorged bodies of other arthropods (e.g., bedbugs) that have fed on a vertebrate. It is but a short step from this kind of habit to tapping directly the body of the vertebrate. Preparasites were perhaps first attracted to waste food, offal and exudations of certain animals, and when conditions encouraged the preparasites to stay with their future host (that is, when the search for food became simplified, or when

it became unnecessary to meet competition by seeking other and more distant sources of food), the preparasites became mess-mates or scavengers, and from this association they became parasites. Evidence for this sort of speculation may be obtained from parasitic insects, among which all gradations from free-living to parasitic forms may be traced. Parallel evolution of the host-seeking instinct and somatic characters seems clear, particularly in such groups as the dipterous Tachinidae (see page 416).

Isopods. Relationships among the isopods are suggested by Figure 25–8.[44] One well-known example of the parasitic members of this group is *Livoneca convexa*, a cymothoid protandrous, hermaphroditic species found on fish. The male isopod is parasitic on the gills of the marine fish, *Chloroscombrus orqueta*, while the female is commensal in the oral cavity of the host. Females are produced from males by a process of metamorphosis, and the relationship between the parasite and host appears to be highly specific.

Copepods. Present knowledge of copepods and their ontogeny is only in an initial stage, and a natural scheme of classification, especially for parasitic species can be only tentative, with many gaps. Since no fossil copepods have been discovered, the question of origins is difficult. The Cyclopoida appear to be the most primitive group, from which the other copepods may have radiated. All parasitic copepods may have a possible origin in the more or less parasitic Ergasilidae,[25] but they possibly exploited this way of life frequently during the course of evolution.

Specialization in parasitic copepods in relation to the distribution of their hosts in time was studied by Leigh-Sharpe,[39] who stated that the lowest phylum of animals in the evolutionary series on which parasitic copepods have been found is the Annulata. Among the copepods on annelids are the most highly modified forms known (e.g., the Saccopsidae), and they are characterized by a loss of appendages and alimentary tract. A striking feature of intimacy with the host is shown by the hermaphroditic *Xenocoeloma*, whose body cavity is continuous with the coelom of its annelid host. Next to this group of highly specialized parasitic copepods on invertebrates, one may find somewhat less specialized species on the more primitive fish,

moderately modified species on less primitive fish, and slightly modified species of parasitic copepods on the most recent fish and on ascidians. In spite of many exceptions to the evolutionary trend suggested above, there

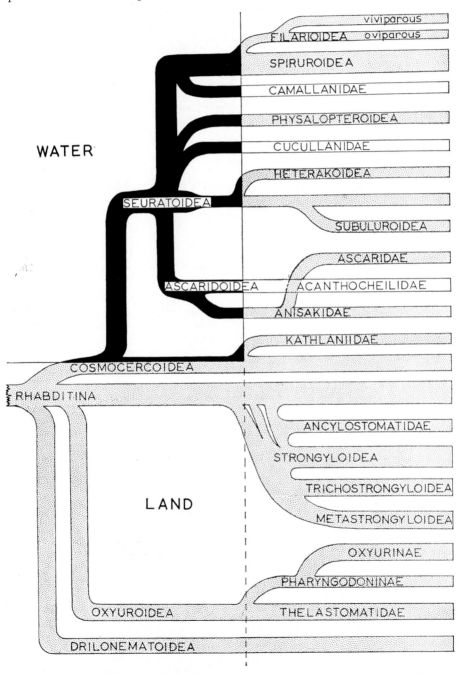

Fig. 25-7. Dendogram illustrating the relationships among the major groups of phasmidian parasites, as postulated by Inglis, 1965. (Courtesy of Blackwell Sci. Pub.)

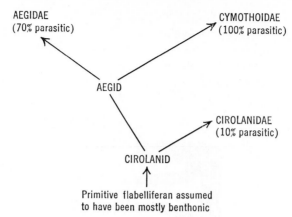

Fig. 25-8. Relationships among isopods. (Modified from Menzies, Bowman, and Alverson, 1955.)

does seem to be a rough correlation between the degree of parasitic modification of copepods and the primitiveness of the host group.

Lice. Lice presumably have been derived from primitive psocid-like ancestors that probably became parasitic first on birds and later on mammals. The Psocoptera (Corrodentia) are known as booklice, and they are not parasitic. They live beneath the bark of trees, and from this habitat ancestral forms could have readily found their way to bird nests and thence to birds themselves. The lice may have passed secondarily from birds to mammals, but the presence of a primitive species of louse on a tree-shrew, a very primitive mammal, suggests a polyphyletic origin for mammal lice.[60] From these beginnings there must have been a considerable amount of parallel evolution.

Among the chewing lice (Mallophaga) of birds, the group known as Amblycera does not possess a spiracular gland, but the Ischnocera do possess the gland, which is similar to the spiracular gland in the sucking lice (Anoplura). The absence of the spiracular gland indicates that the Amblycera were probably well established before the Ischnocera arose. The spiracles of most primitive structures are found in *Eomenacanthus stramineus* of the family Menoponidae. Thus, members of this family are believed to be of more ancient origin than are other Anoplura. The Trichodectidae (Ischnocera) and the Ano-

plura appear to have arisen from a common ischnoceran parasite that migrated from a bird to a primitive insectivore.[58]

A summary of a chronology of relationships among the lice of mammals has been made by Hopkins,[29] as follows:

Here, again, our deductions are greatly hampered by doubt as to whether certain infestations are primary (and very ancient) or secondary (and relatively recent), but it seems possible to arrive at a few conclusions which are beyond reasonable doubt. Distantly related species of *Pediculus* occur on man and the chimpanzee and the genus is so widely separated from any found on non-Primates that its occurrence on the Primates must be ancient—possibly of Miocene date. Turning to the *Anoplura* as a whole, the most important fact is the universal occurrence of Echinophthiriidae on the seals, because this must certainly have originated before the seals adopted a marine life, because they have not been in contact with other mammals since that event, apart from momentary and fatal encounters with Polar bears; seals are definitely known from the Lower Miocene and probably originated in the late Eocene. Infestation of the ungulate branches of the *Ferungulata* is far too nearly universal to be anything but primary, in my opinion, and the fact that seal-lice and the lice of *Camelidae* and *Tayassuidae* have characters annectent between Haematopinidae and Linognathidae seems to confirm that even if the infestation of the *Ferungulata* as a whole does not date back (as I believe it does) to before the ancestors of the ungulate and carnivorous branches (the *Condylarthra* and *Creodonata*) diverged during the lowest Paleocene period (or perhaps the late Cretaceous) it must have spread from one group to the other at a not very much later date. At this point we come to a very serious difficulty—the infestation of the *Glires* with *Anoplura* is also practically universal except for certain Nearctic and Neotropical families of *Hystrichomorpha*. I regard the infestation of the *Glires* with *Anoplura* as at least partly primary, and the absence or rarity of these lice on some *Hystrichomorpha* as the effect of the heavy infestation of these groups with *Mallophaga*, while Vanzolini and Guimarães[58] think that the evidence indicates that *Anoplura* were lacking in South America (and probably everywhere) at the beginning of the Tertiary period and that they were brought in by the Pliocene migration of mammals into South America. If Vanzolini and Guimarães are right, then the *Anoplura* probably arose between the Eocene period and the Pliocene migration and perhaps not long before this migration, whereas if I am right they may be very much older.

Flies. Little is known about the evolution of parasitic Diptera, but one family, Tachinidae, has been studied in some detail[22] and it will be used as one example of the problems encountered in exploring the ancestral histories of insects. The Tachinidae (see p. 416) is a very large assemblage of flies whose classification presents imposing difficulties, and about which there is considerable disagreement among entomologists. They undoubtedly evolved from the family Calliphoridae, a very large group whose larvae are saprophagous or flesh-feeders (e.g., *Sarcophaga*, the fleshfly), or they are parasites of various arthropods. The closest relatives of tachinids among the calliphorids occur in the subfamilies Sarcophaginae and Rhinophorinae. Of these two subfamilies, the Sarcophaginae appears to be closer to the ancestral Tachinidae because of both morphologic and behavioral similarities.

The most primitive subfamily of tachinids is the Dexiinae, and an evolutionary series of subfamilies can be arranged from this group up to the Goniinae. The most primitive method of reaching the host (exhibited in all Dexiinae and also to be found in the Sarcophaginae) is that of active searching by the young larva for its victim. The next step in

the evolution of host-feeding is shown by the Macquartiinae, whose larvae are deposited near and ingested by the host, or the larvae are deposited on the host. In the latter method the larvae are sometimes still encased by the egg shell. A third and much more specialized step, shown by the Phasiinae, is the deposition of parasite eggs on particular hosts, or the injection of the eggs into the host body. The more specialized tachinids are typically parasites of Lepidoptera, whereas the more primitive forms tend more to parasitize a wider range of unrelated hosts.

Fleas. Bird fleas have undoubtedly arisen from mammal fleas. The genus *Ceratophyllus*, for example, is clearly derived from fleas of squirrels and other tree-dwelling rodents whose nests provide conditions not unlike those of bird nests. Most of the relatively few bird fleas are to be found on birds that return to the same nesting site year after year, or that nest on the ground or in holes. Such habits obviously favor the difficult transfer of fleas from a mammal to a bird. *Palaeopsylla*, a genus found on moles and shrews, has been reported from Baltic amber representing the Oligocene. The origins of mammal fleas remain obscure.

Wasps. A study of the parasitic Hymenoptera shows that species feeding on a few kinds of host foods (oligophagous) have arisen repeatedly from polyphagous ancestors. This trend has occurred in spite of the apparent advantages (more available food) of the oligophagous forms. Trends toward specialization seem to be the rule, and toward removing species from competition with one another.

Life histories of wasps and their mites illustrate mutually adaptive modifications, and offer promising subjects for the study of evolution. The eumenine wasp, *Ancistrocerus antilope*, and the larval mite, *Ensliniella trisetosa*, live together throughout the whole life of each member of the alliance.

Male wasps must emerge from the nest bearing propodeal loads of deutonymphs. On the other hand, female wasps must emerge from the nest devoid of deutonymphs, even though possessing nearly identically developed acarinaria, and then acquire deutonymphs from the males at mating. Thus the loss of deutonymphs from adult male wasps must in large part take place by transfer to female wasps. In turn, losses of deutonymphs from female wasps must be chiefly (? or only partly) by deposition in the cells of the new nest during its construction or provisioning. Since adult male wasps must emerge from their cells with full deutonymphal loads, and females must emerge without deutonymphs, it would seem that whatever factors determine that a particular egg will be fertilized, hence diploid and giving rise to a daughter, must also decide that normally deutonymphs will not be released to the cell containing a diploid egg.[14]

An adaptative behavioral pattern of this sort is not without parallel among other nest parasites. But the remarkable transfer of parasites from the host male genital chamber to the female genital chamber has no parallel yet described. "Possibly the mite has somehow become as fully necessary to the wasp as the wasp is to the mite," and it is highly probable that "the life cycles of the two organisms have become inextricably interlaced through an exacting evolutionary co-adaptation."[14]

Mites and Ticks. Among the Acarina parasitism seems to have developed primarily from a predatory habit, but also from the scavenger habit and from phoresy. Little work has been done on the evolution of mites and ticks, but some recent speculation on the phylogeny of trombiculid mites suggests that from essentially reptile-specific groups four lines of development took place. The larvae (chiggers) of these mites are very small, and instead of preying on smaller arthropods, as do the adult mites, they feed parasitically on large arthropods and on vertebrates "by sampling instead of consuming." The first group to arise from ancestral forms was a small number of species that parasitized arthropods. The second group became adapted to living in the skins of amphibians. The third group managed to isolate itself in the lungs of a sea snake, while the fourth, and by far the largest group, spread out over all the mammals, and they developed many genera. The more primitive species of these

mites appear to be distributed over the open home ranges (e.g., fields) of their hosts, whereas other, more specialized species are more restricted in their relationships with their hosts. See also, Hughes[30] and Woolley.[65]

GENERAL RULES AND PRINCIPLES

1. The two basic kinds of changes that normally take place in an organism before it becomes a dependent internal parasite are: (a) physiologic—in which the powers of synthesis of some essential nutritional elements are lost, and (b) morphologic—in which superfluous organs are lost. Such changes, of course, also occur in free-living organisms.

2. Parasitism probably arose in different ways in different groups of animals.

3. To become a parasite, an organism must undergo physiologic adaptations to its host.

4. If a species of parasite is found in a number of different kinds of hosts, the parasite probably has recently been introduced to the area. In other words, parasite-host specificity is normally achieved only after long association.

5. Many factors in addition to length of association play significant roles in determining the incidence and degree of pathogenicity of parasitic infections in new hosts.

6. The terms "degenerate," "regressive," and "retrogressive" are anthropocentric, and they should be used with caution, if at all.

7. We may assume that the first preparasite that entered a host was either pre-adapted to life within the host, or it was able to change rapidly so as to become tolerant of the host.

8. Even a "mild" parasite carries on an assault against its host, and the host is a responsive environment. For these reasons natural selection limits the numbers of successful parasites.

9. Parasites usually evolve more slowly than do their hosts.

10. The advantages of parasitism have tended to preserve large numbers of mutations.

11. Parasites as a whole are worthy examples of the inexorable march of evolution into blind alleys.

REFERENCES

1. Baer, J. G.: Etude Monographique du Groupe des Temoncéphales. Bull. Biol. France-Belg., *65*:1–57, 1931.
2. ———: Ecology of Animal Parasites. Urbana, Ill., Univ. Illinois Press, 1951.
3. Baker, J. R.: The evolution of parasitic Protozoa. *In* Taylor, A. E. R. (ed.): Evolution of Parasites. pp. 1–27. Oxford, Blackwell Sci. Pub., 1965.
4. Ball, G. H.: Parasitism and evolution. Amer. Natur., *77*:345–364, 1943.
5. Bray, R. S.: The exo-erythrocytic phase of malaria parasites. Int. Rev. Trop Med., *2*: 41–47, 1963.
6. Burnet, F. M.: The theories of antibody production. *In* Najjar, V. A. (ed.): Immunity and Virus Infection. New York, John Wiley & Sons, 1959.
7. Bütschli, O.: Studien über die ersten Entwickelungsvorgänge der Eizelle, die Zellteilung, und die Conjugation der Infusorien. Abh. Senck. Ges., *10*:213–464, 1876.
8. Bychowsky, B. E.: Monogenetic Trematodes, Their Systematics and Phylogeny. Acad. Nauk. SSSR. (Eng. editor, W. J. Hargis; trans. by P. C. Oustinoff.) Washington, D.C., Amer. Inst. Biol. Sci., 1957.
9. Cameron, T. W. M.: Host specificity and the evolution of helminthic parasites. *In* Dawes, B. (ed.): Advances in Parasitology. Vol. 2, pp. 1–34. New York, Academic Press, 1964.
10. Chabaud, A. G.: Essai d'Interprétation Phylétique des Cycles Évolutifs Chez les Nématodes Parasites de Vertébrés, Conclusions Taxonomiques. Ann. Parasit. Hum. Comp., *30*:83–126, 1955.
11. Chauhan, B. S.: Studies on the trematode fauna of India. Parts 1 through 4. Rec. Indian Mus., *51*:113–393, 1954.
12. Cheissin, E. M.: [The Taxonomic System of Sporozoa (Class Sporozoa, Phylum Protozoa).] Zoological J., Acad. Sci. USSR, *35*:1281–1298, 1956. (Zool. Zhurnal, Akad. Nauk SSSR.) (In Russian)
13. Clay, T.: Some problems in the evolution of a group of ectoparasites. Evolution, *3*:279–299, 1949.
14. Cooper, K. W.: Venereal transmission of mites by wasps, and some evolutionary problems arising from the remarkable association of *Ensliniella trisetosa* with the wasp *Ancistrocerus antilope*. Biology of eumenine wasps II. Trans. Amer. Entom. Soc., *80*:119–174, 1955.
15. Corliss, J. O.: Comments on the systematics and phylogeny of the Protozoa. Systematic Zool., *8*:169–190, 1959.
16. ———: The Ciliated Protozoa. Elmsford, N. Y., Pergamon Press, 1961.
17. Davenport, C. B.: The animal ecology of Cold Spring Sand Spit, with remarks on the theory of adaptation. Univ. Chicago Publ., *10*:1–22, 1903.
18. Davenport, D.: Specificity and behavior in symbiosis. Quart. Rev. Biol., *30*:29–46, 1955.
19. Dodson, E. O.: A Textbook of Evolution. Philadelphia, W. B. Saunders, 1952.
20. Dougherty, E. C.: Evolution of zoöparasitic groups in the phylum Nematoda, with special reference to host-distribution. J. Parasitol., *37*:353–378, 1951.
21. Dubinina, M. N.: The morphology of Amphilinidae (Cestodaria) in relation to their position in the system of flatworms. Doklady Biol. Sci. Sect. (Doklady Akademii Nauk SSSR), *135*:501–504, 1960. Trans. by the Amer. Inst. Biol. Sci., p. 943–945, 1961.
22. Emden, F. I. van: Evolution of Tachinidae and their parasitism (Diptera) Proc XVth Internat Congr. Zool., London, pp. 664–666, 1958.
23. Fülleborn, F.: On the larval migration of some parasitic nematodes in the body of the host and its biological significance. J. Helminthol., *7*:15–26, 1929.
24. Golvan, Y. J.: Le Phylum des Acanthocephala. Première note. Sa place dans l'eschelle zoologique. Ann. Parasit. Hum. Comp., *33*:538–602, 1958.
25. Gurney, R.: British Fresh-Water Copepoda. 3 volumes. London, The Ray Society, 1931–1933.
26. Hoare, C. A.: Intraspecific biological groups in pathogenic protozoa. Refuah Veterinarith (Quart. Israel Vet. Med. Assoc.), *12*:263–258, 1955.
27. ———: The Transmission of Trypanosomes and its Evolutionary Significance. *In* Biological Aspects of the Transmission of Disease. London, Oliver and Boyd, 1957.
28. Hollande, A.: L'évolution des Flagellés Symbiotiques, Hôtes du *Cryptocercus* et des Termites Inferieurs. Tijdsch. v. Entomol., *95*: 18–110, 1952.
29. Hopkins, G. H. E.: The distribution of Phthiraptera on mammals. *In* First Symposium on Host Specificity Among Parasites of Vertebrates. pp. 64–87. Inst. Zool., Univ. Neuchâtel, 1957.
30. Hughes, T. E.: Mites; or, the Acari. Univ. of London, The Athlone Press, 1959.
31. Huxley, J.: Evolution the Modern Synthesis. London, Geog. Allen and Unwin Ltd., 1947.
32. Hyman, L. H.: The Invertebrates: Platyhelminthes and Rhynchocoela. The Acoelomate Bilateria. Vol. II. McGraw-Hill, 1951.
33. ———: The Invertebrates: Acanthocephala,

Aschelminthes, and Entoprocta. The Pseudo-coelomate Bilateria. Vol. III. New York, McGraw-Hill, 1951.

34. Inglis, W. G.: Patterns of evolution in parasitic nematodes. *In* Taylor, A. E. R. (ed.): Evolution of Parasites. pp. 79–124. Oxford, Blackwell Sci. Pub., 1965.

35. Janiszewska, J.: Parasitogenetic rules. Janicki rule. Zool. Polon., *5*:31–34, 1949.

36. Jones, A. W.: Introduction to Parasitology. Reading, Mass., Addison-Wesley Pub. Co., 1967.

37. La Rue, G. R.: The classification of digenetic trematoda: A review and a new system. Exper. Parasitol., *6*:306–344, 1957.

38. Lederberg, J.: Cell genetics and hereditary symbiosis. Physiol. Rev., *32*:403–430, 1952.

39. Leigh-Sharpe, W. H.: Degeneracy in parasitic copepoda in relation to the distribution of their hosts in time. Parasitology, *20*:421–426, 1928.

40. Llewellyn, J.: The larvae of some monogenetic trematode parasites of Plymouth fishes. J. Mar. Biol. Assoc. U.K., *36*:243–259, 1957.

41. ———: The evolution of parasitic helminths. *In* Taylor, A. E. R. (ed.): Evolution of Parasites. pp. 47–48. Oxford, Blackwell Sci. Pub., 1965.

42. Lyons, K. M.: The chemical nature and evolutionary significance of monogenean attachment sclerites. Parasitology, *56*:63–100, 1966.

43. Mattingly, P. F.: The evolution of parasite-arthropod vector systems. *In* Taylor, A. E. R. (ed.): Evolution of Parasites. pp. 29–45. Oxford, Blackwell Sci. Pub., 1965.

44. Menzies, R. J., Bowman, T. E., and Alverson, F. G.: Studies of the biology of the fish parasite *Livoneca convexa* Richardson (Crustacea, Isopoda, Cymothoidae). Wasman J. Biol., *13*:277–295, 1955.

45. Nicholas, W. L.: The biology of the Acanthocephala. *In* Dawes, B. (ed.): Advances in Parasitology. Vol. II, pp. 205–237. New York, Academic Press, 1967.

46. Noble, G. A., and Noble, E. R.: Entamoebae in farm mammals. J. Parasitol., *38*:571–595, 1952.

47. Osche, G.: Morphological, biological, and ecological considerations in the phylogeny of parasitic nematodes. *In* Dougherty, E. C. (ed.): The Lower Metazoa, Comparative Biology and Phylogeny. pp. 283–302. Berkeley, Univ. of California Press, 1963.

48. Pigulevski, S. V.: [The Phylogeny of Flat-worms.] Papers on Helminthology Presented to Academician K. I. Skryabin on his 80th Birthday. Izdatelstve Akad. Nauk. SSSR, pp. 263–270, 1958. (In Russian)

49. Rothschild, M., and Clay, T.: Fleas, Flukes & Cuckoos. London, Wm. Collins Sons & Co., 1952.

50. Sprent, J. F. A.: Parasitism. Immunity and evolution. pp. 149–165. *In* The Evolution of Living Organisms. A Symposium of the Roy. Soc. Victoria, 1959.

51. Stunkard, H W: The physiology, life cycles and phylogeny of the parasitic flat-worms. Amer. Mus. Nat. Hist., Amer. Mus. Novitates, *908*:1–27, 1937.

52. ———: The life-history and systematic relations of the Mesozoa. Quart. Rev. Biol., *29*:230–244, 1954.

53. ———: The organization, ontogeny, and orientation of the Cestoda. Quart. Rev. Biol., *37*:23–34, 1962.

54. Szidat, L.: Geschichte Anwendung und finige Folgerungen aus parasitogenetischen Regeln. Zeit. Parasitenk., *17*:237–268, 1956.

55. Thorne, W. H.: Further studies on pre-imaginal olfactory conditioning in insects. Proc. Roy. Soc., B: Biol. Sci., *127*:424–433, 1939.

56. Thorpe, W. H., and Jones, F. G. W.: Olfactory conditioning in a parasitic insect and its relation to the problem of host selection. Proc. Roy. Soc., B, *124*:56–81, 1937.

57. Van Cleave, H. J.: Some host-parasite relationships of the Acanthocephala, with special reference to the organs of attachment. Exper. Parasitol., *1*:305–330, 1952.

58. Vanzolini, P. E., and Guimarães, L. R.: Lice and the history of South American land mammals. Rev. Brasil. Ent., *3*:13–46, 1955.

59. Wardle, R. A., and McLeod, J. A.: The Zoology of Tapeworms. Minneapolis, Univ. Minn. Press, 1952.

60. Webb, J. E.: Spiracle structure as a guide to the phylogenetic relationships of the Anoplura (biting and sucking lice) with notes on the affinities of the mammalian host. Proc. Zool. Soc. London, *116*:49–119, 1946.

61. Wenrich, D. H.: A new protozoan, *Amphileptus branchiarum*, n. sp., on the gills of tadpoles. Trans. Amer. Micr. Soc., *43*:191–199, 1924.

62. ———: Host-parasite relations between parasitic protozoa and their hosts. Proc. Amer. Philos. Soc., *75*:605–650, 1935.

63. Wilson, M., and Hindle, E.: Discussion on heteroecism. Proc. Linn. Soc. London, *162*: 4–8, 1950.

64. Wright, C. A.: Relationships between trematodes and molluscs. Ann. Trop. Med. Parasitol., *54*:1–7, 1960.

65. Woolley, T.: A review of the phylogeny of mites. Ann. Rev. Entomol., *6*:263–284, 1961.

Glossary

Many of the words in this list apply also to organisms other than parasites but the definitions are given with special reference to parasitism.

Allopatric species. Species having geographic ranges that do not overlap.
Anthroponoses. Human diseases that can be transmitted to animals.
Axenic. Free from other organisms, such as occurs in a "pure" cultivation medium.

Biocoenosis. A community of living organisms whose interests are integrated by requirements imposed by a circumscribed habitat and by mutual interactions. Sometimes called a "species network."

Cleptoparasite. A parasite that develops on the prey of its host.
Commensal. An organism living in close association with another and benefiting therefrom without harming or benefiting the other.
Cryobiosis. Survival at extremes of low temperature and desiccation.
Cyst. A parasite that is surrounded by a resistant wall or membrane. Technically, the wall or membrane constitute the cyst.

Definitive host. A host in which a parasite becomes sexually mature; a host in which a parasite undergoes the sexual phase of its life cycle. Also known as the final host.
Diapause. A temporary suspension in growth and development.
Disease. A specific morbid process that has a characteristic set of symptoms, and that may affect either the entire body or any part of the body. The pathology, etiology, and prognosis may or may not be known.

Ectoparasite. A parasite that lives on the outside of its host.
Endoparasite. A parasite that lives within its host.
Epidemic. A disease that affects a large number of people and spreads rapidly.
Epidemiology. The study or science of epidemics.
Epizoic. Externally parasitic.
Epizoon. An external parasite or commensal.
Epizootic. A disease that affects many animals and spreads rapidly.
Ethology. A branch of ecology that deals with the behavior of animals.
Euryxenous parasite. A parasite that has a broad host range.

Factitious host. A host that normally cannot be invaded in nature because of ecologic barriers. Usually found through laboratory testing.
Facultative parasite. Capable of living apart from a host; i.e., potentially free-living.
Final host. See "Definitive host."

Gnotobiosis.   An environmental state in which strains of known microorganisms have been inoculated into germfree animals.

Habitat.   The specific place where an organism usually lives (e.g., the small intestine).

Halzoun.   An inflammatory condition of the larynx, nasopharynx or eustachian tubes caused by adult liver flukes (*Fasciola*).   It occurs in Syria when people eat raw, infected liver.

Heterogenetic parasite.   A parasite that has alternation of generations.

Heteroxenous parasite.   A parasite that has two or more types of hosts in its cycle.

Host.   Living animal or plant harboring or affording subsistence to a parasite; also a cell in which a parasite lodges ("host cell").

Host specificity.   Restriction of a parasite to one or more kinds of hosts.

Hyperparasitism.   A situation in which a parasite occurs on or in another parasite.

Hyperplasia.   An increase in the number of tissue elements (excluding tumor formation), thereby increasing the mass of the part or organ involved.

Hypertrophy.   An increase in the mass of tissue due to an increase in size, but not to the number of tissue elements.   Sometimes used to denote an increase in size to meet a demand for increased functional activity.

Hypobiosis.   A condition of reduced body functions during such periods as hibernation or aestivation.   Some parasites undergo hypobiosis with their hosts.

Immunity.   Those natural processes which prevent infection, reinfection or superinfection, or which assist in destroying parasites or limiting their multiplication, or which reduce the clinical effects of infection.

Incidence.   The number of cases of infection occurring during a given period of time in relation to the population unit in which they occur (a dynamic measurement).   Not to be confused with *prevalence*.

Infection.   The entrance, establishment, or maintenance of a parasite within a host, usually involving its reproduction as well.   Also, the resultant condition in the host.

Infestation.   The penetration of an endoparasite into the body or the establishment of an ectoparasite on the tegument.   Sometimes limited to ectoparasitic arthropods.

Inoculation.   The active introduction of an organism through the skin by physical or biologic means.

Inquinilism.   An association whereby one species lives within another species, but does not feed entirely at the expense of the host species.   There is no biochemical dependence on the host species.

Intermediate host.   A host in which the parasite undergoes larval or juvenile development.

Ipsefact.   "All those parts of the environment that an individual, colony, population, or species of animal has modified chemically or physically by its own behavior" (e.g., a nest or home, runs of rodents or deer, excrement, beehives, phaeromones). An ipsefact must be a product of behavior.   (From R. Audy)

Landscape epidemiology.   The study of natural foci of infection in relation to the terrain or landscape, especially in relation to the vegetation and the influence of man.

Latent infection.   An infection that is neither visible nor apparent.

Monoxenous parasite.   A parasite that has a single host in its life cycle.

Multiparasitism.   The occurrence of two or more species of parasites within one host.

Mutualism.   An association whereby two species live together in such a manner that their activities benefit each other.

Neoteny.   The persistence of youthful characteristics into maturity or the achievement of sexual maturity by an immature form.

Niche.   The position or status of the organism within its habitat.

Normal host.   The host in which a parasite usually develops in nature.

Nosogeography.   The study of the geographic distribution of disease.

Obligatory parasite.   Unable to live and multiply except as a parasite on or in a host.

Paraneoxenous.   A situation in which intestinal parasites of vertebrates have secondarily taken up residences in helminth parasites (e.g. Microsporida in flukes and cestodes of snakes, *Giardia* in *Nematodirus* of sheep).

Parasite.   An organism that depends on its host for some essential metabolite and with which a reciprocal chemical relationship exists.

Parasite-mix.   See Parasitocenosis.

Parasitemia.   The presence of parasites in the circulating blood.

Parasitism.   An intimate association between two organisms in which the dependence of the parasite on its host is a metabolic one involving mutual exchange of chemical substances.

Parasitocenosis.   The combined populations of organisms, both flora and fauna, that live together in a host organ or in the entire host, or in the host population.

Parasitogenesis.   The evolution of relationships between the parasite and its host.

Parasitoid.   An organism that is parasitic only during its larval stage.

Parasitosis.   Association between two organism in which one injures the other, causing signs of disease.

Paratenesis.   The passage of an infective-stage larva by a series of transport (paratenic) hosts to the definitive host; the larvae do not undergo essential development within the transport hosts, but are maintained in their infective stage from one season of transmission to another.

Paratenic host.   A potential intermediate host that is not essential to the parasite life cycle, and neither favors nor hinders the completion of the cycle.   The basic difference from an intermediate host is the absence of any larval development in the paratenic host.

Parthenogenesis.   Reproduction by the development of an unfertilized egg.

Pathogenic.   That which causes disease or morbid symptoms.

Pedogenesis.   Reproduction by young or larval forms.

Periodicity.   Regularly recurrent, rhythmic changes in vital functions or recurrence of a parasite at regular intervals of time (e.g., nocturnal recurrence of filaria in peripheral blood of infected host, or recurrence of paroxysms in malaria).

Phoresy.   A relationship between host and commensal involving only passive attachment of the commensal to the surface of the host (e.g., barnacle on whale).   The supporting partner must be continuously moving.

Predator.   That which is predatory, living by preying upon other animals.   The predator consumes all or part of its prey, generally killing the latter.

Premunition.   Resistance of a host to superinfection.   This resistance depends upon the survival of parasites within the host and disappears after their elimination. It may be complete or partial resistance.

Progenesis.   Advanced development of genitalia without actual maturation in a larva. This term and *neoteny* are often used synonymously, but neoteny generally involves gonadal maturation in a larval animal.

Protandry.   Production of sperm, and later of ova, by the same gonad (e.g., certain parasitic Crustacea).

Reservoir host.   A host of the parasite being studied other than the one of major interest to the investigator (e.g., antelopes are reservoir hosts of *Trypanosoma rhodesiense*, the causative agent of African sleeping sickness in man).

Stenoxenous parasite.   A parasite having a narrow host range.

Strain.   Population of the same stock descended from a common ancestor or derived from a single source.

Superinfection.   New infection of a host while a previous infection with a parasite of the same species is still present.

Superparasitism.   See Multiparasitism.

Symbiosis.   Living together of two species of organisms.

Symbiot.   The smaller of two organisms living together in symbiosis.

Synergism.   Cooperative action producing an effect greater than the sum of the two effects taken independently.

Tolerance.   In the immunologic sense, an induced state of unresponsiveness to a specific immunogen.

Transport host.   An organism that merely carries the nondeveloping parasite to the next host.

Trophozoite.   The motile stage of Protozoa.

Vector.   An essential intermediate host, usually an arthropod, in which the parasite undergoes a significant change.

Virulence.   The relative infectiousness of a parasite.

Xenodiagnosis.   Method of diagnosing an infection in a suspected vertebrate host by feeding a vector on it and then examining the vector for the presence of the developmental stages of the parasite.

Zoonosis.   A disease or infection that is naturally transferable between animals and man.   In the broad sense it includes any animal but most studies of zoonoses involve only diseases of vertebrates.

General References

Baer, J. G.: Ecology of Animal Parasites. Urbana, Ill., University of Illinois Press, 1951. 224 p.

Burrows, R. B.: Microscopic Diagnosis of the Parasites of Man. New Haven, Yale University Press, 1965. 328 p.

Cameron, T. W. M.: Parasites and Parasitism. New York, John Wiley, 1956. 322 p.

Caullery, M.: Parasitism and Symbiosis. London, Sidgwick and Jackson, 1952.

Faust, E. C., and Russell, P. F.: Craig and Faust's Clinical Parasitology, 8th ed. Philadelphia, Lea & Febiger, 1970. 890 p.

Geiman, Q.M.: Comparative physiology: mutualism, symbiosis, and parasitism. Ann. Rev. Physiol., 26:75–108, 1964.

Hunter, G. W., III, Frye, W. M., and Swartzwelder, J. C. (Eds.): A Manual of Tropical Medicine, 4th ed. Philadelphia, Saunders, 1966. 931 p.

Hyman, L. H.: The Invertebrates. Vol. I: Protozoa through Ctenophora. 726 p. Vol. II: Platy-helminthes. 550 p. Vol. III: Acanthocephala, Aschelminths and Entoprocta. 572 p. New York, McGraw-Hill, 1940–1951.

Jackson, G. J., Herman, R., and Singer, I.: Immunity to Parasitic Animals. 2 vols. New York, Appleton-Century-Crofts, 1969. Vol. I, 292 p. Vol. II, 917 p.

Jeffrey, H. C., and Leach, R. M.: Atlas of Medical Helminthology and Protozoology. Edinburgh, E. & S. Livingstone Ltd., 1966. 121 plates.

Karakashian, S. J., and Siegel, R. W.: A genetic approach to endocellular symbiosis. Exper. Parasitol., 17:103–122, 1965.

Nutman, P. S., and Mosse, B. (eds.): Symbiotic Associations. Thirteenth Symposium of the Society for General Microbiology, Royal Institution, London, April, 1963. 356 p.

Pan American Health Organization: Immunologic Aspects of Parasitic Infections. Washington, D.C., WHO, 1967. 166 p.

———: Selected publications. Available from Pan American Sanitary Bureau, WHO, 525 Twenty-third St., N. W., Washington, D.C., 20037.

Rogers, W. P.: The Nature of Parasitism. New York, Academic Press, 1962. 287 p.

Rothschild, M., and Clay, T.: Fleas, Flukes & Cuckoos. London, Collins, 1957. 305 p.

Schmidt, G. D. (ed.): Problems in Systematics of Parasites. Baltimore, University Park Press, 1969. 131 p.

Smith, J. C.: Bibliography on the biochemistry of endoparasites. Exptl. Parasitol., 22:352–422, 1968.

Soulsby, E. J. L.: Mönnig's Helminths, Arthropods and Protozoa of Domesticated Animals. Baltimore, Williams & Wilkins, 1968. 824 p.

Spencer, F. M., and Monroe, L. S.: The Color Atlas of Intestinal Parasites. Springfield, Ill., Charles C Thomas, 1961. 142 p.

Sprent, J. F. A.: Parasitism. London, Baillière Tindall and Cox, 1963. 145 p.

Stoll, N. R.: This wormy world. J. Parasitol., 33:1–18, 1947.

Symposium on intraspecific variation in parasitic animals. Systematic Zool., 6:2–28, 1957.

Taylor, A. E. R. (ed.): Techniques in Parasitology. First Symposium of the British Society for Parasitology. Oxford, Blackwell Sci. Pub., 1963. 107 p.

———: Host-Parasite Relationships in Invertebrate Hosts. Oxford, Blackwell Sci. Pub., 1964. 134 p.

———: The Pathology of Parasitic Diseases Symposium. Oxford, Blackwell Sci. Pub., 1966. 53 p.

———: Evolution of Parasites. Oxford, Blackwell Sci. Pub., 1965. 136 p.

Taylor, A. E. R., and Baker, J. R.: The Cultivation of Parasites in Vitro. Oxford, Blackwell Sci. Pub., 1968. 377 p.

Van der Hoeden, J. (ed.): Zoonoses. Amsterdam and New York, Elsevier, 1964. 774 p.

Voge, M., and MacInnis, A. J.: Experiments and Techniques in Parasitology. San Francisco, Freeman & Co., 1970.

Von Brand, T.: Biochemistry of Parasites. New York, Academic Press, 1966. 429 p.

Yamaguti, S.: Systema Helminthum. New York, Interscience, 1958–1963. 5 Vols.

Publications Containing Abstracts or Indices

Biological Abstracts, Philadelphia, 1926–
Excerpta Medica, Sect. IV, Medical Microbiology and Hygiene, Amsterdam, 1948–
Helminthological Abstracts, St. Albans, England, 1932–
Index Catalog of Medical and Veterinary Zoology, U.S.D.A. Washington, D.C., 1932–
Index Medicus, Washington, 1879–
Journal of the American Medical Association, Chicago, 1883–
Quarterly Cummulative Index Medicus, Chicago, 1927–
Reviews of Applied Entomology, Series B (Medical and Veterinary) London, 1913–
Tropical Disease Bulletin, London, 1912–
Tropical Veterinary Bulletin, London, 1912–1930
Veterinary Bulletin, Weymouth, 1951–
Zentralblatt für Backteriologie und Parasitologie, I. Abteilung, Original und Referat, Jena, Germany, 1887–
Zoological Record, London, 1864–

Index

Page numbers in *italics* indicate illustrations. Page numbers followed by "t" indicate tables.

C

Cacoecia sorbiana, and *Onchophanes lanceolator*, 383
Caduceia theobromae, 496
Caging, and stress effects, 476
Calandra orizae, and *Moniliformis moniliformis*, 262
Calcareous corpuscles, of cestodes, 213
Caligidae, 352–354
Caligoida, 352–355
Caligus, 352, *352*
  and *Udonella*, 134
  curtus, and cod, 515
Caliperia brevipes, 117, *122*, 123
Callionymus lyra, and Myxosporida, 486
Calliphora, 418
  and myiasis, 423
  vicina, *405*, *406*, *422*
Calliphoridae, 417–421
  and myiasis, 422
  evolution of, 571
Callosobruchus chinensis, and parasite ecology, 491
Calonymphidae, evolution of, 556
Calotte, 445
Camallanus, and fish age, 488
Cambaroides, and *Paragonimus westermani*, 194
Cambarus, and *Paragonimus westermani*, 194
Camel, and *Dicrocoelium dendriticum*, 184
  and *Echinococcus granulosus*, 245
  and *Onchocerca*, 319
  and *Taenia solium*, 239
  and *Trypanosoma*, 35, 36
  distribution of, 513
Camelidae, and lice evolution, 570
Canaries, and malaria, 461
Capillaria, *annulata*, 323–324, *323*
  evolution of, 568
  hepatica, *323*, 324
  talpae, host specificity of, 530
Caprellidea, 359
Capsaloidea, evolution of, 562
Capsule formation, around parasites, 464
Carabao, and schistosomes, 163
Carbonates, in lakes, ecologic aspects of, 504
Carcinonemertes, 446
Cardium edule, and *Gymnophallus*, 186
Cariama cristata, and Eurasian bustard, 518
Carp lice, 357
  and fish hibernation, 487
Carpenter ants, and *Brachylecithum mosquensis*, 465
Caryophyllaeidae, 227
Caryophyllidea, 227
Castration, host, effect upon parasites, 471, 472
  parasitic, 473–474
Catarrhal enteritis, 45
Cat, and *Alaria arisaemoides*, 162
  and amebas, 62
  and Ancylostomatoidea, 288
  and *Brugia*, 316
  and *Ctenocephalides felis*, 397, *398*, *399*
  and *Dibothriocephalus*, 225
  and *Dipylidium caninum*, 231
  and *Dracunculus*, 310
  and *Echinococcus multilocularis*, 249
  and *Echinostoma ilocanum*, 178
  and *Eimeria*, 86

Cat, (*Continued*)
  and *Gnathostoma*, 310
  and Heterophyidae, 197
  and *Isospora felis*, 86
  and *Metagonimus yokogawai*, 198
  and *Moniliformis moniliformis*, 262
  and *Opisthorchis*, 195, 196, 197
  and *Otodectes*, 434
  and *Paragonimus*, 190
  and *Pentatrichomonas*, 46
  and *Strongyloides stercoralis*, 282
  and *Taenia*, 244
  and *Toxascaris*, 298
  and *Toxoplasma gondii*, 105, 107
  and *Trypanosoma*, 31, 36
Catenotaenia dentritica, and *Urosporidium charletyi*, 473
Catostomus commersoni, cestodes of, periodicity of, 492
Cattle, and amebas, 61, 62
  and *Angiostrongylus cantonensis*, 297
  and *Babesia*, 103
  and *Besnoitia*, 107
  and *Boophilus annulatus*, 427
  and *Bovicola*, 370
  and *Bütschlia*, 118
  and *Cooperia punctata*, 294
  and *Demodex*, 439
  and *Dicrocoelium dendriticum*, 184
  and *Echinococcus granulosus*, 245
  and *Eimeria*, 85
  and Entodiniomorphida, 126
  and *Eurytrema pancreaticum*, 184
  and *Fasciola*, 178, 181
  and *Fascioloides magna*, 181
  and *Gonderia*, 103
  and *Gongylonema*, 310
  and *Haemonchus contortus*, 292
  and *Hypoderma*, 415
  and *Multiceps multiceps*, 244
  and *Nematodirus spathiger*, 294
  and *Onchocerca*, 319
  and *Ostertagia*, 294
  and *Paramphistomum microbothriodes*, 181
  and schistosomes, 163
  and *Taeniarhynchus saginatus*, 242
  and *Theileria parva*, 103
  and *Trichomonas*, 48
  and *Trypanosoma*, 31, 34, 35, 36
  defaunation of, 499
  symbiotic protozoa in, 498–500
Cattle biting louse, 370
Cattle tick fever, 103
Caudal alae, of nematodes, 267
Caudate larva, *382*
Caulleryella, and mosquitoes, 410
Ceca, of trematodes, 134
Cell(s), blood, and parasitism, 462–463
  evolution of, and parasitism, 551
Cellubiase, 499
Cellulase, 498, 499
Cellulose, digestion of, and ciliates, 498, 499
Cement glands, and Acanthocephala, *256*, 257
Centriole, 27
Centrorhynchus, host specificity of, 535
Cepedea, 56
Cephalina, 75